Carotenoids in Photosynthesis

Carotenoids in Photosynthesis

Edited by

Andrew Young

School of Biological and Earth Sciences, Liverpool John Moores University, UK

and

George Britton

Department of Biochemistry, University of Liverpool, UK

CHAPMAN & HALL
London · Glasgow · New York · Tokyo · Melbourne · Madras

Published by Chapman & Hall, 2–6 Boundary Row, London SE1 8HN

Chapman & Hall, 2–6 Boundary Row, London SE1 8HN, UK

Blackie Academic & Professional, Wester Cleddens Road, Bishopbriggs, Glasgow G64 2NZ, UK

Chapman & Hall Inc., 29 West 35th Street, New York NY10001, USA

Chapman & Hall Japan, Thomson Publishing Japan, Hirakawacho Nemoto Building, 6F, 1–7–11 Hirakawa-cho, Chiyoda-ku, Tokyo 102, Japan

Chapman & Hall Australia, Thomas Nelson Australia, 102 Dodds Street, South Melbourne, Victoria 3205, Australia

Chapman & Hall India, R. Seshadri, 32 Second Main Road, CIT East, Madras 600 035, India

First edition 1993

Printed in Great Britain by St Edmundsbury Press, Bury St Edmunds, Suffolk

ISBN 0 412 56250 2

A catalogue record for this book is available from the British Library

Library of Congress Cataloging-in-Publication data available

Contents

Contributors

Dr. W.W. Adams III
Department of Environmental, Population and Organismic Biology
University of Colorado
Boulder
CO 80309
USA

Dr. P.M. Bramley
Department of Biochemistry
Royal Holloway and Bedford New College
Egham
Surrey
TW20 0EX
UK

Dr. G. Britton
Department of Biochemistry
University of Liverpool
PO Box 147
Liverpool
L69 3BX
UK

Professor R.J. Cogdell
Department of Botany
Univeristy of Glasgow
Glasgow
G12 8QQ
UK

Dr. B. Demmig-Adams
Department of Environmental, Population and Organismic Biology
University of Colorado
Boulder
CO 80309
USA

Professor H.A. Frank
Deaprtment of Chemistry
University of Connecticut
Storrs
CT 06269-3060
USA

Dr. H. Hashimoto
Department of Applied Physics
Faculty of Engineering
Sugimoto
Sumiyoshi-ku
Osaka 558
Japan

Professor Y. Koyama
Faculty of Science
Kwansei-Gakuin University
Uegahara
Nishinomiya 662
Japan

Dr. A.J. Young
School of Biological and Earth Sciences
Liverpool John Moores University
Liverpool
L3 3AF
UK

Preface

The carotenoids are probably the most widespread of all natural pigments, ocurring in animals, plants and microorganisms. Carotenoids are responsible for many spectacular shows of colour in nature and, as a consequence, their distribution has been thoroughly examined. Carotenoids are also of commercial importance as they have been used for a number of years as colorants and antioxidants. In recent years, great advances have been made in understanding the biological functions of carotenoids. This is especially true for photosynthetic organisms, where the carotenoids serve two important functions, namely light-harvesting and photoprotection. Although the carotenoids are key components of the photosynthetic apparatus and probably have a vital role in the evolution of photosynthetic organisms and tolerance to oxidative conditions. Carotenoids are also essential for the structural organisation of the photosynthetic apparatus, but their importance in photosynthesis has largely been overlooked and considered secondary to that of the chlorophylls. Recent advances in carotenoid photochemistry have been impressive, and clear roles for carotenoids in photosynthesis are now evident.

The main aim of this book is to bring together for the first time key aspects of the biology, biochemistry and chemistry of carotenoids in photosynthetic organisms. The first chapters provide an introduction to carotenoids, their distribution and their biosynthesis in plants, algae and phototrophic bacteria. the value of carotenoid biosynthesis as an effective herbicide target site is also discussed. The bulk of the book, though, is concerned with light-harvesting and protective roles of carotenoids in photosynthetic organisms. The photochemistry of carotenoids in the processes of photosynthesis is dealt with in detail in a key chapter. Other chapters deal with the involvement of carotenoids in the effects of stress and the specific role of the xanthophyll cycle and zeaxanthin in energy dissipation processes in plants and algae. The final two chapters provide an extensive account of the spectroscopic (resonance Raman and NMR) and other techniques that are used to study carotenoids *in situ* and the methods used to isolate, purify and characterise carotenoids from photosynthetic tissues. These chapters are supported by an appendix which lists the physical data (UV/Vis, MS, NMR, CD and IR) for more than 70 different carotenoids that are commonly found in photosynthetic organsisms.

Andrew Young
George Britton

1

Structure and nomenclature of carotenoids

G. Britton

1.1 INTRODUCTION

In the second edition of the "*Key to Carotenoids*", published in 1987, Pfander listed the structures of approximately 550 different carotenoids that have been isolated from natural sources. Since then, the number of known naturally occurring carotenoids has risen to more than 600. Of these, around 150 have been found in photosynthetic organisms, though by no means all of them play a role in photosynthesis.

The carotenoids are often considered simply to be plant pigments, and they occur universally in the chloroplasts of all higher plants and algae. They are most obvious, however, when they occur in non-photosynthetic tissues, where they are responsible for the yellow, orange or red colours of many fruit, flowers and roots. Carotenoids are also microbial pigments, not only in the photosynthetic membranes of phototrophic bacteria and cyanobacteria, but also in the many non-phototrophic bacteria and fungi. It should not be forgotten that carotenoids are also important pigments in animals, especially birds (yellow and red feathers, egg yolk), fish (orange-red skin and flesh) and invertebrate animals where, in the free form or as complexes with protein, they are responsible for a wide variety of integumental colours. These animal carotenoids are, however, of plant origin, being taken in in the diet.

Carotenoids are used extensively as natural, non-toxic colorants in food, drinks, confectionary, cosmetics and animal feeds (Bauernfeind 1981). Pure, synthetic, "nature identical" carotenoids are used for these purposes, especially β-carotene, zeaxanthin, astaxanthin and 8´-apo-β-caroten-8´-al. Some natural extracts (e.g. annatto) have been used for many years, and the use of natural carotenoids, either as extracts or purified products, is increasing rapidly; the biotechnological production of β-carotene and astaxanthin by the unicellular green algae *Dunaliella* and

Haematococcus, respectively, is currently arousing much interest.

ß-Carotene, and other carotenoids with an unsubstituted ß-ring, play a vital nutritional role in humans and other animals as the main dietary source of vitamin A. The provitamin, ß-carotene, is converted into retinal and thence vitamin A itself, retinol, by an intestinal cleavage enzyme. Recent work is providing increasing evidence that ß-carotene and other carotenoids are effective antioxidants which can afford protection against some forms of cancer and other diseases.

The colour and potential antioxidant properties of carotenoids are a consequence of the most striking structural feature of these compounds, the long system of conjugated double bonds. This polyene chain is responsible for the ability of carotenoids to absorb light in the visible region, which gives them their strong coloration properties. It is also an essential structural requirement for the carotenoids to participate in energy transfer processes that are the basis of the major functions of carotenoids in photosynthesis.

The conjugated double bond system is very susceptible to oxidation and free-radical reactions and is therefore reponsible for the general instability of carotenoids, especially in oxidising conditions.

As a group, the carotenoids are extremely hydrophobic, lipophilic substances, even the glycosides being virtually insoluble in water. *In vivo*, however, the carotenoids are frequently located in close and often specific association with protein, which has a stabilising effect, allows them access to an aqueous environment, and may modify their properties in a way that may be crucial to their functioning. The association of carotenoids with chlorophyll and protein in the pigment–protein complexes is a clear example of this, and is vital for the functioning of the carotenoids in photosynthesis.

1.2 STRUCTURES OF CAROTENOIDS

The basic structure of the carotenoids is a symmetrical tetraterpene skeleton formed by tail-to-tail linkage of two C_{20} units. This basic skeleton is illustrated by the acyclic hydrocarbon lycopene (ψ,ψ-carotene, **1**) but, in many carotenoids, the end-groups are modified into six-membered (occasionally five-membered) rings at one end or at both ends of the molecule to give the monocylic and dicylic carotenoids, respectively. Such a structure is illustrated by the dicyclic hydrocarbon ß-carotene (ß,ß-carotene, **2**).

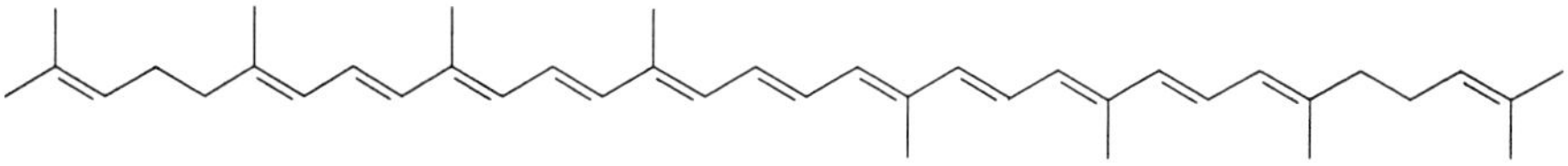

1

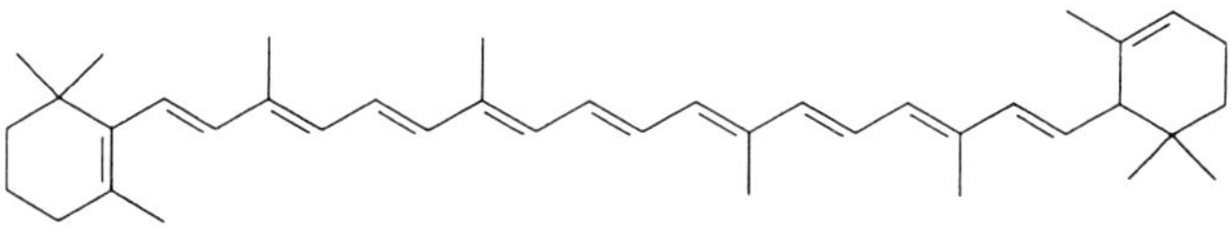

2

The absorption of light by carotenoids is due to a structural feature in the molecule known as the chromophore, which consists mainly or entirely, as in lycopene and ß-carotene, of a conjugated system of C=C double bonds, the polyene chain. For most of the carotenoids that are encountered in photosynthetic systems, the polyene chain consists of 9–13 conjugated double bonds, and this conveys the ability to absorb light in the visible region. The use of ultraviolet (UV)/visible light absorption spectra for the identification and quantitative analysis of carotenoids is described in Chapter 10. In some carotenoids the polyene chain may be modified by the presence of C≡C (acetylenic) or C=C=C (allenic) groups. In some structures, the rings may be modified into aromatic structures which are part of the chromophore.

The hydrocarbon carotenoids are known collectively as **carotenes.** The majority of natural carotenoids, however, contain at least one oxygen function, and are known as **xanthophylls**. The usual oxygen functions that are encountered in organic molecules may be found in the carotenoids, particularly hydroxy-, aldehyde, keto, carboxy-, carbomethoxy-, methoxy-, epoxy- and lactone groups. The hydroxy-groups may also be esterified or glycosylated.

In some carotenoid structures, the C_{40} skeleton has undergone degradation or elimination so that fewer than 40 carbon atoms remain. Compounds which are derived by loss of fragments from one end or both ends of the molecule are known as apo- or diapocarotenoids, respectively. These structures are not commonly seen in photosynthetic systems. Those in which one or more carbon atoms have been eliminated from within the C_{40} skeleton are known as nor-carotenoids. Some nor-carotenoids occur in large amounts in some photosynthetic organisms, e.g. peridinin (5´,6´-epoxy-3,5,3´-trihydroxy-6,7-didehydro-5,6,5´,6´-tetrahydro-10,11,20-trinor-ß,ß-caroten-19´,11´-olide 3-acetate, **3**) in dinoflagellates.

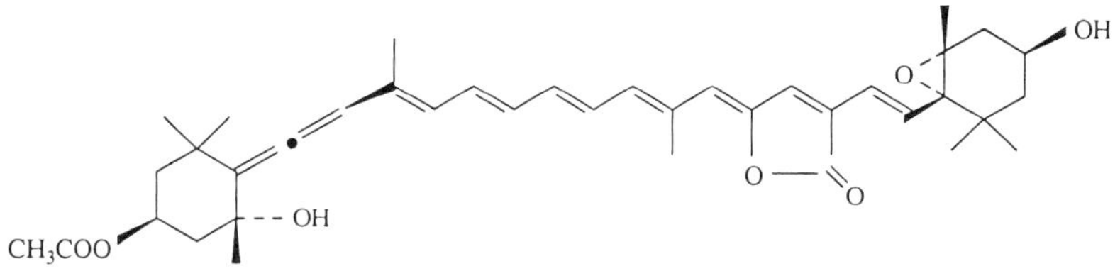

3

1.2.1 Geometrical isomers

Substituted carbon-carbon double bonds can exist in two different configurations depending on the relative disposition of the four substituents. In the case of the carotenoids, a C=C double bond is normally designated *cis* or *trans* on the basis of the disposition of the two substituents that are the continuing parts of the main polyene chain. As illustrated in Fig. 1.1, if these two substituents are located on opposite sides of the axis of the C=C double bond, the configuration is *trans*; if they are on the same side of this axis this gives a *cis* configuration. In recent years, this designation has largely been replaced by the *E/Z* scheme (see Section 1.3.2 (b)) In most cases, *E* and *Z* are synonymous with *trans* and *cis*, respectively, but there are a few cases in which the rules for the specification of configuration (see below) lead to the opposite result.

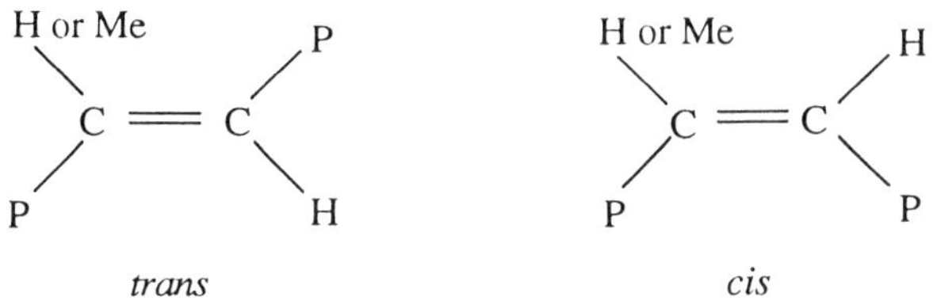

Figure 1.1 Designation of carotenoid double bonds as *trans* or *cis*. P indicates the continuation of the main polyene chain.

Since, in principle, each double bond in the polyene chain of the carotenoid can exist in either of the two forms, a large number of mono-*cis* and poly-*cis* geometrical isomers is theoretically possible for any carotenoid. Fortunately, few of these forms are encountered in practice. This is because the presence of a *cis* double bond creates greater steric hindrance between nearby groups, so that that isomer is rendered less stable than the *trans* form. With some double bonds, i.e. those that are tri-substituted and also the di-substituted C-15,15′ double bond, the steric hindrance is relatively small, so that isomers with *cis* double bonds in these positions are relatively stable and easily formed (e.g. 9-*cis*-, 13-*cis*- and 15-*cis*-ß-carotene). With other di-substituted double bonds, however, a *cis* configuration would result in major steric interference between a methyl group and a hydrogen atom (Fig. 1.2). Such sterically hindered *cis* double bonds are therefore much more difficult to produce and are less stable, so that isomers that contain them (e.g. 7-*cis*- and 11-*cis*-ß-carotene) are rarely encountered.

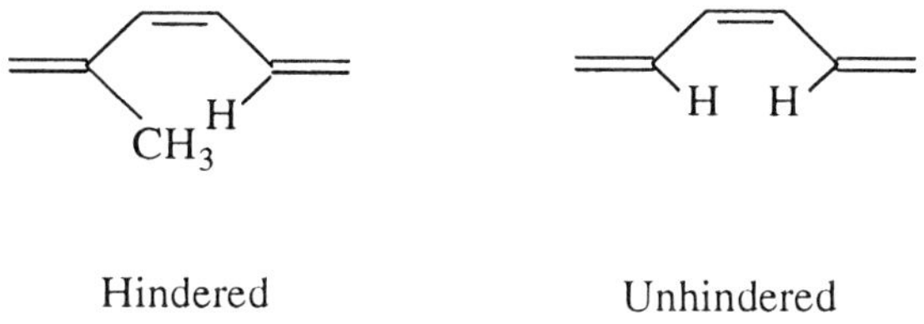

Figure 1.2 Sterically "hindered" and "unhindered" *cis* double bonds in a carotenoid polyene chain.

Although the all-*trans* (all-*E*) isomer is the most stable one, thermodynamically, for most carotenoids, there are some cases in which a *cis* isomer is the preferred, most stable, configuration. Thus, alloxanthin (7,8,7′,8′-tetradehydro-ß,ß-carotene-

3,3´-diol, **4**) occurs naturally in algae as the all-*E* form, but this readily undergoes a virtually irreversible stereomutation to the (9Z,9´Z)-isomer. Adoption of this configuration about the double bonds that are adjacent to the triple bond does not lead to steric interactions that would render the 9*Z* form less stable, as they do in the normal polyene series.

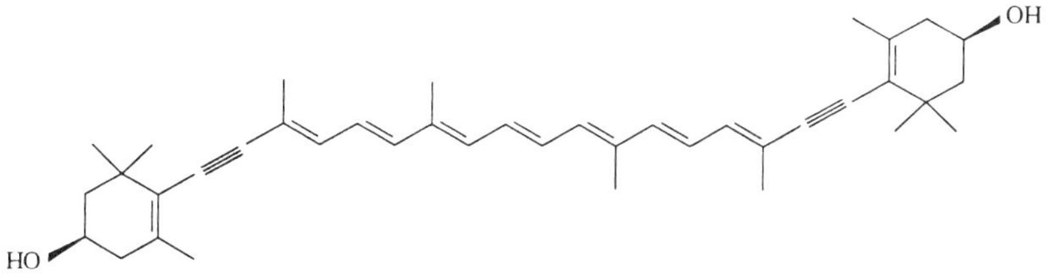

4

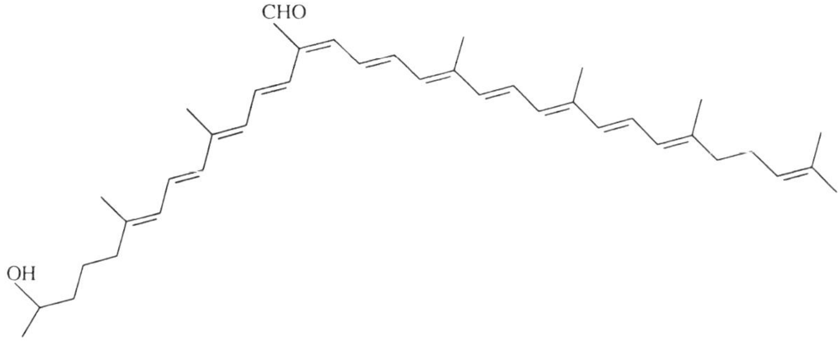

5

An example of naturally occurring stable *cis* isomers is provided by the cross-conjugated aldehydes, e.g. rhodopinal (13-*cis*-1-hydroxy-1,2-dihydro-ψ,ψ-caroten-20-al, **5**) and related compounds that are present in some purple bacteria (see Chapter 2), and exist as the 13-*cis* isomers. N.B. This is an example of a *cis* configuration which is designated *E* by application of the sequence rules (see Section 1.3.2 (b)).

1.2.2 Optical isomers: chirality

Like most groups of natural products, many carotenoids contain chiral centres (in particular, asymmetric carbon atoms, i.e. carbon atoms to which four different atoms or groups are attached). Allenic carotenoids possess a chiral axis. Such compounds can exist in distinct stereoisomeric forms in which the chiral features are not superimposable but bear a mirror-image relationship to one another. Such **optical isomers** or **enantiomers** have identical physical properties except that they can be distinguished by their interactions with polarised light (see Chapter 10). The chirality is designated by the *R,S* convention (see Section 1.3.2 (a)). It is normal for a natural carotenoid to exist in only one of the possible enantiomeric forms, i.e. to be "optically active", but there are examples of the same carotenoid existing in different enantiomeric forms in different organisms, e.g. ε-carotene (ε,ε-carotene, **6**) from higher plants is (6*R*, 6′*R*), whereas the (6*S*, 6′*S*) form has been found in some algae.

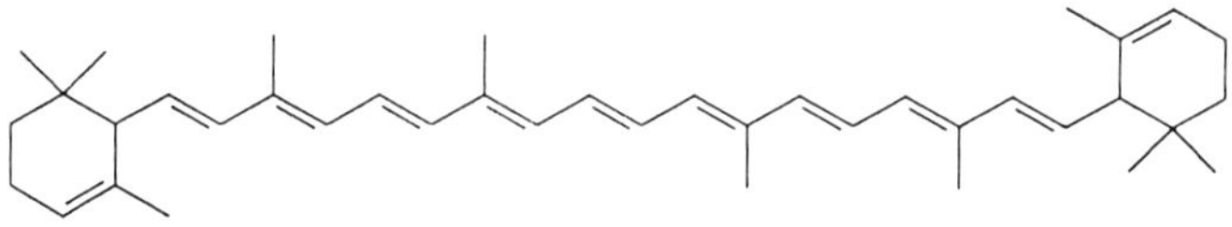

6

1.2.3 Conformation

Geometrical isomers and optical isomers are distinctly different molecular structures. Only a small number of these configurations will be possible for any carotenoid. In addition to this, however, a carotenoid can adopt, in principle, an infinite number of conformations due to the relatively unrestricted rotation that is theoretically possible about any C–C single bond in the polyene chain. In general, the stabilising influence of planarity of the extended π-electron system is such that flexibility is restricted and any degree of twisting is small. In a natural microenvironment, however, the molecular interactions between adjacent carotenoid molecules, or between carotenoid and, for example, protein, can be substantial and can lead to twisting of the entire chromophore into a chiral form. The conformation of a carotenoid chromophore may be important in relation to the functioning of the carotenoid *in vivo*.

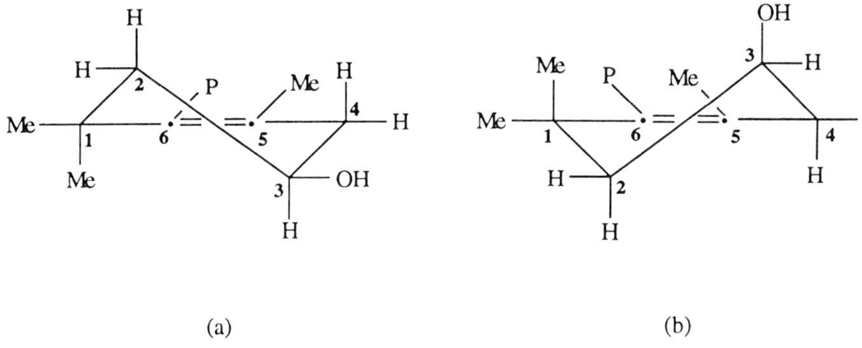

Figure 1.3 Half-chair conformations of the (3*R*)-3-hydroxy-ß-ring: (**a**) with the hydroxy-group equatorial; (**b**) with the hydroxy-group axial.

In carotenoids with cyclic end-groups, the rings are not planar structures (except for five-membered and aromatic rings) and are therefore capable of adopting different conformations. The preferred conformation is normally a "half-chair" (a conventional "chair" conformation modified by the presence of the double bond). In the case of the simple ß-ring, the two "half-chair" forms are of equivalent energy and freely interconvertible, so the molecule exists as a 1:1 equilibrium mixture of the two. With ring systems that contain additional large substituents, such as OH-groups or those in which C-6 is a chiral centre, a preferred conformation will be adopted in which the bulky substituents adopt an equatorial or pseudoequatorial position. Thus, the (3*R*)-3-hydroxy-ß-ring of zeaxanthin exists as the stable form (a), not as (b), Fig. 1.3.

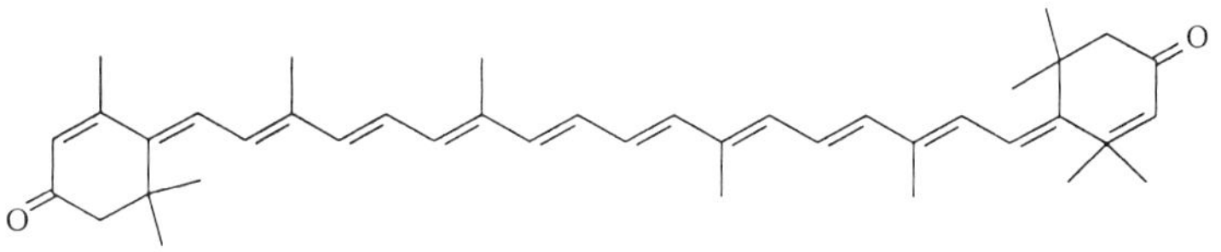

7

In all except the *retro*-carotenoids (e.g. rhodoxanthin, **7**), the end-group ring is linked to the polyene chain by a C-6,7-single bond. In principle, therefore, because of free rotation about this single bond, there can be an infinite number of possible angles of twisting between the ring and the polyene chain. There are thus two extreme structures in which the ring and the polyene chain are coplanar. These, as illustrated in Fig. 1.4, are known as the 6-*s*-*cis* and 6-*s*-*trans* forms. In between these structures an infinite number of conformations in which the ring and polyene chain are twisted out of planarity are theoretically possible. In practice, the preferred conformation is determined by factors such as the need to relieve steric interference between the C-1 and, especially, C-5 methyl substituents of the ring and the C-8 hydrogen atom of the chain and the conflicting stabilising effect of planarity between the chain and ring double bonds. It seems that, in most cases, the carotenoid adopts essentially a 6-*s*-*cis* conformation that is distorted out of planarity to relieve steric interference.

6-*s*-*cis* 6-*s*-*trans*

Figure 1.4 The planar 6-*s*-*cis* and 6-*s*-*trans* conformations of the ß-end-group.

1.3 NOMENCLATURE

Traditionally, carotenoids have been given trivial names, that are usually derived from the name of the biological material from which the carotenoid was first isolated. It is still common practice to use these trivial names, especially in biological work, even though they generally convey no information about the structure of the carotenoid. It would be possible to use the full name for each carotenoid according to the systematic rules of chemical nomenclature, but the names are long and cumbersome and the detailed structural information that they provide can be difficult to interpret. A semi-systematic scheme has therefore been devised which allows any carotenoid to be named unambiguously and in a way that describes and defines its structure. The rules for the nomenclature of carotenoids were issued by the IUPAC Commission on Nomenclature of Organic Chemistry and the IUPAC–IUB Commission on Biochemical Nomenclature, and have been published in *Pure and Applied Chemistry* (the offical journal of IUPAC) and elsewhere. The salient features of the rules that are relevant to the naming of the carotenoids that are found in photosynthetic tissues are outlined here. For full details, the reader should consult the IUPAC reports or the article by Weedon and Moss (1993) concerning carotenoid structures, stereochemistry and nomenclature.

It is accepted that trivial names are well known and will continue to be used, especially in biological and biochemical work, but it is recommended that in any publication the semi-systematic name should be given for each carotenoid that is discussed, for example, when that carotenoid is first mentioned, or in an appendix. This practice will be followed in this book; trivial names will be used throughout, but these are related to the semi-systematic names and structures given in Chapter 2 and in the Appendix.

1.3.1 Semi-systematic nomenclature

(a)Carotenoids: definition

Carotenoids are a class of hydrocarbons and their oxygenated derivatives that consist of eight isoprenoid units joined in such a manner that the arrangement of these units is reversed at the centre of the molecule; the two central methyl sidechains are thus in a 1,6-positional relationship and the remaining non-terminal methyl groups are in a 1,5-positional relationship. All carotenoids may be derived formally from the acyclic $C_{40}H_{56}$ structure of lycopene (**1**) by hydrogenation, dehydrogenation, cyclisation or oxidation, or any combination of these processes. Compounds that arise from certain rearrangements of this carbon skeleton or by the formal removal of part of this structure are also considered as carotenoids.

(b)The carotene stem and end-groups

All specific names are based on the stem name **carotene**, which corresponds to the partial structure and numbering shown in Fig. 1.5, with the long central chain of conjugated double bonds. The broken lines illustrated in the two terminal sections of the molecule represent two "double-bond equivalents".

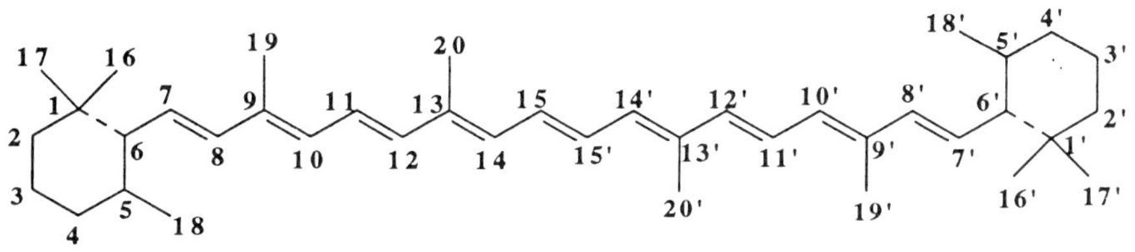

Figure 1.5 General formula and numbering scheme for carotenoids.

Seven different C_9 end-groups are recognised (Fig. 1.6). These may be acyclic, with two double bonds in positions 1,2 and 5,6 (designated the ψ-end-group), or cyclic. The name of a specific carotenoid hydrocarbon is constructed by adding to the stem name "carotene" the two Greek letter prefixes that designate the two C_9 end-groups. The Greek letter prefixes are cited in alphabetical order, that is ß (beta), γ (gamma), ε (epsilon), κ (kappa), ϕ (phi), χ (chi) and ψ (psi).

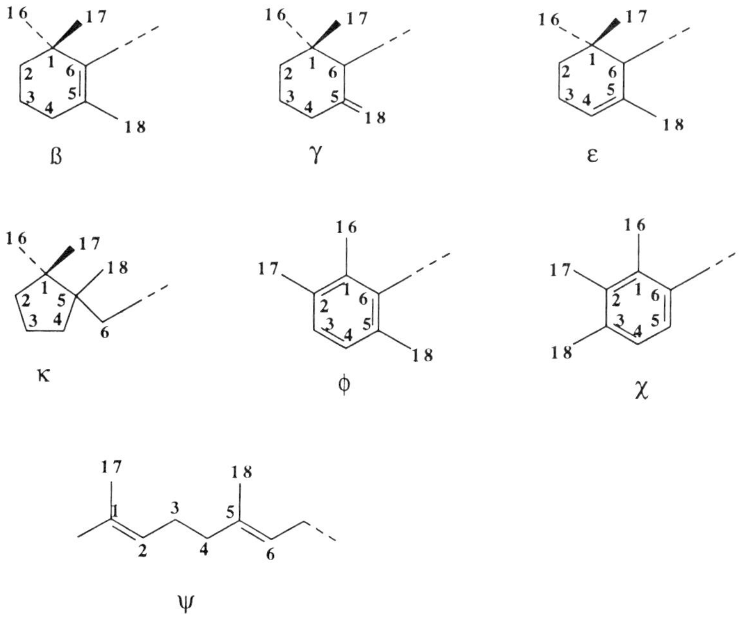

Figure 1.6 The seven carotenoid end-group types.

The basic numbering system is also illustrated in Fig. 1.6. If the two end-groups are dissimilar, lower (unprimed) numbers are given to that end of the molecule which is associated with the Greek letter prefix that is cited first in the name.

(c) Changes in hydrogenation level and substitution

Carotenoid hydrocarbons that differ in hydrogenation level from the parent carotene, as defined above, are named from the latter by use of the prefixes "hydro" and "dehydro", together with the locants (position numbers) that specify the carbon atoms where hydrogen atoms have been added or removed. These prefixes immediately precede the Greek letter end-group prefixes and are cited in the order dehydro before hydro.

(d) Oxygenated derivatives

Oxygenated (and other) derivatives of carotenoid hydrocarbons are named by use of suffixes and prefixes according to the rules of general organic chemical nomenclature. Of the oxygen-containing groups that are present, the one that occurs earliest in the sequence: carboxylic acid – ester of carotenoid acid – aldehyde – ketone – alcohol – ester of carotenoid alcohol, is chosen as the principal group and is cited by the use of a suffix; all other groups are cited as prefixes.

A non-bridging ether group is named by the use of the appropriate alkoxy prefix. Oxygen bridges are indicated by the use of the prefix "epoxy", preceded by the locants of the two carbon atoms that form the bridgeheads of the oxygen bridge. Note that an epoxide that is notionally formed by adding an oxygen atom to a double bond is formally named as an epoxydihydro derivative of the original compound.

Compounds that are derived formally from a carotenoid hydrocarbon by the addition of the elements of water (H, OH) or of methanol (H, OCH_3) to a double bond are named "hydroxydihydro" or "methoxydihydro" derivatives, e.g. hydroxyspheroidene (1′-methoxy-3′,4′-didehydro-1,2,7,8,1′,2′-hexahydro-ψ,ψ-caroten-1-ol, **8**).

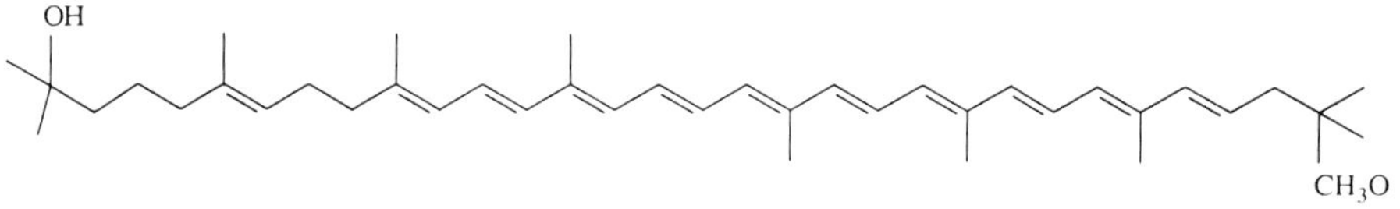

8

If the two C_9 end-groups are dissimilar, the end-group designated by the Greek letter that occurs earlier in the Greek alphabet is cited first and receives unprimed locants. If the two C_9 end-groups of the parent carotene are identical, then the lowest locant possible is assigned to the principal group, which is cited as suffix. Otherwise the principle of "lowest locants" is applied. All unprimed numbers are cited before primed numbers and are considered as lower than all primed numbers.

Note: it is recommended that formulae should normally be drawn so that the end-group with unprimed numbers is to the left. However, when structures are being compared, or biochemical pathways or transformations illustrated, this recommendation is not necessarily followed; some simple structural modifications can change the end-group priorities in closely related compounds (e.g. spheroidene (1-methoxy-3,4-didehydro-1,2,7´,8´-tetrahydro-ψ,ψ-carotene, **9**) and hydroxyspheroidene).

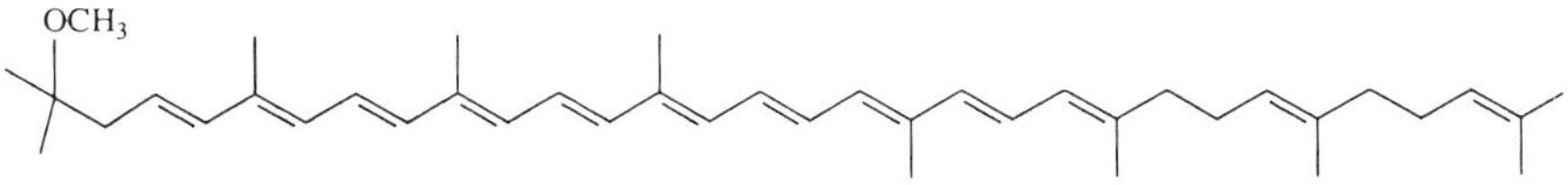

9

(e) Skeletal and polyene chain modifications

Nor-carotenoids

The elimination of one or more carbon atoms from within the carotenoid skeleton is indicated by the prefix "nor", preceded by the locant(s) of the carbon atom(s) that have been eliminated. The basic numbering of the remainder of the carotenoid molecule is retained, e.g. in peridinin (**3**).

Apo-carotenoids

When the carbon skeleton is shortened by the formal removal of fragments from one end or both ends of a carotenoid, the prefix "apo", preceded by a locant, is used to indicate that all of the molecule beyond the carbon atom corresponding to that locant has been replaced by the appropriate number of hydrogen atoms. A sidechain methyl group is considered as a substituent and not as an extension of the chain. The prefix and its locant immediately precede the specific name. If the locant associated with the prefix "apo" is greater than five, no Greek letter end-group designation is given for that end of the molecule. An example is ß-citraurin (3-hydroxy-8´-apo-ß-caroten-8´-al, **10**).

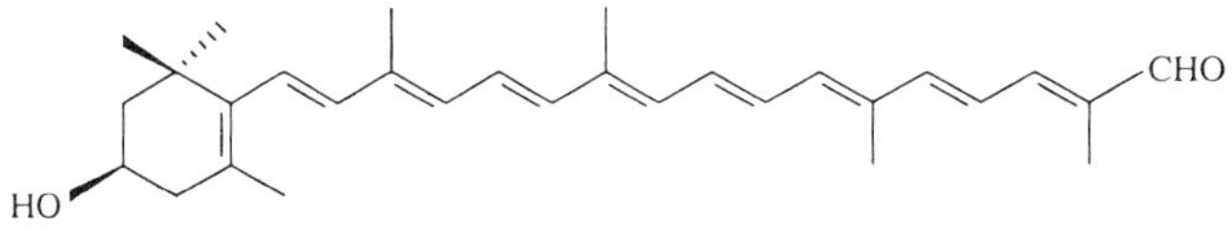

10

Seco-carotenoids

Fission of the bond between two adjacent carbon atoms of a cyclic end-group (other than the 1,6-bond), with formal addition of hydrogen atoms at each terminal group thus created is indicated by the prefix "seco", the original carotenoid numbering being retained, e.g semi-ß-carotenone (5,6-seco-ß,ß-carotene-5,6-dione, **11**).

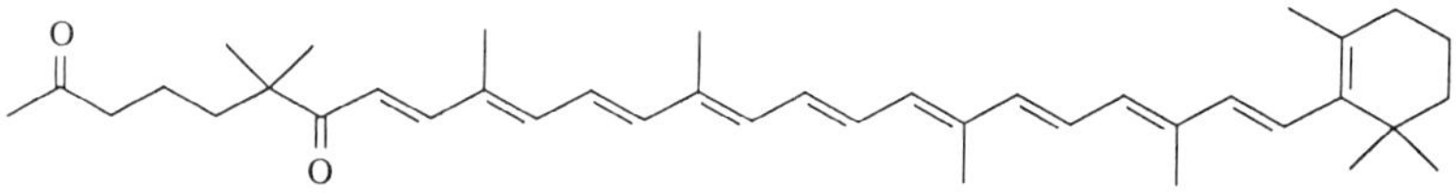

11

Retro-carotenoids

The prefix "*retro*", preceded by a pair of locants, is used to indicate a shift, by one position, of **all** single and double bonds of the conjugated polyene system delineated by the pair of locants. The first locant is that of the carbon atom that has formally lost a proton, the second that of the carbon atom that has gained one, e.g. rhodoxanthin (4´,5´-didehydro-4,5´-*retro*-ß,ß-carotene-3,3´-dione, **7**).

1.3.2 Stereochemistry

(a)Specification of absolute configuration

The absolute configuration of a carotenoid is specified unambiguously by the general *R,S* convention of Cahn, Ingold and Prelog. The main features of the procedure are summarised below, but the application of the rules to carotenoids can, in some cases, be complicated and the reader should consult a convenient introduction to the *R,S* system in general (Cahn 1964) or a more detailed account of the application of the rules to carotenoids (Moss and Weedon 1993) before attempting to assign *R* or *S* chirality to a carotenoid structure.

In order to specify the configuration of any asymmetric carbon atom, two operations must be performed. First, the four ligands that are attached to that carbon atom are ranked a > b > c > d in descending order of priority on the basis of a sequence rule (see below). Secondly, a model of the asymmetric centre is viewed from the side remote from ligand d, i.e. that of lowest priority. If a circle traced from a to b to c then follows a clockwise path, the asymmetric centre is designated *R*; if the path traced from a to b to c is anticlockwise, the designation is *S* (Fig.1.7).

The sequence rule that is used to assign priorities to the ligands consists, in its full form, of four sub-rules, but only one of these needs to be applied to chiral centres in any natural carotenoid. This "sub-rule 1" states that "higher atomic number precedes lower", i.e. O > C > H. It is applied first to the four atoms which directly link the four ligands to the asymmetric carbon atom. If the relative priorities of two or three groups cannot be decided in this way because the first atoms of the ligand are identical, the sub-rule is applied to the next atoms of these ligands, and so on until a decision can be made. If the ligands are complex and contain branched structures, the first atoms of the prior branches (decided by the sequence rule) are first considered and then, if necessary, the first atoms of the second and third branches. The process is repeated, working systematically along the branches, until a distinction is obtained.

a -> b -> c
clockwise
R

a -> b -> c
anticlockwise
S

Figure 1.7 The specification of the absolute configuration of an asymmetric carbon atom as *R* or *S*.

The presence of double and triple bonds in carotenoid molecules introduces a further complication. The convention now used requires that both atoms of a double bond are considered to be duplicated, and the duplicate atoms are considered to be substituted, to quadriligancy, by the appropriate number of "phantom atoms" of atomic number zero. Thus, a C=C double bond (C=C) expands to

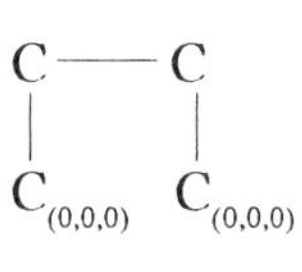

Similar procedures are used to designate the axial chirality of the allene groups in carotenoids such as neoxanthin and fucoxanthin.

(b)Geometrical configuration around double bonds

The stem name carotene implies a *trans* configuration about all double bonds, unless the contrary is indicated. The geometrical configuration of a carotenoid is described by citing the double bond or bonds that have a *cis* configuration, immediately following the designation of absolute configuration. At tri-substituted double bonds the terms *cis* and *trans* refer to the relative position of the two substituents that form parts of the main chain of carbon atoms (see Fig. 1.1).

It is now preferred chemical usage to specify the configuration of a double bond by the *E/Z* convention. For any C=C double bond, the two substituents on each carbon atom are ranked in order of priority according to the sequence rule that is used in the designation of *R* and *S* chirality. If the highest priority substituent on C-1 and that on C-2 are on the same side of the C=C axis, the configuration of the double bond is designated *Z* (from the German "zusammen", the same). If the higher priority substituents on C-1 and C-2 are on opposite sides of the C=C axis the designation is *E* ("entgegen", opposite) (Fig.1.8). For the C=C double bonds of the carotenoid chromophore the remainder of the polyene chain is usually the highest priority substituent at each carbon atom, so the terms *E* and *Z* are usually synonymous with *trans* and *cis*, repectively. If a sidechain methyl group is modified into an oxygen-containing functional group, e.g. CHO or CH_2OH, the order of priority is changed. Thus, for example, 13-*cis*-rhodopin-20-al (**5**) is degnated 13*E*, because of the substitution at C-20.

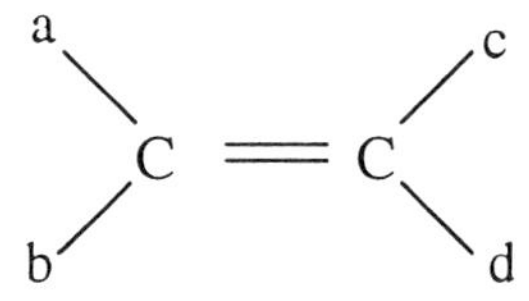

a > b and c > d → Z

a > b and d > c → E

Figure 1.8 Designation of carotenoid double bonds as *E* or *Z*.

In biological work, it may be clearer to retain the terms *trans* and *cis,* in conjunction with trivial names to indicate geometrical configuration in general discussions, but to use the *E/Z* designation when semi-systematic names are defined. This policy will be used throughout the book.

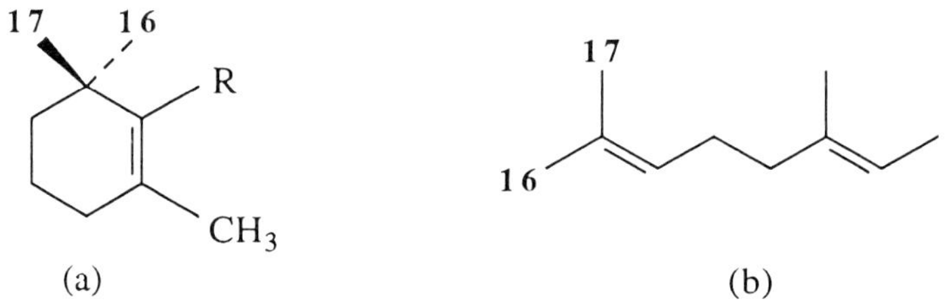

Figure 1.9 Numbering of the C-1 methyl substituents in carotenoid end-groups: (a) cyclic; (b) acyclic.

(c) Numbering of geminal methyl groups at C-1

In an acylic end-group the C-1 substituent that is *trans* to the main skeletal chain is numbered 16 and the one that is *cis* to the chain is numbered 17 (Fig. 1.9). In an end-group of the ß, γ, ε or κ type, the two methyl groups attached to C-1 are non-equivalent. When the potential chirality is as shown in the formula, i.e. with the polyene chain *R* to the right of C-1, the methyl groups below and above the plane of the ring are numbered 16 and 17, repectively.

REFERENCES

Bauernfeind, J.C. (1981) *Carotenoids as Colorants and Vitamin A Precursors. Technological and Nutritional Applications*. Academic Press, New York.

Cahn, R.S. (1964) An introduction to the sequence rule. A system for the specification of absolute configuration. *J. Chem. Educ.*, **41**:116-125.

IUPAC Commission on Nomenclature of Organic Chemistry and IUPAC-IUB Commission on Biochemical Nomenclature (1975) Nomenclature of Carotenoids (Rules approved 1974). *Pure Appl. Chem.*, **41**:407-431.

Pfander, H. (1987) *Key to Carotenoids*. 2nd edn., Birkhäuser Verlag, Basel.

Weedon, B.C.L. and Moss, G.P. (1993) Structures, stereochemistry and nomenclature. In: Britton, G., Pfander, H. and Liaaen-Jensen, S., *Carotenoids*, Vol. 1a. Birkhäuser Verlag, Basel. (In Press)

2

Occurrence and distribution of carotenoids in photosynthetic systems

A.J. Young

2.1 INTRODUCTION

In the past few years the analysis of carotenoids has become much more sensitive with the introduction of high-performance liquid chromatography (HPLC), especially with diode-array detection. This has permitted much more accurate determination of the pigment content and composition of photosynthetic organisms. However, as a note of caution, whilst analytical techniques have improved greatly, they still require very careful extraction of the plant material and subsequent careful handling of the isolated carotenoids. Improper practices at this stage can lead to the formation of a number of artifacts (see Chapter 10). Particularly common is the production of the furanoid derivatives from carotenoids that contain 5,6-epoxide groups. Similarly, many workers continue to employ saponification as an obligatory step during the extraction procedure; at the very least this will cause minor losses of the carotenoids present but it may change the whole nature of the extract by removing xanthophyll esters or destroying some carotenoids.

This chapter is not intended to provide a comprehensive listing of all carotenoids found in all photosynthetic organisms; rather it is intended to act as an overview of this subject and to report the most important trends seen in the distribution of these pigments in photosynthetic organisms. More detailed listings are given by Goodwin (1980), and reports of new observations are generally included in the proceedings of the International Symposia on Carotenoids, which are held every three years.

2.2 HIGHER PLANTS

Carotenoids are biosynthesised and accumulated in the photosynthetic tissues of all higher plants (see Chapters 3 and 4). Both carotenes and xanthophylls have been found in all leaf tissues that have been examined to date (Table 2.1). The cyclic carotenes and xanthophylls found in leaf tissues can have two different ring types, namely ß and ε. Only dicyclic xanthophylls, and particularly those with hydroxyl substituents at C-3 or C-3´, have been found in the photosynthetic tissues of higher plants. Acyclic carotenoids are not normally present, although the application of inhibitors of carotenoid biosynthesis (see Chapter 5) may result in their accumulation. The known occurrence of carotenoids in higher plants has been reviewed by Goodwin (1980), Goodwin and Britton (1988) and Szabolcs (1990).

Table 2.1 Carotenoids of higher plant photosynthetic tissues

Major	*Others*
Antheraxanthin (**5**)	Adonixanthin (**16**)
α-Carotene (**8**)	ß-Carotene-5,6-epoxide (**11**)
ß-Carotene (**1**)	ß-Cryptoxanthin (**9**)
Lutein (**2**)	Lactucaxanthin (**10**)
Neoxanthin (**4**)	Lutein-5,6-epoxide (**7**)
Violaxanthin (**3**)	Lycoxanthin (**17**)
Zeaxanthin (**6**)	Rhodoxanthin (**13**)
	Semi-ß-carotenone (**14**)
	Triphasiaxanthin (**15**)

The same four major carotenoids, namely ß-carotene (ß,ß-carotene, **1**), lutein ((3*R*,3´*R*,6´*R*)-ß,ε-carotene-3,3´-diol, **2**), violaxanthin ((3*S*,5*R*,6*S*, 3´*S*,5´*R*,6´*S*)-5,6,5´,6´-diepoxy-5,6,5´,6´-tetrahydro-ß,ß-carotene-3,3´-diol, **3**) and neoxanthin ((3*S*,5*R*,6*R*,3´*S*,5´*R*,6´*S*)-5´,6´-epoxy-6,7-didehydro-5,6,5´,6´-tetrahydro-ß,ß-carotene-3,5,3´-triol, **4**) have been identified as the major constituents of leaf tissues in all plant species examined to date.

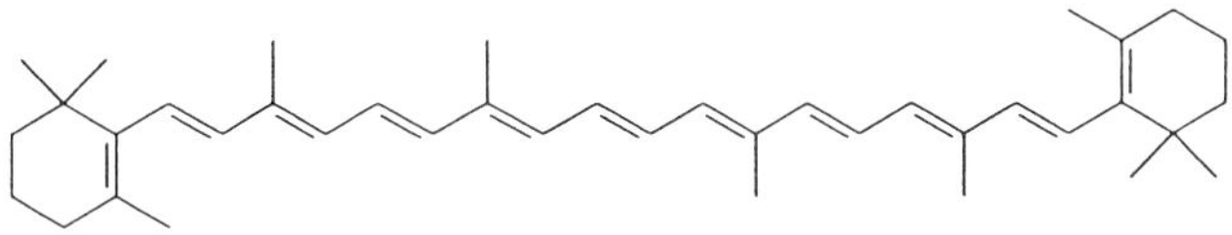

1

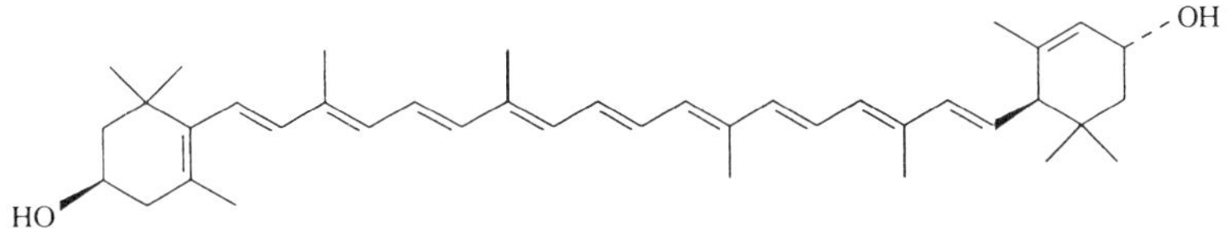

2

These carotenoids are present in leaves taken from plants living in a wide range of habitats. Whilst these four compounds generally account for the bulk of the carotenoid pigment, other carotenoids may also be found in photosynthetic leaf tissues, sometimes in large amounts. These include the other pigments of the xanthophyll cycle, namely antheraxanthin ((3*S*,5*R*,6*S*,3´*R*)-5,6-epoxy-5,6-dihydro-ß,ß-carotene-3,3´-diol, **5**) and zeaxanthin ((3*R*,3´*R*)-ß,ß-carotene-3,3´-diol, **6**).

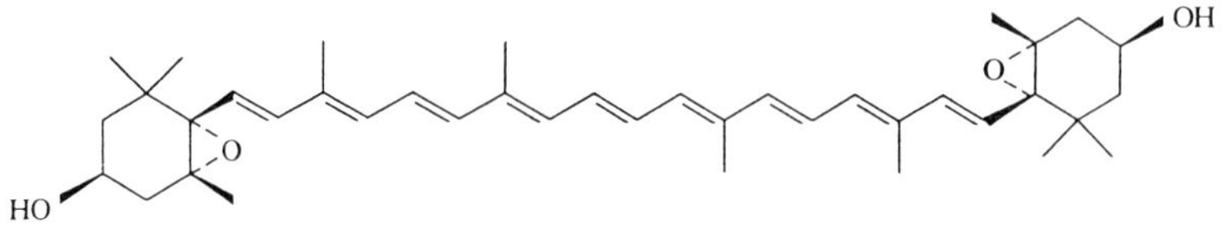

3

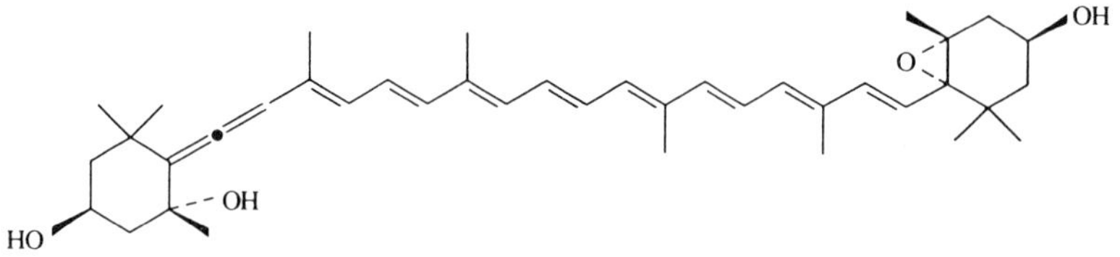

4

The levels of these carotenoids in leaf tissues are greatly affected by environmental and growth/developmental conditions, and may be very high in some plant species (see Chapter 7). Other carotenoids frequently found in leaf tissues are lutein-5,6-epoxide ((3*S*,5*R*,6*S*,3´*R*,6´*R*)-5,6-epoxy-5,6-dihydro-ß,ε-carotene-3,3´-diol, **7**), α-carotene ((6´*R*)-ß,ε-carotene, **8**) and ß-cryptoxanthin ((3*R*)-ß,ß-caroten-3-ol, **9**). The additional presence of lactucaxanthin ((3*R*,6*R*,3´*R*,6´*R*)-ε,ε-carotene-3,3´-diol, **10**) as a major xanthophyll in a number of composites, e.g. lettuce (Siefermann-Harms et al. 1981; Gilmore and Yamamoto 1991) has been confirmed. This is the only report of a carotenoid of the ε,ε-type found in higher plants, although they are found in many algae (see Section 2.3) and fish (Goodwin 1984).

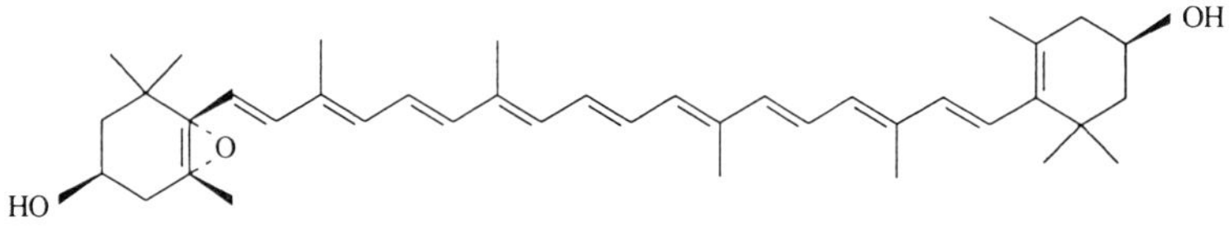

5

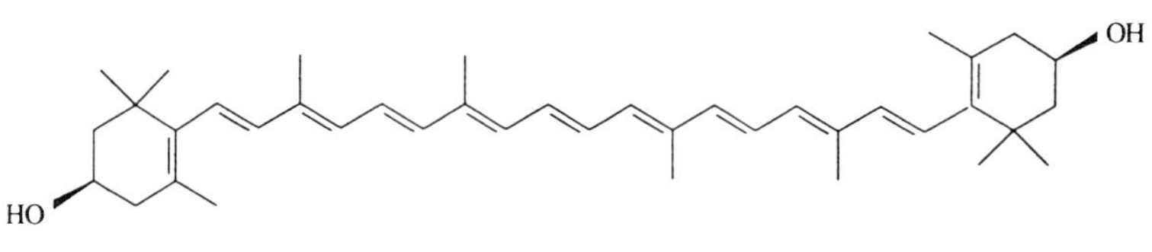

6

The leaves of a large number of plant species, both angiosperms and gymnosperms, contain α-carotene in addition to ß-carotene (Strain 1958; Siefermann-Harms 1985). Levels of α-carotene may be quite high, reaching 25-30% of the total carotene content of the leaves (e.g. in *Calycanthus occidentalis* and *Hedera helix*, Wieckowski and Goodwin 1966). Thayer and Björkmann (1990) observed that, in addition to major changes in the xanthophyll cycle pigments (see Chapter 7), the ratio of α-carotene : ß-carotene was much higher in shade plants than in sun plants. Similarly, those plants containing high levels of α-carotene were shown to have a much higher ratio of α-carotene : ß-carotene when they were grown in low light conditions. Similar data have also been obtained for sun and shade needles of *Picea abies* (Grill and Pfeifhöfer 1985). Young and Britton (1989) showed that, in leaves containing high levels of α-carotene, there was a constant ratio of the two carotenes in all the pigment-protein complexes, suggesting a similar functional role for α-carotene and ß-carotene. α-Carotene is an intermediate in the synthesis of the main light-harvesting carotenoid, lutein, whereas ß-carotene is the precursor for neoxanthin, violaxanthin, antheraxanthin and zeaxanthin.

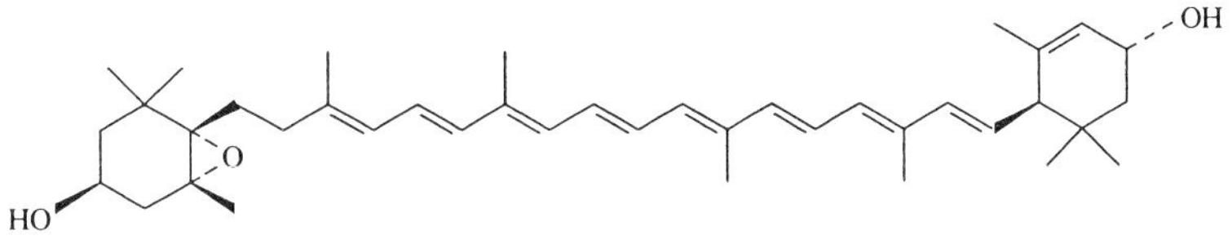

7

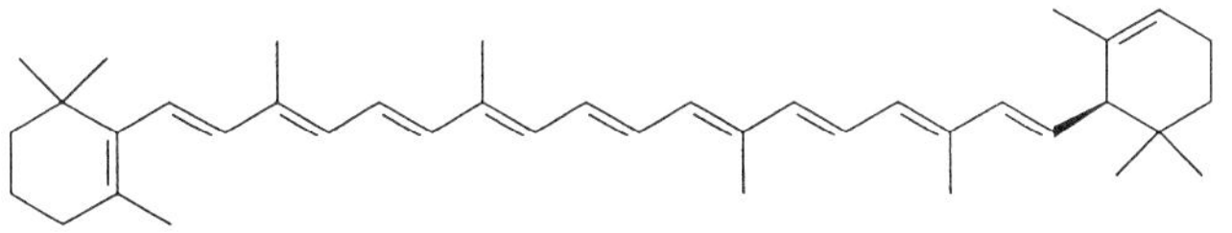

8

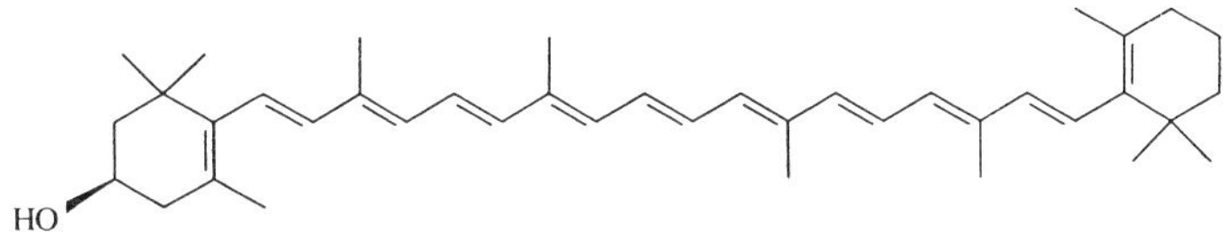

9

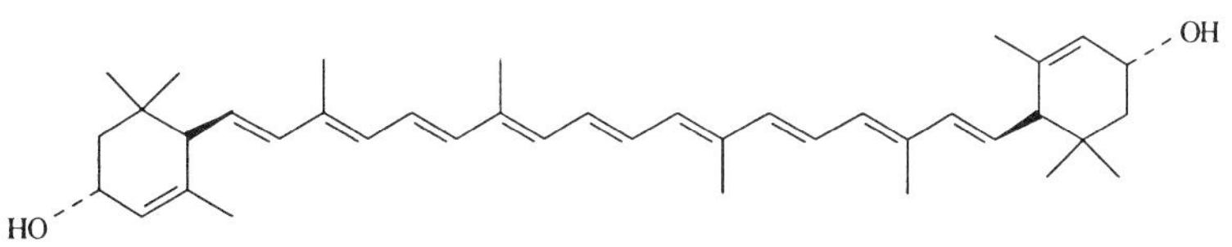

10

The occurrence of ß-carotene-5,6-epoxide (5,6-epoxy-5,6-dihydro-ß,ß-carotene,**11**) in leaves and in isolated chloroplasts or thylakoid membranes has recently been studied. It was first reported by Ashikawa et al. (1987) that this carotenoid was produced by the enzymic epoxidation of ß-carotene when leaves were exposed to high light intensities. Young et al. (1989) showed that ß-carotene-5,6-epoxide was produced by a chemical or photochemical process in leaf tissues (see Chapter 5 for more details). Traces of the 5,8-epoxide (mutatochrome) (5,8-epoxy-5,8-dihydro-ß,ß-carotene, **12**) have frequently been reported in plant tissues, but the formation of this compound is artifactual.

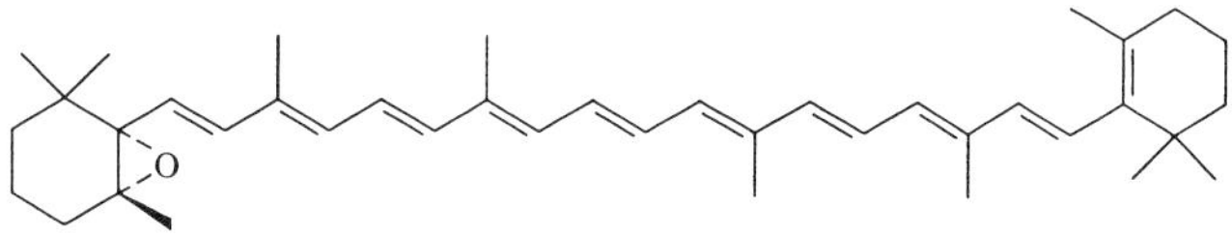

11

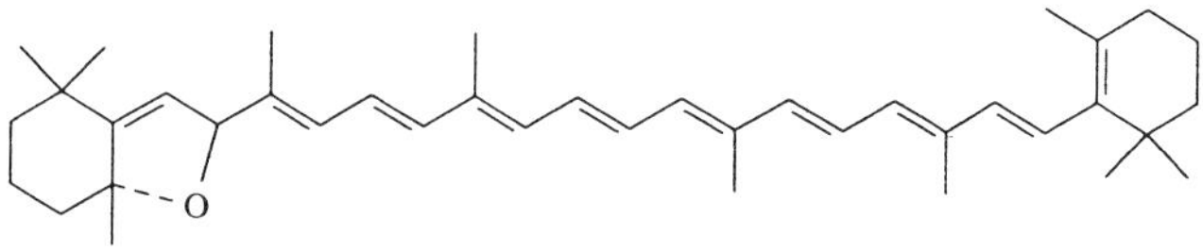

12

Other carotenoids that have been reported in the photosynthetic tissues of higher plants include a number of ketocarotenoids such as rhodoxanthin (4´,5´-didehydro-4,5´-*retro*-ß,ß-carotene-3,3´-dione, **13**) (Goodwin 1980; A. Young and G. Britton, unpublished data), semi-ß-carotenone (5,6-seco-ß,ß-carotene-5,6-dione,**14**; Cardini et al. 1987, 1989) and triphasiaxanthin (3´-hydroxy-5,6-seco-ß,ß-carotene-5,6-dione, **15**; Cardini et al. 1990).

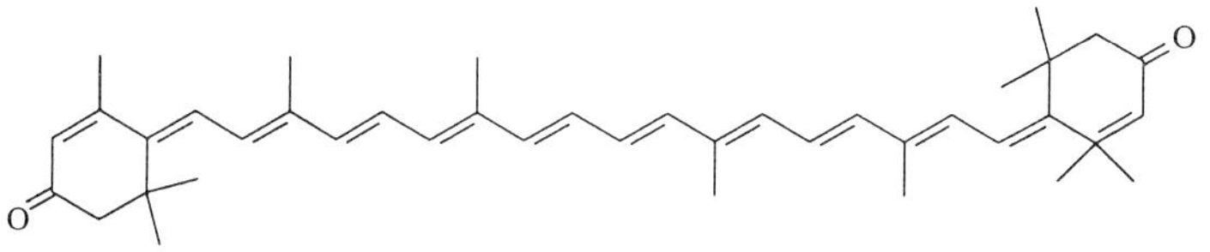

13

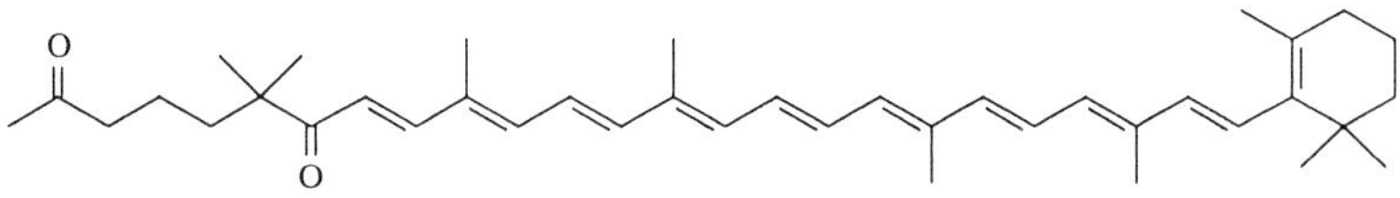

14

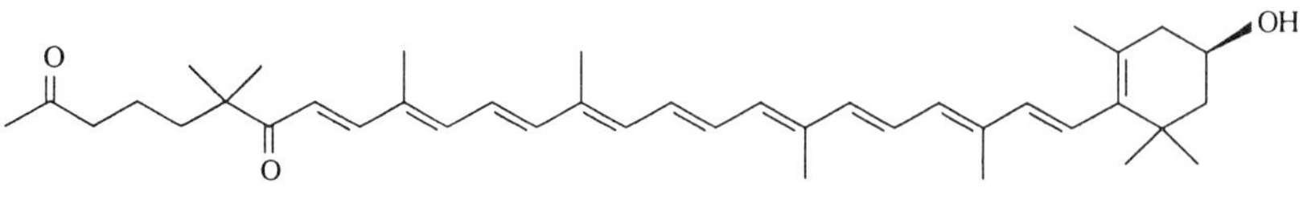

15

It is important to note that, unlike the other carotenoids listed above, these ketocarotenoids are not located in the pigment–protein complexes and that they have no functional role in photosynthesis. Rhodoxanthin is found in a number of Gymnosperms, including members of the Cupressaceae, Taxaceae and Taxodiaceae (e.g. in *Thuja occidentalis* var. Rheingold, where it may account for up to 25%–30% of total carotenoid, A. Young and G. Britton, unpublished data) and it is often responsible for the golden appearance of the autumn leaves or needles (Czeczuga 1986). It has also been reported to be present in leaves of *Potamogeton* spp. and in the aquatic fern *Azolla filiculoides* (A. Tsavalos and A. Young, unpublished data). In the majority of examples the synthesis of rhodoxanthin appears to be regulated by the incident light intensity and its presence is generally accompanied by changes in the morphology of the chloroplast.

Czeczuga (1986) has reported the presence of adonixanthin (3,3´-dihydroxy-ß,ß-caroten-4-one,**16**) in species of the Pinaceae, and also reported that some gymnosperms may contain the unusual carotenoid lycoxanthin (ψ,ψ-caroten-16-ol, **17**).

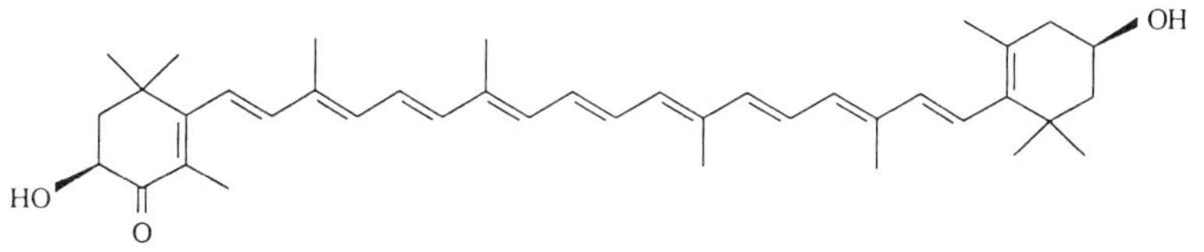

16

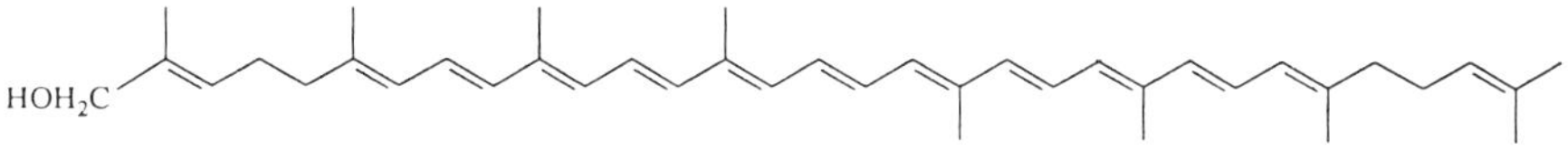

17

The occurrence of the seco-carotenoids, semi-ß-carotenone and triphasiaxanthin has been reported in the young leaflets of two cycads, *Ceratozamia fuscoviridis* and *C. kuesteriana* (Cardini et al. 1987, 1989, 1990). These leaflets accumulate very high levels of these and other, as yet unidentified, seco-carotenoids in oil droplets throughout the plastid.The biosynthesis of carotenoids in green leaves is normally very efficient so that intermediates in the biosynthetic pathway leading to the formation of the normal dicyclic carotenoids are generally absent. The presence of phytoene (7,8,11,12,7´,8´,11´,12´-octahydro-ψ,ψ-carotene, **18**), phytofluene (7,8,11,12,7´,8´-hexahydro-ψ,ψ-carotene,**19**), ζ-carotene (7,8,7´,8´-tetrahydro-ψ,ψ-carotene, **20**), neurosporene (7,8-dihydro-ψ,ψ-carotene, **21**), and lycopene (ψ,ψ-carotene, **22**) may be observed on treatment of plants with inhibitors of carotenogenesis (e.g. diflufenican, aminotriazole (amitrole) – see Chapters 4 and 5 for more details). Some albino mutants of plants and algae may also accumulate these biosynthetic intermediates.

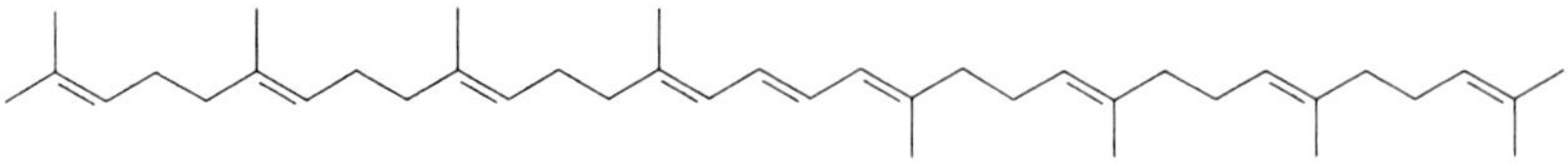

18

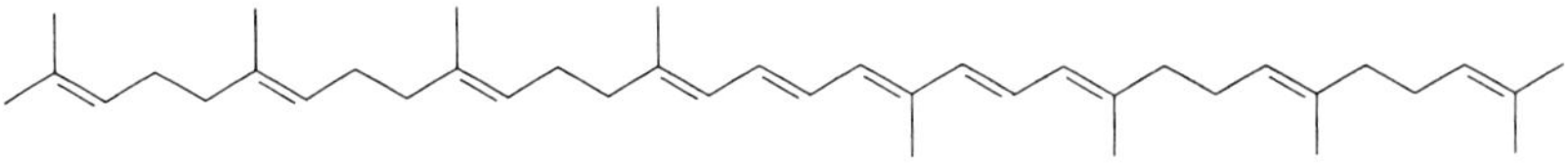

19

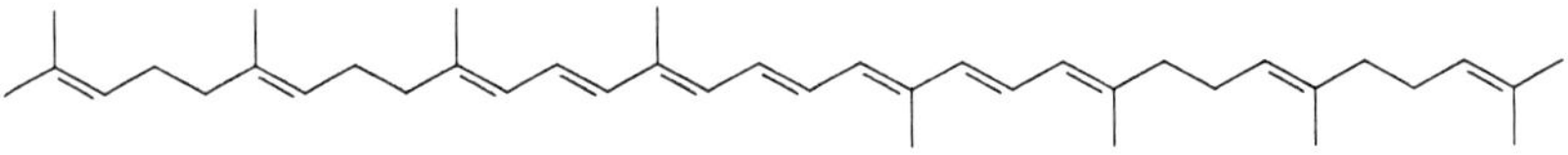

20

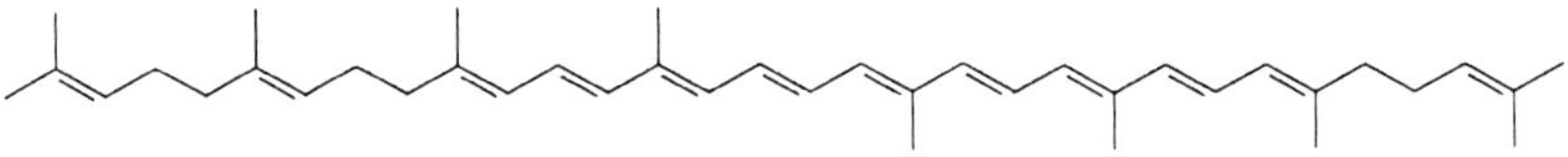

21

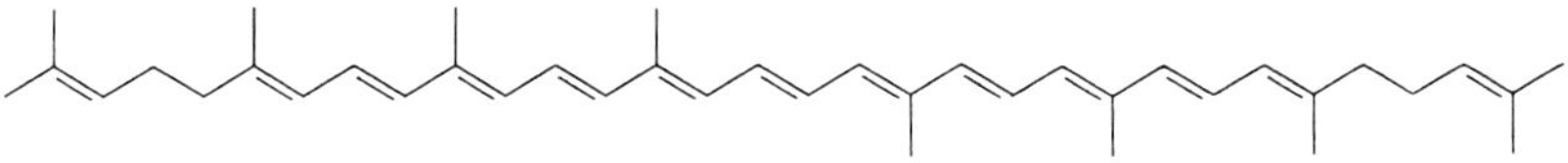

22

Whilst higher plant leaves contain only a limited number of carotenoids, together with chlorophylls *a* and *b*, there is considerable natural variation in the quantitative distribution of these pigments. Developmental and environmental conditions are important factors in determining the stoichiometry of the photosynthetic apparatus and its components, including the carotenoids. Differences or alterations may be seen both in the relative amounts of the individual carotenoids that can be found in the thylakoid membrane (e.g. the relative abundance of α-carotene in plants grown in low or high light) but also in the ratio of total carotenoid : total chlorophyll. Such variation in pigment content may be observed on an inter- and an intra specific basis and may be seen as a direct consequence of environmental conditions, and is discussed in more detail in Chapter 6. The known distribution of carotenoids within the individual pigment–protein complexes of the higher plant photosynthetic apparatus is described in Chapter 3.

2.3 ALGAE AND CYANOBACTERIA

The classification of the eukaryotic algae originally proposed by Christensen (1964) divides the algae into two phyla: Chlorophyta and Chromophyta. The prokaryotic algae belong to the Prochlorophyceae and Cyanophyceae. Problems arise with the classificatory scheme of the eukaryotic algae as they are not monophyletic (Rowan 1989). In particular, the carotenoid composition of the Euglenophyceae sets this class apart from the other Chlorophyceae. This classificatory scheme has, however, been referred to extensively by workers on algal carotenoids, most notably Liaaen-Jensen and co-workers. The classification of eukaryotic algae could be said to be in a state of constant flux, but it is not the intention of this chapter to add to this discussion. The individual classes of algae will be treated separately.

The pigment compositions of the algae have been reviewed extensively in recent years (Goodwin 1971, 1980; Liaaen-Jensen 1977, 1978; Ragan and Chapman 1978; Rowan 1989; Bjørnland 1990a). Other reviews concerning marine carotenoids also contain much information relating to the carotenoids of algae (Liaaen-Jensen 1990, 1991). Stauber and Jeffrey (1988) have perfomed an extensive study by thin layer chromatography (TLC) of the pigments in over 50 species from 13 families of diatoms.

The algae show a wide range of carotenoid structural features. Acetylenic carotenoids, for instance, are restricted to aquatic, mainly marine, animals and algae. There is also much interest in the variations that are encountered in the organisation of the light-harvesting systems and of their pigment contents.

The possible use of the pigment compositions of algae as a taxonomic marker has received a great deal of attention (Bjørnland and Liaaen-Jensen 1989; Rowan 1989). It is in the algae that carotenoids have proved to be the most successful as chemotaxonomic markers (Liaaen-Jensen 1990). In the Chromophyte algae, for example, five of the nine algal classes can be distinguished by their main light-harvesting pigments.

As Bjørnland and Liaaen-Jensen (1989) point out, the carotenoids found in the algae are generally located within the chloroplast and any approach to the use of these pigments in algal systematics should therefore be restricted to photoautotrophic species. The Craspedophyceae, for example, contain only heterotrophic species and cannot be included in any taxonomic treatment based on carotenoids. The examination of the carotenoid composition of many microalgal species which possess few morphological characteristics may prove to be amongst the most useful applications (Liaaen-Jensen 1990).

Of the two main functions for carotenoids in photosynthetic systems, namely light-harvesting and protection against oxidation, Bjørnland and Liaaen-Jensen (1989) and Moore et al. (1990) considered the former of these to be the later evolutionary event.

Liaaen-Jensen (1990) has reported the analysis of 14 algal classes and has assigned the following structural features of carotenoids to them; (i) monocyclic carotenoid glycosides, (ii) common dicyclic xanthophylls, (iii) 4-keto carotenoids, (iv) ε-type end-groups, (v) 5,6-epoxides, (vi) allenes, (vii) acetates, (viii) acetylenic carotenoids, (ix) 8-keto carotenoids, (x) 5,6-diols, (xi) 19-acyloxy, (xii) C_{37}-skeletal butenolides and (xiii) 2-hydroxy-carotenoids. The known distribution of these structural features in the algal classes is shown in Table 2.2 and in Fig. 2.1.

Table 2.2 Distribution of the main structural features of carotenoids found in the algae. See Fig. 2.1 for structures

	Structural feature												
	1	2	3	4	5	6	7	8	9	10	11	12	13
Bacillariophyceae		x			x	x	x	x	x				
Chlorophyceae		x	x	x	x	x					x		x
Cryptophyceae		x		x			x						
Cyanobacteria	x	x	x										
Dinophyceae I		x			x	x	x	x	x		x		
Dinophyceae II		x			x	x	x	x				x	
Euglenophyceae		x	x	x	x	x		x	x		x		
Eustigmatophyceae		x			x	x	?				x		
Phaeophyceae		x			x	x	x		x				
Prasinophyceae		x			x	x			x		x		
Prymnesiophyceae		x			x	x	x	x	x		x		
Raphidophyceae I		x			x	x	x	x		x	x		
Raphidophyceae II		x			x	x	x		x				
Rhodophyceae		x		x									
Xanthophyceae		x			x	x	x	x		x			

In a number of algal classes, the carotenoid compositions of only a limited number of species have been examined, making it difficult to draw any important conclusions concerning the phylogenic value of these pigments (see below). It is very probable that, as the carotenoid compositions of more species of algae are examined, such chemo-taxonomic classificatory schemes will be reviewed and updated, promoting the transfer of some species from one class to another.

Liaaen-Jensen (1990) has pointed out that the two Raphidophyceae pigment types shown in Table 2.2 could be considered to belong to fresh-water and marine environments, respectively. It has been proposed that the first of these two pigment types should belong to the Xanthophyceae and not to the Raphidophyceae (Christensen 1980; Bjørnland and Liaaen-Jensen 1989).

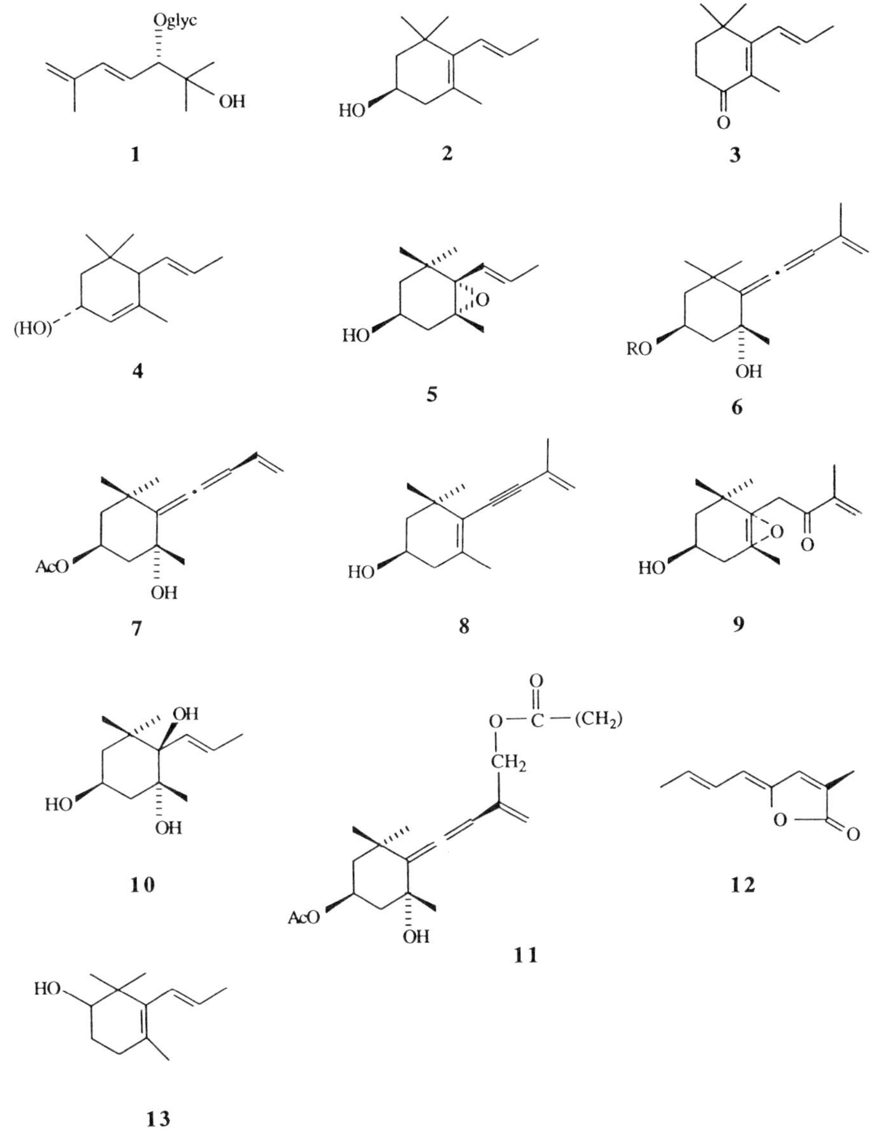

Figure 2.1 Structural features of the main carotenoid end groups found in the algae (see Liaaen-Jensen 1977; Bjørnland and Liaaen-Jensen 1989). 1. Monocylic glucosides; 2. Dicyclic xanthophylls; 3. 4-Keto; 4. ε-Rings; 5. 5,6-epoxides; 6. Allenes; 7. Acetates (e.g. 3-acetate); 8. Acetylenes; 9. 8-Keto; 10. 5,6-Diol; 11. 19-Acyloxy (*n*=2,4); 12. C_{37}-Skeletal butenolides; 13. 2-Hydroxy.

2.3.1 Chlorophyceae

The pigment composition of the Chlorophycean algae is very similar to that found in the photosynthetic tissues of higher plants, i.e. they contain ß-carotene (often with α-carotene), neoxanthin, lutein and the xanthophyll cycle carotenoids, violaxanthin, antheraxanthin and zeaxanthin (Table 2.3). Some species, most notably *Codium* spp., also contain siphonaxanthin (3,19,3´-trihydroxy-7,8-dihydro-ß,ε-caroten-8-one, **23**) (Strain 1951; Kleinig 1969; Walton et al. 1970) and its naturally esterified form, siphonein (**24**).

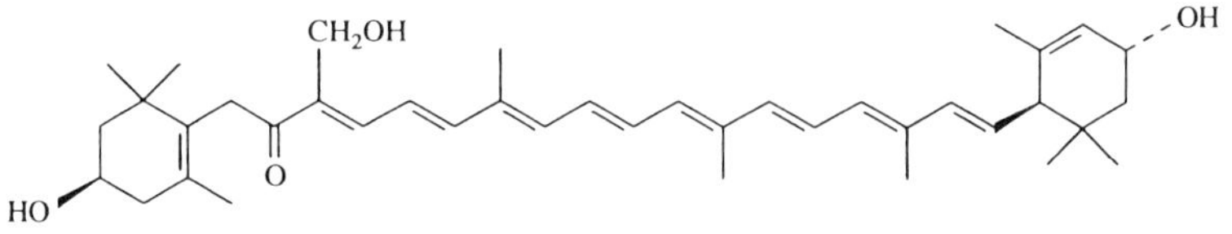

23

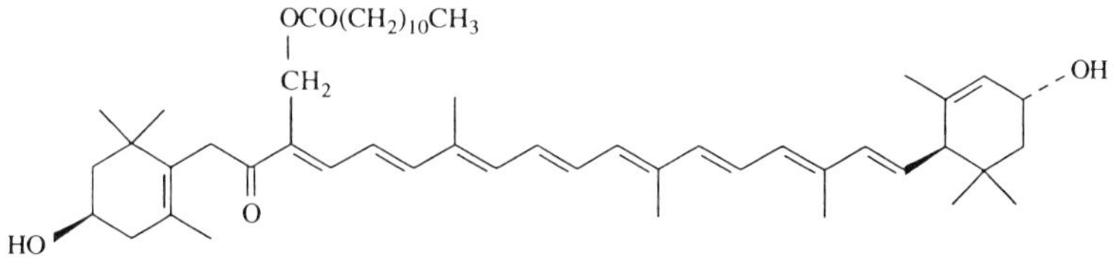

24

Siphonein actually consists of a mixture of molecular species in which siphonaxanthin is esterified with different fatty acids, the main one being lauric acid in *Caulerpa prolifera* (Kleinig and Egger 1967) but a dodecenoic acid in *Codium fragile* (and in the euglenophycean alga *Euglena viridis*). These carotenoids are found in the Derbesidales, Codiales, Caulerpales and Dichotomosiphonales (siphonaxanthin only) and in some species of the Cladophorales and Siphonocladales (Goodwin 1980). Yokohama (1981a,b) also detected siphonein and siphonaxanthin in two species of *Ulva*, and has suggested that their appearance is, in some way, linked to deep water or shade conditions. O´Kelly (1982) showed that members of the Cladophorales contained predominantly high levels of lutein in the intertidal region of the shore, but that this xanthophyll was apparently replaced by siphonaxanthin in species from deeper waters.

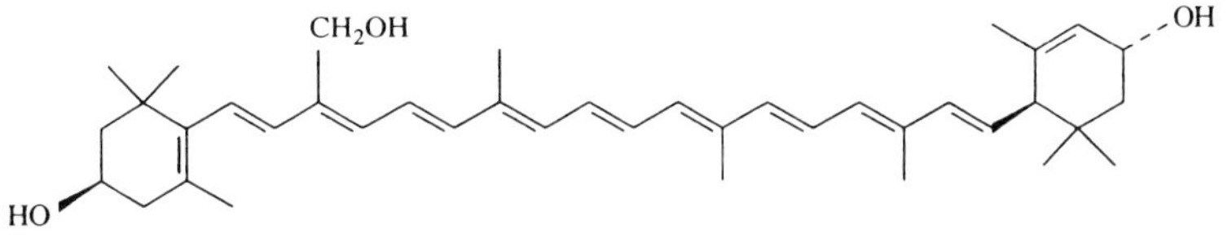

25

Table 2.3 Carotenoids of the Chlorophyceae

Major	*Others*
Antheraxanthin (**5**)	Astaxanthin (**36**)
ß-Carotene (**1**)	α-Carotene (**8**)
Lutein(**2**)	ß,ß-Carotene-2,2'-diol (**26**)
Neoxanthin (**4**)	ß,ε-Caroten-2-ol (**27**)
Siphonaxanthin (**23**)	ε-Carotene (**56**)
Siphonein (**24**)	ß-Cryptoxanthin (**9**)
Violaxanthin (**3**)	2,3-Didehydro-fritschiellaxanthin (**29**)
Zeaxanthin (**6**)	Fritschiellaxanthin (**28**)
	Loroxanthin (**25**)

Loroxanthin ((3*R*,3′*R*,6′*R*)-ß,ε-carotene-3,19,3′-triol, **25**) (= pyrenoxanthin, = trollein in *Chlorella pyrenoidosa,* Aitzetmüller et al. 1969; Francis et al. 1973; Nitsche 1974) has also been detected in many members of the Chlorophyceae, including *Scenedesmus* spp. Like siphonaxanthin, loroxanthin appears to occur in the highest amounts in deep-water algae, in which it possibly replaces lutein in the light-harvesting complexes (Yokohama 1982). Carotenoids with hydroxyl groups at C-2 and/or C-2′ (e.g. ß,ß-carotene-2,2′-diol, **26**, and ß,ε-caroten-2-ol, **27**) have been detected in the air alga *Trentepohlia iolithus* (Buchecker et al. 1974). These have *R*-chirality at C-2.

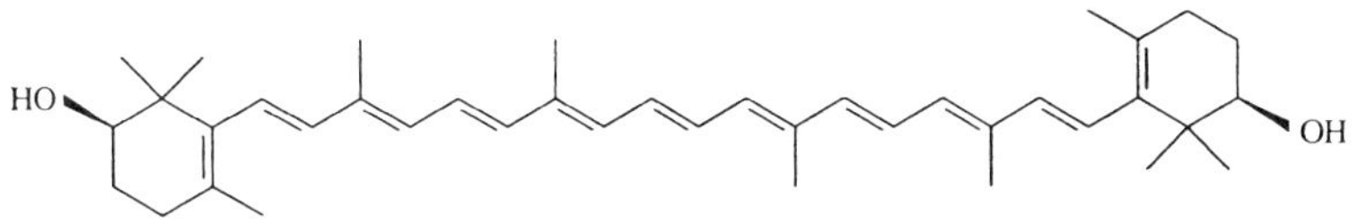

26

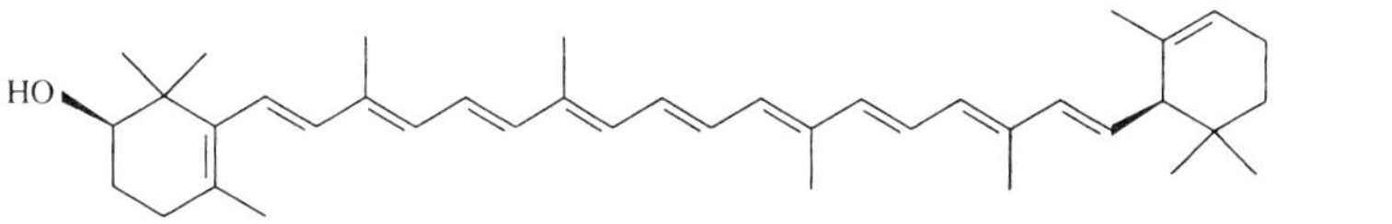

27

Many members of the Chlorophyceae possess the ability to accumulate secondary carotenoids in response to stress conditions. Most notable of these is *Dunaliella salina,* which can accumulate ß-carotene to levels approaching 14% of the cellular dry weight (Borowitzka 1988; Borowitzka and Borowitzka 1988).

First isolated in esterified form from *Fritschiella tuberosa,* fritschiellaxanthin ((3*S*,3′*R*,6′*R*)-3,3′-dihydroxy-ß,ε-caroten-4-one, **28**), a 4-keto derivative of lutein, and 2,3-didehydrofritschiellaxanthin (**29**) have been detected in the cell walls of a number of algae (Burczyk 1987). The latter of these compounds is possibly an artifact.

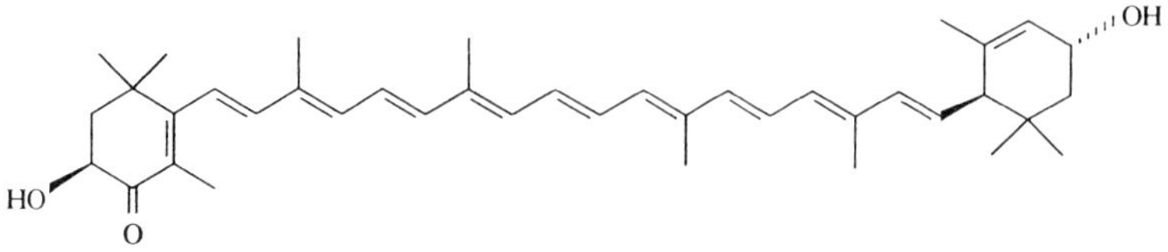

28

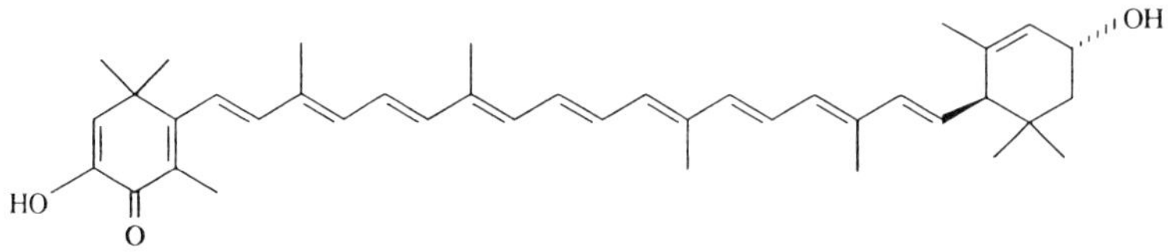

29

Many other algae can synthesise a complex mixture of ketocarotenoids, especially echinenone (ß,ß-caroten-4-one, **30**), monohydroxyechinenone (3´-hydroxy-ß,ß-caroten-4-one, **31**) and canthaxanthin (ß,ß-carotene-4,4´-dione, **32**) (Bernhard 1990). These ketocarotenoids are absent during normal cell growth and an extension of the existing biosynthetic pathway must be introduced to enable their synthesis. The biosynthesis of these secondary carotenoids by algae is discussed in more detail in Chapters 4 and 6.

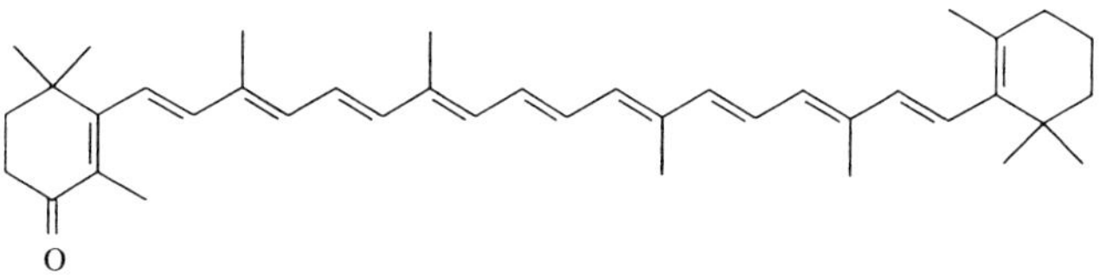

30

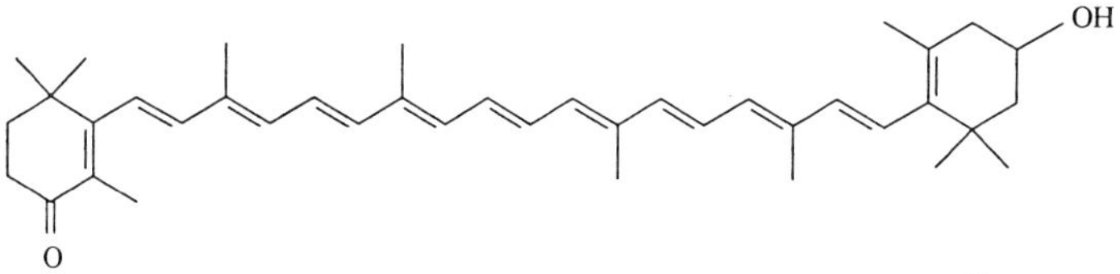

31

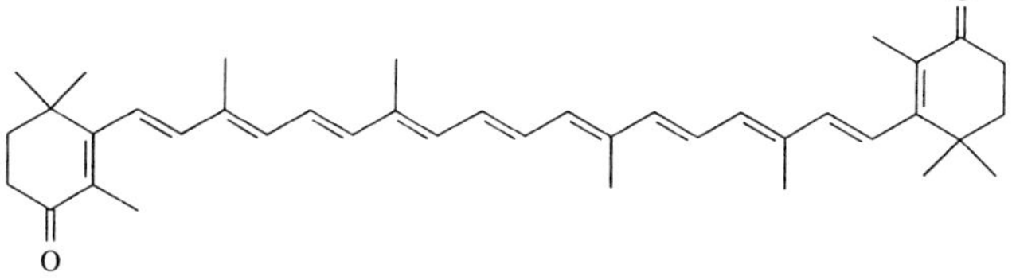

32

A number of pigment mutants of algae have been isolated. Several carotenoid-deficient mutants of *Chlamydomonas reinhardii* are allelic (Iroshnikova and Kvitko

1986). These *lts-1* mutants show anomalies in grana formation, decreased levels of chlorophyll *b*, and a decrease in the ratio of P700: chlorophyll *a* (Ladygin et al. 1982; Maslov and Chunaev 1982). The mutant *pg-101* cannot synthesise carotenoids with ε-rings and therefore lacks α-carotene, lutein and loroxanthin (Eichenberger et al. 1986). Antia and Cheng (1983) have described a strain of *Dunaliella tertiolecta* that lacks the xanthophyll cycle carotenoids and contains high levels of lutein-5,6-epoxide.

A large number of pigment mutants of *C. reinhardii* and *Scenedesmus obliquus* have been used for many years in the study of carotenoid biosynthesis and plastid development (Powls and Britton 1977; Humbeck 1990; see also Chapter 6). Many of these show deficiencies in the normal biosynthetic pathway (see Chapter 4), resulting in the accumulation of biosynthetic intermediates, for example phytoene and ζ-carotene, which are normally absent from photosynthetic tissues.

2.3.2 Prasinophyceae

The pigment composition of these algae is very similar in many respects to that of the Chlorophyceae (Table 2.4, cf. Table 2.3). Ricketts (1970) and Goodwin (1980) described three groups in the Prasinophyceae with the following pigment compositions: (i) the normal higher plant carotenoids plus siphonein and siphonaxanthin (e.g. *Asteromonas propulsa*), (ii) as (i) except that zeaxanthin replaces lutein (e.g. *Pyramimonas olivaceae*), and (iii) algae containing prasinoxanthin (3,6,3´-trihydroxy-7,8-dihydro-γ,ε-caroten-8-one **33**, Foss et al. 1984). Prasinoxanthin (= xanthophyll K) possesses a novel substituted γ-end-group with chirality (3*R*, 3´*R*, 6´*R*) tentatively proposed on biogenetic grounds (Foss et al. 1984) and confirmed by nuclear magnetic resonance (NMR) studies (Foss et al. 1986a, 1987a).

Table 2.4 Carotenoids of the Prasinophyceae

Major	*Others*
Antheraxanthin (**5**)	Astaxanthin (**36**)
ß-Carotene (**1**)	α-Carotene (**8**)
Neoxanthin (**4**)	γ-Carotene (**110**)
Lutein (**2**)	Canthaxanthin (**32**)
Prasinoxanthin (**33**)	Dihydro-prasinoxanthin epoxide (**35**)
Siphonaxanthin (**23**)	Echinenone (**30**)
Siphonein (**24**)	Lutein-5,6-epoxide (**7**)
Violaxanthin (**3**)	Lycopene (**22**)
Zeaxanthin (**6**)	Uriolide (**34**)

Foss et al. (1986a) have identified two further unusual carotenoids in the Prasinophyceae. Uriolide (5,6-epoxy-3,3´-dihydroxy-5,6,7´,8´-tetrahydro-ß,ε-caroten-11´,19´-olide, **34**) is the first reported C_{40} carotenoid to possess an in-chain butenolide group, and it is also the first report of this particular structural feature outside the dinoflagellates. The other newly identified carotenoid is structurally related to prasinoxanthin; it possesses a C-7´,8´ single bond and a 4´,5´-epoxy group (Uebelhart et al. 1986) and has been identified as 7´,8´-dihydroprasinoxanthin-4´,5´-epoxide (**35**). A further revision to the classification of the Prasinophyceae based on their carotenoid compositions may still be necessary (Rowan 1989).

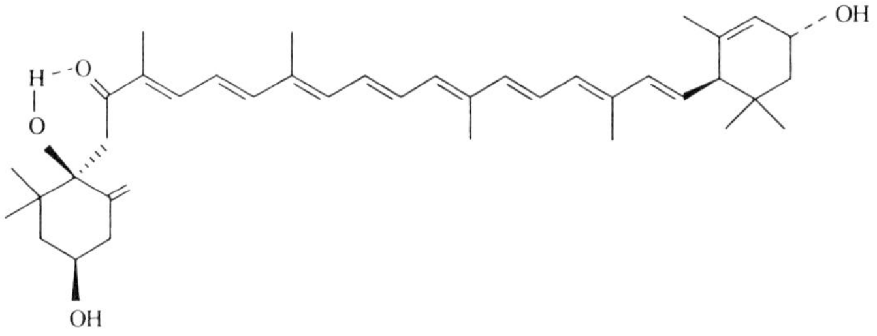

33

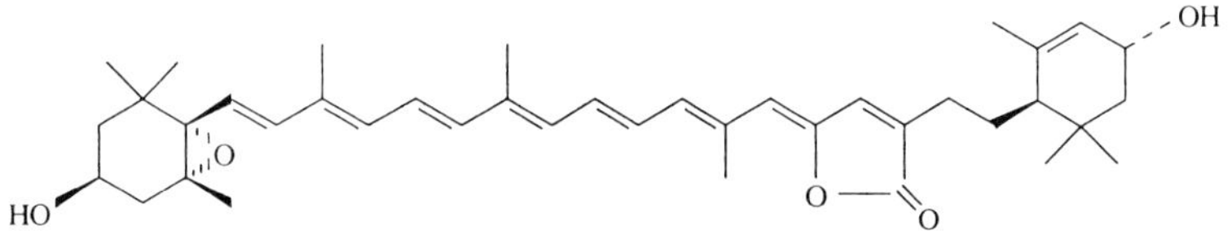

34

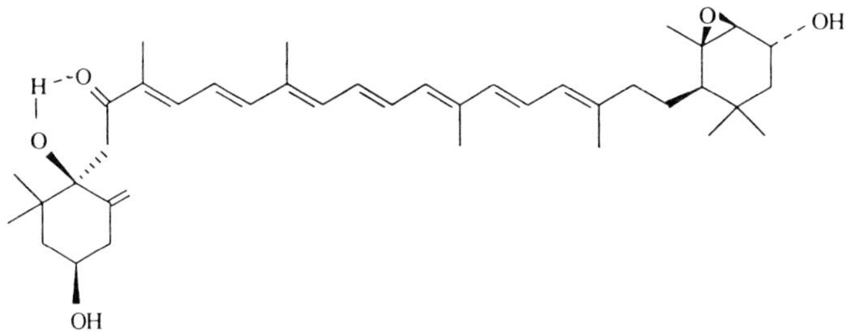

35

Like the Chlorophyceae, many members of the Prasinophyceae have the ability to synthesise very high amounts of ketocarotenoids when grown under stress conditions (usually nutrient deprivation). The most notable example is astaxanthin ((3*S*,3´*S*)-3,3´-dihydroxy-ß,ß-carotene-4,4´-dione, **36**) synthesis in *Haematococcus pluvialis*, which has now been developed into a commercial process with the aim of supplying the aquaculture industry with algal astaxanthin as a pigmenter for salmon and trout flesh (see Chapter 6). In addition to large quantities of astaxanthin (generally a mixture of mono- and di-esters), which are accumulated in this alga when grown under stress conditions, smaller amounts of adonirubin ((3*S*)-3-hydroxy-ß,ß-carotene-4,4´-dione, **37**), usually in an esterified form, are also synthesised.

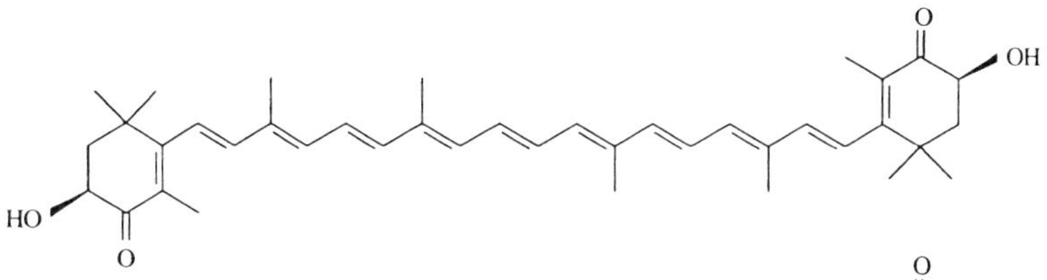

36

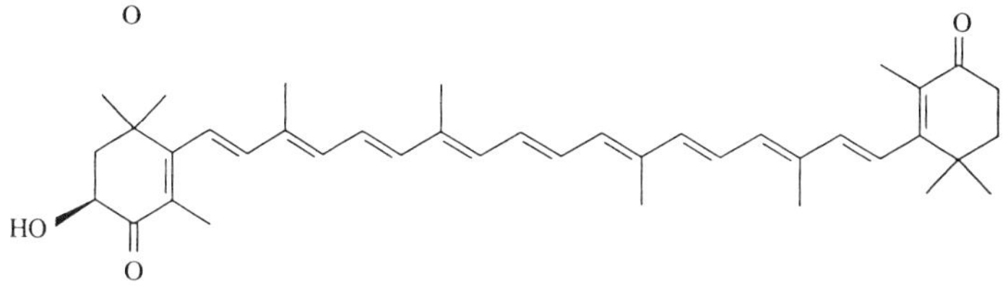

37

2.3.3 Euglenophyceae

Although the Euglenophyceae are included in the Chlorophyta, the pigment composition of these algae more closely resembles that of the Chromophyte algae (Rowan 1989; Table 2.5, cf. Tables 2.3 and 2.7). The Euglenophyceae characteristically contain high levels of diadinoxanthin (5,6-epoxy-7´,8´-didehydro-5,6-dihydro-ß,ß-carotene-3,3´-diol, **38**), with smaller amounts of diatoxanthin (7,8-didehydro-ß,ß-carotene-3,3´-diol, **39**), heteroxanthin (7´,8´-didehydro-5,6-dihydro-ß,ß-carotene-3,5,6,3´-tetrol, **40**; Nitsche 1973) and neoxanthin (Buchecker and Liaaen-Jensen 1977; Bjørnland 1982).

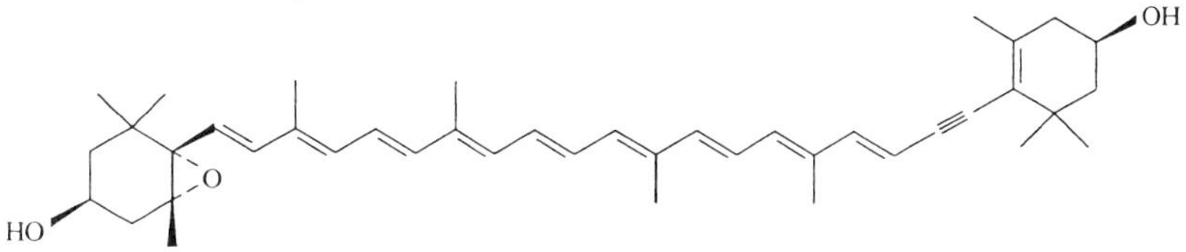

38

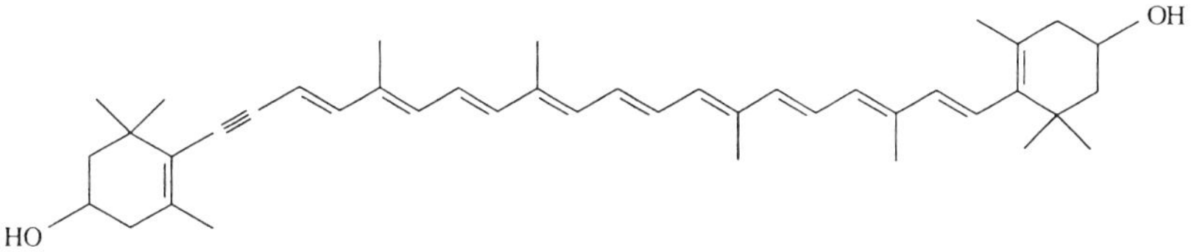

39

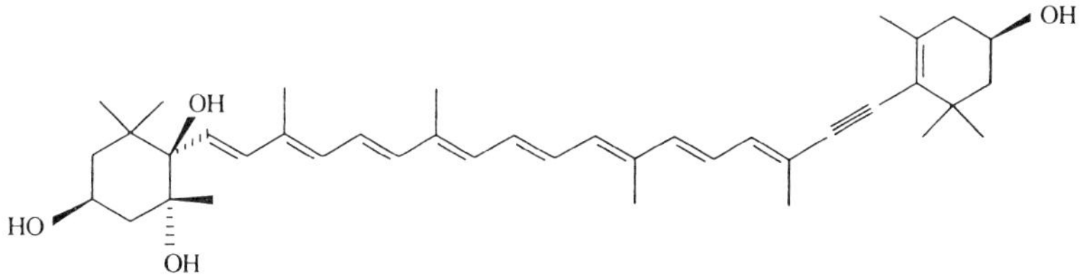

40

The marine alga *Eutreptiella gymnastica* contains siphonein (previously only detected in the Chlorophyceae and Prasinophyceae, Bjørnland 1990a) and two unusual xanthophylls, anhydrodiatoxanthin (3´,4´,7´,8´-tetradehydro-ß,ß-caroten-3-ol, **41**) and eutreptiellanone (3,6-epoxy-3´,4´,7´,8´-tetradehydro-5,6-dihydro-ß,ß-caroten-4-one, **42,** Fiksdahl et al. 1984a). The latter of these, which may be found in quite high levels, is unusual among the algal pigments as it contains a 3,6-oxo-bicycloheptane ring. Two other oxabicycloheptane carotenoids, ß-cryptoeutreptiellanone ((3*S*,5*R*,6*S*,3´*R*)-3´-hydroxy-3,6-epoxy-7´,8´-didehydro-5,6-dihydro-ß,ß-caroten-4-one, **43**) and α-cryptoeutreptiellanone ((3*S*,5*R*,6*S*,3´*R*,6´*R*)-3´-hydroxy-3,6-epoxy-5,6-dihydro-ß,ε-caroten-4-one, **44**) have been isolated in small amounts from this alga. Oxabicycloheptane carotenoids have not been found in other Euglenophycean algae (e.g. *Euglena viridis* and *E. gracilis*).

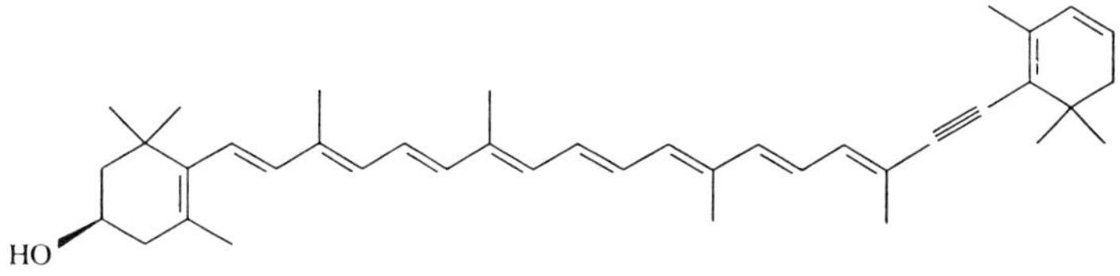

41

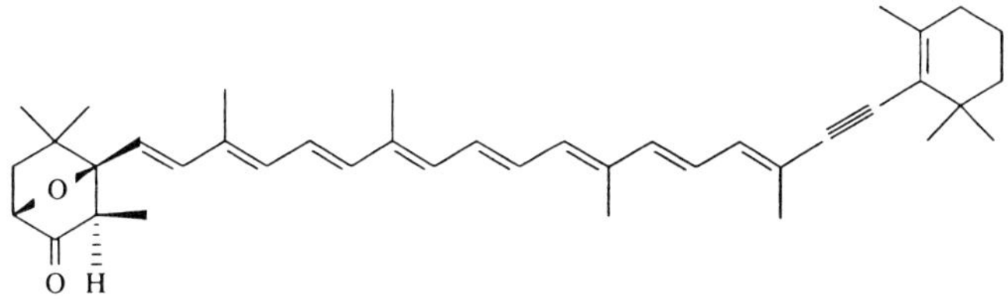

42

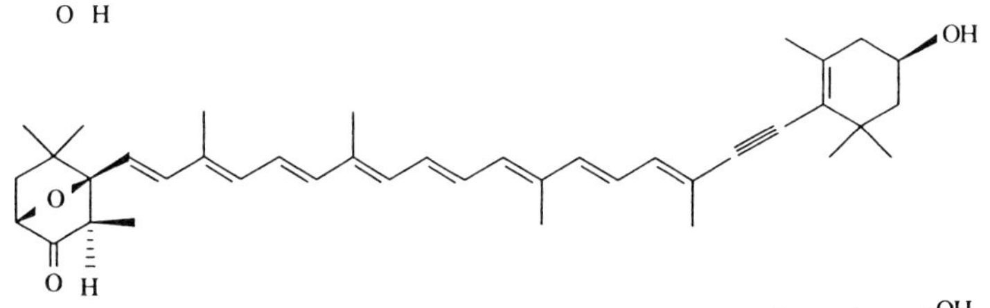

43

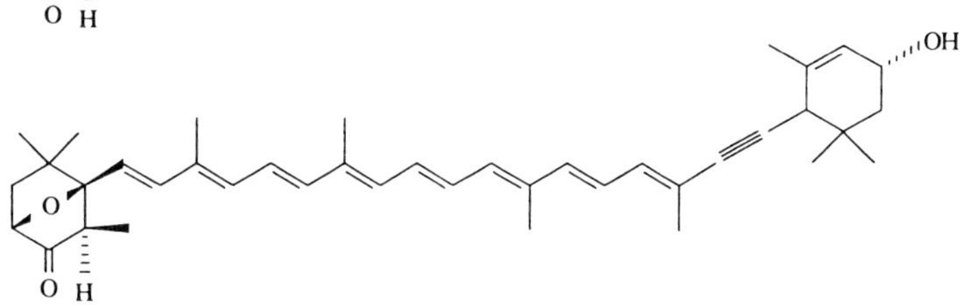

44

Fiksdahl and Liaaen-Jensen (1988) have characterised the acetylenic carotenoids in *E. viridis* and reported the presence of 3,4,7,8,3´,4´,7´,8´-octadehydro-ß,ß-carotene (**45**) and an esterified form of 7,8,3´,4´,7´,8´-hexahydro-ß,ß-caroten-3-ol (**46**). The Euglenophyceae also contain diesters of (3*R*, 3´*R*)-alloxanthin (**47**) with myristic and palmitic acids as the fatty acid moieties. Prior to this observation (Bjørnland 1990a), the bisacetylenic carotenoid alloxanthin (7,8,7´,8´-tetradehydro-ß,ß-carotene-3,3´-diol, **48**) had been considered to be characteristic of the Cryptophycean algae. Preliminary reports of the presence of pectenolone ((3*S*,3´*R*)-

3,3´-dihydroxy-7´,8´-didehydro-ß,ß-caroten-4-one, **49**) in *E. gracilis* have to be confirmed (Bjørnland et al. 1987; Liaaen-Jensen 1990).

Table 2.5 Carotenoids of the Euglenophyceae

Major	*Others*
Diadinoxanthin (**38**)	3',4'-Anhydrodiatoxanthin (**41**)
Diatoxanthin (**39**)	Alloxanthin diester (**47**)
Heteroxanthin (**40**)	Astaxanthin (**36**)
Neoxanthin (**4**)	7,8-Didehydroastaxanthin (**50**)
	Canthaxanthin (**32**)
	α-Cryptoeutreptiellanone (**44**)
	ß-Cryptoeutreptiellanone (**43**)
	Echinenone (**30**)
	Hexadehydro-ß-caroten-3-ol (**46**)
	Octadehydro-ß,ß-carotene (**45**)
	Pectenolone (?) (**49**)
	Pectenolone diester (**51**)
	Siphonein (**24**)

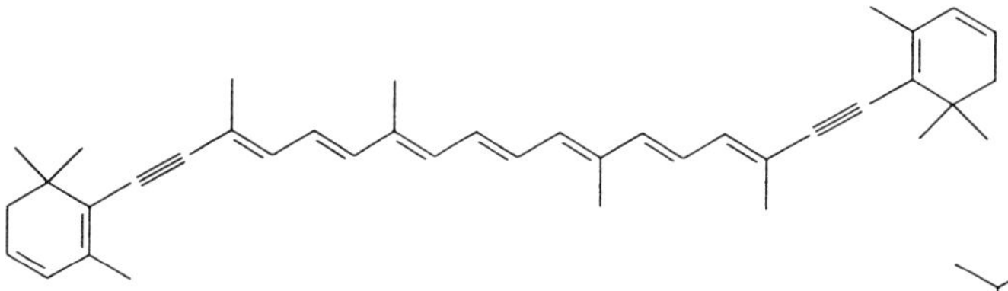

45

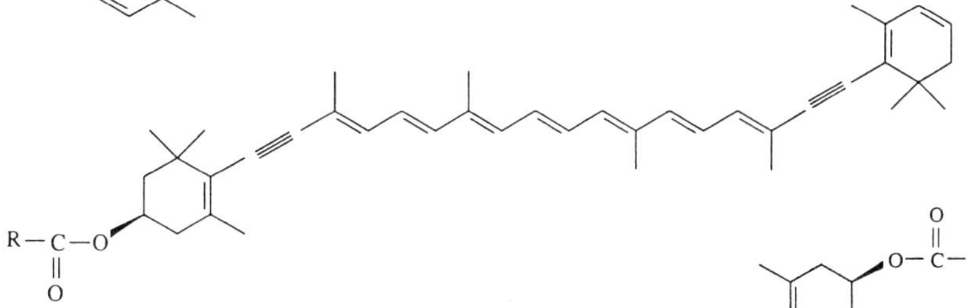

46

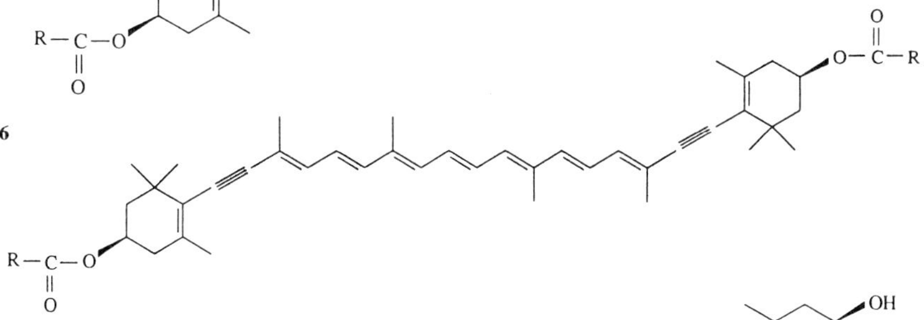

47

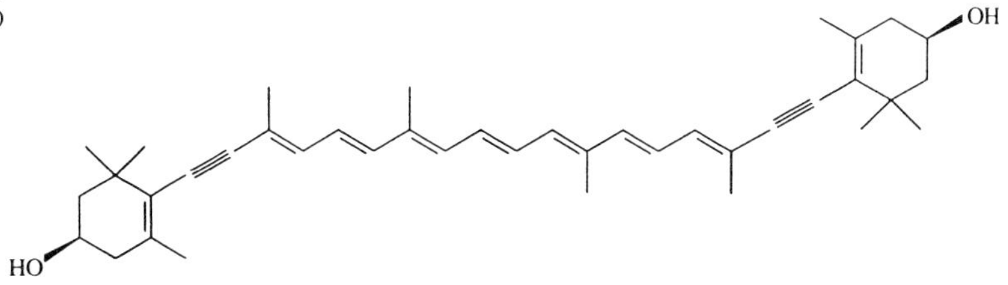

48

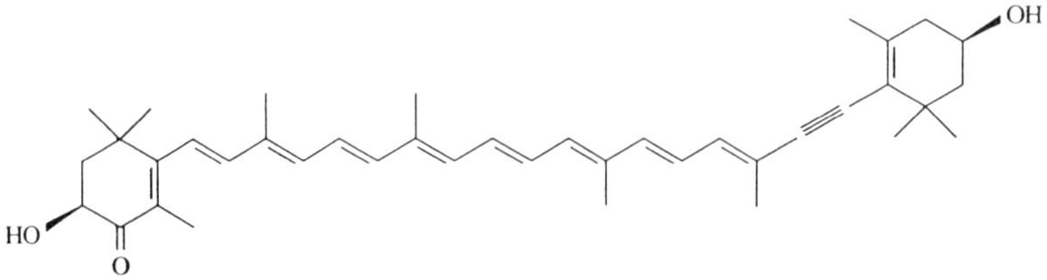

49

Siphonein, isolated recently from two marine species, has been shown to have *n*-2-*trans*-2-dodecenoic acid as its main esterifying acid (Fiksdahl et al. 1984a; Liaaen-Jensen 1985). The Euglenophyceae also contain trace amounts of ketocarotenoids, which are present in the eyespot of the cells. In a recent report,

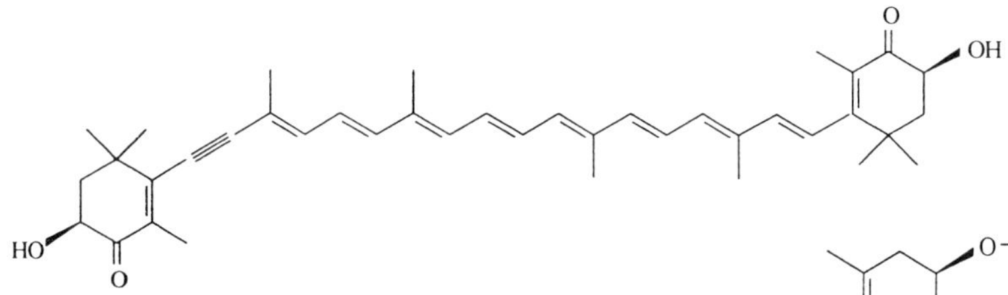

50

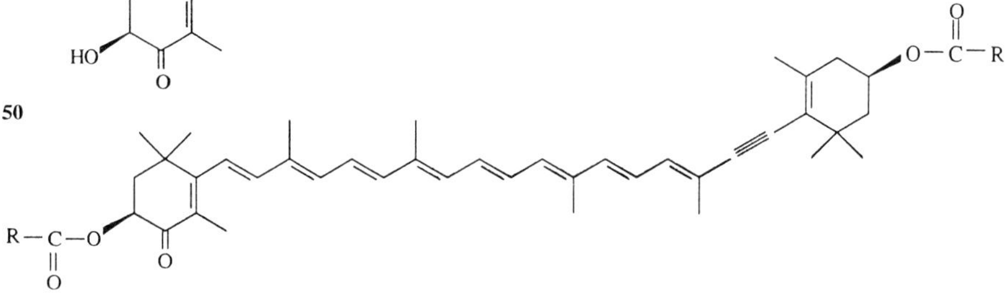

51

Grung et al. (1990) showed that *E. sanguinea* may accumulate secondary carotenoids to 0.7% of the algal dry weight, with astaxanthin diesters making up 91% of total carotenoid. Canthaxanthin and echinenone have been found in cultures of *E. gracilis* (Bjørnland et al. 1987) together with three acetylenic hydroxylated ketocarotenoids. These include mono- and dioxo-derivatives of diatoxanthin, namely pectenolone and 7,8-didehydroastaxanthin ((3*S*,3´*S*)-3,3´-dihydroxy-7,8-didehydro-ß,ß-carotene-4,4´-dione, **50**). Trace amounts of pectenolone diesters (**51**) have also been found in this alga (Bjørnland et al. 1987).

2.3.4 Rhodophyceae

The red algae possess both ß-carotene and α-carotene, together with a range of their simple monohydroxy and dihydroxy derivatives, particularly lutein and zeaxanthin. The overall carotenoid composition of the Rhodophyceae is therefore quite simple (see Table 2.6), but considerable interspecific quantitative differences may be found. Up to nine xanthophylls other than lutein and zeaxanthin have been detected in these algae but it is now apparent that a number of carotenoids that have previously been attributed to this class (e.g. fucoxanthin, 5,6-epoxy-3,3´,5´-trihydroxy-6´,7´-didehydro-5,6,7,8,5´,6´-hexahydro-ß,ß-caroten-8-one 3´-acetate, **52**) may reflect the contamination of the algal thallus by diatoms (Liaaen-Jensen 1978). Great care must therefore be taken when studying the carotenoids of the these and other algae to ensure that all microalgal contamination is removed prior to analysis.

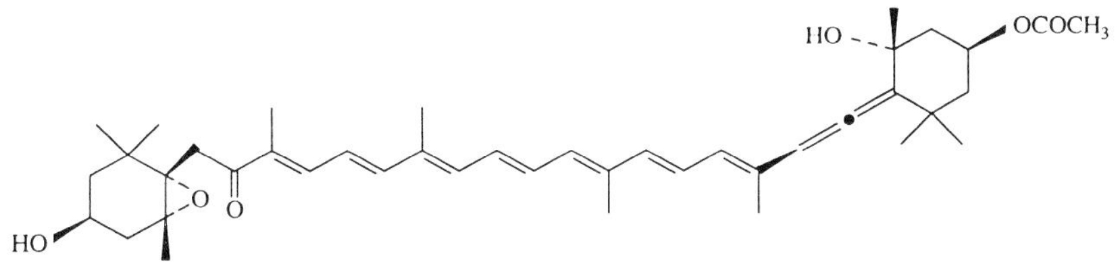

52

Table 2.6 Carotenoids of the Rhodophyceae

Major	*Others*
α-Carotene (**8**)	α-Cryptoxanthin (**53**)
ß-Carotene (**1**)	ß-Cryptoxanthin (**9**)
Lutein (**2**)	Fucoxanthin* (**52**)
Zeaxanthin (**6**)	

* The detection of fucoxanthin is probably the result of contamination by micro-algae (diatoms)

The absolute configurations of (6´*R*)-α-carotene, (3´*R*, 6´*R*)-α-cryptoxanthin (ß,ε-caroten-3´-ol, **53;** see also **53a**), (3*R*, 3´*R*, 6´*R*)-lutein, (3*R*, 3´*R*)-zeaxanthin and all-*trans* (3*S*, 5*R*, 6*S*, 3´*R*)-antheraxanthin from cultured algae, i.e. devoid of possible contamination from other algal species, have been established by ^{1}H-NMR and circular dichroism (CD) (Bjørnland et al. 1986).

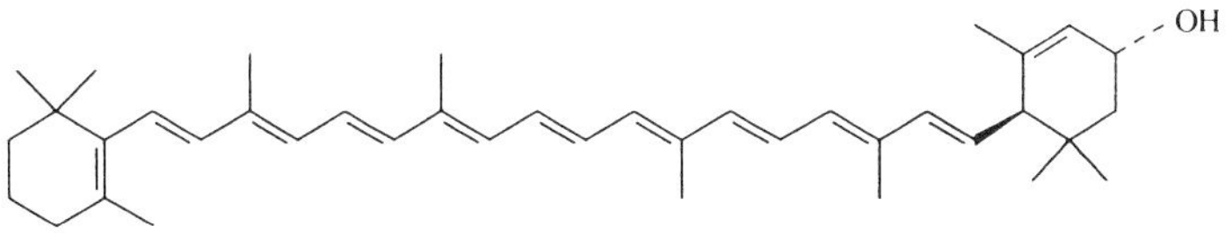

53

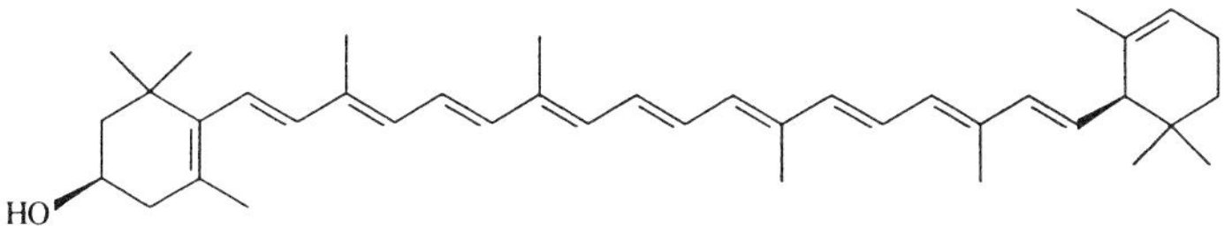

53a

Carotenoid epoxides have rarely been reported for the red algae (Goodwin 1980) although some Rhodophycean algae do appear to be able to synthesise antheraxanthin and lutein-5,6-epoxide (= taraxanthin) (Bjørnland et al. 1984a; Czeczuga and Taylor 1987). Other species appear to lack α-carotene and its xanthophyll derivatives such as lutein (Guillard et al. 1985).

2.3.5 Cryptophyceae

Of all the algal groups, the Cryptophyceae have received the greatest attention from the point of view of their carotenoids. The characteristic pigments of these algae are the ε-ring type and acetylenic carotenoids (Table 2.7), including the bisacetylenic alloxanthin and the monoacetylenic compounds monadoxanthin ((3*R*,3´*R*,6´*R*)-7,8-didehydro-ß,ε-carotene-3,3´-diol, **54**) and crocoxanthin ((3*R*,6´*R*)-7,8-didehydro-ß,ε-caroten-3-ol, **55**) (Liaaen-Jensen 1985; Pennington et al. 1985).

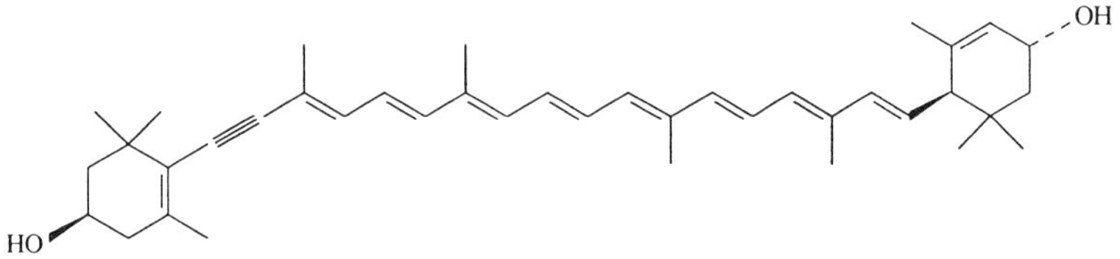

54

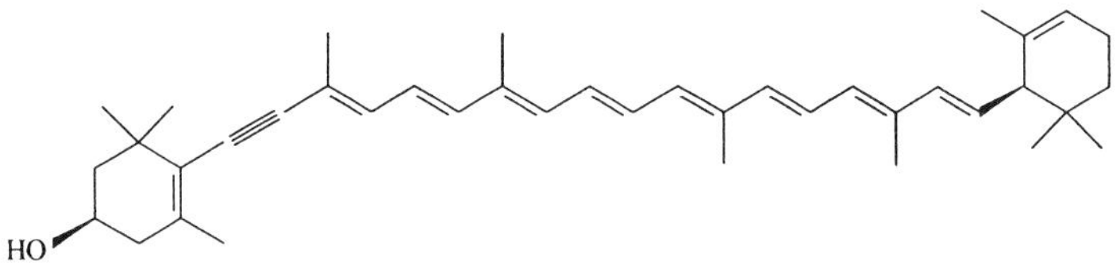

55

ß-Carotene, ε-carotene (ε,ε-carotene, surprisingly the 6*S*,6´*S* isomer, **56**) and (6´*R*)-α-carotene are generally found, although they usually represent only a small percentage of the total carotenoid. Of those carotenoids commonly seen, the bisacetylenic carotenoid alloxanthin (which may account for up to 90% of total carotenoid) is thought to be of major taxonomic significance as it is a major component of many of these algae and is restricted to this particular class. Some species (e.g. *Cryptomonas ovata*) may synthesise zeaxanthin and lycopene (Chapman and Haxo 1963; Pennington et al. 1985). Epoxy-xanthophylls have not been detected in this class.

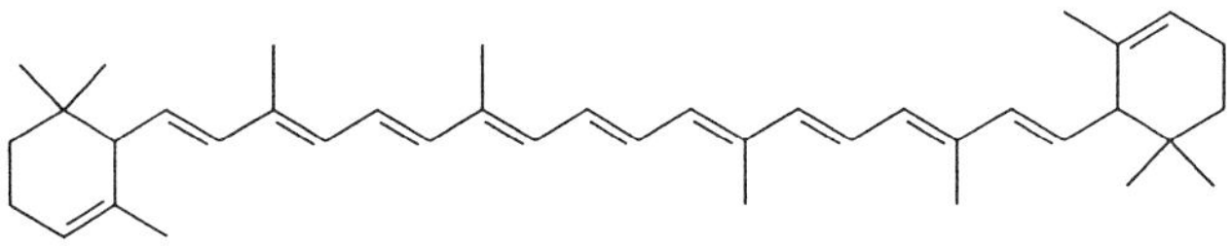

56

Table 2.7 Carotenoids of the Cryptophyceae

Alloxanthin (**48**)
α-Carotene (**8**)
ß-Carotene (**1**)
ε-Carotene (**56**)
Lycopene (**22**)

In contrast to earlier reports by Swift and Millborrow (1981), more recent studies (e.g. Liaaen-Jensen 1985) have failed to detect any carotenoids that could be allenic precursors to these acetylenic carotenoids. Similarly, early reports (e.g. Haxo and Fork 1959; Allen et al. 1964) that ß-carotene was absent from many members of this class have now been largely discounted.

2.3.6 Dinophyceae

Whilst the majority of dinoflagellates are colourless (Bjørnland 1990a), those that are able to photosynthesise produce carotenoids that are structurally more complex than those of the other algal classes (see Table 2.8; Liaaen-Jensen 1985). The unusual C_{37} nor-carotenoid peridinin ((3*S*,5*R*,6*R*,3´*S*,5´*R*,6´*S*)-5´,6´-epoxy-3,5,3´-trihydroxy-6,7-didehydro-5,6,5´,6´-tetrahydro-10,11,20-trinor-ß,ß-caroten-19´,11´-olide 3-acetate, **57**) (Strain et al. 1976; Kjøsen et al. 1976) is characteristic of species belonging to this class, and may represent up to 85% of total carotenoid in certain species. Diadinoxanthin is frequently found in high amounts, but the distribution of diatoxanthin is much more variable, reflecting the epoxidation state of these carotenoids in response to growth conditions.

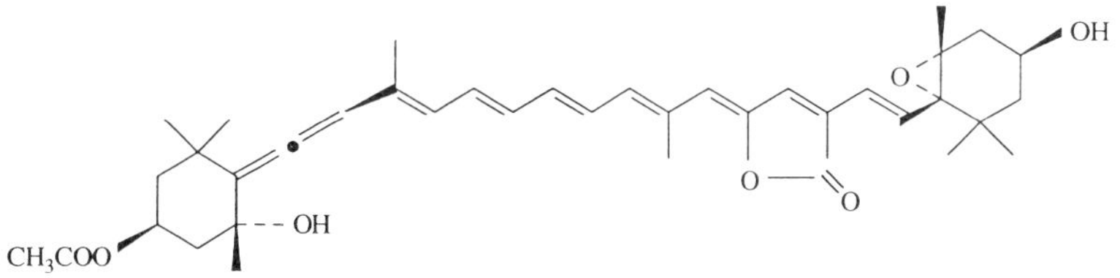

57

These carotenoids are usually accompanied by ß-carotene, pyrrhoxanthin (5´,6´-epoxy-3,3´-dihydroxy-7,8-didehydro-5´,6´-dihydro-10,11,20-trinor-ß,ß-caroten-19´,11´-olide 3-acetate, **58**), dinoxanthin (5´,6´-epoxy-6,7-didehydro-5,6,5´,6´-tetrahydro-ß,ß-carotene-3,5,3´-triol 3-acetate, **59**), pyrrhoxanthinol (5´,6´-epoxy-3,3´-dihydroxy-7,8-didehydro-5´,6´-dihydro-10,11,20-trinor-ß,ß-caroten-19´,11´-olide, **60**) and peridininol (5´,6´-epoxy-3,5,3´-trihydroxy-6,7-didehydro-5,6,5´,6´-tetrahydro-10,11,20-trinor-ß,ß-caroten-19´,11´-olide, **61**). The acetylenic compounds pyrrhoxanthin and pyrrhoxanthinol are also C_{37} carotenoids that are related to peridinin and peridininol, respectively. Johansen et al. (1977) have established the chiralities for peridinin and dinoxanthin.

Table 2.8 Carotenoids of the Dinophyceae

Major	*Others*
Group I	
Diadinoxanthin (**38**)	Astaxanthin (**36**)
Diatoxanthin (**39**)	ß-Carotene (**1**)
Peridinin (**57**)	γ-Carotene (**110**)
	Canthaxanthin (**32**)
	Diadinochrome
	Dinoxanthin (**59**)
	Echinenone (**30**)
	P457 (**62**)
	Peridininol (**61**)
	Pyrrhoxanthin (**58**)
	Pyrrhoxanthinol (**60**)
Major	*Others*
Group II	
Diadinoxanthin (**38**)	19'-Butanoyloxyfucoxanthin (**64**)
Diatoxanthin (**39**)	19'-Butanoyloxyhalocynthiaxanthin 3'-acetate (**69**)
Fucoxanthin (**52**)	α-Carotene (**8**)
19'-Hexanoyloxyfucoxanthin (**63**)	ß-Carotene (**1**)
	Diadinochrome*
	Halocynthiaxanthin 3'-acetate (**67**)
	19'-Hexanoyloxyhalocynthiaxanthin 3'-acetate (**68**)
	19'-Hexanoyloxyparacentrone 3-acetate (**65**)
	Gyroxanthin diester (**66**)

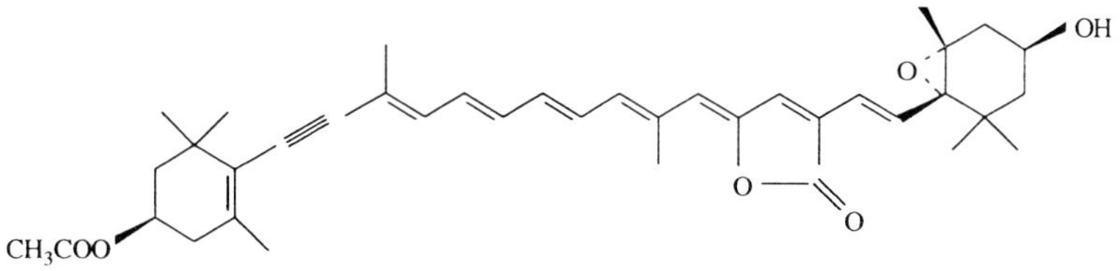

58

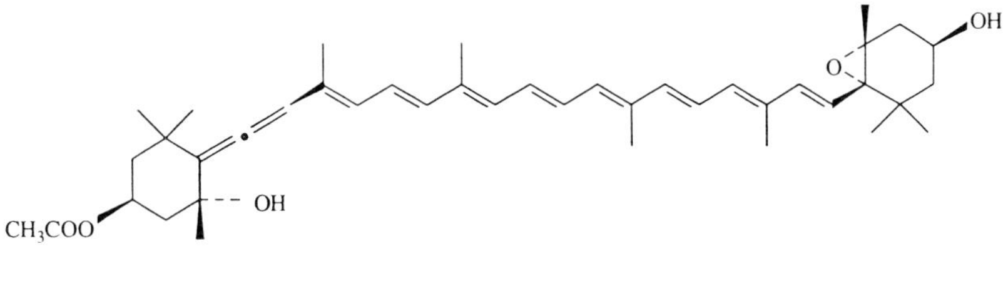

59

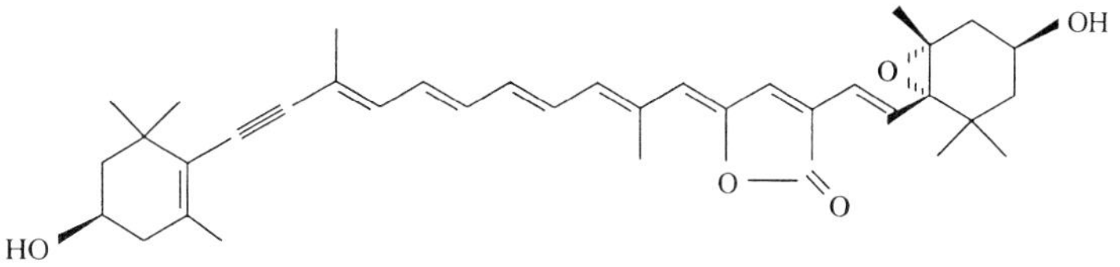

60

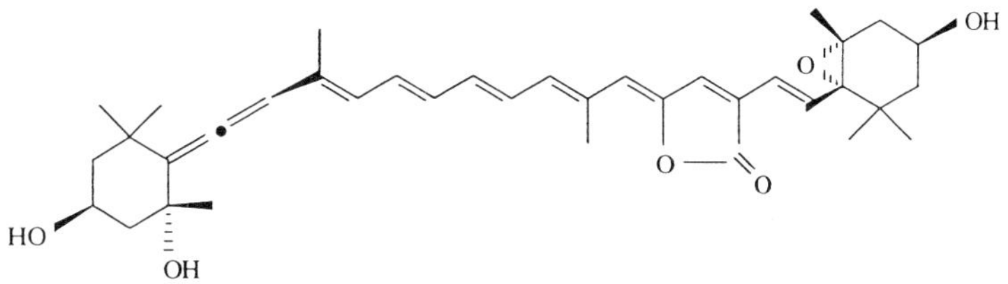

61

A unialgal culture of *Thoracosphera heimii* was found to contain ß-carotene, diatoxanthin, dinoxanthin, pyrrhoxanthin, peridinin, peridininol and the carotenoid glycoside P-457 (Bjørnland 1990b). Various peridinin-synthesising dinoflagellates, including *T. heimii* and *Amphidinium carterae*, have been shown to synthesise P457, which has only recently been shown to be a complex derivative of the C_{40} carotenoid neoxanthin, namely 13´-*cis*-7´,8´-didehydroneoxanthin-20´-al 3´-ß-lactoside (**62**, Aackermann et al. 1990a,b; Liaaen-Jensen 1991). This is the only report of a carotenoid glycoside encountered in algae (other than the cyanobacteria).

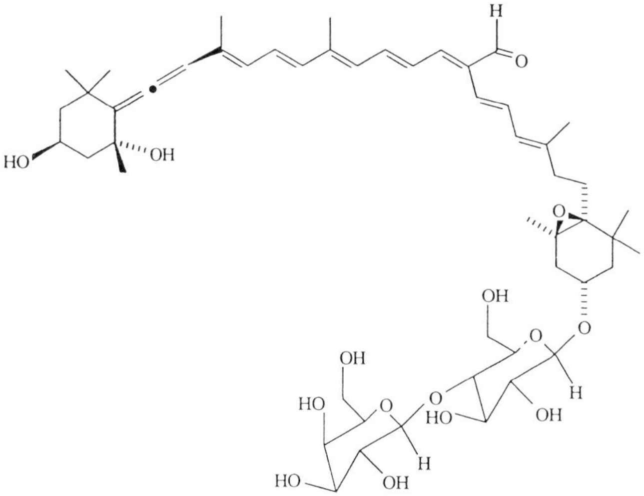

62

Bjørnland (1990b) also reported for the first time the existence of small amounts of the ketocarotenoids echinenone and canthaxanthin in unialgal cultures of a dinoflagellate, although Johansen et al. (1974) had earlier reported the presence of astaxanthin in a *Glenodium* species.

A limited number of species (*Glenodium foliaceum* - Mandelli 1968; *Gymnodinium micrum*, Whittle and Casselton 1969; *Gymnodinium veneficum*, Riley and Wilson 1967; *Peridinium balticum*, Withers et al. 1977) accumulate fucoxanthin, and not peridinin, as their main xanthophyll. In a few other species (e.g. *Gymnodinium galatheanum*, Bjørnland et al. 1987) fucoxanthin may be accompanied by its 19´-hexanoyloxy (**63**) or 19´-butanoyloxy derivatives (**64**). 19´-Hexanoyloxyparacentrone 3-acetate (3,5,19-trihydroxy-6,7-didehydro-5,6,7´,8´-tetrahydro-7´-apo-ß-caroten-8´-one 3-acetate 19-hexanoate, **65**) may also be found in these algae. The occurrence of these 19´-acyloxyfucoxanthins in *Gymnodinium* spp., *Gyrodinium aureolum* (Tangen and Bjørnland 1981) and in *Ptychodiscus brevis* (Bjørnland et al. 1984b) is unusual and may possibly be the result of an endosymbiotic realtionship with a chrysophyte or prymnesiophyte (Bjørnland 1990a). The distribution of the 19´-acyloxyfucoxanthins in the Dinophyceae, Chrysophyceae and Prymnesiophyceae is shown in Table 2.9.

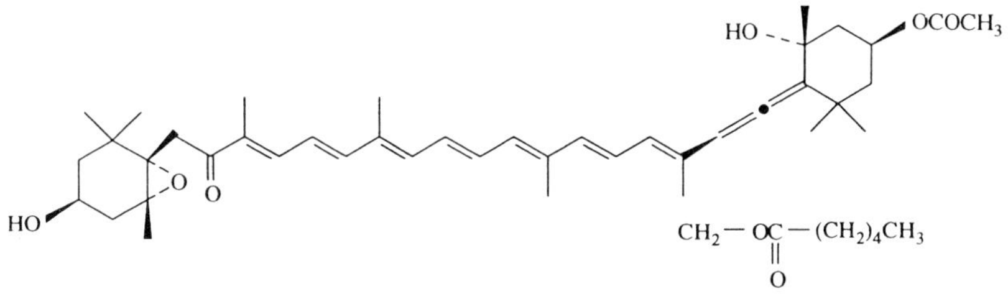

63

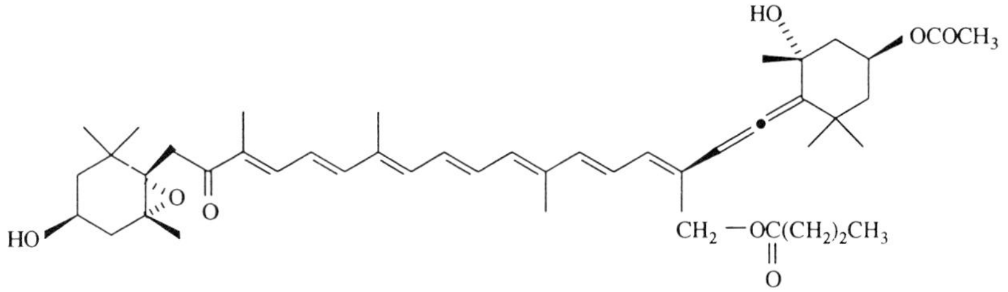

64

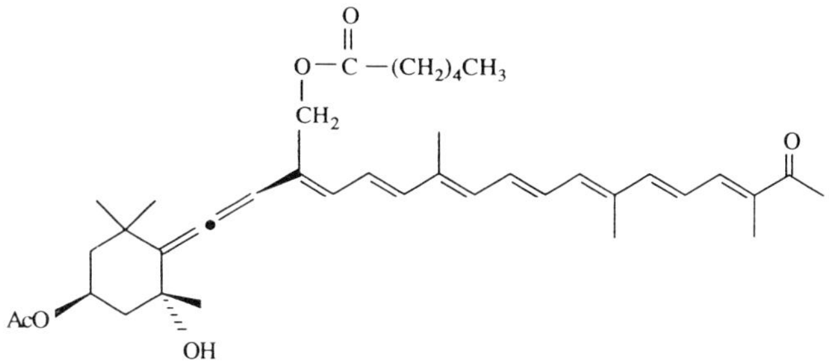

65

Table 2.9 Distribution of 19'-acyloxyfucoxanthins in algae

Chrysophyceae	**1**	**2**	**3**	**Ref.**
Pelagococcus subviridis	x	x		Bjørnland et al. 1984b,1987
P. subviridis	x	x		Vesk and Jeffrey 1987
P. subviridis	x	x	x	Wright and Jeffrey 1987
Dinophyceae				
Gymnodinium galatheanum	x	x	x	Bjørnland et al. 1987
Gyrodinium aureoleum			x	Tangen and Bjørnland 1981
Ptychodiscus brevis	x	x	x	Bjørnland et al. 1984b
Prymnesiophyceae				
Chrysochromulina sp.	x		x	Bjørnland et al. 1987
Corymbellus aureus			x	Gieskes and Kraay 1986
Emiliana huxleyi			x	Arpin et al. 1976
E. huxleyi	x	x	x	Wright and Jeffrey 1987
Phaeocystis pouchetii	x	x	x	Wright and Jeffrey 1987
Phaeocystis sp.	x	x	x	Bjornland et al. 1987

1. Fucoxanthin 2. 19'-Butanoyloxyfucoxanthin 3. 19'-Hexanoyloxyfucoxanthin. Data compiled from Bjørnland (1990a) and Bjørnland and Liaaen-Jensen (1989).

The unusual gyroxanthin diester (5´,6´-epoxy-6,7,7´,8´-tetradehydro-5,6,5´,6´-tetrahydro-ß,ß-carotene-3,5,3´,19´-tetrol 3-acetate 19´-acylate, **66**), which possesses both acetylenic and allenic end-groups, has also been reported (Bjørnland and Liaaen-Jensen 1989) and appears to be exclusive to the dinoflagellates. The allenic end-group of gyroxanthin is comparable to that seen in fucoxanthin, whereas the other end-group (7,8-didehydro and 19-acyloxy) is more unusual. Mass spectrometry of the free and esterified gyroxanthin showed the presence of acetic and dodecanoic acids as the esterifying acids (Bjørnland 1990a). The gyroxanthin diester may account for up to 10% of the total carotenoid found in the dinoflagellates that possess the 19´-acyloxyfucoxanthins. It has been reported to isomerise readily to a 9-*cis* configuration on manipulation and during prolonged storage.

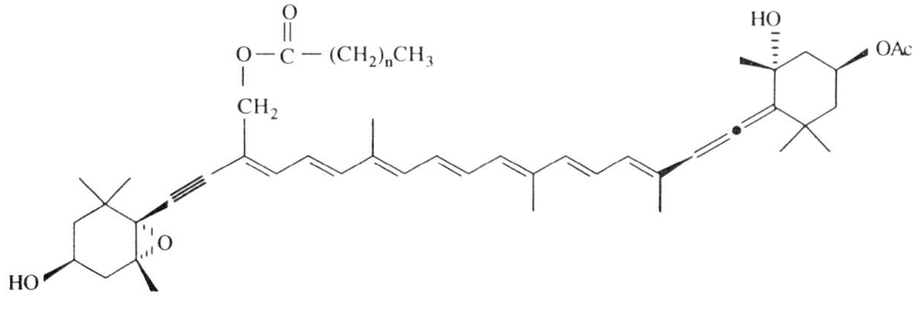

66

Small amounts of halocynthiaxanthin 3´-acetate (5,6-epoxy-3,3´-dihydroxy-

7´,8´-didehydro-5,6,7,8-tetrahydro-ß,ß-caroten-8-one 3´-acetate, **67**), and its 19´-hexanoyloxy (**68**) and 19´-butanoyloxy derivatives (**69**) have been detected in *Gymnodinium galatheanum* (Bjørnland et al. 1987).

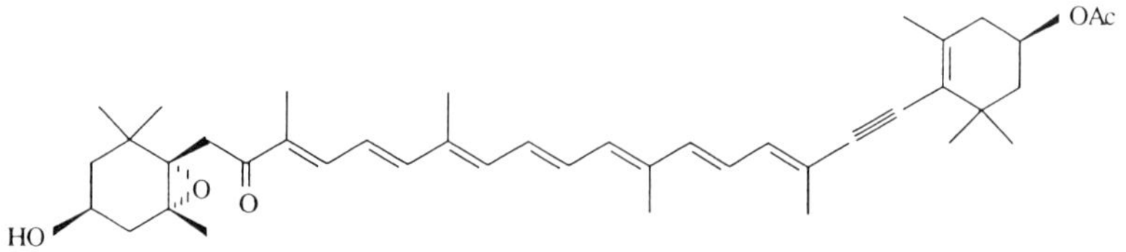

67

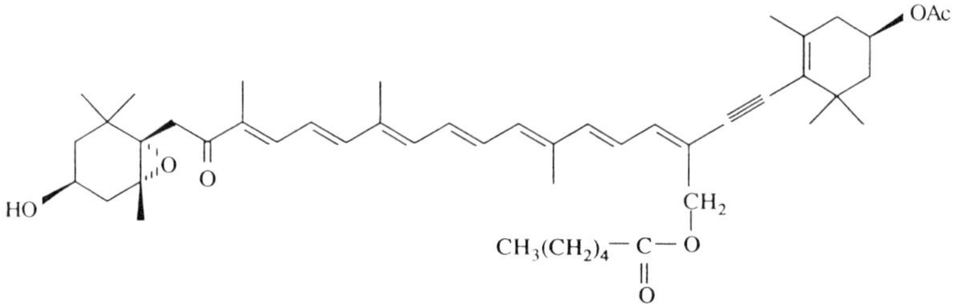

68

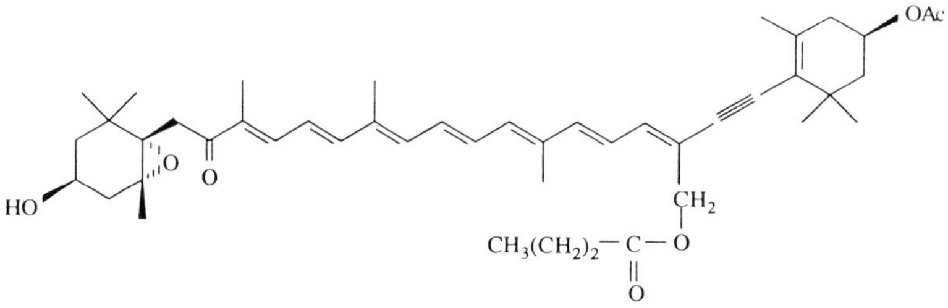

69

2.3.7 Chrysophyceae

The carotenoids that have been detected in species of algae belonging to the Chrysophyceae are listed in Table 2.10. Withers et al. (1981) have examined six species of Chrysophyceae and confirmed the presence of fucoxanthin and the xanthophyll cycle carotenoids violaxanthin, antheraxanthin and zeaxanthin. The acetylenic carotenoids diatoxanthin and diadinoxanthin were more restricted in their distribution. Bjørnland et al. (1984b, 1989) and Vesk and Jeffrey (1987) have established that 19´-butanoyloxyfucoxanthin is present in some species. Indeed the first reports of the occurrence of this carotenoid in marine algae were made for the chrysophycean alga *Pelagococcus subviridis*. Trace amounts of 19´-hexanoyloxyfucoxanthin have also been detected in a limited number of species (Wright and Jeffrey 1987).

Table 2.10 Carotenoids of the Chrysophyceae

Characteristic	*Others*
ß-Carotene (**1**)	Antheraxanthin (**5**)
Fucoxanthin (**52**)	α-Carotene (**9**)
	ε-Carotene (**56**)
	Cryptoxanthin (**9**)
	Cryptoxanthin epoxide (**74**)
	Diadinoxanthin (**38**)
	Diatoxanthin (**39**)
	Neoxanthin (**4**)
	Violaxanthin (**3**)
	Zeaxanthin (**6**)

2.3.8 Raphidophyceae

Bjørnland and Liaaen-Jensen (1989) have proposed that the Raphidophyceae be divided into two groups on the basis of their carotenoid compositions as determined by Fiskdahl et al. (1984a,b). Thus, Group I (*Gonyostomum semen* and *Vacuolaria virescens*), the pigment composition of which resembles that of the Xanthophyceae, contains diadinoxanthin (the major carotenoid), dinoxanthin and neoxanthin. Smaller amounts of the acetylenic tetrol heteroxanthin, vaucheriaxanthin (5´,6´-epoxy-6,7-didehydro-5,6,5´,6´-tetrahydro-ß,ß-carotene-3,5,19,3´-tetrol, **70**) and ß-carotene have been detected (Table 2.11). On the basis of their carotenoid compositions, it has been suggested that these algae belonging to Group I should be moved to the Xanthophyceae (Christensen 1980).

Table 2.11 Carotenoids of the Raphidophyceae

Major	*Others*
Group I	
ß-Carotene (**1**)	Dinoxanthin (**59**)
Diadinoxanthin (**38**)	Heteroxanthin (**40**)
	Neoxanthin (**4**)
	Vaucheriaxanthin-3,19-diacetate (**72**)
Group II	
ß-Carotene (**1**)	Fucoxanthinol (**71**)
Fucoxanthin (**52**)	Violaxanthin (**3**)
	Zeaxanthin (**6**)

In contrast, those algae belonging to Group II, and which now fully represent the Raphidophyceae, synthesise fucoxanthin (the main pigment), fucoxanthinol

((3*S*,5*R*,6*S*,3′*S*,5′*R*,6′*R*)-5,6-epoxy-3,3′,5′-trihydroxy-6′,7′-didehydro-5,6,7,8,5′,6′-hexahydro-ß,ß-caroten-8-one, **71**) and ß-carotene (Table 2.11). The pigment composition of these algae (e.g. *Chattonella japonica*) more closely resembles that of the Chrysophyceae (cf. Table 2.10; see also Section 2.3.7).

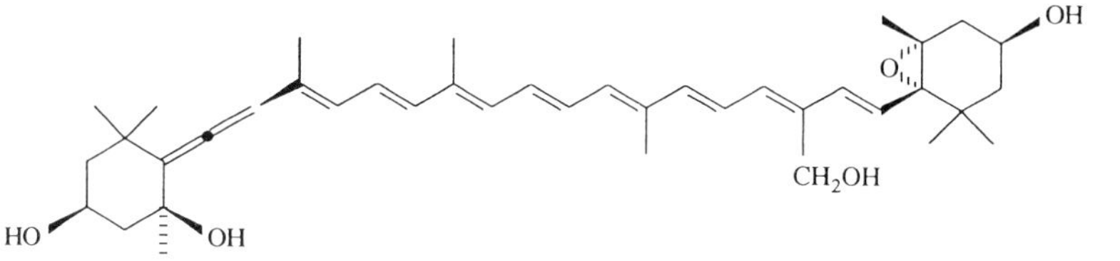

70

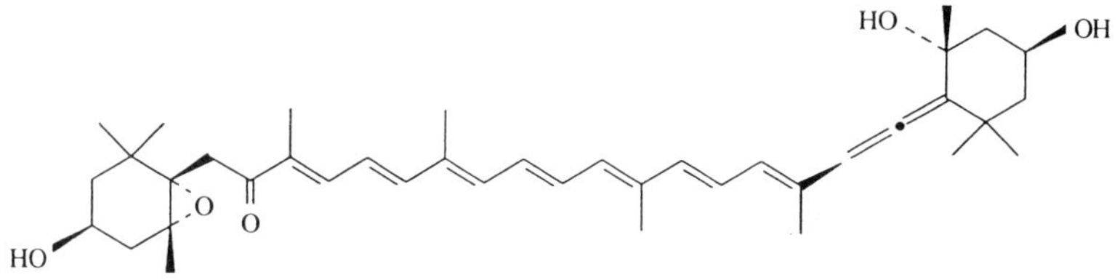

71

2.3.9 Xanthophyceae

Goodwin (1980), Liaaen-Jensen (1978) and Bjørnland and Liaaen-Jensen (1989) have described the carotenoid composition of the Xanthophyceae. Typically these algae synthesise carotenoids with ß-rings, epoxide, acetylenic and allenic groups, acetates, 5,6-glycols and esterified in-chain (19′)-hydroxy groups. The major carotenoids are diadinoxanthin and diatoxanthin, together with heteroxanthin (with 3*S*,5*S*,6*S*,3′*R* chirality, Buchecker et al. 1984), vaucheriaxanthin (as a diacetate **72** - Liaaen-Jensen 1978, or as a 3′-monoacetate **73** as in *Vaucheria sessilis* - (Bjørnland and Liaaen-Jensen 1989)), see Table 2.12.

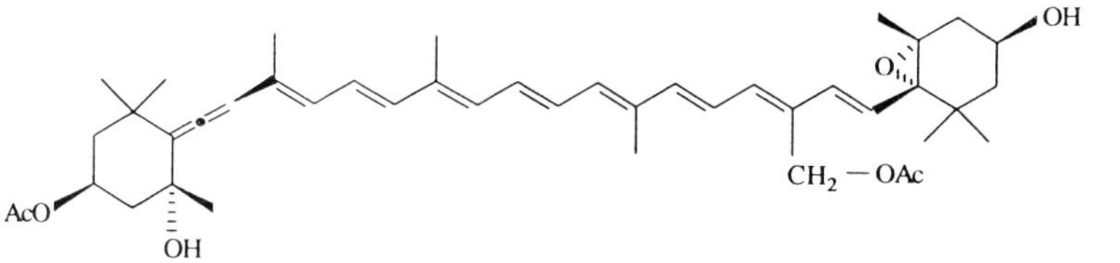

72

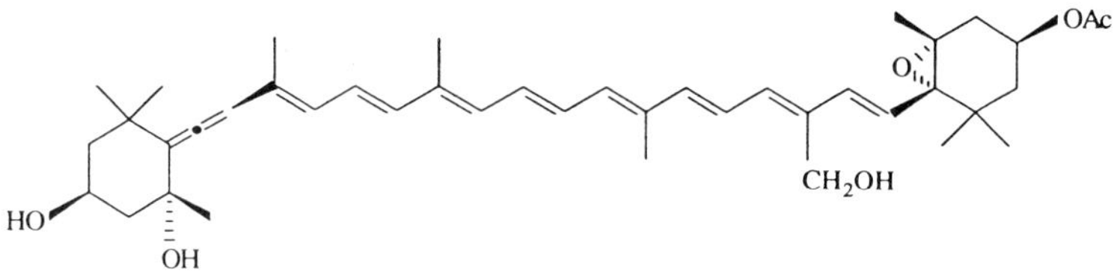

73

Table 2.12 Carotenoids of the Xanthophyceae

Major	*Others*
ß-Carotene (**1**)	ß-Cryptoxanthin-5',6'-epoxide (**74**)
Diadinoxanthin (**38**)	ß-Cryptoxanthin diepoxide (**75**)
Diatoxanthin (**39**)	Neoxanthin (**4**)
Heteroxanthin (**40**)	
Vaucheriaxanthin diester (**72**)	

2.3.10 Eustigmatophyceae

The class Eustigmatophyceae has been formed from species transferred from other algal classes on the basis of their ultrastructural features and carotenoid compositions. These algae appear to lack the ability to synthesise acetylenic carotenoids (Bjørnland and Liaaen-Jensen 1989). The characteristic carotenoid is vaucheriaxanthin, both in the free and esterified forms (Stransky and Hager 1970); with violaxanthin this accounts for the bulk of carotenoid in these algae (Table 2.13; Owens et al. 1987). The presence of ß-carotene, zeaxanthin, antheraxanthin (the levels of the individual xanthophyll cycle carotenoids are probably highly dependent on the culture conditions used) and neoxanthin has been reported for a number of species (Stransky and Hager 1970; Antia et al. 1975), but the presence of small amounts of ß-cryptoxanthin and ß-cryptoxanthin epoxide (5,6-epoxy-5,6-dihydro-ß,ß-caroten-3-ol, **74/74a**) has to be confirmed (Whittle and Casselton 1975). Ketocarotenoids, including astaxanthin and canthaxanthin, are often present as minor components (Norgard et al. 1974; Antia and Cheng 1982). As in many other algae, the synthesis of these ketocarotenoids is probably highly dependent on culture conditions.

Table 2.13 Carotenoids of the Eustigmatophyceae

Major	*Others*
Antheraxanthin (**5**)	Canthaxanthin (**32**)
ß-Carotene (**1**)	ß-Cryptoxanthin (**9**)
Vaucheriaxanthin 3'-acetate (**73**)	ß-Cryptoxanthin-5,6-epoxide (**74**)
Violaxanthin (**3**)	
Zeaxanthin (**6**)	

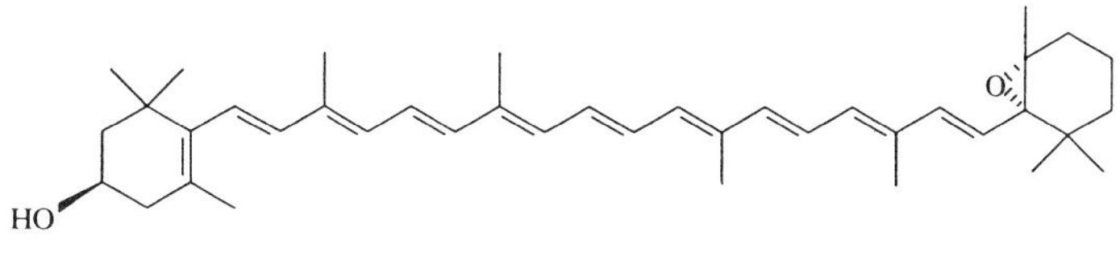

74

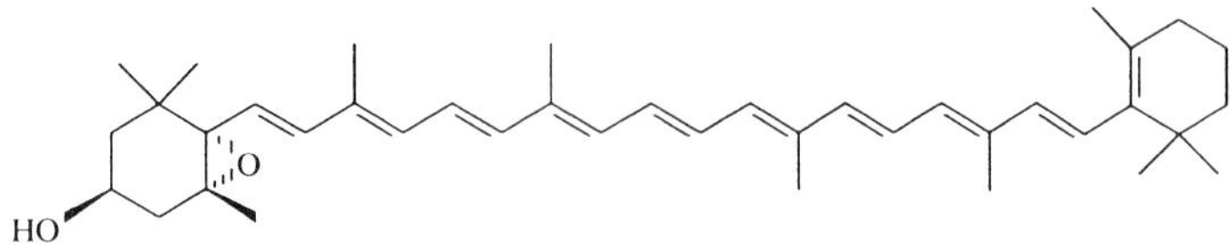

74a

2.3.11 Prymnesiophyceae

Recent studies by Bjørnland et al. (1988), Bjerkeng et al. (1990) and Bjørnland (1990b) have investigated the pigment composition of members of this class of algae. Some species of the Prymnesiophyceae are responsible for the dense phytoplankton blooms which can have severe effects on fish and other marine life. The algae are very similar to the Dinophyceae, but can be distinguished by morphological and ultrastructural features and also by their pigment composition (Bjørnland et al. 1988; Bjørnland 1990a). Members of the Prymnesiophyceae generally contain fucoxanthin and structurally related xanthophylls as their main carotenoids, whereas the Dinophyceae possess peridinin. This distinction has been used to determine whether the alga *Thoracosphera heimii* belongs to either of these two taxonomic groups (see Section 2.3.6). In addition, ß-carotene, diatoxanthin and diadinoxanthin, together with smaller amounts of α-carotene, ß-carotene-5,6-epoxide, ß-cryptoxanthin, ß-cryptoxanthin-5,6,5´,6´-diepoxide (5,6,5´,6´-diepoxy-5,6,5´,6´-tetrahydro-ß,ß-caroten-3-ol, **75**), echinenone, canthaxanthin, fucoxanthinol and deepoxyneoxanthin (6,7-didehydro-5,6-dihydro-ß,ß-carotene-3,5,3´-triol, **76**), are generally found in the Prymnesiophyceae (see Table 2.14; Bjørnland and Liaaen-Jensen 1989).

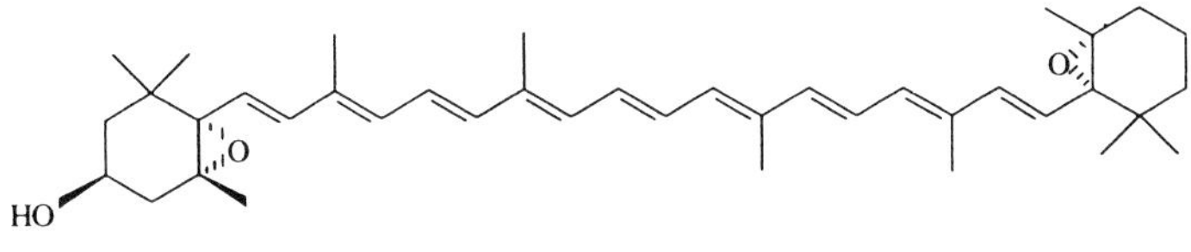

75

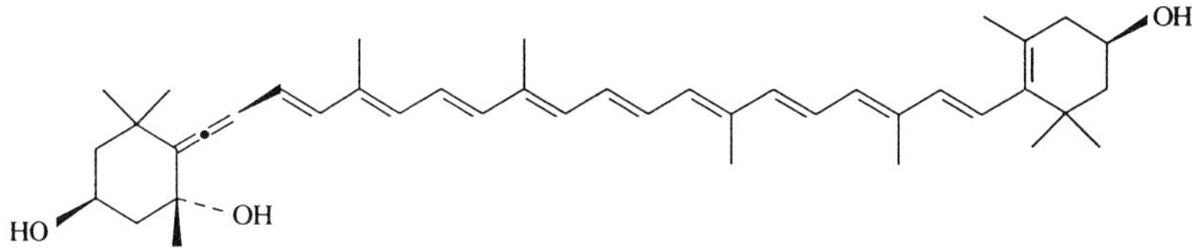

76

Whilst all members contain fucoxanthin, a number also synthesise the structurally related 19´-hexanoyloxy and 19´-butanoyloxy derivatives (see Table 2.9). Bjørnland

and Liaaen-Jensen (1989) have therefore suggested that the Prymnesiophyceae may be subdivided on the basis of their carotenoid composition; i.e. into those containing fucoxanthin alone and those containing fucoxanthin togther with the 19′-acyloxyfucoxanthins. It is important to recognise that such a subdivision does not hold true on the basis of existing algal systematics.

Table 2.14 Carotenoids of the Prymnesiophyceae

Major	*Others*
ß-Carotene (**1**)	Canthaxanthin (**32**)
Diadinoxanthin (**38**)	α-Carotene (**8**)
Diatoxanthin (**39**)	γ-Carotene
Fucoxanthin (**52**)	ß-Carotene epoxide (**11**)
19'-Hexanoyloxyfucoxanthin (**63**)	ß-Cryptoxanthin epoxide (**74**)
	ß-Cryptoxanthin diepoxide (**75**)
	Deepoxyneoxanthin (**76**)
	Dinoxanthin (**59**)
	Echinenone (**30**)
	Fucoxanthinol (**71**)
	19'-Butanoyloxyfucoxanthin (**64**)
	19'-Hexanoyloxyfucoxanthinol
	19-Hexanoyloxyparacentrone 3-acetate (**65**)

Bjerkeng et al. (1990) determined that the following carotenoids were present in a unialgal culture of *Chrysochromulina polylepis*; (3*S*,5*R*,6*S*,3′*S*,5′*R*,6′*S*)-19′-hexanoyloxyfucoxanthin (81% of total carotenoid), fucoxanthin (10%), diadinoxanthin (9%), 19-hexanoyloxyparacentrone 3-acetate (1%). This latter unusual apocarotenoid has been isolated from *Phaeocystis* sp. clone 677-3 as two stereoisomers, tentatively identified as the all-*trans* and 9-*cis* isomers (Bjørnland et al. 1988). Traces of ß-carotene and diatoxanthin were also tentatively identified in these cultures. The authors concluded that the high level of 19′-hexanoyloxyfucoxanthin was characteristic of this alga.

2.3.12 Bacillariophyceae

An extensive analytical study by TLC and high-performance TLC has been performed on members of this class (Stauber and Jeffrey 1988). All species examined contained chlorophylls *a* and c_2 and ß-carotene, fucoxanthin (the main carotenoid), diatoxanthin and diadinoxanthin. One species, *Thalassiothrix heteromorpha* has been found to contain small amounts of a carotenoid provisionally identified as 19′-butanoyloxyfucoxanthin. The close similarity between the diatoms and the Prymnesiophyceae (Table 2.15, cf. Table 2.14) was further illustrated by this study and by that of Pennington et al. (1988) and it was not possible to identify any features of the complex pigment composition that could be used as distinctive taxonomic markers. However, these algae lack any carotenoids containing an ε-ring suggesting the absence of an ε-cyclase (Pennington et al. 1988).

Stauber and Jeffrey (1988) classified three types of pigmentation in the diatoms they examined: (i) species with chlorophylls c_1 and c_2 and fucoxanthin, (ii) species with chlorophylls c_2 and c_3 and fucoxanthin, (iii) species with chlorophylls c_2 and c_3 and derivatives of fucoxanthin.

Table 2.15. Carotenoids of the Bacillariophyceae

Major	*Others*
ß-Carotene (**1**)	ε-Carotene (**56**)
Diadinoxanthin (**38**)	Canthaxanthin (**32**)
Diatoxanthin (**39**)	Echinenone (**30**)
Fucoxanthin (**52**)	Neoxanthin (**4**)

Pennington et al. (1988) examined ten species of diatoms and found a number of monoacetylenic carotenoids (diatoxanthin and diadinoxanthin, comprising 12%–43% of total carotenoid), allenic carotenoids (fucoxanthin and fucoxanthinol, 49%–81%), monoepoxides (diadinoxanthin, fucoxanthin and fucoxanthinol, 79%–91%) and ß-carotene (3%–8%). One species (*Navicula pelliculosa*) also contained traces of canthaxanthin and echinenone.

2.3.13 Phaeophyceae

The main pigment found in these algae is fucoxanthin. ß-Carotene and violaxanthin are also usually present in high amounts, together with up to seven minor carotenoids, including fucoxanthinol, neoxanthin, diatoxanthin, diadinoxanthin and zeaxanthin (Table 2.16; Liaaen-Jensen 1978). Fucoxanthin found in species of the brown algae, and indeed those of all algal classes examined to date, has the (3*S*,5*R*,6*S*,3´*R*,5´*R*,6´*R*) configuration (Bernhard et al. 1976).

Table 2.16 Carotenoids of the Phaeophyceae

Major	*Others*
ß-Carotene (**1**)	Antheraxanthin (**5**)
Fucoxanthin (**52**)	ε-Carotene (**56**)
Violaxanthin (**3**)	Diadinoxanthin (**38**)
	Diatoxanthin (**39**)
	Fucoxanthinol (**71**)
	Neoxanthin (**4**)
	Zeaxanthin (**6**)

Liaaen-Jensen (1990) has recently reported that the neoxanthin present in *Fucus serratus* is the all-*trans* isomer. This is in contrast to some other algae and the majority of higher plants where 9′-*cis* neoxanthin appears to be the natural isomer in photosynthetic tissues (the all-*trans* isomer of neoxanthin is commonly found in non-photosynthetic tissues of higher plants).

2.3.14 Cyanobacteria (blue-green algae)

Goodwin and Britton (1988) have highlighted the following features of carotenogenesis in the cyanobacteria: (i) synthesis of ketocarotenoids, (ii) failure to synthesise ε-rings, (iii) production of 2,3-dihydoxy-ß-derivatives, and (iv) formation of monocyclic and aliphatic carotenoid glycosides (rhamnosides and *O*-methyl pentosides) (Andrewes and Liaaen-Jensen 1984; Rønneberg et al. 1985). Those carotenoids which are known to occur in the cyanobacteria are listed in Table 2.17.

The carotenoid glycosides found in the cyanobacteria may be divided into two sub-groups (Hallenstvet et al. 1979), based on (i) deoxyhexose (chivonose was originally identified as rhamnose by Riesen et al. 1984), (ii) *O*-methyldeoxyhexose (3-*O*-methylfucose Foss et al. 1986b).

Table 2.17 Carotenoids of the Cyanobacteria

Caloxanthin (**79**)
Canthaxanthin (**32**)
ß-Carotene (**1**)
ß-Cryptoxanthin (**9**)
Echinenone (**30**)
Myxoxanthophyll (**77**)
Nostoxanthin (**80**)
Oscillaxanthin (**78**)
Zeaxanthin (**6**)

Riesen et al. (1984) described a species of *Spirulina* which possessed myxol and oscillol α-bound to L-chivonose and L-fucose rather than rhamnose, as suggested above. Foss et al. (1986b) further examined the glycosidic carotenoids of a number of filamentous cyanobacteria and reported an α-bound chivonose moiety for both myxoxanthophyll and oscillaxanthin in *Oscillatoria agardhii* and an *Arthrospora* species. However, in *O. bornetii f. tenus* and in *O. limosa*, ^{1}H NMR studies have established that the sugar bound to myxol was 3-*O*-methylfucose. In general, a single carbohydrate type was thought to be associated with each glycoside. More recently Aackermann et al. (1990b) have shown that mixed glycosides are present in *Spirulina* spp. and in *Oscillatoria* spp. (for example, *S. platensis* has a ratio of myxol α-chivonoside:α-fucoside of 3-4:1). Foss and co-workers (1986b) further demonstrated the possible chemosystematic value of the distribution of carbohydrate moieties in the genus *Oscillatoria*.

Goodwin (1980) listed the following carotenoids as being characteristic of members of the cyanobacteria: ß-carotene, echinenone (= myxoxanthin = aphanin), canthaxanthin, zeaxanthin and the glycosides myxoxanthophyll ((3*R*,2′*S*)-2′-(ß-L-rhamnopyranosyloxy)-3´,4´-didehydro-1´,2´-dihydro-ß,ψ-carotene-3,1´-diol, **77**) (Liaaen-Jensen 1985; Rønneberg et al. 1985) and oscillaxanthin (2,2´-bis(ß-L-rhamnopyranosyloxy)-3,4,3´,4´-tetradehydro-1,2,1´,2´-tetrahydro-ψ,ψ-carotene-1,1´-diol, **78**). Oscillaxanthin probably has the (2*S*,2´*S*) configuration (Liaaen-Jensen 1985), although the (2*R*,2´*R*) configuration was suggested by an earlier study (Rønneberg et al. 1980).

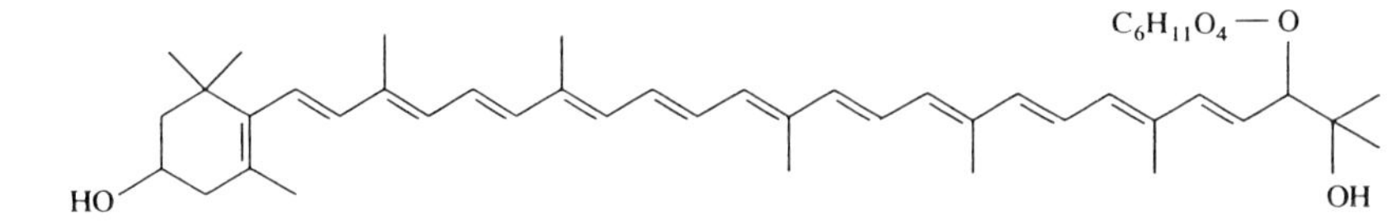

77

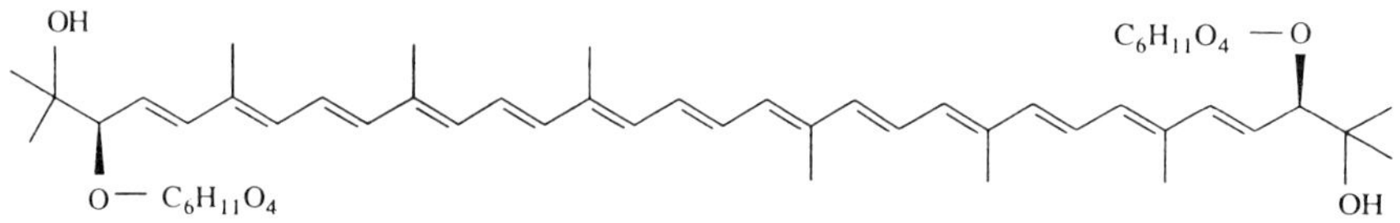

78

The carotenoids caloxanthin ((2*R*,3*R*,3´*R*)-ß,ß-carotene-2,3,3´-triol,**79**) and nostoxanthin ((2*R*,3*R*,2´*R*,3´*R*)-ß,ß-carotene-2,3,2´,3´-tetrol, **80**) are found in a limited number of species, including *Nostoc communa* and *Anacystis nidulans* (Stransky and Hager 1970). These have chirality at C-2 and C-2´ opposite to that of the 2-hydroxycarotenoids that are found in the air-living green alga *Trentepohlia iolithus* (Buchecker et al. 1976).

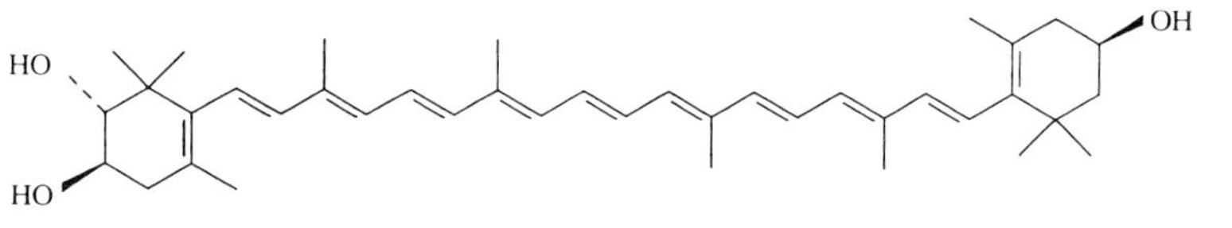

79

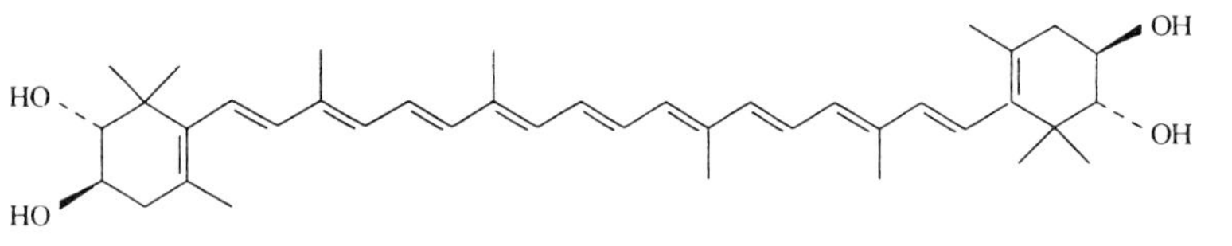

80

2.3.15 Prochlorophyceae

The Prochlorophyceae are similar to the cyanobacteria in that they do not synthesise either ε-rings or epoxides. However, the glycosides that are characteristic of the cyanobacteria have not been detected in members of the Prochlorophyceae (Foss et al. 1987b). Withers et al. (1978a, b) and Foss et al. (1987b) have found high levels of ß-carotene and zeaxanthin, togther with a number of other xanthophylls, generally in smaller amounts, including echinenone and ß-cryptoxanthin. Those carotenoids that have been found in the Prochlorophyceae are listed in Table 2.18.

Table 2.18 Carotenoids of the Prochlorophyceae

Major	*Minor*
ß-Carotene (**1**)	ß-Carotene-5,6-epoxide (**11**)
Zeaxanthin (**6**)	ß-Cryptoxanthin (**9**)
	Echinenone (**30**)

2.4 THE PHOTOTROPHIC BACTERIA

2.4.1 Characteristics and classification

The phototrophic bacteria (the term "phototrophic" is now preferred over the older designation "photosynthetic") are characterised by possessing an anoxygenic photosynthetic system which utilises only one photosystem (PS) akin to PSII of higher plants. The phototrophic bacteria exist in an anaerobic environment but many species can also adopt a chemotrophic (fermentative) lifestyle in aerobic conditions.

There have been several different schemes for the classification of these bacteria, based upon their morphological and metabolic characteristics, and, more recently, upon macromolecular sequences. Thus, changes of name (genus and species) and of family must be borne in mind when older literature is being evaluated. The bacteria were originally collected into the sub-order Rhodobacteriaceae, which was divided into three families: namely, the Thiorhodaceae or purple sulphur bacteria, the Athiorhodaceae or purple non-sulphur bacteria, and the Chlorobacteriaceae or green bacteria. This scheme was in common usage when the majority of analytical studies on the carotenoids of the phototrophic bacteria were carried out.

According to the revised classification that is now in use, the phototrophic bacteria are assigned to the order Rhodospirilleae which is divided into two sub-orders, the Rhodospirillineae (purple bacteria) and the Chlorobiineae. The purple Rhodospirillineae contain three families, the Rhodospirillaceae (non-sulphur) and the Chromatiaceae and Ectothiorhodaceae, which are both purple sulphur bacteria but are separated on the basis of growth requirements and phylogenic considerations. The green bacteria consist of the Chlorobiaceae, sulphur bacteria, and the Chloroflexaceae, filamentous gliding bacteria which have photosynthetic membranes closely resembling those of purple bacteria. Recently, other groups of phototrophic bacteria have been identified, namely, the Heliobacteria and the "aerobic photosynthetic bacteria", but the taxonomic relationship of these families to the main families has not yet been elucidated.

2.4.2 Carotenoid compositions

A diverse range of carotenoids are synthesised by the phototrophic bacteria by a number of different biosynthetic pathways (see Chapter 4). In contrast to their wide distribution in other photosynthetic organisms, ring structures are limited to only a few species, as are primary and secondary hydroxyl groups; the carotenoids of the phototrophic bacteria are therefore primarily acyclic. Schmidt (1978) and Goodwin (1980) provide the most comprehensive reviews of the distribution of carotenoids in the phototrophic bacteria. No such comprehensive survey is appropriate in this chapter, which will describe only the main features.

Schmidt (1978) listed the following characteristics as being typical of the carotenoids found in the phototrophic bacteria: (i) hydration of C-1,2 to yield tertiary hydroxyl groups at C-1, (ii) desaturation of C-3,4 bonds, (iii) saturation of C-1, 2 double bonds, (iv) oxidation of C-2 under aerobic conditions, (v) oxidation of C-4 anaerobically on acyclic end-groups or aerobically and anaerobically in ß-rings, (vi) oxidation of C-20 in conjugation with the polyene chain under anaerobic conditions, (vii) formation of 1,2,3-trimethylphenyl end-groups by aromatization of ß-rings, (viii) formation of 1,2,5-trimethylphenyl end-groups by aromatization of ß-rings, (ix) *O*-methylation.

(a)Rhodospirillaceae

Data are available for many members of the Rhodospirillaceae, and some species have been the subject of intensive investigation, in particular in the classical studies in which the main features of the biosynthetic pathways were elucidated (see Chapter 4). Some confusion is possible, however, as recent reclassifications have led to renaming of several of the organisms and, in some cases, their assignment to different or new genera. The new and old names of some common species are compared in Table 2.19.

Goodwin (1980) lists some 50–60 different carotenoids that have been detected in members of the Rhodospirillaceae grown under normal phototrophic conditions or in the presence of inhibitors of carotenoid biosynthesis (see Table 2.20). These compounds are virtually all variations on the same structural themes and are the products of, or intermediates in, the same few biosynthetic pathways.

Table 2.19 Common species of the Rhodospirillaceae

New name	*Old name*
Rhodobacter capsulatus	*Rhodospeudomonas capsulata*
Rhodobacter sphaeroides	*Rhodospeudomonas spheroides*
Rhodocylcus gelatinosus	*Rhodospeudomonas gelatinosa*
Rhodocyclus tenuis	*Rhodospirillum tenue*
Rhodopila globiformis	*Rhodospeudomonas globiformis*

With very few exceptions, members of the Rhodospirillaceae are unable to biosynthesise cyclic carotenoids. The characteristic carotenoids present are, therefore, acyclic compounds, typically with tertiary hydroxy or methoxy groups at C-1 of one end or both ends of the molecule. Additional desaturation at C-3,4 is also a common feature. Typical carotenoids present are spheroidene (1-methoxy-3,4-didehydro-1,2,7′,8′-tetrahydro-ψ,ψ-carotene, **81**) and spirilloxanthin (1,1′-dimethoxy-3,4,3′,4′-tetradehydro-1,2,1′,2′-tetrahydro-ψ,ψ-carotene, **82**). The number of biosynthetic reactions required for the elaboration of these structures is small, and combinations of these reactions, namely C-1,2-hydration, *O*-methylation and C-3,4-desaturation with variations in their relative rates at the two ends of the precursor molecule, lycopene or neurosporene, can give rise to spirilloxanthin, spheroidene and all possible intermediates in their proposed biosynthetic pathway (see Chapter 4). In the presence of inhibitors such as diphenylamine, other compounds, in particular 1-hydroxy- and 1-methoxy-derivatives of the desaturation intermediates phytoene, phytofluene and 7,8,11,12-tetrahydro-ψ,ψ-carotene (**83**) are also detected.

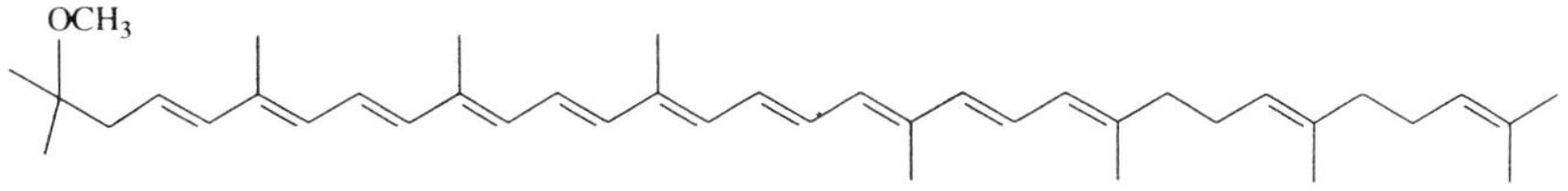

81

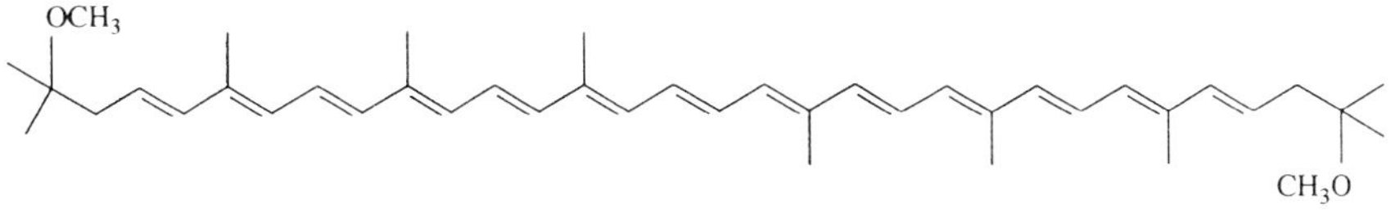

82

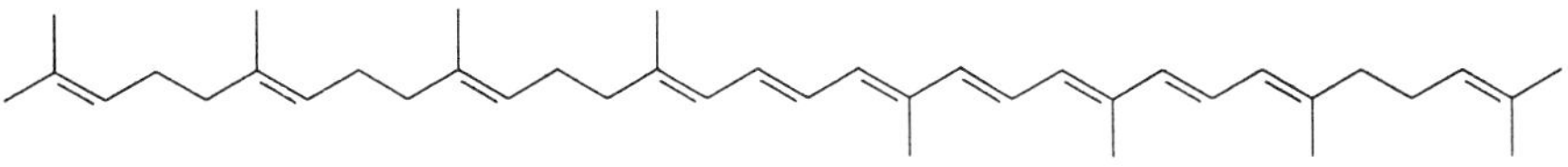

83

Some species apparently are able to carry out the normal desaturation sequence as far as lycopene, and then perform the further reactions at both ends of the molecule. These species include *Rhodospirillum rubrum*, *R. fulvum*, *R. photometricum*, *Rhodopseudomonas acidophila*, *Rps. palustris*, *Rps. cryptolactis*, *Rm. vannielii*, *Rhodocyclus purpureus* and *Rc. tenuis*, and their characteristic major carotenoids are spirilloxanthin, rhodovibrin (1´-methoxy-3´,4´-didehydro-1,2,1´,2´-tetrahydro-ψ,ψ-caroten-1-ol, **84**), anhydrorhodovibrin (1-methoxy-3,4-didehydro-1,2-dihydro-ψ,ψ-carotene, **85**) and rhodopin (1,2-dihydro-ψ,ψ-caroten-1-ol, **86**). *Rps. palustris* also accumulates lycopene and monodemethylspirilloxanthin (1´-methoxy-3,4,3´,4´-tetradehydro-1,2,1´,2´-tetrahydro-ψ,ψ-caroten-1-ol, **87**). *R. fulvum* and *Rps. molischianum* accumulate rhodopin (with its glucoside, **88**, in *R. fulvum*), and lycopene together with small amounts of dihydroxylycopene (1,2,1´,2´-tetrahydro-ψ,ψ-carotene-1,1´-diol, **89**).

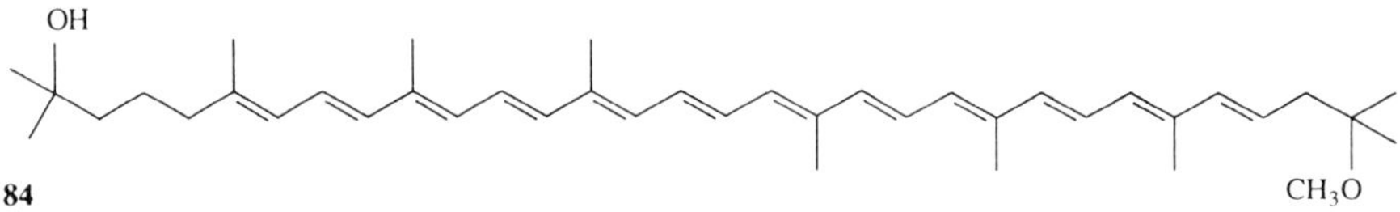

84

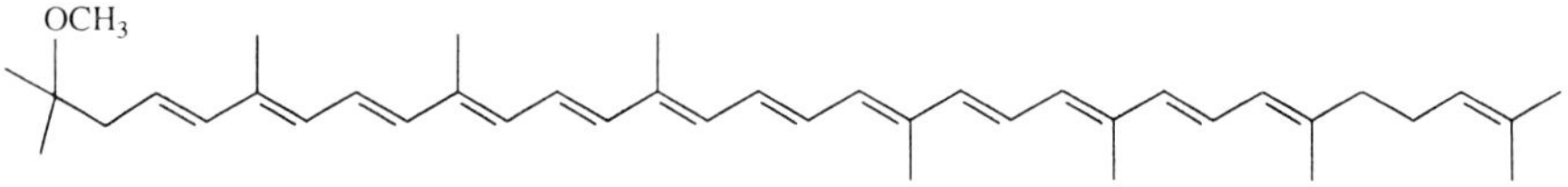

85

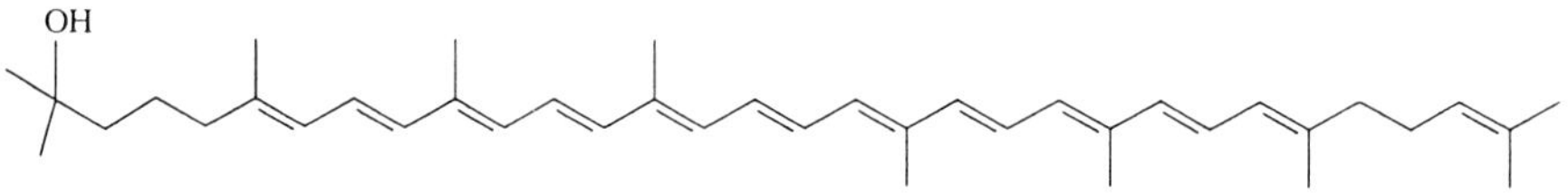

86

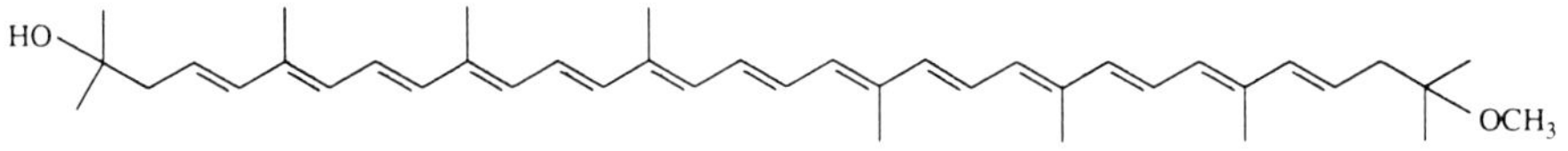

87

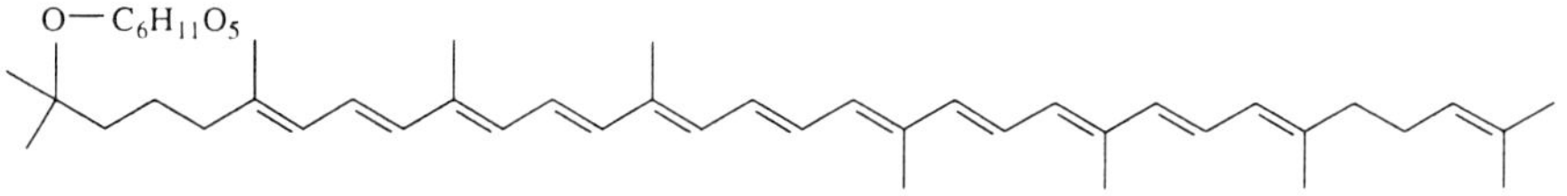

88

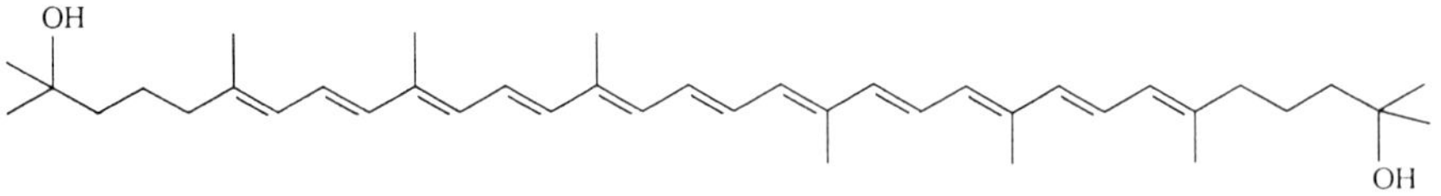

89

Table 2.20. Carotenoids of the Rhodospirillaceae

Anhydrorhodovibrin (**85**)
ß-Carotene (**1**)
ζ-Carotene (**20**)
Chloroxanthin
ß-Cryptoxanthin (**9**)
3',4'-Didehydro-1,2-dihydrolycopen-20(or20')-al
3,4-Didehydrorhodopin
1,2-Dihydroneurosporene
1,2-Dihydrolycopene
1,2-Dihydrophytoene
1,2-Dihydrophytofluene
3,4-Didehydro-1,2-dihydrolycopene
1,2-Dihydro-ζ-carotene
Dihydroxylycopene (**89**)
1,1'-Dimethoxy-1,2,1',2'-tetrahydrolycopen-4-one
1,1'-Dimethoxy-1,2,1',2'-tetrahydrolycopene-4,4'-dione
2,2'-Dioxospirilloxanthin (**93**)
1,2,7,8,11,12-Hexahydrolycopene
Hydroxyspheroidene (**90**)
Hydroxyspheroidenone (**92**)
4-Ketorhodovibrin
Lycopene (**22**)
Lycopen-20-al (**94**)
Methoxyspheroidene
Methoxyspheroidenone
1'-Methoxy-3',4'-didehydro-1',2'-dihydro-ß,ψ-carotene
1-Methoxy-1,2-dihydrolycopen-4-one
1-Methoxy-1'-hydroxy-1,2,1',2'-tetrahydrolycopen-4-one
Monodemethylspirilloxanthin (**87**)
Neurosporene (**21**)
Phytoene (**18**)
Phytofluene (**19**)
Rhodopin (**86**)
Rhodopin glucoside (**88**)
Rhodopin-20-al (**95**)
Rhodopin-20-ol (**97**)
Rhodopin-20-al glucoside (**98**)
Rhodovibrin (**84**)
Spheroidene (**81**)
Spheroidenone (**91**)
Spirilloxanthin (**82**)
1,2,1',2'-Tetrahydroneurosporene
1,2,1',2'-Tetrahydrolycopene

Other species appear to terminate the normal desaturation sequence at the level of neurosporene, which then undergoes further end-group reactions to give spheroidene and hydroxyspheroidene (1′-methoxy-3′,4′-didehydro-1,2,7,8,1′,2′-hexahydro-ψ,ψ-caroten-1-ol, **90**) as the main products. Only five such species have been identified, namely *Rhodobacter capsulatus* and *Rb. sphaeroides*, *Rc. gelatinosus*,

Rps. blastica and *Rps. sulphidophila*. In some cases, subsequent reactions of the second end-group can occur, so that small amounts of rhodovibrin and spirilloxanthin are seen. Earlier discussions of the carotenoid distribution in these organisms often refer to the presence of the normal spirilloxanthin pathway and the spheroidene or alternative spirilloxanthin pathway (see Chapter 4).

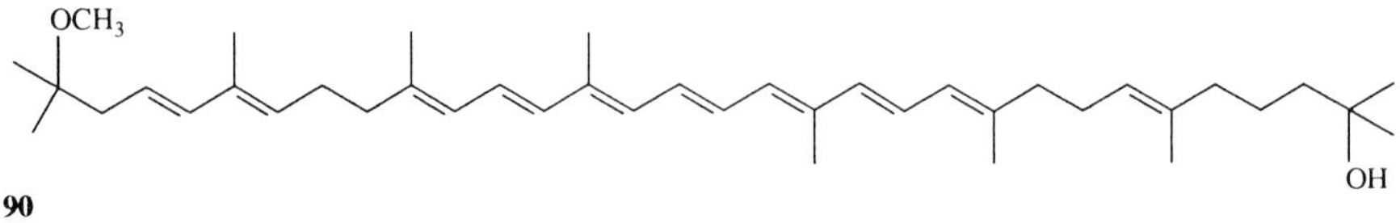

90

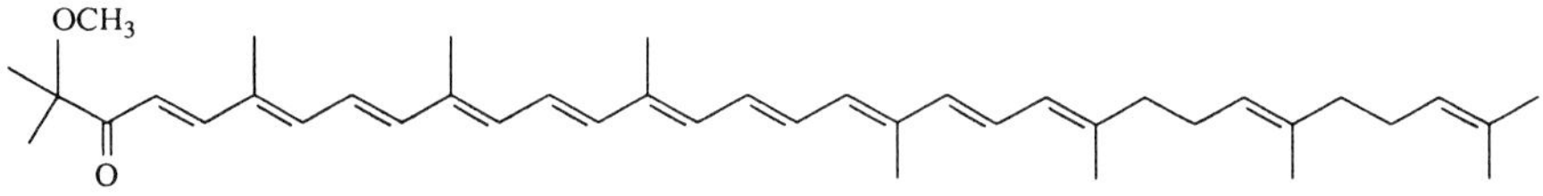

91

Another characteristic of the species that accumulate spheroidene and related compounds is their ability to introduce a ketone group at C-2 on exposure to traces of O_2. Thus, in aerobic conditions, spheroidenone (1-methoxy-3,4-didehydro-1,2,7´,8´-tetrahydro-ψ,ψ-caroten-2-one, **91**) and hydroxyspheroidenone (1´-hydroxy-1-methoxy-3,4-didehydro-1,2,1´,2´,7´,8´-hexahydro-ψ,ψ-caroten-2-one, **92**) become the major carotenoids in *Rb. sphaeroides*. Any carotenoid with a 1-methoxy-3,4-didehydro-1,2-dihydro-ψ end-group is potentially capable of undergoing the reaction, so small amounts of other ketones, including the diketone 2,2´-dioxospirilloxanthin (1,1´-dimethoxy-3,4,3´,4´-tetradehydro-1,2,1´,2´-tetrahydro-ψ,ψ-carotene-2,2´-dione, **93**), may be detected in some species. Generally those organisms that exhibit the efficient formation of ketocarotenoids are able to survive and adapt easily to an aerobic lifestyle. Species that are unable to effect the transformation are frequently strict anaerobes and are killed by exposure to air.

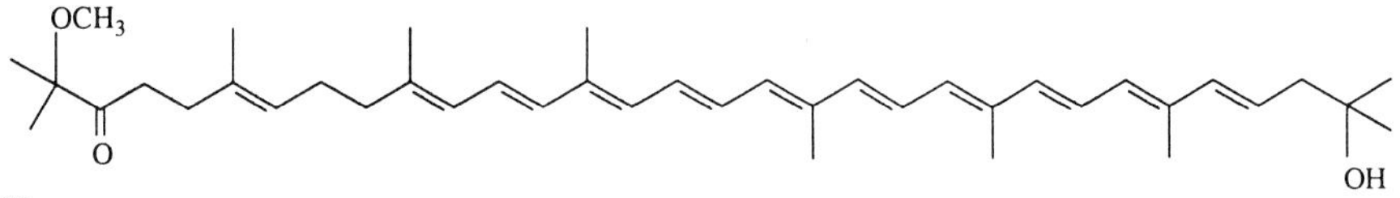

92

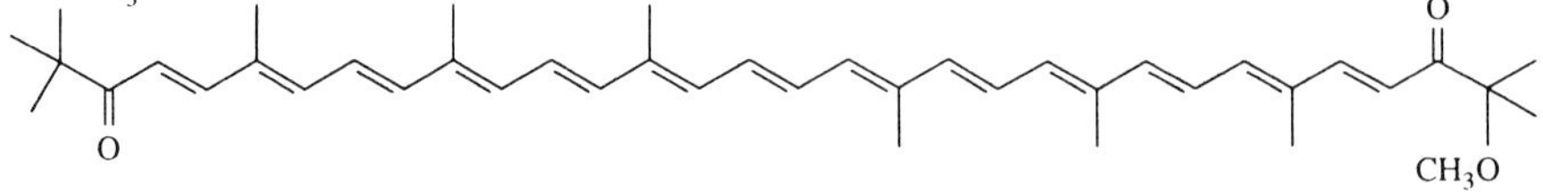

93

Another structural modification that is encountered in several species, notably *Rc. purpureus*, *Rc. tenuis* and *Rps. acidophila* is the oxidation of the C-20 methyl group, usually under anaerobic conditions, to give the cross-conjugated aldehydes, e.g. lycopen-20-al (13-*cis*-ψ,ψ-caroten-20-al, **94**) and rhodopin-20-al (13-*cis*-1-hydroxy-1,2-dihydro-ψ,ψ-caroten-20-al, **95**) and their corresponding alcohols lycopen-20-ol (13-*cis*-ψ,ψ-caroten-20-ol, **96**) and rhodopin-20-ol (13-*cis*-1,2-dihydro-ψ,ψ-carotene-1,20-diol, **97**). Glucosides such as rhodopin-20-al glucoside (**98**) are found in *Rps. acidophila*. This structural feature is considered to be more typical of the Chromatiaceae.

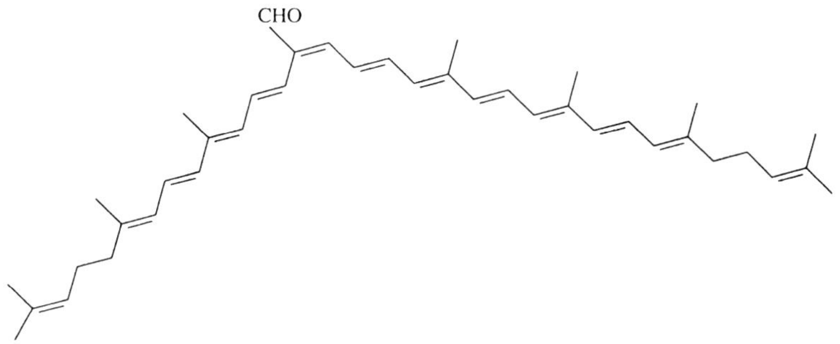

94

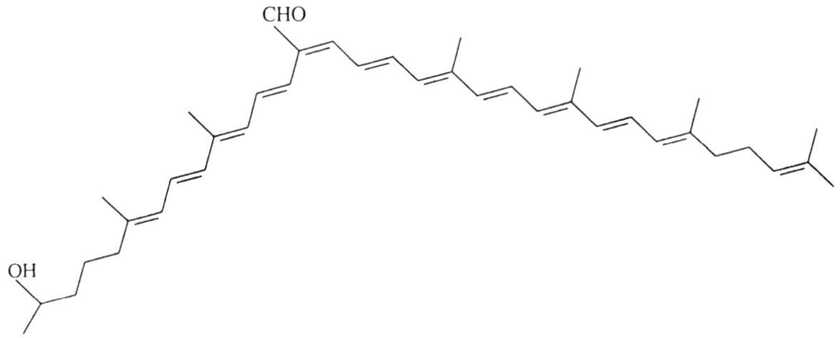

95

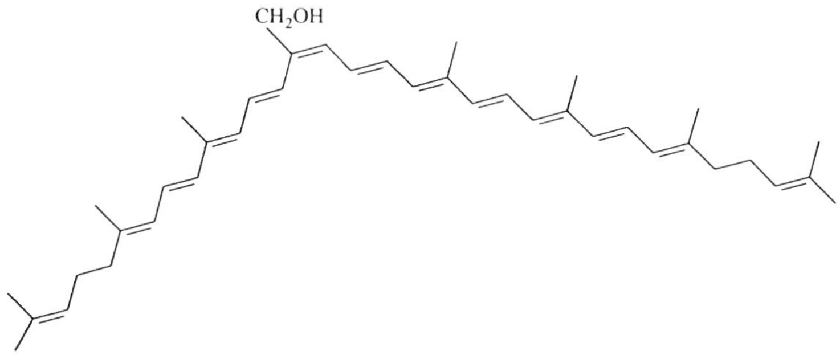

96

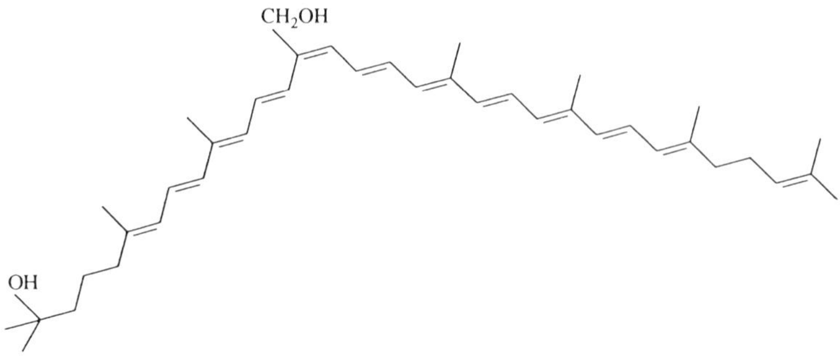

97

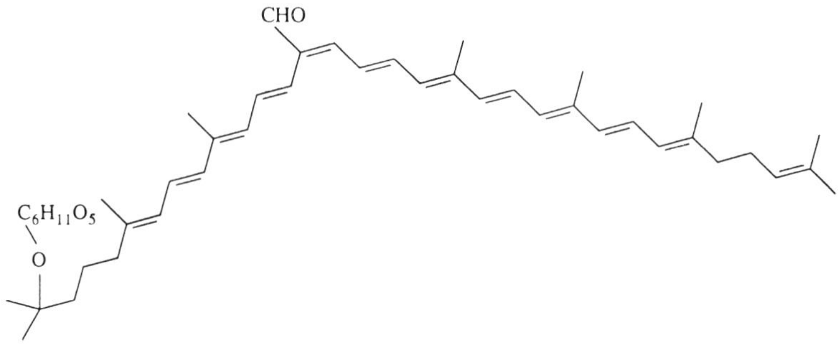

98

Inevitably there are some species which contain additional carotenoids that appear to be unique to one or a very small number of species; for example, there is a significant amount of ß-carotene, and smaller amounts of other ß-ring carotenoids in *Rm. vannielii.*

One notable example, *Rps. viridis*, contains a series of carotenoids that appear to be restricted to this species, i.e. the 1,2-dihydro-derivatives of neurosporene (1,2,7,8-tetrahydro-ψ,ψ-carotene, **99**), and lycopene (1,2-dihydro-ψ,ψ-carotene, **100**) and others in which the terminal double bond is saturated, presumably by the addition of H_2 rather than H_2O across the C-1,2 double bond (Britton et al. 1975, 1976, 1977). Cross-conjugated C-20 aldehyde derivatives of these 1,2-dihydrocarotenes are also present in small amounts. Another unusual species is *Rhodopila globiformis*, the major carotenoids of which include those with the 1-methoxy-1,2-dihydro-4-oxo end-group that occurs in okenone (1′-methoxy-1′,2′-dihydro-χ,ψ-caroten-4′-one, **101**) and other carotenoids that are characteristic of the Chromatiaceae (see Section 2.4.2 (b)) when members of this family are cultured under anaerobic conditions (Schmidt and Liaaen-Jensen 1973).

There has been no systematic survey of the carotenoids of these bacteria by modern HPLC methods, so the potential value of carotenoid compositions as a criterion for systematic classification is not known.

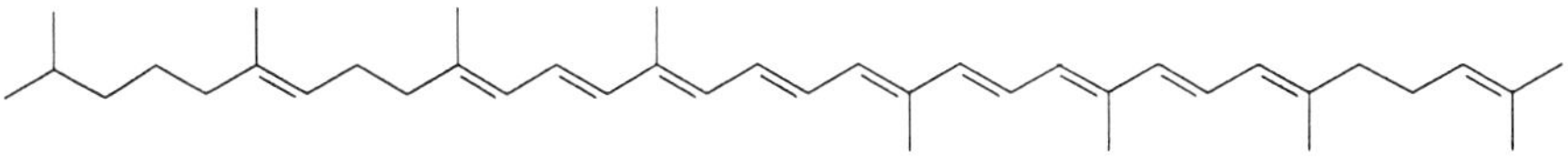

99

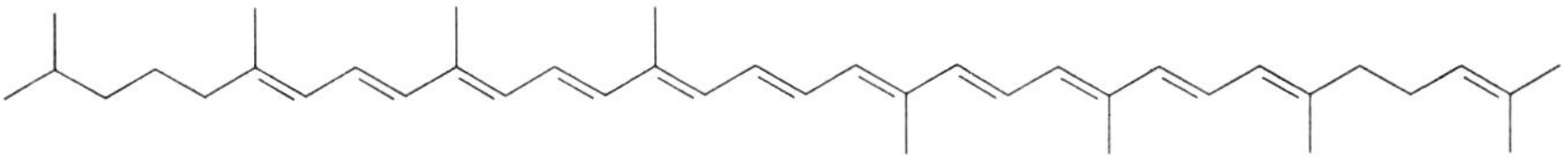

100

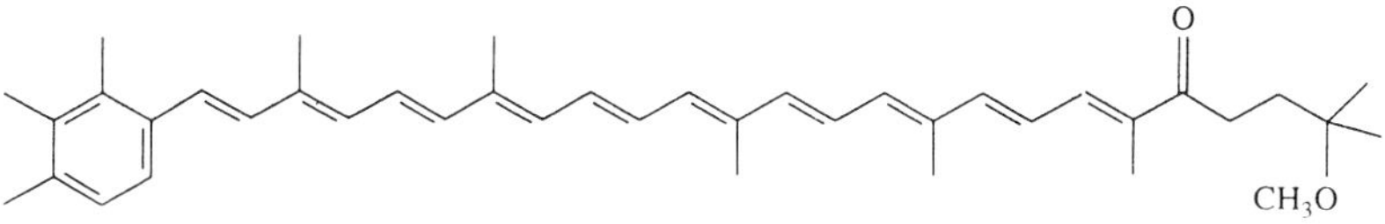

101

(b)Chromatiaceae and Ectothiorhodaceae

The carotenoids of many species of the purple sulphur bacteria are acyclic compounds similar to those of the Rhodospirillaceae. In particular, spirilloxanthin and related compounds are characteristic of *Amoebobacter pendens, A. roseus, Chromatium minutissimum, C. vinosum, Ectothiorhodospira halochloris, E. halophila, E. mobili, E. shaposhnikova* and *Thiospirillum jenense*. These carotenoids often occur together with cross-conjugated C-20 aldehydes of the rhodopinal type, which are characteristic of *C. violascens, C. warmingii, Lamprocystis roseopersicina, Thiocyctis* (formerly *Thiothece*) *violaceae, Thiocapsa pfennigii* and *Thiodictyon bacillosum* and *Thiodictyon elegans*. No species have been found to synthesise the spheroidene group of pigments.

Carotenoids with aromatic rings are typical of many sulphur bacteria (Table 2.21). In members of the Chromatiaceae the characteristic example is okenone, which has the 1,2,3-trimethylphenyl (χ) end-group and is the main carotenoid in the *Chromatium* species *minus, okenii, purpuratum* and *weissei*. The 1-methoxy-1,2-dihydro-4-oxo end-group that forms the second end-group of okenone is also found in the related ß-ring monocyclic and acyclic compounds such as "thiothece-HO-474" that are the main carotenoids in strains of *Thiocystis* and may also be detected in small amounts in okenone-containing species. Carotenoids with this end-group are also present in *Rhodopila globiformis* of the Rhodospirillaceae.

Thiocapsa pfennigii does not synthesise rhodopinal but accumulates high levels of 3,4,3´,4´-tetrahydrospirilloxanthin (1,1´-dimethoxy-1,2,1´,2´-tetrahydro-ψ,ψ-carotene, **102**) together with traces of its C-20 aldehyde derivative (**103**; Aasen and Liaaen-Jensen 1967). In contrast, *Thiodictyon elegans* can synthesise both rhodopin-20-al and 3,4,3´,4´-tetrahydrospirilloxanthin. Lycopen-20-ol, lycopen-20-al and rhodopin-20-ol are frequently found as minor components.

Glucosides such as rhodopin glucoside and 1´-methoxy-1´,2´-dihydrorhodopin glucoside (**104**), may accumulate to very high levels in some species (e.g. *E.*

halochloris Imhoff and Trüper 1977).

Table 2.21 Carotenoids of the Cromatiaceae and Ectothiorhodaceae

Major	*Minor*
Lycopen-20-al (**94**)	Anhydrorhodovibrin (**85**)
Okenone (**101**)	Didehydrorhodopin
Rhodopin (**86**)	Demethylspirilloxanthin (**87**)
Rhodopin-20-al (**95**)	Lycopene (**22**)
Rhodovibrin (**84**)	Lycopen-20-ol (**96**)
3,4,3',4'-Tetrahydrospirilloxanthin (**102**)	Neurosporene (**21**)
	Rhodopin glucoside (**88**)
	Rhodopin-20-ol (**97**)
	Spirilloxanthin (**82**)
	3,4,3',4'-Tetrahydrospirilloxanthinal (**103**)

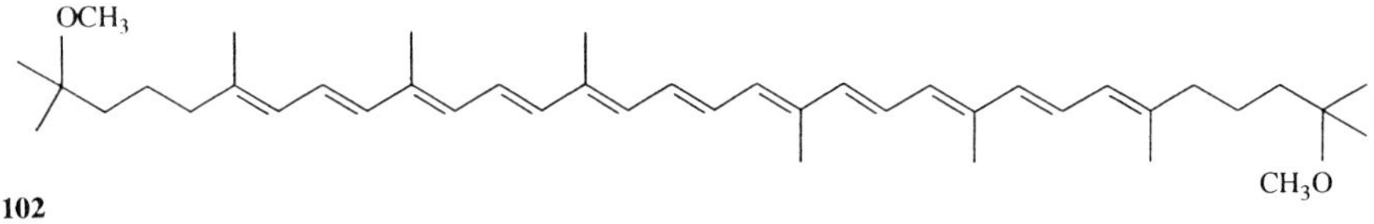

102

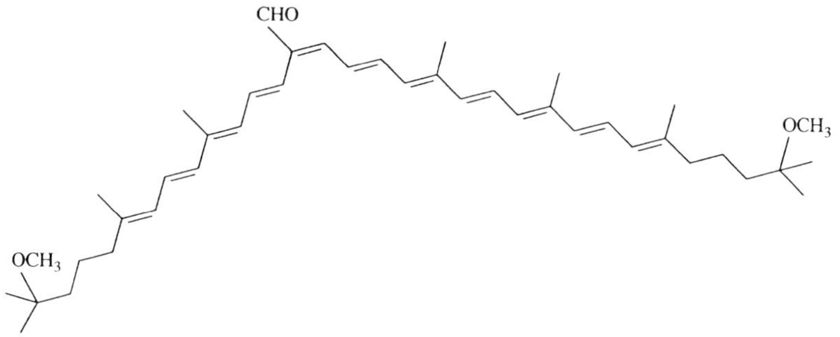

103

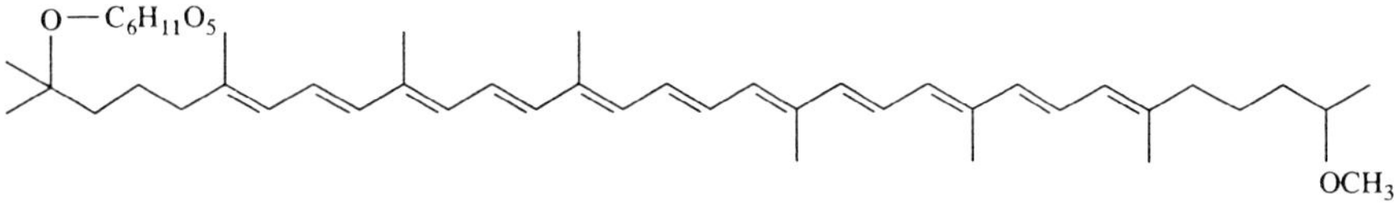

104

(c) *Chlorobiaceae*

Aryl carotenoids are characteristic of this family but, in the case of the Chlorobiaceae, these usually have the 1,2,5-trimethylphenyl (ϕ) end-group, as in chlorobactene (ϕ,ψ-carotene, **105**), hydroxychlorobactene (1´,2´-dihydro-ϕ,ψ-caroten-1´-ol, **106**), isorenieratene (= leprotene, ϕ,ϕ-carotene **107**), and its putative precursor ß-isorenieratene (ß,ϕ-carotene, **108**). The occurrence of hydroxychlorobactene glucoside (**109**) has been reported in *Chlorobium limicola f. thiosulfatophilum* strain 6230 (Schmidt 1980). Schmidt suggested that the occurrence of such glucosides may be much more common than was originally thought, as they are easily obscured by the presence of high amounts of non-glucosidic carotenoids.

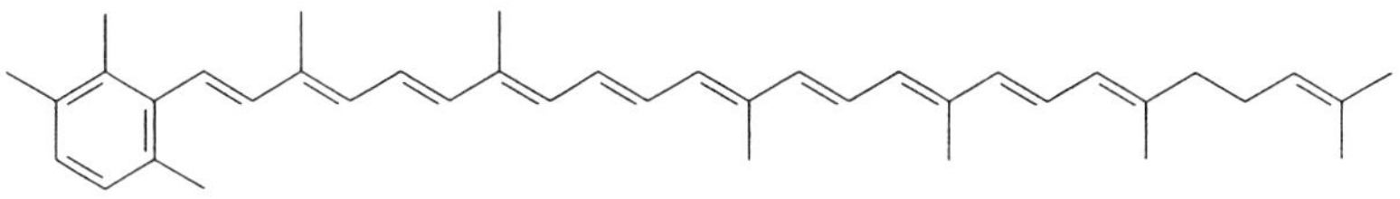

105

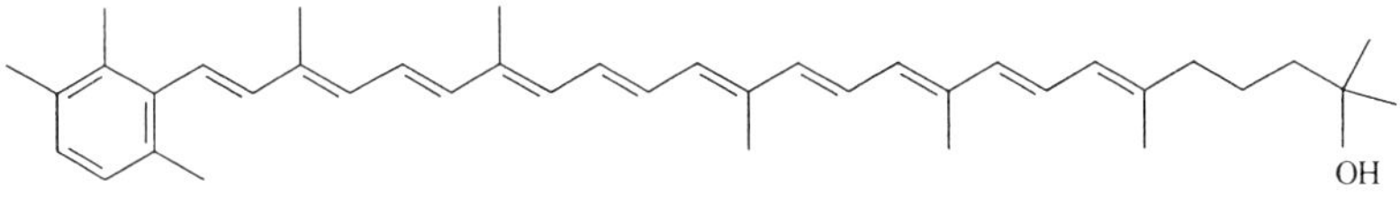

106

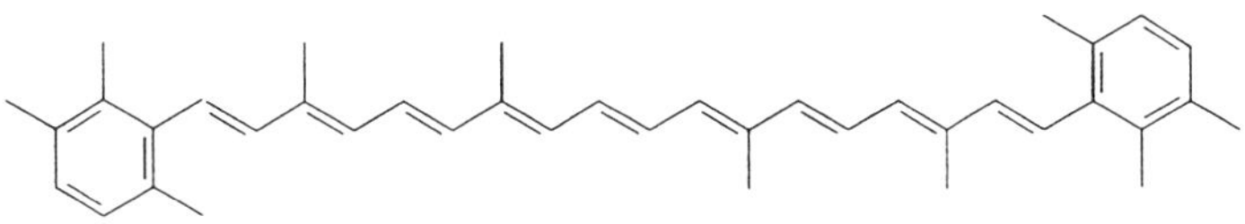

107

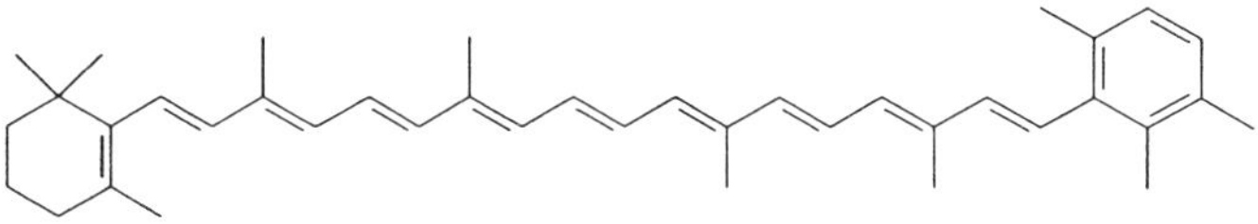

108

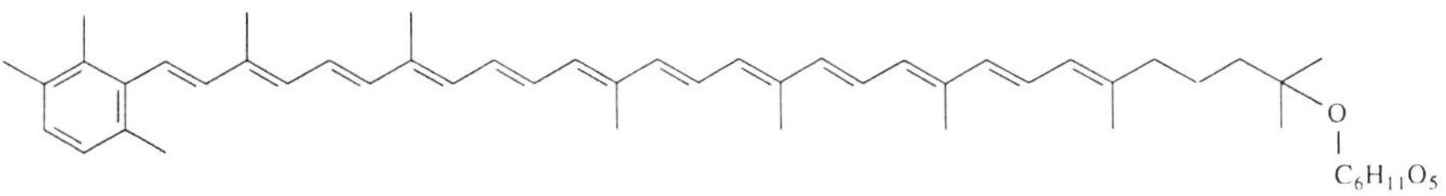

109

The corresponding ß-ring compounds are also often present; for example, chlorobactene is accompanied by γ-carotene (ß,ψ-carotene, **110**) in *Chlorobium* species. In one strain of *Chlorobium limicola*, the 1´,2´-dihydro-analogues of chlorobactene and γ-carotene have been identified, the only known examples of the occurrence of 1,2-dihydrocarotenoids besides *Rps. viridis*. The green strains of the

Chlorobiaceae (e.g. *Chlorobium vibriforme*) accumulate mainly chlorobactene, usually with γ-carotene.

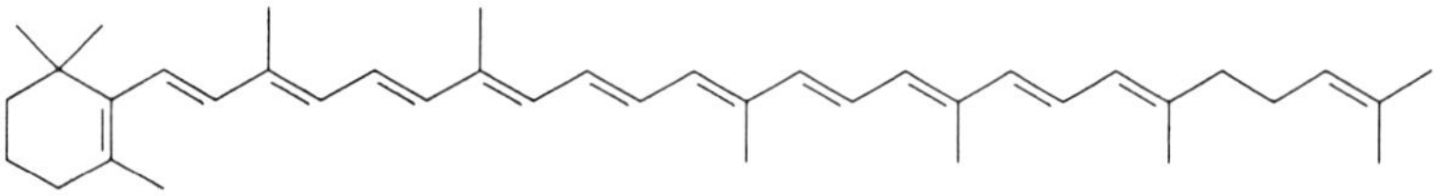

110

Table 2.22 Carotenoids of the Chlorobiaceae

Major	*Minor*
Chlorobactene (**105**)	ß-Carotene (**1**)
Isorenieratene (**107**)	γ-Carotene (**110**)
	ζ-Carotene (**20**)
	Hydroxy-γ-carotene
	Hydroxychlorobactene (**106**)
	ß-Isorenieratene (**108**)
	Lycopene (**22**)
	1'-Methoxy-3',4'-didehydro-1',2'-dihydro-γ-carotene
	Neurosporene (**21**)
	Phytoene (**18**)
	Phytofluene (**19**)
	Rhodopin (**86**)

Hydroxychlorobactene and rhodopin have been shown to be the main pigments of the green sulphur bacterium *Prosthecochloris aestuarii* (Swartoff et al. 1982; Braumann et al. 1986). The carotenoids found in these bacteria are listed in Table 2.22.

(d) Chloroflexaceae

Chloroflexus aurantiacus is a remarkable gliding filamentous organism which contains γ-carotene and several of its derivatives, in particular, 4-oxo-γ-carotene (ß,ψ-caroten-4-one, **111**) and 1´,2´-dihydro-ß,ψ-caroten-1´-ol (**112**) and its glucoside (**113**; Halfen et al. 1972; Schmidt 1978). Small amounts of the ß-carotene derivative, echinenone, are also present (Table 2.23).
In its carotenoids as well as other taxonomic characteristics, *C. aurantiacus* seems to bear little relation to the other phototrophic bacteria.

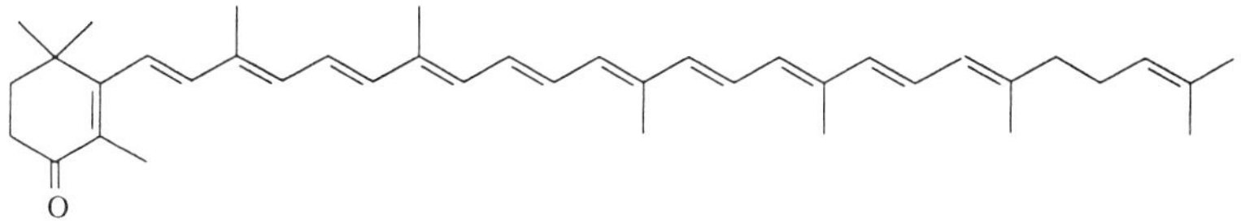

111

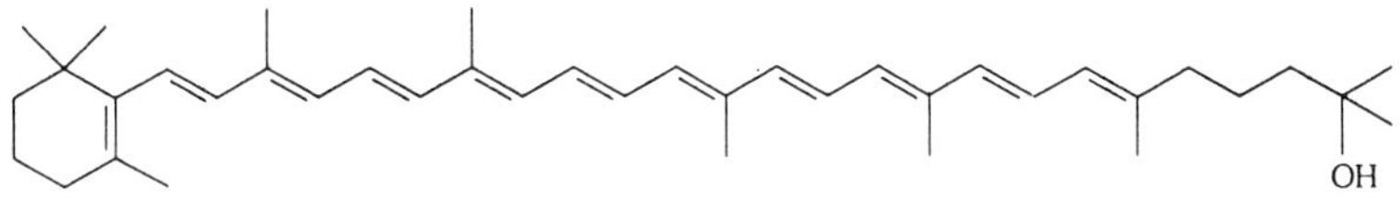

112

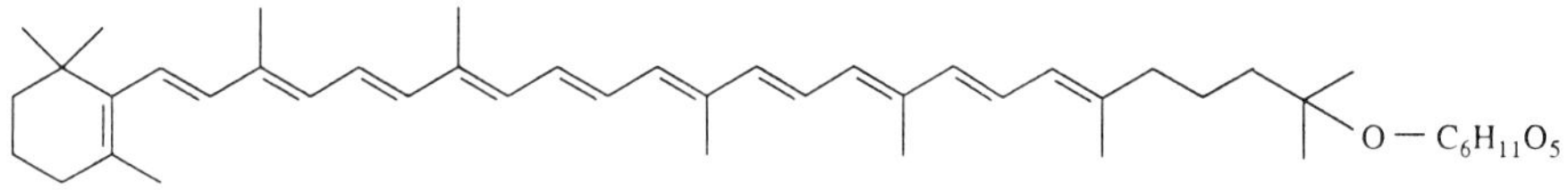

113

Table 2.23 Carotenoids of the Chloroflexaceae

ß-Carotene (**1**)
γ-Carotene (**110**)
Hydroxy-γ-carotene glucoside (**113**)
Didehydro-γ-carotene
Hydroxy-γ-carotene
4-Oxo-γ-carotene (**111**)
4-Oxo-didehydro-γ-carotene glucoside

(e) Aerobic phototrophic bacteria

Although a book has already been published on these recently discovered organisms (Harashima et al. 1989), very few species have been examined. Those that have been described in detail, e.g. *Erythrobacter* species OCh114 and OCh101 (*E. longus*), are organisms that cannot grow under the light, anaerobic conditions that are required or preferred by other phototrophic bacteria. They do, however, utilise light energy well under aerobic conditions and synthesise bacteriochlorophyll *a* and carotenoids even in the dark.

The carotenoid compositions of the two species differ greatly – species OCh114 (now named *Roseobacter denitrificans*) contains spheroidenone as its major carotenoid, with small or trace amounts of other related compounds (Harashima 1989). When dark-grown cells are illuminated under anaerobic conditions, a substantial proportion of the spheroidenone undergoes reduction, the main product isolated being the novel 3,4-dihydrospheroidenone (1-methoxy-1,2,7′,8′-tetrahydro-ψ,ψ-caroten-2-one, **114**; Takaichi et al. 1991a).

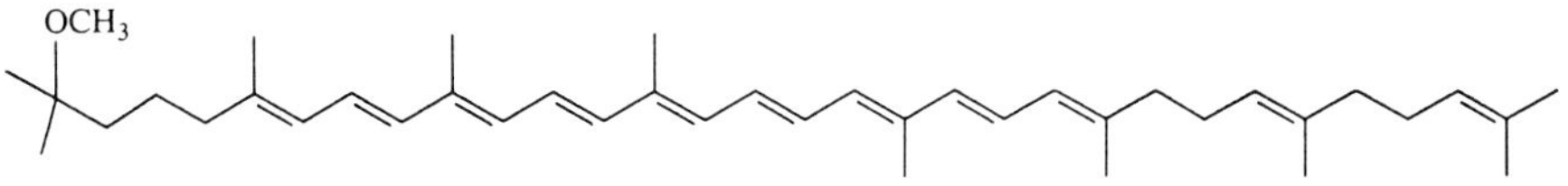

114

In contrast to species OCh114, not all the carotenoid found in *E. longus* is bound to the bacteriochlorophyll–protein complexes (Shimada et al. 1985). The pigment composition of *E. longus* (OCh101) is very complex and more than 20 carotenoids have been observed in aerobic cultures (Takaichi et al. 1988, 1991b), but the structures of many of these, particularly the more polar non-protein bound carotenoids, have yet to be confirmed. The less polar, bound carotenoids, have been well characterised by a range of analytical techniques and are shown in Table 2.24. Spirilloxanthin and related compounds (lycopene and anhydrorhodovibrin) are present, together with several compounds having a 3-hydroxy-ß-ring. These include the novel monocylic compounds (3*R*)-9′-*cis*-1′-methoxy-3′,4′-didehydro-1′,2′-dihydro-ß,ψ-caroten-3-ol (**115**) and (3*R*)-9′-*cis*-3-hydroxy-1′-methoxy-3′,4′-didehydro-1′,2′-dihydro-ß,ψ-caroten-19′-al (**116**) that have been designated "bacteriorubixanthin" and "bacteriorubixanthinal", respectively (Takaichi et al. 1988), but the most surprising finding is the presence of the collection of ß-carotene derivatives ß-cryptoxanthin, zeaxanthin, caloxanthin and nostoxanthin, that are characteristic of the cyanobacteria (Section 2.3.14). Liaeen-Jensen (1991) has pointed out that the 3-OH-ß-ring is rare and that the 19-al feature is a new observation for phototrophic bacteria. As in the cyanobacteria these carotenoids are accompanied by chlorophyll *a*, as well as the expected bacteriochlorophyll.

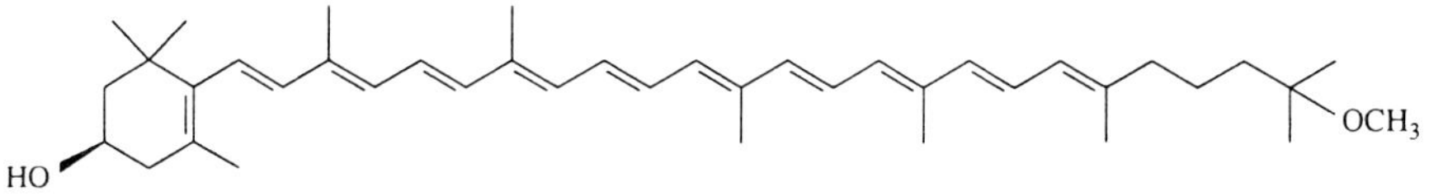

115

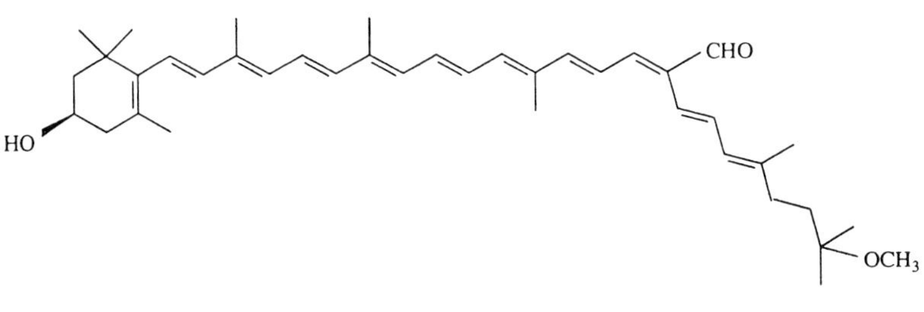

116

Table 2.24 Carotenoids of the aerobic phototrophic bacterium *Erthyrobacter longus*

Major	*Minor*
Bacteriorubixanthinal (**116**)	Anhydrorhodovibrin (**85**)
Caloxanthin (**79**)	Bacteriorubixanthin (**115**)
Nostoxanthin (**80**)	ß-Carotene (**1**)
Zeaxanthin (**6**)	ß-Cryptoxanthin (**9**)
13-*cis*-Zeaxanthin	Caloxanthin sulphate (**118**)
	Erythroxanthin sulphate (**117**)
	Lycopene (**22**)
	Phytoene (**18**)
	Rubixanthin
	Spirilloxanthin (**82**)

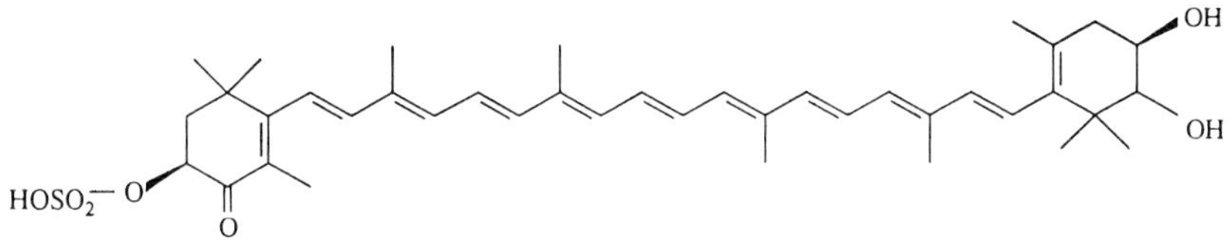

117

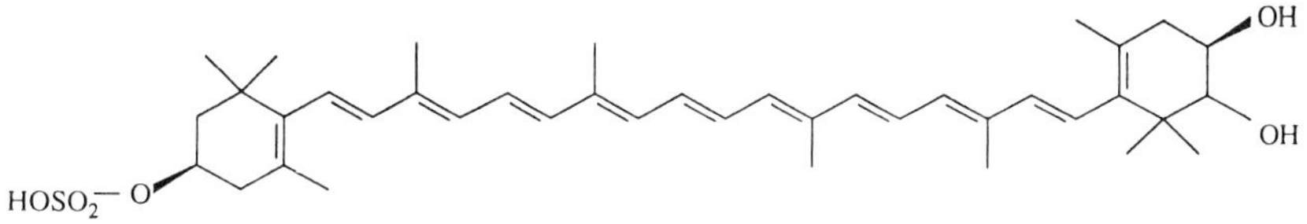

118

Takaichi and co-workers (1991b) have also recently reported the identification in *E. longus* of two carotenoid sulphates, namely (3*S*,2´*R*,3´*R*)-3,2´,3´-trihydroxy-ß,ß-caroten-4-one 3-sulphate (**117**), termed "erythroxanthin sulphate" and (3*R*,2´*R*,3´*R*)-ß,ß-carotene-3,2´,3´-triol 3-sulphate (**118**), termed "caloxanthin sulphate". Carotenoid sulphates have not been found in any other photosynthetic organisms.

REFERENCES

Aakermann, T., Englert, G., Schiedt, K. *et al.* (1990a) Re-isolation and structural elucidation of a cross-conjugated carotenal disaccharide: 13´-*cis*-7´,8´-dihydroneoxanthin-20´-al 3´-ß-lactoside (P457). Abstract, *9th International Symposium on Carotenoids*, Kyoto, p.80.

Aakermann, T., Christensen, B.E., Foss, P. *et al.* (1990b) Analytical studies on carotenoid glycosides and glycosyl esters. Abstract, *9th International Symposium on Carotenoids*,

Kyoto, p.81.

Aasen, A.J. and Liaaen-Jensen, S. (1967) Bacterial carotenoids. XXIV. The carotenoids of Thiorhodaceae 7. Cross-conjugated carotenoids. *Acta Chem. Scand.*, **21**:2185-2204.

Aitzetmüller, K., Strain, H.H., Svec, W.A. *et al.* (1969) Loroxanthin, a unique xanthophyll from *Scenedesmus obliquus* and *Chlorella vulgaris*. *Phytochemistry*, **7**: 525-527.

Allen, M.B., Fries, L., Goodwin, T.W. and Thomas, D.M. (1964) The carotenoids of algae: Pigments from some cryptomonads, a heterokont and some Rhodophyceae. *J. Gen. Microbiol.*, **34**: 259-2647.

Andrewes, A.G. and Liaaen-Jensen, S. (1984) Studies on carotenoid mono and disaccharides. Abstract, *7th International IUPAC Symposium on Carotenoids*, Munich.

Antia, R.S., Bisalputra, T., Cheng, Y.J. and Kalley, J.P. (1975) Pigment and cytological evidence for reclassification of *Nannochloris oculata* and *Monallantus salina* in the Eustigmatophyceae. *J. Phycol.*, **11**: 339-343.

Antia, R.S. and Cheng, Y.J. (1982) The ketocarotenoids of two members of the Eustigmatophyceae. *Br. Phycol. J.*, **17**: 39-50.

Antia, N.J. and Cheng, Y.J. (1983) Evidence for an anomalous xanthophyll composition in a clone of *Dunaliella tertiolecta* (Chlorophyceae). *Phycologia*, **22**: 235-242.

Arpin, N., Svec, W.A. and Liaaen-Jensen, S. (1976) New fucoxanthin-related carotenoid from *Coccolithus huxleyi*. *Phytochemistry*, **15**:529-532.

Ashikawa, I., Kito, M., Satoh, K. *et al.* (1987) All-*trans*-ß-carotene-5,6-epoxide in thylakoid membranes. *Photochem. Photobio*l., **46**: 269-275.

Bernhard, K. (1990) Synthetic astaxanthin. The route of a carotenoid from research to commercialization. In: Krinsky, N.I., Mathews-Roth, M.M. and Taylor, R.F. (eds.), *Carotenoids: Chemistry and Biology* (Plenum Press, New York, pp.337-363.

Bernhard, K., Moss, G.P., Toth, Gy. and Weedon, B.C.L. (1976) Absolute configuration of fucoxanthin. *Tetrahedron Lett.*, 115-118.

Bjerkeng, B., Vernett, M., Nielsen, M.V. and Liaaen-Jensen, S. (1990) Carotenoids of *Chrysochromulina polyepis* (Prymnesiophyceae). *Biochem. Syst. Ecol.*, **18**: 303-306.

Bjørnland, T. (1982) Chlorophylls and carotenoids of the marine red alga *Eutreptiella gymnastica*. *Biochem. Syst. Ecol.*, **11**: 73-76.

Bjørnland, T. (1990a) Carotenoid structures and lower plant phylogeny. In: Krinsky, N.I., Mathews-Roth, M.M. and Taylor, R.F. (eds.), *Carotenoids: Chemistry and Biology*. Plenum Press, New York, pp.21-38.

Bjørnland, T. (1990b) Chromatographic separation and spectrometric characterization of native carotenoids from the marine dinoflagellate *Thoracosphera heimii*. *Biochem. Syst. Ecol.*, **18**: 307-316.

Bjørnland, T. and Liaaen-Jensen, S. (1989) Distribution patterns of carotenoids in relation to chromophyte phylogeny and systematics. In: Green, J.C. and Leadbetter, B.S.C. (eds.), *The Chromophyte Algae: Problems and Perspectives*, Systematics Association Special Vol. No. 38. Clarendon Press, Oxford, pp. 37-60.

Bjørnland, T., Borch, G. and Liaaen-Jensen, S. (1984a) Configurational studies on red algal carotenoids. *Phytochemistry*, **23**:1711-1715.

Bjørnland, T., Pennington, F.C., Haxo, F.T. and Liaaen-Jensen, S. (1984b) Carotenoids of Chrysophyceae and Dinophyceae – "Coc. min. haltenbraken" and *Gymnodinium breve* (Florida red tide). Abstract, *7th. International Symposium on Carotenoids*, Munich, p.21.

Bjørnland, T., Borch, G. and Liaaen-Jensen, S. (1986) Additional oxa-bicyclo-(2.2.1)heptane carotenoids from *Eutreptiella gymnastica*. *Phytochemistry*, **25**: 201-206.

Bjørnland, T., Guillard, R.R.L. and Liaaen-Jensen, S. (1988) *Phaeocystis* sp. clone 677-3 – a typical marine planktonic prymnesiophyte with fucoxanthin and 19´-hexanoyloxyfucoxanthin as chemosytematic markers. *Biochem. Syst. Ecol.*, **16**: 445-452.

Bjørnland, T, Liaaen-Jensen, S. and Thorndsen, J. (1989) Carotenoids of the marine chrysophyte *Pelagococcus subviridis* from the Norwegian Sea. *Phytochemistry*, **28**: 3347-3353.

Bjørnland, T., Fiskdahl, A., Haxo, F.T. and Liaaen-Jensen, S. (1987) New and unusual features of *Euglena gracilis* (Euglenophyceae) Abstract. *8th. International Symposium on Carotenoids* (Boston). P1.

Borowitzka M.A. (1988) Vitamins and fine chemicals from micro-algae. In: Borowitzka M. A., Borowitzka L.J. (eds.), *Micro-algal Biotechnology*. Cambridge University Press, Cambridge, pp. 153-196.

Borowitzka M.A. and Borowitzka L.J. (1988) *Dunaliella*. In: Borowitzka M.A., Borowitzka L.J. (eds.), *Micro-algal Biotechnology*. Cambridge University Press, Cambridge, pp. 27-58.

Braumann, T., Vasmel, H., Grimme, L.H. and Amesz, J. (1986) Pigment composition of the photosynthetic membrane and reaction centre of the green bacterium *Prosthecochloris aestuarii*. *Biochim. Biophys. Acta*, **848**: 83-91.

Britton, G., Singh, R.K., Goodwin, T.W. and Ben-Aziz, A. (1975) The carotenoids of *Rhodomicrobium vannielii* (Rhodospirillaceae) and the effect of diphenylamine on the carotenoid composition. *Phytochemistry*, **14**: 2427-2433.

Britton, G., Malhotra, H.C., Singh, R.K. *et al.* (1976) Methoxyspheroidene and methoxyspheroidenone, two carotenoids from *Rhodopseudomonas sphaeroides*. *Phytochemistry*, **15**: 1971-1972.

Britton, G., Singh, R.K. and Goodwin, T.W. (1977) Carotenoid biosynthesis in *Rhodomicrobium vannielii*. Experiments with nicotine and 2-(4-chloro-phenylthio)triethylammonium chloride (CPTA). *Biochim. Biophys. Acta*, **488**: 475-483.

Buchecker, R. and Liaaen-Jensen, S. (1977) Absolute configuration of heteroxanthin and diadinoxanthin. *Phytochemistry*, **16**: 729-733.

Buchecker, R., Eugster, C.H., Kjøsen, H. and Liaaen-Jensen, S. (1974) Algal carotenoids. IX. Absolute configuration of ß,ε-caroten-2-ol, ß,ß-caroten-2-ol and ß,ß-carotene-2,2′-diol. *Acta Chem. Scand.*, **B28**, 449-452.

Buchecker, R., Liaaen-Jensen, S., Borch, G. and Siegelman, H.W. (1976) Carotenoids of *Anacystis nidulans*, structures of caloxanthin and nostoxanthin. *Phytochemistry*, **15**: 1015-1018.

Buchecker, R., Marti, U. and Eugster, C.H. (1984) Synthese von optisch aktiven Carotinoiden mit 3,5,6-trihydroxy-5,6-dihydro-ß-endgruppen. *Helv. Chim. Acta*, **67**: 2043-2046.

Burczyk, J. (1987) Cell wall carotenoids in green algae that form sporopollenins. *Phytochemistry*, **26**, 121-128.

Cardini, F., Giananneschi, M., Silva, A. and Chelli, M. (1987) Semi-ß-carotenone from the leaves of two cycads. *Phytochemistry*, **26**: 2029-2031.

Cardini, F., Britton, G. and Selva, A. (1989) A seco-carotenoid from the leaves of two cycads. *Phytochemistry*, **28**: 2793-2795.

Cardini, F., Britton, G., Selva, A. and Chelli, M. (1990) Occurrence of the very rare carotenoid triphasiaxanthin in the leaves of two cycads. *Biochem. Syst. Ecol.*, **18**: 317-319.

Chapman, D.J. and Haxo, F.T. (1963) Identity of ε-carotene and ε_1-carotene. *Plant Cell Physiol.*, **4**: 59-63.

Christensen, T. (1964) The gross classification of algae. In: Jackson, D.F. (ed.), *Algae and Man*. Plenum Press, New York, pp. 59-64.

Christensen, T. (1980) *Algae. A Taxonomic Survey*. Fasc. 1 AiO Tryk, Odense.

Czeczuga, B. (1986) Autumn carotenoids in the leaves of some trees. *Biochem. Syst. Ecol.*, **14**: 203-205.

Czeczuga, B. and Taylor, R.F. (1987) Carotenoid content in some species of the brown and red algae from the coastal area of New Zealand. *Biochem. Syst. Ecol.*, **15**: 5-10.

Eichenberger, W., Boschetti, A. and Michel, H.P. (1986) Lipid and pigment composition of a chlorophyll *b*-deficient mutant of *Chlamydomonas reinhardii*. *Physiol. Plant.*, **66**:589-594.

Fiksdahl, A. and Liaaen-Jensen, S. (1988) Diacetylenic carotenoids from *Euglena viridis*. *Phytochemistry*, **27**: 1447-1450.

Fiksdahl, A., Withers, N., Guillard, R.R.L. and Liaaen-Jensen, S. (1984a) Carotenoids of the Raphidophyceae - a chemosystematic contribution. *Comp. Biochem. Physiol.*, **78B**: 265-

271.
Fiksdahl, A., Withers, N. and Liaaen-Jensen, S. (1984b) Carotenoids of *Heterostigma akashiwo*. A chemosystematic contribution. *Biochem. Syst. Ecol.*, **12**: 355-356.
Foss, P., Guillard, R.R.L. and Liaaen-Jensen, S. (1984) Prasinoxanthin. A chemosystematic marker for algae. *Phytochemistry*, **23**: 1629-1633.
Foss, P., Guillard, R.R.L. and Liaaen-Jensen, S. (1986a) Carotenoids from eucaryotic ultraplankton clones (Prasinophyceae). *Phytochemistry*, **25**: 118-124.
Foss, P., Skulberg, O.M., Kilaas, L. and Liaaen-Jensen, S. (1986b) The carbohydrate moieties bound to the carotenoids myxol and oscillol and their chemosystematic applications. *Phytochemistry*, **25**: 1127-1132.
Foss. P., Skjetne, T. and Liaaen-Jensen, S. (1987a) Algal carotenoids. 36. NMR studies of prasinoxanthin - stereochemical analysis. *Acta Chem. Scand.* **B40**: 172-177.
Foss. P., Lewin, R.A. and Liaaen-Jensen, S. (1987b) Carotenoids of *Prochloron* sp. (Prochlorophyta). *Phycologia*, **26**: 142-144.
Francis, G.W., Knutsen, G. and Lein, T. (1973) Loroxanthin from *Chlamydomonas reinhardii*. *Acta Chem. Scand.*, **27**: 3359-3600.
Gieskes, W.W.C. and Kraay, G.W. (1986) Analysis of phytoplankton pigments by HPLC before, during and after mass occurrence of the microflagellate *Corymbellus aureus* during the spring bloom in the open North Sea in 1983. *Mar. Biol.*, **92**:45-52.
Gilmore, A.M. and Yamamoto, H.Y. (1991) Resolution of lutein and zeaxanthin using a nonendcapped, lightly carbon-loaded C-18 high performance liquid chromatographic column. *J. Chromatog.*, **543**:137-145.
Goodwin, T.W. (1971) Algal carotenoids. In: Goodwin, T.W. (ed.), *Aspects of Terpenoid Chemistry and Biochemistry*. London, Academic Press, pp. 315-356.
Goodwin, T.W. (1980) *The Biochemistry of Carotenoids*, 2nd. Ed. Vol. 1. Plants. Chapman and Hall, London.
Goodwin, T.W. (1984) *The Biochemistry of Carotenoids*, 2nd. Ed. Vol. 2. Animals. Chapman and Hall, London.
Goodwin, T.W. and Britton, G. (1988) Distribution and analysis of carotenoids. In: Goodwin, T.W. (ed.), *Plant Pigments*. Academic Press, London, pp. 61-134.
Grill, D. and Pfeifhöfer, H.W. (1985) Carotinoide in fichtennadelen II. Quantitative untersuchungen. *Phyton.*, **25**: 1-15.
Grung, M,, Bjerking, B., Borowitzka, M.A. *et al.* (1990) Alternative sources of astaxanthin, including secondary carotenoids of microalgae. Abstract, *9th. International Symposium on Carotenoids Kyoto*, p 217.
Guillard, R.R.L., Murphy, L.S., Foss, P. and Liaaen-Jensen, S. (1985) *Synechococcus* spp. as likely zeaxanthin-dominant ultraplankton in the North Atlantic. *Limnol. Oceanogr.*, **30**: 412-414.
Halfen, L.N., Pierson, B.K. and Francis, G.W. (1972) Carotenoids of a gliding organism containing bacteriochlorophylls. *Arch. Mikrobiol.*, **82**: 240-246.
Hallenstvet, G. M., Liaaen-Jensen, S. and Skulberg, O.M. (1979) Carotenoids of *Oscillatoria bornetii* f. *tenuis*. *Biochem. Syst. Ecol.*, **7**: 1-2.
Harashima, K. (1989) Carotenoids, quinones and other lipids, In: Harashima, K., Shiba, T. and Murata, N. (eds.), *Aerobic Photosynthetic Bacteria*. Springer-Verlag,Tokyo, pp. 125-148.
Harashima, K., Shiba, T. and Murata, N. (1989) *Aerobic Photosynthetic Bacteria*. Springer-Verlag, Tokyo.
Haxo, F.T. and Fork, D.C. (1959) Photosynthetically active accessory pigments of cryptomonads. *Nature*, **184**: 1051-1052.
Humbeck, K. (1990) Light-dependent carotenoid biosynthesis in mutant C-6D of *Scenedesmus obliquus*. *Photochem. Photobiol.*, **51**: 113-118.
Imhoff, J.F. and Trüper, H.G. (1977) *Ectothiorhodospira halochloris* sp. *nov.*, a new extremely halophilic phototrophic bacterium containing bacteriochlorophyll. *Arch. Mikrobiol.*, **114**:115-121.

Iroshnikova, G.A. and Kvitko, K.V. (1986) Studies on mutations modifying pigmentation in *Chlamydomonas reinhardii* strains with different ploidy IV. Estimation of the linkage group for the gene *Its-1*. *Genetika*, **22**:761-766.

Johansen, J.E., Svec, W.A. and Liaaen-Jensen, S. (1974) Carotenoids of the Dinophyceae. *Phytochemistry*, 13: 2261-2271.

Johansen, J.E., Liaaen-Jensen, S. and Borch, G. (1977) Studies on the absolute configuration of peridinin and dinoxanthin. In: Faulkner, D.J. and Fenical, W.H. (eds.), *Marine Natural Products Chemistry*. Plenum, New York, pp. 225-237.

Kjøsen, H., Norgård, S., Liaaen-Jensen, S. *et al.* (1976) Algal carotenoids. XV. Structural studies on peridinin. Part 2. Supporting evidence. *Acta Chem. Scand.*, **B30**: 157-164.

Kleinig, H. (1969) Carotenoids of siphonous green algae: A chemotaxonomic study. *J. Phycol.*, **5**: 281-284.

Kleinig, H. and Egger, K. (1967) Carotenoide der vaucheriales *Vaucheria* und *Botrydium* (Xanthophyceae). *Z. Naturforsch.*, **22b**, 868-872.

Ladygin, V.G., Semeonova, G.A., Chemerilova, V.I. *et al.* (1982) Structural organisation of chloroplast variability in light-sensitive allelic mutants of *Chlamydomonas reinhardii*. *Tsitologiya*, **24**: 391-399.

Liaaen-Jensen, S. (1977) Algal carotenoids and chemosystematics. In: Faulkner, D.J. and Fenical, W.H. (eds.), *Marine Natural Products Chemistry*. Plenum, New York, pp. 239-259.

Liaaen-Jensen, S. (1978) Marine carotenoids. In: Scheuer, P. (ed.), *Marine Natural Products*. Academic Press, New York, pp.2-73.

Liaaen-Jensen, S. (1985) Carotenoids of lower plants - recent progress. *Pure Appl. Chem.*, **57**: 649-658.

Liaaen-Jensen, S. (1990) Marine carotenoids - selected topics. *New J. Chem.*, **14**, 747-759.

Liaaen-Jensen, S. (1991) Marine carotenoids: recent progress. *Pure Appl. Chem.*, **63**: 1-12.

Mandelli, E.F. (1968) Carotenoid pigments of the dinoflagellate *Glenodium foliaceum* Stein. *J. Phycol.*, **4**: 347-348.

Maslov, V. G. and Chunaev, A.S. (1982) Study of primary photoprocesses in photosystem 2 of *Chlamydomonas* mutants by hole-burning spectroscopy. *Mol. Biol.*, **16**: 604-611.

Moore, T.A., Gust, D. and Moore, A.L. (1990) The function of carotenoid pigments and their possible involvement in the evolution of higher plants. In: Krinsky, N.I., Mathews-Roth, M.M. and Taylor, R.F. (eds.), *Carotenoids: Chemistry and Biology*. Plenum Press, New York, pp.223-228.

Nitsche, H. (1973) Heteroxanthin in *Euglena gracilis*. *Arch. Mikrobiol.*, **90**: 151-155.

Nitsche, H. (1974) Die identität von loroxanthin mit pyrenoxanthin, trollein und trihydroxy-α-carotin. *Arch. Microbiol.*, **95**: 79-90.

Norgård, S., Svec, W.A., Liaaen-Jensen, S. *et al.* (1974) Chloroplast pigments and algal systematics. *Biochem. Syst. Ecol.*, **2**: 3-6.

O'Kelly, C.J. (1982) Chloroplast pigments in selected marine Chaetophoraceae and Chaetosiphonaceae (Chlorophyta): The occurrence and significance of siphonaxanthin. *Bot. Mar.*, **25**: 133-137.

Owens, T.G., Gallagher, J.C. and Alberte, R.S. (1987) Photosynthetic light-harvesting function of violaxanthin in *Nannochloropsis* spp, (Eustigmatophyceae). *J. Phycol.*, **23**:79-85.

Pennington, F.C., Haxo, F.T., Borch, G. and Liaaen-Jensen, S. (1985) Carotenoids of Cryptophyceae. *Biochem. Syst. Ecol.*, **13**, 215-219.

Pennington, F.C., Guillard, R.R.L. and Liaaen-Jensen, S. (1988) Carotenoid distribution patterns in Bacillariophyceae (diatoms). *Biochem. Syst. Ecol.*, **16**: 589-592.

Powls, R. and Britton, G. (1977) A series of mutant strains of *Scenedesmus obliquus*, with abnormal carotenoid compositions. *Arch. Microbiol.*, **113**: 275-280.

Ragan, M.A. and Chapman, D.J. (1978) *A Biochemical Phylogeny of the Protista*. Academic Press, New York.

Ricketts, T.R. (1970) The pigments of the Prasinophyceae and related organisms.

Phytochemistry, **9**:1835-1843.

Riley, J.P. and Wilson, T.R.S. (1967) The pigments of some marine phytoplankton species. *J. Mar. Biol. Ass. U.K.*, **47**: 351-362.

Riesen, R., Pfander, H. and Schiedt, K. (1984) New carotenoid glycosides from *Spirulina* sp. Abstract, *7th. International Symposium on Carotenoids Munich*, P. 40.

Rønneberg, H., Foss, P., Ramdahl, T. *et al.* (1980) Occurrence and chirality of oscillaxanthin. *Phytochemistry*, **19**: 2167-2170.

Rønneberg, H., Andrewes, A.G., Borch, G. *et al.* (1985) CD correlation of C-2´ substituted monocyclic carotenoids. *Phytochemistry*, **24**: 309-319.

Rowan, K.S. (1989) *Photosynthetic Pigments of Algae*. Cambridge University Press, Cambridge.

Schmidt, K. (1978) Biosynthesis of carotenoids. In: Clayton, R.K. and Sistrom, W.R. (eds.), *The Photosynthetic Bacteria*. Plenum Press, New York, pp. 729-750.

Schmidt, K. (1980) A comparative study on the composition of chlorosomes (*Chlorobium* vesicles) and cytoplasmic membranes from *Chloroflexus aurantiacus* strain Ok-70-fl and *Chlorobium limicola* f. *thiosulfatophilum* strain 6230. *Arch. Mikrobiol.*, **124**: 21-31.

Schmidt, K. and Liaaen-Jensen, S. (1973) Bacterial carotenoids. XLII. New ketocarotenoids from *Rhodopseudomonas globiformis* (Rhodospirillaceae). *Acta Chem. Scand.*, **27**: 3040-3052.

Shimada, K., Hayashi, H. and Tasumi, M. (1985) Bacteriochlorophyll-protein complexes of aerobic bacteria, *Erythrobacter longus* and *Erythrobacter* species OCh114. *Plant Cell Physiol.*, **29**: 861-865.

Siefermann-Harms, D. (1985) Carotenoids in photosynthesis. I. Location in photosynthetic membranes and light-harvesting function. *Biochim. Biophys. Acta*, **811**: 325-355.

Siefermann-Harms, D., Hertzberg, S., Borch, G. and Liaaen-Jensen, S. (1981) Lactucaxanthin, an ε,ε-carotene-3,3´-diol from *Lactuca sativa*. *Phytochemistry*, **20**:85-88.

Stauber, J.L. and Jeffrey, S.W. (1988) Photosynthetic pigments in fifty-one species of marine diatoms. *J. Phycol.*, **24**:158-172.

Strain, H.H. (1951) The pigments of algae In: Smith, G.M. (ed.), *Manual of Phycology*. Chronica Botanica, Waltham, MA., pp. 243-262.

Strain, H.H. (1958) Chloroplast pigments and chromatographic analysis. *32nd Annual Priestly Lecture*. PennsylvaniaState University Press, University Park, PA.

Strain, H.H., Svec, W.A., Wegfahrt, P. *et al.* (1976) Algal carotenoids. XIV. Structural studies on peridinin. Part 1. Structural elucidation. *Acta. Chem. Scand.*, **B30**: 109-120.

Stransky, H. and Hager, A. (1970) Das carotinoidmuster und die verbreitung des lichtinduzierten xanthophyllcyclus in verschiedenen algenklassen. II. Xanthophyceae. *Arch. Mikrobiol.*, **71**:164-190.

Swartoff, T., Kramer, H.J.M. and Amesz, J. (1982) Thin-layer chromatography of pigments of the green photosynthetic bacterium *Prosthecochloris aestuarii*. Biochim. Biophys. Acta, 681:354-358.

Swift, I.A. and Millborrow, B.V. (1981) Stereochemistry of allene biosynthesis and the formation of the acetylenic carotenoids diadinoxanthin and peridinin from neoxanthin. *Biochem. J.*, **199**: 69-74.

Szabolcs, J. (1990) Plant carotenoids. In: Krinsky, N.I., Mathews-Roth, M.M. and Taylor, R.F. (eds.), *Carotenoids: Chemistry and Biology*. Plenum Press New York, pp.39-58.

Takaichi, S., Shimada, K. and Ishidsu, G. (1988) Monocyclic cross-conjugated carotenals from an aerobic photosynthetic bacterium, *Erythrobacter longus*. *Phytochemistry*, **27**: 3605-3609.

Takaichi, S., Furihata, K. and Harashima, K. (1991a) Light-induced changes of carotenoid pigments in anaerobic cells of the aerobic photosynthetic bacterium, *Roseobacter denitrificans* (*Erythrobacter* species OCh-114) -reduction of spheroidenone to 3,4-dihydrospheroidenone. *Arch. Microbiol.*, **155**: 473-476.

Takaichi, S., Furihata, K., Ishidsu, J. and Shimada, K. (1991b) Carotenoid sulphates from the aerobic photosynthetic bacterium, *Erythrobacter longus*. *Phytochemistry*, **30**:3411-3415.

Tangen, K. and Bjørnland, T. (1981) Observations on pigments and morphology of *Gyrodinium aureolum* Hulbert, a marine dinoflagellate containing 19´-hexanoyloxyfucoxanthin as the main carotenoid. *J. Plankton Res.*, **3**: 389-401.

Thayer, S.S. and Björkman, O. (1990) Leaf xanthophyll content and composition in sun and shade determined by HPLC. *Photosyn. Res.*, **23**: 331-343.

Uebelhart, R., Baumelor, A., Haag, A. *et al.* (1986) Optisch aktive 4,5-epoxy-4,5-dihydro-α-ionone und syntheses der stereoisomeren 4,5,4´,5´-diepoxy-4,5,4´,5´-tetrahydro-ε,ε-carotine und der sterische verlauf ihrer hyrolyse. *Helv. Chim. Acta.*, **69**: 816-834.

Vesk, M. and Jeffrey, S.W. (1987) Ultrastructure and pigments of two strains of the picoplankton alga *Pelagococcus subviridis* (Chrysophyceae). *J. Phycol.*, **23**:322-336.

Walton, T.J., Britton, G., Goodwin, T.W. *et al.* (1970) The structure of siphonaxanthin. *Phytochemistry*, **9**: 2545-2552.

Whittle, S.J. and Casselton, P.J. (1969) Peridinin as the major xanthophyll of the Dinophyceae. *Br. Phycol. Bull.*, **3**:602-603.

Whittle, S.J. and Casselton, P.J. (1975) The chloroplast pigments of the algal classes Eustigmatophyceae and Xanthophyceae. I. Eustigmatophyceae. *Br. Phycol. J.*, **10**:179-191.

Wieckowski, S. and Goodwin, T.W. (1966) Assimilatory pigments in cotyledons of 4 species of pine seedling grown in darkness and in light. *Phytochemistry*, **5**:1345-1348.

Withers, N.W., Cox, E.R., Tomas, R. and Haxo, F.T. (1977) Pigments of the dinoflagellate *Peridinium balticum* and its photosynthetic endosymbiont. *J. Phycol.*, **13**: 354-358.

Withers, N.W., Alberte, R.S., Lewin, R.A. *et al.* (1978a) Photosynthetic unit size, carotenoids and chlorophyll-protein composition of *Prochloron* sp., a prokaryotic green alga. *Proc. Natl. Acad. Sci. USA*, **75**: 2301-2305.

Withers, N.W., Vidaver, W. and Lewin, R.A. (1978b) Pigment composition, photosynthesis and fine structure of a non-blue green prokaryotic algal symbiont (*Prochloron* sp.) in a didemnid ascidian from Hawaiian waters. *Phycologia*, **17**: 167-171.

Withers, N.W., Fiksdahl, A., Tuttle, R.C. and Liaaen-Jensen, S. (1981) Carotenoids of the Chrysophyceae. *Comp. Biochem. Physiol.*, **63B**: 343-349.

Wright, S.W. and Jeffrey, S.W. (1987) Fucoxanthin pigment markers of marine phytoplankton analyzed by HPLC and HPTLC. *Mar. Ecol. Progr. Ser.*, **38**: 259-266.

Yokohama, Y. (1981a) Distribution of the green light-absorbing pigments, siphonaxanthin and siphonein in marine green algae. *Bot. Mar.*, **24**: 637-640.

Yokohama, Y. (1981b) Green light-absorbing pigments in marine green algae, their ecological significance and systematic significance. *Jpn. J. Phycol.*, **29**: 209-222.

Yokohama, Y. (1982) The distribution of lutein and its derivatives in marine green algae. *Jpn. J. Phycol.*, **30**: 311-317.

Young, A.J. and Britton, G. (1989) The distribution of α-carotene in the photosynthetic pigment-protein complexes of higher plants. *Plant Sci.*, **64**: 179-183.

Young, A.J., Barry, P. and Britton, G. (1989) The occurrence of ß-carotene-5,6-epoxide in the photosynthetic apparatus of higher plants. *Z. Naturforsch.*, **44C**: 959-965.

3

Carotenoids in pigment-protein complexes

A.J. Young

3.1 HIGHER PLANTS

3.1.1 Structure and composition of pigment-protein complexes

A number of different terminologies have been and are used in the description of the pigment-protein complexes found in higher plants. To avoid any confusion, those that are preferred by Thornber and co-workers will be used in this article wherever possible. These designations and the alternatives used by many other groups are given in Table 3.1, together with some of the characteristics of the individual complexes.

Each photosystem consists of two multi-protein binding complexes, namely the core complex or reaction centre and the antenna pigments organised into light-harvesting complexes (LHC). The light-harvesting complexes, particularly that associated with photosystem II (PSII), are rich in both chlorophylls and xanthophylls. Of the approximately 250 chlorophyll molecules found per P680 in PSII, for example, it has been estimated that ~80% is found in the chlorophyll *a/b*–xanthophyll complexes of LHCII (Ghirardi et al. 1986). The carotenoid compositions of the two photosystems are discussed in Sections 3.1.1 (a) and (b). Information concerning the polypeptide composition of these complexes is kept to a minimum, as this is outside the scope of this chapter, and readers are referred to many of the recent reviews or research papers on this subject: e.g. Siefermann-Harms 1985; Antonopoulou and Akoyunoglou 1986; Green 1988; Thornber et al. 1988, 1989; Melis 1991; Peter and Thornber 1991a,b;.

The analysis of the pigment composition of the individual pigment–protein complexes has always been hampered by the high levels of carotenoid (and often chlorophyll) that are found in the free pigment band seen on polyacrylamide gels after the use of detergents such as sodium dodecyl sulphate (SDS). More recently, and especially in the laboratory of Thornber and co-workers, the resolution of the pigment-protein complexes through the judicous use of glycosidic surfactants followed by Deriphat-polyacrylamide gel electrophoresis (PAGE), has enabled a much more accurate determination of their carotenoid content and composition. Indeed, recent reports (e.g. Vainstein et al. 1989; Peter and Thornber 1991b) show that there is no release of free pigment when individual complexes are resolved.

Table 3.1 The pigment-protein complexes of higher plants (from Peter and Thornber 1991a)

Complex	*Alternative name(s)*	*Chlorophyll* a/b	*% Total chlorophyll*
PSI		6.0	38
CCI	CP1, Chl*a*-P1	*a* only	20
LHC1a	LHCP1a	1.4	
LHC1b	LHCP1b	2.3	18
(others)			
PSII		2.0	63
CCII-RC		*a* + phaeophytin	1
CCIIa	CPIII, Chl*a*-P2, CP47	*a* only	10
CCIIb	CPIV, Chl*a*-P3, CP43	*a* only	
LHCIIa	Chl*a/b*-P1, CP29	2.5–3.5	3.3
LHCIIb	Chl*a/b*-P2, CP11	1.0–1.33	42
LHCIIc	CP27, CP26	1.8	3.3
LHCIId	CP24	0.83–1.0	3.3
LHCIIe		1.4	2.0

(a)Photosystem I

The pigment content and composition of PSI is not as well documented as that of PSII (see Section 3.1.1 (b)), largely because there are still difficulties in obtaining clean and relatively uncontaminated preparations of the individual complexes. PSI isolated from higher plants contains a core complex (CCI) which is enriched in, and probably contains only ß-carotene (often with small amounts of α-carotene) and chlorophyll *a*. The reports of xanthophylls such as lutein and violaxanthin being present in CCI probably result from contamination by the light-harvesting antenna complex, LHCI.

The light-harvesting complex of PSI, LHCI, has been isolated and shown to contain 80–120 molecules of chlorophyll. On the whole, LHCI has a higher ratio of chlorophyll *a*:*b* than LHCII at ~3.5:1 (see Table 3.1). The absorption spectra of three separate components of LHCI (LHCIa, LHCIb and LHCIc) isolated from bundle sheath cells of the C_4 plant, maize, are shown in Fig. 3.1. LHCIb has a ratio of chlorophyll *a*:*b* of 2–2.5:1, whilst LHCIa and LHCIc contain much smaller proportions of chlorophyll *b*, and have a ratio of 4–5:1 (Vainstein et al. 1989).

The carotenoid compositions of the different pigment–protein complexes of LHCI are not well documented, although it is known that, as in LHCII, carotenes (i.e. α- and ß-carotene) are absent and only xanthophylls such as lutein, neoxanthin and violaxanthin are present. It is interesting to note that Vainstein et al. (1989) have observed that LHCIa probably contains a much higher ratio of carotenoid:chlorophyll than do other higher plant pigment–protein complexes. The "xanthophyll cycle" (i.e. the enzymic de-epoxidation of violaxanthin into antheraxanthin and zeaxanthin, and the subsequent epoxidation reactions) is known to operate in LHCI (G. Noctor, A. Young and P. Horton, unpublished data; see also Chapter 7).

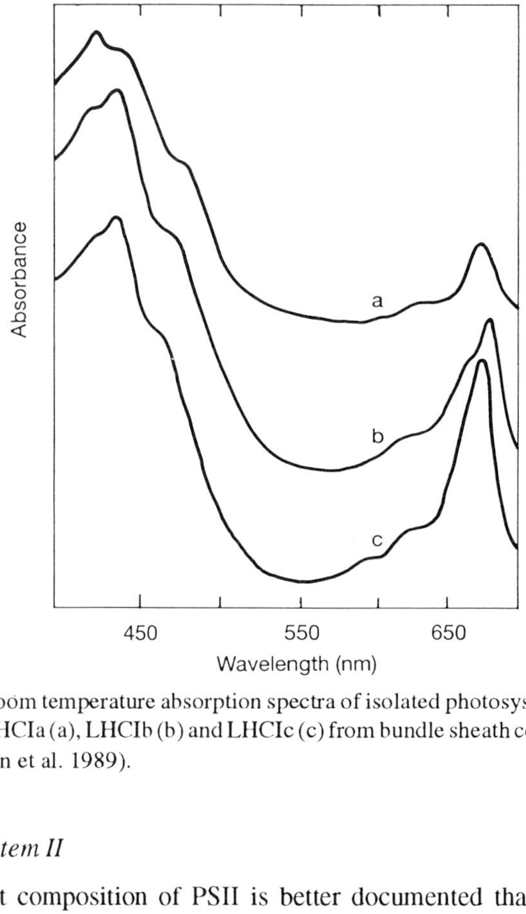

Figure 3.1 Room temperature absorption spectra of isolated photosystem I light harvesting complexes, LHCIa (a), LHCIb (b) and LHCIc (c) from bundle sheath cells of maize (redrawn from Vainstein et al. 1989).

(b) Photosystem II

The pigment composition of PSII is better documented than that of PSI. The absorption spectra of the isolated D1/D2/Cyt-*b*559 complex and of the oxygen-evolving core complex are shown in Fig. 3.2 (Newell et al. 1988). As in PSI, the reaction centre of PSII (CCII) contains only ß-carotene, phaeophytin and chlorophyll *a*. The first report on the pigment composition of the isolated D1/D2/Cyt-*b*559 complex was that of Nanaba and Satoh (1987), who found proportions of 5 chlorophyll *a*:2 phaeophytin:1 ß-carotene in the complex isolated from spinach.

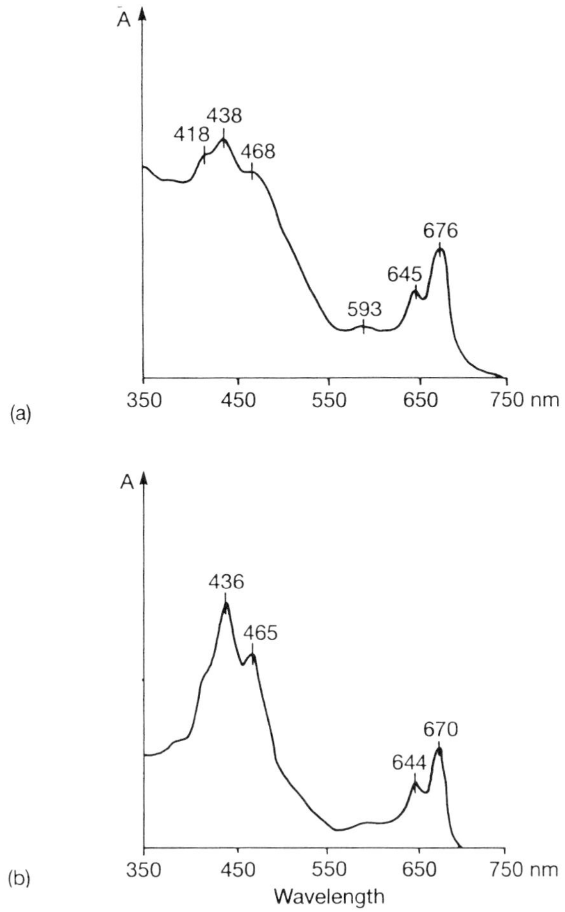

Figure 3.2 Absorption spectra of (a) D1/D2/Cyt *b*-559 complex, and (b) oxygen-evolving core complex of PSII isolated from pea. Spectra were measured in 50 mM Tris-Cl, pH 7.2 (chlorophyll = 5 $\mu g ml^{-1}$; reproduced from Newell et al. 1988)

The proportions of these pigments in the complex from pea was shown to be 4:2:1 (Barber et al. 1987). Other, more recent, reports give the proportions of pigments as 5–6:2:2 per reaction centre (Gounaris et al. 1990; Kobayashi et al. 1990; Telfer et al. 1991). This is in contrast to the earlier reports (e.g. Nanaba and Satoh 1987) where only one ß-carotene molecule per reaction centre (i.e. per two phaeophytin) was detected. This complex has also been isolated from the cyanobacterium *Synechocystis* 6803, and proportions of 16:2:2 have been reported (Gounaris et al. 1989). Although this complex was found to have a high ratio of chlorophyll *a*:phaeophytin, the levels of phaeophytin and ß-carotene are in agreement with those reported from the studies on the D1/D2/Cyt-*b*559 complex from higher plants.

The pigment stoichiometry of the D1/D2/Cyt-*b*559 complex is quite different from that of the reaction centre complex from phototrophic bacteria, in which molar proportions of 4 bacteriochlorophyll:2 bacteriophaeophytin:1 carotenoid have been reported (see Section 3.3). The reaction centre from higher plants, therefore, probably contains an additional two molecules of chlorophyll *a* and one of ß-carotene, compared to the bacterial complex. The role of these pigments in the higher plant complex is not known.

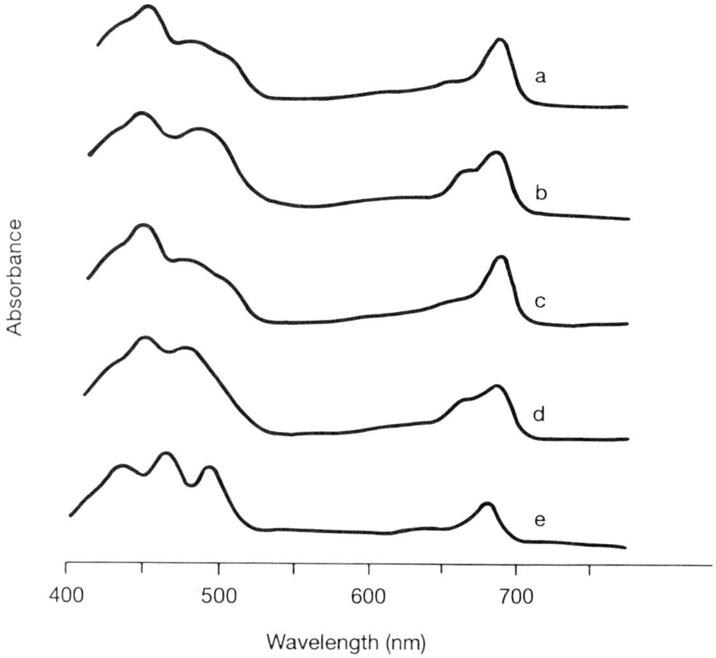

Figure 3.3 Room temperature absorption spectra of purified LHCII pigment–protein complexes; LHCIIa (a), LHCII b (b); LHCII c (c), LHCIId (d) and LHCIIe (e). Spectra were recorded of polyacrylamide gel strips containing each complex (redrawn from Peter and Thornber 1991b).

There are contradictory reports concerning the orientation of the ß-carotene molecules in the reaction centre of PSII. van Dorssen et al. (1987) have reported the presence of two distinct forms of ß-carotene, oriented perpendicular to each other, whereas Newell et al. (1991) observed only one form of carotene molecule.

The light-harvesting complexes associated with PSII show characteristic pigment compositions, and a detailed study of the carotenoid content of the five known complexes of LHCII has recently been carried out by Peter and Thornber (1991b). The absorption spectra of these five complexes are shown in Fig. 3.3, and the distribution of the xanthophylls, expressed on the basis of chlorophyll *a*, is given in Table 3.2. As predicted, ß-carotene has been shown to be completely absent from

LHCII and only xanthophylls are found in association with chlorophylls *a* and *b*. Even when Deriphat-gels are used in order to minimise the levels of free pigment, not all the violaxanthin has been found to be directly associated with these complexes, and indeed this is the only pigment found in the free pigment fraction of the gels. It could therefore be concluded that violaxanthin is probably the only photosynthetic pigment in the thylakoids that is not directly associated with protein (Peter and Thornber 1991b). It should be noted that violaxanthin is one of the main pigments of the plastid envelope (see Section 3.1.1 (d)), which may contribute to contamination of thylakoid preparations. The distribution of the other xanthophyll cycle carotenoids, antheraxanthin and zeaxanthin, when formed either through *de novo* synthesis or via the xanthophyll cycle, on these improved gels has not been reported, although preliminary data suggest that zeaxanthin may also be released into the free pigment, as it is in the older and less efficient gel systems (A. Pascal, A. Young and P. Horton, unpublished data).

Table 3.2 Pigment composition of the main light-harvesting complex of photosystem II, LHCII isolated from barley (from Peter and Thornber 1991b). The values are expressed per molecule of chlorophyll *a*

	Neoxanthin	*Violaxanthin*	*Lutein*	*Chl.* b
LHCIIa	0.11	0.22	0.22	0.44
LHCIIb	0.125	0.0625	0.25	0.75
LHCIIc	0.14	0.07	0.28	0.57
LHCIId	0.05	0.05	0.25	1.10
LHCIIe	0.5	1.5	1.0	0.7

The enrichment of xanthophylls in LHCIIe, recently identified by Thornber and co-workers as a major component of LHCII, is particularly interesting, although the role of the complex is unclear. Peter and Thornber (1991b) have suggested that this complex may have a photoprotective role and that it is not necessarily related structurally to PSII. The high levels of violaxanthin found in LHCIIe (see Table 3.2) are of particular interest due to their possible involvement in the xanthophyll cycle (see Chapter 7).

The technique of isoelectric focussing can provide particularly impressive separation of the complexes from BBY particles, with the minor complexes such as CP24 and CP26 being well resolved. The carotenoid composition of IEF-resolved complexes has been determined for both dark and light-pre-treated BBY in order to examine the distribution of the xanthophyll cycle pigments under these conditions. Preliminary results showing a heterogeneous organisation of carotenoids in these complexes confirm the findings of Thornber. Neoxanthin is present in all complexes and probably has a vital role in the organisation of these LHC. The xanthophyll cycle carotenoids are found in all complexes, although they are present in relatively higher levels (as a percentage of total chlorophyll) in the minor LHC compared to the main light-harvesting complex, LHCIIb (Ruban, A.V., Pascal, A., Horton, P. and Young,

A.J., unpublished data).

(c) Photosystem stoichiometry

Whilst it is outside the scope of this chapter to go into any detail concerning the stoichiometry of PSI and PSII in higher plants, the reader should be aware of some of the factors that can regulate this stoichiometry and, hence, the photosynthetic pigment content and composition. The photosystem stoichiometry of different photosynthetic organisms has been determined (see Melis 1991, for a review). Higher plant chloroplasts have a ratio of PSII:PSI of 1.8:1.0, whereas green algae have a ratio of 1.4:1.0 and cyanobacteria and red algae (Rhodophyceae) a ratio of 0.3–0.7:1.0. The ratio of PSII:PSI seen in developing higher plant chloroplasts (under intermittent light) and in a range of chlorophyll *b*-less mutants is generally > 2.0:1.0 (Melis 1991).

The carotenoid compositions of a series of *chlorina* (chlorophyll *b*-less) mutants of *Hordeum vulgare* have been determined (Simpson et al. 1985). The majority of the *chlorina* f2 mutants are characterised by a high level of ß-carotene and much reduced levels of all the xanthophylls. Only one mutant (*chlorina* 106) possessed an unusual carotenoid composition in that it had a much higher ratio of xanthophylls (lutein + neoxanthin + violaxanthin) to chlorophyll *a*. The location of these xanthophylls in the individual pigment-protein complexes of these mutants is not known. Rock and Zeevaart (1991) have reported the occurrence of a mutant of *Arabidopsis thaliana* that is deficient in violaxanthin and neoxanthin, but possesses higher levels of zeaxanthin than the wild type (see Chapter 7 for full details). Again, it has not been determined which particular complexes may be deficient in these carotenoids and it will be of interest to see how the absence of neoxanthin in these complexes affects their organisation as it has always been thought that this particular carotenoid has an important role in the assembly of the LHC. Should sufficient plant material be available from these mutants (which are difficult to grow), it may be possible to conduct reconstitution experiments investigating carotenoid binding in the LHC from *Arabadopsis*. Such experiments may reveal a great deal of information concerning the role of carotenoids in the organisation of LHC in higher plants.

(d)The plastid envelope

The biochemistry of plastid envelopes has recently been reviewed by Douce and Joyard (1990) and Joyard et al. (1991), in whose laboratory much of the work on envelopes has been carried out. The envelope may account for 1%–2% of plastid protein and its area has been estimated as 5.9×10^{-11} m^2 per chloroplast; i.e. about 10–20 times less than the thylakoids from a "typical" C_3 plant (Joyard et al. 1991).

The biosynthesis of the carotenoids found in leaves is located within the plastid itself and it is generally assumed that biosynthesis takes place within the thylakoid membrane (see Chapter 4). A role for the plastid envelope in the biosynthesis of these pigments has been suggested, particularly as non-green plastids are also capable of synthesising many, if not all, of the carotenoids found in photosynthetic tissues (Douce and Joyard 1990). Enzyme activity for carotenoid biosynthesis has been located within isolated envelope preparations; Lütke-Brinkhaus et al. (1982) detected phytoene dehydrogenase activity and Costes et al. (1979) have reported zeaxanthin epoxidase activity. Siefermann-Harms et al. (1978) have also demonstrated that the xanthophyll cycle operates in the chloroplast envelope. Much work is needed, however, before any positive role in carotenoid biosynthesis is assigned to the plastid envelope. It is important to note that during the course of plastid development (see Chapter 6), the nature of the envelope changes very little (Douce and Joyard 1990).

Envelope membranes isolated from both green and non-green plastids are yellow, due to the presence of carotenoids. The presence of chlorophyll in these preparations is usually taken as evidence of contamination by thylakoid membranes. All membrane preparations made to date are highly enriched in violaxanthin (which generally accounts only for 15%–25% of total carotenoid in higher plant thylakoids), although the other "normal" chloroplast carotenoids are also present (Jeffrey et al. 1974; Douce and Joyard 1979; Douce et al. 1985). ß-Carotene levels are very much reduced and the ratio of total xanthophyll:ß-carotene has generally been estimated as approximately 6:1 in chloroplast envelopes and nearer 3:1 in the thylakoid membranes (Table 3.3; Joyard et al. 1991).

Table 3.3 The carotenoid content of thylakoid and envelope fractions of isolated spinach chloroplasts (K.E. Pallett and J. Joyard, unpublished data). The corrected data for the envelopes account for possible contamination of the envelope preparation by chlorophyll and other thylakoid pigments. Figures in parentheses represent % total carotenoid

	Thylakoid (μg/mg protein)	*Envelope (μg/mg protein)*	*Envelope (μg/mg protein) corrected*
ß-Carotene	8.7 (26.1)	1.3 (10.3)	0.9 (8.0)
Violaxanthin	6.9 (20.7)	7.4 (58.7)	7.1 (63.4)
Antheraxanthin	0.6 (1.8)	0.4 (3.2)	0.3 (2.7)
Lutein	12.8 (38.4)	2.8 (22.2)	2.3 (20.5)
Neoxanthin	4.3 (12.9)	0.7 (5.6)	0.6 (5.4)
Total carotenoid	33.3	12.6	11.2
Total chlorophyll	151.8	5.7	—

Envelope preparations from green (e.g. spinach) and non-green tissues (e.g. sycamore amyloplasts) have very similar pigment compositions. The absorption spectra of envelopes isolated from spinach chloroplasts, cauliflower proplastids and potato tuber amyloplasts are very similar (R. Douce, J. Joyard and P. Barry, unpublished data). These workers found that the overall carotenoid compositions of these preparations were very similar, with the exception that the presence of (9Z)-

neoxanthin was restricted to envelopes obtained from green tissues and that the all-*E* isomer was found in the non-green plastids. This observation fits in with current ideas concerning the general distribution of xanthophyll isomers in photosynthetic and non-photosynthetic tissues, i.e. it is the 9*Z* isomer of neoxanthin that predominates in photosynthetic tissues. Block and co-workers (1983) have reported that both the inner and outer membranes contain carotenoids, although the outer membrane may be slightly richer in neoxanthin. It is not known whether the carotenoids present in envelopes are free or bound to protein in pigment–protein complexes, as in the thylakoid.

The cytoplasmic membranes from the cyanobacterium *Anacytis nidulans* have been shown to be preferentially enriched in zeaxanthin, whereas the thylakoids are rich in ß-carotene; the ratio of xanthophyll:ß-carotene is therefore higher in the membranes than in the thylakoid (Omata and Murata 1983). This pigment composition is very similar to that seen in higher plants if violaxanthin (not detected in the cyanobacteria) is replaced by zeaxanthin.

Carotenoids may now be considered to be genuine components of the plastid envelope and Joyard et al. (1991) have put forward a strong argument for an active role for these membranes in carotenogenesis and, possibly, in chlorophyll biosynthesis.

3.2 ALGAE

3.2.1 Organisation of the photosynthetic apparatus in algae

The photosynthetic apparatus of the green algae is very similar to their higher plant counterparts. However, those of the Chromophytes, or chlorophyll *c*-containing algae, are very different, both in terms of structure and in pigment composition. As in the higher plants all the pigments that are active in photosynthesis in algae are bound to proteins. The pigment compositions of the individual pigment–protein complexes have been studied in many marine and fresh water algae, although the resolution of the complexes and their pigments is not as well defined as for higher plants (see Section 3.1). Improvements are being made all the time (Berkaloff et al. 1990), especially with the use of newer detergents (e.g. in the resolution of the fucoxanthin–chlorophyll *a/c* complex in the chromophyte algae), and progress in recent years has been impressive. Many studies still report high levels of carotenoid in the free-pigment fraction of gels and, as a result, there is a degree of uncertainty concerning the origin of this "free-pigment".

(a)The core complexes

In comparison to the light-harvesting complexes, little attention has been paid to the pigments of PSI isolated from algae. Several workers have isolated PSI-enriched fractions from a range of algae by use of sucrose density gradients. As in higher plants (see Section 3.1), the core complexes of PSI and PSII isolated from a range of algae contain largely ß-carotene (or α-carotene in the case of *Codium fragile*; Benson and Cobb 1983) and chlorophyll *a*. Different isolation procedures, especially choice of detergent, have marked effects on the final pigment content of isolated CCI and CCII and it is therefore difficult to provide any definitive figures for the composition of these core complexes. Anderson and Barrett (1986) reported that, in addition to P700, CCI contains 40–60 molecules of chlorophyll *a* and 5-10 molecules of ß-carotene bound to a protein of 68 000 M_r. Trace quantities of xanthophylls have been reported in some preparations, although these are likely to result from contamination by the light-harvesting complexes, LHCI and LHCII (Anderson and Barrett 1986).

(b) The light-harvesting complexes

The diverse carotenoid structures that are found in the algae (see Chapter 2) are primarily xanthophylls that are associated specifically with the light-harvesting complexes (see Table 3.4). There has been much interest, therefore, in isolating and characterising these light-harvesting complexes. Wilhelm (1990) has published a comprehensive review of the nature of the light-harvesting systems of the green algae and of the chromophytes, in which he identified four distinct light-harvesting systems in the chlorophyll *b*-containing algae (although more may exist):

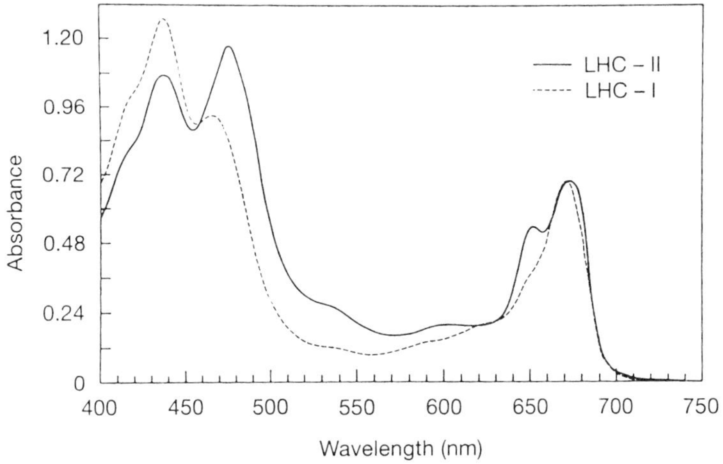

Figure 3.4 Absorption spectra of the siphonaxanthin-chlorophyll *a/b* protein of photosystems I and II isolated from the green alga, *Codium* (reproduced from Chu and Anderson 1985).

(i) A chlorophyll *a/b*–lutein antenna system similar to that seen in higher plants (see Section 3.1), found in many members of the Chlorophyceae (e.g. *Chlorella vulgaris*, Menger et al. 1984; *C. fusca,* Grimme et al. 1987; Knoetzel et al. 1988; *Chlamydomonas reinhardii*, Knoetzel et al. 1988; *Dunaliella tertiolecta*, Sukenik et al. 1987).

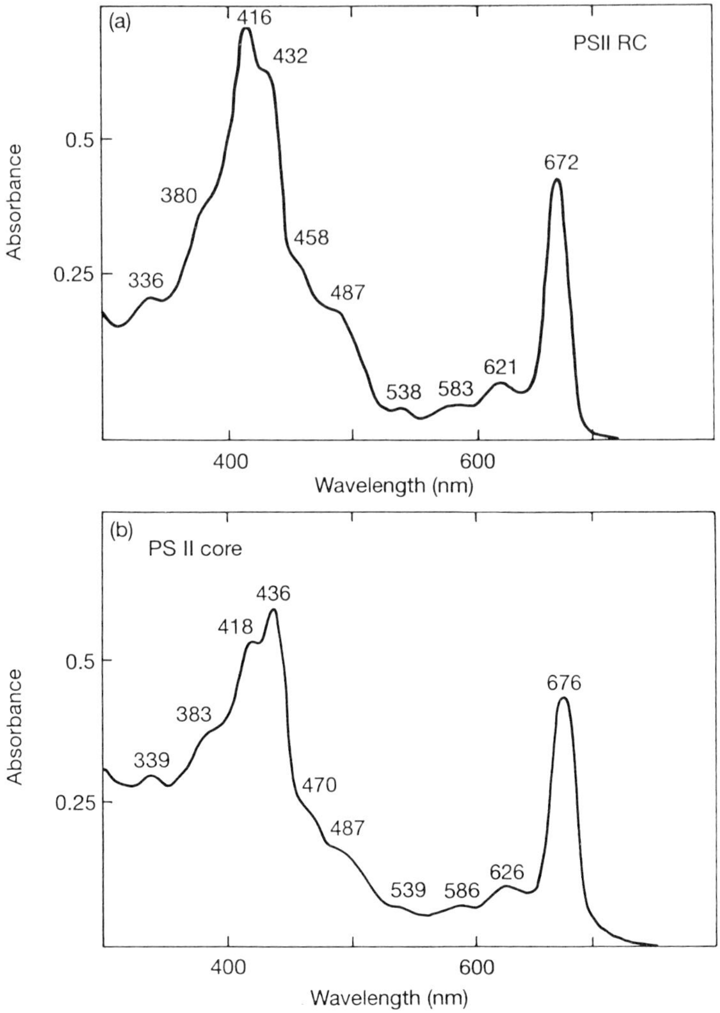

Figure 3.5 Absorption spectra of (a) inact cells of *Micromonas* sp. (Prasinophyceae), and (b) the prasinoxanthin-chlorophyll *a/c* light-harvesting complex (LHCI) of *Micromonas* sp. (reproduced from Wilhelm et al. 1986).

(ii) A chlorophyll *a/b*–siphonaxanthin–siphonein system (see Fig. 3.4), typically found in a range of algae inhabiting deep waters, e.g. *Bryopsis maxima* (Nakayama et al. 1986) and *Acetabularia mediterranea* (Green 1988). It has been reported that these algae possess a higher ratio of xanthophyll (i.e. siphonaxanthin and siphonein):chlorophyll *a* than those algae that have a chlorophyll *a/b*–lutein antenna system.

(iii) A chlorophyll *a/b/c*–prasinoxanthin system (see Fig. 3.5), typical of the Prasinophyceae; the proportions of chlorophyll *a*:*b*:*c*:total xanthophyll =

100:115:20:50 (Wilhelm et al. 1986; Wilhelm and Lenartz-Weiler 1987). The xanthophyll composition of the isolated light-harvesting complexes is dependent on the light regime experienced during growth (Wilhelm 1990).

(iv) A chlorophyll *a/b*–diadinoxanthin system, as seen in *Euglena*; the molar ratio of total xanthophyll:chlorophyll *a* in *Euglena* has been reported to be 30:100 (Cunningham and Schiff 1985). This value does vary and appears to be dependent largely on the ontogenic stage of the cells. Thus, Brandt and Wilhelm (1990) reported values ranging between 40 molecules of diadinoxanthin:100 molecules of chlorophyll *a* and 90:100 in cells at the onset of the light period and at the middle of the light phase, respectively.

Table 3.4 Light-harvesting pigment–protein complexes from chlorophyll *b*-containing algae (from Wilhelm 1990). The pigment content is expressed per 1000 chlorophyll *a*. 1 Kan & Thornber (1976); 2 Menger et al. (1984);3 Wilhelm and Lenartz-Weiler (1987); 4 Delepaire and Chua (1979); 5 Nakayama et al. (1986); 6 Nakayama & Okada (1990); 7 Chu & Anderson (1985); 8 Cunningham & Schiff (1985)

Chl. *a/b*–lutein LHCII
Chl *a* (1000), lutein (120), violaxanthin (100), neoxanthin (60) (1)
Chl *b* (900), lutein (120), violaxanthin (40), neoxanthin (80), loroxanthin (160) (2)
Chl *b* (666), lutein (67), violaxanthin (33), neoxanthin (44), loroxanthin (119) (3)
Chl *a/b* - lutein LHCI
Chl b (270), carotenoids not determined(4)
Chl *a/b* - siphonaxanthin LHCII
Chl *b* (1160), siphonaxanthin (510), siphonein (152), neoxanthin (278) (5)
Chl *b* (1333), siphonaxanthin (333), siphonein (166), neoxanthin (166) (6)
Chlorophyll *a/b* - siphonaxanthin LHCI
Chl *b* (555), carotenoids not determined (7)
Chlorophyll *a/b* - diadinoxanthin LHCII
Chl *b* (385), neoxanthin (66), diadinoxanthin (225) (8)
Chlorophyll *a/b/c*- prasinoxanthin
Chl *b* (1150), Chl *c* (175), neoxanthin (107), violaxanthin (90), prasinoxanthin (222) (3)

In the chlorophyll *c*-containing algae, the range of carotenoids and light-harvesting polypeptides is much greater than in the chlorophyll *b*-containing algae described above (Wilhelm 1990). Of the carotenoids identified in these algae, fucoxanthin and its 19′-acyloxy derivatives (19′-hexanoyloxyfucoxanthin and 19′-butanoyloxyfucoxanthin) are probably the most widely distributed, although diadinoxanthin (e.g. in *Pleurochloris meiringensis*, Wilhelm et al. 1988), antheraxanthin (e.g. in *Chrysosphera* sp., Alberte and Anderson 1986; *Cryptomonas rufescens*, Lichtlé et al. 1987), violaxanthin (e.g. in *Nannochloropsis salina*, Fig. 3.6; Brown 1986) and, especially, peridinin (e.g. in *Glenodinium* sp. and *Gonyaulax polyhedra*, Fig. 3.7; Prézelin and Haxo 1976), have also been identified as major light-harvesting pigments in some species of algae (see Table 3.5).

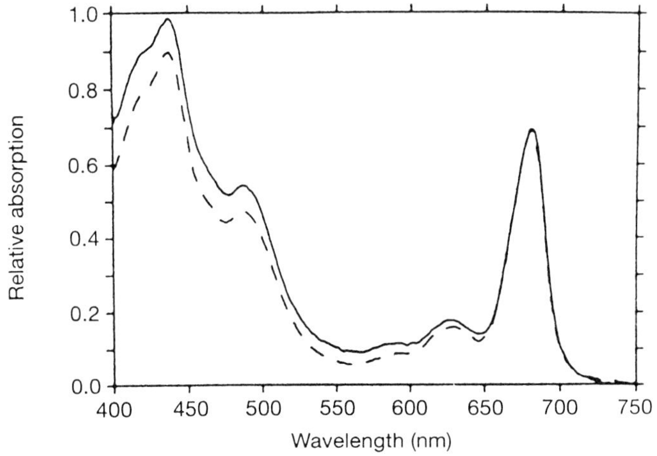

Figure 3.6 Absorption spectrum of cells of *Nannochloropsis salina* clone GSB Sticho (——) and *Nannochloropsis* sp. CAR A2 (- - - - -). These strains contain violaxanthin (~ 60% of total carotenoid) and vaucheriaxanthin as their major light-harvesting carotenoids. Strain CAR A2 lacks canthaxanthin (Owens et al. 1987).

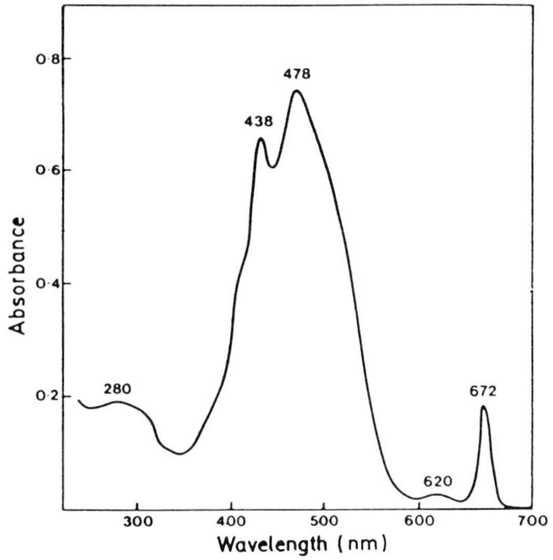

Figure 3.7 Absorption spectrum of the peridinin–chlorophyll *a* protein from *Glenodinium* sp. (Reproduced from Prézelin and Haxo 1976).

The fucoxanthin–chlorophyll *a*/*c* protein complex has attracted much interest in recent years but there has been only partial success in the isolation of this complex by using detergents. In the diatom *Phaeodactylum tricornutum* the pigments have been reported to exist in the proportions 1.0:0.09:0.28:2.22 (chlorophyll *a*:*c1*:*c2*:fucoxanthin) in this complex (Owens and Wold 1986). The absorption spectrum of a fucoxanthin–chlorophyll *a/c* complex from *Pavlova lutherii* (Hiller et al. 1988) is shown in Fig. 3.8. The spectroscopic properties of this complex and complexes obtained from similar studies were changed with respect to those in intact cells, i.e. there was a pronounced blue-shift in the isolated complex. It has become clear from recent work by Mimuro and Katoh (1991) that this blue-shift is directly related to the dissociation of the complex.

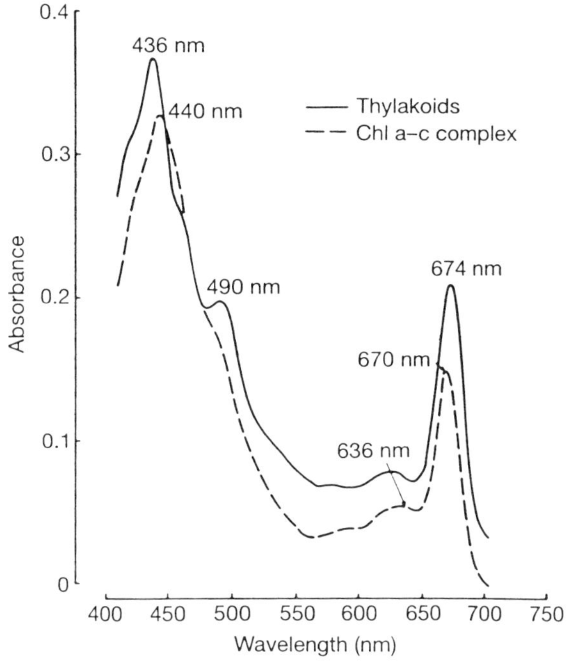

Figure 3.8 Absorption spectra of thylakoids (——) and the fucoxanthin–chlorophyll *a/c* complex (- - - - -)from *Pavlova lutherii* (Reproduced from Hiller et al. 1988).

Using a different detergent (decanoyl sucrose) Katoh et al. (1989) isolated a brown-orange complex of 590 000 M_r from the phaeophycean alga *Dictyota dichotoma* which they have termed the fucoxanthin–chlorophyll *a/c* assembly (Mimuro and Katoh 1991). The pigment composition per dimer in *D. dichotoma* was estimated to be 13:3:10:1 for chlorophyll *a*:chlorophyll *c*:fucoxanthin:violaxanthin (Katoh et al. 1989; Mimuro et al. 1990). Mimuro et al. (1990) identified two functionally active forms of fucoxanthin and one inactive species in the complex. A similar supramolecular complex was found in *Petalonia fascia* (Phaeophyceae) and was shown to consist of seven units of a 4.8 S fucoxanthin–chlorophyll *a/c*–protein and contained 128 molecules of chlorophyll *a*, 27 of chlorophyll *c*, 69 of fucoxanthin and 8 of violaxanthin (Katoh and Ehara 1990). The absorption spectrum of the intact supramolecular complex is shown in Fig. 3.9. The similarity between these two algae suggests that the assembly of the fucoxanthin–chlorophyll *a/c* complex may be very similar in all brown algae.

The analysis of complexes from widespread macro-algae such as *Laminaria* spp. and *Fucus* spp. has been greatly hampered both by their polysaccharide walls and the general lability of these complexes. In general, the pigment composition of the light-harvesting complexes from these algae shows an enrichment of chlorophyll *c*, fucoxanthin and violaxanthin and a depletion in chlorophyll *a* relative to the chloroplasts (Passaquet et al. 1991). The pigment composition of the complexes from these and other brown algae is shown in Table 3.5.

Table 3.5 Light-harvesting pigment–protein complexes from chlorophyll *c*-containing algae (after Wilhelm 1990). The pigment content is expressed per 1000 chlorophyll *a*. 1 . Prézelin & Haxo 1976; 2 Boczar 1985; 3 Boczar & Prézelin 1986; 4 Gugliemelli 1984; 5 Friedman & Alberte 1984;6 Owens & Wold 1986;7Caron & Brown 1987;8 Katoh et al. 1989; 9 Mimuro et al. 1990;10 Katoh & Ehara 1990;11 Passaquet et al. 1991; 12 Wilhelm et al. 1988; 13 Brown 1986.

Chl *a*-peridinin
Peridinin (4000) (1)
Chl *a/c*2-protein
Chl *c* (400) (2)
Chl *a/c*2-peridinin
Chl *c* (200), carotenoids not determined (3)
Chl *a/c*-fucoxanthin
Chl *c* (400), fucoxanthin (2000) (4)
Chl *c* (500), fucoxanthin (2500) (5)
Chl *c* (370), fucoxanthin (2200) (6)
Chl *c* (312), fucoxanthin (1250), diadinoxanthin (217) (7)
Chl *c* (231), fucoxanthin (770), violaxanthin (77) (8,9)
Chl *c* (211), fucoxanthin (539), violaxanthin (63) (10)
Chl *c* (180–300), fucoxanthin (610–1070), violaxanthin (60–300), ß-carotene (20–40) (11)
Chl *a/c*-diadinoxanthin
Chl *c* (224), diadinoxanthin (264), heteroxanthin (148) (12)
Chl *a*-violaxanthin
Neoxanthin (20), violaxanthin (260), vaucheriaxanthin (250) (13)

(c)Other membrane-bound carotenoids

The three membrane systems that are found in the cyanobacteria (the thylakoid membrane, the cytoplasmic membrane and the outer membrane) all possess carotenoids, but only the thylakoid membrane contains chlorophyll *a*. A carotenoprotein (containing 3′-hydroxyechinenone) has been identified in aqueous extracts from *Spirulina maxima*, *Anabaena flosaquae* and *Microcystis aeruginosa* (Holt and Krogmann 1981). This was shown to have a stoichiometry of 20–40 carotenoid molecules per mol protein (16 000 M_r). Bullerjhan and Sherman (1986) identified a carotenoid-binding protein (45 000 M_r) in the cytoplasmic membrane of *Synechocystis* PCC 6714. The outer membrane from this alga contained zeaxanthin, myxoxanthophyll, two carotenoid glycosides, echinenone and an unidentified polar

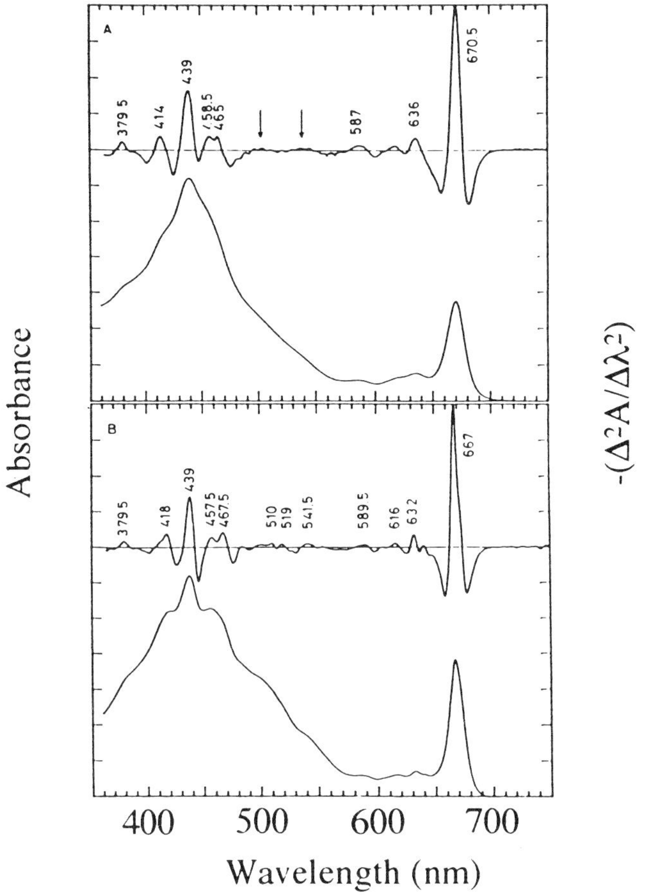

Figure 3.9 Absorption spectra and their second derivatives of the intact fucoxanthin–chlorophyll *a/c* complex (FCPA) isolated from *Dictyota dichotoma*: Spectra are shown at (A) room temperature and (B) at 77 K. For clarity, the second derivative spectra are displayed upside down (reproduced from Katoh et al. 1989).

carotenoid, with the first four of these accounting for more than 75% of total carotenoid from the outer membrane (Omata and Murata 1983, 1984; Jürgens and Weckesser 1985; Jürgens et al. 1985). The function of these carotenoids in non-photosynthetic membranes is not known, although a structural role is most likely.

The orientation of carotenoids in the outer membrane of *Synechococcus* PCC 6307 and *Synechocystis* PCC 6714 has been examined by linear dichroism, which has shown a predominant orientation of the carotenoid transition moments perpendicular to the *x*, *y*-plane (Jürgens and Mäntele 1991). The absorption spectra of these membrane-bound carotenoids are shown in Fig. 3.10.

Berkaloff (1977) observed that the extrachloroplastic pigmented globules that accumulate in *Protosiphon botryoides* during encystment contain lutein- and canthaxanthin-carotenoproteins. Berkaloff suggested that these proteins originated from the degenerating chloroplast and may be used as a storage pool, allowing the rapid regreening of the cells.

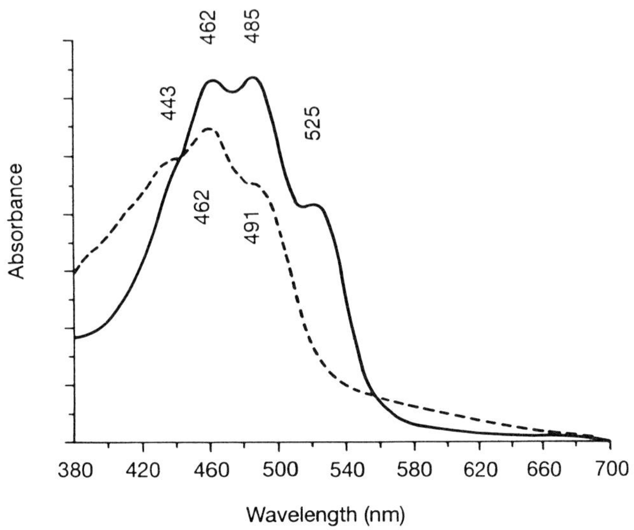

Figure 3.10 Absorption spectra of the aqueous outer membrane suspensions from *Synechocystis* PCC 6714 (——) and *Synechocystis* PCC 6307 (- - - -). Redrawn from Jürgens and Mäntele (1991).

3.3 PHOTOTROPHIC BACTERIA

A wide variety of carotenoids can be found in the chromatophores of the phototrophic bacteria (see Chapter 2). The pigment composition of the pigment–protein complexes found in phototrophic bacteria has been reviewed by Cogdell and Thornber (1980), Thornber et al (1983), Drews (1985), Siefermann-Harms (1985) and Thornber (1986). Many of the details referring to the organisation of the carotenoids in the complexes from phototrophic bacteria can also be found in Chapter 8, and this chapter therefore only serves as a brief introduction to this topic.

Under anaerobic conditions the purple bacteria develop photosynthetic membranes which contain high levels of pigment–protein complexes in the intracytoplasmic (chromatophore) membranes (Fig. 3.11). The vast bulk of pigment is associated with the antenna complexes, which serve to harvest light energy and transfer the excitation energy to the reaction centres. The reactions involved in the energy transfer process in these complexes are discussed in detail in Chapter 8.

In comparison to the reaction centres (see below), the structures of the light-harvesting complexes are much less well understood. These complexes are defined on the basis of the near-infrared absorbance peak or peaks of the bacteriochlorophyll–

proteins with the prefix "B" indicating that the bulk of the pigment absorbs at this wavelength. Three types of antenna complex have been identified, namely B875/B890 of *Rhodobacter sphaeroides* and B800–850 and B800–820 of *Rhodospirillum rubrum* (Thornber et al. 1983). The core antenna are closely associated, and are in fixed stoichiometry, with the reaction centre. The B875 complex, which is found in a large number of species, has a ratio of bacteriochlorophyll:carotenoid of 1:1 (e.g. in *Rb. sphaeroides* and *Rb. capsulatus*). The B890 complex of *R. rubrum* and *Chromatium vinosum* has a ratio of bacteriochlorophyll:carotenoid (spirilloxanthin in *R. rubrum*) of 2:1. The thermophilic purple sulphur bacterium, *Chromatium tepidium*, possesses two light-harvesting complexes, a B920 antenna complex and a B800–850 complex (Garcia et al. 1987).

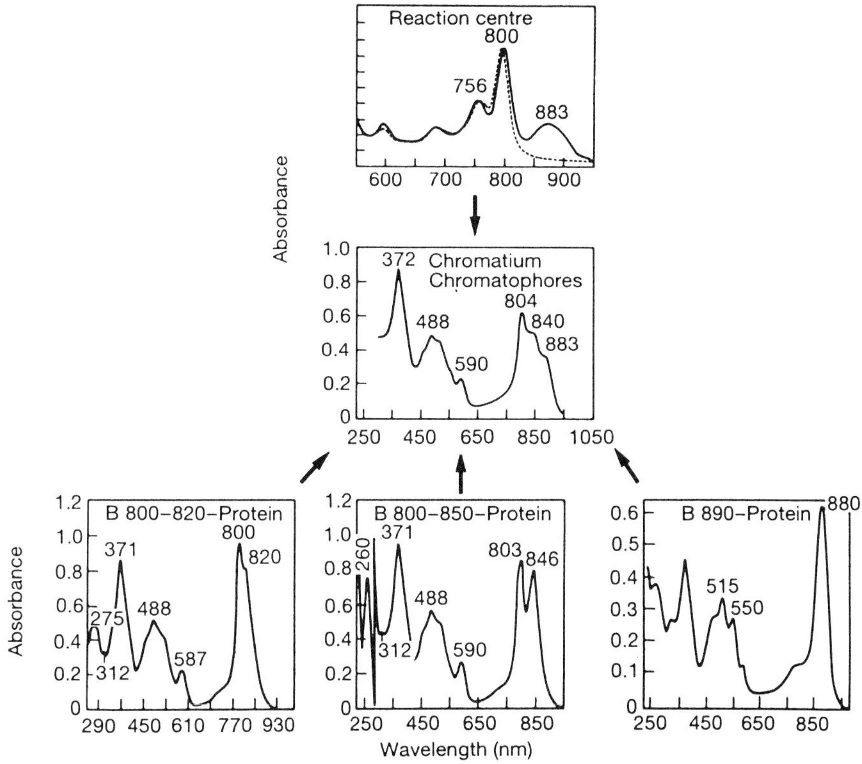

Figure 3.11 Absorption spectra of the pigment-protein complexes isolated from the purple phototrophic bacterium, *Chromatium vinosum*. Three light-harvesting bacteriochlorophyll complexes are present in this species; B800-820, B800-850 and B890 (reproduced from Thornber 1986).

In common with the majority of photosynthetic organisms, a great deal of variation is encountered in the carotenoid content and composition of the light-harvesting complexes, as opposed to the relatively conserved nature of the reaction centres. Unlike in higher plants and algae, both xanthophylls and carotenes can be bound to the bacterial reaction centre proteins and can also be found in the light-harvesting complexes. The complexes are reported to show little specificity regarding the carotenoid which is bound and the composition appears to reflect the carotenoid composition of the species/strain from which the pigment–protein complexes are isolated (Cogdell and Crofts 1978). Carotenoids with longer chromophores appear, however, to be primarily associated with reaction centres, which also show a slight specificity for carotenoids possessing methoxy end-groups.

Each light-harvesting complex is known to consist of oligomeric pairs of transmembrane proteins to which bacteriochlorophyll and carotenoids are non-covalently attached (Drews 1985; Brunisholz et al. 1986). In the B890 antenna, two bacteriochlorophyll molecules are bound to each α/ß-polypeptide pair (Broglie et al. 1980; Brunisholz et al. 1986). The ratio of bacteriochlorophyll:carotenoid in the B890 complex of *R. rubrum* has been shown to be 2:1 (Cogdell et al. 1982; Picorel et al. 1983; Ueda et al. 1985).

Many species of phototrophic bacteria also possess a second type of pigment–protein complex of which the B800–850 complex found in *Rb. sphaeroides* is an example. In the B800–850 complexes, three bacteriochlorophyll molecules are bound (Sauer and Austin 1978; Broglie et al. 1980). There is, however, a certain amount of conflicting data concerning the ratio of bacteriochlorophyll:carotenoid found in complexes of the B800–850 type. Cogdell and Crofts (1978) originally reported a ratio of 3:1 in *Rb. sphaeroides*. Later studies, including those of Ueda et al. (1985), reported a ratio of 2:1 for *Rb. sphaeroides*. In *Rb. capsulatus,* ratios of 2:1 and 3:1 have been reported by Franck and Drews (1978) and Shiozawa et al. (1982), respectively. A recent study by Evans et al. (1988) determined a ratio of 2:1 by using a number of published methods for the isolation of the B800–850 complex from *Rb. sphaeroides* (in this study, the B890 complex was also found to have a bacteriochlorophyll:carotenoid ratio of 2:1).

In some species of phototrophic bacteria the light-harvesting proteins of the B800–850 type may be absent and they may therefore contain only one type of antenna complex (e.g. *R. rubrum*). In other species, the growth conditions of the bacteria, especially incident light intensity, may affect the synthesis of a particular light-harvesting protein, usually of the B800–850 type. One of the most notable features of some phototrophic bacteria (e.g *Rb. sphaeroides*) is their ability, when the bacteria are exposed to even traces of oxygen, to convert spheroidene into spheroidenone while these carotenoids are *in situ*, i.e. bound to the complexes.

Although the reaction centres of these bacteria contain only a fraction of the total pigment found in the chromatophores, they have been isolated and studied extensively. The reaction centres from a number of species of purple bacteria have been isolated and studied by X-ray crystallography. These studies have shown clearly that in *Rhodopseudomonas viridis*, *Rhodobacter sphaeroides* R.26 and wild type strain 2.4.1, four molecules of bacteriochlorophyll, two bacteriophaeophytin and one carotenoid are bound to the reaction centre (Deisenhofer et al. 1985;

Deisenhofer and Michel 1988, 1989a,b; Allen et al. 1988).

The nature of the carotenoid bound to the reaction centre and light-harvesting proteins varies greatly between species and can depend greatly on the culture conditions used to grow the organism. Lists of the main carotenoids that have been detected in members of the Rhodospirillineae and in other classes of phototrophic bacteria are given in Chapter 2.

REFERENCES

Alberte, R. and Anderson, R.A. (1986) Antheraxanthin, a light-harvesting carotenoid found in a chromophyte alga. *Plant Physiol.*, **86**: 583-587.

Allen, J. P., Feher, G., Yeates, T. O. *et al.* (1988) Structure of the reaction centre from *Rhodobacter sphaeroides* R-26 and 2.4.1. In: Breton, J. and Vermeglio, A. (eds.), *The Photosynthetic Bacterial Reaction Centre*. NATO ASI Series A: Life Sciences. Plenum, New York, pp.149:5-11.

Anderson, J.M. and Barrett, J. (1986) Light-harvesting pigment–protein complexes of algae. In: Staehelin, L.A. and Arntzen, C.J. (eds.), *Encyclopedia of Plant Physiology*, Vol. 19 (2nd. Ed). Springer-Verlag, Berlin, pp. 144-157.

Antonopoulou, P. and Akoyunoglou, G. (1986) Pigment and polypeptide composition of the light-harvesting complex of the photosystem I unit. In: Akoyunoglou, G. and Senger, H. (eds.), *Regulation of Chloroplast Differentiation*. Alan R. Liss, New York, pp. 267-272.

Barber, J., Chapman, D.J. and Telfer, A. (1987) Characterisation of a photosystem II reaction centre isolated from the chloroplasts of *Pisum sativum*. *FEBS Lett.*, **220**:647-73.

Benson, E.E. and Cobb, A.H. (1983) Pigment/protein complexes of the intertidal alga *Codium fragile*. *New Phytol.*, **95**: 581-594.

Berkaloff, C. (1977) Carotenoproteins in extrachloroplastic structures in the green alga *Protosiphon botryoides*. *Plant Sci. Lett.*, **10**: 45-48.

Berkaloff, C., Caron, L. and Rousseau, B. (1990) Subunit organization of PSI particles from brown algae and diatoms. *Photosynth. Res.*, **23**:181-193.

Block, M.A., Dorne, A.-J., Joyard, J. and Douce, R. (1983) Preparation and characterisation of membrane fractions enriched in inner and outer envelope membranes from spinach chloroplasts. II - Biochemical characterisation. *J. Biol. Chem.*, **258**: 13281-13286.

Boczar, B. A. (1985) Functional organisation of chlorophyll in chlorophyll *c*-containing marine phytoplankton. Ph.D. Thesis, University of California, Santa Barbara.

Boczar, B. A. and Prézelin, B.B. (1986) Light- and $MgCl_2$-dependent characteristics of four chlorophyll protein complexes from the marine dinoflagellate, *Glenodinium* sp. *Biochim. Biophys. Acta*, **850**: 300-309.

Brandt, P. and Wilhelm, C. (1990) The light-harvesting system of *Euglena gracilis* during the cell cycle. *Planta*, **180**: 293-296.

Broglie, R.M., Hunter, C.N., Delepelaire, P. *et al.* (1980) Isolation and characterisation of the pigment–protein complexes of *Rhodopseudomonas sphaeroides* by lithium dodecyl sulfate/polyacrylamide gel electrophoresis. *Proc. Natl. Acad. Sci. USA*, **77**:87-91.

Brown, J. (1986) Functional organisation of chlorophyll *a* and carotenoids in the alga *Nannochloropsis salina*. *Plant Physiol.*, **83**: 434-437.

Brunisholz, R.A., Zuber, H., Valentine, J. *et al.* (1986) The membrane location of the B–890 complex from *Rhodospirillum rubrum* and the effect of carotenoid on the conformation of its 2 apoproteins exposed at the cytoplasmic surface. *Biochim. Biophys. Acta*, **849**:295-303.

Bullerjhan, G. S. and Sherman, L.A. (1986) Identification of a carotenoid-binding protein in the cytoplasmic membrane from the heterotrophic cyanobacterium *Synechocystis* sp.

strain PCC6714. *J. Bacteriol.*, **167**: 396-399.

Caron, L. and Brown, J. (1987) Chlorophyll-carotenoid protein complexes from the diatom *Phaeodactylum tricornutum*: spectrophotometric, pigment and polypeptide analysis. *Plant Cell Physiol.*, **28**: 775-785.

Chu, Z. and Anderson, J. (1985) Isolation and characterisation of a siphonaxanthin–chlorophyll *a/b*–protein complex of photosystem I from a *Codium* species. *Biochim. Biophys. Acta*, **806**: 154-160.

Cogdell, R. J. and Crofts, A.R. (1978) Analysis of the pigment content of an antenna pigment–protein complex from three strains of *Rhodopseudomonas sphaeroides*. *Biochim. Biophys. Acta*, **502**:409-416.

Cogdell, R. J. and Thornber, J. P. (1980) Light-harvesting pigment–protein complexes of purple photosynthetic bacteria. *FEBS Lett.*, **122**:1-8.

Cogdell, R.J., Lindsay, J.G., Valentine, J. and Durant, I. (1982) The isolation and partial purification of a light-harvesting pigment–protein component of *Rhodopseudomonas acidophila*. *Biochim. Biophys. Acta*, **502**:409-416.

Costes, C., Burghofer, C., Joyard, J. *et al.* (1979) Occurrence and biosynthesis of violaxanthin in isolated spinach chloroplast envelope. *FEBS Lett.*, **103**: 17-21.

Cunningham, F.X. and Schiff, J. (1985) Chlorophyll–protein complexes from *Euglena gracilis* and mutants deficient in chlorophyll *b*. *Plant Physiol.*, **80**: 223-230.

Deisenhofer, J. and Michel, H. (1988) The crystal structure of the photosynthetic reaction centre from *Rhodopseudomonas viridis*. In: Breton, J. and Vermeglio, A. (eds.), *The Photosynthetic Bacterial Reaction Centre*, NATO ASI Series A: Life Sciences. Plenum, New York, 149:1-3.

Deisenhofer, J. and Michel, H. (1989a) The photosynthetic reaction centre from the purple bacterium *Rhodopseudomonas viridis*. *Science*, **245**:1463-1473.

Deisenhofer, J. and Michel H. (1989b) The photosynthetic reaction centre from the purple bacteria *Rhodopseudomonas sphaeroides* and *Rhodospirillum rubrum*. *Chemica Scripta*, **29**:205-220.

Deisenhofer, J., Epp, O., Miki, K. *et al.* (1985) Structure of the protein subunits in the photosynthetic reaction centre of *Rhodopseudomonas viridis* at 3 angstrom resolution. *Nature*, **318**:618-624.

Delepelaire, P. and Chua, N-H., (1979) Lithium-dodecylsulfate-polyacrylamide gel electrophoresis of thylakoid membranes at 4°C: characterization of two additional chlorophyll–protein complexes. *Proc. Natl. Acad. Sci. USA*, **76**: 111-115.

van Dorssen, R.J., Breton, J., Plijter, J.J., *et al.* (1987) Spectroscopic properties of the reaction centre and of the 47 kDa chlorophyll protein of photosystem II. *Biochim. Biophys. Acta*, **893**: 267-274.

Douce, R. and Joyard, J. (1979) Structure and function of the plastid envelope. *Adv. Bot. Res.*, **7**:1-116.

Douce, R. and Joyard, J. (1990) Biochemistry and function of the plastid envelope. *Annu. Rev. Plant Physiol.*, **6**: 173-216.

Douce, R., Block, M.A., Dorne, A.-J. and Joyard, J. (1985) The plastid envelope membranes: Their structure, composition, and role in chloroplast biogenesis. *Subcell. Biochem.*, **10**: 1-86.

Drews, G. (1985) Structure and functional organisation of light-harvesting complexes and photochemical reaction centres in membranes of phototrophic bacteria. *Microbiol. Rev.*, **49**:59-70.

Evans, M.B., Cogdell, R.J. and Britton, G. (1988) Determination of the bacteriochlorophyll :carotenoid ratios of the B890 antenna complex of *Rhodospirillum rubrum* and the B800-850 complex of *Rhodobacter sphaeroides*. *Biochim. Biophys. Acta*, **935**:292-298.

Friedman, A.L. and Alberte, R.S. (1984) A diatom light-harvesting pigment–protein complex. *Plant Physiol.*, **76**:483-489.

Garcia, D., Parot, P. and Verméglio, A. (1987) Purification and characterisation of the photochemical reaction centre of the thermophilic purple sulfur bacterium *Chromatium tepidium*. *Biochim. Biophys. Acta*, **894**:379-385.

Ghirardi, M.L., McCauley, S.W. and Melis, A. (1986) Photochemical apparatus organisation in the thylakoid membrane of *Hordeum vulgare* wild type and chlorophyll *b*-less *chlorina* f2 mutant. *Biochim. Biophys. Acta*, **851**:331-339.

Gounaris, K., Chapman, D.J. and Barber, J. (1989) Isolation and characterisation of the D1/D2/cytochrome *b*-559 complex from *Synechocystis* 6803. *Biochim. Biophys. Acta*, **973**:296-301.

Gounaris, K., Chapman, D.J., Booth, P. *et al.* (1990) Comparison of the D1/D2/cytochrome *b*559 reaction centre complex photosystem two isolated by two different methods. *FEBS Lett.*, **265**: 88-92.

Green, B.R. (1988) The chlorophyll–protein complexes of higher plant photosynthetic membranes or just what green band is that? *Photosynth. Res.*, **15**: 3-32.

Grimme, L.H., Damm, I., Steinmetz, D. and Scheffczyk, B. (1987) Pigment-protein complexes of algal thylakoid membranes: variations in pattern, pigment composition and reaction centre II types during the cell cycle of *Chlorella fusca* and after adaptation to low light intensities. In: Biggins, J. (ed.), *Progress in Photosynthesis Research*, Vol. II. Nijhoff, Dordrecht, pp. 347-350.

Gugliemelli, L. (1984) Isolation and characterization of pigment–protein particles from light-harvesting complex of *Phaeodactylum tricornutum*. *Biochim. Biophys. Acta*, **766**:45-50.

Hiller, R., Larkum, A.W.D. and Wrench, P. (1988) Chlorophyll–proteins of the Prymnesiophyte *Pavlova lutherii* comb. nov.: identification of the main light-harvesting pigment complex. *Biochim. Biophys. Acta*, **932**:223-231.

Holt, T.K. and Krogmann, D.W. (1981) A carotenoid protein from cyanobacteria. *Biochim. Biophys. Acta*, **637**:408-414.

Jeffrey, S.W., Douce, R. and Benson, A.A. (1974) Carotenoid transformations in the chloroplast envelope. *Proc. Natl. Acad. Sci. USA.*, **71**: 807-810.

Joyard, J., Block, M.A. and Douce, R. (1991) Molecular aspects of plastid envelope biochemistry. *Eur. J. Biochem.*, **199**: 489-509.

Jürgens, U.J. and Mäntele, W. (1991) Orientation of carotenoids in the outer membrane of *Synechocystis* PCC 6714 (cyanobacteria). *Biochim. Biophys. Acta*, **1067**:208-212.

Jürgens, U.J. and Weckesser, J. (1985) Carotenoid-containing outer membranes of *Synechocystis* PCC 6714. *J. Bacteriol.*, **164**:384-389.

Jürgens, U.J., Golecki, J.R. and Weckesser, J. (1985) Characterisation of the cell wall of the unicellular cyanobacterium (blue-green alga) *Synechocystis* PCC 6714. Arch. Microbiol., 142:168-174.

Kan, K. -S. and Thornber, J.P. (1976) The light-harvesting chlorophyll *a/b* protein complex of *Chlamydomonas reinhardtii*. *Plant Physiol.*, **57**:47-52.

Katoh, T. and Ehara, T. (1990) Supramolecular assembly of fucoxanthin–chlorophyll–protein complexes isolated from a brown alga, *Petalonia fascia*: Electron microscope studies. *Plant Cell Physiol.*, **31**:439-448.

Katoh, T., Mimuro, M. and Takaichi, S. (1989) Light-harvesting particles isolated from a brown alga, *Dictyota dichotoma*: a supramolecular assembly of fucoxanthin–chlorophyll–protein complexes. *Biochim. Biophys. Acta*, **976**:233-240

Knoetzel, J., Braumann, T. and Grimme, L.H. (1988) Pigment–protein complexes of green algae: improved methodological steps for the quantification of pigments in pigment–protein complexes derived from the green algae *Chlorella* and *Chlamydomonas*. *J. Photochem. Photobiol.*, **1B**:475-491.

Kobayashi, M., Maeda, H., Wantanabe, T. *et al.* (1990) Chlorophyll *a* and ß-carotene content in the D1/D2/cytochrome *b*-559 reaction centre complex from spinach. *FEBS Lett.*, **260**:138-140.

Lichtlé, C., Duval, J-C. and Lemoine, Y. (1987) Comparative biochemical, functional and ultrastructural studies of photosystem particles from a Cryptophyceae: *Cryptomonas rufescens*; isolation of an active phycoerythrin particle. *Biochim. Biophys. Acta*, **894**:79-90.

Lütke-Brinkhaus, F., Liedvogel, B., Kreuz, K. and Kleinig, H. (1982) Phytoene synthase and phytoene dehydrogenase associated with envelope membranes from spinach chloroplasts. *Planta*, **156**: 176-180.

Melis, A. (1991) Dynamics of photosynthetic membrane composition and function. *Biochim. Biophys. Acta*, **1058**:87-106.

Menger, W., Knötzel, J., Braumann, T. and Grimme, L. (1984) Function of chlorophylls and carotenoids in thylakoid membranes: pigment composition of LHCP-complexes of spinach and *Chlorella*. *Adv. Photosynth. Res*., **2**:145-148.

Mimuro, M. and Katoh, T. (1991) Carotenoids in photosynthesis: absorption, transfer and dissipation of light energy. *Pure Appl. Chem*., **63**:123-130.

Mimuro, M., Katoh, T. and Kawai, H. (1990) Spatial arrangement of pigments and their interaction in the fucoxanthin–chlorophyll *a/c* protein assembly (FCPA) isolated from *Dictyota dichotoma*: analysis by means of polarized spectroscopy. *Biochim. Biophys. Acta*, **1015**:450-456.

Nakayama, K. and Okada, M. (1990) Purification and characterisation of light-harvesting chlorophyll *a/b*–protein complexes of PSII from the green alga, *Bryopsis maxima*. *Plant Cell Physiol*., **31**: 253-260.

Nakayama, K., Itagaki, T. and Okada, M. (1986) Pigment composition of chlorophyll–protein complexes isolated from the green alga *Bryopsis maxima*. *Plant Cell Physiol*., **27**:311-317.

Nanba, O. and Satoh, K. (1987) Isolation of a photosystem II reaction centre consisting of D-1 and D-2 polypeptides and cytochrome *b*-559. *Proc. Natl. Acad. Sci. USA*, **84**: 109-112.

Newell, W.R., van Amerongen, H., van Grondelle, R. *et al.* (1988) Spectroscopic characterisation of the reaction centre of photosystem II from higher plants. *FEBS Lett*., **228**: 162-166.

Newell, W.R., van Amerongen, H., Barber, J. and van Grondelle, R. (1991) Spectroscopic characterisation of the reaction centre of photosystem II using polarized light: evidence for ß-carotene excitons in PSII reaction centres. *Biochim. Biophys. Acta*, **1057**:232-238.

Omata, T. and Murata, N. (1983) Isolation and characterization of the cytoplasmic membranes from the blue-green alga (Cyanobacterium) *Anacystis nidulans*. *Plant Cell Physiol*., **24**:1101-1112.

Omata, T. and Murata, N. (1984) Isolation and characterisation of 3 types of membranes from the cyanobacterium (blue-green alga) *Synechocystis* PCC 6714. *Arch. Microbiol*., **139**:113-116.

Owens, T.G. and Wold, E.R. (1986) Light-harvesting function in the diatom *Phaeodactylum tricornutum*. *Plant Physiol*., **80**:732-738.

Owens, T.G., Gallagher, J.C. and Alberte, R.S. (1987) Photosynthetic light-harvesting function of violaxanthin in *Nannochloropsis* spp. *J. Phycol*., **23**:729-738.

Passaquet, C., Thomas, J.C., Caron, L. *et al.* (1991) Light-harvesting complexes of brown algae. Biochemical characterization and immunological relationships. *FEBS Lett*., **280**:21-26.

Peter, G.F. and Thornber, J. P. (1991a) Electrophoretic procedures for fractionation of photosystems I and II pigment–proteins of higher plants and for determination of their subunit composition. In: *Methods in Plant Biochemistry*, Vol. 5. Academic Press, London, pp. 195-210.

Peter, G.F. and Thornber, J.P. (1991b) Biochemical composition and organization of higher plant photosystem II light-harvesting pigment-proteins. *J. Biol. Chem*., **266**:16745-16754.

Picorel, R., Belanger, G. and Gingras, G. (1983) Antenna holochrome B880 of *Rhodospirillum rubrum* S1. Pigment, phospholipid and polypeptide composition. *Biochemistry*, **22**:2491-2494.

Prézelin, B. and Haxo, F. (1976) Purification and characterisation of peridinin-chlorophyll *a*-proteins from the marine dinoflagellate *Glenodinium* sp. and*Gonyaulax polyedra*. *Planta*, **128**:133-141.

Rock, C.D. and Zeevaart, J.A.D. (1991) The *aba* mutant of *Arabadopsis thaliana* is impaired in epoxy-carotenoid biosynthesis. *Proc. Natl. Acad. Sci. USA*, **88**:7496-7499.

Sauer, K. and Austin, L.A. (1978) Bacteriochlorophyll–protein complexes from the light-harvesting antenna of photosynthetic bacteria. *Biochemistry*, **17**:2011-2019.

Shiozawa, J.A., Welte, W., Hodapp, N. and Drews, G. (1982) Studies on the size and composition of the isolated light-harvesting B800–850 pigment–protein complexes of *Rhodopseudomonas capsulata*. *Arch. Biochem. Biophys.*, **213**:473-485.

Siefermann-Harms, D. (1985) Carotenoids in photosynthesis. I. Location in photosynthetic membranes and light-harvesting function. *Biochim. Biophys. Acta*, **811**: 325-355.

Siefermann-Harms, D., Joyard, J. and Douce, R. (1978) Light-induced changes of the carotenoid levels of chloroplast envelopes. *Plant Physiol.* **61**: 530-533.

Simpson, D.J., Machold, O., Høyer-Hansen, G. and von Wettstein, D. (1985) *Chlorina* mutants of barley (*Hordeum vulgare* L.). *Carlsberg Res. Commun.*, **50**: 223-238.

Sukenik, A., Wyman, K.D., Bennet, J. and Falkowsky, P. (1987) A novel mechanism for regulating the excitation of photosystem II in a green alga. *Nature*, **327**: 704-707.

Telfer, A., De Las Rivas, J. and Barber, J. (1991) ß-Carotene within the isolated photosystem II reaction centre: Photooxidation and irreversible bleaching of this chromophore by oxidised P680. *Biochim. Biophys. Acta*, **1060**: 106-114.

Thornber, J.P. (1986) Biochemical characterisation and structure of pigment–proteins of photosynthetic organisms. In: Staehelin, L.A. and Arntzen, C.J. (eds.), *Encyclopedia of Plant Physiology*, Vol. 19 (2nd. ed). Springer-Verlag, Berlin, pp. 98-142.

Thornber, J. P., Cogdell, R. J., Pierson, B. K. and Seftor, R. E. B. (1983) Pigment–protein complexes of purple photosynthetic bacteria. *J. Cell. Biochem.*, **23**:159-169.

Thornber, J.P., Peter, G.F., Chitnis, P.R. *et al.* (1988) The light-harvesting complex of photosystem II of higher plants. In: Stevens, S.E. and Bryant, D.A. (eds.), *Light Energy Transduction in Photosynthesis*. The American Society for Plant Physiologists, New York, pp. 137-154.

Thornber, J.P., Peter, G.F., Chitnis, P.R. and Vainstein, A. (1989) Molecular and cellular biology of the major light-harvesting pigment–protein (LHCII*b*) of higher plants. In: Singhal, G.S., Barber, J., Dilley, R.A. *et al.* (eds.), *Photosynthesis. Molecular Biology and Bio-energetics*. Springer-Verlag, London, pp. 373-387.

Ueda, T., Morimoto, Y., Sato, M. *et al.* (1985) Isolation, characterisation and composition of a ubiquitous pigment–protein complex consisting of a reaction centre and light-harvesting bacteriochlorophyll protein present in purple photosynthetic bacteria. *J. Biochem.*, **98**:1487-1498.

Vainstein, A., Peterson, C. and Thornber, J.P. (1989) Light-harvesting pigment proteins of photosystem I in maize. *J. Biol. Chem.*, **264**: 4058-4063.

Wilhelm, C. (1990) The biochemistry and physiology of light-harvesting processes in chlorophyll *b*- and chlorophyll *c*-containing algae. *Plant Physiol. Biochem.*, **28**:293-306.

Wilhelm, C. and Lenartz-Weiler, I. (1987) Energy transfer and pigment composition in three chlorophyll *b*-containing light-harvesting complexes isolated from *Mantoniella squamata* (Prasinophyceae), *Chlorella fusca* (Chlorophyceae) and *Sinapis alba*. *Photosynth. Res.*, **13**: 125-141.

Wilhelm, C., Lenartz-Weiler, I., Wiedermann, I. and Wild, A. (1986) The light-harvesting system of *Micromonas* spec. Butcher (Prasinophyceae): the combination of three different chlorophyll species in one single chlorophyll–protein complex. *Phycologia*, **25**:304-312.

Wilhelm, C., Büchel, C. and Rousseau, B. (1988) Isolation and characterisation of a chlorophyll *a*/*c*–heteroxanthin/diadinoxanthin light-harvesting complex from *Pleurochloris meiringensis* (Xanthophyceae). In, Scheer, H. and Schneider, S. (eds.), *Photosynthetic Light-Harvesting Systems*. de Gruyter, Berlin, pp. 167-174.

4

Biosynthesis of carotenoids

G. Britton

4.1 INTRODUCTION

The carotenoids that are found in the photosynthetic pigment–protein complexes of higher plants, algae and phototrophic bacteria, including cyanobacteria, are C_{40} tetraterpenes. They are biosynthesised by a specialised branch of the isoprenoid or terpenoid pathway which is also used for the biosynthesis of a wide variety of other important compounds. All isoprenoid compounds are built up from the C_5 "isoprene unit" the carbon skeleton of which is clearly seen in the important intermediate isopentenyl diphosphate (IDP). A series of diphosphate intermediates with carbon chains consisting of multiples of five carbon atoms is then built up from this isoprene unit intermediate. In the first step of this process, IDP undergoes a double bond isomerisation to give dimethylallyl diphosphate (DMADP). The prenyl transferase enzymes then build up the isoprenoid chain from these two intermediates. The first condensation, between DMADP and IDP, gives the C_{10} compound geranyl diphosphate (GDP), which is the precursor of monoterpenes. Further additions of IDP then extend the chain to produce, successively, the C_{15} farnesyl diphosphate (FDP), precursor to the sesquiterpenes, triterpenes and sterols, and the C_{20} geranylgeranyl diphosphate (GGDP), which gives rise to the carotenoids and to the diterpenes including phytol, which provides the isoprenoid sidechain of the chlorophylls and, in most cases, the bacteriochlorophylls. The general outline of the isoprenoid pathway, illustrating the biosynthesis of carotenoids in relation to that of other isoprenoid compounds, is given in Fig. 4.1

The overall pathway of carotenoid biosynthesis can be considered conveniently as a series of stages, as illustrated in Fig. 4.2. The reactions involved are discussed below. Further details can be obtained from a number of review articles, e.g.: Britton 1976, 1988, 1990a,b, 1991; Davies 1980; Goodwin 1971; Spurgeon and Porter

1983.

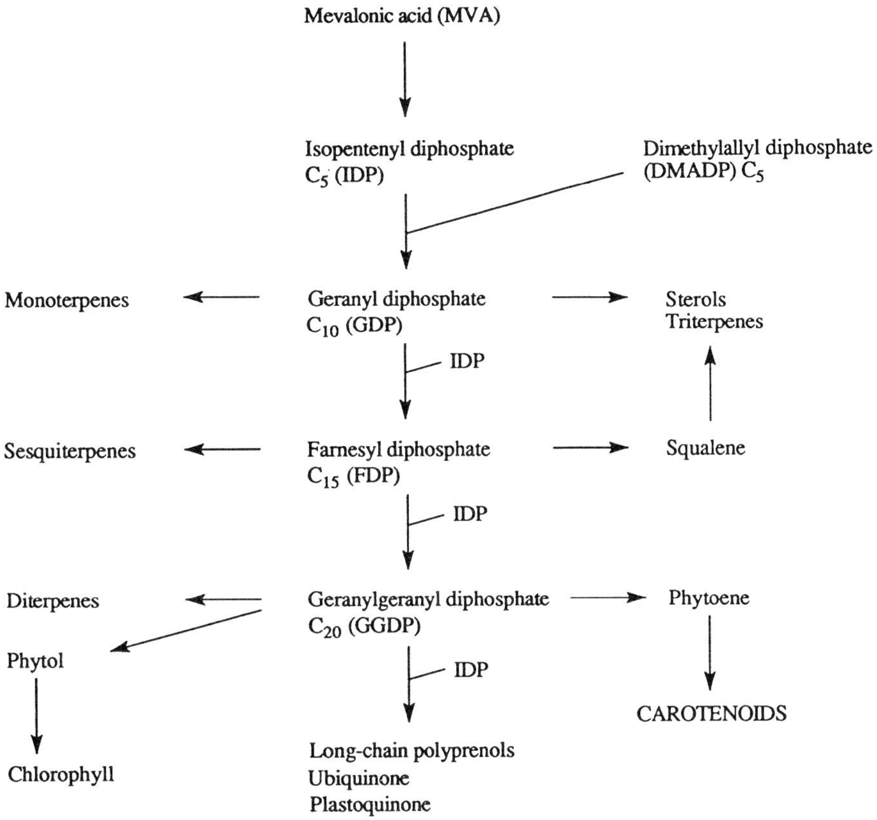

Figure 4.1 Summary of the biosynthesis of isoprenoid compounds.

4.2 REACTIONS OF CAROTENOID BIOSYNTHESIS

4.2.1 Plants and algae

(a)Early stages: origin of isoprenoid precursors

It is well known that, in plants, the carotenoids that accumulate and function in the chloroplast are biosynthesised within that organelle. The enzymes that are specific for carotenoid biosynthesis are encoded in the nucleus and are synthesised in cytoplasmic ribosomes; they are then transported into the chloroplast. There are still arguments, however, about whether the early isoprenoid precursors, especially mevalonate (MVA) and IDP, have to be imported into the chloroplast or are

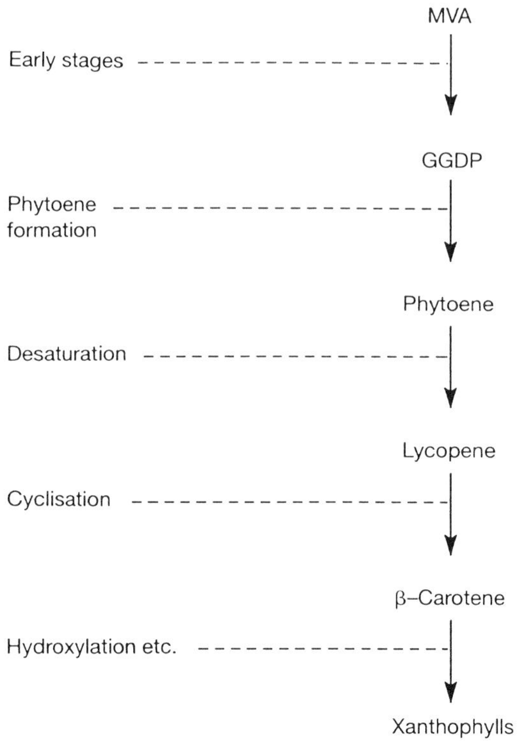

Figure 4.2 Summary of the stages of carotenoid biosynthesis. For abbreviations, see Fig. 4.1.

synthesised there (Kleinig 1989). Some authors (e.g. Kreuz and Kleinig 1981) have concluded that the chloroplast must import IDP. Other workers believe that the chloroplast is a biosynthetically autonomous organelle and can biosynthesise carotenoids from CO_2 via acetyl-coenzyme A (CoA) and MVA. A chloroplast HMGCoA reductase (for the conversion of 3-hydroxy-3-methylglutaryl-CoA into MVA) has been reported (Grumbach and Bach 1979). Recent work by Heintze et al. (1990) may have resolved the controversy. This work revealed that the biosynthetic capability of chloroplasts changes as the chloroplasts develop or the leaves mature. Thus, in the leaf base segment of seedlings of barley (*Hordeum vulgare*) photosynthetically fixed CO_2 was converted into ß-carotene and other isoprenoids by a pathway, via pyruvate, acetyl CoA and MVA, that was located entirely within the chloroplast. The leaf base contains developing chloroplasts with incomplete thylakoid stacking but a considerable rate of photosynthetic CO_2 fixation. The leaf tip, however, contains mature chloroplasts with maximal photosynthetic activity, and these chloroplasts were unable to utilise CO_2 for isoprenoid synthesis, and had to import extra-plastidic IDP. The change was attributed to a drastic decrease in pyruvate decarboxylase-dehydrogenase activity as the chloroplasts developed and matured. These findings are in agreement with earlier reports that very young tissues of spinach and other plants can utilise photosynthetically fixed CO_2 very efficiently to synthesise chloroplast isoprenoids (Rogers et al. 1966).

(b)Formation of the C_{20} precursor, GGDP

The reactions by which the isoprenoid chain (as GGDP) is constructed are illustrated in Fig. 4.3. The enzymes and the reaction mechanisms have generally been studied in non-carotenogenic systems, particularly in relation to the biosynthesis of lower terpenoids and sterols, but the details are considered to be the same in the biosynthesis of all isoprenoid compounds. These reactions are not discussed further in this chapter, but details may be obtained from the general review articles cited above.

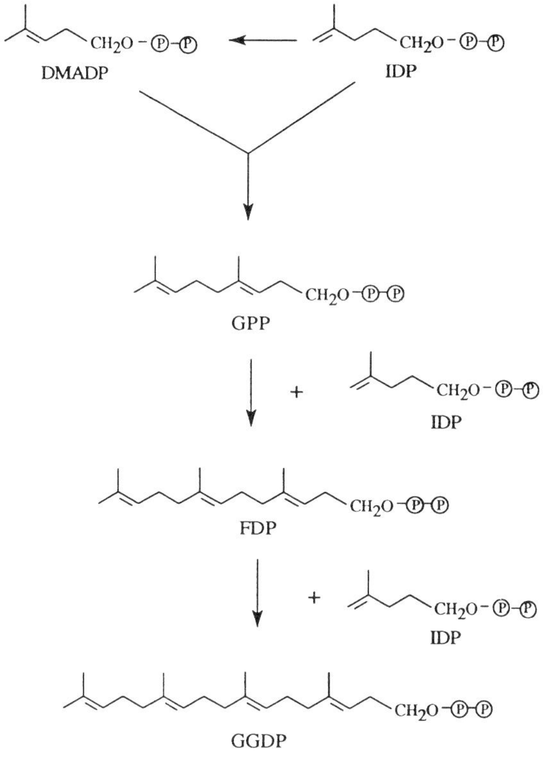

Figure 4.3 Formation of geranylgeranyl diphosphate (GGDP) from isopentenyl diphosphate (IDP). For abbreviations, see Fig. 4.1.

(c) Formation of phytoene

The first reaction that is specific to carotenoid biosynthesis is that by which the first C_{40} hydrocarbon, phytoene, is formed form two molecules of GGDP. This reaction proceeds via a C_{40} intermediate, prephytoene diphosphate (PPDP), and, in all photosynthetic systems that have been studied, the phytoene product has been shown to be the 15*Z* (15-*cis*) isomer. The stereochemistry of hydrogen loss from GGDP in the formation of phytoene has been established (Gregonis and Rilling

1974), and a mechanism has been proposed that is consistent with the stereochemical course of formation of (15*Z*)-phytoene from GGDP via the natural (1*R*,2*R*,3*R*) isomer of PPDP (Fig. 4.4).

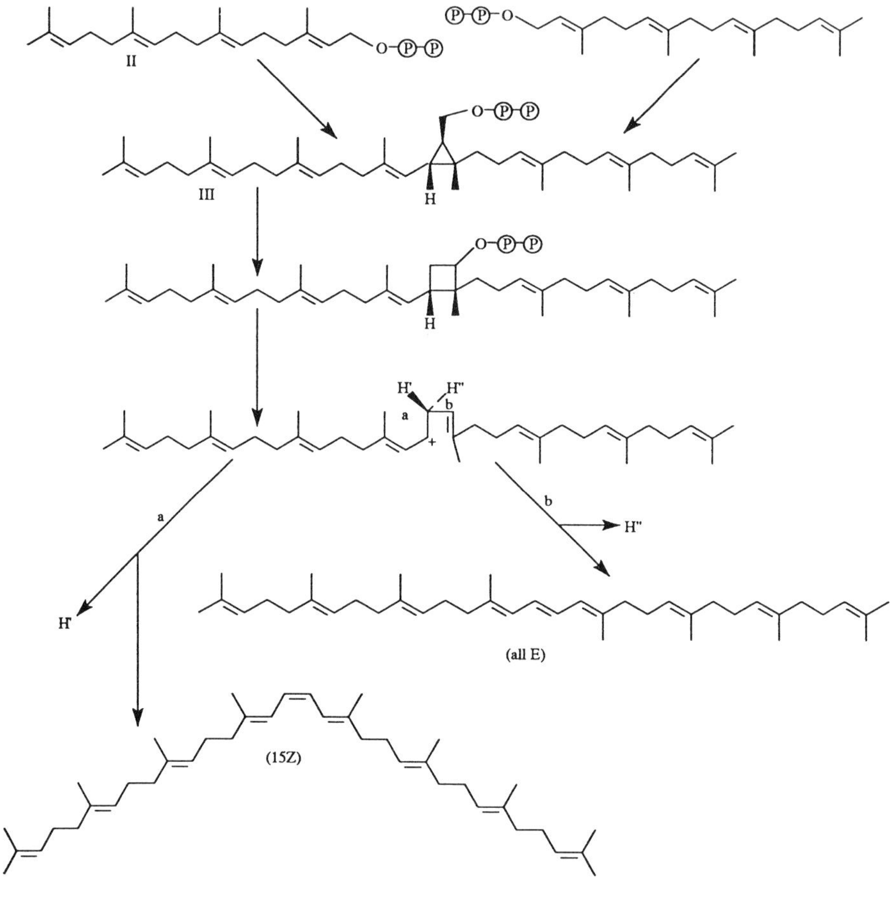

Figure 4.4 Mechanism and stereochemistry of the formation of (all-*E*)- and (15*Z*)-phytoene.

(d) Desaturation

Phytoene, with a chromophore of only three conjugated double bonds, is colourless. Its conversion into the coloured carotenoids that are normally present in photosynthetic systems involves the sequential extension of this chromophore by a series of desaturation reactions, each of which removes two hydrogen atoms by *trans* elimination to introduce a new double bond (Fig. 4.5; McDermott et al. 1973).

In plants and algae, the desaturations occur on alternate sides of the polyene chromophore and each extends the chromophore by two conjugated double bonds to give, successively, phytofluene, ζ-carotene, neurosporene and lycopene (Fig. 4.6).

Figure 4.5 Stereochemistry of desaturation. (H_{5R}, H_{5S}, H_{2R}, H_{2S}) are the hydrogen atoms derived from the 5-*pro-R*, 5-*pro-S*, 2-*pro-R* and 2-*pro-S* hydrogens of mevalonic acid, respectively.

The final coloured products of carotenoid biosynthesis generally have an all-*trans* polyene chromophore, although the phytoene that can be isolated is the 15-*cis* (15*Z*)-isomer. Isomerization of the C-15,15′ double bond from *Z* to *E* (*cis* to *trans*) must therefore occur at some stage in the biosynthetic sequence. This is discussed further below (Section 3.2) when recent work on the genes and enzymes of carotenoid biosynthesis is considered.

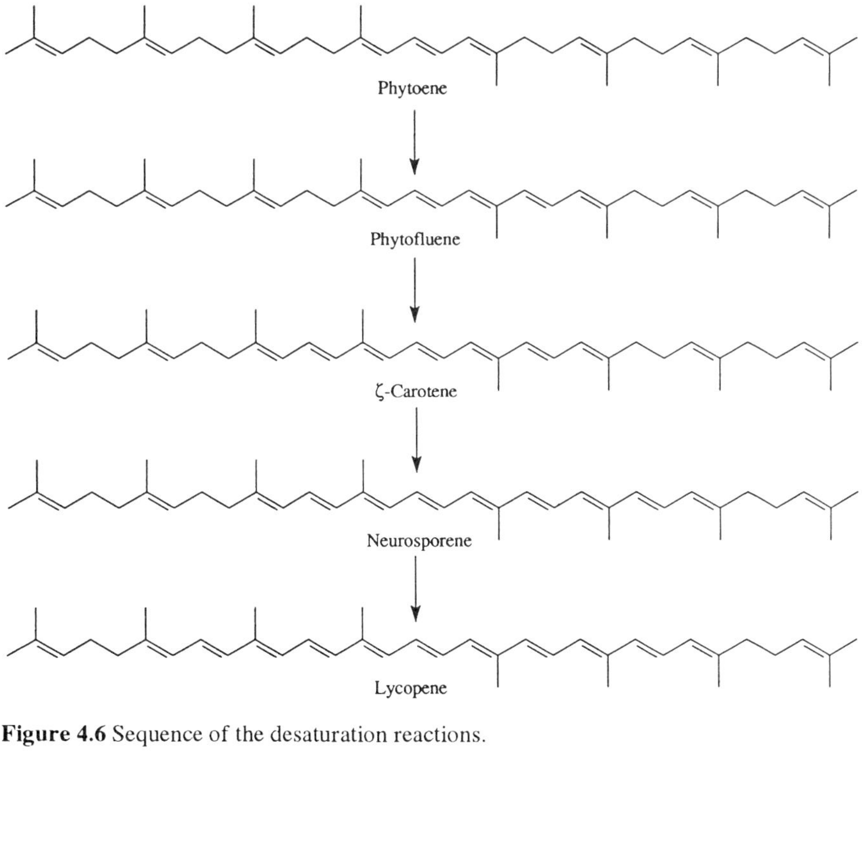

Figure 4.6 Sequence of the desaturation reactions.

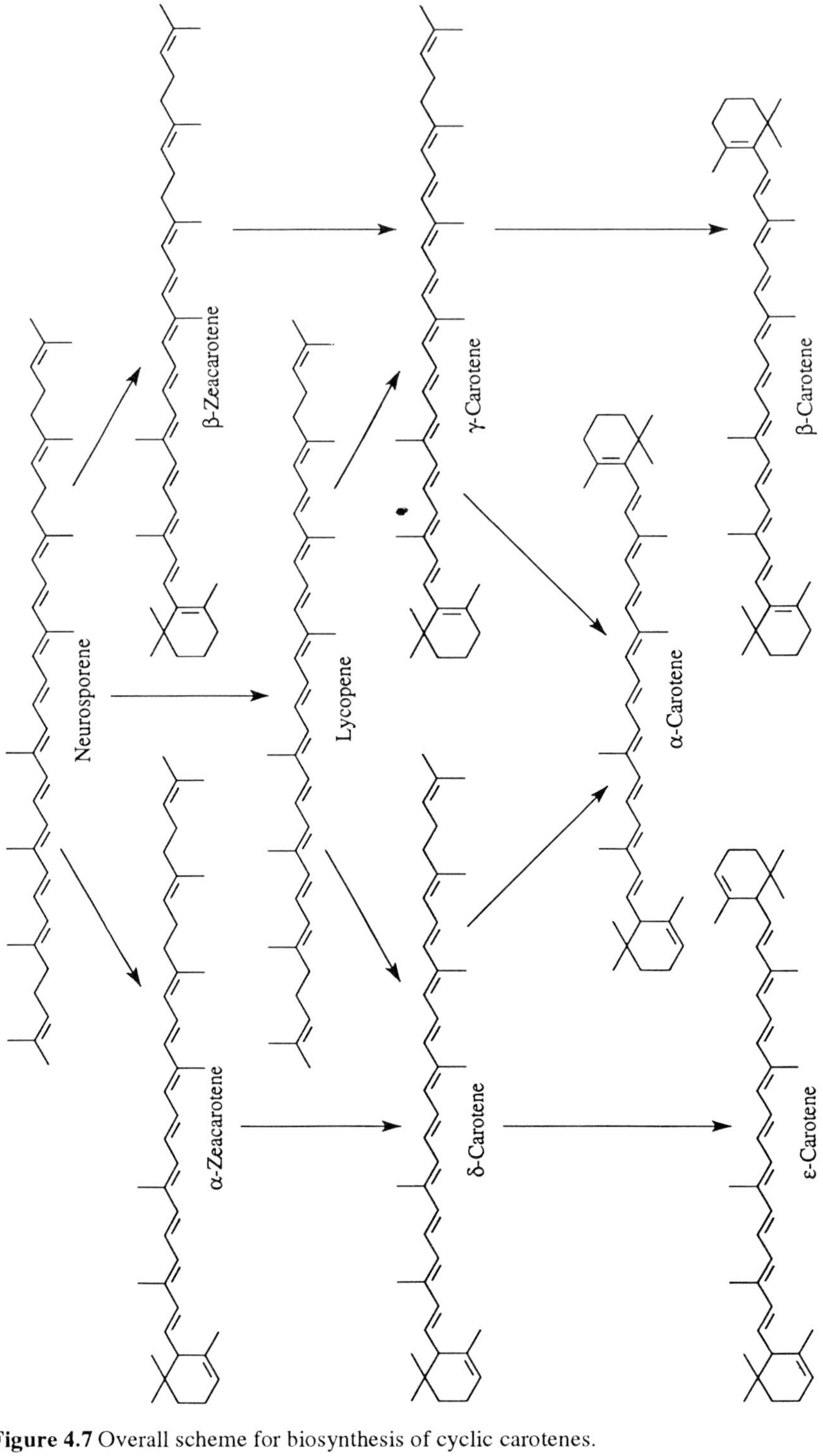

Figure 4.7 Overall scheme for biosynthesis of cyclic carotenes.

(e) Cyclisation

Because of the rigid polyene chain, in the carotenoids cyclisation is limited to the ends of the molecule, where formation of a six-membered ring is common. A scheme illustrating the "alternative routes" for the formation of the dicyclic compounds α-carotene, ß-carotene and ε-carotene is given in Fig. 4.7. There are still arguments about whether the first intermediate that cyclises is neurosporene or lycopene. The actual cyclisation reaction, however, would be the same in both cases and it is more satisfactory to consider that the requirement for cyclisation to occur is that one end-group, i.e. one half of the carotenoid molecule, must reach the lycopene level of desaturation.

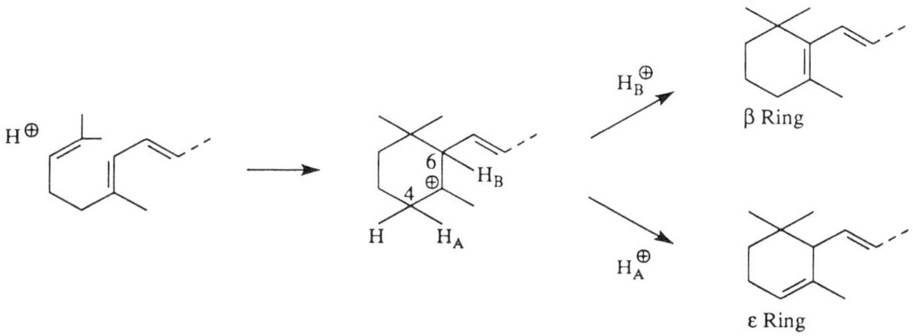

Figure 4.8 General mechanism of cyclisation.

It is now well established that cyclisation is initiated by proton attack at C-2 of a suitably folded acyclic end-group. Alternative loss of a proton from C-6 or C-4 of the proposed carbenium ion intermediate then gives rise to the ß- or ε-ring, respectively (Fig. 4.8; Williams et al. 1967). The stereochemical course of these cyclisation rections has been determined with a mutant strain of the green alga *Scenedesmus obliquus* (Britton and Mundy 1980; Britton 1985). On illumination of the dark-grown algae, ζ-carotene that accumulated was converted, *in situ*, into the normal cyclic carotenoids. When this conversion was allowed to proceed in ^2H_2O, a deuterium atom was introduced at C-2 of the cyclic carotenes and xanthophylls with the stereochemistry shown in Fig. 4.9. The stereochemical behaviour of the C-1 methyl substituents was established by labelling experiments with ^2H and ^{13}C (Britton 1985, 1990b). Possible schemes for the detailed stereochemistry of the cyclisation reactions are illustrated in Fig. 4.10.

(f) Hydroxylation

The introduction of oxygen functions and other structural modifications by which the great diversity of carotenoid structures, especially in algae, is elaborated is

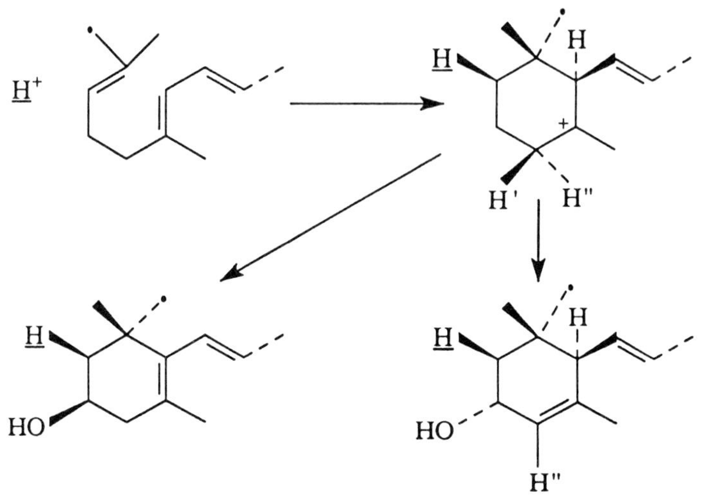

Figure 4.9 Stereochemistry of cyclisation to form the ß- and ε-rings in zeaxanthin and lutein.

considered to occur as the final processes in carotenoid biosynthesis. The reaction that has been studied most is the introduction of hydroxy-groups at C-3 of the ß- and ε-rings to give the common xanthophylls, zeaxanthin and lutein. Labelling studies have shown that these hydroxylations occur by stereospecific direct replacement of hydrogen by OH (Walton et al. 1969; Milborrow et al. 1982). This gives the hydroxylated products of opposite chirality at C-3 in the ß- and ε-rings (though both are designated *R*) as illustrated in Fig. 4.11.

(g)Later modifications

The reversible enzymic process by which epoxy-groups are introduced at C-5,6 and C-5′,6′ of zeaxanthin to give violaxanthin as part of the "violaxanthin cycle" ("xanthophyll cycle") is discussed in detail in Chapter 7. It is not known whether the enzyme that catalyses this reaction in the original biosynthesis of violaxanthin is the same as the epoxidase that is involved in the cyclic interconversions.

Little is known about the other late reactions, especially end-group modifications, by which the many different individual carotenoids that are found in plant and algal chloroplasts are elaborated. Pathways have been proposed for the formation of several different end-groups from the 3-hydroxy-5,6-epoxy-ß-end-group as in violaxanthin. For example, formation of the allenic end-group of neoxanthin can be explained by the simple rearrangement reaction of the 3-hydroxy-5,6-epoxy-5,6-dihydro-ß end-group of violaxanthin (Fig 4.12), but there is no direct evidence to support this chemically reasonable hypothesis.

Milborrow has reported some work on the biosynthesis of some of the structurally complex carotenoids found in algae. The conversion of zeaxanthin into the acetylenic diadinoxanthin, the allenic neoxanthin and the C_{37} carotenoid

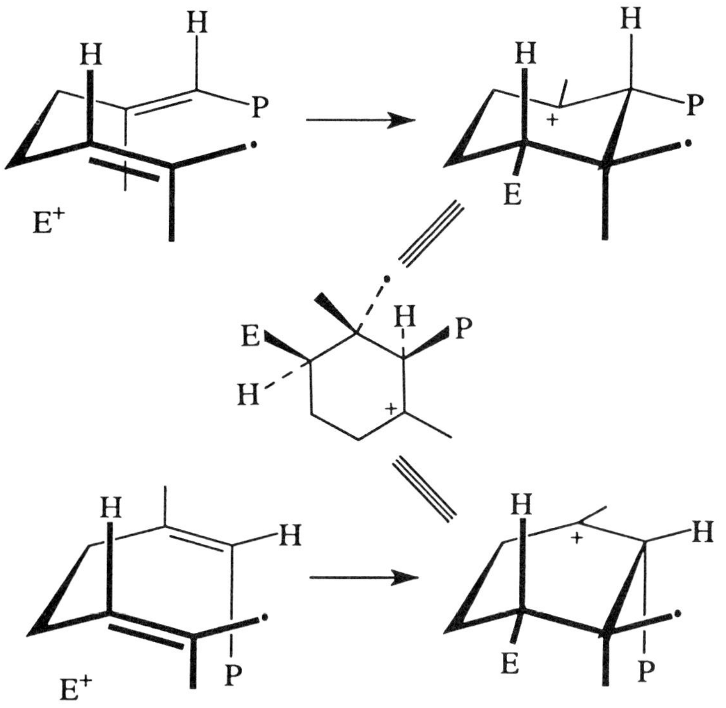

Figure 4.10 Alternative possibilities for the stereochemistry of folding of the acyclic end-group during carotenoid cyclisation.

peridinin in the dinoflagellate *Amphidinium carterae* were described (Swift and Milborrow 1981, Milborrow 1982, Swift et al. 1982). The conversion of the allenic into the acetylenic end-group was proposed, and a mechanism postulated for the excision of a C_3 fragment to form the C_{37} skeleton of peridinin (Fig. 4.13).

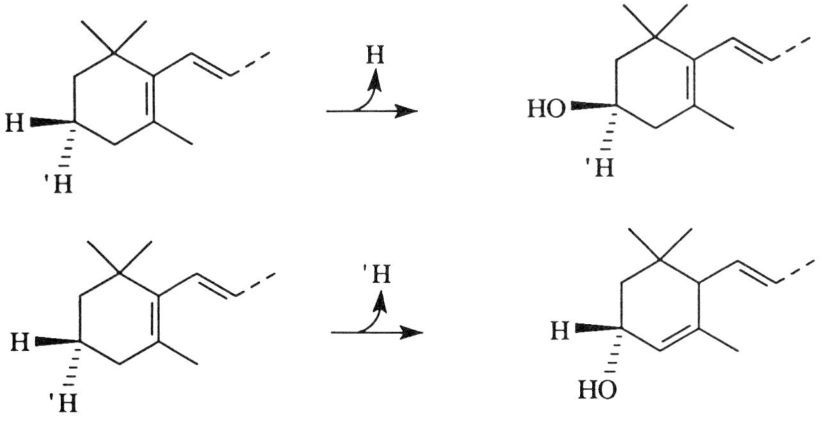

Figure 4.11 Stereochemistry of hydroxylation at C-3 in the ß- and ε-rings.

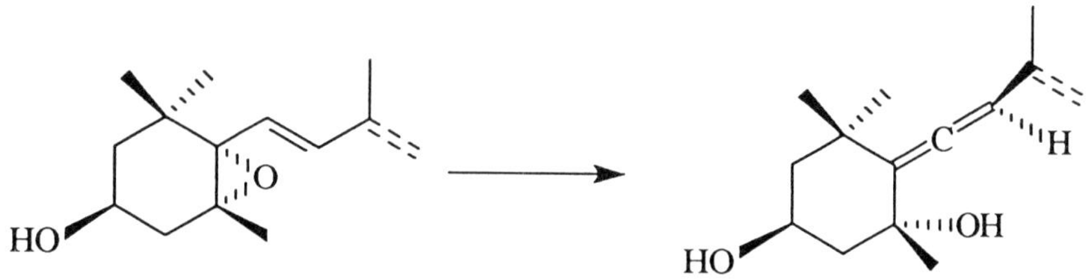

Figure 4.12 Scheme for the possible formation of the allenic end-group from a 5,6-epoxy-5,6-dihydro-ß-ring.

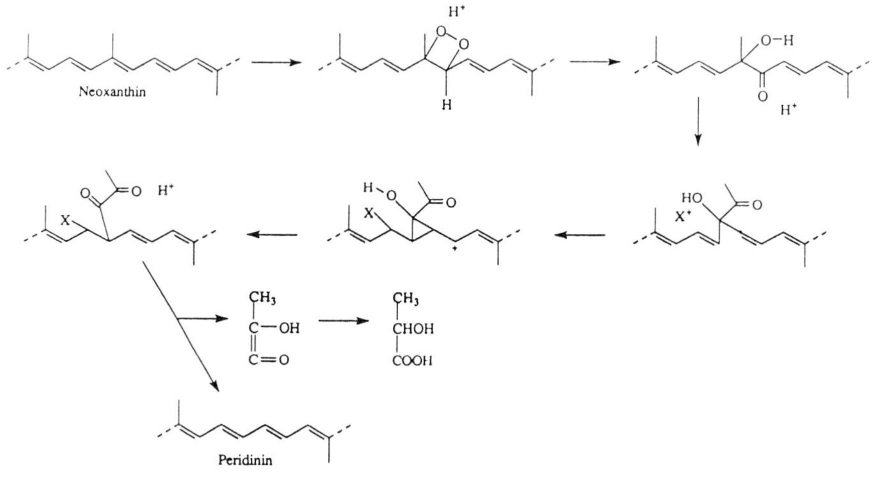

Figure 4.13 Mechanism proposed by Milborrow (1982) for the extrusion of a C_3 unit in the biosynthesis of the C_{37} norcarotenoid peridinin.

4.2.2 Phototrophic bacteria

(a) General features

As outlined in Chapter 2, the purple and green phototrophic bacteria very rarely contain any of the monocyclic and dicyclic carotenoids that are normally found in plants, algae and cyanobacteria. Carotenoids that contain the aryl φ and χ end-groups, e.g. chlorobactene and okenone are, however, frequently encountered in the sulphur bacteria. The characteristic carotenoids in the phototrophic bacteria are acyclic compounds which have tertiary hydroxy and methoxy substituents in the C-1 and C-1′ positions, e.g. spheroidene and spirilloxanthin. The biosynthesis of these carotenoids in the purple bacteria of the Rhodospirillaceae has become particularly topical in view of recent major advances in molecular genetics (see Section 4.2.1).

(b) Main Pathways

The pathway of carotenoid biosynthesis in these bacteria is, in principle, similar to that described above for plants and algae, but it differs in detail. It is generally assumed that the early stages, up to GGDP, are the same as in plants but there is little direct evidence to confirm this hypothesis. The incorporatation of acetate and IDP into carotenoids has been achieved, but there are doubts about whether the pathway in the purple bacteria involves MVA. The condensation of two molecules of GGDP to give phytoene appears to be the same as in higher plants, with the main product being the (15*Z*)-isomer. The sequence of desaturation reactions is, however, different. The reaction mechanism may be similar, but the first two desaturations take place in the same half of the molecule to give 7,8,11,12-tetrahydrolycopene, with its unsymmetrically located chromophore, in place of the symmetrical ζ-carotene. Also, in many bacteria, e.g. *Rhodobacter* spp., desaturation does not proceed beyond neurosporene. In other cases, however, e.g. *Rhodomicrobium vannielii* or *Rhodospirillum rubrum,* four desaturations do occur to give lycopene.

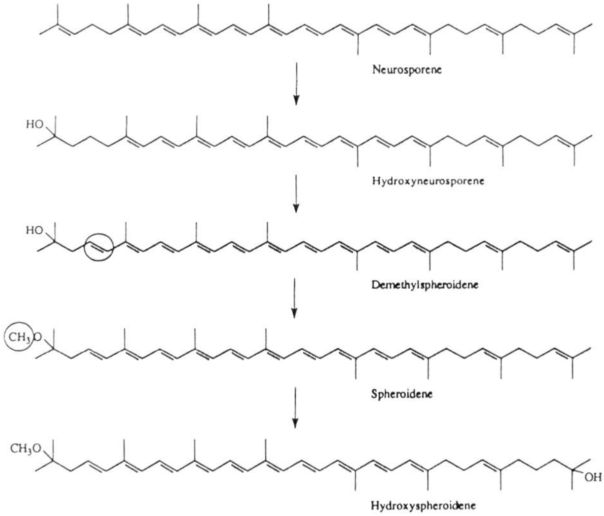

Figure 4.14 Pathway for the biosynthesis of hydroxyspheroidene from neurosporene via spheroidene.

Further modifications to the carotenoid molecule are then limited, and largely restricted to the end-groups. Characteristic reactions are the addition of water across the terminal C-1,2 double bond, the *O*-methylation (with *S*-adenosylmethionine) of the tertiary C-1 hydroxy-group thus introduced, additional desaturation of the C-3,4 bond and, under aerobic conditions, introduction of a conjugated ketone group at C-2. The genes relating to these reactions have been identified and are discussed later in this chapter. On the basis of these reactions, the pathways illustrated in Figs 4.14 and 4.15 have been proposed for the biosynthesis of hydroxyspheroidene in *Rhodobacter*, and spirilloxanthin in *Rhodospirillum rubrum*, respectively. Many related compounds, including 1-hydroxy- and 1-methoxy-derivatives of phytoene, phytofluene and 7,8,11,12-tetrahydrolycopene, have been isolated, usually in small amounts, from normal or inhibited cultures. These compounds could be considered to be intermediates in alternative pathways, but their presence can be explained by slight variations in the relative rates of the reactions or by lack of absolute substrate specificity of the enzymes, which are able to accept any carotenoid that has the required structural feature, irrespective of the rest of the molecule.

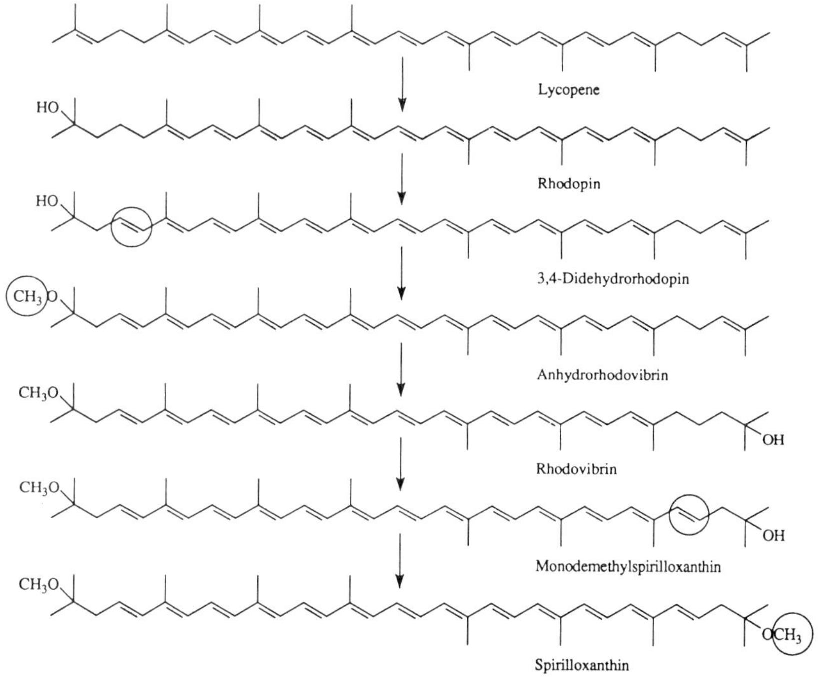

Figure 4.15 Pathway for the biosynthesis of spirilloxanthin from lycopene.

The C-1 hydroxylation is believed to occur by a hydration reaction that is initiated by H^+ attack at C-2 and is therefore analogous or mechanistically similar to the cyclisation reaction. Formation of the 1,2-dihydrocarotenes that are characteristic of *Rhodopseudomonas viridis* is thought to occur by a similar reaction involving the addition of H_2 across the C-1,2 double bond (Fig. 4.16).

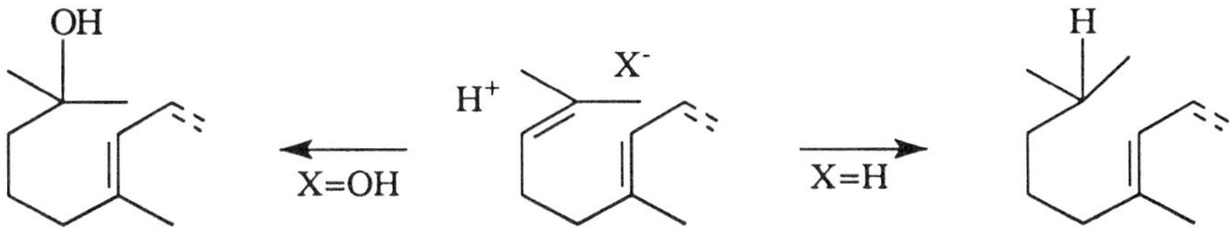

Figure 4.16 Hydroxylation and hydrogenation at C-1,2 as alternatives to cyclisation.

Nothing is known about the reactions by which other structural modifications are elaborated, e.g. the introduction of an oxo-group at C-4 (in okenone) under strictly anaerobic conditions, and the oxidation of the C-20 methyl group to produce cross-conjugated aldehydes such as rhodopin-20-al.

(c)Aromatisation

It is assumed that the 1,2,5-trimethylphenyl (ϕ) and 1,2,3-trimethylphenyl (χ) end-groups are derived biosynthetically by dehydrogenation–rearrangement processes from the ß-ring. The formation of the 1,2,5-trimethylphenyl ring has been investigated by incubating *Chloropseudomonas ethylica* with radioactively labelled MVA and then chemically degrading the product, chlorobactene (Moshier and Chapman 1973). This showed that the methyl group that remains at C-1 is that which arises from C-2 of MVA, so it is the methyl group derived from the C-3′ methyl group of MVA that migrates from C-1 to C-2 of the carotenoid ring during the aromatisation. The stereochemical disposition of the C-1 methyl groups in the proposed ß-ring precursor has not been established directly but, if it is the same as in the ß-ring of zeaxanthin in plants, algae and non-phototrophic bacteria (Fig. 4.9), the aromatisation must follow the stereochemical course illustrated in Fig. 4.17.

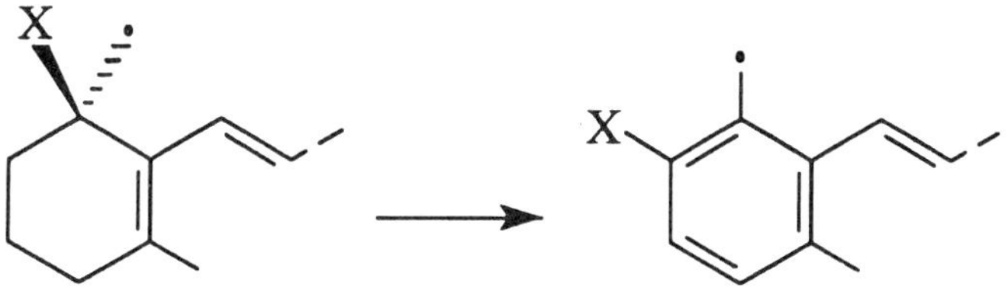

Figure 4.17 Mechanism and stereochemistry of rearrangement to form the 1,2,5-trimethylphenyl (ϕ) end-group of chlorobactene.

The series of rearrangements that would be required for the subsequent conversion of the 1,2,5-trimethylphenyl into the 1,2,3-trimethylphenyl end-group must be considerably more complex, and no experimental information is yet available about the reaction. A "Ladenburg prism" mechanism has been proposed (Fig. 4.18) and provides a good working hypothesis (Liaaen-Jensen 1969).

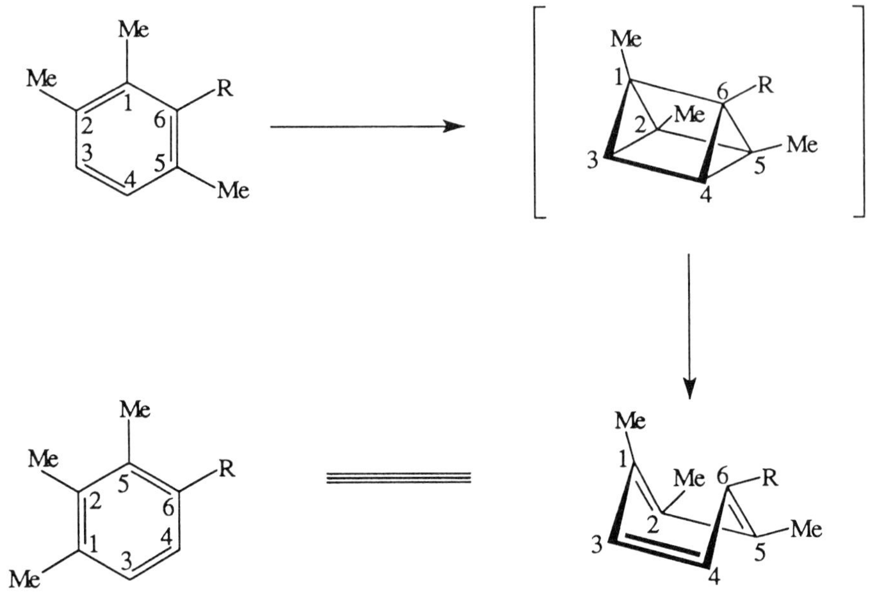

Figure 4.18 Proposed Ladenburg prism mechanism for the rearrangment to form the 1,2,3-trimethylphenyl (χ) end-group.

4.3 ENZYMES OF CAROTENOID BIOSYNTHESIS

The pioneering work on the biosynthesis of carotenoids in cell-free systems was performed in Porter's laboratory (summarised by Spurgeon and Porter 1983), with acetone powders made from chromoplasts of tomato fruit (*Lycopersicon esculentum*). Over a period of several years, acetone powders were prepared from a range of different strains of tomato, and were used to demonstrate first the incorporation of IDP into phytoene, lycopene and cyclic carotenes and then the direct conversion of precursor carotenoids into products. The biosynthesis of carotenoids *in vitro* has been reviewed by Bramley (1985).

4.3.1 Phytoene synthase

The synthesis of phytoene was studied in detail. A soluble enzyme system was isolated from the acetone powders and could synthesise phytoene from IDP. Substantial purification (350-fold) yielded a 200 000 M_r complex that contained all the enzyme activities necessary to convert IDP into phytoene (Maudinas et al. 1977). The complex was unstable and could not be purified to homogeneity. During ion-exchange chromatography, dissociation occurred; the ability to synthesise phytoene was lost, but the preparation still contained a 40 000 M_r protein which actively converted IDP into PPDP (Islam et al. 1977). This protein seems remarkably small, considering that it contains three different enzyme activities, but it has been purified to homogeneity. It showed a requirement for Mn^{2+} which could partly be replaced by Mg^{2+}. For phytoene synthesis also by the undissociated complex, Mn^{2+} was required and ATP was markedly stimulatory though it is not directly involved in the reaction.

More recently, a crude enzyme system from chromoplasts of red pepper (*Capsicum annuum*) has been used in Camara's laboratory (Camara and Monéger 1982; Camara et al. 1982; Camara 1985). The soluble fraction from the chromoplasts was able to incorporate IDP into phytoene, though later steps occurred exclusively in the membrane fraction. A 190 000 M_r phytoene synthase complex similar to that from tomato has been isolated (Camara 1985). The synthesis of phytoene from IDP by this complex was stimulated by polyethylene glycol and inhibited by allylic diphosphates.

An affinity chromatography method has now been used to purify to homogeneity the phytoene synthase from *Capsicum* chromoplasts (Dogbo et al. 1988). The enzyme proved to be a single monomeric protein of 47 500 M_r that carried the two enzyme activities that catalyse the formation of PPDP from two molecules of GGDP and also the subsequent conversion of PPDP into phytoene. The enzyme activity, which was strictly dependent on Mn^{2+}, was inhibited by inorganic pyrophosphate and by hydroxyphenylglyoxal, a reagent that is specific for arginine residues. The conclusion that phytoene synthase is a bifunctional enzyme was supported by the finding that in no instance were the two reactions kinetically uncoupled.

In daffodil chromoplasts, the enzymes which catalyse the conversion of IDP into phytoene form an association that is peripheral to the intraplastidic membranes, whereas the later desaturase and cyclase enzymes are integral membrane proteins (Kreuz et al. 1982).

4.3.2 Phytoene desaturase

The acetone powders from tomato chromoplasts also provided the first crude cell-free preparations with which the direct desaturation of phytoene was demonstrated. With such systems from normal red tomatoes, several incorporations were achieved, notably of (15*Z*)-phytoene, and of (15*Z*)-phytofluene, (all-*E*)-phytofluene, and (all-*E*)-ζ-carotene into neurosporene, lycopene and monocylic and dicyclic carotenoids (Kushwaha et al. 1970; Qureshi et al. 1974). Anaerobic conditions were used and a requirement for $NADP^+$, FAD and Mn^{2+} was reported. The desaturation of phytoene to phytofluene and ζ-carotene by the acetone powder preparation from *Capsicum* chromoplasts has also been reported (Camara and Monéger 1982), but the desaturase enzyme system was not studied in detail.

Major progress has been made recently by Kleinig, Beyer and co-workers, who have elucidated details of the desaturation sequence and the desaturase enzymes in daffodil chromoplasts (Beyer and Kleinig 1990). This impressive work showed that the desaturase (and cyclase) enzymes are integral membrane proteins which can be solubilised by the zwitterionic detergent CHAPS. The enzyme activity is lost as a result of the solubilisation but is restored when the solubilised enzymes are reconstituted in proteoliposomes. The multi-component system required at least a terminal electron acceptor in addition to the individual enzymes (Mayer et al. 1990). With this system, at least two segments or phases of the desaturation sequence could be identified, with subsequent cyclisation constituting a third phase (Beyer et al. 1989). In the first phase of desaturation, the conversion of (15*Z*)-phytoene into (15*Z*)-ζ-carotene occurred in the dark, with O_2 as an essential requirement. The reaction was independent of cofactors such as $NADP^+$ and NADPH but was at least stimulated by an oxidoreductase. (15*Z*)-ζ-Carotene could not be desaturated further, but when this was isomerised by light or during isolation, the ζ-carotene isomeric mixture that was thus formed then underwent further desaturation to give lycopene, not as the (all-*E*) form but as the (7*Z*,9*Z*,7′*Z*,9′*Z*)-isomer, prolycopene. It is assumed that the photoisomerisation produces (all-*E*)-ζ-carotene which is then used as substrate in the subsequent desaturations, but it would be difficult to distinguish between (all-*E*)- and (9*Z*,9′*Z*)-ζ-carotene by chromatography and light absorption spectrum. The only significant work on phytoene desaturation in photosynthetic systems has come from the laboratory of Sandmann. The approach adopted has been to use a coupled assay technique (Bramley 1990) in which a soluble fraction from a strain of the mould *Phycomyces blakesleeanus* is included in the incubation with IDP to generate *in situ* labelled phytoene that becomes the substrate for the later, membrane-bound enzymes. Because it requires the addition of a complex and undefined extract to the detergent-solubilised membrane preparation, this approach is obviously much less satisfactory than the direct use of exogenous labelled phytoene as substrate and the results obtained must be evaluated with caution. Nevertheless, this approach has produced much information about carotenoid biosynthesis, especially desaturation, in the cyanobacteria *Aphanocapsa* and *Anacystis*. The efficient formation of desaturated carotene products by the mixed system has been achieved (Sandman and Bramley 1985a). Desaturation of phytoene to lycopene was O_2-dependent but, in contrast to the results with daffodil chromoplasts, an enhancement by NAD^+ or $NADP^+$ (but not their reduced forms) was seen (Sandman and Kowalczyk 1989).

4.3.3 Cyclases

In the classical work of Porter's group, acetone powders from the "high beta" and "high delta" strains of tomato were used to demonstrate the light-dependent incorporation of radioactive lycopene into a range of monocyclic and dicyclic carotenes, especially α-carotene, ß-carotene, γ-carotene and δ-carotene (Kushwaha et al. 1969; Papastephanou et al. 1973). The reported requirement for FAD is surprising because the cyclisation should not involve any oxidative step. In other work the incorporation of labelled lycopene into ß-carotene and other cyclic carotenoids *in vitro* by preparations from bean leaf chloroplasts (Hill et al. 1971) and spinach chloroplasts (Kushwaha et al. 1970) was reported. All these studies, however, used only crude preparations; no purification of the enzymes was achieved.

More recently, preparations with lycopene cyclase activity have been obtained from chromoplasts of *Capsicum annuum* by solubilisation of the chromoplast membranes with Tween 80 or via an acetone powder. These preparations converted lycopene into ß-carotene, but no formation of the monocyclic γ-carotene was detected. There was no requirement for $NADP^+$ or FAD.

Important new information has come from the laboratory of Beyer and Kleinig (Beyer et al. 1989). With the solubilised membrane preparation from daffodil chromoplasts, as described above, the third stage of biosynthesis to be investigated was the cyclisation of lycopene to ß-carotene. Although the desaturation reactions required O_2, the subsequent cyclisation occurred only under strictly anaerobic conditions. (all-*E*)-Lycopene was not a substrate, even though the products of cyclisation, ß-carotene etc., were exclusively the (all-*E*)-isomers. The cyclisation was shown to proceed efficiently only with prolycopene (7Z,9Z,7´Z,9´Z) or (7Z,7´Z)-lycopene as substrate. In contrast to this, however, it was reported that, in the mutant strain C-6D of the green alga *Scenedesmus*, desaturation occurs by a poly-*cis* pathway in the dark to give prolycopene but cyclisation occurs only after light-catalysed isomerisation of this to the all-*trans*-isomer, (Humbeck 1990; Sandmann 1991).

4.3.4 Enzymes for hydroxylation and other later reactions

Almost nothing is known about the later enzymes of carotenoid biosynthesis, and little progress has been achieved in demonstrating carotenoid hydroxylation and other transformations in cell-free systems.

(a)Hydroxylation

The coupled incubation system of Sandmann and Bramley has been used to achieve the formation of ß-cryptoxanthin by membrane preparations of the cyanobacterium *Aphanocapsa* (Sandmann and Bramley 1985a) and, later, the direct incorporation of ß-carotene into ß-cryptoxanthin by the *Aphanocapsa* preparation was reported (Sandmann and Bramley 1985b). The conversion was dependent on O_2, stimulated by NADPH and sensitive to CO and other inhibitors, i.e. it showed characteristics typical of a monooxygenase reaction.

(b) Epoxidation

Enzymes responsible for the epoxidation of zeaxanthin to antheraxanthin and violaxanthin and for the reverse of this process have been obtained from plant chloroplast membranes, but it is likely that these processes occur in relation to the functioning of the violaxanthin cycle, and these enzymes may not be involved in the initial biosynthetic reactions (Yamamoto 1979).

(c) Other reactions

Antheraxanthin and violaxanthin may be incorporated into capsanthin and capsorubin, respectively, by preparations of *Capsicum* chromoplasts (Camara and Monéger 1982) and zeaxanthin into neoxanthin, diadinoxanthin and peridinin by a homogenate of a dinoflagellate, *Amphidinium carterae* (Swift et al. 1982).

4.4 GENETICS OF CAROTENOID BIOSYNTHESIS

Genetic studies of carotenoid biosynthesis can be considered to have had two phases: the traditional classical genetics, mainly with tomatoes; and the recent advances in molecular genetics.

4.4.1 Classical genetics

Evidence from classical genetics (summarised by Kirk and Tilney-Bassett (1978)) shows that, in higher plants, all genes responsible for carotenoid biosynthesis are nuclear encoded. Mutations involving desaturation and cyclisation have been identified, and phytoene-, ζ-carotene- and lycopene-accumulating mutants of maize, wheat and various other plant species have been produced.

The most extensive work was carried out in the 1940s to 1960s with mutants of tomato. Many mutants were produced and their phenotypes identified. In particular, genes were identified that control the total amount of carotenoid formed (r^+/r), the geometry of the lycopene product (t^+/t), the formation of ß- and ε- rings (b^+/b and del^+/del, respectively) as well as various moderator genes. On the basis of these identifications the Porter–Lincoln scheme, which first set out the main features of the pathway of carotenoid biosynthesis was proposed. Goodwin (1971, 1980) gave a concise account of the main findings.

4.4.2 Molecular genetics

In recent years, rapid progress has been made on the molecular genetics of carotenoid biosynthesis in phototrophic bacteria. This work will now be intensified and will lead to major advances in the genetics of carotenoid biosynthesis in higher plants and algae. It is this approach, when applied in combination with traditional chemical and biochemical procedures, that will bring the elucidation of details of the enzymes and their organisation, the reaction mechanisms and the regulation of carotenoid biosynthesis during the development of the photosynthetic apparatus in plants, algae and bacteria.

(a) Phototrophic bacteria

Work on the genetics of carotenoid biosynthesis in the phototrophic bacteria follows from the pioneering studies of Marrs some ten years ago. Marrs (1982)

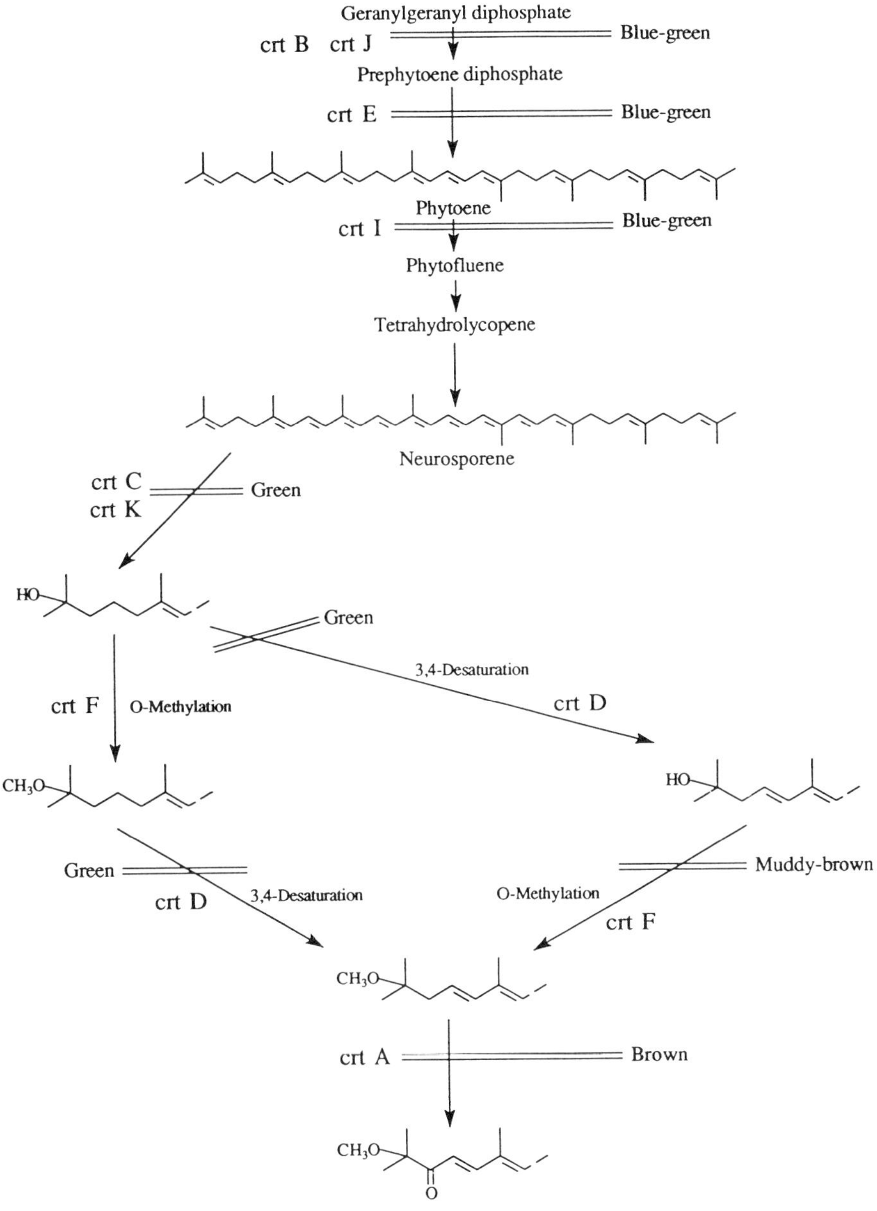

Figure 4.19 Sequence of reactions in the biosynthesis of carotenoids in *Rhodobacter* spp. The reactions catalysed by the known gene products are shown, and the colour phenotypes given by mutations in the various *crt* genes are indicated.

discovered that the biosynthesis of all the main components of the pigment–protein complexes of *Rhodobacter capsulatus*, namely peptides, bacteriochlorophyll and carotenoids, is encoded on a tight gene cluster which can be inserted into a plasmid and is amenable to molecular genetic manipulations such as transposon or interposon mutagenesis. Since then, a great deal of exciting progress has been made in the laboratories of Marrs, Scolnik, Giuliano and Hearst with *Rb. capsulatus* and of Harding and Pemberton, and Hunter with *Rb. sphaeroides*. The main findings of this work are summarised below.

In the case of *Rb. capsulatus*, all the known carotenoid genes lie within a 46 kilobase photosynthetic gene cluster which is derived from the bacterial chromosome and includes at least nine carotenoid (*crt*) and eleven bacteriochlorophyll (*bch*) genes. A variety of transposon, interposon and point mutants have been generated and characterised by their altered pigmentation, as determined by rigorous identification of the carotenoids present, by high-performance liquid chromatography (HPLC), UV-Vis and mass spectra. The *crt* genes, introduced into plasmid pRPS404, have been mapped by genetic and physical techniques, and biochemical functions have been assigned to the gene products by identification of the biosynthetic intermediates that accumulate in the mutants (Giuliano et al. 1988; Armstrong et al. 1989, 1990). The sites in the biosynthetic pathway where the various gene products act and the colour phenotype given by a mutation in each *crt* gene are illustrated in Fig. 4.19.

A segment of contiguous DNA from the gene cluster has now been cloned and sequenced. This segment contains eight of the nine *crt* genes that have been recognised, namely *crtA, crtB, crtC, crtD, crtE, crtF, crtI,* and *crtK*. The sequence of 11039 base pairs has been published and the sequence of each *crt* gene deduced (Armstrong et al. 1989, 1990; Bartley and Scolnik 1989). The ninth gene, *crtJ,* has been detected elsewhere in the cluster, among the *bch* genes. The location of the *crt* genes in the cluster is given in Fig. 4.20. Not all *crt* genes are transcribed in the same direction.

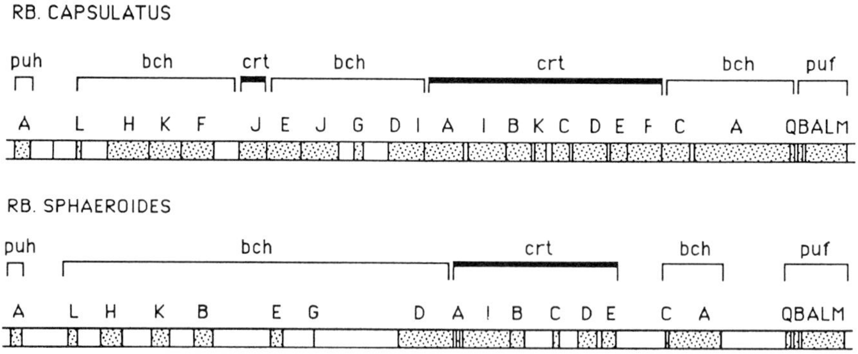

Figure 4.20 Maps of the photosynthetic gene clusters of (a) *Rhodobacter capsulatus*, (b) *Rb. sphaeroides*.

From the base sequence, and from a consideration of the usage of codons in *Rb. capsulatus*, which shows a strong bias against codons that have A (adenosine) or T (thymine) in the third position, sites of translational starts have been proposed for the *crt* genes, some possible ribosome-binding sites have been identified, and possible transcription termination signals have been detected.

From the base sequences, the amino acid sequences, molecular masses and mean hydropathy values have been deduced for the *crt* gene products, the biosynthetic enzymes. Confirmation of the sequences and properties awaits the isolation of the proteins themselves. Except for CrtK, the proteins appear not to be particularly hydrophobic. A remarkably high degree of amino acid homology was found for the enzymes CrtI and CrtD, which catalyse 7,8- and 11,12- desaturation (CrtI) and 3,4-desaturation (CrtD), respectively.

Results from different laboratories are broadly in agreement, but the regulation of the genes is still an area of argument and speculation. It appears that all the *crt* genes may be organised as single transcriptional units (Giuliano et al. 1988), though Young et al. (1989) have obtained evidence of superoperonal organisation of the photosynthesis genes of *Rb. capsulatus*. Oxygen is certainly a major regulatory factor; mRNA levels for at least three genes, *crtA, C* and *E*, increase substantially following transfer from aerobic to anaerobic photosynthetic growth conditions, when a considerable increase in carotenoid content occurs.

For *Rb. sphaeroides* a cluster of six carotenoid biosynthesis genes (*crtA, B, C, D, E, F*) was described in 1986 by Pemberton and Harding (1986). Hunter and coworkers (Coomber et al. 1990) have used transposon mutagenesis to characterise a 45 kilobase photosynthetic gene cluster of *Rb. sphaeroides* which is similar in size and composition to that of *Rb. capsulatus*.

From the number of genes that have been identified, it is clear that there is simply one enzyme for each reaction. For example, for the C-7,8, C-11,12, and C-11′,12′ desaturations there is only one gene, *crtI*, not different genes for the desaturation of phytoene, phytofluene, and ζ-carotene (or 7,8,11,12-tetrahydrolycopene). The enzyme molecules and their reaction mechanisms must, therefore, be the same. Similarly there is only one gene, *crtC*, for the C-1,2 hydration reaction and hence there is only one type of hydration enzyme, not different enzymes for the two ends of the molecule or one for each possible substrate. 1-Hydroxy-1,2-dihydrophytoene is present along with phytoene in desaturation (*crtI*) mutants, showing that the enzyme that normally adds water across the C-1,2 double bond of neurosporene will also hydrate phytoene if this is present in place of neurosporene. This indicates that the enzymes of carotenoid biosynthesis show specificity towards a particular end-group or structural feature rather than towards a particular substrate molecule. It also shows that the genes for hydration (*crtC*) and desaturation (*crtI*) are transcribed independently. Although the earlier biosynthetic step, desaturation, is blocked by the mutation in *crtI*, the enzyme (CrtC) for the later reaction is still present.

Subsequent work in which the carotenoid biosynthesis genes of the non-phototrophic bacterium *Erwinia* (*herbicola* and *uredovora*) have been cloned in

Escherichia coli provides information which may also be relevant to the biosynthesis of carotenoids in photosynthetic systems, particularly the organisation of the carotenogenic enzymes (Misawa et al. 1990). These enzymes are generally believed to be located in or bound to membranes as an organised multi-enzyme complex. The work with the *Erwinia* genes has shown that these enzymes can be expressed and correctly organised in *E. coli*, an organism that is normally incapable of carotenoid biosynthesis. The significance of this may be far-reaching, since a model is indicated in which at least one of the carotenogenic enzymes is able to recognise and attach to a protein-binding site on the *E. coli* membrane. It also suggests that the other enzymes then aggregate into a self-assembling multi-enzyme complex which is able to synthesise carotenoids efficiently without the need of additional binding of regulatory proteins or any separate isomerase enzymes. The organisation of the enzymes in the correct location and orientation in a functional multi-enzyme complex is essential for carotenoid biosynthesis to take place and is a critical factor in determining which carotenoid is made.

A simple model such as this may be applicable to bacteria but it seems likely that, in more complex organisms such as plants and algae, the arrangement, organisation and regulation of the carotenogenic enzymes may be much more complicated, especially if the involvement of geometrical isomers and isomerisation, as indicated for the daffodil system, is substantiated. Some very recent work shows that some progress is at last beginning to be made on the molecular genetics of carotenoid biosynthesis in plants.

(b) Plants and algae

Attempts to use the bacterial genes as heterologous probes to identify and clone the plant genes have proved unsuccessful. This is not now surprising in view of the lack of homology between the bacterial genes and those plant genes that have been isolated (Pecker et al. 1992). Progress has been made, however, by using genes from the cyanobacterium *Synechococcus* PCC7942 as probes, or via cDNA clones associated with the ripening of fruit, especially tomatoes. The genes *psy* and *pds*, which encode phytoene synthase and phytoene desaturase, respectively, from *Synechococcus* have been cloned. The gene *pds* has been used as a heterologous probe for identifying and cloning cDNA from tomato and soyabean (*Glycine max*; Bartley et al. 1991; Pecker et al. 1992). There was substantial homology between the cyanobacterial gene and part of the plant genes, but the product of the plant gene is a substantially larger polypeptide which appears to contain a transit peptide region that serves to facilitate import of the protein into the plastids. The importation of tomato and soyabean phytoene desaturase into isolated pea chloroplasts has been demonstrated. The pre-enzyme is processed into a functional enzyme of 52 000 Mr, the same size as that in the cyanobacterium.

A cDNA clone from tomato, pTOM5, that is differentially expressed during ripening of the fruit (Bird et al. 1991), has been shown to be homologous with *psy*, the phytoene synthase gene of *Synechococcus* (Chamovitz et al. 1992). The earlier gene, coding for GGDP synthase, from *Capsicum annuum* has recently been cloned (Romer S., unpublished data).

Evidence is beginning to accumulate to indicate that regulation at the transcriptional level is important in carotenoid biosynthesis in ripening fruit, and that biosynthesis in fruit and leaves is regulated differentially (Bird et al. 1991; Pecker et al. 1992).

These studies are at an early stage and have not yet yielded any appreciable information about the enzymes of carotenoid biosynthesis in the chloroplast and their regulation. Now that a start has been made, however, progress in this area is likely to be extremely rapid.

4.5 REGULATION OF CAROTENOID BIOSYNTHESIS

4.5.1 As part of chloroplast development

Carotenoid biosynthesis is an essential part of the construction of the pigment–protein complexes in the thylakoids, and its regulation is closely linked to that of other components such as chlorophylls, proteins and lipids. The whole process of chloroplast development, including expression of the genes for carotenogenesis, is under direct nuclear control. There must, however, be other control mechanisms operating on the individual biosynthetic pathways in order to regulate the supply of the various components, including the carotenoids, as required during the assembly of the pigment–protein complexes and thylakoids.

Because there are several different carotenoids which have different locations and functions within the pigment–protein complexes it is an oversimplification to think of the biosynthesis of the total carotenoid in leaf or chloroplast as a single uniform process. The factors which regulate the biosynthesis of each individual carotenoid and determine the carotenoid composition of each pigment–protein complex are of major importance.

Some experimental information relevant to these points has been obtained. Labelling studies have shown that ß- and ε-rings are not interconverted (Williams et al. 1967), so the final stages of lutein and ß-carotene synthesis must be independent. When young seedlings are grown either in the light or in the dark and then incubated, in the light, with [^{3}H]-MVA and CO_2 simultaneously, different ^{3}H:^{14}C ratios are observed for ß-carotene and lutein, indicating that the time courses for synthesis of these two carotenoids are different (Sergeant and Britton 1984). However, the stage at which the two pathways diverge is not known. Lutein and ß-carotene could be made by two distinct enzyme assemblies. Alternatively, the divergence may occur at a later stage, to utilise different final cyclisation and hydroxylation reactions after phytoene synthesis and desaturation have taken place on a common enzyme assembly. Sergeant and Britton (1984), also showed that there were no great differences in either total incorporations or ^{3}H:^{14}C ratios for any individual carotenoid in the different thylakoid fractions.

There must be delicate regulatory mechanisms which determine the specific destination and location, within the thylakoid pigment–protein complexes, of each carotenoid molecule that is made. It may be significant that carotenoid molecules that are labelled with various substrates in a short incubation lose their labelling much more quickly than ones that are labelled in a longer incubation. This suggests that newly formed carotenoid molecules may enter a labile pool from which they are either taken for incorporation into the pigment–protein comlexes or rapidly degraded.

Several different phases or variations of carotenoid biosynthesis in chloroplasts must also be considered:

(i) Synthesis of the bulk of the pigment during the initial construction of the photosynthetic apparatus.

(ii) Synthesis that continues in mature chloroplasts as part of turnover.

(iii) In greening etiolated plants, continuing synthesis by the enzymes that were active in the etiolated leaves, especially in the early stages of greening before new enzyme synthesis becomes established.

(iv) Synthesis to modify carotenoid compositions as a response to, or a consequence of, changes in environmental conditions, especially light intensity.

It is also likely that chloroplast differentiation and its associated carotenoid biosynthesis do not follow the same course universally. Substantial differences may be expected between different plant species and even between different varieties of the same species. Growth conditions and the age of the plant tissue under investigation may also be very important, and cotyledons and true leaves, even of the same plant, may behave quite differently and are not strictly comparable. Greening etiolated seedlings provide a favoured and amenable experimental system, but it must be remembered that the course of chloroplast differentiation and its associated carotenoid biosynthesis in a greening etiolated plant is not the same as it is in a plant growing under a normal light–dark regime. Extreme caution should be exercised, therefore, when comparing systems, and especially when extrapolating from one system to another. The regulation of carotenoid biosynthesis is thus extremely complicated. It seems likely that control at the genetic and transcriptional level will be important. This promises to be one of the most interesting areas of study that will be opened up by developments in molecular genetics in the near future.

4.5.2 Carotenoid transformations in greening etiolated plants

Although it is not the normal course of plant growth and chloroplast development (under normal conditions of a day–night light regime, chloroplasts develop direct from proplastids), the greening of etiolated seedlings and the associated transformation of etioplast to chloroplast provide an extremely convenient experimental system in which to study the incorporation of carotenoids into the developing chloroplasts. Etiolated seedlings generally contain substantial amounts of carotenoids, although the distribution of the individual pigments differs considerably from that in the green plant. In particular, etiolated leaves and etioplasts normally contain little or no carotene and a modified collection of xanthophylls; lutein and violaxanthin are predominant but a substantial amount of antheraxanthin is usually present and there is little or no neoxanthin (Bahl 1977; Barry et al. 1991; see Chapter 6).

In the etioplasts, the carotenoids are reported to be located in prothylakoids, prolamellar bodies, and the etioplast envelope. The enzymes of carotenoid biosynthesis are also present in the etioplast, the phytoene synthase complex in a soluble form in the stroma and the subsequent enzymes in membrane fractions (Lütke-Brinkhaus and Kleinig 1987).

When dark-grown etiolated plants are illuminated, the prolamellar bodies break down and the prothylakoids grow to form thylakoids and, eventually, grana stacks (Virgin et al. 1963; Henningsen and Boynton 1970). Over a period of 24–48 hr the characteristic carotenoid/chlorophyll composition of the chloroplast is attained.

A series of experiments from the author's laboratory has compared the effects of different light intensities on carotenoid biosynthesis and carotenoid compositions in different sections of the leaves of greening etiolated barley seedlings (Barry et al. 1991). In general, it took 24h longer in low intensity (1000 lux) than in high intensity light (10 000 lux) for the normal chloroplast pigment composition and content to be attained. The rate of pigment synthesis was, as expected, greatest in the rapidly growing bottom zone of the shoot, where carotenoid levels doubled, but virtually no increase was seen in the shoot tip. The levels of ß-carotene and neoxanthin increased several-fold, and that of antheraxanthin (and eventually that of zeaxanthin) decreased markedly.

This work provided no information on whether the carotenoid molecules that were present in the etioplast could be used and incorporated into the chloroplast and its pigment–protein comlexes. A subsequent series of experiments (P. Barry and G. Britton, unpublished results), allowed the etiolated seedlings to become green while being watered with H_2O/D_2O (1:1). The carotenoids were isolated from three zones of the shoot (bottom, middle and tip) and analysed by mass spectrometry. Under these conditions, existing molecules, synthesised in the etiolated seedlings, contain no deuterium, whereas molecules synthesised during greening in D_2O show substantial levels of deuterium incorporation. Patterns characteristic of synthesis from photosynthetically fixed CO_2 or from existing reduced carbon substrates can also be distinguished. From this work, the following conclusions could be drawn:

(i) The lutein and violaxanthin that were present in the etiolated seedlings were retained during greening and were incorporated into the chloroplasts and thylakoids.

(ii) ß-Carotene was synthesised almost entirely *de novo*, initially from existing reduced carbon substrates, later from CO_2, especially in the leaf base region.

(iii) Although the level of neoxanthin increased approximately five-fold during greening, no incorporation of deuterium was seen. The neoxanthin appears to be produced from antheraxanthin or violaxanthin that was present in the etioplasts.

4.5.3 Regulation of carotenoid biosynthesis in phototrophic bacteria

Although many phototrophic bacteria are obligate anaerobes, some can adopt several different lifestyles and can grow under anaerobic or aerobic conditions in the light or dark. It is not surprising, therefore, that light and oxygen are the main factors that affect carotenoid content and composition. In those organisms that can live either phototrophically under anaerobic conditions or heterotrophically in air e.g. *Rhodobacter* spp., transfer from aerobic to anaerobic conditions causes a considerable increase in carotenoid content. Oxygen is the major regulatory factor and Giuliano et al. (1988) have shown that mRNA levels for three genes, *crtA, C* and *E*, increase substantially following the transfer. With *crtE*, a transcript appears that is almost undetectable under aerobic conditions.

When *Rhodobacter* spp. growing under phototrophic conditions are exposed to even traces of oxygen this is sufficient to cause the rapid transformation of the main carotenoids spheroidene and hydroxyspheroidene into their 2-oxo-derivatives, spheroidenone and hydroxyspheroidenone. These changes apparently occur while the carotenoids are in position in the pigment–protein complexes.

Changes in light intensity also lead, albeit more slowly, to changes in carotenoid content. In particular, transfer to lower light intensity leads to an increase in the size of the light-harvesting antenna complexes, especially the B800–850 type, so the overall carotenoid content increases substantially.

4.6 CONCLUSIONS

Broad details of the pathways and reactions of carotenoid biosynthesis have been known for many years. Progress on the isolation and study of the enzymes, however, has been very slow. The difficulty of such work has usually been attributed to a requirement for the enzymes to be associated and organised as a multi-enzyme complex which is membrane bound and which loses its activity when dissociated. The carotenogenic enzymes are of considerable practical importance as a target site for herbicide action (see Chapter 5) and also in relation to the biotechnological production of natural carotenoids, especially ß-carotene and astaxanthin, as natural colorants for food or feed, or as health-promoting substances.

The recent rapid advances in the molecular genetics of carotenoid biosynthesis in bacteria are now leading to progress also with plants and algae. Imaginative use of these powerful techniques, in combination with rigorous biochemical work and an appreciation of the chemistry and properties of the carotenoids themselves will surely lead to major advances in knowledge and understanding of the genes, enzymes and reactions of carotenoid biosynthesis, of the organisation of the enzymes into functional complexes and of the intricate and vital regulatory mechanisms that must operate, particularly in plants, to control carotenoid biosynthesis as part of the development and maintenance of functional photosynthetic apparatus. The future for carotenoid biosynthesis is exciting.

REFERENCES

Armstrong, G.A., Alberti, M., Leach, F. and Hearst, J.E. (1989) Nucleotide sequence, organisation, and nature of the protein products of the carotenoid biosynthesis gene cluster of *Rhodobacter capsulatus*. *Molec. Gen. Genet.*, **216**:254-268.

Armstrong, G.A., Schmidt, A., Sandmann, G. and Hearst, J.E. (1990) Genetic and biochemical characterisation of carotenoid biosynthesis mutants of *Rhodobacter capsulatus*. *J. Biol. Chem.*, **265**:8329-8338.

Bahl, J. (1977) Chlorophyll, carotenoid and lipid content in *Triticum sativa* L. Plastid envelopes, prolamellar bodies, stroma lamellae and grana. *Planta*, **136**:21-24.

Barry, P., Young, A.J. and Britton, G. (1991) Accumulation of pigments during the greening of etiolated seedlings of *Hordeum vulgare*. *J. Exp. Bot.*, **42**:229-234.

Bartley, G.E. and Scolnik, P.A. (1989) Carotenoid biosynthesis in photosynthetic bacteria. *J. Biol. Chem.*, **264**:13109-13113.

Bartley, G.E., Viitanen, P.V., Pecker, I. *et al.* (1991) Molecular cloning and expression in photosynthetic bacteria of a soybean cDNA coding for phytoene desturase, an enzyme of the carotenoid biosynthesis pathway.*Proc. Natl. Acad. Sci. USA*, **88**:6532-6536.

Beyer, P. and Kleinig, H. (1990) On the desaturation and cyclisation reactions of carotenes in chromoplast membranes. In: Krinsky, N.I., Mathews-Roth, M.M. and Taylor, R.F. (eds.), *Carotenoids: Chemistry and Biology*. Plenum, New York, pp. 195-206.

Beyer, P., Mayer, M. and Kleinig, H. (1989) Molecular oxygen and the state of geometric isomerism of intermediates are essential in the carotene desaturation and cyclization reactions in daffodil chromoplasts. *Eur. J. Biochem.*, **184**:141-150.

Bird, C.R., Ray, J.A., Fletcher, J.D. *et al.* (1991) Using antisense RNA to study gene function: inhibition of carotenoid biosynthesis in transgenic tomatoes. *Bio/Technology*, **9**:635-639.

Bramley, P.M. (1985) The in vitro biosynthesis of carotenoids. *Adv. Lip. Res.*, 21:243-279.

Bramley, P.M. (1990) The use of cell-free systems in studies of carotenoid biosynthesis. In: Krinsky, N.I., Mathews-Roth, M.M. and Taylor, R.F. (eds.), *Carotenoids: Chemistry and Biochemistry*. Plenum Press, New York, pp. 185-194.

Britton, G. (1976) Biosynthesis of carotenoids. In: Goodwin, T.W. (ed.), *Chemistry and Biochemistry of Plant Pigments*, 2nd Ed., Vol. 1. Academic Press, New York, pp. 262-327.

Britton, G. (1985) Stable isotopes in carotenoid biochemistry. *Pure Appl. Chem.*, **57**:701-708.

Britton, G. (1988) Biosynthesis of carotenoids. In: Goodwin, T.W. (ed.), *Plant Pigments*. Academic Press, New York, pp. 133-182.

Britton, G. (1990a) Carotenoid biosyntheis – an overview. In: Krinsky, N.I., Mathews-Roth, M.M. and Taylor, R.F. (eds.), *Carotenoids: Chemistry and Biology*. Plenum Press, New York and London, pp. 167-184.

Britton, G. (1990b) The stereochemistry of carotenoid biosynthesis. In: Atta-ur-Rahman (ed.), *Studies in Natural Products Chemistry*, Vol. 7. Elsevier, Amsterdam, pp. 317-367.

Britton, G. (1991) The biosynthesis of carotenoids: a progress report. *Pure Appl. Chem.*, **63**:101-108.

Britton, G. and Mundy, A.P. (1980) The use of deuterium labelling from deuterium oxide to study the biosynthesis of chloroplast carotenoids. *Dev. Plant Biol.*, **6**:345-350.

Camara, B. (1985) Carotenogenic enzymes from *Capsicum* chromoplasts. *Pure Appl. Chem.*, **57**:675-677.

Camara, B. and Monéger, R. (1982) Biosyntheic capabilities and localization of enzymatic activities in carotenoid metabolism of *Capsicum annuum* isolated chloroplasts. *Physiol. Végétale*, **20**:757-773.

Camara, B., Bardat, F. and Monéger, R. (1982) Sites of biosynthesis of carotenoids in *Capsicum* chromoplasts. *Eur. J. Biochem.*, **127**:255- 258.

Chamovitz, D., Misawa, N., Sandmann, G. and Hirschberg, J. (1992) Molecular cloning and expression in *E.coli* of a cyanobacterial gene coding for phytoene synthase, a carotenoid biosynthesis enzyme. *FEBS lett.*, **296**:305-310.

Coomber, S.A., Chaudhri, M., Connor, A. *et al.* (1990) Localised transposon Tn*5* mutagenesis of the photosynthetic gene cluster of *Rhodobacter sphaeroides*. *Molec. Microbiol.*, **4**:977-989.

Davies, B.H. (1980) Carotenoid biosyntheis. In: Czygan, F-C. (ed.), *Pigments in Plants*, 2nd Ed. Gustav Fischer, Stuttgart, pp. 31-56.

Dogbo, O., Laferriere, A., d'Harlingue, A. and Camara, B. (1988) Carotenoid biosynthesis: Isolation and characterisation of a bifunctional enzyme catalyzing the synthesis of phytoene. *Proc. Natl. Acad. Sci. USA*, **85**:7054-7058.

Giuliano, G., Pollock, D., Stapp, H. and Scolnik, P.A. (1988) A genetic-physical map of the *Rhodobacter capsulatus* carotenoid biosynthesis gene cluster. *Molec. Gen. Genet.*, **213**:78-83.

Goodwin, T.W. (1971) Biosynthesis. In: Isler, O. (ed.), *Carotenoids*. Birkhäuser, Basel, pp. 577-636.

Goodwin, T.W. (1980) Carotenoids in higher plants. In: Goodwin, T.W. (ed.), *The Biochemistry of the Carotenoids*, 2nd Ed, Vol. 1 Plants. Chapman and Hall, London, pp. 143-203.

Gregonis, D.E. and Rilling, H.C. (1974) The stereochemistry of *trans*-phytoene synthesis. Some observations on lycopersene as a carotene precursor and a mechanism for the synthesis of *cis*- and *rans*- phytoene. *Biochemistry*, **13**:1538-1542.

Grumbach, K.H. and Bach, T.J. (1979) The effect of PSII herbicides, amitrole and SAN6706 on the activity of 3-hydroxy-3-methylglutaryl coenzyme A reductase and the incorporation of [2-^{14}C]-acetate and DL-[2-^{3}H]-mevalonate into chloroplast pigments of radish seedlings. *Z. Naturforsch.*, **34c**:941-943.

Heintze, A., Gorlach, J., Leuschner, C. *et al.* (1990) Plastid isoprenoid synthesis during chloroplast development. Change from metabolic autonomy to a division-of-labour stage. *Plant Physiol.*, **93**:1121-1127.

Henningsen, K.W. and Boynton, J.E. (1970) Macromolecular physiology of plastids. VIII. Pigment and membrane formation in plastids of barley greening under low light intensity. *J. Cell Biol.*, **44**:290-304.

Hill, H.M., Calderwood, S.K. and Rogers, L.J. (1971) Conversion of lycopene to ß-carotene by plastids isolated from higher plants. *Phytochemistry*, **10**:2051-2058.

Humbeck, K. (1990) Photoisomerisation of lycopene during carotenogenesis in mutant C-6D of *Scenedesmus obliquus*. *Planta*, **182**:204-210.

Islam, M.A., Lyrene, S.A., Miller, E.M. and Porter, J.W. (1977) Dissociation of prelycopersene pyrophosphate synthetase from phytoene synthetase complex of tomato fruit plastids. *J. Biol. Chem.*, **252**:1523-1525.

Kirk, J.T.O. and Tilney-Bassett, R.A.E. (1978) *The Plastids: their Chemistry, Structure, Growth and Inheritance*, 2nd Ed. Elsevier– North Holland, Amsterdam.

Kleinig, H. (1989) The role of plastids in isoprenoid biosynthesis. *Annu. Rev. Plant Physiol. Plant Molec. Biol.*, **40**:39-59.

Kreuz, K. and Kleinig, H. (1981) Synthesis of prenyllipids in cells of spinach leaf. Compartmentation of enzymes for formation of isopentenyl diphosphate. *Eur. J. Biochem.*, **141**:531-535.

Kreuz, K., Beyer, P. and Kleinig, H. (1982) The site of carotenogenic enzymes in chromoplasts from *Narcissus pseudonarcissus* L. *Planta*, **154**:66-69.

Kushwaha, S.C., Subbarayan, C., Beeler, D.A. and Porter, J.W. (1969) The conversion of lycopene-15,15′-^{3}H to cyclic carotenes by soluble extracts of higher plant plastids. *J. Biol. Chem.*, **244**:3635-3642.

Kushwaha, S.C., Suzue, G., Subbarayan, C. and Porter, J.W. (1970) The conversion of phytoene-^{14}C into acyclic, monocyclic and dicyclic carotenes and the conversion of lycopene-15,15′-^{3}H to mono- and dicyclic carotenes by soluble enzyme systems obtained from the plastids of tomato fruit. *J. Biol. Chem.*, **245**:4708-4717.

Liaaen-Jensen, S. (1969) Selected examples of structure determination of natural carotenoids. *Pure Appl. Chem.*, **20**:421-448.

Lütke-Brinkhaus, F. and Kleinig, H. (1987) Carotenoid and chlorophyll biosynthesis in isolated plastids from mustard seedling cotyledons (*Sinapis alba* L.) during etioplast–chloroplast conversion. *Planta*, **170**:121-129.

Marrs, B.L. (1982) Genetic analysis of carotenogenesis in *Rhodopseudomonas capsulata*. In: Britton, G, and Goodwin, T.W. (eds.), *Carotenoid Chemistry and Biochemistry*. Pergamon, Oxford, pp. 273-277.

Maudinas, B., Bucholtz, M.L., Papastephanou, C. *et al.* (1977) The partial purification and properties of a phytoene-synthesizing enzyme system. *Arch. Biochem. Biophys.*, **180**:354-362.

Mayer, M., Beyer, P. and Kleinig, H. (1990) Quinone compounds are able to replace molecular oxygen as terminal electron acceptor in phytoene desaturation in chromoplasts of *Narcissus pseudonarcissus* L. *Eur. J. Biochem.*, **191**:359-363.

McDermott, J.C.B., Britton, G. and Goodwin, T.W. (1973) Carotenoid biosynthesis in a *Flavobacterium* sp.: Stereochemistry and hydrogen elimination in the desaturation of phytoene to lycopene, rubixanthin and zeaxanthin. *Biochem. J.*, **134**:1115-1117.

Milborrow, B.V. (1982) Stereochemical aspects of carotenoid biosynthesis. In: Britton, G. and Goodwin, T.W. (eds.), *Carotenoid Chemistry and Biochemistry*. Pergamon, Oxford, pp. 279-295.

Milborrow, B.V., Swift, I.E. and Netting, A.G. (1982) Stereochemistry of hydroxylation of the carotenoid lutein in *Calendula officinalis*. *Phytochemistry*, **21**:2853-2857.

Misawa, N., Nakagawa, M., Kobayashi, K. *et al.* (1990) Elucidation of *Erwinia uredovora* carotenoid biosynthesis pathway by functional analysis of gene products expressed in *Escherichia coli*. *J. Bacteriol.*, **172**:6704- 6712.

Moshier, S.E. and Chapman, D.J. (1973) Biosynthetic studies on aromatic carotenoids. *Biochem. J.*, **136**:395-404.

Papastephanou, C., Barnes, F.J., Briedis, A.V. and Porter, J.W. (1973) Enzymatic synthesis of carotenes by cell-free preparations of fruit of several genetic selections of tomatoes. *Arch. Biochem. Biophys.*, **157**:415-425.

Pecker, I., Chamowitz, D., Linden, H. *et al.* (1992) A single polypeptide catalyzing the conversion of phytoene to ζ-carotene is transcriptionally regulated during tomato fruit ripening. *Proc. Natl. Acad. Sci. USA*. 89:4926-4966.

Pemberton, J.M. and Harding, C.M. (1986) Cloning of carotenoid biosynthesis genes from *Rhodopseudomonas sphaeroides*. *Curr. Microbiol.*, **14**:25-29.

Qureshi, A.A., Andrewes, A.G., Qureshi, N. and Porter, J.W. (1974) The enzymatic conversion of *cis*-[^{14}C]-phytofluene, *trans*-[^{14}C]- phytofluene and *trans*-[^{14}C]-ζ-carotene to more unsaturated acyclic, monocyclic and dicyclic carotenes by a cell-free preparation of red tomato fruits. *Arch. Biochem. Biophys.*, **162**:93-107.

Rogers, L.J., Shah, S.P.J. and Goodwin, T.W. (1966) Compartmentation of terpenoid biosynthesis in green plants. *Biochem. J.*, **114**:395-405.

Sandmann, G. (1991) Light-dependent switch from formation of poly-*cis* carotenes to all-*trans* carotenoids in the *Scenedesmus* mutant C-6D. *Arch. Microbiol.*, **155**:229-233.

Sandmann, G. and Bramley, P.M. (1985a) Carotenoid biosynthesis by *Aphanocapsa* homogenates coupled to a phytoene-generating system from *Phycomyces blakesleeanus*. *Planta*, **164**:259-263.

Sandmann, G. and Bramley, P.M. (1985b) The *in vitro* biosynthesis of ß-cryptoxanthin and related xanthophylls with *Aphanocapsa* membranes. *Biochim. Biophys. Acta*, **843**:73-77.

Sandmann, G. and Kowalczyk, S. (1989) *In vitro* carotenogenesis and characterisation of the phytoene desaturase reaction in *Anacyctis*. *Biochem. Biophys. Res. Commun.*, **163**:916-921.

Sergeant, J.M. and Britton, G. (1984) Observations on the biosynthesis of chloroplast carotenoids. In: Sybesma, C. (ed.), *Advances in Photosynthesis Research*, Vol. IV.

Martinus Nijhoff/Dr W. Junk, Dordrecht, The Netherlands, pp. 779-782.
Spurgeon, S.L. and Porter, J.W. (1983) Biosynthesis of carotenoids. In: Porter, J.W. and Spurgeon, S.L. (eds.), *Biosynthesis of Isoprenoid Compounds*, Vol. 2. Wiley, New York, pp. 1-122.
Swift, I.E. and Milborrow, B.V. (1981) Stereochemistry of allene biosynthesis and the formation of the acetylenic carotenoids diadinoxanthin and peridinin (C_{37}) from neoxanthin. *Biochem. J.*, **199**:69-74.
Swift, I.E., Milborrow, B.V. and Jeffrey, S.W. (1982) Formation of neoxanthin, diadinoxanthin and peridinin from [^{14}C] zeaxanthin by a cell-free system from *Amphidinium carterae. Phytochemistry,* **21**:2859-2864.
Virgin, H.I., Kahn, A. and Von Wettstein, D. (1963) The physiology of chlorophyll formation in relation to structural changes in the chloroplast. *Photochem. Photobiol.*, **2**:83-91.
Walton, T.J., Britton, G. and Goodwin, T.W. (1969) Biosynthesis of xanthophylls in higher plants. Stereochemistry of hydroxylation at C- 3. *Biochem. J.*, **112**:383-385.
Williams, R.J.H., Britton, G. and Goodwin, T.W. (1967). The biosynthesis of cylic carotenes. *Biochem. J.*, **105**:99-105.
Yamamoto, H.Y. (1979) Biochemistry of the violaxanthin cycle in higher plants. *Pure Appl. Chem.*, **51**:639-648.
Young, D.A., Bauer, C.A., Williams, J.C. and Marrs, B.L. (1989) Genetic evidence for superoperonal organisation of genes for photosynthetic pigments and pigment-binding proteins in *Rhodobacter capsulata. Molec. Gen. Genet.*, **218**:1-12.

5

Inhibition of carotenoid biosynthesis

P.M. Bramley

5.1 INTRODUCTION

The inhibition of carotenoid biosynthesis has been studied for some 40 years. The earliest investigations, using compounds which cause the accumulation of less unsaturated carotenoids, particularly phytoene, were aimed at the elucidation of the carotenoid biosynthetic pathway (e.g. see Goodwin 1952; Goodwin and Land 1956). Since that time, a massive number of compounds have been reported to inhibit not only desaturation reactions, but also the formation of cyclic carotenes and xanthophylls. During the 1960s and 1970s such inhibitors not only facilitated studies on the pathways, but were used also to investigate the mechanisms of desaturation and cyclisation. In more recent years, however, there has been a change of emphasis in studies with carotenoid inhibitors. Over the past decade, most investigations have centred upon inhibitors which can act as bleaching herbicides. Since carotenoids are essential components of photosynthetic tissues, it follows that inhibition of their formation leads to cell death, primarily due to photooxidation (Kunert and Dodge 1989). There has been a substantial effort, therefore, to understand the molecular mechanisms of inhibition, to pinpoint the precise target sites of inhibitors and, as a result, to design new compounds which are specific and potent inhibitors of carotenogenesis.

The aim of this chapter is to review the modes of action of inhibitors of carotenoid formation, particularly of compounds which affect photosynthetic organisms, and to explain the experimental approaches which are used in such investigations. The biosynthetic pathways leading to the formation of the coloured carotenoids found in photosynthetic tissues and referred to in this chapter have been discussed in detail in Chapter 4, as have the properties of the enzymes that catalyse the reactions. The principal target sites for inhibitors of carotenoid biosynthesis are outlined in Fig. 5.1. Other compounds, especially those that inhibit or divert electron transport in photosynthesis are also known to affect the carotenoid content and composition of photosynthetic organisms. The effect of such inhibitors is not discussed in this chapter and readers are instead invited to consult Chapter 6. The trivial names of inhibitors are used throughout the text; their corresponding systematic names are listed in Table 5.1.

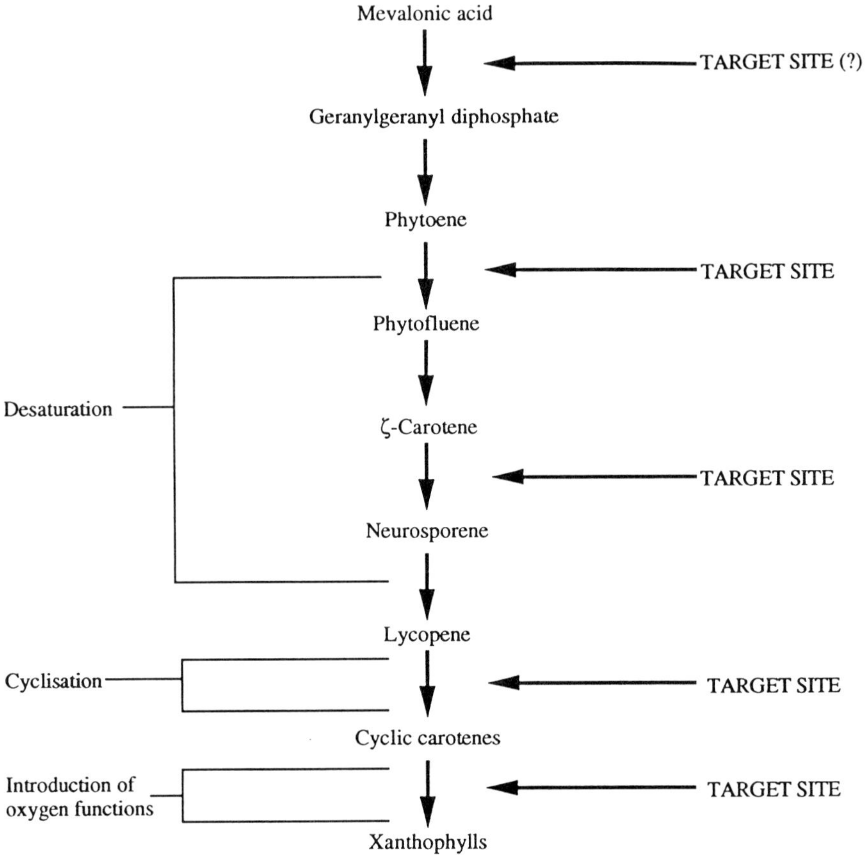

Figure 5.1 Proposed target sites of inhibitors of carotenoid biosynthesis.

Table 5.1 Trivial and systematic names of carotenoid inhibitors

Trivial Name	*Systematic nomenclature*
Amitrole	3-Amino-*s*-triazole
AMO-1618	(4-Hydroxy-5-isopropyl-2-methylphenyl) trimethylammonium chloride, 1-piperidine-carboxylate
CDEB	4-Chloro-[ß-(diethylamino)-ethyl]benzoate
CGA 22867	1-Chlorobenzyl-4-(*N,N*-dimethylamino-ethylenamino)-3-methylphenyl ether
CPTA	2-(4-Chlorophenylthio)triethylamine
Cyclic ureas	1,3-Diaryl-2-pyrimidones
Cycocel	(2-Chloroethyl)trimethylammonium chloride
Dichlormate	3, 4-Dichlorobenzyl methylcarbamate
Diflufenican (MB 38544)	*N*-(2,4-Difluorophenyl)-2-(3-trifluoro-methoxyphenoxy)-3-pyridine carboxyamide
Difunone (EMD-IT 5914)	5-(Dimethylaminomethylene)-2-oxo-4-phenyl-(1*H*)-pyridone -2,5-dihydrofuran-carbonitrile
Dimethazone, (Clomazone, FMC 57020)	2-(2-Chlorophenyl)methyl-4,4-dimethyl-3-isoxazolidinone
DTP	1,3-Dimethyl-4-(2,4-dichlorobenzoyl)-5-hydroxypyrazole-3-(2*H*)-one
Fluometuron	1,3-Dimethyl-3-(3-trifluoromethylphenyl) urea
Fluorochloridone (R-40244)	1-(*m*-Trifluoromethylphenyl)-3-chloro-4-chloromethyl-2-pyrrolidone
Fluridone (EL171)	1-Methyl-3-phenyl-5-[3-(triflurometyl)-phenyl]-4-(1*H*)-pyridone
Flurtamone	5-Methylamino-2-phenyl-4-(3-trifluoromethylphenyl)-3-(2*H*)-furanone
Fomesafen	5-[2-Chloro-4-(trifluoromethyl)phenoxy]-2-nitro-*N*-methylsulphonyl-benzoate
Haloxydine	3,5-Dichloro-2,6-difluoro-4-hydroxypyridine
J334	2-Propyl-5-(2-chloro)-benzyl-6-methyl pyrimidine
J852	4-Isobutoxy-2-isopropylamino-6-methylpyrimidine
KM 143-958	2-Isopropylamino-4-(2-chloroallyloxy)-6-methyl-pyrimidine
LAB 117682	3-(1,2,4-Triazolyl-1)-1-(4-chlorophenyl)-4, 4-dimethyl-pentan-1-one
LS 80707	Ethyl-*cis*-5-methyl-6-ethyl-2-phenyl-5,6-dihydropyran-4-one-3-carboxylate
LS 80717	Ethyl-*trans*-5-methyl-6-ethyl-2-phenyl-5,6-dihydropyran-4-one-3-carboxylate
LS 780091	Ethyl-*cis*-5-methyl-6-methyl-2-(4-fluorophenyl)-5,6-dihydropyran-4-one-3-carboxylate
Methyoxyphenone, Kayaphenon (NK094)	3, 3′-Dimethyl-4-methoxybenzophenone
MPTA	2-(4-Methylphenoxy)-triethylamine

Pyrichlor	2,3,5-Trichloro-4-pyridinol
RE-40885	5-Methylamino-2-phenyl-4-(α,α,α-trifluoro-*m*-tolyl)furan-3(2*H*)-one
SAN 6706, metflurazon	4-Chloro-5-(dimethylamino)-2-(α,α,α-trifluoro-*m*-tolyl)-3-(2*H*)-pyridazonone
SAN 9789, norflurazon	4-Chloro-5-(methylamino)-2-(α,α,α-trifluoro-*m*-tolyl)-3-(2*H*)-pyridazinone
SAN 9785	4-Chloro-5-(dimethylamino)-2-phenyl-3-(2*H*)-pyridazinone
SAN 380H	4-(2-Chloro-2-propenoxy)-2-isopropylamino-6-methylpyrimidine
SC-0051(ICIA 0051)	2-[2´-Chloro-4´-(methanesulphonyl)benzoyl]-5,5-dimethylcyclohexa-1,3-dione
S-3422	3-(2, 5-Dimethylphenoxy)-*N*-ethylbenzamide
Tetcyclasis (LAB 102883)	5-(4-Chlorophenyl)-3,4,5,9,10-pentazotetracyclo-5,4,02,6,08,11-dodeca-3, 9-diene
WL 110547	1-(3´-Trifluoromethylphenyl)-5-phenoxy-tetrazole

5.2 EXPERIMENTAL APPROACHES TO STUDIES OF INHIBITORS

A number of different, but often complementary, techniques are used to study the inhibition of carotenogenesis. In general, they aim to pinpoint target sites, to discover which functional groups are prerequisites for inhibitory activity, and to establish the physiological effects that result from the inhibition of carotenoid formation.

5.2.1 Choice of tissue

A wide variety of organisms, including photosynthetic and non-photosynthetic tissues, has been used in studies of carotenoid inhibitors. The choice is governed by several factors. Earlier studies, aimed at elucidating the biosynthetic pathways, tended to use microorganisms which could be cultured easily in large quantities, whose growth was not affected by the inhibitor, and which were permeable to the compound under investigation. More recent investigations, aimed primarily at bleaching herbicides, have used photosynthetic microorganisms such as green algae and cyanobacteria in order to relate, as far as possible, laboratory experiments to the effects on crop and weed species in the field. It must be said, however, that these choices are also influenced by the personal preferences of different research groups.

Phototrophic bacteria of the Rhodospirillaceae such as *Rhodospirillum rubrum* (Davies 1970a,b; Davies and Aung Than 1974), *Rhodobacter sphaeroides* (then named *Rhodopseudomonas spheroides*; Maudinas et al. 1972), *Rps. viridis* (Britton et al. 1977a,b) and *Rhodomicrobium vannielii* (Britton et al. 1975) have been widely used with inhibitors such as diphenylamine and nicotine in studies on the carotenoid biosynthetic pathway. When cultured anaerobically in the light, or aerobically in darkness, these bacteria remain viable in the absence of desaturated carotenoids. In addition, the inhibitor can easily be removed by washing the cells, which can then be resuspended in sterile medium where reversal of the inhibitory effect can be monitored. Cultures of cyanobacteria have been used in order to demonstrate the inhibition of carotenogenesis by bleaching herbicides (e.g. see Clarke et al. 1985; Sandmann et al. 1985, 1990a,b; Linden et al. 1990) and non-herbicidal compounds (Jüttner and Bogenshütz 1983) and also to establish cellular I_{50} values (Clarke et al. 1985; Sandmann et al. 1990a).

In order to ascertain that inhibitors, especially herbicides, exert their prime mode of action on the formation of carotenoids, rather than by the photo-oxidation of the pigments (including chlorophylls) it is necessary to grow plants or culture algae in darkness. This serves to preclude the differential light effects on carotenoids and chlorophylls after the addition of inhibitors. This can be achieved reliably with heterotrophic cultures of green algae such as *Scenedesmus*, which can be grown in darkness with glucose as a carbon source, yet still possess a fully developed photosynthetic apparatus. This alga has been utilised extensively by Böger and Sandmann in Konstanz (e.g. Sandmann et al. 1981; Lambert and Böger 1983; Sandmann and Böger 1983; Sandmann and Albrecht 1990). Such a model system can distinguish between direct inhibition of the pathway and oxidation of existing molecules (Sandmann and Böger 1983). Other advantages of using *Scenedesmus* to assay herbicidal bleaching have been described (Sandmann et al. 1979). In particular, the ease of maintenance of controlled growth conditions in defined media for a rapidly growing population of cells in which the uptake, accumulation and excretion of herbicides can be followed is attractive. A less well-exploited alga is *Chlorella fusca*, which can be grown with sub-lethal concentrations of herbicides under heterotrophic conditions (Kummel and Grimme 1975; Tantawy and Grimme 1982), but the growth of *Chlorella vulgaris* is severely affected by carotenoid inhibitors, and cannot be used for such studies (Goodwin 1980). *Euglena gracilis* has also been cultured with herbicides, especially to monitor light-induced chloroplast development (Schiff et al. 1982).

Higher plant seedlings have been used in many laboratory experiments with bleaching herbicides. Once again, a wide variety of species has been tested, under a range of lighting conditions from darkness (etiolated seedlings), to full white, red or blue light. Typically, a light/dark cycle is used to mimic field conditions. The herbicides can be applied in different ways, e.g. in the potting compost (pre-emergent) (e.g. see Ben-Aziz and Koren 1974), during watering (e.g. see Grumbach and Britton 1983), direct to intact leaves (post-emergent) or leaf discs (Lay et al. 1985; Rogers et al. 1987) or by soaking the seeds prior to germination (e.g. see Dahlin and Ryberg 1986). An unusual protocol has been described whereby the herbicide is applied to flower buds (Hlousek-Radojcic and Ljubesic 1988). Application of bleaching herbicides in different ways can provide some information about the best route of uptake and hence application procedure for the inhibitor (Wightman and Haynes 1985). The subcellular distribution of carotenoids following growth with bleaching herbicides has been reported for radish (Grumbach and Britton 1983) and wheat seedlings (Dahlin and Ryberg 1986). Radish seedlings contained phytoene in both thylakoid and plastid envelope fractions, whereas in norflurazon-treated wheat, the plastoglobuli were found to be the main location of phytoene.

Studies on intact cells or plants do provide important information on the potency of inhibitors and may pinpoint which biosynthetic step(s) is affected. Additionally, the results reflect factors such as uptake, translocation and metabolism of the inhibitor. They do not, however, enable accurate measurements of intracellular concentrations to be made and so cannot provide definitive values for inhibition constants (K_i), nor can they distinguish between the various possible target sites (Table 5.2). Such biochemical criteria must be obtained from enzyme preparations, preferably by use of pure proteins (see Section 5.2.4).

Table 5.2 Possible target sites for inhibitors of carotenoid biosynthesis

1.	Gene expression: transcription or translation
2.	Bind to the active site of the enzyme
3.	Bind to other parts of the enzyme, reversibly or irreversibly
4.	Affect the association of a multi-enzyme complex
5.	Affect membrane organisation
6.	Inhibit the electron transport chain
7.	Affect cofactor availability
8.	Affect transport of the enzyme into the plastid
9.	Prevent targeting onto the plastid membrane
10.	Affect carrier proteins involved in substrate transfer or transport to pigment–protein complexes
11.	Inhibit post-translation modifications, e.g. isoprenylation

5.2.2 Carotenoid analyses

In order to establish that a compound inhibits carotenoid biosynthesis, it is necessary to extract, identify and quantify the carotenoids from control and treated tissues. This has been achieved traditionally by lipid extraction followed by chromatographic separation, identification and quantitation from absorption spectra and mass spectra. Such protocols are described in detail in several reviews (Britton and Goodwin 1971; Davies 1976; Britton 1985, 1991; Goodwin and Britton 1988; see also Chapter 10). Over the past few years this approach has been streamlined and simplified by using high-performance liquid chromatography (HPLC) diode-array detection, a technique that combines high chromatographic resolution with simultaneous and instant spectral analysis (reviewed by Bramley 1992, and in Chapter 10). This enables routine screening of carotenoids from treated and untreated tissues to be carried out in a single operation (Britton et al. 1987, 1989; Young et al. 1989; Barry and Pallett 1990).

5.2.3 Electron Microscopy

Electron micrographs of higher plants and algae following treatment with bleaching herbicides reveal changes in the ultrastructure of cells (particularly plastids) resulting from the inhibition of carotenogenesis (Bartels and Hyde 1970; Vaisberg and Schiff 1976; Khan et al. 1979; Schiff et al. 1982; Duke and Paul 1986; Hlousek-Radojcic and Ljubesic 1988). They can also indicate any redistribution of lipids after such treatment, e.g. the increased presence of plastoglobuli containing phytoene in SAN 9789-treated wheat seedlings (Dahlin and Ryberg 1986). Electron microscopy is most usefully carried out in conjuction with carotenoid analyses (Section 5.2.2), especially of organelle fractions. In isolation, electron microscopy cannot elucidate the primary mode of action of inhibitors of carotenoid biosynthesis.

5.2.4 Enzyme assays

In order to investigate the molecular mechanisms of action of inhibitors and to prove that the target site is a particular carotenogenic enzyme, it is necessary to assay the enzyme *in vitro*, preferably using pure protein. This ideal is not yet possible with membrane-bound carotenogenic enzymes, since only the soluble phytoene synthase from *Capsicum* chromoplasts has been purified (Dogbo et al. 1988). In addition, chloroplast preparations often have very low carotenogenic activities (see Bramley 1985, 1990) and so are unsuitable for inhibition studies. Consequently, alternative *in vitro* preparations have been used as model systems, in particular thylakoid or membrane preparations from cyanobacteria such as *Aphanocapsa* and *Anacystis*, chromoplasts from *Narcissus* and *Capsicum*, and crude cell extracts of the fungus *Phycomyces blakesleeanus* (Table 5.3).

Table 5.3 Cell extracts used to study the inhibition of carotenogenesis *in vitro*

Organism/extract	*Inhibitor*	*Reference*
Anacystis thylakoids	Substituted tetrahydropyrimidines	Babczinski et al. 1990
	Norflurazon	Sandmann et al. 1989
	Flurtamone	Sandmann et al. 1990a
Aphanocapsa membranes	Miscellaneous herbicides	Clarke et al. 1983,1985, Clarke 1984; Sandmann et al. 1984, 1985; Zöbelin 1985; Kostal 1986
Capsicum chromoplasts	WL 110547 Un-and tridecyldiethylamines	Kerr and Whitaker 1987; Camara et al.1985
Cheiranthus chromoplasts	2-(4-Chloro-2-nitrobenzoyl)-5,5-dimethylcyclohexane-1,3-dione	Sandmann et al. 1990b
Chenopodium thylakoids	CPTA, J852, difunone	Kostal 1986; Sandmann 1988

Corn chloroplasts	2-(4-Chloro-2-nitrobenzoyl)-5,5-dimethylcyclohexane-1,3-dione	Sandmann et al. 1990b
Narcissus chromoplasts	Pyridazinones, fluridone, J334	Mayer et al. 1989
	Dimethazone	Lutzow et al. 1990
Phycomyces blakesleeanus crude extract	Norflurazon	Sandmann et al. 1980
	Diphenyl compounds	Clarke et al. 1983; Bramley et al. 1984
Spinach chloroplasts	Dimethazone	Böger and Sandmann 1990
Synechococcus PCC7942 thylakoids	2-(4-Chloro-2-nitrobenzoyl)-5,5-dimethylcyclohexane-1,3-dione	Sandmann et al. 1990b

A further complication in the use of cell extracts is the complexity of carotenogenic enzyme assays. Catalytic activities are measured by the incorporation of radioactivity, preferably ^{14}C, from substrate into product. Such ^{14}C-labelled compounds cannot be purchased and must be prepared in the laboratory (see Bramley 1985, 1992). One method for estimating phytoene desaturase activity is to use a coupled assay in which ^{14}C-labelled phytoene is synthesised *in situ* from [2-^{14}C]mevalonic acid (MVA) with a cell extract of a phytoene-accumulating mutant of *Phycomyces*. Once the [^{14}C]-phytoene has been formed, the extract under investigation is added, with and without inhibitor, and the incubation continued in order to estimate the desaturation of phytoene (Sandmann and Bramley 1985; Fraser et al. 1991). Presenting the lipophilic substrate to the enzyme in this way avoids the use of surfactants. Variations on this theme have been used to assay other carotenogenic enzymes (Lütke-Brinkhaus et al. 1982; El-Jack et al. 1987). In any radiochemical assay, it is vital to purify exhaustively the products of the reaction in order to demonstrate unequivocally carotenoid formation *in vitro* (Britton 1985, 1991; Goodwin and Britton 1988).

Provided that these constraints are recognised, cell-free systems are most valuable for studying the interactions between enzyme and inhibitor. Incubations containing different concentrations of substrate, enzyme and inhibitor yield data which, when plotted graphically, establish inhibition constants (K_i), the type of inhibition and also whether inhibition is reversible or irreversible (Sandmann and Böger 1987a; Sandmann et al. 1989). A comparison of an *in vivo* I_{50} value and an *in vitro* K_i for the phenoxybenzamide S3422 with *Aphanocapsa* is shown in Fig. 5.2. In this example, the two values are closely similar, indicating that there is no intracellular activation or deactivation of the herbicide, nor any barrier to its uptake by the cells. Such close agreements between cellular and *in vitro* inhibition constants are common, but some exceptions have been reported. For example, inhibition by CGA 22867 shows an I_{50} much lower than the K_i value for phytoene desaturase (Sandmann et al. 1985). This may indicate that the herbicide is metabolised *in vivo* to a more potent derivative, or that its primary target site is not the inhibition of phytoene desaturase. Similarly, 2-(4-chloro-2-nitrobenzoyl)-5,5-

dimethylcyclohexane-1,3-dione is ineffective *in vitro*, but inhibits carotenoid desaturation in seedlings when applied to the roots (Sandmann et al. 1990b). In this case, it is thought that this compound is activated during uptake and/or translocation in the plant, especially as it is not inhibitory when applied to leaves or single cells.

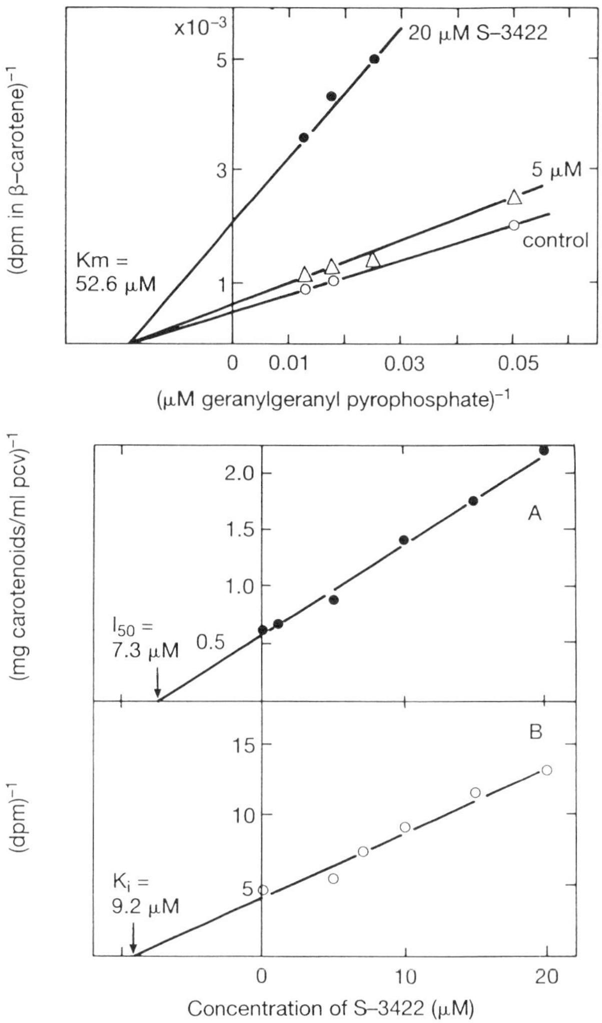

Figure 5.2 Determination of cellular I_{50} (A) and cell free K_i (B) and K_m values for the inhibition of carotenogenesis by S3422 in *Aphanocapsa*. The geranylgeranyl diphosphate concentration in B was 30 mM. (From Clarke et al. 1985; reproduced with permission).

Cell extracts can also be used to demonstrate inhibition of carotenogenic enzyme activities with organisms that are impermeable to the molecule. Thus, phytoene desaturase of *Phycomyces* is inhibited *in vitro* by diphenylamine and related biphenyl compounds, but the intact cells are impermeable to these molecules.

Inhibition is non-competitive and reversible (Clarke et al. 1983). In contrast, however, diphenylamine, fluorenone and 2-hydroxybiphenyl were ineffective with a soluble extract of tomato fruit. This led Bucholtz et al. (1977) to conclude that the compounds affect gene expression or a pathway essential to carotenoid formation.

5.2.5 Quantitative Structure–Activity Relationships (QSAR)

QSAR studies can be carried out on intact cells, whole plants or with enzyme preparations. The purpose of such investigations is to gain information on which functional moieties of an inhibitor are necessary for activity, with a view to designing a more potent molecule. Typically, a series of analogues is tested with respect to inhibitory activity, and the effectiveness of each compound is related to physicochemical parameters of the molecule, using constants available in the literature (Hansch et al. 1973a,b). Values for lipophilicity (π), electronic parameters (σ_μ and σ_π), steric bulk (MR), Verloop's length parameter (*L*) and STERIMOL parameter B_3 have been derived for bleaching herbicides and can then be used to derive regression equations for the series. Such studies have not been reported for non-herbicidal inhibitors of carotenogenesis.

(a)Phenylpyridazinones

The phenylpyridazinones are a multi-functional group of compounds, which can be assigned by cluster analysis to five functional groups (St John et al. 1984), one of which contains inhibitors of phytoene desaturase.

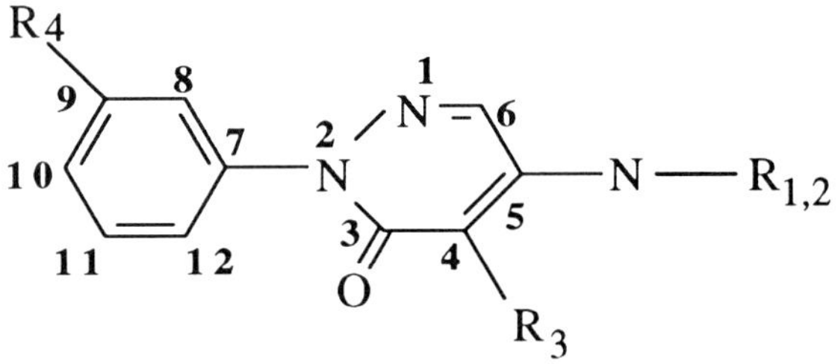

A comparison of seven different phenylpyridazinones by Urbach and coworkers (1976) found that an *m*-trifluoromethyl substituent on the phenyl ring and a methyl-substituted 4-amino group are favourable for inhibitory activity. A later study compared inhibitory activity of phenylpyridazinones substituted at different positions on the heterocyclic ring. Substituents at C-4 enhanced inhibition with increasing σ_μ values, but substituents at C-5 by groups with low π values were favourable as electrons are shifted towards the pyridazinone ring (Sandmann et al. 1989).

Furthermore, substituents at the *meta* position of the phenyl ring influenced activity in the same way as did substituents at C-4. Therefore the maximum inhibition of carotenogenesis by phenylpyridazinones requires ligands which stabilise a resonance structure, leading to a positive charge on N-2, and a negative charge on the oxygen at C-3.

A QSAR correlation for substituted 2-phenylpyridazinones has been established, with an equation in which pI_{50} is directly dependent upon σ_{μ} and π. Steric factors did not influence inhibitory activity (Sandmann and Böger 1982). Cell-free inhibition data obtained with *m*-substituted phenylpyridazinones have also been reported (Sandmann 1987). Five compounds with very similar electronic parameters but different π values were chosen. A plot of log inhibition values versus π showed a linear relationship (Fig. 5.3), in accordance with the QSAR equation obtained with intact cells.

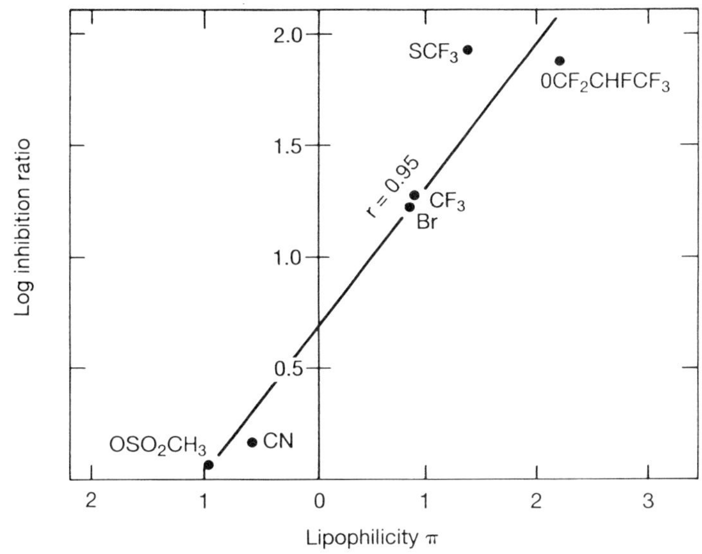

Figure 5.3 Relationship between inhibition by phenylpyridazinones and lipophilicity (π). The pyridazinones were *m*-phenyl substituted as indicated. See Fig. 5.5 for structures. (From Sandmann 1987; reproduced with permission).

(b) Phenoxybenzamides

Phenoxyphenyl compounds interact with plant metabolism in multiple ways. The essential part of the molecule for inhibition of phytoene desaturation is the benzamide structure of substituted diphenyl ethers (Sandmann 1987). Therefore, these herbicides, typified by S-3422, are called phenoxybenzamides. The 3-phenoxy group can be replaced by CF_3 to produce an increased inhibitory effect. Systematic modification of *m*-phenoxybenzamides indicates that substituents that lower the partial positive charge on the carbonyl carbon enhance the inhibitory effect. It is interesting to note that the introduction of a nitro group at C-6 alters the mode of action from a carotenoid inhibitor to a peroxidative herbicide (Sandmann et al. 1984).

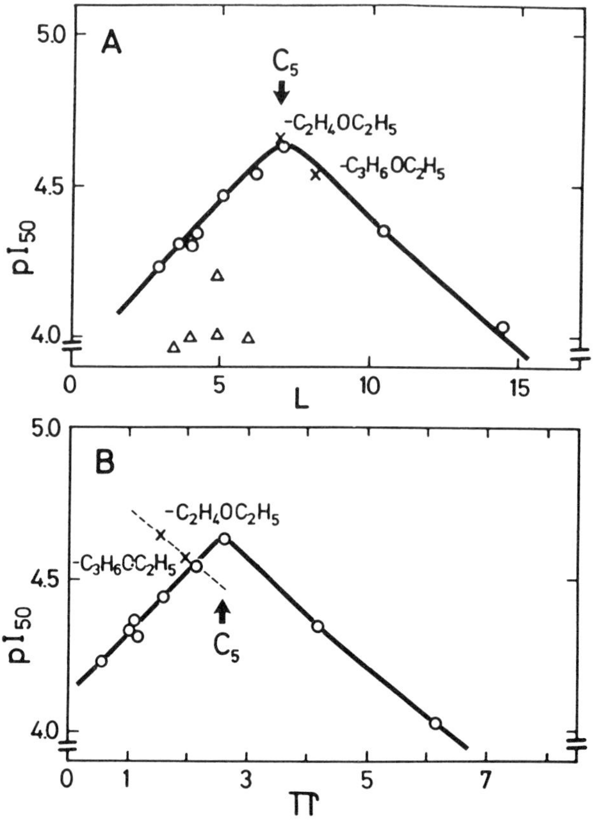

Figure 5.4 Plot of pI_{50} for cellular inhibition of carotenoid biosynthesis versus STERIMOL parameter, *L*(A) or lipophilicity, π (B) of branched (-Δ-), unbranched (-o-), or unbranched ethoxy (-x-)*N*-alkyl *m*-phenoxybenzamides. (From Sandmann 1987; reproduced with permission).

It is also possible to increase the *N*-alkyl chain length of these compounds. The presence of a branched chain caused a decrease in inhibitory activity, when compared to a straight chain with the same number of carbon atoms (Clarke et al. 1985). A C_5 linear alkyl chain was the most effective, with derivatives up to C_{12} being less inhibitory. Thus, plotting pI_{50} versus Verloop's length parameter (*L*) or lipophilicity (π), shows a parabolic curve (Fig. 5.4). In order to distinguish between the effects of π and *L*, compounds were synthesised with approximately the same *L* values, but lower values of π, by replacing a CH_2 group in the alkyl chain with an ether bridge. The pI_{50} values for ethoxy derivatives closely fit on the curve of the unbranched alkyl chains in the plot of pI_{50} versus *L* (Fig. 5.4), but in the plot of pI_{50}

versus π, the data do not match the descending branch of the curve. It was concluded that the chain length, rather than lipophilicity, is important for interaction of the phenoxybenzamide with phytoene desaturase (Sandmann 1987). The STERIMOL parameter, B_3, gave the best description of the additional steric effect of branched alkyl chains *in vivo*. QSAR analysis was also performed with *in vitro* inhibition data for branched and unbranched *N*-alkylphenoxybenzamides. In this case, a significant contribution of B_3 could not be observed for the branched alkyl chains. Sandmann (1987) concludes that this difference is caused by the swelling of the membranes containing phytoene desaturase during preparation procedures, leading to an alteration of the target site.

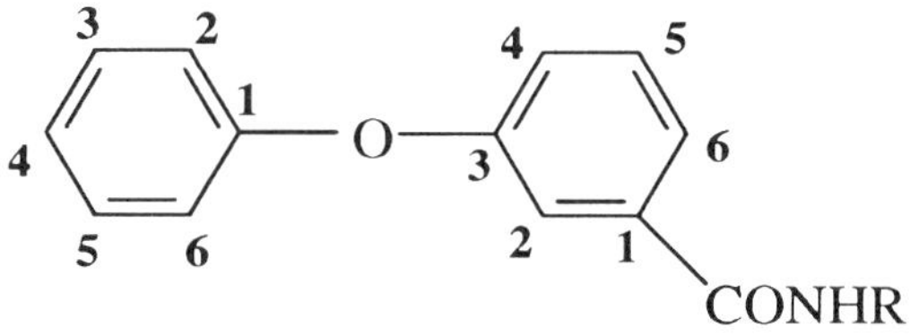

(c) Phenoxynicotinamides

The phenoxynicotinamides, typified by diflufenican (Fig. 5.5) show some structural similarities to phenoxybenzamides but, in contrast to the latter, the nicotinamides exhibit very high inhibition as *o*-phenoxy derivatives. They possess higher affinities for phytoene desaturase, with I_{50} and K_i values at least one order of magnitude lower than those of *m*-phenoxybenzamides (Sandmann 1987).

The effects of substituents on inhibitory activity have been studied *in vitro* (Sandmann 1987). The nicotinamide with a *p*-fluorophenyl substitution showed high inhibition at 5 μM. An additional methyl substituent on the nitrogen significantly decreased inhibition, whilst an even more marked decrease in inhibitory activity was found with a derivative containing an ethylene group between the nitrogen and the phenyl moiety.

5.3 TARGET SITES

The inhibition of carotenoid biosynthesis could be caused by several effects. Although the most direct, and obvious, is the inhibition of enzyme activity, several other possibilities should be considered, as shown in Table 5.2. The advent of active cell extracts has enabled the inhibition of enzyme activity to be measured, but our ignorance of many aspects of carotenogenesis and its regulation, and also of chloroplast development, unfortunately prevent the experimental study of several of the alternative target sites.

In this section, the current evidence for the target sites and modes of action of inhibitors are discussed, as well as data for those inhibitors with multiple or uncertain target sites. Several reviews on this topic have been published (Ridley 1982; Spurgeon and Porter 1983; Sandmann and Böger 1987a, 1989; Britton et al. 1989; Böger and Sandmann 1990; Bramley 1991; Pallett 1991; Young 1991).

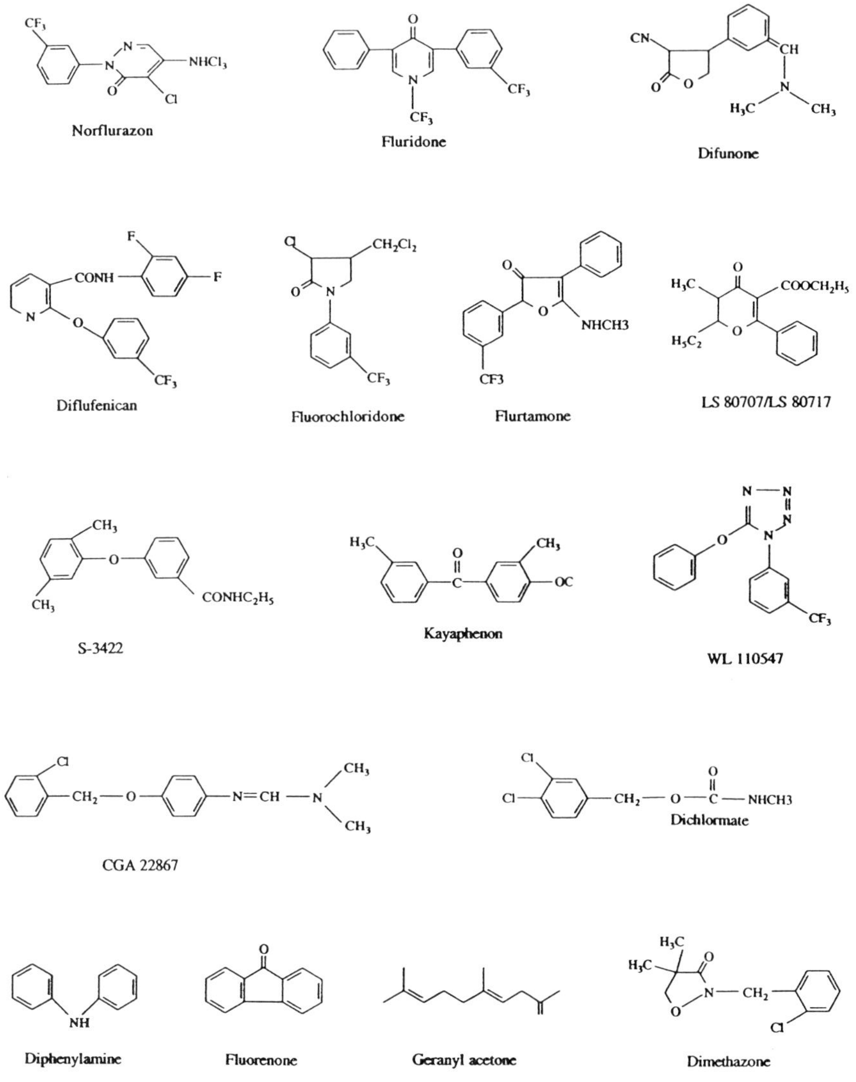

Figure 5.5 Inhibitors of desaturation reactions.

5.3.1 Phytoene Synthesis

The formation of phytoene from two molecules of geranylgeranyl diphosphate (GGDP), via prephytoene diphosphate (PPDP) is the first specific step in carotenoid biosynthesis (see Chapter 4). This two-step reaction has a mechanism different from that of the superficially analogous reaction of squalene synthesis from farnesyl diphosphate (Gregonis and Rilling 1973) and should, therefore, be a carotenoid-specific target site for bleaching herbicides.

There have been no reports, however, of a phytoene synthase-specific herbicide. Norflurazon, which is generally regarded as an inhibitor of phytoene desaturase (see Section 5.3.2) does inhibit phytoene synthesis in cell extracts of *P. blakesleeanus* (Sandmann et al. 1980) and also in wheat seedlings at concentrations <0.05 μM (Wilkinson 1985). The effect on wheat can be partly reversed by several analogues of norflurazon, such as desmethyl norflurazon, although the analogues are not inhibitors themselves. They may compete for binding to the target site on the phytoene synthase protein (Wilkinson 1987a). The kaurene synthetase inhibitor 2-chloro-*N*,*N*-bis-2-propylenacetamide (CPAA) also partially reverses the effect of norflurazon on phytoene synthesis (Wilkinson 1987b).

Since carotenoids are formed as a branch of the isoprenoid pathway, inhibitors of the early steps from acetyl-coenzyme A (CoA) to GGDP will have an effect on carotenoid levels. Thus, the potent inhibitor of HMG-CoA reductase, mevinolin, inhibits both sterol and carotenoid synthesis in *Phycomyces* (P. D. Fraser and P. M. Bramley, unpublished results). In plants, however, mevinolin affects only cytoplasmic reactions, presumably because it cannot enter the plastid (Bach 1987; Bach and Lichtenthaler 1987). Irreversible inhibitors of IDP isomerase have also been reported (Poulter 1990).

Dimethazone (Clomazone, FMC 57020) was introduced in 1984 as a selective pre-emergence/pre-plant-incorporated herbicide that causes bleaching in susceptible plants, leading to a decrease in the level of carotenoids and chlorophylls (Duke and Kenyon 1986). The results of studies since then indicate that dimethazone affects an early step in isoprenoid synthesis, rather than carotenoid biosynthesis specifically. Phytoene did not accumulate in dimethazone-treated leaves (Duke et al. 1985) and carotenogenic enzymes in *Aphanocapsa* cell extracts were not inhibited by the herbicide (Sandmann and Böger 1987b). Furthermore, the inhibition of other branches of the isoprenoid pathway, e.g. phytol formation (Duke et al. 1985; Duke and Kenyon 1986; Sandmann and Böger 1986), α-tocopherol synthesis (Sandmann and Böger 1986) and gibberellin biosynthesis (Duke et al. 1985; Sandmann and Böger 1986) by dimethazone also indicate that an early, common step in the isoprenoid pathway is the target site. Precisely which reaction is inhibited, however, is still uncertain. *In vitro* assays, using *Aphanocapsa* cell extracts with several substrates, have shown that 20 μM dimethazone inhibits reactions from mevalonic acid to GGDP, but the strongest inhibition (90%) was on the conversion of MVA to IDP (Böger and Sandmann 1990). An earlier study from the same group compared I_{50} values *in vivo* and *in vitro* for dimethazone acting on isoprenoid biosynthesis. They found that the target enzyme exhibited different sensitivity to the inhibitor depending upon its intracellular location, with the intra-plastidic reactions being some 45-fold more sensitive than those in the cytoplasm (Sandmann and Böger 1987b).

In contrast to these results, however, millimolar concentrations of dimethazone did not inhibit the activities *in vitro* of the plastidic enzymes of mustard and daffodil which convert IDP into GGDP (Lutzow et al. 1990). It may be, therefore, that the primary target site is within the series of reactions from MVA to IDP, and not later in the isoprenoid pathway.

5.3.2 Desaturation reactions

Compounds which interfere with the desaturation steps from phytoene to lycopene (see Chapter 4 for details of the pathway) are the most prominent group of inhibitors of carotenoid biosynthesis, and include a great number of chemically unrelated compounds (Fig. 5.5). The majority of these are herbicides, which cause the accumulation of phytoene or, less commonly, ζ-carotene in the treated tissues. This difference in the more saturated carotene that is accumulated after such treatment has led to the use of the terms "phytoene desaturase" and "ζ-carotene desaturase" inhibitors (e.g. Sandmann and Böger 1989; Böger and Sandmann 1990). It should be stressed, however, that at present there is no biochemical evidence that demonstrates unequivocally that two distinct enzymes exist, since there are no reports of their purification. It is more accurate, therefore, to refer to the inhibitors in this category as those causing the accumulation of either phytoene or ζ-carotene.

Table 5.4 Non-herbicidal inhibitors that cause the accumulation of more saturated carotenoids in higher plants and photosynthetic organisms

Compound	*Organism*	*Reference*
Diphenylamine	*Rhodospirillum rubrum*	Davies 1970a,b; Malhotra et al. 1970; Davies and Aung Than 1974
	Rhodomicrobium vannielii	Britton et al. 1975
	Rhodopseudomonas globiformis	Schmidt and Liaaen-Jensen 1973
	Chlorobium sp.	Goodwin and Land 1956
	Chromatium sp.	Fuller and Anderson 1958
	Aphanocapsa[a]	Clarke 1984; Sandmann et al. 1984
2-Hydroxybiphenyl	*Rhodospirillum rubrum*	Maudinas et al. 1973
	Rhodopseudomonas sphaeroides	Maudinas et al. 1972
Geranyl derivatives	*Synechococcus* PCC6911	Jüttner and Bogenschütz 1983

[a] *In vitro*

(a) Compounds that cause the accumulation of phytoene

Before the advent of commercial bleaching herbicides some 15–20 years ago, the only compounds documented to inhibit the desaturation of phytoene were bis-phenyl derivatives such as diphenylamine and 2-hydroxybiphenyl (Table 5.4). These were particularly effective in studies on the biosynthesis of carotenoids in phototrophic bacteria, algae and fungi (see Section 5.2.1 and reviews by Davies 1973, 1975; Goodwin 1980; Spurgeon and Porter 1983).

A study of the relative inhibitory activities of a range of bis-phenyl compounds and geranylacetone on carotenogenesis in cell extracts of *Aphanocapsa*, showed that, of the 13 compounds tested, methoxybenzophenone was the most inhibitory, whilst bis-phenyl compounds with two atoms between the phenyl ring, or with only one phenyl ring were only very weak inhibitors (Bramley et al. 1984). Structural similarities between the C-12 to C-8′ portion of the phytoene molecule and both bisphenyl compounds as well as alkylphenoxybenzamides provide some support for the hypothesis that such inhibitors compete for the active site of the desaturase (Rilling 1965; Herber et al. 1972; Bramley et al. 1984). Certainly diphenylamine does inhibit the enzymic activity rather than the synthesis of phytoene desaturase (Clarke et al. 1983).

A large number of commercial bleaching herbicides exert their primary mode of action by inhibiting carotenoid biosynthesis and causing the accumulation of phytoene (Table 5.5a). The oldest and best known compounds are the phenylpyridazinones such as norflurazon (SAN 9789) and its corresponding dimethylamino-derivative metflurazon (SAN 6706). Both compounds have been shown to cause the accumulation of phytoene *in vivo* and *in vitro*. More recently introduced herbicides such as fluridone, fluorochloridone, the *m*-phenoxybenzamides and diflufenican are also powerful inhibitors of carotenogenesis at the level of phytoene desaturation.

Table 5.5 Herbicides that cause the accumulation of more saturated carotenoids

Compound	*Organism*	*Reference*
(a) Phytoene accumulation		
CGA 22867	*Aphanocapsa*[a]	Sandmann et al. 1984
Cyclic ureas	*Anacystis*[a]	Babczinski et al. 1990
Diflufenican	Crop and weed species	Wightman and Haynes 1985
	Aphanocapsa[a]	Sandmann et al. 1985
Difunone	*Ankistrodesmus braunii*	Urbach et al. 1976;
	Aphanocapsa[a]	Clarke et al. 1982b
DTP	*Raphanus sativus*	Soeda and Uchida 1987
Fluorochloridone	*Zea mays*	Lay and Niland 1983
Fluridone	*Triticum aestivum*	Bartels and Watson 1978; Ridley 1982
Flurtamone	*Anacystis*[a]	Sandmann et al. 1990a

Compound	Organism	Reference
m-Phenoxybenzamides	*Scenedesmus*	Lambert and Böger 1983
	Aphanocapsa[a]	Clarke et al. 1985
RE-40885	Weed and crop species	Rogers et al. 1987
SAN 6706	*Triticum aestivum*	Bartels and McCullough 1972; Ben-Aziz and Koren 1974
	Chlorella fusca	Kummel and Grimme 1975; Tantawy and Grimme 1982
	Narcissus pseudonarcissus[a]	Beyer et al. 1980
	Raphanus sativus	Grumbach and Britton 1983
SAN 9789	*Euglena gracilis*	Vaisberg and Schiff 1976
	Triticum aestivum	Bartels and Watson 1978; Wilkinson 1985; Dahlin and Ryberg 1986
	Scenedesmus	Sandmann et al. 1981; Sandmann and Albrecht 1990
	Dunaliella bardawil	Ben Amotz et al. 1987
	Anacystis R2[a]	Sandmann et al. 1989
	Synechococcus	Linden et al. 1990
SC-0051	*Glycine max*	Mayonado et al. 1989

Compound	*Organism*	*Reference*
(b) ζ-Carotene accumulation		
Amitrole[b]	*Triticum aestivum*	Burns et al. 1971
Dichlormate[c]	*Triticum, Hordeum*	Burns et al. 1971; Ridley 1982
Haloxydine[c]	*Avena*	Ridley 1982
KM 143-958	*Scenedesmus, Lemna,* grasses and dicots	Chollet et al. 1990
LS 80717	*Aphanocapsa*[a]	Sandmann et al. 1985
LS 780091	*Aphanocapsa*[a]	Sandmann and Böger 1989
Methoxyphenone	*Echinochloa frumentacea*	Fujii et al. 1977
6-Methylpyrimidines	*Hordeum*	Ridley 1982
Pyrichlor[c]	*Triticum aestivum*	Burns et al. 1971
WL 110547	*Capsicum annuum* [a]	Kerr and Whitaker 1987

[a] *In vitro*
[b] Also has other modes of action. See Section 3.5
[c] Phytoene and phytofluene also accumulated

Chlorosis-inducing herbicides such as the phenylpyridazinones cause a decrease in both carotenoids and chlorophylls (Ridley 1982) with a concurrent accumulation of phytoene (Urbach et al. 1976; Pardo and Schiff 1980). Early studies were designed to determine whether such inhibitors affected the synthesis of both carotenoids and chlorophyll, or whether the disappearance of the latter was a secondary effect due to photooxidation. Bartels and Hyde (1970) showed that the inhibition of carotenogenesis by metflurazon in wheat seedlings was independent of light conditions, but the decrease in chlorophyll was higher with increasing light intensities. Experiments with dark-grown heterotrophic cultures of *Scenedesmus* showed that carotenoid formation was inhibited by norflurazon, but that the chlorophyll content was unaffected (Sandmann and Böger 1983). In addition, with the development of *in vitro* assays for phytoene desaturation (see Table 5.3), it was shown that both metflurazon and norflurazon inhibit the activity, rather than the synthesis, of phytoene desaturase (Beyer et al. 1980; Clarke et al. 1982a; Mayer et al. 1989). These studies have led to the generally accepted conclusion that the *N*-phenylpyridazinones inhibit phytoene desaturase activity and that the decrease in chlorophyll occurs by $^1O_2$-mediated photooxidation due to the absence of coloured carotenoids, which are known to scavenge or prevent the formation of $^1O_2$.

The advent of cell-free systems, from *Aphanocapsa, Anacystis* and *Narcissus* in particular, has resulted in several recent studies on the kinetics of inhibition by *N*-phenylpyridazinones, leading to the graphical analyses of data described in Section 5.2.4. With all three *in vitro* systems, norflurazon exhibited non-competitive inhibition of phytoene desaturase (Zöbelin 1985; Mayer et al. 1989; Sandmann et al. 1989). Dissociation constants (K_i) for the desaturase–norflurazon complex were 0.39 and 0.09 μM for the *Narcissus* and *Anacystis* enzymes, respectively (Mayer et al. 1989; Sandmann et al. 1989). The latter K_i is the lowest yet determined for any herbicidal inhibitor of phytoene desaturation. Reversibility of the binding was shown either by washing the membranes to remove the inhibitor (Mayer et al. 1989), or by displacement of ^{14}C-labelled norflurazon with unlabelled norflurazon (Sandmann et al. 1989). Inhibition of carotenogenesis in wheat seedlings by norflurazon is reversed when non-herbicidal derivatives are applied simultaneously (Wilkinson 1987a). It has been suggested that norflurazon binds to an allosteric site on the desaturase molecule and mimics an unsaturated carotene which normally causes feedback inhibition by binding to that site (Sandmann and Böger 1989). Some structural similarity between the carotenes and the inhibitor (norflurazon) molecules might be expected for this to occur. Unsaturated carotenes have been shown to inhibit phytoene desaturation *in vitro* with lycopene being the most effective of the group tested (Zöbelin 1985; Sandmann and Kowalczyk 1989). Zöbelin (1985) showed that replacement of the 3-CF_3 group of norflurazon by -SCF_3 resulted in irreversible non-competitive inhibition of the desaturase.

It is pertinent to note that certain phenylpyridazinones, such as SAN 9785, inhibit the desaturation of linoleic to linolenic acid in plants (St John 1976; Khan et al. 1979; Murphy et al. 1980). At high concentrations norflurazon also inhibits linoleic acid desaturation in wheat (St John 1976), whilst metflurazon has a similar effect on the desaturation of linoleate associated with monogalactosyldiacylglycerol (Khan et al. 1979) and the desaturation of palmitate to 3-hexadecenoate (Davies and Harwood 1983). Thus, a minor change in the pyridazinone molecule (e.g. the removal of the 3-trifluoromethyl group on the phenyl ring) changes an inhibitor of phytoene desaturation into one that is directed against linoleic acid desaturase.

Fluorochloridone, a selective herbicide, inhibits carotenoid biosynthesis (Devlin et al. 1979; Lay and Niland 1983) and the formation of polar membrane lipids (St John 1985). Its primary mode of action is the former, and inhibition has been demonstrated *in vitro* with a K_i of 4.5 μM (Mayer et al. 1989). Selectivity appears to be due to differences in the accumulation of the herbicide: it is found in leaves of susceptible plants, but is restricted to the roots of tolerant species (Lay and Niland 1983). Other herbicides which cause the accumulation of phytoene both *in vivo* and *in vitro* are fluridone (Bartels and Watson 1978; Mayer et al. 1989), difunone (Urbach et al. 1976; Clarke et al. 1982b) and tetrahydropyrimidinones (Babczinski et al. 1990).

All the herbicides discussed so far, when tested *in vitro*, have inhibited the activity of phytoene desaturase, and generally at concentrations very similar to those used with intact plants or algae. There are, however, some herbicides which do not fit this pattern, although they cause the accumulation of phytoene in susceptible tissues. For example, a comparison of I_{50} (whole cell) and K_i (cell-free) values for the herbicide CGA 22867 with *Aphanocapsa* shows that the former value is considerably lower than the *in vitro* inhibition constant (Sandmann et al. 1985). This difference could be due to enhanced uptake, metabolism to a more active derivative or enrichment at the intracellular target site in the intact cell.

Recently, two other herbicides have been reported to show major differences in their effects on carotenogenesis when used *in vivo* and *in vitro*. A comparison of norflurazon with the experimental herbicide SC-0051 on soybean plants has shown that both compounds caused the accumulation of phytoene, with a reduction in unsaturated carotenoids and chlorophylls (Mayonado et al. 1989). However, these workers noted that the two herbicides caused differences in the bleaching symptoms of leaves. Norflurazon bleached primarily the old (unifoliate) leaves, whereas SC-0051 bleached new (trifoliate) leaves. They suggest that norflurazon inhibits carotenogenesis involved in replacement or turnover of pigments in mature chloroplasts, whilst SC-0051 interferes with carotenoid formation in developing plastids. No studies have been reported on isolated plastids incubated with SC-0051.

The chemically related 2-(4-chloro-2-nitrobenzoyl)-5,5-dimethylcyclohexane-1,3-dione also shows a differential effect on new and old leaves. When this compound was added to germinating cress or soybean, the levels of the coloured carotenoids of primary leaves were decreased and phytoene accumulated. However, when the cotyledons of 3-day-old seedlings were soaked with the herbicide, the pigments of subsequent primary leaves were unaffected (Sandmann et al. 1990b). Similarly, soybean cell suspension cultures and algae were hardly affected by the herbicide at 10^{-5} M. *In vitro* preparations also showed no inhibition of phytoene desaturation. Parallel experiments with norflurazon gave the expected inhibition. The authors conclude that metabolic modification of cyclohexanediones in general is required for inhibitory activity and that this is dependent upon uptake and/or translocation in higher plants.

A further consequence of the accumulation of phytoene *in vivo* by bleaching herbicides has been demonstrated by several groups. The advent of rapid HPLC separations of carotenoids from crude extracts (see Section 5.2.2) has revealed that a number of hydroxy- and epoxy-phytoenes and, to a lesser extent, phytofluenes accumulate in intact tissues under these conditions. Although it was first thought that these may be intermediates in a two-step mechanism of hydroxylation followed by dehydration (Britton et al. 1987), later studies have discounted this hypothesis and suggest that the compounds are produced either non-enzymically or by non-specific hydroxylases (Barry and Pallett 1990; Sandmann and Albrecht 1990).

(b) Compounds that cause the accumulation of ζ- carotene

Compared to the large number of inhibitors that cause the accumulation of phytoene (see Section 5.3.2 (a); Table 5.5a), relatively few compounds inhibit only the desaturation of ζ-carotene (Table 5.5b; structures, Fig. 5.5). Pyrichlor, dichlormate (Burns et al. 1971), methoxyphenone (Kayaphenon; Fujii et al. 1977), haloxidine (Ridley 1982) and KM 143-958 (Chollet et al. 1990) all cause the accumulation of ζ-carotene, together with phytoene and phytofluene, in light-grown plants. *In vitro* studies on these compounds have not been reported, so it is not known whether they inhibit enzyme activity. Amitrole has also been shown to inhibit ζ-carotene metabolism *in vivo* (Burns et al. 1971), but it may have other modes of action, as discussed in Section 5.3.5.

Other inhibitors of ζ-carotene desaturation have been investigated in more detail, by the use of cell-free systems. The 6-methylpyrimidines, such as J852, J334 and J004, have all been shown to be inhibitory *in vitro*, J852 with extracts of *Aphanocapsa* and *Chenopodium* (Sandmann and Böger 1989), and the other two with *Narcissus* plastid membranes (Mayer et al. 1989). J334 and J004 also inhibited phytoene desaturase, but with I_{50} values some two-fold higher than those for ζ-carotene desaturation. J852 shows the same effect *in vivo*, with barley seedlings (Ridley 1982) as well as with *Triticum medicago* and *Aphanocapsa* (Kostal 1986).

Similar studies have been reported with the dihydropyrone LS 80707. ζ-Carotene, phytoene and traces of phytofluene accumulate in *Aphanocapsa* cells grown with LS 80707, and the corresponding preparation *in vitro* is affected in the same fashion, showing similar I_{50} values of about 5mM (Sandmann et al. 1985). The analogue LS 780091 has been shown to be an irreversible non-competitive inhibitor that binds to the active site of carotene desaturase (Sandmann and Böger, 1989), but LS 80717 (the same structure as LS 80807, but with alkyl chains at C-5 and C-6 in the *trans* configuration) causes the accumulation of phytoene but not that of ζ-carotene (Sandmann et al. 1989).

5.3.3 Cyclisation reactions

Nicotine and CPTA (Fig. 5.6) have been shown to inhibit carotene cyclisation in bacteria, fungi and higher plants (Table 5.6). This inhibition generally results in the accumulation of lycopene, and lesser amounts of monocyclic carotenoids. In a number of microorganisms it is possible to convert the accumulated lycopene into cyclic carotenoids after removal of the inhibitor (reviewed by Spurgeon and Porter 1983).

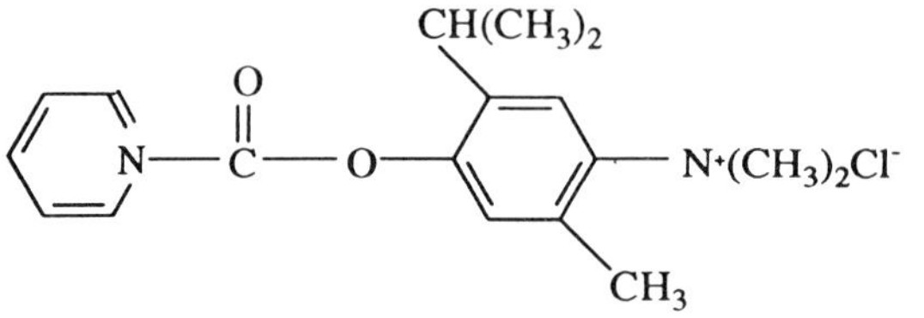

AMO-1618

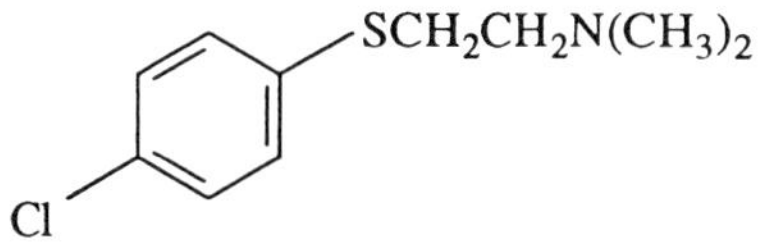

CPTA

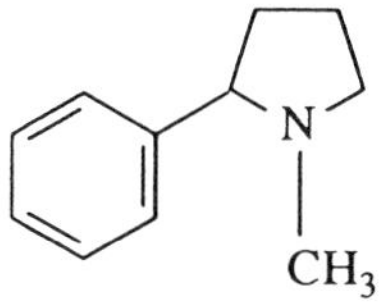

Nicotine

Figure 5.6 Inhibitors of cyclisation reactions.

Table 5.6 Inhibitors that cause the accumulation of lycopene or neurosporene in photosynthetic bacteria and higher plants

Compound	*Organism*	*Reference*
Nicotine	*Rhodopseudomonas* spp	McDermott et al. 1973; Singh et al. 1973
	Rhodomicrobium vannielii	Britton et al. 1977a; Leutwiler and Chapman 1978
	Chlorobium limicola	Leutwiler and Chapman 1979, 1981
	Cucurbita ficafolia,	Howes 1974
	Cucurbita pepo	Howes 1974
CPTA	*Rhodopseudomonas viridis*	Britton et al. 1977a
	Rhodomicrobium vannielii	Britton et al. 1977b; Rabinowich and Rudich 1972
	Tomato fruit and cells	Fosket and Radin 1983
	Grapefruit	Yokoyama et al. 1972; Coggins et al. 1970
	Lemon fruit	Valadon and Mummery 1978; Coggins et al. 1970
	Orange fruit	Coggins et al. 1970; Yokoyama et al. 1972
	Peach fruit	Coggins et al. 1970
	Prune fruit	Coggins et al. 1970
	Apricot fruit	Coggins et al. 1970

	Sinton citrangequat fruit	Coggins et al. 1970
	Carrot root	Coggins et al. 1970
	Sweet potato root	Coggins et al. 1970
	Capsicum annuum fruit	Camara 1984
	Cucurbita moschata cotyedons	Seyama and Splittstoesser 1975;
CDEB	Lemon and oranges	Coggins et al. 1970
Cycocel	*Cucurbita pepo* cotyledons	Knypl 1969
Undecyldiethylamine,	*Nicotiana tabacum*	Camara et al. 1985
tridecyldiethylamine	Soybean cultures	Camara et al. 1985
	Helianthus annuus	Camara et al. 1985
	Pisum sativum	Camara et al. 1985
4-[α-(Diethylamino) -ethoxy]benzophenone	*Rhodospirillum rubrum*	Hayman and Yokoyama 1976
Phenoxytriethylamines	Spinach, pumpkin	Wright et al. 1980
AMO 1618	*Aphanocapsa* .	Sandmann et al. 1985

The inhibition by CPTA has been studied in some detail in *Aphanocapsa*. Lycopene, together with small amounts of ζ-carotene, is accumulated in whole cells, cultured with 10 to 100mM CPTA (Sandmann and Böger 1985; Zöbelin 1985), with a concurrent decrease in the ß-carotene content. Lycopene cyclisation was non-competitively inhibited, but no stimulation of lycopene synthesis occurred. *In vivo* and *in vitro* I_{50} values of 45 and 35 μM, respectively, were obtained (Zöbelin 1985). These results contrast with previous *in vitro* studies. Bucholtz and coworkers (1977) were unable to demonstrate significant inhibition of cyclic carotenoid formation by CPTA in a tomato plastid system. Similarly, only slight inhibition of cyclisation was found with a *Narcissus* chromoplast preparation (Beyer et al. 1980).

In citrus fruits a range of onium compounds related to CPTA (with a -$CH_2N(C_2H_5)_2$ group) regulate the formation of all-*trans* carotenoids, in that not only is cyclisation blocked, but also the formation of lycopene is stimulated. The latter effect requires protein synthesis (Hsu et al. 1972). Indeed, the tertiary amine MPTA causes gene expression and translation of poly(A)$^+$RNA on 80 S ribosomes (Yokoyama et al. 1984). Clearly the mode of action of CPTA and related compounds is complex, and may lead to multiple effects in plants.

Other inhibitors of cyclisation include AMO1618, CDEB, Cycocel and decyldiethylamines (Table 5.6). Detailed biochemical studies with these compounds have not been reported.

5.3.4 Xanthophyll formation

Several reactions of xanthophyll formation are catalysed by mixed function oxidases (see Chapter 4). Although several inhibitors of similar reactions in sterol (Benveniste 1986) and gibberellin (Graebe 1987) biosynthesis have been developed, there are no reports of herbicides which inhibit hydroxylations in carotenogenesis. The bioregulators tetcyclasis and LAB 11682 (Fig. 5.7), do inhibit the formation of ß-cryptoxanthin *in vitro* when used at high concentrations (Sandmann and Bramley 1985). Also, nicotine prevents the hydroxylation at C-1 of acyclic carotenes in phototrophic bacteria (McDermott et al. 1973; Britton et al. 1977a) probably because the mechanism of this hydroxylation is similar to that of cyclisation (see Section 5.3.3).

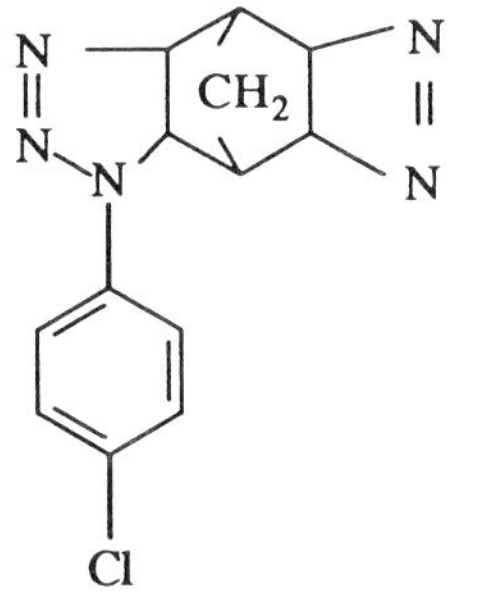

Tetcyclasis

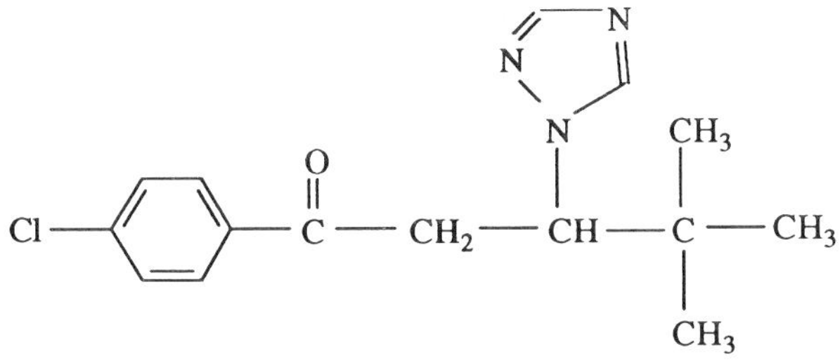

LAB 117682

Figure 5.7 Inhibitors of xanthophyll formation.

5.3.5 Multiple target sites

Fluometuron (Fig. 5.8) is primarily an inhibitor of photosynthetic electron transport (Appleman 1952), but it also inhibits carotenoid formation. A comparison of the I_{50} values for each inhibitor effect indicates that it is a more potent inhibitor of electron flow through photosystem II (I_{50} = 2 µM) than of carotenogenesis (I_{50} = 10 µM) (Sandmann and Böger 1989). Similarly, fomesafen (Fig. 5.8) initiates peroxidation, but also inhibits carotenoid biosynthesis (Sandmann and Böger 1989).

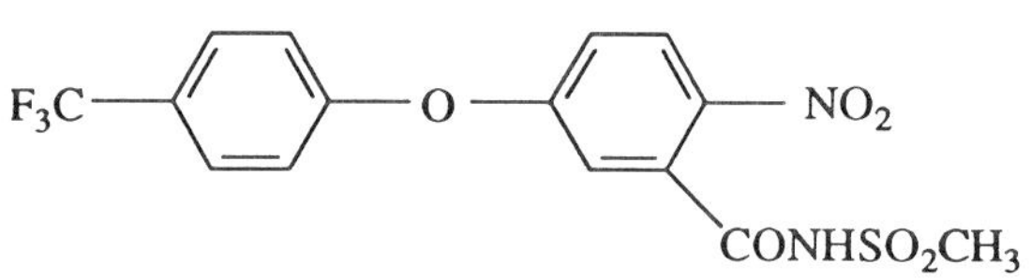

Fomesafen

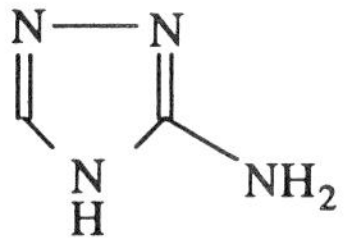

Amitrole

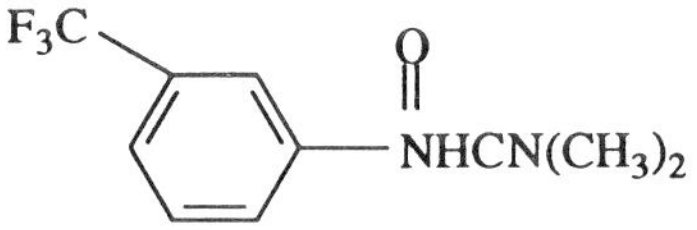

Fluometuron

Figure 5.8 Inhibitors with multiple target sites.

Compounds such as AMO-1618, CPTA and nicotine that inhibit carotene cyclisation also have multiple actions, as described in Section 5.3.3. Treatment of plants with amitrole causes the accumulation of ζ-carotene (Burns et al. 1971), but more recent studies suggest that it may have multiple modes of action. Although it is effective *in vivo*, with *Aphanocapsa* or *Chenopodium*, no inhibitory activity could be detected with cell extracts of these species, even at a concentration of 50 μM (Kostal 1986). In addition, there are reports that amitrole causes the accumulation of lycopene (Rüdiger et al. 1976) and phytoene (Grumbach 1984) in a range of higher plant species.

5.4 RESISTANCE TO INHIBITORS

Resistance to inhibitors of carotenogenesis because of impaired uptake by plants and microorganisms has been reported occasionally. Most herbicides are effective against only a restricted number of plant species, with the other plants being tolerant to the dosage used in the field. This is attributed largely to poor absorption or translocation of the molecule into the plant tissues (as exemplified by the uptake of SAN 6706 and 9789 by cotton, corn and soybean; Strang and Rogers 1974), but it may also be due to the application of the inhibitor to plants at a stage of their development when the herbicide is ineffective (e.g. SC-0051, see Section 5.3.2 (a)). The metabolism of bleaching herbicides to non-toxic derivatives is not thought to be a cause of resistance.

Despite the fact that bleaching herbicides have been used for weed control for some 20 years, there have been no reports of resistant biotypes of weeds. In order to obtain herbicide-resistant strains therefore, laboratory mutagenesis techniques have been successfully employed. Christianson and Deal (1984) were able to produce, by γ-irradiation, cell cultures of *Nicotiana tabacum* which were resistant to norflurazon. Three patterns of pleiotropic interaction with herbicides were found: one group of norflurazon-resistant cells with altered sensitivity to fluridone, another group cross-resistant to difunone and pyrichlor, and a third cross-resistant only to pyrichlor. No group was solely resistant to norflurazon.

Chemical mutagenesis of *Synechococcus* PCC7942 has generated cells resistant to norflurazon (Linden et al. 1990). Several mutant strains with different degrees of resistance were isolated. Assays of carotenoid biosynthesis *in vivo* and *in vitro* have shown that the binding of norflurazon to phytoene desaturase is decreased. One strain was also resistant to flurochloridone but not to other bleaching herbicides. The herbicide-resistant gene from one mutant has now been cloned by genetic complementation of the resistant trait into wild type cells (Chamovitz et al. 1990). Difunone- and flurtamone-resistant *Anacystis* mutants have also been isolated, but studies on cross-resistance have yet to be reported (Sandmann et al. 1990a).

5.5 CONCLUSIONS AND FUTURE PROSPECTS

Our inadequate understanding of the properties of carotenogenic enzymes as well as the regulatory mechanisms which control their synthesis, turnover and targeting within the cell precludes the absolute categorisation of carotenogenic inhibitors with respect to their modes of action. Although it is possible to list potential target sites (Table 5.2), our ability to test such hypotheses experimentally is still some distance into the future.

Some significant progress in recent years has been made. The development of precise *in vitro* assays for many carotenogenic enzymes in higher plant and cyanobacterial cells (Table 5.3) has revealed that many inhibitors of the carotene desaturase reactions bind to the enzyme(s) non-competitively and reversibly (Sections 5.2.4 and 5.3.2). The recent availability of herbicide-resistant cells (Section 5.4) shows that individual inhibitors probably bind to different sites of the targeted protein. The use of pure enzymes in such detailed biochemical analyses should be a major objective in the near future. Until it is shown that a particular inhibitor binds to the pure enzyme, and thus inhibits catalytic activity, it can still be argued that the effects shown in crude preparations *in vitro* are influenced by other reactions in the cell extract. The availability of radioactive herbicide derivatives which covalently bond to the target sites would be beneficial to such investigations.

The cloning and sequencing of carotenoid genes has gained considerable momentum over the past two years, with the characterisation of genes from *Rhodobacter* (Armstrong et al. 1989), *Neurospora* (Nelson et al. 1989; Schmidhauser et al. 1990), *Erwinia* (Misawa et al. 1990), *Aphanocapsa* (Schmidt and Sandmann 1990), *Synechococcus* (Chamovitz et al. 1991), soybean (Bartley et al. 1991) and tomato (Bird et al. 1991). Several of these genes have been expressed in *E. coli*, and if the subsequent proteins can be reconstituted to fully active enzymes, with the same properties as the proteins found in the original species, then they can be used to provide large amounts of pure enzymes for detailed biochemical analyses or the routine screening of novel inhibitors which target carotenogenic enzymes.

Apart from studies on the enzymes themselves, it is as important to understand more about the development of the chloroplast, the transport and integration of the enzymes into the plastid and the movement of carotenoids themselves into pigment–protein complexes. At the present time, we do not have the requisite molecular probes to investigate such processes. The availability of antibodies to the enzymes would be useful for immunocytology studies, but it must first be ensured that the antibody is specific to the protein under investigation. It is possible that antibodies to enzymes expressed in bacteria are not sufficiently specific for this purpose.

We are entering an exciting area of studies of carotenogenesis in plants and microorganisms. The knowledge that is gained in the future should enable a more rational design of commercial bleaching herbicides to be attempted, rather than the current approach of random screening of many chemicals.

ACKNOWLEDGEMENTS

I thank Dr Gerhard Sandmann for providing Figs 5.3 and 5.4, and Miss Justine Baker for typing the manuscript.

REFERENCES

Appleman, D. (1952) Catalase–chlorophyll relationship in barley seedlings. *Plant Physiol.*, **27**: 613-621.

Armstrong, G.A., Alberti, M., Leach, F. and Hearst, J.E. (1989) Nucleotide sequence, organization and nature of the protein products of the carotenoid biosynthesis gene cluster of *Rhodobacter capsulatus*. *Molec. Gen. Genet.*, **216**: 254-268.

Babczinski, P., Blunck, M., Sandmann, G. *et al.*(1990) Substituted tetrahydropyrimidones. A new class of herbicidal compounds inducing chlorosis by inhibition of phytoene desaturation. *Pestic. Sci.*, **30**: 339-342.

Bach, T.J. (1987) Synthesis and metabolism of mevalonic acid in plants. *Plant Physiol. Biochem.*, **25**: 163-178.

Bach, T.J. and Lichtenthaler, H.K. (1987) Plant growth regulation by mevinolin and other sterol biosynthesis inhibitors. In: Fuller, G. and Nes, W.D. (eds.), *Ecology and Metabolism of Plant Lipids*. American Chemical Society, Washington, DC, pp. 109-115.

Barry, P. and Pallett, K.E. (1990) Herbicidal inhibition of carotenogenesis detected by HPLC. *Z. Naturforsch.*, **45c**: 492-497.

Bartels, P.G. and Hyde, A. (1970) Chloroplast development in Sandoz 6706 - treated wheat seedlings. A pigment, ultrastructural and ultracentrifugal study. *Plant Physiol.*, **45**: 807-810.

Bartels, P.G. and McCullough, C. (1972) A new inhibitor of carotenoid synthesis in higher plants: 4-chloro-5-(dimethylamino)-2,α,α,α(trifluoro-*m*-tolyl)-3(2H)-pyridazinone (Sandoz 6706). *Biochem. Biophys. Res. Commun.*, **48**: 16-22.

Bartels, P.G. and Watson, C.W. (1978) Inhibition of carotenoid synthesis by fluridone and norflurazon. *Weed Sci.*, **26**: 198-203.

Bartley, G.E., Viitanen, P.V., Pecker, I. *et al.* (1991) Molecular cloning and expression in photosynthetic bacteria of a soybean cDNA coding for phytoene desaturase, an enzyme of the carotenoid biosynthetic pathway. *Proc. Natl. Acad. Sci.* USA, **88**: 6532-6536.

Ben-Amotz, A., Gressel, J. and Avron, M. (1987) Massive accumulation of phytoene induced by norflurazon in *Dunaliella bardawil* (Chlorophyceae) prevents recovery from photoinhibition. *J. Phycol.*, **23**: 176-181.

Ben-Aziz, A. and Koren, E. (1974) Interference in carotenogenesis as a mechanism of action of the pyridazinone herbicide Sandoz 6706. *Plant Physiol.*, **54**: 916-920.

Benveniste, P. (1986) Sterol biosynthesis. *Annu. Rev. Plant Physiol.*, **37**: 275-308.

Beyer, P., Kreuz, K. and Kleinig, H. (1980) ß-Carotene synthesis in isolated chromoplasts from *Narcissus pseudonarcissus*. *Planta*, **150**: 435-438.

Bird, C.R., Ray, J.A., Fletcher, J.D. *et al.* (1991) Using antisense RNA to study gene function: Inhibition of carotenoid biosynthesis in transgenic tomatoes. *Biotechnology*, **9**: 635-639.

Böger, P. and Sandmann, G. (1990) Modern herbicides affecting typical plant processes. In: Bowers, W.S., Ebing, W., Martin, D. and Wegler, R. (eds.), *Chemistry of Plant Protection*, Vol. 6. Springer, Berlin, pp. 173-216.

Bramley, P.M. (1985) The *in vitro* biosynthesis of carotenoids. *Adv. Lip. Res.*, **21**: 243-279.

Bramley, P.M. (1990) The use of cell-free systems in studies of carotenoid biosynthesis. In: Krinsky, N.I., Mathews-Roth, M.M. and Taylor, R.F. (eds.), *Carotenoids: Chemistry and Biology*. Plenum Press, New York, pp. 185-194.

Bramley, P.M. (1991) Carotenoid biosynthesis. In: Kirkwood, R.F. (ed.), *Target Sites of Herbicides Action*. Plenum Press, New York, pp. 95- 122.

Bramley, P.M. (1992) Analysis of carotenoids by HPLC and diode-array detection. *Phytochemical Anal.*, **3**:97-104.

Bramley, P.M. (1993) Carotenoid biosynthesis. In: Lea, P.J. (ed.), *Methods in Plant Biochemistry*, Vol. 9. Academic Press, London, pp. 281-297.

Bramley, P.M., Clarke, I.E., Sandmann, G. and Böger, P. (1984) Inhibition of carotene biosynthesis in cell extracts of *Phycomyces blakesleeanus*. *Z. Naturforsch.*, **39c**: 460-463.

Britton, G. (1985) General carotenoid methods. *Methods Enzymol.*, **111**: 113-149.

Britton, G. (1991) Carotenoids. In: Charlwood, B.V. and Banthorpe, D.V. (eds.), *Methods in Plant Biochemistry*, Vol. 7. Academic Press, London, pp. 473-518.

Britton, G. and Goodwin, T.W. (1971) Biosynthesis of carotenoids. *Methods Enzymol.*, **18c**: 654-701.

Britton, G., Singh, R.K., Goodwin, T.W. and Ben-Aziz, A. (1975) The carotenoids of *Rhodomicrobium vannielii* (Rhodospirillaceae) and the effect of diphenylamine on the carotenoid composition. *Phytochemistry*, **14**: 2427-2433.

Britton, G., Singh, R.K., Malhotra, H.C. *et al.* (1977a) Biosynthesis of 1,2-dihydrocarotenoids in *Rhodopseudomonas viridis*: experiments with inhibitors. *Phytochemistry*, **16**: 1561-1566.

Britton, G., Singh, R.K. and Goodwin, T.W. (1977b) Carotenoid biosynthesis in *Rhodomicrobium vannielii*. Experiments with nicotine and CPTA. *Biochim. Biophys. Acta*, **488**: 475-483.

Britton, G., Barry, P. and Young, A.J. (1987) The mode of action of diflufenican: Its evaluation by HPLC. *Proc. Brit. Crop Prot. Conf. (Weeds)*: 1015-1022.

Britton, G., Barry, P. and Young, A.J. (1989) Carotenoids and chlorophylls: Herbicidal inhibition of pigment biosynthesis. In: Dodge, A.D. (ed.), *Herbicides and Plant Metabolism*. Cambridge University Press, Cambridge, pp. 51-72.

Bucholtz, M.L., Maudinas, B. and Porter, J.W. (1977) Effects of *in vivo* inhibitors of carotene biosynthesis on the synthesis of carotenes by a soluble tomato plastid system. *Chem-Biol. Interactions*, **17**: 359- 362.

Burns, E.R., Buchanan, G.A. and Carter, M.C. (1971) Inhibition of carotenoid synthesis as a mechanism of action of amitrole, dichlormate and pyrichlor. *Plant Physiol.*, **47**: 144-148.

Camara, B. (1984) Terpenoid metabolism in plants, sites of phytoene synthetase activity and synthesis in plant cells. *Plant Physiol.*, **74**: 112-116.

Camara, B., d'Harlingue, A., Dogbo, O. *et al.* (1985) Highly potent inhibitors of plant terpenoid biosynthesis. *C.R. Acad. Sci.*, **Ser III**,: 297-300.

Chamovitz, D., Pecker, I., Sandmann, G. *et al.* (1990) Cloning a gene coding for norflurazon resistance in cyanobacteria. *Z. Naturforsch.*, **45c**: 482-486.

Chamovitz, D., Pecker, I. and Hirschberg, J. (1991) The molecular basis of resistance to the herbicide norflurazon. *Plant Mol. Biol.*, **16**: 967- 974.

Chollet, R., Sandmann, G., Diethelm, R. *et al.* (1990) ζ-Carotene accumulation and bleaching by new pyrimidine compounds. *Pestic. Sci.*, **30**: 326-329.

Christianson, A.C. and Deal, L.M. (1984) Tobacco plants resistant to inhibitors of phytoene desaturation. *Genetics*, **107**: S20.

Clarke, I.E. (1984) Inhibitors of carotenogenesis in cell-free systems of *Aphanocapsa* and *Phycomyces*. PhD thesis, University of London.

Clarke, I.E., Sandmann, G., Bramley, P.M. and Böger, P. (1982a) Carotene biosynthesis with isolated photosynthetic membranes. *FEBS Lett.*, **140**: 203-206.

Clarke, I.E., Bramley, P.M., Sandmann, G. and Böger, P. (1982b) Herbicidal action on carotenogenesis in a photosynthetic cell-free system. In: Wintermans, J.F.G.M. and Kuiper, P.J.C. (eds.), *Biochemistry and Metabolism of Plant Lipids*. Elsevier, Amsterdam, pp. 549-554.

Clarke, I.E., de la Concha, A., Murillo, F.J. *et al.* (1983) The effect of diphenylamine on carotenogenesis in *Phycomyces blakesleeanus*. *Phytochem*istry, **22**: 435- 439.

Clarke, I.E., Sandmann, G., Bramley, P.M. and Böger, P. (1985) Phytotoxicity of *m*-phenoxybenzamides: inhibition of cell-free phytoene desaturation. *Pestic. Biochem. Physiol.*, **23**: 335-340.

Coggins, C.W., Henning, G.L. and Yokoyama, H. (1970) Lycopene accumulation induced by CPTA. *Science*, **168**: 1589-1590.

Dahlin, C. and Ryberg, H. (1986) Accumulation of phytoene in plastoglobuli of SAN 9789 (norflurazon)-treated dark-grown wheat. *Physiol. Plant.*, **68**: 39-45.

Davies, A.O. and Harwood, J.L. (1983) Effect of substituted pyridazinones on chloroplast structure and lipid metabolism in greening barley leaves. *J.Exp. Bot.*, **34**: 1089-1100.

Davies, B.H. (1970a) A novel sequence for phytoene dehydrogenation in *Rhodospirillum rubrum*. *Biochem. J.*, **116**: 93-99.

Davies, B.H. (1970b) Alternative pathways of spirilloxanthin biosynthesis in *Rhodospirillum rubrum*. *Biochem. J.*, **116**: 101-110.

Davies, B.H. (1973) Carotene biosynthesis in fungi. *Pure Appl. Chem.*, **35**: 1-28.

Davies, B.H. (1975) Carotenoids - aspects of biosynthesis and enzymology. *Ber. Deutsch. Bot. Ges.*, **88**: 7-25.

Davies, B.H. (1976) Carotenoids. In: Goodwin, T.W. (ed.), *Chemistry and Biochemistry of Plant Pigments*, Vol. 2, 2nd Edn. Academic Press, New York, pp. 38-162.

Davies, B.H. and Aung Than. (1974) Monohydroxylated carotenoids from diphenylamine inhibited cultures of *Rhodospirillum rubrum*. *Phytochemistry*, **13**: 209-219.

Devlin, R.M., Kisiel, M. and Kostusiak, A.S. (1979) Influence of R-40244 on pigment content of wheat and corn. *Weed Res.*, **19**: 59-61.

Dogbo, O., Laferrière, A., d'Harlingue, A. and Camara, B. (1988) Carotenoid biosynthesis: isolation and characterisation of a bifunctional enzyme catalysing the synthesis of phytoene. *Proc. Natl. Acad. Sci. USA*, **85**: 7054-7058.

Duke, S.O. and Kenyon, W.H. (1986) Effects of dimethazone (FMC 57020) on chloroplast development. II. Pigment synthesis and photosynthetic function in cowpea (*Vigna unguiculata*) primary leaves. *Pestic. Biochem. Physiol.*, **25**: 11-18.

Duke, S.O. and Paul, R.N. (1986) Effects of dimethazone (FMC 57020) on chloroplast development I. Ultrastructural effects in cowpea (*Vigna unguiculata*) primary leaves. *Pestic. Biochem. Physiol.*, **25**: 1-10.

Duke, S.O., Kenyon, W.H. and Paul, R.N. (1985) FMC 57020 effects on chloroplast development in morning glory (*Ipomea lacurrosa*) cotyledons. *Weed Sci.*, **33**: 786-794.

El-Jack, M., Bramley, P.M. and Mackenzie, A. (1987) Carotene biosynthesis by a cell extract of *Aspergillus giganteus* mut. *alba*. *Phytochemistry*, **26**: 2525-2529.

Fosket, D.E. and Radin, D.N. (1983) Induction of carotenogenesis in cultured cells of *Lycopersicon esculentum*. *Plant Sci. Lett.*, **30**: 165- 175.

Fraser, P.F., de las Rivas, J., Mackenzie, A. and Bramley, P.M. (1991) *Phycomcyces blakesleeanus carB* mutants: Their use in assays of phytoene desaturase. *Phytochemistry*, **30**: 3971-3976.

Fujii, Y., Kurokawa, T., Inoue, Y. *et al.* (1977) Inhibition of carotenoid biosynthesis as a possible mode of herbicidal action of 3,3´-dimethyl-4´-methoxybenzophenone (NK-049). *J. Pestic. Sci.*, **2**: 431-437.

Fuller, R.C. and Anderson, I.C. (1958) Suppression of carotenoid synthesis and its effects on the activity of photosynthetic bacterial chromatophores. *Nature*, **181**: 252-254.

Goodwin, T.W. (1952) Studies on carotenogenesis. 3. Identification of the minor polyene components of the fungus *Phycomyces blakesleeanus* and a study of their synthesis under various cultural conditions. *Biochem. J.*, *50*: 550-558.

Goodwin, T.W. (1980) *The Biochemistry of Carotenoids*, Vol I: Plants. 2nd Edn. Chapman and Hall, London.

Goodwin, T.W. and Britton, G. (1988) Distribution and analysis of carotenoids. In: Goodwin, T.W. (ed.), *Plant Pigments*. Academic Press, London, pp. 61-131.

Goodwin, T.W. and Land, D.G. (1956) Carotenoids of some species of *Chlorobium*. *Biochem. J.*, **62**: 553-556.

Gregonis, D.E. and Rilling, H.C. (1973) Photoinduced prephytoene pyrophosphate synthesis in a *Mycobacterium* sp. *Biochem. Biophys. Res. Commun.*, **54**: 449-454.

Graebe, J.E. (1987) Gibberellin biosynthesis and control. *Annu. Rev. Plant Physiol.*, 38: 419-465.

Grumbach, K.H. (1984) Herbicides which interfere with the biosynthesis of carotenoids and their effect on pigment excitation, chlorophyll fluorescence and pigment composition of the thylakoid membrane. *Z. Naturforsch.*, **39c**: 455-458.

Grumbach, K.H. and Britton, G. (1983) Distribution of carotenoids in subcellular and subplastidic fractions of radish seedlings (*Raphanus sativus*) grown in the presence of bleaching herbicides. *Phytochemistry*, **22**: 1937-1940.

Hansch, C., Leo, A., Unger, S.H. *et al.* (1973a) Aromatic substituent constants for structure – activity correlations. *J. Med. Chem.*, **16**: 1207-1216.

Hansch, C., Unger, S.H. and Forsythe, A.B. (1973b) Strategy in drug design. Cluster analysis as an aid in the selection of substituents. *J. Med. Chem.*, **16**: 1217-1222.

Hayman, E.P. and Yokoyama, H. (1976) Effects of 4[ß-(diethylamino)- ethoxy] benzophenone upon carotenogenesis in *Rhodospirillum rubrum*. *J. Bacteriol.*, **127**: 1030-1031.

Herber, R., Maudinas, B. and Villoutreix, J. (1972) Influence de différents composés chimiques sur la caroténogenèse de *Mucor hiemalis*. *Phytochemistry*, **11**: 3461-3464.

Hlousek-Radojcic, A. and Ljubesic, N. (1988) The development of daffodil chromoplasts in the presence of herbicides SAN 9789 and SAN 9785. *Z. Naturforsch.*, **43c**: 418-422.

Howes, C.D. (1974) Nicotine inhibition of carotenoid cyclisation in *Cucurbita ficifolia* cotyledons. *Phytochemistry*, **13**: 1469-1471.

Hsu, W.J., Yokoyama, H. and Coggins, C.W. (1972) Carotenoid biosynthesis in *Blakeslea trispora*. *Phytochemistry*, **11**: 2985-2990.

Jüttner, F. and Bogenschütz, O. (1983) Geranyl derivatives as inhibitors of carotenogenesis in *Synechococcus* PCC 6911 (Cyanobacteria). *Z. Naturforsch.*, **38c**: 387-392.

Kerr, M.W. and Whitaker, D.P. (1987) The mode of action of the herbicide WL 110547. *Proc. Brit. Crop Prot. Conf. (Weeds)*: 1005- 1013.

Khan, M.U., Lem, N.W., Chandorkar, K.B. and Williams, J.M. (1979) Effects of substituted pyridazinones (SAN 6706, 9774 and 9785) on glycerolipids and their associated fatty acids of *Vicia faba* and *Hordeum vulgare*. *Plant Physiol.*, **64**: 300-305.

Knypl, J.S. (1969) Accumulation of lycopene in detached cotyledons of pumpkins treated with CPTA. *Naturwissenschaften*, **56**: 572.

Kostal, H. (1986) Charakterisierung der ζ-Carotindesaturase als Angriffsstelle Herbizidaler Inhibitoren. Thesis, Universität Konstanz.

Kummel, H.W. and Grimme, L.H. (1975) The inhibition of carotenoid biosynthesis in green algae by SANDOZ 6706: Accumulation of phytoene and phytofluene in *Chlorella fusca*. *Z. Naturforsch.*, **30c**: 333-336.

Kunert, K.J. and Dodge, A.D. (1989) Herbicide-induced radical damage and antioxidative systems. In Böger, P. and Sandmann, G. (eds.), *Target Sites of Herbicide Action*. CRC Press, Boca Raton, FL, pp. 45-78.

Lambert, R. and Böger, P. (1983) Inhibition of carotenogenesis by substituted diphenyl ethers of the *m*-phenoxybenzamide type. *Pestic. Biochem. Physiol.*, **20**: 183-187.

Lay, M.M. and Niland, A.M. (1983) The herbicidal mode of action of R- 40244 and its absorption by plants. *Pestic. Biochem. Physiol.*, **19**: 337-343.

Lay, M.M., Henstrand, J.M., Lawrence, S.R. and Chromartie, T.H. (1985) Studies on the mode of action of the herbicide flurochloridone. *Proc. Brit. Crop Prot. Conf. (Weeds)*: 179-186.

Leutwiler, L.S. and Chapman, D.J. (1978) Biosynthesis of carotenoids in *Rhodomicrobium vannielii*. *FEBS Lett.*, **89**: 248-252.

Leutwiler, L.S. and Chapman, D.J. (1979) Biosynthesis of aryl carotenoids: Inhibitor studies of chlorobactene biosynthesis in *Chlorobium limicola* F,. *thiosulfatophilum*. *Arch. Microbiol.*, **123**: 267-273.

Leutwiler, L.S. and Chapman, D.J. (1981) Biosynthesis of aryl carotenoids: Studies on the nature of the precyclisation intermediate. *Plant Cell Physiol.*, **22**: 781-787.

Linden, H., Sandmann, G., Chamovitz, D. *et al.* (1990) Biochemical characterisation of *Synechococcus* mutants selected against the bleaching herbicide norflurazon. *Pestic. Biochem. Physiol.*, **36**: 46-51.

Lütke-Brinkhaus, F., Liedvogel, B., Kreuz, K. and Kleinig, H. (1982) Phytoene synthase and phytoene dehydrogenase associated with envelope membranes from spinach chloroplasts. *Planta*, **156**: 176-180.

Lutzow, M., Beyer, P. and Kleinig, H. (1990) The herbicide Command does not inhibit the prenyl diphosphate-forming enzymes in plastids. *Z. Naturforsch.*, **45c**: 856-858.
Malhotra, H.C., Britton, G. and Goodwin, T.W. (1970) The occurrence of hydroxy derivatives of phytoene and phytofluene in diphenylamine-inhibited cultures of *Rhodospirillum rubrum*. *FEBS Lett.*, **6**: 334- 336.
Maudinas, B., Herber, R. and Villoutreix, J. (1972) Action de différentes substances chimiques sur la caroténogenése de *Rhodopseudomonas sphaeroides*. *Chem. Biol. Interactions,* **5**: 341-350.
Maudinas, B., Oelze, J., Villoutreix, J. and Reisinger, O. (1973) The influence of 2-hydroxybiphenyl on membranes of *Rhodospirillum rubrum*. *Arch. Microbiol.*, **93**: 219-228.
Mayer, M.P., Bartlett, D.L., Beyer, P. and Kleinig, H. (1989) The *in vitro* mode of action of bleaching herbicides on the desaturation of 15-*cis*--phytoene and *cis*-ζ-carotene in isolated daffodil chromoplasts. *Pestic. Biochem. Physiol.* **34**: 111-117.
Mayonado, D.J., Hatzios, K.K., Orcutt, D.M. and Wilson, H.P. (1989). Evaluation of the mechanism of action of the bleaching herbicide SC- 0051 by HPLC analysis. *Pestic. Biochem. Physiol.* **35**: 138-145.
McDermott, J.B.C., Ben-Aziz, A., Singh, R.K., Britton, G. and Goodwin, T.W. (1973) Recent studies of carotenoid biosynthesis in bacteria. *Pure Appl. Chem.*, **35**: 29-45.
Misawa, N., Nakagama, M., Kobayashi, K. *et al.*(1990) Elucidation of the *Erwinia uredovora* carotenoid biosynthetic pathway by functional analysis of gene products expressed in *E. coli*. *J. Bacteriol.*, **172**: 6704-6712.
Murphy, D.J., Harwood, J.L., Lee, K.A. *et al.* (1980) Differential responses of a range of photosynthetic tissues to a substituted pyridazinone, Sandoz 9785. Specific effects on fatty acid desaturation. *Phytochemistry*, **24**: 1923-1929.
Nelson, M.A., Morelli, G., Carattoli, A. *et al.* (1989) Molecular cloning of a *N. crassa* carotenoid biosynthesis gene (albino-3) regulated by blue light and the products of the white collar genes. *Molec. Cell Biol.*, **9**: 1271-1276.
Pallett, K.E. (1991) Other primary target sites of herbicides. In: Kirkwood, R.F. (ed.), *Target Sites for Herbicide Action*. Plenum Press, New York, pp. 123-167.
Pardo, A.D. and Schiff, J.A. (1980) Plastid and seedling development in SAN 9789 - treated etiolated bean seedlings. *Can. J. Bot.*, **58**: 25-35.
Poulter, C.D. (1990) IDP to squalene – enzymology and inhibition. In: Kuhn, P.J., Trinci, A.P.J., Jung, M.J. *et al.* (eds.), *Biochemistry of Cell Walls and Membranes in Fungi*. Springer, New York, pp. 169-188.
Rabinowitch, H.D. and Rudich, J. (1972) Effects of ethephon and CPTA on color development of tomato fruits at high temperatures. *Hortic. Sci.*, **7**: 76-77.
Ridley, S.M. (1982) Carotenoids and herbicide action. In: Britton, G. and Goodwin, T.W. (eds.), *Carotenoid Chemistry and Biochemistry*. Pergamon Press, New York, pp. 353-369.
Rilling, H.C. (1965) A study of inhibition of carotenoid synthesis. *Arch. Biochem. Biophys.*, **110**: 39-46.
Rogers, D.D., Kirby, B.W., Hulbert, J.C. *et al.* (1987) RE-40885: A new broadleaf herbicide in cotton, peanut, *Sorghum* and sunflower. *Proc. Brit. Crop Prot. Conf. (Weeds)*: 69-75.
Rüdiger, W., Benz, J., Lempart, U. *et al.* (1976) Hemmung der Phytol Akkumulation mit Herbiziden. Geranylgeraniol- and dihydrogeranylgeraniol-haltiges Chlorophyll aus Weizenkeimlinger. *Z. Pflanzenphysiol.*, **80**: 131-135.
Sandmann, G. (1987) Structure and activity of herbicidal inhibitors of phytoene desaturase. In: Greenhalgh, R. and Roberts, T.R. (eds.), *Pesticide Science and Technology*. Proc. 6th IUPAC Congr. Pestic. Chem., pp. 43-48.
Sandmann, G. (1988) *In vitro* carotenoid biosynthesis in *Aphanocapsa*. *Methods Enzymol.*, **167**: 329-335.
Sandmann, G. and Albrecht, M. (1990) Accumulation of colourless carotenes and derivatives

during the interaction of bleaching herbicides with phytoene desaturation. *Z. Naturforsch.*, **45c**: 487-491.

Sandmann, G. and Böger, P. (1982) Bleaching activity of new 2-phenylpyridazinones: Structure–activity relationship. *Z. Naturforsch.*, **37c**: 1092-1094.

Sandmann, G. and Böger, P. (1983) Comparison of the bleaching activity of norflurazon and oxyfluorfen. *Weed Sci.*, **31**: 338-341.

Sandmann, G. and Böger, P. (1986) Interference of dimethazone with the formation of terpenoid compounds. *Z. Naturforsch.*, **41c**: 729-732.

Sandmann, G. and Böger, P. (1987a) Herbicides affecting plant pigments. *Proc. Brit. Crop Prot. Conf. (Weeds)*: 139-148.

Sandmann, G. and Böger, P. (1987b) Interconversion of prenyl pyrophosphates and subsequent reactions in the presence of FMC 57020. *Z. Naturforsch.*, **42c**: 803-807.

Sandmann, G. and Böger, P. (1989) Inhibition of carotenoid biosynthesis by herbicides. In: Böger, P. and Sandmann, G., (eds.), *Target Sites of Herbicide Action.* CRC Press, Boca Raton, FL, pp. 25-44.

Sandmann, G. and Bramley, P.M. (1985) Carotenoid biosynthesis by *Aphanocapsa* homogenates coupled to a phytoene-generating system from *Phycomyces blakesleeanus. Planta*, **164**: 259-263.

Sandmann, G. and Kowalcyck, S. (1989) *In vitro* carotenogenesis and characterisation of the phytoene desaturase reaction in *Anacystis. Biochem. Biophys. Res. Commun.*, **163**: 916-921.

Sandmann, G., Kunert, K.J. and Böger, P. (1979) Biological systems to assay herbicidal bleaching. *Z. Naturforsch.*, **34c**: 1044-1046.

Sandmann, G., Bramley, P.M. and Böger, P. (1980) The inhibitory mode of action of the pyridazinone herbicide norflurazon on a cell-free carotenogenic system. *Pestic. Biochem. Physiol.*, **14**:185-191.

Sandmann, G., Kunert, K.J. and Böger, P. (1981) Bleaching activity and chemical constitution of phenylpyridazinones. *Pestic. Biochem. Physiol.*, **15**:288-293.

Sandmann, G., Clarke, I.E., Bramley, P.M. and Böger, P. (1984) Inhibition of phytoene desaturase – the mode of action of certain bleaching herbicides. *Z. Naturforsch.*, **39c**:443-449.

Sandmann, G., Bramley, P.M. and Böger, P. (1985) New herbicidal inhibitors of carotene biosynthesis. *J. Pestic. Sci.*, **10**:19-24.

Sandmann, G., Linden, H. and Böger, P. (1989) Enzyme kinetic studies on the interaction of norflurazon with phytoene desaturase. *Z. Naturforsch.*, **44c**:787-790.

Sandmann, G., Ward, C.E., Lo, W.C. *et al.* (1990a) Bleaching herbicide flurtamone interferes with phytoene desaturase. *Plant Physiol.*, **94**:476-478.

Sandmann, G., Böger, P. and Kumita, I. (1990b) Atypical inhibition of phytoene desaturation by 2-(4-chloro-2-nitrobenzoyl)-5,5-dimethyl-cyclohexane-1,3-dione. *Pestic. Sci.*, **30**:353-355.

Schiff, J.A., Cunningham, F.X. and Green, M.S. (1982) Carotenoids in relation to chloroplasts and other organelles. In: Britton, G. and Goodwin, T.W. (eds.), *Carotenoid Chemistry and Biochemistry.* Pergamon Press, Oxford, pp. 329-338.

Schmidhauser, T.J., Lauter, F.R., Russo, V.E.A. and Yanofsky, C. (1990) Cloning, sequencing and photoregulation of *al-1*, a carotenoid biosynthetic gene of *N. crassa. Mol. Cell Biol.*, **10**:5064-5070.

Schmidt, K. and Liaaen-Jensen, S. (1973) Bacterial carotenoids. XLII. New keto carotenoids from *Rhodopseudomonas globiformis. Acta Chem. Scand.*, **27**:3040-3052.

Schmidt, A. and Sandmann, G. (1990) Cloning and nucleotide sequence of the *crtI* gene encoding phytoene dehydrogenase from the cyanobacterium *Aphanocapsa* PCC 6714. *Gene*, **91**:113-117.

Seyama, N. and Splittstoesser, W.E. (1975) Pigment synthesis in *Cucurbita moschata* cotyledons as influenced by CPTA and several inhibitors. *Plant Cell Physiol.*, **16**: 13-19.

Singh, R.K., Ben-Aziz, A., Britton, G. and Goodwin, T.W. (1973) Biosynthesis of spheroidene and hydroxyspheroidene in *Rhodopseudomonas sphaeroides*: Experiments with nicotine as inhibitor. *Biochem. J.*, **132**:649-652.

Soeda, T. and Uchida, T. (1987) Inhibition of pigment synthesis by 1,3- dimethyl-4-(2,4-dichlorobenzoyl)-5-hydroxypyrazole, norflurazon and new herbicidal compounds in radish and flatsedge plants. *Pestic. Biochem. Physiol.*, **29**:35-42.

Spurgeon, S.L. and Porter, J.W. (1983) The biosynthesis of carotenoids. In: Porter, J.W. and Spurgeon, S.L. (eds.), *Biosynthesis of Isoprenoids*, Vol. 2. Wiley, New York, pp. 1-122.

St John, J.B. (1976) Manipulation of galactolipid fatty acid composition with substituted pyridazinones. *Plant Physiol.*, **57**:38-40.

St John, J.B. (1985) Action of R-40244 on chloroplast pigments and polar lipids. *Pestic. Biochem. Physiol.*, **23**:13-18.

St John, J.B., Schirmer, U., Ritting, F.R. and Bleiholder, H. (1984) Substituted pyridazinone herbicides: Structural requirements for action on membrane lipids. In: Magee, P.S., Kohn, G.K. and Menn, J.J. (eds.), *Pesticide Synthesis Through Rational Approaches*. ACS Symposium Series 255. American Chemical Society, Washington DC, pp. 145-162.

Strang, R.H. and Rogers, R.L. (1974) Behavior and fate of two phenylpyridazinone herbicides in cotton, corn and soybean. *J. Agr. Food Chem.*, **22**:1119-1125.

Tantawy, M.M. and Grimme, L.H. (1982) The bleaching induced by the phenylpyridazinone herbicide metflurazon in *Chlorella* is a metabolic and reversible process. *Pestic. Biochem. Physiol.*, **18**: 304-314.

Urbach, D., Suchanka, M. and Urbach, W. (1976) Effect of substituted pyridazinone herbicides and of difunone (EMD-IT 5914) on carotenoid biosynthesis in green algae. *Z. Naturforsch.*, **31c**: 652- 655.

Vaisberg, A.J. and Schiff, J.A. (1976) Events surrounding the early development of *Euglena* chloroplasts. 7. Inhibition of carotenoid biosynthesis by the herbicide SAN 9789 and its developmental consequences. *Plant Physiol.*, **57**: 260-269.

Valadon, L.R.G. and Mummery, R.S. (1978) Effects of two triethylamines on the carotenogenesis of Turkish lemons and oranges. *Z. Pflanzenphysiol.*, **90**: 11-19.

Wightman, P. and Haynes, C. (1985) The mode of action and basis of selectivity of diflufenican in wheat, barley and selected weed species. *Proc. Brit. Crop Prot. Conf. (Weeds)*: 171-178.

Wilkinson, R.E. (1985) Carotenogenic inhibition by norflurazon in wheat. *Pestic. Biochem. Physiol.*, **23**: 370-377.

Wilkinson, R.E. (1987a) Reversal of norflurazon carotenogenesis inhibition by isomers. *Pestic. Biochem. Physiol.*, **28**: 381-388.

Wilkinson, R.E. (1987b) 2-Chloro-*N,N*-di-2-propylene-acetamide reversal of carotenogenic inhibition by low concentrations of norflurazon. *Pestic. Biochem. Physiol.*, **29**: 146-151.

Wright, B.J., Baillie, A.C., Wright, K. *et al.* (1980) Synthesis of potential herbicides designed to uncouple photophosphorylation. *Phytochemistry*, **19**: 61-65.

Yokoyama, H., De Benedict, C., Coggins, C.W. and Henning, G.L. (1972) Induced color changes in grapefruit and orange. *Phytochemistry*, **11**: 1721-1724.

Yokoyama, H., Hsu, W.J., Hayman, E. and Poling, S. (1984) Bioregulation of plant constituents. *Recent Adv. Phytochemistry*, **18**: 231-250.

Young, A.J. (1991) Inhibition of carotenoid biosynthesis. In: Baker, N.R. and Percival, M.P. (eds.), *Herbicides, Topics in Photosynthesis*, Vol. 10. Elsevier, London, pp. 130-171

Young, A.J., Britton, G. and Musker, D. (1989) A rapid method for the analysis of the mode of action of bleaching herbicides. *Pestic. Biochem. Physiol.*, **35**: 244-250.

Zöbelin, V. (1985) Inhibitorstudien an zellfreier Carotin biosynthese. Thesis, Universität Konstanz.

6

Factors that affect the carotenoid composition of higher plants and algae

A.J. Young

6.1 CAROTENOIDS IN HIGHER PLANTS

The photosynthetic tissues of higher plants contain a range of pigments, including chlorophylls *a* and *b* and a number of carotenoids (carotenes and xanthophylls). These pigments are bound to proteins in the pigment–protein complexes of the thylakoid membrane (see Section 6.1.1), where they have a direct and essential role in photosynthesis (see Chapters 3 and 8). Small amounts of carotenoids may also be located in the plastid envelope (see Chapter 3), perhaps reflecting a role for this membrane in carotenoid biosynthesis. Under certain conditions (for example, reflecting the development of leaves), some carotenoids (often in high amounts) may be found in oil droplets or osmiophilic globules either inside or outside the plastid (e.g. the synthesis of the *retro*-carotenoid rhodoxanthin in bronzed leaves of *Thuja* spp.). As these carotenoids have no apparent function in photosynthesis or in photoprotection they will not be considered. Thus, except where stated, this chapter refers to the function and fate of those carotenoids which are specifically bound to the photosynthetic pigment–protein complexes, and probably have an active role in photosysnthesis.

6.1.1 Plastid development

The carotenoid composition of the photosynthetic tissues of higher plants is very highly conserved (see Chapter 2), especially in comparison with other photosynthetic organisms, including the algae. ß-Carotene, often with α-carotene (Goodwin 1980), is found in the reaction centres from all higher plants examined to date. A number of xanthophylls are generally found in the light-harvesting complexes of higher plants (a complete listing is given in Chapter 2) and a description of the carotenoid composition of the individual complexes is given in Chapter 3. Lutein is the predominant xanthophyll, with high levels of neoxanthin and the xanthophyll cycle carotenoids, violaxanthin, antheraxanthin and zeaxanthin. Smaller amounts of lutein-5,6-epoxide are frequently found in thylakoid preparations. Whilst the overall carotenoid composition does not vary greatly from one species to another, the relative levels of these carotenoids in leaf tissues can be affected by a number of environmental and developmental factors, for example nutrient deprivation and leaf (needle) age. The pigment content of needles of *Picea abies* (Norway spruce) and other trees has been shown to be highly dependent on the age class of the needle, with younger needles generally containing much lower levels of pigments. The ratio of total carotenoid:chlorophyll in these different age classes has, however, been shown to be constant (Lange et al. 1987).

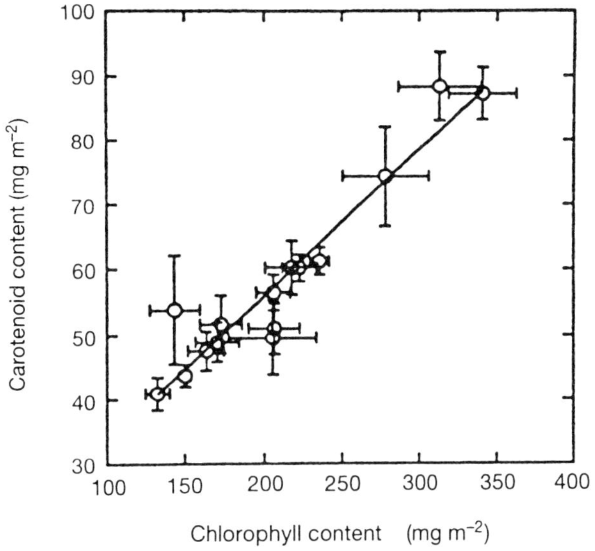

Figure 6.1 Relationship between chlorophyll and total carotenoid content in the leaves of higher plants, expressed on a unit area basis (G. Johnson, A. Young and P. Horton, unpublished data). Each point represents an individual plant species, from either sun or shade populations.

The most important factor in determining the levels of the main carotenoids in photosynthetic tissues of higher plants is the unit size of the photosynthetic apparatus (see Figs. 6.1 and 6.2). It is important, however, to recognise that different parts of a leaf, and leaves or needles in different areas of a tree canopy, can experience gradients in light quality and quantity which may, in the longer term, affect photosystem stoichiometry and the balance of photosynthetic pigments.

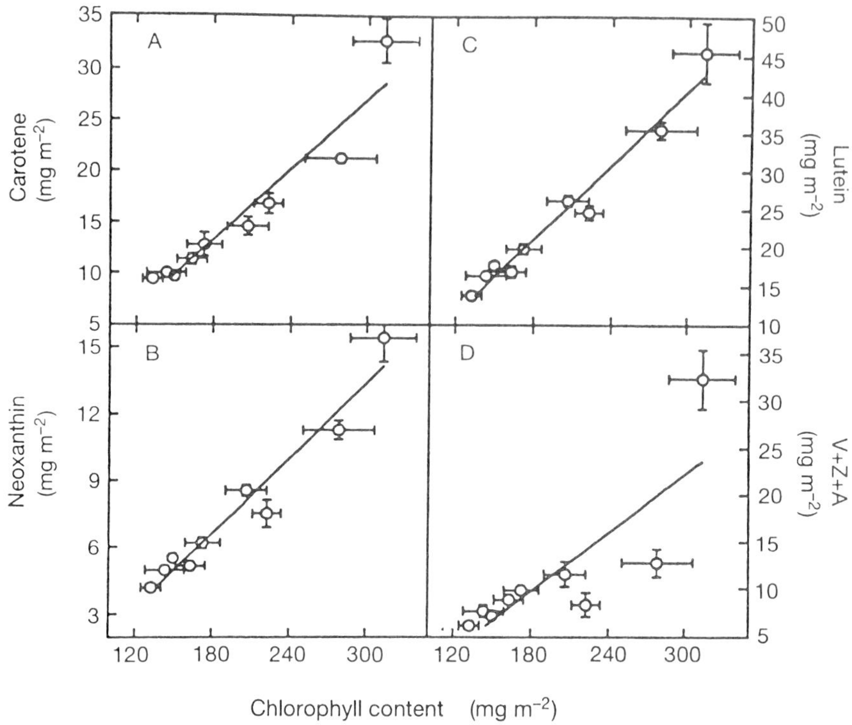

Figure 6.2 Relationship between chlorophyll and individual carotenoid contents in leaves of higher plants, expressed on a unit area basis. A: total carotene (α-carotene + ß-carotene); B: neoxanthin; C: lutein; D: xanthophyll cycle carotenoids (violaxanthin + antheraxanthin + zeaxanthin). (G. Johnson, P. Horton and A. Young, unpublished data). Each point represents an individual plant species, from either sun or shade populations.

(a)Etiolation and greening

The etioplasts of angiosperms are characterised by being devoid of chlorophyll and having a much reduced level of total carotenoid. In particular, the proportion of ß-carotene found in etioplasts is much lower than that found in fully developed chloroplasts (Bahl 1977). The carotenoids are generally thought to be located in the prolamellar bodies within the etioplasts (Sandelius et al. 1981).

The size and composition of the carotenoid pool in etiolated seedlings or in isolated etioplasts has been determined for a number of plant species including barley (Lichtenthaler 1969a; Grumbach and Lichtenthaler 1975; Kleugden and Grumbach 1983; Barry et al. 1991), milo (Malhotra et al. 1982; Oelmüller and Mohr 1985), mustard (Lütke-Brinkhaus and Kleinig 1987) and radish (Buschmann et al. 1980). Table 6.1 shows the carotenoid composition of etiolated seedlings of barley (Barry et al. 1991).

A number of common conclusions about the carotenoid content of etiolated seedlings can be drawn from the above work:

Table 6.1 The carotenoid compositions of etiolated, greened and light-grown seedlings of *Hordeum vulgare*. The values are expressed as a percentage of total carotenoid and reflect changes in the tip of the first leaf. Considerable variation can be seen in different regions of the leaf as a result of greening (Barry et al. 1991)

Carotenoid	*Etiolated*	*Greened 48hrs*[1]	*Greened 24hrs*[2]	*Light-grown*[2]
Neoxanthin	3.6	8.9	7.6	12.4
Violaxanthin	35.7	29.7	25.8	20.1
Antheraxanthin	11.7	1.6	5.2	1.1
Lutein	44.4	41.7	37.8	38.0
Zeaxanthin	1.3	—	3.0	—
ß-Carotene	3.2	18.2	20.6	28.4
Ratio xanthophyll: ß-carotene	22.5:1	—	3.85:1	2.52:1

[1] 15–25 $\mu mol\ m^{-2}\ s^{-1}$
[2] 150 $\mu mol\ m^{-2}\ s^{-1}$

(i) The level of ß-carotene is low and the ratio of total xanthophyll to ß-carotene ranges from approx. 10:1 to at least 20:1 (in contrast to light-grown tissues which usually have a ratio of 2–3:1).

(ii) The predominant xanthophylls are violaxanthin, lutein and antheraxanthin (levels of neoxanthin are generally very low).

(iii) The amount of carotenoid varies considerably, not only with the age of the tissue but also with the species studied.

The greening of dark-grown seedlings has been used for many years in order to study the assembly of the chloroplast thylakoid membrane. When dark-grown seedlings are illuminated, the etioplasts present develop into fully functional chloroplasts. This sequence of development involves the breakdown of the prolamellar bodies, with concomitant growth of the prothylakoids to form thylakoids and eventually grana stacks (Henningsen and Boynton 1970). During this process the pigment–protein complexes are assembled and the pigments accumulated (Shimada et al. 1990). Although carotenoids are clearly synthesised in the dark, the rate of synthesis is greatly accelerated in the presence of light. Lichtenthaler (1975) and Lichtenthaler and Becker (1975) demonstrated that over a period of 36 h illumination the rate of synthesis was five to six times greater than that seen in seedlings retained in the dark. Grumbach and Lichtenthaler (1975) demonstrated an 18-fold increase in ß-carotene level, a 9-fold increase in neoxanthin level and a 3-fold increase in the level of lutein. There was, however, little overall change in the level of violaxanthin in those plants transferred from dark to light. In all plants examined to date, the largest increase in carotenoid levels in greening tissues is seen for ß-carotene (Fig. 6.3).

The presence of zeaxanthin has been reported in etiolated seedlings of barley

(Table 6.1; Barry et al. 1991). Its subsequent presence in greened tissues is largely dependent on the light intensity to which the plants are exposed. Thus, zeaxanthin may be rapidly lost when plants are greened at low light, but significant increases may be seen when the plants are greened at much higher light intensities. This latter observation could be the result of one or more processes: (i) direct synthesis of zeaxanthin *de novo*, (ii) induction of the xanthophyll cycle and the de-epoxidation of violaxanthin (see Chapter 7), and (iii) transformation from the large antheraxanthin pool. A combination of these is likely.

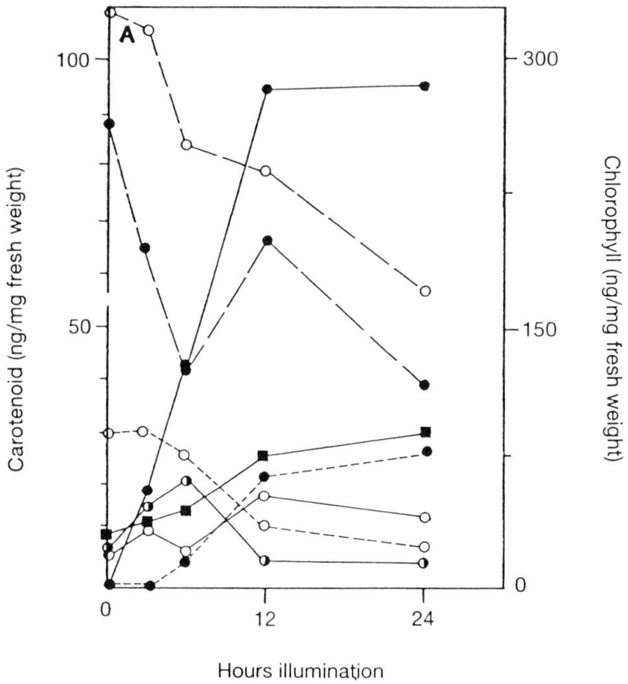

Figure 6.3 Changes in the carotenoid composition of the tips of etiolated seedlings of *Hordeum vulgare* (barley) greened in white light (150 μmol m^{-2} s^{-1}). Chlorophyll *a* (●–●), chlorophyll *b* (●- - -●), ß-carotene (■——■), neoxanthin (○——○), lutein (○– –○), violaxanthin (●——●), antheraxanthin (○- - - - -○), zeaxanthin (○——○). The middle and basal regions of the developing leaves accumulated pigments at a faster rate, although the overall pattern was similar (Barry et al. 1990).

Axelsson et al. (1981) reported that a flash of light could induce chlorophyll synthesis in etiolated wheat leaves that had been pre-treated with norflurazon, which inhibits the biosynthesis of the normal cyclic carotenoids.

It is outside the scope of this chapter to consider the carotenoid content of chromoplasts which can be formed as the next stage in development of the plastid, as is the case, for example, in many fruit (see Goodwin 1980).

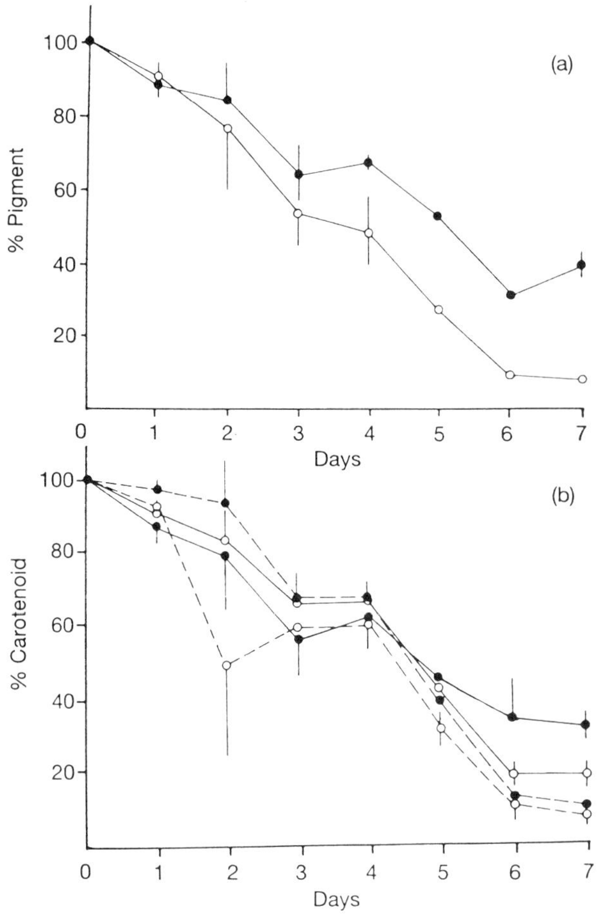

Figure 6.4 Changes in the levels of (a): total chlorophyll (○—○), and total carotenoid (●—●), and, (b) levels of the individual carotenoids : (neoxanthin (○--○), violaxanthin (●--●), lutein (○—○) and ß-carotene (●—●)) during the senescence of leaves from seedlings of *Hordeum vulgare* (Young et al. 1991).

(b) Senescence

Senescence of whole plants or leaves is a natural process that occurs in all plants as part of the normal developmental sequence. Senescence may also be considered to be one of the most likely consequences of severe and prolonged stress. Among the most obvious symptoms of senescence is the bleaching of leaves, especially during the later stages of tissue degradation. In many deciduous trees a yellowing of leaves is often observed prior to leaf fall in the autumn months. This change in appearance of the leaves is usually interpreted as being the result of the rapid degradation of the chlorophylls to expose the carotenoid pigments, which appear to be less susceptible to damage in this particular case. Although the precise nature of the oxidising species involved is not known (Halliwell 1991), peroxidase activity has been suggested to be involved in the degradation of pigments during senescence.

The pattern of pigment loss in senescent leaves of barley and wheat has been examined by Young et al. (1991). In leaves of these plants, the expected patterns of pigment loss were observed, i.e. the carotenoids were lost at a slower rate than were the chlorophylls. Chlorophyll *a* was slightly more susceptible to these conditions than was chlorophyll *b*, resulting in a reduction in the ratio of chlorophyll *a:b*. Similar patterns have been observed in leaves of *Festuca pratensis*, where the gradual yellowing of the leaves was accompanied by a severe decrease in the ratio of chlorophyll *a:b* (Gut et al. 1987). Gut and co-workers (1987) also observed that ß-carotene was the least sensitive of all the chloroplast carotenoids to the destructive events associated with senescence. This pattern of pigment loss is quite different from that seen during photooxidation (brought about by exposure to high light, or herbicide action). In these examples, ß-carotene is generally the most rapidly destroyed of all the chloroplast pigments (see Section 6.2; Young and Britton 1990a) and the carotenoids are lost at a much faster rate than are the chlorophylls.

The pattern of carotenoid loss during senescence may, however, differ between species. Young et al. (1991) have reported that the carotenoids in senescent leaves of wheat were degraded in a pattern similar to that seen in *F. pratensis* (Gut et al. 1987), i.e. ß-carotene > neoxanthin > lutein > violaxanthin. In contrast, the degradation of the carotenoids in senescent leaves of barley followed the pattern, neoxanthin > violaxanthin > lutein > ß-carotene (Fig. 6.4). One of the more obvious features of senescence, the formation of xanthophyll acyl esters (discussed in detail in Section 6.2.5), does not occur in the senescent leaves of all species.

The carotenoid composition of thylakoids obtained from highly senescent leaves of wheat and barley was identical with that seen in control tissues, i.e. there was no evidence of any specific pigment destruction in any complex. This indicates that the difference in the susceptibility of ß-carotene seen in these species may reflect differences in the behaviour of the carotenoids once they have been released into the cytosol. Once they have been released from the pigment–protein complexes, some of the pigments may become incorporated into plastoglobuli, which may serve to increase their relative stability with respect to oxidation (see Section 6.2.5).

6.1.2 Sun /shade adaptation

The photosynthetic apparatus of higher plants is influenced directly by the intensity of light available during development and growth. The pigment composition, structure and function of the plastid may be affected by the light environment. In general, chloroplasts from plants grown in high light have fewer thylakoid membranes and smaller grana stacks than plants grown under low light (Lichtenthaler et al. 1982a,b). Photophosphorylation activity, photosynthetic rates and photochemical capacity of photosystem (PS) I and PSII are reduced in plants maintained in low light conditions (Chow and Anderson 1987; El Sebba et al. 1987; De la Torre and Burkey 1990).

A number of studies have examined differences in the pigment content and composition of plants grown under low or high light, or changes upon the transfer of plants from one light environment to another (e.g. Grumbach and Lichtenthaler 1982; Lichtenthaler et al. 1981, 1982a,b, 1983). The data shown in Figs. 6.1 and 6.2 show that, irrespective of growth in either sun or shade conditions, the overall carotenoid composition of higher plant leaves is largely dependent on the unit size of the photosynthetic apparatus (Johnson et al. 1992; G. Johnson, A. Young and P. Horton, unpublished data). The level of chlorophyll, expressed on a unit area basis, shows considerable variability in response to growth habitat. Plants from a high light environment generally have a higher level of chlorophyll and a higher ratio of chlorophyll *a:b* than those from low light (see also De la Torre and Burkey 1987, 1990). Anderson et al. (1988) observed that plants native to shaded habitats have lower ratios of chlorophyll *a:b* than those grown in full sun. Increases in this ratio reflect less light-harvesting complex (LHC) II and more core chlorophyll *a* proteins in PSI and PSII.

Some early reports (e.g. Lichtenthaler et al. 1982b, 1983) recorded changes in the levels of the carotenes and of the xanthophylls in response to light intensity, so that the ratio of total xanthophylls:total carotene ((neoxanthin + lutein + xanthophyll cycle carotenoids): (α- + ß-carotene)) increased in plants grown in low light, i.e. sun leaves were found to contain much lower levels of xanthophylls than shade leaves. An increase in the amounts of xanthophyll and chlorophyll *b* in shade plants would improve the efficiency for light capture, and the increased grana stacking would help to maintain efficiency in low irradiances by reducing spillover from PSII to PSI. Anderson et al. (1988) also showed that the ratio of xanthophyll:ß-carotene is higher in shade than in sun chloroplasts.

These studies did not report any alterations to the ratio of α-carotene:ß-carotene (see below), although the thin layer chromatographic separation techniques used may not have been effective in resolving these two carotenes. Conflicting data (see below and Chapter 7) suggest that these observations, which were restricted to only a limited number of species, do not follow for all plants.

More recent reports have shown that shade plants generally have a much lower ratio of xanthophyll:carotene than do sun plants (Grumbach and Lichtenthaler 1982; De la Torre and Burkey 1990; Thayer and Björkman 1990; Demmig-Adams and Adams 1992; G. Johnson, A. Young and P. Horton, unpublished data). These alterations in xanthophyll content in sun leaves compared to shade leaves can nearly all be accounted for by increases in the levels of the xanthophyll cycle carotenoids, violaxanthin, antheraxanthin and zeaxanthin; levels of lutein also increase slightly in a high light environment, but neoxanthin levels were found not to vary greatly (Thayer and Björkman 1990). Thayer and Björkman (1990) observed that the ratio of lutein:chlorophyll *b* was consistently higher in sun than in shade leaves and that the ratio of neoxanthin:chlorophyll *b* remained fairly constant in all light regimes. In addition to these changes in the xanthophyll content, the levels of ß-carotene are generally higher in sun leaves. More recently G. Johnson, A. Young and P. Horton (unpublished data) have shown a clear inverse relationship between the lutein and zeaxanthin content of leaves from sun and shade plants (data not shown); thus, leaves from shade-loving plants were found to contain higher levels of lutein than did sun plants. These results would suggest that these two structurally very similar carotenoids (lutein has one ß- and one ε-ring, compared to the two ß-rings of zeaxanthin) have the ability to occupy similar binding sites in the pigment–protein complexes. When taken in conjunction with the changes in the α-carotene levels reported by Thayer and Björkman (1990), these results strengthen the suggestion put forward by Thayer and Björkman that the level of incident light may exert some control on carotenoid biosynthesis at the point of cyclisation of lycopene (see Chapter 4).

Table 6. 2 The α-carotene content of sun and shade leaves of higher plants (Thayer and Bjørkmann 1989). The α-carotene content is expressed as a percentage of total carotene

Species	*α-Carotene*	*Species*	*α-Carotene*	
Shade only			*Shade*	*Sun*
Asarum	40	*Arbutus*	<1	<1
Calycanthus	49	*Citrus*	29	15
Monstera	50	*Cyrtomium*	36	2
Spathiphyllum	76	*Gossypium*	<4	<1
		Hibiscus	16	0.7
		Toxicodendron	17	4
Sun only				
Ailanthus	<1			
Cucurbita	<1			
Gaura	<1			
Helianthus	<1			
Lycopersicon	<1			
Phytolacca	<1			
Xanthium	<1			

The pigment content and composition of leaves from a number of sun- and shade-adapted species of higher plants has been examined by Thayer and Björkmann (1990). The latter authors observed that, in addition to major changes in the xanthophyll cycle pigments (see Chapter 7), the ratio of α-carotene:ß-carotene differed in sun and shade leaves. α-Carotene levels were found to be much higher in shade plants grown in the shade or in the sun. In contrast, little or no α-carotene was found in sun leaves. Similarly, those plants containing high levels of α-carotene were shown to have a much higher ratio of α-carotene:ß-carotene when grown in low light conditions (Table 6.2). Figure. 6.5 shows a fairly good correlation between light intensity and the α-carotene content of leaves of *Hibiscus tiliaceus*. Grill and Pfeifhöfer (1985) showed that needles of *Picea abies* from the shade had a much higher proportion of total carotene as α-carotene than did sun-grown needles. Similar results have been obtained for a number of bromeliads (e.g. *Guzmania* spp.)

when grown under low light (approx. 100 mmol m^{-2} s^{-1}) or high light (approx. 400 mmol m^{-2} s^{-1}; K. Maxwell, H. Griffiths and A. Young, unpublished data). Not all shade-loving or shade-tolerant plants show the ability to accumulate α-carotene in their leaves. In a recent examination of more than 15 sun and shade species G. Johnson, A. Young and P. Horton (unpublished data) found that only two accumulated α-carotene in low light.

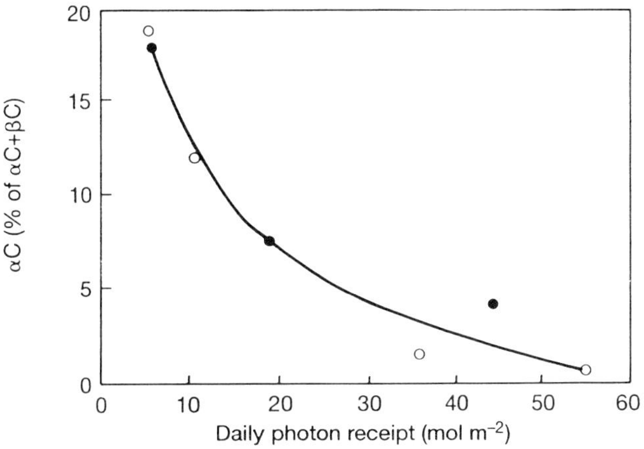

Figure 6.5 Effect of light regime on the α-carotene content of leaves of *Hibiscus tiliaceus*. Closed and open symbols represent artificial and natural light regimes, respectively. The total carotene content (α-carotene + ß-carotene) of the leaves was 87 ± 8 mmol mol^{-1} chlorophyll *a* + *b* (Thayer and Björkman 1990).

Young and Britton (1989) showed that in leaves containing high levels of α-carotene there was a constant ratio of these two hydrocarbons in all the pigment–protein complexes where they were detected, suggesting similar functional roles for α-carotene and ß-carotene. The accumulation of α-carotene and ß-carotene in shade leaves may therefore have little or no functional significance and may simply reflect the contents of lutein and the other xanthophylls, the levels of which also vary considerably in these plants (G. Johnson, P. Horton and A. Young, unpublished data).

6.2 PHOTOBLEACHING PROCESSES

The general photoprotective properties of carotenoids in photosynthetic systems (quenching of triplet state chlorophyll or bacteriochlorophyll molecules and scavenging for singlet oxygen) are discussed in detail in Chapter 8. A number of reviews have been published in recent years which have also covered this topic very well: Siefermann-Harms 1985, 1987; Cogdell 1988; Cogdell and Frank 1988; Frank et al. 1991; Truscott 1990; Koyama 1991.

Whilst the light-harvesting carried out by the carotenoids of the antenna complexes is a useful process, the protection of the photosynthetic apparatus by carotenoids is recognised as being essential to the survival of the plant. This is particularly well illustrated by the behaviour of various mutants of higher plants and microorganisms which are deficient in carotenoids. In the absence of their normal complement of chloroplast carotenoids, these organisms are readily subject to photooxidative damage, in the presence of light and oxygen. Such damage can be serious enough to kill the organism (Krinsky 1979). Similar photooxidative damage can also be seen when inhibitors of carotenoid biosynthesis are used on photosynthetic organisms, preventing the formation of their normal carotenoids and resulting in the accumulation of acyclic presursors such as phytoene (see Chapter 5), which are much less effective in photoprotection.

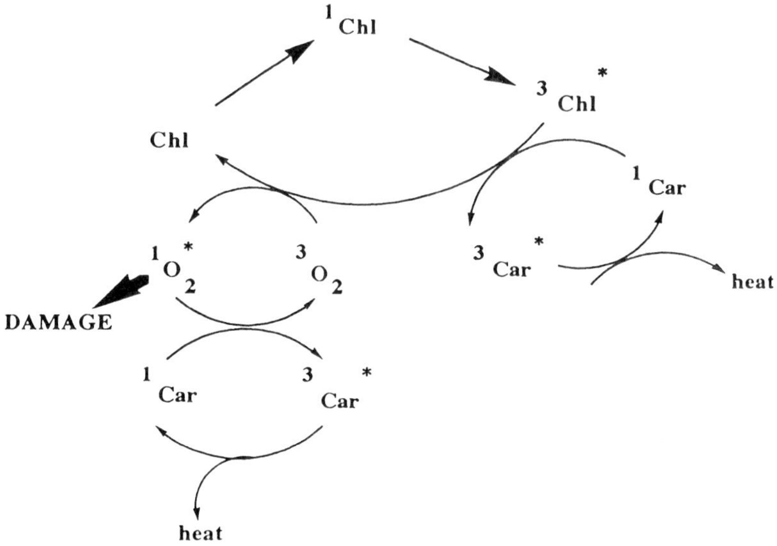

Figure 6.6 A schematic representation of the photoprotective role of carotenoids in photosynthetic tissues.

The mechanisms whereby carotenoids are able to protect the photosynthetic apparatus from oxidative damage are described in detail in Chapter 8 and can be summarised by the information contained in Fig. 6.6. The ability of carotenoids to photoprotect, both by the quenching of triplet states and by the scavenging of $^1O_2$ and other reactive oxygen species is governed largely by the number of conjugated double bonds in the carotenoid chromophore (see Fig. 6.7). Thus, carotenoids that possess nine or more conjugated double bonds are effective quenchers of $^3Chl^*$ and $^1O_2$. Carotenoids can be very effective in protecting chlorophyll from photobleaching in solution (Fig. 6.7). A precise structural arrangement between the carotenoid molecule and either the $^3Chl^*$ or $^1O_2$ molecule is also essential for rapid and efficient quenching (see Chapter 8 for a full discussion).

Whilst some precursors that accumulate as the result of inhibitor action (see Chapter 5), for example lycopene, have a sufficient degree of conjugation to enable them to scavenge for $^1O_2$ *in vitro*, the lack of the normal cyclic carotenoids may prevent the precise structural assembly of the complexes required for this process. The precursors are probably not accumulated in the complexes to any great extent and are therefore not in the correct site to quench either ^{3}Chl* or $^1O_2$. As a result of such ineffective photoprotection, illumination by even moderate light intensities of photosynthetic organisms treated with inhibitors of carotenoid biosynthesis induces rapid bleaching of chlorophyll and other cellular components arising from $^1O_2$ production via ^{3}Chl*.

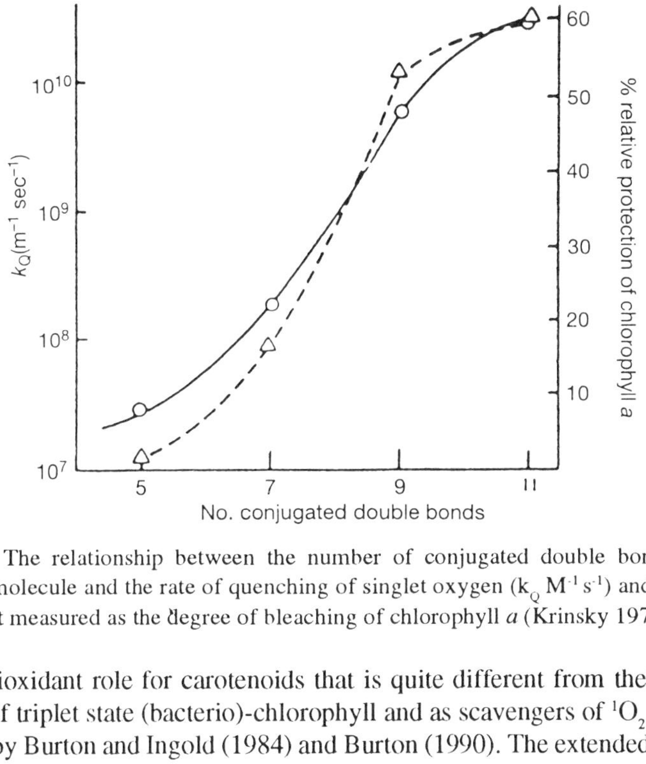

Figure 6.7 The relationship between the number of conjugated double bonds in the carotenoid molecule and the rate of quenching of singlet oxygen (k_Q $M^{-1}\ s^{-1}$) and ability to photoprotect measured as the degree of bleaching of chlorophyll *a* (Krinsky 1979).

An antioxidant role for carotenoids that is quite different from their role as quencher of triplet state (bacterio)-chlorophyll and as scavengers of $^1O_2$ has been identified by Burton and Ingold (1984) and Burton (1990). The extended series of conjugated double bonds seen in the structure of the ß-carotene molecule and in many other carotenoids makes them susceptible to attack by peroxy radicals. In the presence of oxygen a new chain-carrying peroxy radical species of ß-carotene is formed (reactions 1 and 2). ß-Carotene was found readily to undergo autooxidation, which was found to be highly dependent on the partial pressure of oxygen. Thus, ß-carotene would be much more effective at preventing the oxidation of cellular components at low partial pressures of oxygen and may indeed complement the role of vitamin E, which is effective at high oxygen pressures (Burton 1990). This would therefore be of greatest benefit in the conditions found in many mammalian cells where low oxygen concentrations prevail, but the effectiveness of this chain-breaking property in photosynthetic tissues is not known. The carbon-centred ß-carotene radical produced in reaction 6.1 would be highly reactive and would lead to the formation of other radical species.

$$\beta - Carotene + ROO^{\bullet} \rightarrow \beta - Car^{\bullet} \quad (6.1)$$

$$\beta - Car^{\bullet} + O_2 \leftrightarrow \beta - CarOO^{\bullet} \quad (6.2)$$

It has also been proposed that the ß-carotene radical (ß-car·) can be removed through interaction with another peroxyl radical (reaction 6.3) (Burton and Ingold 1984). This termination reaction is thought to proceed very rapidly and efficiently. The overall effect of these reactions is to prevent a series of reactions involving lipid peroxidation by promoting what would be much less serious reactions with the more expendable carotenoid molecules (Burton 1990). Importantly, from the point of view of photosynthetic systems where a complex mixture of carotenoids may be present, these unusual properties of ß-carotene are also seen in other carotenoid molecules which possess a series of conjugated double bonds.

$$\beta - Car^{\bullet} + ROO^{\bullet} \rightarrow inactive \quad products \quad (6.3)$$

It is clear that, *in vitro*, carotenoids have the ability to scavenge for $^1O_2$ and other highly reactive and potentially lethal oxygen species. The main photoprotective role of these pigments in photosynthetic systems is undoubtedly through the direct quenching of the triplet-state sensitiser (chlorophyll or bacteriochlorophyll; Borland et al. 1989). It has also been suggested that were the role of the carotenoids *in vivo* to be restricted simply to the scavenging of $^1O_2$ then a significant proportion of (bacterio-) chlorophyll molecules would be irreversibly photooxidised. Furthermore, Cogdell and Frank (1988) suggested that only partial photoprotection would be achieved *in vivo*, as the efficiency of the scavenging would depend on the proximity of molecular oxygen to carotenoid, and the carotenoid would be competing for the $^1O_2$ energy with other substrates. Direct quenching of the triplet-state chlorophyll molecule, thus preventing the generation of $^1O_2$ in the first place, provides a much more efficient and effective protective strategy (see Chapter 8). The ability of membrane-bound carotenoids to scavenge for exogenous sources of $^1O_2$ and free radicals, such as those produced in the presence of atmospheric pollutants, must be severely limited. On their own, carotenoids are not sufficient to protect the photosynthetic apparatus (Siefermann-Harms 1990), and should be considered as important, and probably essential, contributors to the photoprotection of the photosynthetic apparatus alongside other antioxidants such as ascorbate, a-tocopherol and superoxide dismutase.

The bleaching of photosynthetic tissues is one of the more obvious symptoms of photooxidative damage to the chloroplast. The following sections describe the fate of carotenoids under conditions that lead to such bleaching. In some of these examples the carotenoids may form part of a "first line of defence" against oxidation (e.g. photoinhibition), in others (e.g. in the presence of atmospheric pollutants) they may play no direct protective role and may be destroyed along with the other components of the photosynthetic apparatus.

The photoprotective role afforded by carotenoids in photosynthetic systems is highly complex and may be based on a number of different processes. Although the mechanistics of many of these processes are not fully understood, it is quite clear that carotenoids play an essential role in protecting photosynthetic systems from photooxidation.

6.2.1 Photoinhibition

Under optimum conditions the carotenoids associated with the photosynthetic apparatus serve to protect the chloroplast from photooxidative damage through the processes described above (see also Chapters 7 and 8). In many cases these protective systems become overloaded, resulting in damage to the components of the chloroplast. Such photooxidative damage may occur when leaves, or isolated chloroplasts, are exposed to high light intensities in aerobic conditions, especially when they are exposed to some additional form of stress (e.g. low temperatures).

Exposure to high light intensities may initially induce some subtle changes in the carotenoid content and composition of photosynthetic tissues. These changes may involve the photoinduced isomerisation of certain carotenoids and the induction of the de-epoxidation part of the xanthophyll cycle, resulting in alterations to the levels of violaxanthin, antheraxanthin and zeaxanthin (see Chapter 7). Prolonged illumination of leaf tissues, chloroplasts or thylakoids, with high light intensities and under aerobic conditions, will generally result in the rapid bleaching of pigments, accompanied by the degradation of other cellular components.

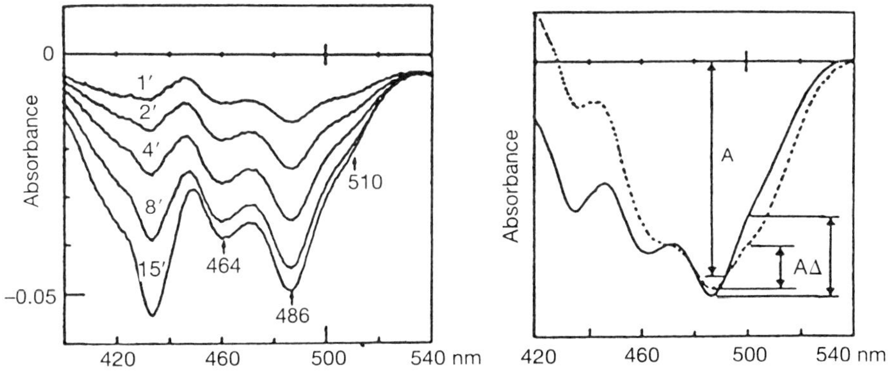

Figure 6.8 Irreversible absorption changes (light–dark) in isolated PSII reaction centres illuminated with 600 μmol $m^{-2} s^{-1}$ red light in the presence of 100μM DMBIB (Telfer et al. 1992).

Barber and colleagues have examined some of the events associated with photoinhibition of the isolated PSII reaction centre complex. This complex binds two ß-carotene molecules, two phaeophytin *a*, six chlorophyll and one cytochrome *b*559 molecule (see Chapter 3). The isolated PSII reaction centre complex is highly vulnerable to damage when illuminated under aerobic conditions in the absence of additional electron acceptors. Such damage is initially characterised by the bleaching of the P680 chlorophyll and conformational changes in the reaction centre (Telfer et al. 1991; Barber, 1992), but will ultimately lead to the breakdown of the D1/D2 proteins. Under these conditions, ß-carotene is not oxidised by $P680^+$ and therefore not bleached. In the presence of dibromethymoquinone (DBMIB) or another electron acceptor, exposure of the isolated complex to photoinhibitory conditions causes the oxidation and subsequent irreversible bleaching of ß-carotene within the reaction centre (Fig. 6.8; Telfer et al. 1991, 1992). If a donor such as Mn^{2+} is present, pigment bleaching occurs only under aerobic conditions. By acting as an electron donor, ß-carotene is able to prevent the accumulation of $P680^+$ and thereby protect PSII from photoinhibitory damage.

Damm et al. (1990) have studied the bleaching of pigments in both the isolated PSI reaction centre and the antenna complex, LHCI. Illumination of the isolated reaction centre induced a rapid bleaching of pigments, especially of ß-carotene

The bleaching observed as a result of photoinhibitory conditions reflects oxidative damage to these photosynthetic systems (Satoh 1970) and the photobleaching of chlorophyll in the light-harvesting complexes, for example, can be considered to be a self-sensitised type II reaction (Asada et al. 1982). Foote (1991) has recently stressed the need for careful definition of type I and type II photosensitised oxidations. Type II, with which we are concerned here, involves the reaction between an excited sensitiser (possibly a singlet, though nearly always a triplet-state) and oxygen, leading to the formation of $^1O_2$ (and in limited cases superoxide) through energy transfer. In contrast, type I involves a reaction between the sensitiser and a substrate to yield radicals or radical ions (e.g. hydroxyl radicals).

The bleaching processes observed in leaves and chloroplasts as a result of exposure to high light and many herbicides (see Section 6.2.3) are highly dependent on the presence of oxygen. The extent of bleaching can be controlled through the action of various scavengers of singlet oxygen and other active oxygen species (e.g benzoate will scavenge hydroxyl radicals). The bleaching may also be promoted by certain conditions, e.g. by the presence of several compounds which serve to extend the lifetime of $^1O_2$. Under anaerobic or reducing conditions bleaching of pigments is usually minimal, i.e. it is the combination of light and oxygen that is lethal. Thus, bleaching can be stimulated in the presence of D_2O (which can extend the lifetime of $^1O_2$ up to 10-fold) and prevented by the addition of ascorbate (Ridley 1977; Barry et al. 1990). An identical pattern of pigment loss was seen with D_2O as in the illuminated control tissues, suggesting that a similar oxidative process, probably involving $^1O_2$, is involved (A. Young and G. Britton, unpublished data).

Solvent-based mixtures of chlorophylls and carotenoids *in vitro* are generally highly unstable and rapid bleaching of these pigments occurs in the presence of light and oxygen. The stability of these pigments when bound to apoproteins in the pigment–protein complexes is greatly increased (Siefermann-Harms 1990). The sensitivity of these bound photosynthetic pigments to oxidation may be enhanced by modification of the organisation of the complexes (e.g through the addition of

Triton X-100).

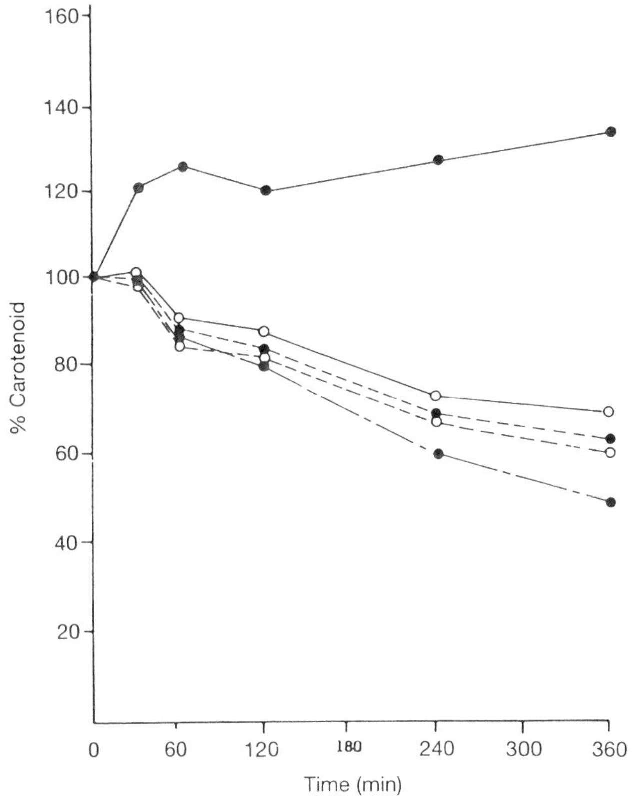

Figure 6.9 Photodestruction of individual carotenoids in chloroplasts of *Hordeum vulgare* subjected to a light intensity of 250 μmol $m^{-2} s^{-1}$. Neoxanthin (○---○), violaxanthin (●---●), lutein (○—○), ß-carotene-5,6-epoxide (●——●), ß-carotene (●---●). The data are expressed as percentage of the individual carotenoid level at the start of the experiment (Barry et al. 1990).

The photobleaching of pigments has generally been found to occur in a well-defined order when leaves or isolated chloroplasts or thylakoids are exposed to high light intensities. The carotenoids are generally destroyed at a faster rate than are the chlorophylls, with the reaction centre-bound ß-carotene being particularly susceptible. Neoxanthin is also very susceptible to these conditions and is lost much more readily than the other xanthophylls (Fig. 6.9). Lutein, which is the major carotenoid found in the chloroplast, is, under most conditions, the most stable to photooxidative degradation. The degree of epoxidation of the xanthophyll cycle carotenoids does not affect the pattern of bleaching and, indeed, these xanthophylls are bleached at similar rates *in vivo*. Zeaxanthin does not appear to be intrinsically any more resistant to oxidative conditions than any other xanthophyll. Although zeaxanthin is readily formed from violaxanthin (and antheraxanthin) in high light in intact leaves, exposure to light *in vitro* (e.g. isolated chloroplasts and thylakoids) will not normally induce the xanthophyll cycle unless additional ascorbate is present. Allowing for alterations to the epoxidation state of the xanthophyll cycle carotenoids, identical patterns of bleaching are seen *in vivo* and *in vitro*. The analysisby high-performance liquid chromatography (HPLC) of pigments from untreated control chloroplasts and from chloroplasts that have been exposed to very high light intensities is shown in Figs. 6.10 and 6.11.

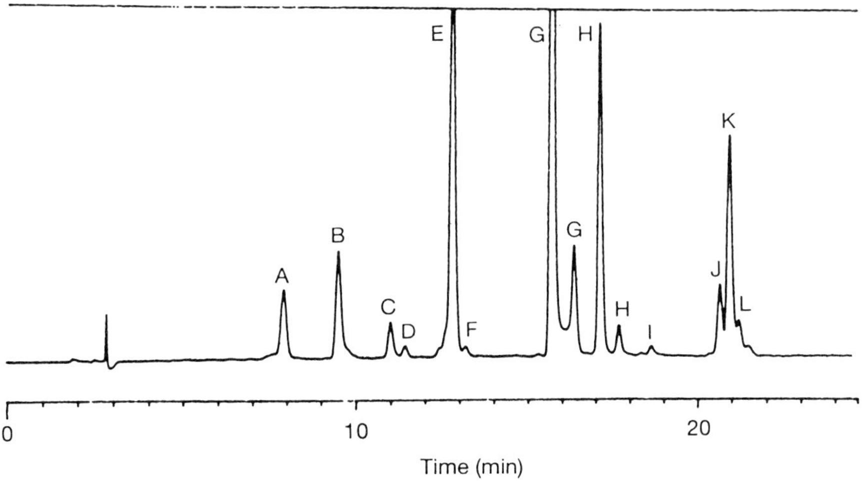

Figure 6.10 Reversed-phase high-performance liquid chromatography of carotenoids and chlorophylls from leaves of *Digitalis purpurea*. Peak identifications: A. neoxanthin; B. violaxanthin; C. lutein-5,6-epoxide; D. antheraxanthin; E. lutein; F. zeaxanthin; G. chlorophyll *b*; H. chlorophyll *a*; I. ß-carotene-5,6-epoxide; J. α-carotene; K. all-*trans*-ß-carotene; L. *cis*-ß-carotene. The chromatogram was monitored at 447 nm (A. Young and P. Horton, unpublished data).

Barry et al. (1990) examined the site(s) of pigment loss in chloroplasts and showed by SDS-polyacrylamide gel electrophoresis (PAGE) that high light caused photodestruction of pigments in all pigment-protein complexes. These conditions reduced the relative pigment content of the LHCII/CPa complex (containing the PSII core) to a greater extent than that of any other complex. Pigment levels in the CP1/CP1a complex (i.e. core complex (CC) I, the PSI core) were also greatly reduced by exposure to high light intensities. The degradation of the complexes in these experiments resulted in greatly increased levels of chlorophylls and carotenoids in the free-pigment micelles compared to those in control (dark-adapted) tissues. A similar increase in free-pigment levels was obtained for chloroplasts obtained from greened etiolated seedlings, although the extent and rate of pigment loss were even greater (Barry et al. 1990; Wieckowski and Majewska 1990). An almost identical pattern of pigment photobleaching for the individual complexes was seen in the presence of the electron transport inhibitors, DCMU (chloroplasts; Barry et al. 1990) and monuron (intact leaves; A. Young and G. Britton, unpublished data). There is a need to repeat these studies using the much more sensitive and accurate Deriphat-PAGE method, as reported in Chapter 3.

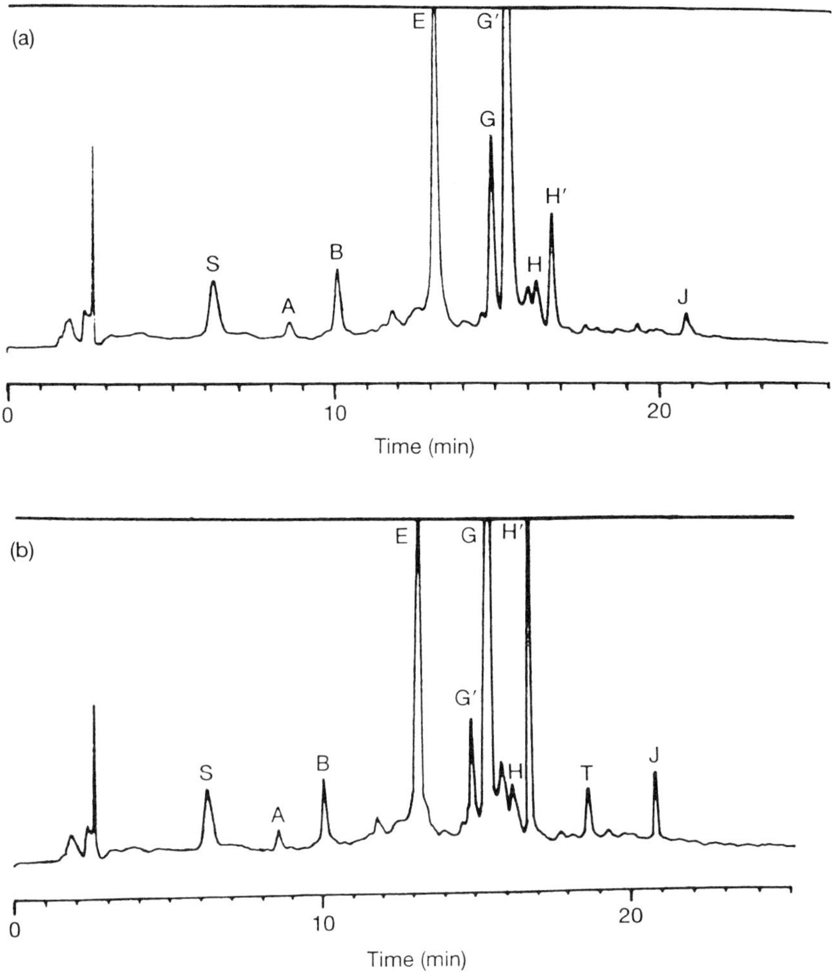

Figure 6.11 Chromatograms showing the effects of (A) exposure of isolated chloroplasts to high light intensities, and (B) the effect of 1.0 mM paraquat (methyl viologen) on pigments in intact leaves of *Hordeum vulgare*. Peak S represents an unidentified chlorophyll breakdown product; other peak identifications as in Fig. 6.9 (A. Young and G. Britton, unpublished data).

Damm et al. (1987) compared the patterns of bleaching in PSI particles that had been fractionated into LHCI and CCI. In high light treated LHCI, chlorophylls and ß-carotene were rapidly degraded, with the various xanthophylls remaining much more stable to oxidation. These workers suggested that the kinetics of ß-carotene destruction, which was greater than that for either of the chlorophylls, indicated a major protective role for this carotenoid in the antenna complex. The low level of oxidative damage seen with the xanthophylls (lutein, violaxanthin and neoxanthin) led Damm et al. (1987) to suggest that these pigments primarily provide a structural rather than a photoprotective role. The isolated CCI, exposed to similar conditions, showed very rapid losses of chlorophyll *b*, compared to ß-carotene, chlorophyll *a* or the remaining xanthophylls. The high stability of chlorophyll *a* in CCI was probably due to the very high levels of ß-carotene in this complex. The low xanthophyll levels in CCI (a result of contamination by LHCI) were thought to afford little protection, if any, to chlorophyll *b*. Miller and Carpentier (1991) examined the kinetics of photobleaching in isolated thylakoids and in PSI and PSII preparations and reported that carotenoids were bleached to a greater extent in PSI preparations. The bleaching was characterised initially by a slow rate of loss, indicating the presence of energy traps, which serve to protect the bulk pigments. Miller and Carpentier also observed that the core–antenna complexes were more sensitive to illumination than were the peripheral complexes.

The importance of carotenoids in maintaining the structural integrity of the pigment–protein complexes has recently been shown in a series of reconstitution studies by Schmidt and co-workers (Plumley and Schmidt 1987; Cammarata et al. 1990; Paulsen et al. 1990). They have shown that there is an absolute requirement for xanthophylls, especially the allenic carotenoid neoxanthin, in the successful reassembly of isolated LHCII. The effectiveness of many bleaching herbicides (e.g norflurazon) which function through the inhibition of carotenogenic enzymes (resulting in the accumulation of acyclic precursors such as phytoene and phytofluene in place of the normal cyclic carotenoids) may be due to the lack of the structural integrity that is provided by the normal chloroplast carotenes and xanthophylls, especially in PSII. Studies using greening mustard cotyledons have shown that carotenoids are required for PSII but not PSI activity (Markgraff and Oelmüller 1991).

6.2.2 Formation of ß-carotene-5,6-epoxide

In a series of experiments *in vitro*, S.W.M. Lee and G. Britton (unpublished data) showed that the formation of ß-carotene-5,6-epoxide (5,6-epoxy-5,6-dihydro-ß,ß-carotene) was a reliable indicator of oxidative damage to ß-carotene. Although it has been reported to be widely present in flowers and fruit this epoxide has not yet been detected as a natural component of photosynthetic tissues, and has been found only as a consequence of oxidative damage to leaves or isolated chloroplasts. Ashikawa et al. (1987) suggested that a correlation existed between the observed photobleaching of ß-carotene in thylakoid membranes and the occurrence of all-*trans*-ß-carotene monoepoxide. They further suggested that the appearance of this epoxide in spinach thylakoids and in the cyanobacterium *Synechococcus vulcanus* was largely the result of an enzymic epoxidation/de-epoxidation cycle comparable with the xanthophyll cycle. The production of ß-carotene-5,6-epoxide was thought to be based on a diurnal cycle, i.e. levels of the epoxide were low following a dark period and rose following exposure to natural daylight. This is, in fact, the reversal of the events in the xanthophyll cycle in which it is the de-epoxidation reactions which are promoted in the light.

In a more recent study (Young et al. 1989), neither ß-carotene-5,6-epoxide nor its furanoid derivative ß-carotene-5,8-epoxide (mutatochrome) were detected in the leaves of a number of higher plant species or algae, except under conditions of oxidative stress. These studies also confirmed by spectroscopic analysis that the epoxide produced in these reactions was optically inactive; it was therefore not produced enzymically but must have been derived from ß-carotene by purely photochemical or chemical reactions to give a mixture of the (5*RS*,6*RS*) isomers.

Conflicting data exist concerning the location of the ß-carotene-5,6-epoxide produced in chloroplasts or thylakoids. Ashikawa et al. (1987) reported that the epoxide was preferentially located in PSII particles (presumably in CCII), whereas Young et al. (1989) observed high levels of the epoxide in the PSI core complexes. This latter study could only detect the epoxide in the PSII complexes after very harsh treatment of thylakoids with the herbicide DCMU (see Section 6.2.3(a)), or through exposure to very high light intensities (Fig. 6.9), i.e. conditions that promoted rapid and extensive bleaching of pigments.

More recently, Kennedy and Liebler (1991) have identified ß-carotene-5,6-epoxide as one of the products formed from the action of peroxyl radicals on ß-carotene *in vitro*. This supports the hypothesis that the epoxide can be formed in leaves or chloroplasts as a result of oxidative processes.

6.2.3 Effects of herbicides

The action of certain herbicides on photosynthetic tissues provides one of the best illustrations of the need for carotenoids to protect plants from photooxidation. Herbicides can act upon the chloroplast pigments of plants in two ways: either they can (i) destroy the existing pigments, generally through the production of oxidative species such as singlet oxygen and superoxide (e.g., through the action of DCMU and paraquat, respectively), or (ii) they can interfere with the biosynthesis of these pigments (e.g. diflufenican, aminotrizole; see Chapter 5). A limited number of herbicidal compounds, especially some of the diphenyl ethers, may act by both inhibiting biosynthesis and promoting oxidation. Damage to the chloroplast and its components can be brought about both by a lack of energy dissipation when the electron flow through PSII or cyclic PSI becomes limited and, with certain compounds, by the direct production of toxic oxygen species (e.g. the action of paraquat). Halliwell (1991) has reviewed the role of toxic oxygen species in the action of herbicides and the role that various antioxidants, including the chloroplast carotenoids, may play in their control.

(a) Inhibition of carotenoid biosynthesis

The inhibition of carotenogenesis in higher plants and in other photosynthetic organisms is covered in detail in Chapter 5. A number of reviews have also recently been published concerning the mode of action of those compounds that inhibit carotenoid biosynthesis (Ridley 1977; Britton et al. 1989; Bramley 1991; Young 1991) and it is not the intention to go into any detail in this chapter. The vital role that carotenoids play in photoprotection is well illustrated in two *in vivo* systems: (i) a range of mutants (microorganisms and higher plants) deficient in their normal carotenoid composition, and (ii) the effect of inhibiting the biosynthesis of carotenoids in photosynthetic systems.

The inhibition of carotenoid biosynthesis has two main effects. First, the synthesis of the normal cyclic carotenoids (e.g. ß-carotene, lutein, etc.) is prevented through the inhibition of the desaturase and cyclisation reactions in the biosynthetic pathway, resulting in the accumulation of a number of precursors (e.g. phytoene). Secondly, the absence of these cyclic carotenoids prevents the normal assembly of the photosynthetic apparatus so that any plastids that may be formed following treatment with these inhibitors are generally either inactive or show poor photosynthetic performance; the effect of these inhibitors on chloroplast ultrastructure can readily be seen by electron microscopy. Reconstitution studies by Plumley and Schmidt (1987), for instance, have shown an absolute requirement for xanthophylls in the successful reassembly of isolated LHCII therefore the absence of these carotenoids is likely to lead to incomplete assembly. Markgraff and Oelmüller (1991) have shown, through the inhibition of carotenoid biosynthesis by norflurazon, that carotenoids are required for functional PSII but not for PSI in mustard seedling cotyledons. On transfer of the seedlings from the herbicide to water, lutein levels were seen to rise greatly in line with PSII activity, whereas the levels of the carotenes and other xanthophylls remained largely unaltered. Markgraff and Oelmüller (1991) therefore proposed that lutein synthesis was one of the main rate-limiting steps in the formation of a functional photosynthetic system. Plumley and Schmidt (1987) observed that the incorporation of neoxanthin into LHCII was especially important for reassembly to take place.

A number of studies have shown that plants treated with inhibitors of carotenoid biosynthesis are readily susceptible to photooxidation (e.g. Grumbach 1984; Ridley 1982). These inhibitors, unlike those that interfere with electron transport processes, do not have a requirement for light for their action and after herbicide treatment both light-grown and dark-grown plants are usually bleached white or, in the case of inhibitors of cyclisation (e.g. aminotriazole), yellow.

(b)Interference with electron transport processes

Many herbicidal compounds will affect the photosynthetic pigments of treated tissues through the generation of highly reactive oxygen species (e.g. superoxide generation by paraquat) which may result in extensive chlorosis and necrosis of leaf tissues (Ridley 1982; Halliwell 1991). The effect of several compounds representing a number of different classes of herbicides on chloroplast pigments *in vivo* and *in vitro* has recently been undertaken (Barry et al. 1990; A. Young and G. Britton, unpublished results). Inhibitors of electron transport at PSII have an effect similar to those of exposure to high light (see Fig. 6.8) both in isolated chloroplasts (DCMU) and in intact leaves (monuron). This agrees with current thinking that, in a manner similar to the exposure to high light intensities, these compounds lead to greatly increased levels of singlet oxygen in the chloroplast. The levels of ß-carotene-5,6-epoxide produced in leaves in the presence of these compounds is often much higher than is seen with other examples of photooxidation. An additional effect is seen with DCMU and monuron as they both inhibit the de-epoxidation of violaxanthin into zeaxanthin *in vivo*. The patterns of pigment destruction seen in leaf tissues that have been subjected to a range of photooxidative conditions are shown in Table 6.3.

The application of paraquat to intact leaves in the light, conditions which will result in highly elevated levels of superoxide, causes very rapid and severe bleaching of tissues. The pattern of pigment loss is similar to that seen with DCMU or high light except that the presence of paraquat appears to induce violaxanthin de-epoxidase at low light intensities, leading to the accumulation of zeaxanthin (A. Young and G. Britton, unpublished data). It is generally recognised that carotenoids are able to scavenge for a number of active oxygen species, including superoxide, albeit with poor kinetics and there may be some interaction between the carotenoids and oxidative species produced by paraquat action.

The application of compounds which have actions other than the inhibition of PSI and PSII, e.g. the diphenyl ethers (e.g. acifluorfen, fomasafen) may produce a quite different pattern of pigment destruction. This suggests that different oxidising species are involved (e.g. lipid peroxides) and/or that these oxidising species are produced at different sites either inside or outside the chloroplast. It is important to remember that the membrane-bound carotenoids in the chloroplast are only effective in scavenging for oxidising species that are produced in relatively close proximity to them, e.g. $^1O_2$ produced by triplet chlorophyll molecules in the pigment–protein complexes. Thus, the carotenoids bound to pigment–protein complexes may ultimately be affected by a range of oxidising conditions, but in many cases they are not in a position to scavenge for these oxidising molecules effectively *in situ,* as the latter are probably more likely to react readily with other substrates, i.e. other cellular components.

This large group of diphenyl ether herbicides was originally thought to act through an initial and direct interaction with membrane-bound carotenoids in the thylakoid membrane, resulting in the sensitisation of the herbicide molecule. This would then lead to the production of oxidising species, most probably in the form of $^1O_2$ and lipid peroxides. It is now known that the carotenoids are not directly involved in the action of these compounds, and that the diphenyl ethers cause the accumulation of protoporphyrin IX, which may act as a sensitiser of $^1O_2$ formation. Like many other herbicides, the diphenyl ether compounds cause rapid bleaching of plant tissues, although close examination shows that the kinetics of bleaching for carotenoids and chlorophylls are quite different (see Table 6.3; A. Young and G. Britton, unpublished data). In contrast to the bleaching processes described above

Table 6.3 Pigment composition of barley leaves as determined by reversed-phase HPLC following exposure to a range of conditions that lead to bleaching of photosynthetic tissues (A. Young and G. Britton, unpublished data)

Carotenoid	*Control*	*High light (>2000 μmol m⁻² s⁻¹)*	*Paraquat (100 μM 16 h)*	*Monuron (1.0mM 8 h)*
Neoxanthin	12.4	13.4	16.1	12.4
Violaxanthin	20.1	7.4	10.8	17.6
Antheraxanthin	1.1	0.9	2.5	0.7
Lutein	38.0	49.5	67.2	55.0
Zeaxanthin	ND	8.9	ND[a]	1.7
ß-Carotene-5,6-epoxide	ND	1.7	ND[b]	1.7
ß-Carotene	28.4	18.4	3.4	12.6
Total xanthophyll acyl esters	ND	ND	ND	ND
Ratio lutein: ß-carotene	1.3:1	2.7:1	19.8:1	4.4:1
Ratio chlorophyll *a*:*b*	2.3:1	1.5:1	1.2:1	1.4:1
Ratio carotenoid: chlorophyll	0.5:1	0.3:1	0.2:1	0.3:1
	Senescence	*Rose-bengal (96 hrs)*	*Acifluorfen (10μM 72 h)*	*Fomasafen (10μM 72h)*
Neoxanthin	3.5	4.6	2.8	3.1
Violaxanthin	4.7	9.9	4.8	6.9
Antheraxanthin	ND	3.1	2.4	2.6
Lutein	18.0	13.8	16.2	18.3
Zeaxanthin	ND	2.0	1.6	2.5
ß-Carotene-5,6-epoxide	ND	ND	ND	ND
ß-Carotene	20.2	23.3	27.1	16.9
Total xanthophyll acyl esters	54.4	42.9	44.9	49.6
Ratio lutein: ß-carotene	0.9:1	0.6:1	0.6:1	1.1:1
Ratio chlorophyll *a*:*b*	1.4:1	2.9:1	4.5:1	3.5:1
Ratio carotenoid: chlorophyll	2.5:1	0.9:1	0.7:1	0.8:1

[a] Zeaxanthin formed in initial exposure to paraquat

[b] ß-Carotene-5,6-epoxide formed in initial exposure to paraquat

ND Not detected

for those compounds that interfere with PSI and PSII electron transport, the diphenyl ethers acifluorfen and fomasafen were found to bleach chlorophylls *a* and *b* at a much faster rate than the carotenoids. In this example, the xanthophylls, especially violaxanthin and neoxanthin, were much more readily destroyed than ß-carotene: this suggests that when bound to the reaction centre ß-carotene is not as greatly involved in photoprotection under these conditions as it is in photoinhibitory conditions or in the presence of certain other herbicides.

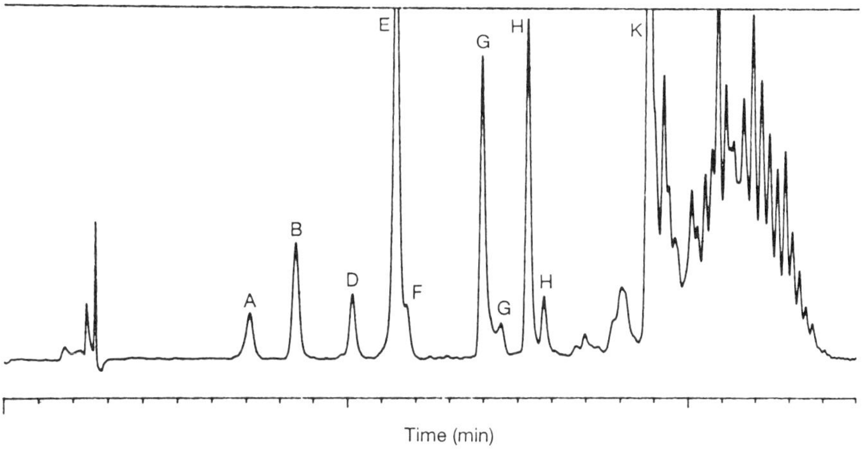

Figure 6.12 Chromatogram showing the formation of very high levels of xanthophyll acyl esters in leaves of *Hordeum vulgare* following treatment with 10 μM acifluorfen for 48 hours. Peak identifications as in Fig. 6.9; with the exception of ß-carotene (peak K) all peaks with a retention time greater than 18 min are individual xanthophyll acyl esters (A. Young and G. Britton, unpublished data).

One of the most noticeable features of the action of the diphenyl ethers is the production of large amounts of xanthophyll acyl esters (see Section 6.2.5) following prolonged treatment. This is illustrated in Fig. 6.12, which is a chromatogram showing very high levels of xanthophyll acyl esters, accounting for more than 60% of total carotenoid, from acifluorfen-treated leaves of barley. These esters are normally absent from healthy leaves and are observed only under quite specific oxidising conditions. Treated tissues also accumulate large numbers of plastoglobuli as the chloroplasts degrade. These can be isolated by sucrose density centrifugation and analysed by HPLC (see Fig. 6.13; A. Young and G. Britton, unpublished data). Under certain oxidative conditions the levels of esters in the plastoglobuli may be much higher than those illustrated in Fig. 6.13, and, in some cases, the bulk of carotenoid in the plastoglobuli may be in the form of xanthophyll acyl esters (A. Young and G. Britton, unpublished data).

The application of some diphenyl ethers to leaf tissues also induces the de-epoxidation side of the xanthophyll cycle so that small amounts of zeaxanthin may be found in treated tissues, although it is not known whether the diphenyl ethers (or indeed paraquat) can either directly promote the de-epoxidation of violaxanthin or whether they have a secondary effect that will ultimately affect the epoxidation state of the leaves (e.g. alteration of the trans-thylakoid pH gradient).

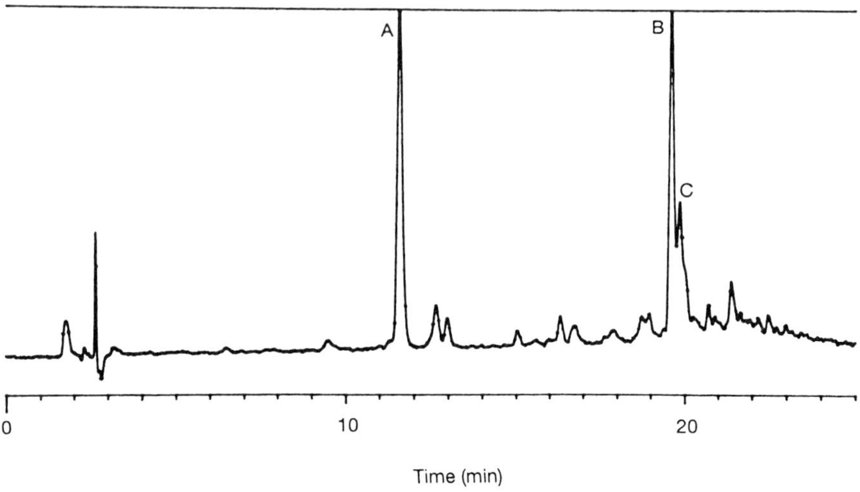

Figure 6.13 Chromatogram of pigments extracted from plastoglobuli obtained from sucrose density gradient centrifugation of chloroplasts from senescing leaves of *Hordeum vulgare*. Peak identifications: A, lutein; B, ß-carotene; C, violaxanthin acyl ester (A. Young and G. Britton, unpublished data).

6.2.4 Other bleaching processes

Two distinct patterns of pigment bleaching are found in plant tissues. The majority of stress treatments (exposure to high light intensities, herbicide treatment etc., see Sections 6.2.1–3) ultimately result in an increase in the ratio of total carotenoid:chlorophyll. In many of these examples the carotenoids are probably in the first line of defence as quenchers of triplet-state chlorophyll and scavengers of $^1O_2$ produced in the thylakoid membrane. Another commonly observed pattern of pigment loss that may be observed shows the chlorophylls degraded at a faster rate than the carotenoids (e.g. the diphenyl ether herbicides). This may be due partly to enzymic degradation, although purely oxidative conditions may produce similar results (see Section 6.2.4). A number of specific examples of environmental conditions or treatments that lead to bleaching of chloroplast pigments are described below.

(a) Atmospheric pollutants

There is still conflicting evidence as to the cause of the forest decline that is prevalent in various parts of the world, including Western Europe. Direct photooxidative damage to photosynthetic tissues by atmospheric pollutants such as SO_2 and O_3 is thought to be one of the major factors responsible for needle loss in conifers. Chlorosis of photosynthetic tissues and subsequent needle fall are among the main features of this process; this has led to determinations of the pigment content and composition of leaves and needles from a number of plant species that have been subjected to a range of pollutants. The series of events involved in such damage has not been fully elucidated, although it is clear that photooxidative destruction of chloroplast components occurs in many cases (Elstner et al. 1985). The direct effects of a number of pollutants (O_3, NO_2 and SO_2) on the carotenoid composition of leaves and needles have recently been investigated.

Table 6.4 Pigment composition of (a) green, (b) partially chlorotic, and (c) partially necrotic needles of *Picea abies* taken from trees at a site in the Bavarian forest that exhibited characteristic yellowing of previous year and older needles for 3.5 years. The data shown are the means of at least 3 separate determinations by reversed-phase HPLC analysis (standard errors not shown). Considerable differences in the pigment content and composition of different age classes of green needles were also observed (Young et al. 1991)

Carotenoid	*Green needles (current year)*	*Partially chlorotic (previous year)*	*Partially necrotic (3 and 4 year-old)*
Neoxanthin	12.1	6.2	6.4
Violaxanthin	11.8	7.1	0.9
Lutein-5,6-epoxide	1.5	2.1	0.2
Antheraxanthin	0.6	0.5	0.2
Lutein	46.4	39.0	53.8
Zeaxanthin	ND	0.5	3.9
α-Carotene	3.7	6.0	2.9
ß-Carotene	23.6	38.1	31.7
Ratio total carotenoid: chlorophyll	0.39:1	0.44:1	1.03:1
Ratio lutein: total carotene	1.69:1	0.88:1	1.55:1
Ratio chlorophyll *a*:*b*	2.69:1	3.72:1	0.93:1

ND Not detected

The direct effects of photooxidative processes on plant tissues can readily be confused with the effects of nutrient deprivation (also recognised as a major contributory factor in forest decline) and, although observations have been made for the leaves and needles of a number of species, it may be very difficult in many natural populations of plants to differentiate between the effects of these processes simply by examining the pigment composition of photosynthetic tissues.

Damage to photosynthetic tissues is readily observed as the result of O_3 treatment of leaves and needles, and probably involves the production of hydroxyl radicals (Asada and Takahashi 1987). A detailed examination of the changes in pigment composition in young barley seedlings as a result of exposure to O_3 and NO_2 under controlled conditions has recently been undertaken (A. Young, A. Price, P. Beckett, G. Britton and P.J. Lea, unpublished data; Price et al. 1990). The visible symptoms of fumigation with these pollutants were observed to be quite different; O_3 resulted in the appearance of chlorotic patches all over the leaves, whereas NO_2 caused necrotic lessions to appear specifically at the edge of the leaf. Detailed analysis of the pigment composition of treated leaves showed differences also in the pattern of pigment bleaching with these two pollutants. The destruction of chlorophylls *a* and *b* proceeded at similar rates in the presence of O_3, but NO_2 treatment resulted in an increase in the ratio of chlorophyll *a*:*b*. The pattern of carotenoid destruction was quite similar with both pollutants (Table 6.4); neoxanthin levels were greatly reduced whereas ß-carotene levels were least affected, which is a pattern quite different from that seen for many other oxidative treatments (e.g. exposure to high light intensities). In these experiments, the main xanthophyll, lutein, was seen to be much less sensitive to NO_2 treatment than to O_3. Unfortunately it is not known whether these changes in carotenoid content and composition are changes that occur whilst the carotenoid is bound in the pigment–protein complexes (as with high light treatment, Section 6.2.1) or simply changes that occurred once the carotenoids were released into the cytosol after the degradation of the complexes. Provisional results that need to be confirmed suggest the latter (A. Young, P.J. Lea and G. Britton, unpublished data). Specific and highly focused or localised loss of carotenoids in the pigment–protein complexes is likely to be the result of $^1O_2$, or radical production, sensitised via triplet chlorophyll.

Traces of xanthophyll acyl esters were detected in leaves after prolonged treatment with both of these pollutants. Earlier reports (Sakaki et al. 1983; Senser et al. 1987) indicated that exposure to O_3 resulted in the rapid and preferential bleaching of ß-carotene. The preferential loss of ß-carotene in these tissues would be predicted from earlier studies on high-light-mediated bleaching (see Section 6.2.1). Other studies (Price et al. 1990; Senser et al. 1990; Young et al. 1990) have clearly shown that this is not the case and that leaves or needles exposed to O_3 show large reductions in the levels of xanthophylls and that ß-carotene levels, although subject to an initial reduction, may stabilise (Table 6.4).

Short-term exposure to 100% O_2 has been shown to lead to very rapid degradation of photosynthetic tissues and pigments (Mikulska et al. 1987). ß-Carotene, violaxanthin and zeaxanthin levels were particularly reduced whereas chlorophylls *a* and *b*, lutein and neoxanthin were comparatively stable. On transfer from 100% O_2 back to ambient conditions these plants continued to show substantial losses of ß-carotene and violaxanthin, although levels of zeaxanthin recovered, probably as a result of *de novo* synthesis. Bortitz (1964) has reported that exposure of a number of plant species to elevated levels of SO_2 resulted in an increase in the ratio of ß-carotene to chlorophyll.

As bleaching of photosynthetic tissues is one of the main consequences of pollution, Wolfenden et al. (1988) have proposed the use of changes in the pigment content and composition of leaves and needles as diagnostic indicators of forest decline. These workers analysed the pigment content of leaves taken from trees in field sites where O_3 was thought to be the major pollutant and suggested that long-term exposure to low concentrations of O_3 may serve to promote the epoxidation of zeaxanthin to antheraxanthin and violaxanthin. Lichtenthaler and Buschmann (1984) reported quite sensitive changes in the pigment composition of needles taken from German forests. Photoinduced bleaching or yellowing of the needles of sun-exposed conifer branches was frequently observed, whereas needles on the lower, shaded, side of the branch remained green and healthy. Again, this emphasises the requirement for light in bleaching. In needles of *P. abies* taken from field sites, the levels of chlorophyll *a* and ß-carotene were much more severely reduced than those of chlorophyll *b* or the xanthophylls, suggesting that the reaction centres were being degraded prior to the light-harvesting complexes. In a series of experiments carried out under controlled conditions in open-top chambers, Senser and colleagues (1990) observed an increase in the violaxanthin content of young spruce trees when exposed to a combination of O_3 and acid mist. The correlation between the violaxanthin content of the needles and O_3-treatment was especially evident when the data were expressed per unit chlorophyll. Such changes were proposed to form part of a regulatory process leading to protection from light-mediated stress. The ratios of the various chloroplast pigments have been used to predict whether the observed damage to photosynthetic tissues was primarily due to metabolic events or to photooxidation (Kandler et al. 1987; Senser et al. 1990).

In a recent study (A. Young and P. Putwain, unpublished data) the potential for using the photosynthetic pigment content of leaves or needles as predictors of damage was examined. The pigment content and composition of needles taken from young Norway spruce and Scots pine which had been subjected to elevated levels of O_3 and SO_2 for a number of years at an experimental field fumigation site were determined. Trees at this site lacked any visible signs of photobleaching and the needles taken from these trees showed no appreciable changes in the carotenoid or chlorophyll content which may have been of use as a possible indicator of the early stages of photodamage.

Koestner et al. (1990) have followed seasonal alterations to the pigment content and composition of Norway spruce in an area where SO_2 was recognised as being the main pollutant. The synthesis of chlorophylls in needles followed a distinct pattern, with a sharp decline in the levels of chlorophylls in the winter and immediately prior to bud break. The levels of carotenes and xanthophylls did not correlate with either the chlorosis of needles or peak SO_2 levels. Koestner et al. (1990) proposed that the observed chlorosis was a result of nutritional deficiencies.

Young et al. (1990) recorded considerable variation in the violaxanthin and zeaxanthin contents of damaged needles of *P. abies*, particularly with respect to needle age. The levels of zeaxanthin relative to the other carotenoids increased with increasing visible signs of damage (i.e. bleaching) and were highest in necrotic needles (Table 6.4). In another study, Price et al. (1990) could not, however, observe any alteration in the epoxidation state of these xanthophylls in barley seedlings treated with high levels of O_3.

(b)Nutrient deprivation

Deficiencies in certain nutrients (e.g. N, K, P, S, Cl and a number of micronutrients (e.g. Mn; Wilkinson and Ohki 1988)) can lead to alterations in the carotenoid composition of photosynthetic tissues (Goodwin 1980). Fabian-Galan (1972), for example, reported that sunflower leaves deficient in K^+ have a lowered total carotenoid content compared to control plants. Further analysis showed that the ß-carotene content of these leaves was increased relative to the xanthophylls. The apparent increase in the levels of lutein seen in this study are possibly a reflection of an increase in zeaxanthin levels as a result of the de-epoxidation of violaxanthin (lutein and zeaxanthin are difficult to separate chromatographically).

Perhaps the best studied example of the effects of nutrient deprivation is that of iron deficiency in a number of crop plants. The chlorotic effects of iron deficiency are well known (Terry and Abadia 1986) and the effects on the pigment composition of leaves from sunflower, peanut (Monge et al. 1987), pea (Abadia et al. 1989) and sugar beet (Morales et al. 1990) have been reported. In pea, where a pattern of development-related bleaching of leaves occurs, a deficiency in iron was found to result in a decrease in the light-harvesting complexes and a relative enrichment in the PSII reaction centre. In the most chlorotic leaves (showing up to 80% reduction in pigment levels) the proportion of xanthophylls increased from only 12% of total pigment (carotenoid + chlorophyll) to over 60%. This was found to be due largely to an increase in the levels of the xanthophyll cycle carotenoids, and the bleaching of the chlorophylls and ß-carotene. The ratios of chlorophyll *a*:*b* and total carotenoid:total chlorophyll were much higher in the deficient leaves than in the control leaves. In sugar beet, iron deficiency resulted in an increase in the ratio of lutein:chlorophyll *a* and of the xanthophyll cycle pool:chlorophyll *a* (Morales et al. 1990).

Monge et al. (1987) found that of all the normal chloroplast pigments lutein was the least affected by iron deficiency. They suggested that this could be due to the presence of large numbers of plastoglobuli in iron-deficient leaves, which are preferentially enriched in lutein (and ß-carotene). Young and Britton (1990a) have suggested that the "entrapment" of carotenoids within plastoglobuli results in an increase in their stability to further oxidative processes.

(c)Effects of extremes of temperature

Low temperature (chilling)

The combination of low temperature and light may be particularly damaging, resulting in photoinhibition in leaves (Björkman 1987). Chlorophyll bleaching in conifers during the winter has been shown to be much more pronounced in sun-exposed than in shade-adapted needles (Tranquillini 1964). Öquist et al. (1987) suggested that two distinct processes may account for the increased incidence of bleaching of photosynthetic tissues in the winter months: (i) a preferential photooxidation of chlorophyll *a* in the antennae, and (ii) a catabolic process resulting in the breakdown of the light-harvesting complex. Jin et al. (1990) have shown that carotenoid synthesis can take place in the needles of conifers during the winter months, whereas chlorophyll biosynthesis is very much reduced. This could explain, at least in part, the observed chlorosis of these needles in winter-stressed pines (see also Section 6.2.4 (a); Koestner et al. 1990).

Chilled *Cucumus sativus* that had been exposed to light showed pigment photooxidation with a pattern of pigment loss similar to that seen in other photoinhibited tissues (see Section 6.2.1), i.e. ß-carotene was most sensitive to these conditions (van Hasselt 1972). van Hasselt and van Berlo (1980) further reported that an atmosphere of N_2 completely prevented pigment loss when plant material was exposed to similar light and temperature regimes. An atmosphere enriched in O_2 served to increase injury at the low temperatures.

Changes in the activities of active oxygen scavengers, including carotenoids, in spinach plants under conditions of chilling and high light have been investigated (Schöner et al. 1990). Only minor alterations to the pigment bed occurred as a result of these conditions. Whilst levels of the xanthophyll cycle carotenoids were greatly altered, the levels of ß-carotene and of neoxanthin were largely unchanged compared to those in the control plants. Low temperatures can also directly affect the de-epoxidation of violaxanthin into zeaxanthin in some species (see Chapter 7).

In an earlier study Öquist (1986) reported that the ß-carotene content of the PSI reaction centre was increased in the thylakoids of winter-stressed pine needles compared to that in trees in the summer months. Öquist concluded that such changes were the result of long-term acclimatisation or adaptation to winter conditions and they allow the increased proportion of ß-carotene present in PSI to act as a sink for the dissipation of excess energy. Unfortunately no information was gathered in this study concerning possible alterations to the pool of xanthophyll cycle carotenoids.

High temperature

Little information is available concerning the effects of high temperature on the bleaching of photosynthetic pigments. The inhibition of various metabolic functions on exposure to high temperature (e.g. CO_2 fixation, O_2 evolution and phosphorylation) in turn leads to very rapid degradation of the chloroplast and its components, including the chlorophylls and carotenoids. The effects of a combination of heat stress, oxygen and high light are especially damaging, whereas under reducing conditions (e.g. in the presence of a nitrogen atmosphere or in the dark) bleaching may be minimal.

Williams et al. (1986) showed that, although PSII activity was greatly reduced in higher plant chloroplasts by high temperature, the pigments associated with PSI showed a greater susceptibility to photobleaching. In contrast to the photobleaching pattern seen in cells exposed to high light intensities at room temperature (Section 6.2.1), at higher temperature the chlorophylls were destroyed to a greater extent than were the carotenoids (see also Young and Britton 1990b).

(d) Exogenous sources of singlet oxygen

A number of photosensitiser dyes can be used to produce $^1O_2$ in the presence of light and oxygen. The action of one of these sensitisers, rose bengal, on the photooxidative breakdown of photosynthetic tissues has been investigated by Knox and Dodge (1984, 1985) and by Young and Britton (1990c). Illumination in the presence of rose bengal causes the very rapid bleaching of pigments *in vitro* or in leaf discs. Such treatment increased the ratio of chlorophyll *a:b* and of total carotenoid:total chlorophyll in leaf discs. The ratio of lutein:ß-carotene was severely reduced, indicating a relatively high stability of ß-carotene compared to the xanthophylls and, indeed, to the chlorophylls (see Table 6.4 and Fig. 6.14). Exposure of leaves to rose bengal in the light also results in the formation of very high amounts of xanthophyll acyl esters (see Section 6.2.5). Although the action of rose bengal as a sensitiser leading to the formation of $^1O_2$ is well recorded, a comparison of the pattern of photobleaching in leaf tissues in the presence of this sensitiser with that seen as a result of other photooxidative conditions where $^1O_2$ is also thought to be an important oxidative agent (e.g exposure to high light intensities, Section 6.2.1) shows that quite different processes are taking place. The changes in pigment composition of leaf tissues as a result of rose bengal treatment suggest that although high levels of $^1O_2$ may be produced, the effects on the chloroplast are quite different from that seen when $^1O_2$ is produced *in situ* through triplet chlorophyll-mediated sensitisation of molecular oxygen in the pigment–protein complexes.

6.2.5 Formation of xanthophyll acyl esters

The occurrence of xanthophyll acyl esters in photosynthetic tissues has only recently been reported. These compounds have not been detected in leaves maintained under optimum conditions and their presence is restricted to tissues subjected to oxidative damage, e.g. treatment of leaves with the $^1O_2$ sensitiser rose bengal (see Figs. 6.12 and 6.14), or with certain of the diphenyl ether herbicides (e.g. acifluorfen; Young and Britton 1990c), drought-stressed leaves of barley (Barry et al. 1992), O_3 and NO_2-treated leaves (Price et al. 1990), and in senescent leaves (Young et al. 1991). They are not, however, found universally in all bleached and photooxidised tissues, and have not been detected following prolonged exposure to high light intensities or to the PSI or PSII herbicides DCMU (or monuron) and paraquat (see Tables 6.3 and 6.4). The production of these xanthophyll acyl esters is not light-dependent as they can be formed in senescing leaves in the dark. This suggests that their formation is promoted under certain specific oxidative conditions, perhaps as a result of lipid peroxidation. Unsaturated fatty acids in the thylakoid membrane are particularly prone to attack by free radical species and the resulting peroxidation leads to disruption of the membrane, releasing lipids into the cytosol. The free fatty acids liberated from these will then be available to esterify the free xanthophylls that have been released from the degraded pigment–protein complexes. The synthesis of xanthophyll acyl esters in leaf tissues at the expense of the unesterified, "free", xanthophylls (lutein + neoxanthin + xanthophyll cycle pool) is shown in Fig. 6.15.

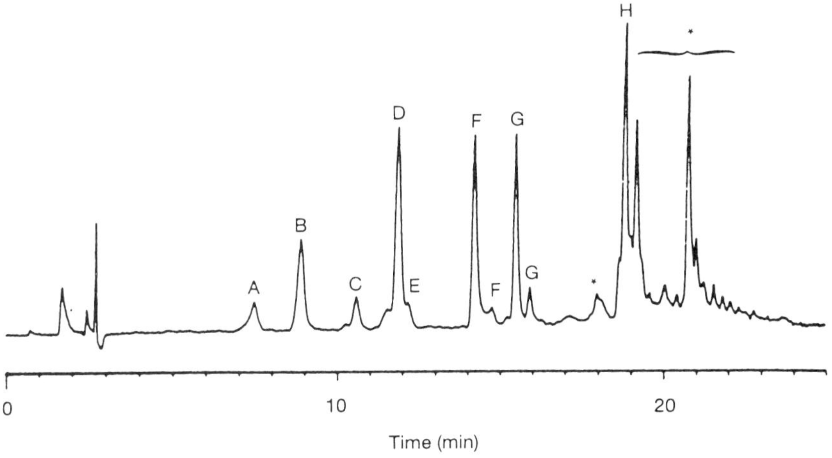

Figure 6.14 Chromatogram showing the formation of xanthophyll acyl esters in leaves of *Hordeum vulgare* following exposure to the singlet-oxygen-sensitiser rose bengal. Peak identifications: A, neoxanthin; B, violaxanthin; C, antheraxanthin; D, lutein; E, zeaxanthin; F, chlorophyll *b*; G, chlorophyll *a*; H, ß-carotene. With the exception of peak H. (ß-carotene) all peaks with a retention time greater than 17 min are xanthophyll acyl esters (A.J. Young and G. Britton, unpublished data).

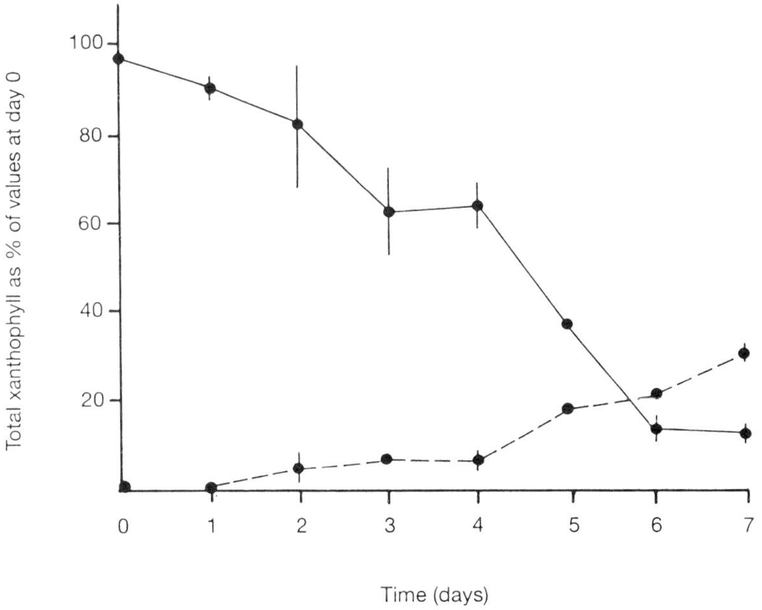

Figure 6.15 Production of xanthophyll acyl esters in senescent leaves of *Hordeum vulgare* as determined by HPLC analysis: total xanthophyll acyl esters (●----●), total free (unesterified) xanthophylls (●——●) (A. J. Young and G. Britton, unpublished data).

Esters of all major chloroplast xanthophylls have been detected in stressed tissues, and the single main component has been identified as lutein bis-linolenate (Barry et al. 1992). Data also indicated that the xanthophylls were esterified in approximately the same proportions as they were originally in the thylakoid membrane.

A large amount of the esterified carotenoid is found in the numerous plastoglobuli (see Fig. 6.13), which are seen in senescent tissues and which also contain free, unesterified, carotenoids (e.g. ß-carotene and lutein) (Lichtenthaler 1969b; Young et al. 1991) and possibly chlorophyll (Trevini and Steinmuller 1985).

Young and Britton (1990c) have suggested that two factors are important in the production of these xanthophyll acyl esters in photosynthetic tissues, namely the nature of the oxidising species involved and its site of production. They tentatively proposed that an extra-chloroplastic source of oxidising agents, involving lipid peroxidation, may be responsible for their formation. Clearly, more work needs to be undertaken in this area.

6.3 CAROTENOIDS IN ALGAE AND CYANOBACTERIA

In comparison to the information gained concerning the environmental regulation of carotenoid content and composition in higher plants, very little is known concerning the behaviour of these pigments in the algae and cyanobacteria. The chlorophycean algae possess a similar pigment composition to higher plants (see Chapter 2) and, as far as is known, behave in a similar manner with respect to changes in their light environment, exposure to herbicides that interfere with photosynthetic electron transport, etc. Much less is known about modifications to the carotenoid composition in species from other classes of algae in which the pigment composition differs from that of higher plants. More work needs to be directed at the function and behaviour of carotenoids in these important and widely distributed organisms.

6.3.1 Effect of light on the carotenoid content and composition of algae

Anderson and Barrett (1986) listed four factors which may modulate light-harvesting processes in the Chlorophyta and, probably, the Cryptophyta. These are: (i) total amounts of pigments associated with each photosystem; (ii) the ratio of LHCI and LHCII to CCI and CCII; (iii) the antenna unit sizes of both photosystems; and (iv) the ratio of P690:P700. Some of these will be reflected in alterations to the carotenoid content and composition of the algae.

Kohata and Wantanabe (1988) have reported diel changes in the pigment content and composition of the raphidophycean (Group II) alga *Chattonella antiqua* (responsible for red tides in the seas off Japan). The observed changes could be closely related to cell division in this alga, which takes place once a day. The major observations were: (i) the synthesis of all pigments was stimulated in the light; (ii) there was a correlation between chlorophyll *a*, chlorophyll *c* and fucoxanthin levels and cellular carbon, levels reaching their maximum in the dark part of the cycle; (iii) diel variation in ß-carotene and violaxanthin levels, with maximum levels occurring in the light, was quite distinct from the behaviour of fucoxanthin; and (iv) the synthesis of the separate carotenoids occurred at different stages of the cell division cycle, for example the synthesis of violaxanthin was restricted to the beginning of the light period and that of fucoxanthin towards the end.

The behaviour of fucoxanthin and violaxanthin in *C. antiqua* supports the theory that these carotenoids are synthesised by different biosynthetic routes (Chapter 4; Liaaen-Jensen 1978) and incorporated into different pigment–protein complexes. Similar behaviour has also been observed in other raphidophycean algae and in *Pyramimonas parkeae* (Prasinophyceae), which synthesises siphonaxanthin and violaxanthin.

Kohata and Wantanabe (1988) have suggested that each pigment accumulated in *C. antiqua* during cell division has a discrete function in light-harvesting and photoprotection. It is interesting to note that neither zeaxanthin nor antheraxanthin was detected in this alga during the study, the authors concluding that violaxanthin itself plays an important role in photoprotection rather than in light-harvesting. Brandt and Wilhelm (1990) have reported that the ratio of diadinoxanthin:chlorophyll *a* increased greatly during the cell division cycle of *Euglena gracilis*.

The response of the red alga *Porphyridium cruentum* to changes in light intensity and spectral quality has been investigated by Cunningham et al. (1989, 1990). Growth of this alga in high light led to a decrease in the unit sizes of PSI and PSII, a corresponding three-fold reduction in chlorophyll levels and a two-fold decrease in ß-carotene levels. As the levels of zeaxanthin did not vary with respect to growth irradiance, the authors proposed that zeaxanthin had a role other than as a major light-harvesting pigment in *P. cruentum*. The carotenoid content and composition (zeaxanthin ~50%, ß-carotene ~40%, cryptoxanthin ~5%) of this alga was little affected by the spectral quality of incident light during growth.

The effects of spectral quality and intensity on the carotenoid content of two cyanobacteria, namely *Spirulina platensis* and *Oscillatoria agardhii*, have been studied (Olaizola and Duerr 1990, and Millie et al. 1990, respectively). Whilst ß-carotene and echinenone are the most abundant carotenoids in *Spirulina* under white light, exposure to red and blue light reduced the myxoxanthophyll content and served to increase the levels of ß-carotene. Levels of other carotenoids were unaffected. In *Oscillatoria*, an increasing photon flux density caused a decrease in levels of chlorophyll *a* and *c*-phycocyanin, but an increase in the relative levels of myxoxanthophyll and zeaxanthin and decreases in the relative levels of ß-carotene and echinenone. In this organism, exposure to blue wavelengths of light had little effect, whereas red light increased the relative levels of echinenone and zeaxanthin.

The quality and intensity of light during the growth of some species of microalgae also plays an important role in the synthesis of secondary carotenoids. This is discussed in Section 6.3.2.

(a)Regulation of development

In comparison with the higher plants, a more diverse range of carotenoids is found in the algae, particularly in the xanthophylls associated with light-harvesting complexes. The pigmentation of the core complexes of PSI and PSII is much more highly conserved and usually restricted to chlorophyll *a* and ß-carotene (see Chapter 3). The distribution of these carotenoids in the various classes of algae is described in Chapter 2.

A number of algae will grow under heterotrophic conditions, making them suitable for studies on the biogenesis of the photosynthetic apparatus and its pigments on transfer to light (see Chapter 4). The best characterised of these is the alga *Scenedesmus*, for which a large number of mutants are available (Britton et al. 1977; Powls and Britton 1977; Humbeck 1990). Many of the mutants that have been isolated grow only under heterotrophic conditions and die when exposed to light under aerobic conditions. Of particular interest is the PG1 strain, which, when grown in the dark, accumulates ζ-carotene as its main carotenoid in place of the normal range of cyclic carotenes and xanthophylls. In the light, the alga develops into the normal dark-green colour of the wild type and is able to photosynthesise. The pigment composition of the light-grown PG1 has been shown to be identical with that seen in the wild type (Britton et al. 1977).

The C-6D mutant of *Scenedesmus obliquus* has been studied extensively. This mutant lacks the ability to synthesise the normal chloroplast pigments when grown heterotrophically and accumulates the acylic precursors lycopene, neurosporene, ζ-carotene and ß-zeacarotene in place of the cyclic carotenes and xanthophylls (Humbeck 1990; Humbeck et al. 1990). When C-6D is transferred to light, the rapid synthesis of the normal cyclic carotenoids (ß-carotene, neoxanthin, lutein, violaxanthin, antheraxanthin, zeaxanthin and loroxanthin) is seen, together with a reduction in the levels of the precursors. This overall pattern is quite similar to that observed in higher plants (see Section 6.1.1), i.e. the initial synthesis of ß-carotene, presumably direct from the original lycopene pool via cyclisation, is particularly rapid. The xanthophylls, with the exception of lutein, exhibit a lag phase in their synthesis. The regulation of the biosynthesis of lutein appears to be under control that is different from that of the other xanthophylls, as levels of lutein also rise rapidly on transfer of the alga to light. The changes in pigment composition closely accompany the formation of fully functional photosynthetic apparatus (Senger and Strassberger 1978; Humbeck et al. 1989, 1990; Humbeck 1990).

6.3.2 Secondary carotenoid synthesis

The regulatory factors governing secondary carotenoid production by microalgae are quite different from those that regulate the synthesis of the normal chloroplast carotenoids (which are often referred to as primary carotenoids). A limited number of algal species, most notably *Dunaliella* spp., are reported to be accumulators of ß-carotene under stress conditions (Goodwin 1980). However, many more microalgae may synthesise large amounts of ketocarotenoids, usually as a complex mixture of several carotenoids (e.g. astaxanthin, adonirubin, echinenone and canthaxanthin) with varying degrees of esterification, in response to similar stress conditions.

The synthesis of secondary carotenoids in microalgae appears to be dependent upon a large number of environmental and developmental factors. Of particular importance are depletion of the nitrogen source in the growth medium (often simply reflected by the age of the culture), exposure to high light and exposure to high salinity (e.g. in *Dunaliella*). In general, the production of secondary carotenoids by algae has been reported in cultures that have been subjected to a combination of reduced nitrate levels and high light intensities. Interestingly, Takada et al. (1990) have recently reported that a number of strains of *Chlorella* and *Scenedesmus* accumulated small amounts of ketocarotenoids when subjected to salt or osmotic stress (0.5 M NaCl or equivalent). In some algae carotenoid synthesis may be independent of light (e.g. *Trentepohlia aurea*, Czygan and Kalb 1966), whereas in others (e.g. *Dunaliella* spp.) the degree of secondary carotenoid formation is highly dependent on exposure to high light intensities.

(a) β-Carotene

The best-documented case of secondary carotenoid synthesis in any microalga is that of ß-carotene accumulation in *Dunaliella* (Ben-Amotz and Avron 1983, 1989, 1990; Borowitzka et al. 1984, 1986, 1990; Ben-Amotz 1987; Curtain et al. 1987; Moulten et al. 1987; Borowitzka 1988; Borowitzka and Borowitzka 1988, 1989; ; Lers et al. 1990; Schlipalius 1990).

The vast majority of algae contain ß-carotene as an integral component of the photosynthetic apparatus (see Chapter 3) although only a limited number of algal species can accumulate ß-carotene in high amounts. These include *Dunaliella* spp., *Trentepohlia aurea* and *Protosiphon botryoides* (see Goodwin 1980). A large number of other terrestrial algae (e.g. *Hormidium* spp.) are also reported to accumulate carotenoid (probably ß-carotene) in red-coloured oil droplets in the cells during drought conditions. In *Dunaliella,* ß-carotene accumulates in membrane-free osmiophilic globules located in the interthylakoid spaces (Ben-Amotz et al. 1982). Berland et al. (1989) have described a halotolerant strain of the cyanobacterium *Aphanothece* sp. which accumulates ß-carotene to 70%–80% of its total carotenoid content.

Not all species of *Dunaliella* have the potential to accumulatelarge amounts of ß-carotene (e.g. *D. parva 19/9*) and one has been reported to accumulate canthaxanthin under stress conditions (Massyuk and Radchenko 1973). *Dunaliella bardawil* and *D. salina* (there is continuing confusion concerning the exact taxonomic position of these species) have been reported to accumulate ß-carotene to levels approaching 14% of cell dry weight (Ben-Amotz et al. 1982; Ben-Amotz and Avron 1990) and are therefore used commerically.

Dunaliella spp. have been shown to produce a mixture of all-*trans* and 9-*cis*-ß-carotene (synthetic ß-carotene is >99% all-*trans*). Ben-Amotz et al. (1988) and Ben-Amotz and Avron (1989) have shown that the proportion of ß-carotene in the 9-*cis* form accumulated by *D. bardawil* increases with light intensity, so that the ratio of 9-*cis* to all-*trans* rises from 0.2:1 in low light to over 1.5:1 when the alga is grown at high light intensities (Fig. 6.16). When *Dunaliella* is grown in the presence of the potent inhibitor of carotenoid biosynthesis norflurazon, the ratio of *cis*- (tentatively identified as 9-*cis*):all-*trans* phytoene is also dependent on the light intensity.

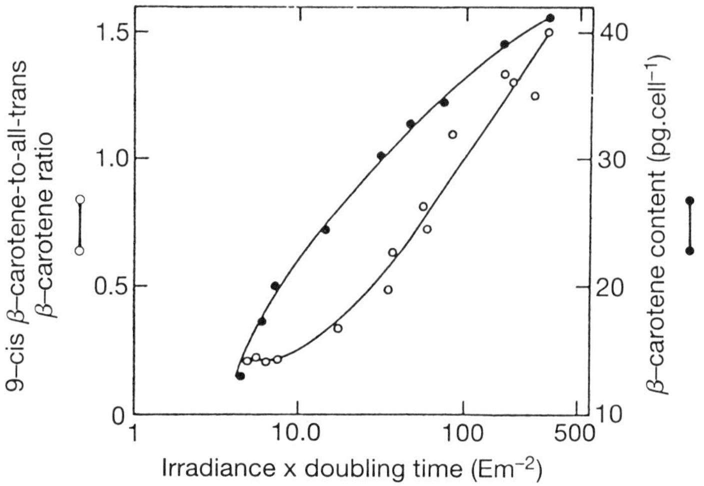

Figure 6.16 Dependence of the ß-carotene content and ratio of 9-*cis*- to all-*trans*-ß-carotene in *Dunaliella bardawil* on the irradiance per division cycle (Ben-Amotz et al. 1988).

The function of the accumulated ß-carotene and, indeed, other carotenoids that are synthesised in very high amounts (e.g. astaxanthin) when microalgae are exposed to stress conditions is not known, although Ben-Amotz et al. (1987, 1989) have observed that ß-carotene-rich *Dunaliella* are substantially more resistant to photoinhibitory stress than are *Dunaliella* that are low in ß-carotene. Ben-Amotz and co-workers have proposed that the accumulated ß-carotene effectively protected *Dunaliella* by screening irradiation in the blue region of the absorption spectrum.

(b) Astaxanthin and other carotenoids

Studies by Czygan (1968a,b, 1970), Goodwin (1980), Borowitzka (1988) and Nelis and De Leenheer (1990) have showed that a wide range of microalgae have the ability to synthesise high levels of astaxanthin or other ketocarotenoids when exposed to stress conditions (see Table 6.5). The astaxanthin accumulated in higher plants and in algae occurs as the (3*S*,3´*S*) isomer (Renstrøm et al. 1981a,b).

It is quite clear though that the synthesis of ketocarotenoids under stress conditions is much more widely distributed amongst algal taxa than was reported in these reviews and there are many microalgal cultures that turn red or orange as they age due to the accumulation of ketocarotenoids (A. Tsavalos, J. Day and A.

Young, unpublished data). Heterotrophically grown *Haematococcus pluvialis* accumulates secondary carotenoids to levels approaching that of the light-grown cultures, especially on an acetate-based media, albeit at a much slower rate. Thus, although light is not essential for astaxanthin synthesis it is stimulatory (Goodwin and Jamikorn 1954; Kobayashi et al. 1991), as it is in *Chlorococcum wimmeri* (Brown et al. 1967).

Table 6.5 The secondary carotenoid composition of selected microalgae. The data were compiled from [a]Grung et al. 1990; [b]Grung et al. 1989;[c]Kobayashi et al. 1991; [d]Renstrøm et al. 1981

Carotenoid	*Botryococcus braunii* [a,b]		*Euglena sanguinea*[a]	*Haematoccus pluvialis* [c,d]			*Haematococcus* sp.[a]	
ß-Carotene	1*	6*	—	—	—	1	5	3
Echinenone	28	20	—	—	—	—	4	—
Canthaxanthin	46	46	—	—	—	—	4	—
Adonirubin esters	—	—	—	6	9	3	—	3
Astaxanthin monoesters	—	—	—	74	69	76	46	71
Astaxanthin bisesters	—	—	91	10	20	7	34	16
Astaxanthin (free)	—	—	—	—	—	1	1	3
3'-Hydroxyechinenone	10	—	—	—	—	—	—	—
Lutein/zeaxanthin	5	17	1	9	1	7	6	4
Violaxanthin	—	10	—	—	—	2	—	1
Neoxanthin	—	—	—	—	—	1	—	1
Diatoxanthin	—	—	6	—	—	—	—	—
Unidentified	—	—	2	—	—	—	—	—
Carotenoid (% dry weight)	0.25-0.30		0.7	—	—	—	0.7	2.0

The alga that has received the most attention for the production of astaxanthin is *Haematococcus,* although there are very few published reports of carotenoid synthesis and its regulation. Cultures may contain as much as 6%–8% dry weight as astaxanthin (Tsavalos et al. 1992). As with many other algae, when maintained under optimum growth conditions the alga remains green but following nutrient deprivation, generally nitrogen starvation, the cultures turn red as astaxanthin is accumulated. Morphological changes usually accompany the alterations in the pigment composition of these cells, and large, deep-red, encysted cells are often formed, although astaxanthin may also be accumulated in high amounts in vegetative cells in *H. pluvialis* (astaxanthin accumulation and encystment may, however, be seperated).The formation of these cysts may be accompanied by a high degree of bleaching of the cells (Kobayashi et al. 1991; M. Harker, A. Tsavalos and A. Young, unpublished data).

Unlike *Dunaliella* spp. in which the accumulated ß-carotene is deposited in oil droplets in the interthylakoid space, those algae that synthesise ketocarotenoids are thought to localise them in cytoplasmic oil droplets. Vechel et al. (1991) have examined the distribution of secondary carotenoids in the pigment–protein complexes of the alga *Eremosphaera viridis*. Under conditions of nitrogen depletion, this alga accumulates relatively small amounts of canthaxanthin, echinenone and astaxanthin. Traces of these pigments were found to be associated with LHCI. Similar trends are seen for other prymnesiophycean algae, and for members of the Chlorophyceae (e.g. *Chlamydomonas nivalis*) and Euglenophyceae (e.g. *E. sanguinea*). Grung et al. (1989) have shown that *B. braunii* accumulates high levels of echinenone and canthaxanthin during the stationary phase of growth (see Table 6.5). *Chlorella fusca* also accumulates these two ketocarotenoids at high levels (Goodwin 1980). In the cyanobacteria, members of the Nostocales (e.g. *Phormidium favelorarum*) and Chroococcales can produce high levels of echinenone. In many of these algae, large amounts of canthaxanthin may also be accumulated.

REFERENCES

Abadia, A., Lemoine, Y., Trémolières, A. *et al.* (1989) Iron deficiency in pea: Effects on pigment, lipid and pigment–protein complex composition of thylakoids. *Plant Physiol. Biochem.*, **27**: 679-687.

Anderson, J.M. and Barrett, J. (1986) Light-harvesting pigment–protein complexes of algae. In: Staehelin, L.A. and Arntzen, C.J. (eds.), *Encyclopedia of Plant Physiology*, Vol. 19 2nd. Ed. Springer-Verlag, Berlin, pp. 144-157.

Anderson, J.M., Chow, W.S. and Goodchild, D.J. (1988) Thylakoid membrane organisation in sun/shade acclimation. *Aust. J. Plant Physiol.*, **15**:11-26.

Asada, K. and Takahashi, M. (1987) Production and scavenging of active oxygen in photosynthesis. In: Kyle, D.J., Osmond, C.B. and Arntzen, C.J. (eds.), *Photoinhibition*. Elsevier, Amsterdam, pp. 227-287.

Asada, K., Takahashi, M., Nakano, Y. and Shimada, S. (1982) Active oxygen in chloroplasts: Suppression of singlet oxygen formation and scavenging of hydrogen peroxide. In: Nozaki, M., Yamamoto, S., Ishimura, Y. *et al.* (eds.), *Oxygenases and Oxygen Metabolism*. Academic Press, London, pp. 255-266.

Ashikawa, I., Kito, M., Satoh, K. *et al.* (1987) All-*trans*-ß-carotene-5,6-epoxide in thylakoid membranes. *Photochem. Photobiol.*, **46**: 269-275.

Axelsson, L., Klockare, B., Ryberg., H. and Sandelius, A.S. (1981) The function of carotenoids during chloroplast development. II. Photostability and organisation of early forms of chlorophyll(ide). In: Akoyunoglou, G. (ed.), *Photosynthesis*, Vol. V. Balaban International Science Services, Philadelphia, pp. 285-293.

Bahl, J. (1977) Chlorophyll, carotenoid and lipid content in *Triticum sativa* L. Plastid envelopes, prolamellar bodies, stroma lamellae, and grana. *Planta*, **136**: 21-24.

Barry, P., Young, A.J. and Britton, G. (1990) Photodestruction of pigments in higher plants by herbicide action. I. The effect of DCMU (diuron) on isolated chloroplasts. *J. Exp. Bot.*, **41**: 123-129.

Barry, P., Young, A.J. and Britton, G. (1991) Accumulation of pigments during the greening of etiolated seedlings of *Hordeum vulgare* L. *J. Exp. Bot.*, **42**: 229-234.

Barry, P., Evershed, R.P., Young, A.J.., Prescott, M.C. and Britton, G. (1992) Characterisation of carotenoid acyl esters produced in drought-stressed barley seedlings. *Phytochemistry*, **31**:3163-3168.

Ben-Amotz, A. (1987) Effect of irradiance and nutrient deficiency on the chemical composition of *Dunaliella bardawil* Ben-Amotz and Avron (Volvocales, Chlorophyta). *J. Plant Physiol.*, **131**: 479-487.

Ben-Amotz, A. and Avron, M. (1983) On the factors which determine massive ß-carotene accumulation in the halotolerant alga *Dunaliella bardawil*. *Plant Physiol.*, **72**: 593-597.

Ben-Amotz, A. and Avron, M. (1989) The wavelength dependence of massive carotene synthesis in *Dunaliella bardawil* (Chlorophyceae). *J. Phycol.*, **25**: 175-178.

Ben-Amotz, A. and Avron, M. (1990) The biotechnology of cultivating the halotolerant alga *Dunaliella*. *Trends Biotech.*, **8**: 121-126.

Ben-Amotz, A., Katz, A. and Avron, M. (1982) Accumulation of ß-carotene in halotolerant algae: Purification and characterization of ß-carotene-rich globules from *Dunaliella bardawil* (Chlorophyceae). *J. Phycol.*, **18**: 529-537.

Ben-Amotz, A., Gressel, J. and Avron, M. (1987) Massive accumulation of phytoene induced by norflurazon in *Dunaliella bardawil* (Chlorophyceae) prevents recovery from photoinhibition. *J. Phycol.*, **23**: 176-181.

Ben-Amotz, A., Lers, A. and Avron, M. (1988) Stereoisomers of ß-carotene and phytoene in the alga *Dunaliella bardawil*. *Plant Physiol.*, **86**: 1286-1291.

Ben-Amotz, A., Shaish, A. and Avron, M. (1989) Mode of action of the massively accumulated ß-carotene of *Dunaliella bardawil* in protecting the alga against damage by excess irradiation. *Plant Physiol.*, **91**: 1040-1043.

Berland, B., Le Champion, T. and Campos Baeta Neves, M.H. (1989) Interaction de la salinité et de la temperature sur la morphologie, la croissance et la composition cellulaire d'une cyanobacteria halotolerante (*Aphanothece* sp.). *Bot. Mar.*, **32**: 317-329.

Björkman, O. (1987) Low-temperature chlorophyll fluorescence in leaves: Its relation to photon yield of photosynthesis in photoinhibition. In: Kyle, D.J., Osmond, C.B. and Arntzen, C.J. (eds.), *Photoinhibition*, Topics in Photosynthesis, Vol. 9. Elsevier, Amsterdam, pp. 123-144.

Borland, C.F., Cogdell, R.J., Land, E.J. and Truscott, T.G. (1989) Bacteriochlorophyll *a* triplet state and its interactions with bacterial carotenoids and oxygen. *J. Photochem. Photobiol.*, **3**: 237-245.

Borowitzka, M.A. (1988) Vitamins and fine chemicals from micro-algae. In: Borowitzka, M.A. and Borowitzka, L.J. (eds.), *Micro-algal Biotechnology*. Cambridge University Press, Cambridge, pp. 153-196.

Borowitzka, M.A. and Borowitzka, L.J. (1988) *Dunaliella*. In: Borowitzka, M.A. and Borowitzka, L.J. (eds.), *Micro-algal Biotechnology*. Cambridge University Press, Cambridge, pp. 27-58.

Borowitzka, L.J. and Borowitzka, M.A. (1989) ß-Carotene (provitamin A) production with algae. In: Vandamme, E.J. (ed.), *Biotechnology of Vitamins, Pigments and Growth Factors*. Elsevier Applied Science, London, pp. 15-26.

Borowitzka, L., Borowitzka, M.A. and Moulton, T.P. (1984) The mass culture of *Dunaliella salina* for fine chemicals: From laboratory to pilot plant. *Hydrobiology*, **116**: 115-134.

Borowitzka, M.A., Moulton, T.P. and Borowitzka, L.J. (1986) Salinity and the commercial production of ß-carotene from *Dunaliella salina*. *Nova Hedwig.*, **83**: 224-229.

Borowitzka, M.A., Borowitzka, L.J. and Kessley, D. (1990) Effects of salinity increase on carotenoid accumulation in the green alga *Dunaliella salina*. *J. Appl. Phycol.*, **2**: 111-119.

Bortitz, S. (1964) Physiologische und biochemische beitrage zur rauchschades-forschung: 1. Mitteilung. *Biologische Zentralblatt*, **83**: 501-513.

Bramley, P.M. (1991) Carotenoid biosynthesis. In: Kirkwood, R.F. (ed.), *Target Sites of Herbicides Action*. Plenum Press, New York, pp. 95-122.

Brandt, P. and Wilhelm, C. (1990) The light-harvesting system of *Euglena gracilis* during the cell cycle. Planta, 180: 293-296.

Britton, G., Powls, R. and Schulze, R. M. (1977) The effect of illumination on the pigment composition of the ζ-carotenic mutant, PG1, of *Scenedesmus obliquus*. *Arch. Microbiol.*, **113**: 281-284.

Britton, G., Barry, P. and Young, A.J. (1989) Carotenoids and chlorophylls: Herbicide inhibition of carotenoid biosynthesis. In: Dodge, A.D. (ed.). *Herbicides and Plant*

Metabolism. Cambridge University Press, Cambridge, pp. 51-72.

Brown, T.E., Richardson, F.L. and Vaughn, M.L. (1967) Development of red pigmentation of *Chlorococcum wimmeri* (Chlorophyta: Chlorococcales). *Phycologia,* **6**:167-184.

Burton, G.W. (1990) Antioxidant properties of carotenoids. *J. Nutr.*, **119**:109-111.

Burton, G.W. and Ingold, K.U. (1984) ß-Carotene: An unusual type of antioxidant. *Science*, **224**: 569-573.

Buschmann, C., Grumbach, K.H. and Bach, T.J. (1980) Herbicides which inhibit photosystem II or produce chlorosis and their effect on production and transformation of pigments in etiolated radish seedlings (*Raphanus sativus*). *Physiol. Plant.*, **49**:455-458.

Cammarata, K., Plumley, F.G. and Schmidt, G.F. (1990) Reconstitution of light-harvesting complexes: A single apoprotein binds Chl*a*, Chl*b*, and xanthophylls. In: Baltscheffsky, M. (ed.), *Current Research in Photosynthesis*, Vol. II. Kluwer Academic Publishers, Dordrecht, pp. 341-344.

Chow, W.S. and Anderson, J.M. (1987) Photosynthetic responses of *Pisum sativum* to an increase in irradiance during growth. II. Thylakoid membrane components. *Aust. J. Plant Physiol.*, **14**:9-19.

Cogdell, R. (1988) The functions of pigments in chloroplasts. In: Goodwin, T.W. (ed.), *Plant Pigments*. Academic Press, London, pp. 183-230.

Cogdell, R.J. and Frank, H. (1988) How carotenoids function in photosynthetic bacteria. *Biochim. Biophys. Acta*, **895**:63-79.

Cunningham, F.X., Dennenberg, R.J., Mustardy, L. *et al.* (1989) Stoichiometry of photosystem I, photosystem II, and phycobilisomes in the red alga *Porphyridium cruentum* as a function of growth irradiance. *Plant Physiol.*, **91**:1179-1187.

Cunningham, F.X., Dennenberg, R.J., Jursinic, L. and Gantt, E. (1990) Growth under red light enhances photosystem II relative to photosystem I and phycobilisomes in the red alga *Porphyridium cruentum*. *Plant Physiol.*, **93**:888-895.

Curtain, C.C., West, S.M. and Schlipalius, L. (1987) Manufacture of ß-carotene from the salt lake alga *Dunaliella salina*: The scientific and technical background. *Aust. J. Biotechnol.* **1**:51-57.

Czygan, F.-C. (1968a) Sekunder-carotenoide in grünalgen. I. Chemie, vorkommen und faktoren, welche die bildung dieser polyene beeinflussen. *Arch. Mikrobiol.*, **61**:81-102.

Czygan, F.-C. (1968b) Sekunder-carotenoide in grünalgen. II. Untersuchungen zur biogenese. *Arch. Mikrobiol.*, **62**:209-236.

Czygan, F.-C. (1970) Blutregen und blutschnee: Stickstoffmangel-zellen von *Haematococcus pluvialis* und *Chlamydomonas nivalis*. *Arch. Mikrobiol.* **74**: 69-76.

Czygan, F.-C. and Kalb, K. (1966) Untersuchungen zur biogenese der carotenoide in *Trentepohlia aurea*. *Z. Pflanzenphysiol.*, **55**: 59-64.

Damm, I., Knoetzel, J. and Grimme, L.H. (1987) On the protective role of carotenoids in the PSI reaction centre and LHCI complexes of the thylakoid membrane. In: Biggins, J.W. (ed.), *Progress in Photosynthesis Research*. Vol. II. Kluwer Academic Publishers, Dordrecht, pp. 607-610.

De la Torre, W. R. and Burkey, K.O. (1987) Adaptation of barley thylakoid membranes to changing light intensity. In: Biggins, J. (ed.), *Progress in Photosynthesis Research,* Vol. 2. Martinus Nijhoff Publishers, Dordrecht, pp. 343-346.

De la Torre, W. R. and Burkey, K.O. (1990) Acclimation of barley to changes in light intentsity: Chlorophyll organisation. *Photosynth. Res.*, **24**:117-125.

Demmig-Adams, B. and Adams, W.W. (1992). Carotenoid composition in sun and shade leaves of plants with different life forms. *Plant Cell Environ.*, **15**:411-419.

Elstner, E.F., Osswald, W. and Youngman, R.J. (1985) Basic mechanisms of pigment bleaching and loss of structural resistance in spruce (*Picea abies*) needles: advances in phytochemical diagnostics. *Experientia*, **41**:591-597.

Fabian-Galan, G. (1972) Influence of K and P deficiency on carotenoids from sunflower leaves. *Rev. Roum. Biol. Ser. Bot.*, **17**:263-269.

Foote, C.S. (1991) Definition of Type I and Type II photosensitized oxidation. *Photochem.*

Photobiol., **54**:5.

Frank, H., Violette, C.A., Trautman, J.K., Shreve, A.P., Owens, T.G., and Albrecht, A.C. (1991) Carotenoids in photosynthesis: structure and photochemistry. *Pure Appl. Chem.*, **63**:109-114 .

Goodwin, T.W. (1980) *The Biochemistry of Carotenoids*, 2nd Ed., Vol. 1, Plants. Chapman and Hall, London.

Goodwin, T.W. and Jamikorn, M. (1954) Studies in carotenogenesis. II. Carotenoid synthesis in the alga *Haematococcus pluvialis*. *Biochem. J.*, **57**: 376-381.

Grill, D. and Pfeifhöfer, H.W. (1985) Carotinoide in fichtennadelen. II. Quantitative untersuchungen. *Phyton*, **25**:1-15.

Grumbach, K.H. (1984) Biosynthesis of acylic and cyclic carotenoids in higher plants and the control by phytochrome. *Physiol. Plant.*, **60**:389-394.

Grumbach, K.H. and Lichtenthaler, H.K. (1975) Kinetics of lipoquinone and pigment synthesis during red light-induced thylakoid formation in etiolated barley seedlings. *Z. Naturforsch.*, **30c**:337-341.

Grumbach, K.H. and Lichtenthaler, H.K. (1982) Chloroplast pigments and their biosynthesis in relation to light intensity. *Photochem. Photobiol.*, **35**:209-212.

Grung, M., Metzger, P. and Liaaen-Jensen, S. (1989) Primary and secondary carotenoids in two races of the green alga *Botryococcus braunii*. *Biochem. Syst. Ecol..*, **17**:263-269.

Grung, M., Bjerkeng, B., Borowitzka, M.A., Skulberg, O. and Liaaen-Jensen, S. (1990) Alternative sources of astaxanthin, including secondary carotenoids of microalgae. Abstract, *9th.,International Symposium on Carotenoids,* (Kyoto), 217.

Gut, H., Rutz, C., Matile, R. and Thomas, H. (1987) Leaf senescence in a non-yellowing mutant of *Festuca pratensis*: Degradation of carotenoids. *Physiol. Plant.*, **70**:659-663.

Halliwell, B. (1991) Oxygen radicals: Their formation in plant tissues and their role in herbicide damage. In: Baker, N.R. and Percival, M.P. (eds.), *Herbicides*, Topics in Photosynthesis, Vol. 10. Elsevier Science Publishers, Amsterdam, pp 88-129.

Hasselt, P.R. van (1972) Photo-oxidation of leaf pigments in *Cucumis* leaf discs during chilling. *Acta Bot. Neerl.*, **21**: 539-548.

Hasselt, P.R. van and Berlo, H.A.C. van (1980) Photooxidative damage to the photosynthetic apparatus during chilling. *Physiol. Plant.*, **50**:52-56.

Henningsen, K.W. and Boynton, J.E. (1970) Macromolecular physiology of plastids of barley greening under low light intensity. *J. Cell Biol.*, **44**:290-304.

Humbeck, K. (1990) Light-dependent carotenoid biosynthesis in mutant C-6D of *Scenedesmus obliquus*. *Photochem. Photobiol.*, **51**:113-118.

Humbeck, K., Römer, S. and Senger, H. (1989) Evidence for an essential role of carotenoids in the assembly of an active photosystem I. *Planta*, **179**:242-250.

Humbeck, K., Römer, S. and Senger, H. (1990) The role of carotenoids in light-induced assembly of photosystem II in mutant C-6D of *Scenedesmus obliquus*. In: Baltscheffsky, M. (ed.), *Current Research in Photosynthesis,* Vol. 1. Kluwer Academic Publishers, Dordrecht, pp. 655-658.

Jin, Y.H., Tao, D.L. and Du, Y.J. (1990) Freezing tolerance, pigments and SOD of five conifers in Shenyang (China). *Acta Bot. Sinica*, **32**:702-706.

Johnson, G., Young, A.J., Horton, P. and Scholes, J. (1992) Variation in the carotenoid content of leaves from different plant species: relation to growth in sun and shade. Abstract, *AFRC Meeting on Photosynthesis*, Wye College. p.43.

Kandler, O., Miller, W. and Ostner, R. (1987) Dynamik der'akuten verglbung' der fichte. Epidemiologische und physiologische befunde. *Allge. Forstzeitsch.*, **27-29**:715-723.

Kennedy, T.A. and Liebler, D.C. (1991) Peroxyl radical oxidation of ß-carotene: Formation of ß-carotene epoxides. *Chem. Res. Toxicol.*, **4**:290-295.

Kleudgen, H.K. and Grumbach, K.H. (1983) Differences in phytochrome action on the formation of carotenoids and quinones during chloroplast development in seedlings of barley (*Hordeum vulgare*). *Physiol. Plant.*, **57**: 363-366.

Knox, J.P. and Dodge, A.D. (1984) Photodynamic damage to plant leaf tissue by rose bengal.

Plant Sci. Lett., **37**:3-8.

Knox, J.P. and Dodge, A.D. (1985) Singlet oxygen and plants. *Phytochemistry*, **24**: 889-896.

Kobayashi, M., Kakizono, T. and Nagai, S. (1991) Astaxanthin production by a green alga, *Haematococcus pluvialis,* accompanied with morphological changes in acetate media. *J. Ferment. Bioeng.*, **71**: 335-339.

Koestner, B., Czygan, F.-C. and Lange, O.L. (1990) An analysis of needle yellowing in healthy and chlorotic Norway spruce (*Picea abies*): In a forest decline area of the Fichtelgebirge (northeast Bavaria) Germany: 1. Annual time-course changes in chloroplast pigments for five different needle age classes. *Trees*, **4**: 55-67.

Kohata, K. and Wantanabe, M. (1988) Diel changes in the composition of photosynthetic pigments and cellular carbon and nitrogen in *Chattonella antiqua* (Raphidophyceae). *J. Phycol.*, **24:** 58-66.

Koyama, Y. (1991) Structures and functions of carotenoids in photosynthetic systems. *J. Photochem. Photobiol. B*, **9**:265-280.

Krinsky, N.I. (1979) Carotenoid protection against oxidation. *Pure Appl. Chem.*, **51**:649-660.

Lange, O.L., Zellner, H., Gebel., J. *et al.* (1987) Photosynthetic capacity, chloroplast pigments and mineral content of the previous year's spruce needles with and without the new flush: analysis of the forest decline phenomenon of needle bleaching. *Oecology*, **73**:351-357.

Lers, A., Biener, Y. and Zamir, A. (1990) Photoinduction of massive ß-carotene accumulation by the alga *Dunaliella bardawil.* Kinetics and dependence on gene activation. *Plant Physiol.*, **93**:389-395.

Liaaen-Jensen, S. (1978) Marine carotenoids. In: Scheuer, P.J. (ed.), *Marine Natural Products*, Vol. 2. Academic Press, New York, pp. 1-73.

Lichtenthaler, H.K. (1969a) Light-stimulated synthesis of plastid quinones and pigments in etiolated barley seedlings. *Biochim. Biophys. Acta*, **184**:164-172.

Lichtenthaler, H.K. (1969b) Die plastoglobuli von spinat; ihre große, isolierung und lipochinonzusammensetzung. *Protoplasm*, **68**:65-77.

Lichtenthaler, H.K. (1975) Regulation of prenyl chain synthesis in etiolated *Hordeum* seedlings by far-red and white light. *Physiol. Plant.*, **33**:241-244.

Lichtenthaler, H.K. and Becker, R. (1975) The influence of continuous far-red and white light on prenyl chain synthesis in plastids of *Raphanus* seedlings. *Planta*, **122**:252-258.

Lichtenthaler, H.K. and Buschmann, C. (1984) Photooxidative changes in pigment composition and photosynthetic activity of air-polluted spruce needles (*Picea abies* L.). In: Sybesma, C. (ed.), *Advances in Photosynthesis Research*, Vol. 4. Martinus Nijhoff/ Dr W. Junk, The Hague, pp. 245-250.

Lichtenthaler, H.K., Buschmann, C., Doll, M. *et al.* (1981) Photosynthetic activity, chloroplast ultrastructure and leaf characteristics of high and low light plants from sun and shade leaves. *Photosyn. Res.*, **2**: 117-128.

Lichtenthaler, H.K., Kuhn, G., Prenzel, U. *et al.* (1982a) Chlorophyll–protein levels and degree of thylakoid stacking in radish chloroplasts from high-light, low-light and bentazon-treated plants. *Physiol. Plant.*, **56**:183-188.

Lichtenthaler, H.K., Kuhn, G., Prenzel, U. *et al.* (1982b) Adaptation of chloroplast ultrastructure and of chlorophyll protein levels to high-light and low-light growth conditions. *Z. Naturforsch.*, **37c**:464-477.

Lichtenthaler, H.K., Burgstahler, R., Buschmann, C., Meier, D., Prenzel, U. and Schöntal, A. (1983) Effects of high light and high light stress on composition, function and structure of the photosynthetic apparatus. In: Moralle, R. and Clijsters, H. (eds.), *Stress Effects on Photosynthesis*. Dr. W. Junk Publishers, The Hague, pp. 353-370.

Lütke-Brinkhaus, F. and Kleinig, H. (1987) Carotenoid and chlorophyll biosynthesis in isolated plastids from mustard seedling cotyledons (*Sinapis alba* L.) during etioplast–chloroplast conversion. *Planta*, **170**:121-129.

Malhotra, K., Oelze-Karow, H. and Mohr, H. (1982) Action of light on accumulation of carotenoids and chlorophylls in the milo shoot (*Sorghum vulgare* Pers.). *Planta*, **154**:361-370.

Markgraff, T. and Oelmüller, R. (1991) Evidence that carotenoids are required for the accumulation of a functional photosystem II, but not photosystem I in the cotyledons of mustard seedlings. *Planta*, **185**:97-104.

Massyuk, N.P. and Radchenko, M.I. (1973) New taxons of *Dunaliella* Teod., genus III. *Ukranskya Botanichnya Zhournal*, **30**:468-471.

Mikulska, M., Siedlecka, J.W., Poskuta, S. and Maleeszewski, S. (1987) Effects of long-term action of high oxygen concentration on photosynthetic apparatus in french bean leaves. I. Changes in net photosynthesis rate and in chlorophylls and carotenoids contents. *Photosynthetica*, **21**:175-178.

Miller, N. and Carpentier, R. (1991) Energy dissipation and photoprotection mechanisms during chlorophyll photobleaching in thylakoid membranes. *Photochem. Photobiol.*, **54**:465-472.

Millie, D.F., Ingram, D.A. and Dionigi, C.P. (1990) Pigment and photosynthetic responses of *Oscillatoria agardhii* (Cyanophyta) to photon flux density and spectral quality. *J. Phycol.*, **26**:660-666.

Monge, E., Val, J., Heras, L. and Abadia, J. (1987) Photosynthetic composition of higher plants grown under iron stress. In: Biggins, J. (ed.), *Progress in Photosynthesis Research*, Vol. IV. Martinus Nijhoff Pubishers, Dordrecht, pp. 201-204.

Morales, F., Abadia, A. and Abadia, J. (1990) Characterization of the xanthophyll cycle and other photosynthetic pigment changes induced by iron deficiency in sugar beet (*Beta vulgaris* L.). *Plant Physiol.*, **94**:607-613.

Moulton, T.P., Borowitzka, M.A. and Vincent, D.J. (1987) The mass culture of *Dunaliella salina* for ß-carotene: From pilot plant to production plant. *Hydrobiology*, **151/152**:99-105.

Nelis, H.J. and De Leenheer, P. (1989) Microbial production of carotenoids other than ß-carotene. In: Vandamme, E.J, (ed.), *Biotechnology of Vitamins, Pigments and Growth Factors*. Elsevier, London, pp. 43-80.

Oelmüller, R. and Mohr, H. (1985) Carotenoid composition in milo (*Sorghum vulgare*) shoots as affected by phytochrome and chlorophyll. *Planta*, **164**:390-395.

Olaizola, M. and Duerr, E.O. (1990) Effects of light intensity and quality on the growth rate and photosynthetic pigment content of *Spirulina platensis*. *J. Appl. Phycol.*, **2**: 97-104.

Öquist, G. (1986) Effects of winter stress on chlorophyll organisation and functions in Scots pine. *J. Plant Physiol.*, **122**:169-179.

Öquist, G., Greer, D.H. and Ogren, E. (1987) Light stress at low temperature. In: Kyle, D.J., Osmond, C.B. and Arntzen, C. J. (eds.), *Photoinhibition*, Topics in Photosynthesis, Vol. 9. Elsevier, Amsterdam, pp. 67-88.

Paulsen, H., Rümler, U. and Rüdiger, W. (1990) Reconstitution of pigment-containing complexes from light-harvesting chlorophyll *a/b*-binding protein overexpressed in *Escherichia coli*. *Planta*, **181**:204-211.

Powls, R. and Britton, G. (1977) A series of mutant strains of *Scenedesmus obliquus*, with abnormal carotenoid compositions. *Arch. Microbiol.*, **113**: 275-280.

Plumley, F.G. and Schmidt, G.W. (1987) Reconsititution of chlorophyll *a/b* light-harvesting complexes: Xanthophyll-dependent assembly and energy transfer. *Proc. Natl. Acad. Sci. USA.*, **84**:146-150.

Price, A., Young, A.J., Beckett, P. *et al.* (1990) The effect of ozone on plant pigments. In: Baltscheffsky, M. (ed.) *Current Research in Photosynthesis*, Vol IV. Kluwer Academic Publishers, Dordrecht, pp. 595-598.

Renstrøm, B., Borch, G., Skulberg, O. and Liaaen-Jensen, S. (1981a) Optical purity of (3*S*,3´*S*)-astaxanthin from *Haematococcus pluvialis*. *Phytochemistry*, **20**: 2561-2564.

Renstrøm, B., Berger, H. and Liaaen-Jensen, S. (1981b) Esterified, optically pure (3*S*, 3´*S*)-astaxanthin from flowers of *Adonis annua*. *Biochim. Syst. Ecol.*, **9**: 249-257.

Ridley, S.M. (1977) Interaction of chloroplasts with inhibitors. Induction of chlorosis by

diuron during prolonged illumination *in vitro*. *Plant Physiol.*, **59**: 724-732.
Ridley, S.M. (1982) Carotenoids and herbicide action. In: Britton, G. and Goodwin, T.W. (eds.), *Carotenoid Chemistry and Biochemistry*. Pergamon, Oxford, pp. 353-369.
Sakaki, T., Kondo, N. and Sugahara, K. (1983) Breakdown of photosynthetic pigments and lipids in spinach leaves with ozone fumigation: role of active oxygen. *Physiol. Plant.*, **59**: 28-34.
Sandelius, A.S., Axelsson, L., Klockare, B. *et al.* (1981) The function of carotenoids during chloroplast development. IV. Protection of galactolipids from photodecomposition, sensitised by early forms of chlorophyll. In: Akoyunoglou, G. (ed.), *Photosynthesis*, Vol. V. Balaban International Science Services, Philadelphia, pp. 305-309.
Satoh, K. (1970) Mechanisms of photoinactivation in photosynthetic systems. II. The occurrence and properties of two different types of photoinactivation. *Plant Physiol.*, **11:** 29-38.
Schlipalius, L. (1990) ß-Carotene from algae. *Food Aust.*, **41**:742-745.
Schöner, S., Foyer, C., Lelandais, M. and Krause, G.H. (1990) Increase in activities of scavengers for active oxygen in spinach related to cold acclimation in excess light. In: Baltscheffsky, M. (ed.), *Current Research in Photosynthesis,* Vol. II. Kluwer Academic Publishers, Dordrecht, pp. 483-486.
El Sebba, D., Prioul, J.L. and Brangeon, J. (1987) Acclimation of adult *Lilium multiflorum* leaves to changes in irradiance: Effect on leaf photosynthesis and chloroplast ultrastructure. *J. Plant Physiol.*, **127**:431-441.
Senger, H. and Strassberger, G. (1978) Development of the photosystems in greening algae. In: Akoyunoglou, G. (ed.), *Chloroplast Development*. Elsevier, Amsterdam, pp. 367-378.
Senser, M., Höpker, K.-A., Peuker, A. and Glashagen, B. (1987) Wirkungen extremer ozonkonzentrationen auf koniferen. *Allge. Forstzeitsch.*, **27-29**:709-714.
Senser, M., Kloos, M. and Lütz, C. (1990) Influence of soil substrate and ozone plus acid mist on the pigment content and composition of needles from young spruce trees. *Env. Poll.*, **64**: 295-312.
Shimada, Y., Tanaka, A., Tanaka, Y., Takabe, T., Takabe, T. and Tsuji, H. (1990) Formation of chlorophyll–protein complexes during greening. 1. Distribution of newly synthesised chlorophyll among apoproteins. *Plant Cell Physiol.*, **31**: 639-647.
Siefermann-Harms, D. (1985) Carotenoids in photosynthesis. I. Location in photosynthetic membranes and light-harvesting function. *Biochim. Biophys. Acta*, **811**: 325-355.
Siefermann-Harms, D. (1987) The light-harvesting and protective functions of carotenoids in photosynthetic membranes. *Physiol. Plant.*, **69:** 561-568.
Siefermann-Harms, D. (1990) Protective function of the apoprotein of the light-harvesting chlorophyll–*a*/*b*–protein complex in pigment photo-oxidation. *J. Photochem. Photobiol.*, **4B**: 283-295.
Takada, S., Okuda, M. and Morishita, H. (1990) Carotenoid formation in unicellular green algae under salt stress. Abstract, *9th. International Symposium on Carotenoids* (Kyoto), p. 141.
Telfer, A., De Las Rivas, J. and Barber, J. (1991) ß-Carotene within the isolated photosystem II reaction centre: photooxidation and irreversible bleaching of this chromophore by oxidised P680. *Biochim. Biophys. Acta*, **1060**: 106-114.
Telfer, A., De Las Rivas, J., Woollin, C. and Barber, J. (1992) Dimeric behaviour of the two ß-carotenes present in isolated photosystem II reaction centres. Abstract, *AFRC Meeting on Photosynthesis*, Wye College, p. 21.
Terry, N. and Abadia, J. (1986) Function of iron in chloroplasts. *J. Plant Nutr.*, **9**: 609-646.
Thayer, S.S. and Björkman, O. (1990) Leaf xanthophyll content and composition in sun and shade determined by HPLC. *Photosynth. Res.*, **23**:331-343.
Tranquillini, W. (1964) The physiology of plants at high altitudes. *Annu. Rev. Plant Physiol.*, **15**:345-360.
Trevini, M. and Steinmuller, D. (1985) Composition and function of plastoglobuli. II. Lipid

composition of leaves and plastoglobuli during beech leaf senescence. *Planta*, **163**: 91-96.
Truscott, T.G. (1990) The photophysics and photochemistry of the carotenoids. *J. Photochem. Photobiol. B*, **6**:359-371.
Tsavalos, A.T, Harker, M., Daniels, M. and Young, A.J. (1992) Secondary carotenoid synthesis in microalgae. In: (Murata, N. ed.), *Research in Photosynthesis*, Vol. III, Kluwer, Dordrect. pp.47-51.
Vechel, B., Pistorious, E.K. and Ruppel, H.G. (1991) Occurrence of secondary carotenoids in photosystem I complexes isolated from *Eremosphaera viridis* De Bary (Chlorophyceae). *Z. Naturforsch.*, **47c**: 51-56.
Wieckowski, S. and Majewska, G. (1990) Chlorophyll photobleaching in thylakoid membranes isolated from cucumber cotyledons at various stages of greening. *J. Plant Physiol.*, **136**: 701-704.
Wilkinson, R.E. and Ohki, K. (1988) Influence of manganese deficiency and toxicity on isoprenoid synthesis. *Plant Physiol.*, **87**:841-846.
Williams, W.P., Sen, A. and Fork, D.C. (1986) Selective photobleaching of PSI-related chlorophylls in heat-stressed pea chloroplasts. *Photosynth. Res.*, **10**:75-92.
Wolfenden, J., Robinson, D.C., Cape, J.N. *et al.* (1988) Use of carotenoid ratios, ethylene emissions and buffer capacities for the early diagnosis of forest decline. *New Phytol.*, **109**:85-95.
Young, A.J. (1991) Inhibition of carotenoid biosynthesis. In: Baker, N.R. and Percival, M.P. (eds.), *Herbicides*, Topics in Photosynthesis, Vol. 10. Elsevier, Amsterdam, pp. 130-171.
Young, A.J. and Britton, G. (1989) The distribution of α-carotene in the photosysnthetic pigment–protein complexes of higher plants. *Plant Sci.*, **64**:179-183.
Young, A.J. and Britton, G. (1990a) Carotenoids and stress. In: Alscher, R.G. and Cumming, J.R. (eds.), *Stress Responses in Plants: Adaptation and Acclimation Mechanisms.* Wiley-Liss, New York, pp. 87-112.
Young, A.J. and Britton, G. (1990b) Photobleaching in the unicellular green alga *Dunaliella parva* 19/9. *Photosynth. Res.*, **25**:129-136.
Young, A.J. and Britton, G. (1990c) Carotenoids and oxidative stress. In: Baltscheffsky, M. (ed.), *Current Research in Photosynthesis*, Vol. IV. Kluwer Academic Publishers, Dordrecht, pp. 587-590.
Young, A.J., Barry, P. and Britton, G. (1989) The occurrence of ß-carotene-5,6-epoxide in the photosynthetic apparatus of higher plants. *Z. Naturforsch.*, **44c**:959-965.
Young, A.J., Britton, G. and Senser, M. (1990) Carotenoid composition of needles of *Picea abies* L. showing signs of photodamage. *Z. Naturforsch.*, **45c**:1111-1116.
Young, A.J., Wellings, R. and Britton, G. (1991) The fate of chloroplast pigments during the senescence of primary leaves of *Hordeum vulgare* and *Avena sativum*. *J. Plant Physiol.*, **137**:701-705.

7

The xanthophyll cycle

B. Demmig-Adams and W.W. Adams III

7.1 LIGHT-DEPENDENT INTERCONVERSIONS OF XANTHOPHYLLS WITHIN THE PHOTOSYNTHETIC MEMBRANE

Among all the carotenoids which are present in the photosynthetic membranes of higher plants, only three undergo rapid, light-induced changes in their concentration. These are the three xanthophylls of the xanthophyll cycle and they undergo interconversions induced by changes in light intensity. Furthermore, the acclimation of photosynthetic organs to high light involves a strong increase in the total sum of these three xanthophylls. Both the interconversions among the components of the xanthophyll cycle and the increase of the size of the total xanthophyll cycle pool occur in response to light stress. Plants are subjected to light stress in the field in high light habitats, as well as in habitats where they encounter other environmental stress factors in combination with light. The characteristics of the xanthophyll cycle are discussed here in relation to its specific association with high light stress and its recently described function in a photoprotective process of fundamental importance.

7.1.1 Occurrence of reversible de-epoxidation reactions in different taxonomic groups

It was first reported by Sapozhnikov et al. (1957; cf. Sapozhnikov 1973) that the violaxanthin content of a leaf could be reversibly decreased by exposing that leaf to light and that this process was reversed by returning the leaf to darkness. A similar finding was reported shortly thereafter by Blass et al. (1959). Yamamoto et al. (1962) showed that violaxanthin was converted into zeaxanthin (Fig. 7.1) upon exposure of leaves to light. In the following decades several groups of workers elucidated many details of the xanthophyll cycle, including the regulatory factors governing the reactions. Higher plants, ferns, mosses, and green and brown algae all possess this xanthophyll cycle (Table 7.1), which involves the conversion of violaxanthin into antheraxanthin, and of antheraxanthin into zeaxanthin (forward, de-epoxidation sequence), as well as the reverse, epoxidation sequence (zeaxanthin into antheraxanthin into violaxanthin; Fig. 7.1). This is a cycle, and not a simple equilibrium reaction, since the forward and back reaction sequences are catalysed by two different enzymes (violaxanthin de-epoxidase and zeaxanthin epoxidase, respectively), and the factors controlling each are quite different (for reviews, see Siefermann-Harms 1977; Yamamoto 1979; Hager 1980). The xanthophyll cycle found in higher plants has also been referred to as the violaxanthin cycle (e.g. Yamamoto 1979), primarily because there is another xanthophyll cycle involving only two xanthophylls (diatoxanthin and its monoepoxide diadinoxanthin) which occurs in several algal taxa (Table 7.1). The cycle involving violaxanthin, antheraxanthin, and zeaxanthin is referred to as the xanthophyll cycle in this chapter because it is not the epoxidised form violaxanthin but instead the de-epoxidised forms, and in particular zeaxanthin, which appear(s) to be the most functionally important component(s) of the cycle. The function of these xanthophylls, in a photoprotective process, is discussed in Section 7.2.

Table 7.1 Photosynthetic taxa which can form zeaxanthin but which do not contain xanthophyll epoxides nor exhibit any cyclic changes in xanthophyll composition (Group 1), photosynthetic taxa which contain the diadinoxanthin-diatoxanthin cycle (Group 2), and photosynthetic taxa which contain the xanthophyll cycle (Group 3) consisting of cyclic changes in violaxanthin, antheraxanthin, and zeaxanthin. Modified after Hager (1980)

No epoxide cycle, but zeaxanthin may be present.	*Diadinoxanthin* ↓↑ *Diatoxanthin*	*Violaxanthin* ↓↑ *Antheraxanthin* ↓↑ *Zeaxanthin*
Group 1	**Group 2**	**Group 3**
Rhodophyceae	Bacillariophyceae	Higher plants
Cryptophyceae[a]	=Diatomeae	Ferns
Cyanophyceae	Chrysophyceae	Mosses
	Xanthophyceae	Phaeophyceae
	Chloromonadophyceae	Chlorophyceae and
	Dinophyceae	individual species
	Euglenophyceae	of Rhodophyceae,
		Chrysophyceae, and
		Xanthophyceae[b]

[a]only *Cryptomonas* analysed by Hager and Stransky (1970).
[b]Aihara and Yamamoto 1968; Hager and Stransky 1970; Stransky and Hager 1970a,b; Brown and McLachlan 1982.

Blue-green algae or cyanobacteria, red algae (Rhodophyceae), and

Cryptophyceae are generally free from xanthophylls with epoxide groups. However, this is not an exclusive pattern. The presence of antheraxanthin was reported for two species of red algae (Aihara and Yamamoto 1968). Another, extensive study on the genus *Gracilaria* (Rhodophyceae) revealed the frequent occurrence of antheraxanthin, and in many cases violaxanthin was also present (Brown and McLachlan 1982). Among the Xanthophyceae and Chrysophyceae, there have also been reports of at least one species within each family possessing violaxanthin, antheraxanthin and zeaxanthin, not diadinoxanthin and diatoxanthin (Hager and Stransky 1970; Stransky and Hager 1970a; Hager 1980). However, even in those taxa which do not possess any epoxide cycle, zeaxanthin can often be found in substantial quantities. Thus, the presence of zeaxanthin is not restricted to those taxa that have the xanthophyll cycle. Zeaxanthin is among the major carotenoids frequently reported in red algae (the other major ones being α-carotene, ß-carotene, and lutein; e.g. Bjørnland and Aguilar-Martinez 1976; Liaaen-Jensen 1978), and there are also frequent reports of zeaxanthin in the blue-green algae, where this carotenoid may be a major component of some species (Stransky and Hager 1970b).

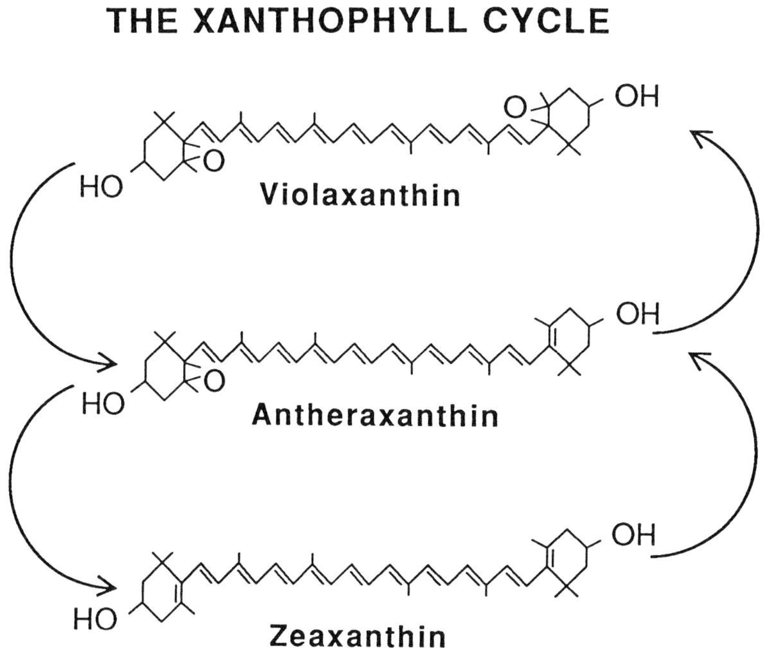

Figure7.1 The reactions that constitute the xanthophyll cycle, namely the conversion (de-epoxidation) of violaxanthin into antheraxanthin to zeaxanthin, as well as the reverse reactions (epoxidation). One enzyme is responsible for both de-epoxidation steps, and one enzyme is responsible for both epoxidation steps.

We have demonstrated that zeaxanthin can be present in lichens with blue-green algae as symbionts (Demmig-Adams et al. 1990a). Blue-green algal lichens were found to contain similar amounts of zeaxanthin as green algal lichens in habitats where they presumably experienced an excess of light (W.W. Adams, B. Demmig-Adams and O.L. Lange, unpublished data). Under no circumstances, however, were the xanthophylls that contain epoxide groups present in the blue-green algal lichens. Furthermore, zeaxanthin was not formed rapidly, at least not over the course of several hours, in these blue-green algal lichens (B. Demmig-Adams and W.W. Adams, unpublished data). In conclusion, the ability to form zeaxanthin is extremely common among aerobic photosynthetic organisms but the possession of an epoxide cycle appears to be related to the ability to form and remove zeaxanthin rapidly, i.e. within minutes instead of days.

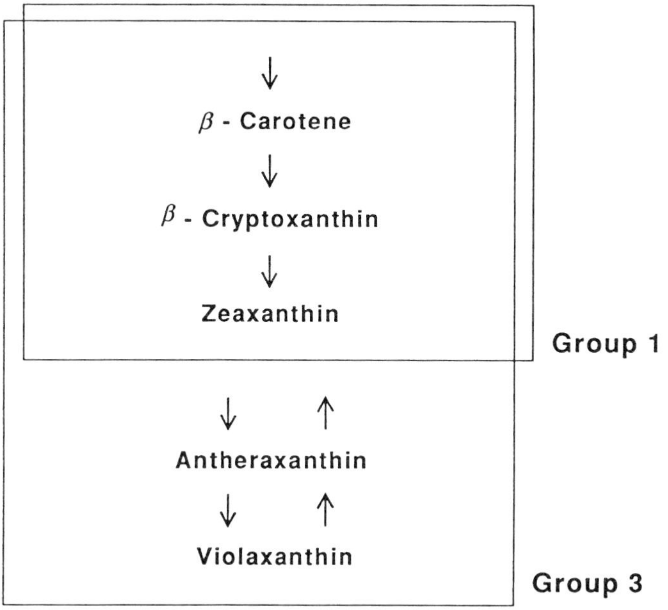

Figure 7.2 Presumed biosynthetic pathway leading to zeaxanthin in the Groups 1 and 3 organisms (see Table 7.1) as well as the xanthophyll cycle found in Group 3 organisms.

The occurrence of zeaxanthin in organisms which lack the xanthophyll cycle is consistent with the presumed general biosynthetic pathway leading to the formation of zeaxanthin, antheraxanthin and violaxanthin in all organisms (Fig. 7.2). Zeaxanthin is presumably formed first, from ß-carotene, with antheraxanthin and violaxanthin arising from the epoxidation of zeaxanthin. All xanthophylls of the xanthophyll cycle are derivatives of ß-carotene. This aspect of the biosynthetic pathway (Fig. 7.2) has recently been indirectly confirmed with a mutant of *Arabidopsis thaliana* (Duckham et al. 1991; Rock and Zeevaart 1991). This mutant accumulates greater amounts of zeaxanthin and ß-carotene than does the wild type, and it contains very low amounts of violaxanthin (all-*trans*-violaxanthin) and neoxanthin (9′-*cis*-neoxanthin). From this phenotype it was postulated that the mutant is deficient in the zeaxanthin epoxidase.

Although the xanthophyll cycle itself occurs in the thylakoid membranes of chloroplasts, the biosynthesis of the components of the cycle has been suggested to occur in the chloroplast envelope (see Chapter 3; Schultz et al. 1992). This includes not only the *de novo* synthesis of zeaxanthin, but also the generation of antheraxanthin and violaxanthin, as a zeaxanthin epoxidase has been reported to be present in the chloroplast envelope as well (Costes et al. 1979). As outlined in Fig. 7.2, these final steps leading to the synthesis of zeaxanthin, antheraxanthin and violaxanthin would therefore include: (precursors) → ß-carotene → ß-cryptoxanthin → zeaxanthin → antheraxanthin → violaxanthin (Jones and Porter 1986) in the envelope membranes, presumably followed by a transfer of violaxanthin to the thylakoid membranes (Siefermann-Harms et al. 1978). Violaxanthin de-epoxidase is found only in the thylakoid membranes, i.e. only these membranes contain the entire xanthophyll cycle.

The progress in the characterisation of the xanthophyll cycle has been influenced strongly by the development of appropriate methodology for the separation of the various carotenoids (see Chapter 10). Several thin layer chromatography (TLC) techniques may be used (e.g. Hager and Meyer-Bertenrath 1966; see also Demmig-Adams et al. 1990b). However, the recent development of separation methods using high-performance liquid chromatography (HPLC) has allowed more rapid progress. The first chromatographic methods used in the study of the cycle did not allow for the separation of lutein (usually present in large amounts) and zeaxanthin (present in variable amounts from zero to substantial), which often resulted in increases in zeaxanthin content being mistaken for an increase in lutein. More recently a partial separation of lutein and zeaxanthin by HPLC (Braumann and Grimme 1981; Prenzel and Lichtenthaler 1982; Mantoura and Llewellyn 1983; Krinsky and Welankiwar 1984; Wright and Shearer 1984; Lemoine et al. 1987; Siefermann-Harms 1988) was replaced by a baseline separation of these two xanthophylls (see Chapter 10; Thayer and Björkman 1990; Gilmore and Yamamoto 1991c; Rock and Zeevaart 1991).

The kinetics of the interconversion of the components of the xanthophyll cycle within chloroplasts or leaves are best monitored by non-intrusive optical methods. The conversion of violaxanthin into zeaxanthin (via antheraxanthin), or the reverse sequence, leads to an absorbance change in the 505 nm wavelength region of the spectrum. This technique of measuring the absorbance change was used extensively in isolated chloroplasts by Siefermann and Yamamoto (1974a, 1975), but more recently has been applied to intact leaves in which absorbance changes at about 505, 515 and 535 nm were characterised in detail (Bilger et al. 1989; Bilger and Björkman 1990, 1991). It was shown by Bilger and co-workers that the absorbance change at 505 nm and part of the absorbance change at 535 nm (the light-scattering signal) were associated with the conversion of violaxanthin into zeaxanthin, since they were absent (at 505 nm) or decreased (at 535 nm) in dithiothreitol (DTT)-treated leaves which did not undergo the conversion of violaxanthin into zeaxanthin. However, the rapid absorbance change at 515 nm was the same, independent of whether or not de-epoxidation occurred. A linear relationship exists between the absorbance change at 505 nm in leaves and the epoxidation state of the xanthophyll cycle, expressed as the ratio of (violaxanthin + 0.5x antheraxanthin):(violaxanthin + antheraxanthin + zeaxanthin) (Bilger and Björkman 1991). The potential for the use of these spectral shifts in the remote sensing of the state of the xanthophyll cycle is being explored (Gamon et al. 1990).

7.1.2 Relationship between photosynthesis, irradiance, and violaxanthin de-epoxidation in leaves

In Section 7.1.1 it was mentioned that the conversion of violaxanthin to zeaxanthin *in vivo* does not occur in darkness but only in the light. The degree to which violaxanthin is converted into zeaxanthin varies with the light level, i.e. the photon flux density (PFD), and/or with changes in the rate of photosynthesis, and this is examined in greater detail in this section. However, the interconversion of the xanthophyll cycle components does not respond to these changes in photosynthesis or PFD *per se*. Rather, the xanthophyll cycle responds to changes in the ratio of photosynthesis:PFD, i.e. the balance between the rate at which photons are absorbed by chlorophyll and the rate at which that captured energy is utilised through photosynthesis. Thus, the conversion of violaxanthin into zeaxanthin is induced by an excess of light, i.e. the absorption of more light by chlorophyll than can be utilised through the process of photosynthesis. Of course a change in the ratio of photosynthesis to PFD results from changes in either photosynthesis or PFD. Conditions where excess light is absorbed can potentially result in damage to the photosynthetic system. In Section 7.2 the association of the xanthophyll cycle, or rather its de-epoxidised state, with a photoprotective process is discussed.

(a) PFD response of violaxanthin de-epoxidation

The relationship between incident PFD, photosynthesis, the ratio of PFD:photosynthesis, and the levels of violaxanthin, antheraxanthin, and zeaxanthin in *Spinacia oleracea* (spinach) are shown in Fig. 7.3. As long as photosynthesis increases linearly with increases in PFD, there is no change in the ratio of photosynthesis:PFD (it remains maximal; for every increase in absorbed light energy there is a proportional increase in the rate of the utilisation of that absorbed light through photosynthesis), and the absorbed light is not excessive. There is also no accumulation of zeaxanthin in this portion of the light-response curve. Thus, the xanthophyll cycle is shifted predominantly towards violaxanthin under conditions of non-excessive light. As soon as photosynthesis no longer increases in proportion to the increases in PFD, more light is being absorbed than can be utilised via photosynthesis, i.e. the light becomes excessive. The ratio of photosynthesis:PFD decreases, or the ratio of PFD:photosynthesis (which is an approximate measure of excess light) increases, and violaxanthin is converted into antheraxanthin and zeaxanthin.

Similar PFD responses of the zeaxanthin level were obtained also for a variety of other species including *Helianthus annuus* (sunflower), *Monstera deliciosa* (split-leaf philodendron), and *Hedera helix* (English ivy; Demmig-Adams et al. 1989c), *Lemna gibba* (Siefermann-Harms 1977), and sun and shade leaves of *Hedera canariensis* (Algerian ivy; Bilger and Björkman 1990) (see also the response of the xanthophyll cycle to natural changes in PFD in the field in Section 7.3.1).

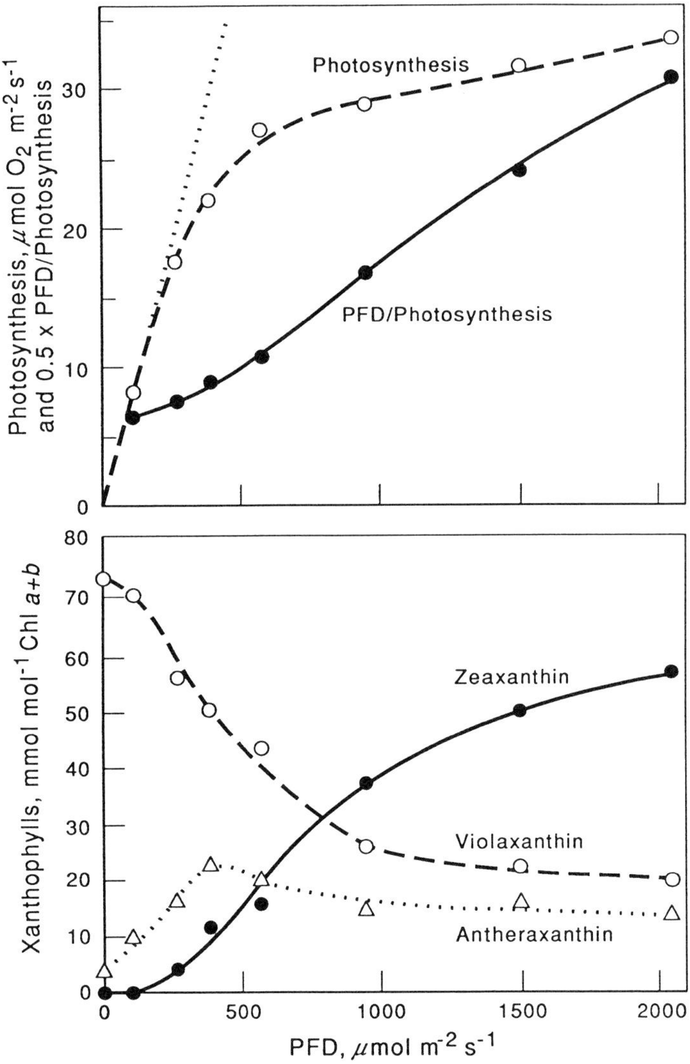

Figure 7.3 Changes in the rate of photosynthetic oxygen evolution, the ratio of PFD:photosynthesis, and the levels of violaxanthin, antheraxanthin, and zeaxanthin in leaves of spinach (*Spinacia oleracea*) exposed to different PFD. These leaves were grown in a naturally lit glasshouse with maximal PFD of 1700 µmol m^{-2} s^{-1}. (W.W. Adams and B. Demmig-Adams, unpublished data.)

Over a range of PFDs, the zeaxanthin content increases almost linearly with PFD. However, the zeaxanthin content can follow the increase in excessive light only until the violaxanthin pool is exhausted. The maximal potential for zeaxanthin formation depends therefore on the total size of the xanthophyll cycle pool. The maximal degree of conversion of violaxanthin into zeaxanthin is typically reached at PFDs exceeding the growth PFD (Demmig-Adams et al. 1989c) in leaves subjected to high light stress under controlled conditions in the laboratory. Furthermore, the maximal extent of de-epoxidation that is possible with the given amount of violaxanthin can vary as well. Earlier it was reported that typically no more than 66% of the violaxanthin present became converted into zeaxanthin (Yamamoto 1979; Hager 1980). It was therefore suggested that there may be significant pools of violaxanthin which are not available for de-epoxidation. However, more recently it has been reported that much higher percentages of violaxanthin could be converted into zeaxanthin, particularly in sun-grown leaves (Demmig et al. 1987; Thayer and Björkman 1990). Under natural conditions in the field, high degrees of de-epoxidation were also observed at noon in full sunlight in various shrubs and vines (Demmig-Adams and Adams 1992a). In these plants violaxanthin represented only 5%–12% of the total xanthophyll cycle pool at noon (see Section 7.3.1).

The absolute PFD at which zeaxanthin begins to accumulate varies among leaves and corresponds to the PFD at which the light-response curve of photosynthesis departs from linearity (Fig. 7.4). In leaves which are acclimated to shade or low light, photosynthesis saturates at relatively low PFDs, and light becomes excessive above these relatively low PFDs (e.g. English ivy in Fig. 7.4). On the other hand, leaves which developed in high light generally have much higher capacities for photosynthesis, and do not exhibit saturation of photosynthesis until relatively high PFDs are reached (e.g. sunflower in Fig. 7.4). Therefore, shade leaves tend to form and accumulate zeaxanthin at PFDs much below that which would result in a comparable accumulation of zeaxanthin in a sun leaf. This demonstrates that the accumulation of zeaxanthin is associated with the ratio of PFD:photosynthesis (see also the shade/sun acclimation of the pool of the xanthophyll cycle components in Section 7.3.3).

So far we have considered only the zeaxanthin levels which are present in leaves under steady-state conditions at various levels of PFD. One can also, however, consider the kinetics of zeaxanthin formation on sudden transfer of a leaf to high PFD. Such a transfer, using a leaf from the xerophyte *Nerium oleander*, is illustrated in Fig. 7.5. Although considerable amounts of zeaxanthin were formed in this leaf in a relatively short period of time, the de-epoxidation of violaxanthin can occur even faster (e.g. a complete conversion of the available violaxanthin into zeaxanthin within three min in spinach; Demmig-Adams et al. 1990c).

(b) Violaxanthin de-epoxidation in response to changes in photosynthesis rate at a constant PFD

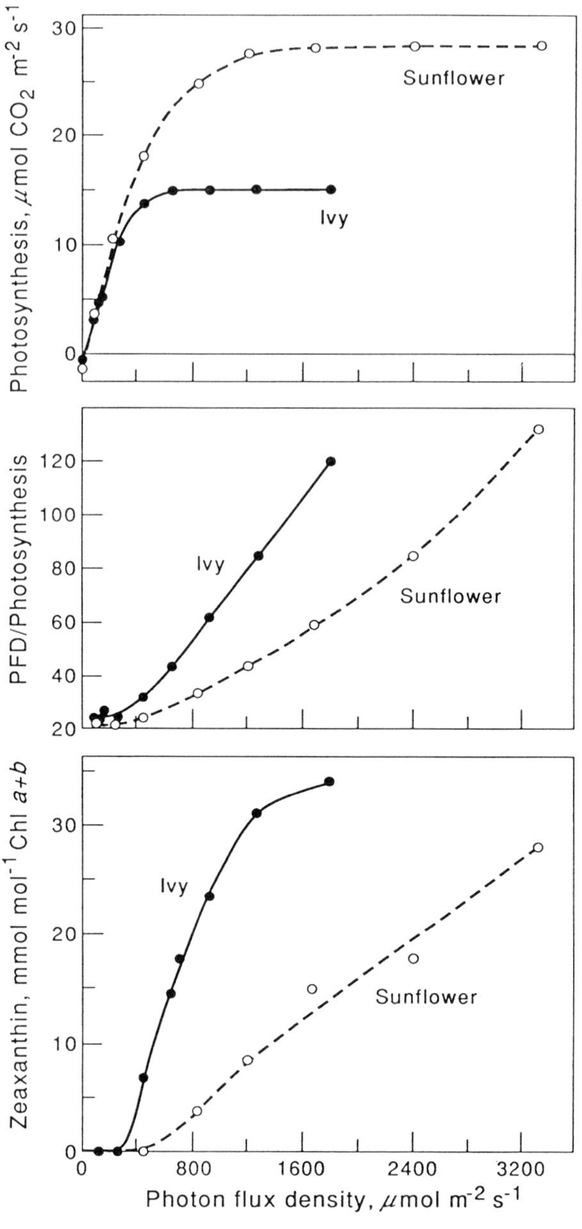

Figure 7.4 Changes in photosynthetic CO_2 exchange, the ratio of PFD:photosynthesis, and the level of zeaxanthin in leaves of ivy (*Hedera helix*) and sunflower (*Helianthus annuus*) exposed to different PFDs. (Data from Demmig-Adams et al. 1989c.)

In this section the response of the xanthophyll cycle to changes in photosynthesis is described for a constant PFD under conditions that could be created only in the laboratory, i.e. changes in the gas composition around the leaf. However, a similar phenomenon can occur under natural conditions in the field. Any stress factor (such as low temperature) which lowers photosynthesis rates will alter the ratio of PFD:photosynthesis, even at a constant PFD.

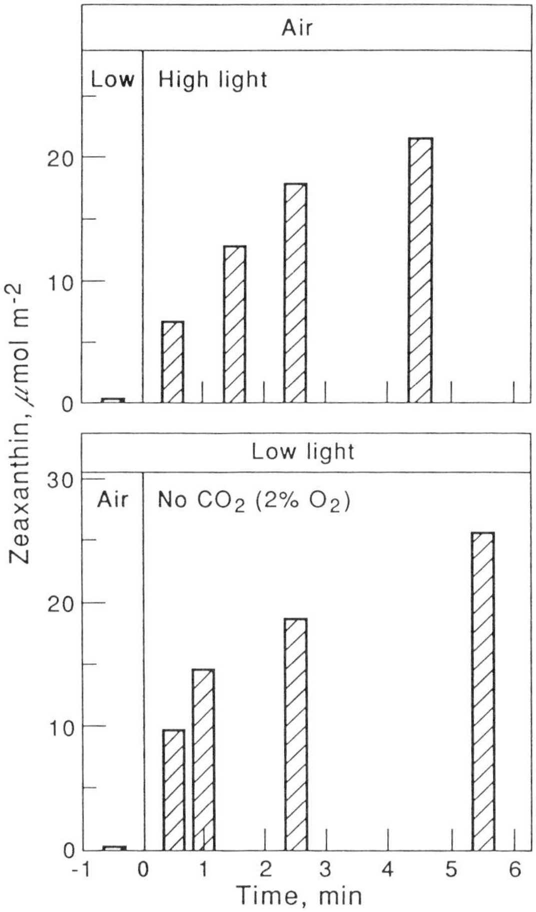

Figure 7.5 Kinetics of the accumulation of zeaxanthin in leaves of *Nerium oleander* upon transfer from low light (200 μmol photons m^{-2} s^{-1}) to high light (1800 μmol photons m^{-2} s^{-1}) in air and upon transfer from air to an atmosphere of 2% O_2, 0% CO_2 in constant low light (200 μmol photons m^{-2} s^{-1}). B. Demmig-Adams, unpublished data.

At a relatively low incident PFD of 200 μmol m^{-2} s^{-1} in the presence of air, photosynthetic electron transport proceeds at a rate which is sufficient to utilise all of the absorbed light energy, and there is no zeaxanthin present in such a leaf (Fig. 7.5). However, when the composition of the atmosphere surrounding a leaf is altered so that there is no CO_2 available for photosynthesis and only enough O_2 to allow respiration but no photorespiration, the same phenomenon that occurs upon an increase in PFD in the presence of air is observed: i.e. there is a large increase in the level of zeaxanthin in the leaf. From measurements of the zeaxanthin levels alone it is not possible to determine whether there has been an increase in the level of absorbed light energy or a decrease in the utilisation of the absorbed energy through photosynthesis (Fig. 7.5). Thus, it is clear that it is not the absolute level of photosynthesis, nor the absolute level of PFD, which is important in controlling the formation of zeaxanthin in the xanthophyll cycle, but rather the ratio of photosynthesis:PFD.

7.1.3 Regulation of the enzymic de-epoxidation and epoxidation reactions

Many aspects of the regulation of the xanthophyll cycle reactions were characterised by Yamamoto, Hager and Siefermann-Harms and their co-workers in the 1960s and 1970s (Siefermann-Harms 1977; Yamamoto 1979; Hager 1980). This characterisation was rather thorough, but left the function of the cycle (see Section 7.2.2) unresolved. The major regulatory factors of the xanthophyll cycle are summarised in Fig. 7.6. As mentioned above, there are two enzymes involved in the cycle; a de-epoxidase responsible for removing either one of the epoxide groups from violaxanthin (forming antheraxanthin) or both (forming zeaxanthin), and an epoxidase responsible for converting zeaxanthin into antheraxanthin and then into violaxanthin. Both enzymes are present in the thylakoid membranes of the chloroplast. The de-epoxidation reaction (zeaxanthin formation) is generally more rapid than the epoxidation reaction (Yamamoto 1979).

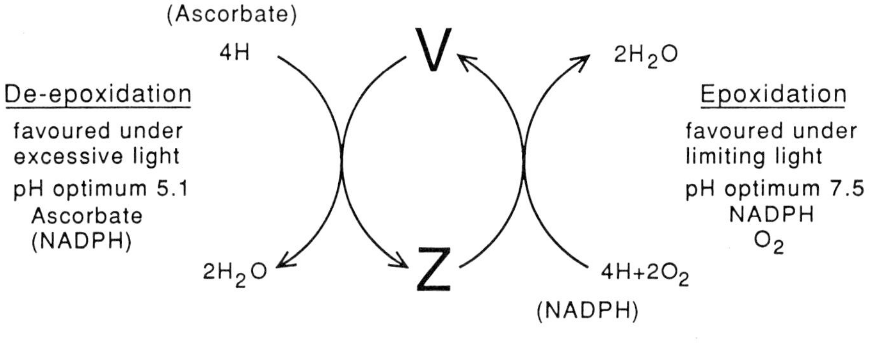

Figure 7.6 Diagram of the regulatory factors and substrates involved in the xanthophyll cycle. (Modified after Hager 1980.)

The primary regulatory factors of the xanthophyll cycle include: (i) **De-epoxidation** — a low pH optimum for the de-epoxidase (Hager 1969) and the involvement of ascorbate (and ultimately NADPH) as a reductant for the formation of zeaxanthin (Siefermann and Yamamoto 1974b). It was also postulated that the availability of violaxanthin for de-epoxidation is regulated. (ii) **Epoxidation** — a high pH optimum for the epoxidase, and the utilisation of O_2 and NADPH as substrates for the formation of violaxanthin.

Given the proper manipulation of these various factors, the biochemistry of the xanthophyll cycle can proceed in darkness. Incubation of isolated chloroplasts (Hager 1969) or leaves (Bilger et al. 1989) at low pH led to a slow formation of zeaxanthin in darkness. However, *in vivo,* light is involved in the regulation of the cycle through its effects on the pH gradient (ΔpH) across the thylakoid membrane, the supply of reductant, and perhaps generally the reduction state of the photosynthetic electron transport chain. These regulatory factors were elucidated through studies on the effects of uncouplers, electron transport inhibitors, etc., on the xanthophyll cycle in isolated chloroplasts (Hager 1969; Yamamoto 1979; Hager 1980; see also Gilmore and Yamamoto 1991a,b,d).

Several of these factors, including ΔpH (or in particular the pH on the lumen side of the thylakoid membrane) and the level of reduction of the electron transport chain which controls the supply of reducing equivalents such as NADPH, are all products of the photosynthetic electron transport reactions. All of these can thus be indicators for changes in the ratio of photosynthesis:PFD, i.e. changes in the balance between the absorption of light and the utilisation of the products of electron transport for photosynthesis. When fewer ATP and NADPH molecules are consumed than are produced, an increase in the ΔpH, and in the reduction state of the electron transport chain and thus in the supply of NADPH, is usually the result.

(a)pH and localisation

A low pH is required for a high degree of de-epoxidation leading to the formation of antheraxanthin and zeaxanthin (Yamamoto 1979; Hager 1980). The de-epoxidase has a pH optimum at 5.1, whereas the epoxidase has a pH optimum at 7.5. The de-epoxidation and epoxidation reactions can, however, proceed simultaneously in the light *in vivo*. Based upon the different pH optima for the two enzymes, a "trans-membrane model" for their localisation in the thylakoid membrane was proposed which postulated that the de-epoxidase was on the lumen side of the thylakoid membrane and the epoxidase was located on the stroma side (Yamamoto 1979; Hager 1980). It has been emphasised, however, that there is no direct evidence for this trans-membrane model (Yamamoto 1979). Although the de-epoxidase has been partially purified from spinach (Yamamoto et al. 1974) and *Lactuca sativa* (lettuce; Hager and Perz 1970), the epoxidase has not been isolated and is presumed to be bound tightly within the thylakoid membrane (Yamamoto 1979). The de-epoxidase was classified as a lipoprotein containing monogalactosyldiacylglyceride (MGDG), the major lipid component of chloroplasts, and requiring MGDG in order to function (Yamamoto et al. 1974). The de-epoxidase has been reported to have an M_r of 54 000–60 000 (Hager and Perz 1970; Yamamoto and Higashi 1978). Since DTT reversibly inhibits its activity, violaxanthin de-epoxidase presumably contains a disulphide group critical for its activity (Yamamoto and Kamite 1972; see also Section 7.2.5). The violaxanthin de-epoxidase is a mono de-epoxidase which is specific for 3-hydroxy-5,6-epoxy carotenoids in a 3*S*,5*R*,6*S* configuration and all *trans* (all-*E*) in the polyene chain (Yamamoto and Higashi 1978).

The precise localisation of violaxanthin, antheraxanthin, and zeaxanthin within the thylakoid membranes has also been uncertain. A problem confounding the localisation of the xanthophyll cycle components within the thylakoid membranes is that they are apparently easily lost from pigment–protein complexes during isolation. In the early studies, a very large portion of these xanthophylls was present in the "free pigment fraction". This finding gave rise to speculation about their being free in the membrane *in vivo*. But with gentler isolation procedures a high percentage (80%–100%) of these xanthophylls can be found to be bound to the isolated complexes (S.S. Thayer and O. Björkman, unpublished data). Violaxanthin has been found to be present in similar amounts (on a chlorophyll *a* basis) in stromal and granal lamellae of lettuce chloroplasts (Siefermann and Yamamoto 1976), which suggested that violaxanthin, i.e. the xanthophyll cycle, was associated with both photosystem (PS) II and PSI. This was directly confirmed in a study on the pigment composition of isolated PSII and PSI chlorophyll–protein complexes from *Gossypium hirsutum* (cotton; S.S. Thayer and O. Björkman, unpublished data). Similarly large amounts of violaxanthin + antheraxanthin + zeaxanthin were associated with PSI and with PSII (see also Noctor et al. 1991). Experiments in which the chlorophyll–protein complexes were isolated before and after cotton leaves were exposed to excessive light show that violaxanthin de-epoxidation to zeaxanthin occurs in each of these photosystems. Further confirmation for the operation of the xanthophyll cycle within PSI comes from a study on the pigment composition of bundle-sheath cells of C_4 plants (S.S. Thayer, H.Y. Yamamoto and O. Björkman, unpublished data). The bundle-sheath cells of C_4 plants (of the NADP-malic enzyme type) contain large amounts of PSI and are highly deficient in PSII. Even though bundle-sheath cell preparations isolated from leaves of *Zea mays* (maize; NADP-malic enzyme type) contained only trace amounts of PSII, the amount of violaxanthin + antheraxanthin + zeaxanthin (V+A+Z) (on a chlorophyll basis) equalled or exceeded that found in the whole leaf. The response of the de-epoxidation reaction within isolated bundle-sheath cells to factors such as pH, ascorbate and the inhibitor DTT were indistinguishable from those found in chloroplasts and leaves from C_3 plants. These results clearly show that the xanthophyll cycle is associated with both PSI and PSII.

(b) Ascorbate

Ascorbate does not readily cross membranes, and the movement of ascorbate into and out of the chloroplast stroma is facilitated by a carrier in the chloroplast envelope (Anderson et al. 1983; Beck et al. 1983). Once within the chloroplast, ascorbate appears to bind to the thylakoid membrane (Aristarkov et al. 1987). Given the purported localisation of the de-epoxidase on the lumen side of the thylakoid membrane, Foyer et al. (1992) have suggested that ascorbate may be a limiting factor in the operation of the de-epoxidase if the accessibility of ascorbate is restricted by its ability to cross the thylakoid membrane to act as a reductant for this enzyme. However, as pointed out above, the localisation of the two enzymes of the xanthophyll cycle may not be as simple as initially postulated. Alternatively, another (membrane-bound) redox component might be involved after ascorbate, as was discussed by Yamamoto (1979).

The re-reduction of the oxidised forms of ascorbate requires NADPH. Monodehydroascorbate is reduced by an NADPH-dependent monodehydroascorbate reductase (Asada and Takahashi 1987). Dehydroascorbate is reduced by a dehydroascorbate reductase which is dependent on reduced glutathione, which is in turn reduced by NADPH (cf. Foyer et al. 1992). Siefermann-Harms (1977) suggested that the availability of reduced ascorbate may increase with increasing PFD, as the ratio of photosynthesis:PFD decreases and NADPH becomes consumed by photosynthesis less rapidly than it is formed through electron transport. However, Foyer et al. (1991) have shown that the redox state of the ascorbate pool remains virtually constant over a wide range of PFDs. This, nonetheless, begs the question of the absolute rates of turnover of the ascorbate/dehydroascorbate pair. A constant redox state of the pool does not *per se* exclude different turnover rates at low and high PFD.

Ascorbate is involved not only in the reductive de-epoxidation of violaxanthin but also in the removal of H_2O_2, which can be formed when O_2 acts as an electron acceptor under excess light (cf. Foyer et al. 1992; see also Neubauer and Yamamoto 1991). In view of these functions under excess light, it is very interesting that the total pool of ascorbate has been found to increase over a period of days upon transfer of the plant to a higher PFD, and that in a canopy of maize plants the size of the total ascorbate pool was severalfold larger in the most exposed leaves as compared to the most shaded leaves (Foyer et al. 1992).

(c) NADPH

In addition to reducing equivalents for the de-epoxidation reaction, there is also a requirement for reducing equivalents (NADPH) for the epoxidation reaction. The reconversion of zeaxanthin into violaxanthin (epoxidation) proceeds more rapidly in low light than in darkness (Hager 1967; Yamamoto et al. 1967). The dependence on NADPH is a likely cause for this stimulation by light.

(d) Other factors

It was suggested by Siefermann and Yamamoto (1974a) that the maximum extent of violaxanthin de-epoxidation is regulated and might depend on the redox state of some component of the electron transport chain (see also Siefermann-Harms 1977; Yamamoto 1979). They speculated that the availability of violaxanthin to the de-epoxidase might in some manner be affected by the redox state of this particular electron carrier. This effect saturated at low PFD and therefore would not seem to be responsible for the dependence of violaxanthin de-epoxidation on excessive light (Siefermann-Harms 1977). It has also been suggested that the availability of violaxanthin, and thus the maximum extent of violaxanthin de-epoxidation, is controlled by the bulk pH of the thylakoid lumen (Pfündel et al. 1991).

(e) Feedback effects on the regulating factors

All of the above factors are subject to feedback regulation by the de-epoxidation state of the xanthophyll cycle. As discussed in detail in Section 7.2, zeaxanthin (or the de-epoxidised state of the xanthophyll cycle) is associated with the removal (dissipation) of excess excitation energy directly within the photochemical system. When excess excitation energy begins to accumulate, ΔpH, the reduction state of the electron transport chain, and thus NADPH availability, all begin to increase, and this consequently induces an increase in thermal energy dissipation. This increase in zeaxanthin-associated thermal energy dissipation, in turn, presumably leads to a decrease in ΔpH, the reduction state of the electron transport chain, and the availability of reducing equivalents. This has been shown for the redox state of PSII and PSI (mostly through measurements of chlorophyll fluorescence; cf. Weis and Berry 1987; Demmig-Adams et al. 1990c; Bilger and Björkman 1990; Harbinson et al. 1990) and for the redox state of the stromal compounds (Dujardyn and Foyer 1989). The redox state of the photosynthetic electron transport chain, as monitored by the redox state of PSII, is maintained at quite a low level in leaves over a range of PFDs. In normally photosynthesising leaves or cladodes in the field, PSII rarely becomes fully reduced and typically remains as low as 20%–30% reduced throughout the day (Adams et al. 1989). Direct measurements of the ΔpH in leaves as a function of PFD, however, are not available.

7.2 FUNCTION OF THE XANTHOPHYLL CYCLE

7.2.1 Excess light and its removal

The characteristics of the de-epoxidation of violaxanthin into zeaxanthin via antheraxanthin were described in Section 7.1; its regulation by light, and in particular its association with an excess of light. In addition to this formation of zeaxanthin (and antheraxanthin), another process has been found to undergo quantitatively similar changes in response to PFD, or rather in response to an excess of PFD. This second process is the thermal dissipation of excessive excitation energy directly within the photochemical apparatus. In this process the excess of excitation energy that cannot be utilised for photosynthesis is removed in a controlled and "safe" fashion immediately upon the absorption of this (excess) light. Linear relationships have been obtained between the zeaxanthin content of leaves and their energy dissipation activity for a wide variety of species and under many different conditions where excessive light was present. As is discussed in Section 7.2.5, there is strong evidence that zeaxanthin (or the de-epoxidised state of the xanthophyll cycle) is involved in this thermal energy dissipation process, protecting the photosynthetic apparatus against the adverse effects of excessive light.

An excess of excitation energy within the photochemical apparatus could potentially result in the accumulation of excitation energy. This, in turn, could lead to the formation of various destructive O_2 species, such as superoxide (O_2^-) and hydrogen peroxide (H_2O_2), as a result of the reduction of O_2 through PSI, or singlet oxygen ($^1O_2$) in the chlorophyll pigment bed (cf. Asada and Takahashi 1987; see also Chapter 6). However, the accumulation of excess excitation energy within the photochemical system is apparently counteracted: when more light is absorbed than can be used in photosynthesis, the photochemical system is much less reduced than would be expected from the ratio of photosynthesis:PFD. This is because thermal dissipation of excess excitation energy occurs directly within the light-capturing and/or primary photochemical system through a regulated removal of the excess energy.

7.2.2 Correlations between zeaxanthin and energy dissipation activity

The level of thermal energy dissipation is currently measured from changes in the yield of chlorophyll fluorescence. A high activity of thermal energy dissipation causes the chlorophyll fluorescence yield to be low, i.e. causes quenching of chlorophyll fluorescence (see Section 7.2.3). Figure 7.7 shows typical examples of the concomitant changes in these two parameters with changes in incident PFD in two species. These are the same English ivy and sunflower leaves which were examined in Section 7.1, with the level of energy dissipation depicted by the rate constant for thermal energy dissipation in the chlorophyll pigment bed, k_D. As long as the rate of photosynthesis increases linearly with PFD, no zeaxanthin is formed and no thermal energy dissipation occurs in either of the leaves. This is as would be expected, given that the light which is being absorbed is not excessive. When photosynthesis becomes light saturated, there is an increase in both zeaxanthin level and energy dissipation activity, in a quantitatively similar fashion. A close correlation exists between these two parameters in both ivy and sunflower. In the ivy leaf which developed in open shade, photosynthesis saturates at lower PFD than in sun-grown sunflower and reaches lower maximal rates. Since at a given high PFD ivy exhibits a lower photosynthesis rate than does sunflower, the same level of light absorbed by the ivy leaf represents a greater excess than that absorbed by the sunflower leaf. More energy dissipation would thus be required to remove this greater excess of light energy. That is precisely what is observed when the ivy leaf is compared with the sunflower leaf (Fig. 7.7): in ivy, much higher values of zeaxanthin per chlorophyll $a + b$ as well as of k_D are reached at PFDs approaching those of full sunlight. The energy dissipation activity at 1200 µmol m^{-2} s^{-1} is four times greater in the ivy leaf than in the sunflower leaf. The shape of the response of energy dissipation activity and zeaxanthin content to PFD is also different in ivy compared to sunflower. In ivy, both parameters increase strongly over a narrow range of PFDs, approaching saturation at half of full sunlight. Such PFD levels were never encountered by these ivy leaves in their natural growth habitat. In contrast the sunflower leaves frequently experienced full sunlight (2000 µmol m^{-2} s^{-1}, at noon) where they were growing. In sunflower a gradual, more linear increase in zeaxanthin levels and in k_D were observed, up to as much as 1.5 times full sunlight, which is far above the PFD at which photosynthesis rates saturate in these leaves. At this PFD the capacity to form zeaxanthin, and to develop thermal energy dissipation, have not yet saturated. Similar correlations between the level of zeaxanthin (or the de-epoxidation state of the xanthophyll cycle) and the activity of thermal energy dissipation have been shown for the PFD response of other species such as spinach (Demmig-Adams et al. 1990c) and Algerian ivy (Bilger and Björkman 1990).

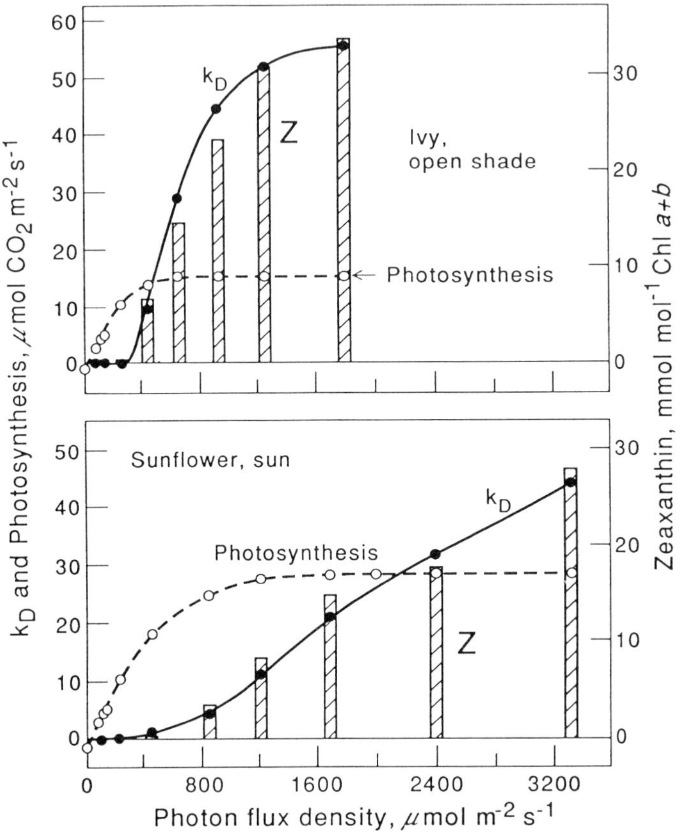

Figure 7.7 Changes in the rate constant for thermal energy dissipation in the chlorophyll (Chl) pigment bed (k_D), in the rate of photosynthetic CO_2 exchange, and in the level of zeaxanthin (Z) in response to PFD in leaves of ivy (*Hedera helix*) and of sunflower (*Helianthus annuus*). (Data from Demmig-Adams et al. 1989c.)

Another condition under which a correlation was found between the level of zeaxanthin and the energy dissipation activity was during the transition period between steady-state conditions after transfer from low to high (excessive) light. Such correlations have been reported for leaves of ivy (Demmig-Adams et al. 1989b; Bilger and Björkman 1990), and for isolated thylakoids of *Pisum sativum* (pea) (Gilmore and Yamamoto 1991d). Similar correlations have also been obtained during the transfer of a leaf to an excess of light, without any change in PFD, through a removal of the primary photosynthetic electron acceptors (e.g. Fig. 7.5) resulting in a strong decrease in the rate of photosynthetic electron transport (Demmig-Adams et al. 1989b,d).

In leaves of *Nerium oleander* plants which were subjected to slowly developing water stress, resulting in a gradual decrease in photosynthesis rates over a number of days, the levels of zeaxanthin and thermal energy dissipation were determined at the end of each night over a period of 12 days (see also Section 7.3.2 (a); Demmig et al. 1988; cf. also Demmig-Adams et al. 1989b,d). Both parameters showed a pronounced and quantitatively similar increase over this period. This means that close correlations can be observed between zeaxanthin and thermal energy dissipation when both parameters undergo sustained increases.

Finally, a correlation has been found recently between the zeaxanthin level (or de-epoxidation state of the xanthophyll cycle, as estimated from the 505 nm absorbance change) and thermal energy dissipation activity in isolated thylakoids that were treated with different concentrations of DTT so that they contained different zeaxanthin levels at one single PFD (Gilmore and Yamamoto 1991d; see also Section 7.2.5).

7.2.3 Estimation of thermal energy dissipation activity from the yield of chlorophyll fluorescence

Thermal dissipation of excitation energy within the photochemical apparatus is, at present, frequently measured from changes in the yield of chlorophyll fluorescence from PSII. The portion of such chlorophyll fluorescence quenching which is indicative of thermal energy dissipation is "non-photochemical fluorescence quenching". This non-photochemical fluorescence quenching has been studied quite intensively during the past few years and has been shown to depend on the ΔpH (Krause et al. 1982); it also depends on the presence of zeaxanthin, as is discussed in Section 7.2.5 (cf. Demmig-Adams 1990; Bilger and Björkman 1990; Gilmore and Yamamoto 1991d).

The interpretation of the changes in chlorophyll fluorescence yield is based upon the relationships depicted in Fig. 7.8. Fluorescence emission from the excited singlet state of chlorophyll is always a small fraction of the total excitation energy which has been absorbed. However, this fluorescence signal can serve as an indicator of the rate of photosynthesis and also the rate of alternative (thermal) energy dissipation. The fluorescence yield is low when either the rate of photochemistry or the rate of thermal energy dissipation is high. When such a low fluorescence yield (fluorescence quenching) is caused by a high rate of photosynthesis (photochemistry) it is referrred to as "photochemical quenching". When caused by a high activity of thermal energy dissipation it is called "non-photochemical quenching". The effect of photosynthesis can be excluded, either by measuring chlorophyll fluorescence at very low temperatures (77 K) or by using the more recent saturation pulse technique (non-intrusive, room temperature fluorescence; see also Fig. 7.9) by which the photosynthetic reaction centres are transiently closed (reduced; i.e. they are unable to undergo another charge separation) by a high intensity flash of light.

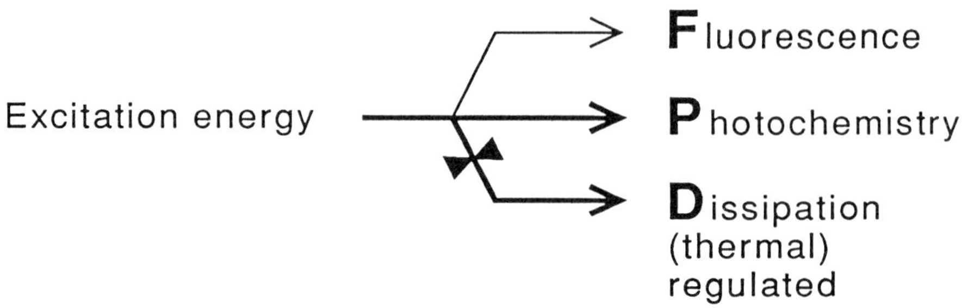

Figure 7.8 Diagram of the possible fates of excitation energy once light has been absorbed by chlorophyll. A fraction of the excitation energy is always re-emitted in the form of chlorophyll fluorescence, although it represents only a small portion of the total energy absorbed by chlorophyll. If there is a high level of either photochemistry or thermal energy dissipation, the yield of fluorescence is lowered. Photochemistry is the "normal" fate of excitation energy leading to photosynthesis. Thermal energy dissipation occurs through a regulated process whenever more energy accumulates than can be utilised through photochemistry.

Until a few years ago it was not recognised that an alternative pathway for energy dissipation existed within the photochemical apparatus, and thus there was some controversy over the interpretation of the fact that fluorescence yield could be low when photosynthesis rates were high or low. The recognition of this thermal dissipation process in the photochemical system (Weis and Berry 1987; Demmig and Björkman 1987) provided the explanation for this phenomenon.

Decreases in the yield of both the level of instantaneous fluorescence, F_o, and the maximum level of fluorescence, F_m, are typically observed upon exposure of leaves (Demmig and Björkman 1987; Demmig-Adams et al. 1989c; Bilger and Björkman 1990) or chloroplasts (Demmig-Adams et al. 1990c; Rees et al. 1989; Gilmore and Yamamoto 1991d) to an excess of light. This response has been suggested to result from a thermal energy dissipation process in the chlorophyll pigment bed (Kitajima and Butler 1975; Demmig and Björkman 1987; Horton 1990). Such a process would divert excess excitation energy away from the photochemical reaction centres.

The term F_o represents the situation in which all of the reaction centres (or acceptors of PSII) are fully oxidised (i.e. no charge separation has occurred and the potential for photochemistry is maximal at this instant); F_m represents the situation in which all of the reaction centres (or acceptors of PSII) are fully reduced (i.e. charge separation has occurred and no further photochemistry is possible at this instant). The yields of F_o and F_m have therefore been defined as

$$F_0 = \frac{k_F}{k_F + k_D + k_P} \quad (7.1)$$

$$F_m = \frac{k_F}{k_F + k_D} \quad (7.2)$$

where k_F, k_D and k_P are the rate constants for fluorescence, energy dissipation and photochemistry, respectively.

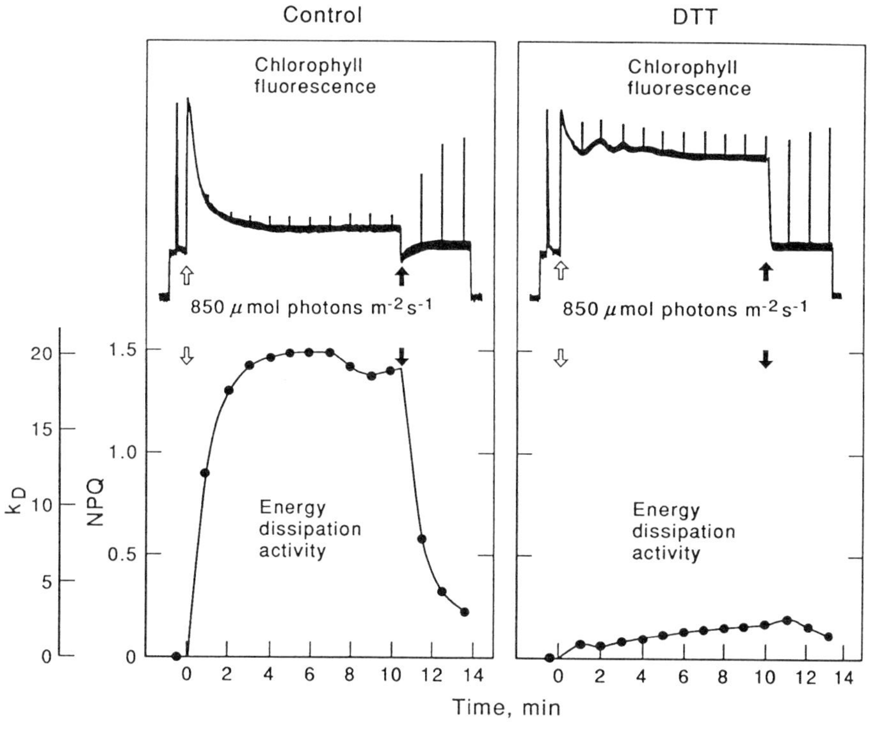

Figure 7.9 Kinetics of the changes in the chlorophyll fluorescence yield and in thermal energy dissipation represented by the rate constant for thermal energy dissipation (k_D) and in non-photochemical quenching (NPQ), both of the latter calculated from the decrease in the yield of maximal fluorescence (see text for calculations), in a control leaf of *Oxalis* sp. and in one pretreated with DTT (dithiothreitol). The control level of instantaneous fluorescence (F_o), obtained in darkness, is represented by the initial fluorescence trace, and the control level of maximal fluorescence (F_m), also obtained in darkness, is represented by the first spike in fluorescence (obtained with a saturating flash of light) at the left in each figure. Following these determinations, the actinic light (850 μmol m^{-2} s^{-1}) was switched on and maximal

fluorescence determined at one minute intervals during illumination (F_m'). After 10 min of illumination the actinic light was switched off (at the black arrows) to determine the level of instantaneous fluorescence as close as possible to that during actinic illumination (F_o'). (W.W. Adams and B. Demmig-Adams, unpublished data.)

The lowering of fluorescence yield (of both F_o and F_m) occurs in both PSII and PSI, as judged from the similarity of the changes in 77 K chlorophyll fluorescence characteristics from both photosystems (Demmig and Björkman 1987). Furthermore, the xanthophyll cycle apparently operates in both PSII and PSI (Siefermann and Yamamoto 1976; S.S. Thayer, H.Y. Yamamoto and O. Björkman, unpublished data). The occurrence of these responses in more than one site is one reason why energy dissipation activity is referred to as occurring in the "chlorophyll pigment bed" rather than in a specific pigment–protein complex.

The same type of change in the chlorophyll fluorescence yield (lowering of the yield of F_o and F_m) in PSII can furthermore be induced (in isolated chloroplasts) by conditions of either cyclic or linear electron flow (Gilmore and Yamamoto 1991d). The relationship between zeaxanthin and the energy dissipation activity was the same under both conditions. Thus, the induction of zeaxanthin formation/energy dissipation (measured in PSII) could be afforded similarly by both electron transport systems.

Until recently it was thought (Krause 1988) that there were two different processes with the same fluorescence characteristics, presumably both energy dissipation processes, one of which operated in the short term and the other over longer periods. This was the accepted interpretation because after a short-term exposure to high PFD the yield of chlorophyll fluorescence increases very rapidly upon return to non-excessive light (within minutes; this was termed "high-energy state quenching, q_E" or "pH-dependent quenching"). After longer exposures, or when a combination of high light and (long-term) water stress was involved in causing the decrease in the fluorescence yield, recovery of the fluorescence yield occurred more slowly. (Some authors referred to this slowly reversible effect as "photoinhibitory quenching, q_I"; cf. Krause 1988.) However, more recently it has become clear that one and the same process (causing a lowering of the yield of F_o and F_m) can have widely different relaxation kinetics and can apparently be responsible for both types of change in the fluorescence yield (q_E and part of q_I; cf. Demmig-Adams 1990). One line of evidence for this is that the same linear correlations were observed between the zeaxanthin content of leaves and changes in the fluorescence yield when the decrease in fluorescence yield was rapidly reversible *and* when it was slowly reversible (Demmig-Adams et al. 1989b).

There are certainly, however, cases in which slowly reversible changes in chlorophyll fluorescence characteristics do not involve zeaxanthin or thermal energy dissipation alone, but instead involve a slowly reversible effect on photochemistry which is presumably caused by "damage" to the system. The reversal of this response is thought to be due to repair of such damage (Greer et al. 1986). This damage appears to be involved particularly during the first exposure of plants grown in low light to a high PFD, and is associated with fluorescence characteristics that differ from those that are associated with increased thermal energy dissipation (Demmig and Björkman 1987; Franklin et al. 1992).

Two slightly different approaches are currently used to derive the energy dissipation activity from the degree to which chlorophyll fluorescence yield is lowered, both of which have provided quantitatively similar results. The first linear relationships between dissipation activity and zeaxanthin content were reported by use of a measure for energy dissipation activity developed by Kitajima and Butler (1975), the rate constant k_D for radiationless energy dissipation in the chlorophyll pigment bed (cf. Fig. 7.7). This arose from $F_m = k_F/(k_F+k_D)$ (see eq. 7.2), solved for k_D:

$$k_D = \frac{F_m(unquenched)}{F'_m(quenched) \bullet 0.074} - 1 \quad (7.3)$$

The factor 0.074 is the control value for a maximally high yield of F_m in a leaf with a high value of F_v/F_m of 0.855, which is typical for unstressed leaves (Björkman and Demmig 1987). For $F_v/F_m = 0.855$ the following values were assumed: $k_F = 1$, $k_P = 80$ and $k_D = 12.5$. It then follows that $F_o = 1/(1+80+12.5) = 0.0107$ and $F_m = 1/(1+12.5) = 0.074$.

The second approach to estimating energy dissipation activity from the yield of chlorophyll fluorescence, using the Stern–Volmer expression (F_m/F_m' - 1), was suggested by Bilger and Björkman (1990). A comparison of the two expressions shows that they are equivalent and will vary proportionally (see Demmig-Adams 1990; this is provided that only changes in k_D are considered, i.e. if the value of 12.5 for controls is subtracted from all values of k_D.) The expression (F_m/F_m'—1) has been referred to as NPQ, for non-photochemical quenching (see Bilger and Björkman 1990, 1991) or SV_N (Stern–Volmer quenching; Gilmore and Yamamoto 1991d).

The xanthophyll levels in leaves have also been expressed in various ways. Zeaxanthin has been related to energy dissipation on a leaf area, chlorophyll *a* + *b*, and chlorophyll *a* basis, or expressed as the ratio of zeaxanthin:(V+A+Z). Furthermore, an empirically derived correlation between the epoxidation state (EPS) of the xanthophyll cycle and the expression ((photosynthesis:PFD)x(V+A+Z)) has also been demonstrated for a variety of species exposed to high light (Thayer and Björkman 1990). The EPS is equal to the ratio of the epoxidised to the de-epoxidised components of the xanthophyll cycle (=(V+0.5A)/(V+A+Z)) with a factor of 0.5 for antheraxanthin, since it has one epoxide group. There are reports that EPS also yields close correlations with the level of thermal energy dissipation (Bilger and Björkman 1991). This might indicate that antheraxanthin possesses half the activity of zeaxanthin in thermal energy dissipation. In contrast, the ratio of zeaxanthin:(V+A+Z) yields closer correlations with thermal energy dissipation than does EPS in other situations (e.g. when rather large amounts of antheraxanthin are present; W.W. Adams and B. Demmig-Adams, unpublished data). However, more work is needed to resolve whether both products of de-epoxidation, zeaxanthin and antheraxanthin, are involved in thermal energy dissipation, i.e. whether antheraxanthin can also lower the yield of chlorophyll fluorescence.

7.2.4 The presence of zeaxanthin in non-energised thylakoid membranes

There are conditions under which zeaxanthin is present in the thylakoid membranes even when no excessive light is present. Such conditions exist, for example, following a high light exposure when PFD is low and non-excessive, but the reconversion of zeaxanthin into violaxanthin has not been completed. Under such conditions the presence of zeaxanthin (or the de-epoxidised state of the xanthophyll cycle) is often not associated with thermal energy dissipation (Demmig-Adams et al. 1989d). This occurs frequently upon a return to low light subsequent to only a short exposure to high light; the recovery of the yield of chlorophyll fluorescence is then typically more rapid than the reconversion of zeaxanthin into violaxanthin. Upon a second exposure of the same leaf to high light, before all the zeaxanthin has been reconverted into violaxanthin, the yield of chlorophyll fluorescence decreases much more rapidly than during the first exposure to high light (Demmig-Adams et al. 1989d; Gilmore and Yamamoto 1991d). This association between existing levels of zeaxanthin and a particularly rapid development of thermal energy dissipation upon exposure to an excess of light does suggest a requirement for zeaxanthin for such energy dissipation.

There is a second situation where zeaxanthin has been observed to be present without actual thermal energy dissipation (i.e. without a low yield of chlorophyll fluorescence). When leaves were incubated in an acidic medium in darkness, this resulted in the conversion of violaxanthin into zeaxanthin in the absence of an excess of (or any) light. In this case the yield of fluorescence also remained high, at least as long as leaves remained in darkness or at low (non-excessive) PFDs (Bilger et al. 1989).

7.2.5 Evidence for a causal relationship between zeaxanthin and thermal energy dissipation

Various conditions have been identified under which violaxanthin de-epoxidation is inhibited or slow (see Table 7.2). Under such conditions it was determined whether or not thermal energy dissipation was present. The most important progress was obtained by the employment of DTT treatments of whole excised leaves (Bilger et al. 1989). By allowing leaves to take up DTT through the transpiration stream under non-excessive PFDs it was possible to prevent completely the de-epoxidation of violaxanthin in these leaves during a subsequent exposure to excessive PFD. It was found that DTT inhibited not only the formation of zeaxanthin, but also the development of thermal energy dissipation, i.e. it prevented the types of change in the fluorescence yield that are associated with zeaxanthin, in leaves of spinach (Adams et al. 1990; Demmig-Adams et al. 1990c), Algerian ivy (sun and shade leaves; Bilger and Björkman 1990), cotton and *Malva parviflora* (Bilger and Björkman 1991), as well as *Cucumis sativus* (cucumber), *Cucurbita pepo* (pumpkin), sunflower, *Oxalis* sp. and *Philodendron caudatum* (W.W. Adams and B. Demmig-Adams, unpublished data). A typical time course of changes in the fluorescence yield upon tranfer of a leaf to high PFD is shown in Fig. 7.9. In the control leaf, the yield of (maximal) fluorescence decreased strongly within a few minutes under high PFD. In the lower portion of the graph the corresponding energy dissipation activity (expressed either as k_D or NPQ) is shown. Fluorescence yield rapidly increased again upon return to darkness. This type of change in the fluorescence yield has been referred to as "q_E". In DTT-treated leaves the strong decrease of fluorescence yield under high PFD does not occur, i.e. the increase in thermal energy dissipation fails to develop in DTT-treated zeaxanthin-free leaves. The example depicted in Fig. 7.9 is from a shade-grown leaf of *Oxalis*. The same phenomena have also been observed in sun leaves of other species (see above). The maximal extent of k_D or NPQ is less in this shade leaf than is typically observed in sun leaves.

Table 7.2 Summary of evidence for the involvement of zeaxanthin in thermal energy dissipation and protection of the photosynthetic apparatus against damage by excessive energy

Presence (+Z) or absence (-Z) of zeaxanthin and photosynthetic system	*Thermal energy dissipation (decreases in both F_o and F_m)*	*Redox state (i.e. reduction state of PS II)*	*Photoprotection (rapid recovery of photosynthesis)*
+Z Control leaves[a]	Yes	Low	Yes
+Z Control green-algal lichens[b]	Yes	Low	Yes
-Z DTT-treated leaves[a]	No	High	No
-Z DTT-treated green algal lichens[b]	No	High	ND
-Z Blue-green algal lichens[b]	No	High	No
-Z Leaves at chilling temperatures[c]	No (slow)	High	No
+Z DTT-treated leaves[d] (pre-exposed)	Yes	ND	ND
+Z Blue-green algal lichens[b] (from the field)	Yes	Low	Yes
+Z Leaves at chilling temperatures[c] (pre-exposed)	Yes	ND	Yes

ND = not determined
[a]Bilger et al. 1989; Bilger and Björkman 1990, 1991; Adams et al. 1990; Demmig-Adams et al. 1990c
[b]Demmig-Adams et al. 1990a,b,d
[c]Demmig-Adams et al. 1989e
[d]W. Bilger and O. Björkman, unpublished work

Another approach to test whether thermal energy dissipation is related to zeaxanthin involved the comparison of blue-green algal systems, which lack the xanthophyll cycle, with green algal systems which possess the cycle. We used lichens (associations of fungal and algal partners) with either a green or a blue-green algal partner. Due to the absence of the xanthophyll cycle, or rather the epoxides of the xanthophyll cycle, blue-green algal lichens cannot form zeaxanthin rapidly. They can, however, form it slowly, as discussed in Section 7.1. Interestingly the blue-green algal lichens which we have used contained very few phycobilipigments, in contrast to free-living blue-green algae grown in low light. The use of lichens has therefore provided a convenient system for comparing the green and blue-green algae. Studies of the chlorophyll fluorescence characteristics of green and blue-green algal lichens revealed that green algal lichens showed evidence of zeaxanthin-associated thermal energy dissipation similar to that in leaves of higher plants whereas (zeaxanthin-free) blue-green algal lichens did not (Demmig-Adams et al. 1990a). In fact the fluorescence response in zeaxanthin-free blue-green algal lichens was similar to that observed in DTT-treated zeaxanthin-free green algal lichens or leaves.

There are also other differences between control leaves/green algal lichens on one hand, and DTT-treated leaves/green algal lichens, and blue-green algal lichens on the other hand, which would be predicted from the absence of thermal energy dissipation. The reduction state of PSII is higher in DTT-treated systems and in the blue-green algal lichens which do not contain zeaxanthin or exhibit thermal energy dissipation. Such photosynthetic systems were also characterised by an increased sensitivity to high light stress (see Section 7.2.6).

Another condition under which zeaxanthin formation can be prevented (for some time) is a leaf temperature in the chilling range. At such temperatures, formation of zeaxanthin through violaxanthin de-epoxidation is very slow, as one would expect given that this is an enzymic reaction. Leaves at chilling temperatures also exhibit a very slow decrease of the fluoresence yield, i.e. a slow increase in thermal energy dissipation under excess light (Demmig-Adams et al. 1989e).

In summary, under all these different conditions, where de-epoxidation is inhibited by DTT, where de-epoxidation does not occur naturally, and where de-epoxidation is extremely slow, thermal energy dissipation (of the type associated with zeaxanthin) is also absent or develops slowly. Furthermore, the effect of the presence of zeaxanthin in blue-green algal lichens and the effect of the simultaneous presence of zeaxanthin and DTT in leaves has also been investigated. Treatment of leaves with DTT in the light when zeaxanthin had already been formed did not cause inhibition of thermal energy dissipation (W. Bilger and O. Björkman, unpublished data). This indicates that the inhibition of thermal energy dissipation by DTT occurs through the inhibition of zeaxanthin formation. (This assumes that the effect of DTT on thermal energy dissipation is the same when DTT is added in low or in high light.)

Furthermore, blue-green algal lichens collected in the field sometimes contained zeaxanthin if the lichens experienced excessive light in the moist state. These thalli exhibited the same types of change in the fluorescence yield (indicative of thermal energy dissipation) which were otherwise observed in (zeaxanthin-forming) green algal lichens, but not in zeaxanthin-free blue-green algal lichens (Demmig-Adams et al. 1990a). Zeaxanthin-containing blue-green algal lichens also exhibited a lowered PSII reduction state and a reduced sensitivity to high light stress relative to zeaxanthin-free blue-green algal lichens. In zeaxanthin-containing blue-green algal lichens the fluorescence yield can nevertheless be high in darkness (end of night), and thermal energy dissipation develops only upon exposure to excess light. This is similar to what is observed in leaves subsequent to high light treatments when zeaxanthin could still be present but the fluorescence yield had already recovered (see Section 7.2.4). It has been suggested that even in the presence of zeaxanthin an activation of thermal energy dissipation through excessive light occurs both in blue-green algal lichens and in leaves (cf. Demmig-Adams 1990).

In conclusion, zeaxanthin (and possibly antheraxanthin) is involved in the energy dissipation process which occurs in the chlorophyll pigment bed of the chloroplasts.

7.2.6 Evidence for the involvement of zeaxanthin in photoprotection

When the ability to develop thermal energy dissipation was compared in the various zeaxanthin-containing versus zeaxanthin-free photosynthetic systems, two other parameters relating to photoprotection were also determined (Table 7.2). These were the reduction state of PSII under excessive light, and the effect of prolonged exposure to highly excessive PFD on photosynthesis rates. It was predicted that, without the type of thermal energy dissipation associated with zeaxanthin, excess excitation energy should accumulate within the photochemical reaction centres, resulting in a higher reduction state of these centres. That is what was observed in those systems which did not form zeaxanthin (Table 7.2) in the DTT-treated versus control leaves and green algal lichens, and in zeaxanthin-free blue-green algal lichens versus green algal lichens. Furthermore, the blue-green algal lichens which contained zeaxanthin were also able to maintain a low reduction state of PSII. This is consistent with a function of zeaxanthin in removing excess excitation energy immediately after the latter is absorbed by chlorophyll. In another study (Khamis et al. 1990) it was reported that leaves of maize, grown with low nitrogen availability, contained more zeaxanthin, exhibited a higher level of thermal energy dissipation, and also possessed a lower reduction state of PSII at high PFD than comparable leaves grown with high nitrogen availability. This was so in spite of the fact that the leaves from plants grown in low nitrogen possessed lower photosynthesis rates, i.e. a lower capacity to utilise the absorbed light through photosynthesis.

A high reduction state (i.e. a strong accumulation of excessive excitation energy) should result in a high sensitivity to high light stress or, more specifically, to the adverse effects of high PFD on the photosynthetic process. The parameter which has been used to assess the sensitivity of leaves or other photosynthetic organs to high light stress is the rate of photosynthesis subsequent to a high light treatment. The photosynthesis rate in limiting light, i.e. the efficiency of photosynthetic energy conversion, was determined before, and at different times after, exposure to high light in leaves and other photosynthetic organs which were able to form zeaxanthin rapidly upon absorbing excessive PFD versus others in which zeaxanthin formation was prevented or naturally did not occur. Upon prolonged exposure, all systems in which no zeaxanthin was present or formed also consistently experienced sustained depressions of the photosynthesis rate (see Table 7.1). In the control systems (those which contained or formed zeaxanthin), there was a transient decrease in the photosynthesis rate which was, however, rapidly reversible. This transient decrease is the expected side-effect of a process (thermal energy dissipation) which diverts energy away from the photochemical reaction centres, i.e. away from photochemistry.

Another line of evidence also indicates a photoprotective role of thermal energy dissipation. Treatment of chloroplasts with various uncouplers inhibited the decrease of the fluorescence yield by excess light (i.e. removed q_E) and thus thermal energy dissipation. The treated chloroplasts, with a high fluorescence yield, experienced a greater depression of PSII- and PSI-driven electron transport during a high light treatment than did control suspensions with a low fluorescence yield (q_E present; Krause and Behrend 1986). From this study it is, however, not known whether all the thermal energy dissipation (q_E) was of the type associated with zeaxanthin.

7.2.7 Possible mechanisms for a zeaxanthin-associated energy dissipation process

The evidence summarised in Section 7.2.5 indicates that zeaxanthin is involved in the thermal dissipation of excess excitation energy in the chlorophyll pigment bed. The precise mechanism of this involvement is not known. On the grounds that the lowering of the yield of chlorophyll fluorescence is associated with the activity of zeaxanthin in energy dissipation, Demmig-Adams and co-workers have suggested that this thermal energy dissipation process consists of a de-excitation of the excited singlet state of chlorophyll that gives rise to fluorescence (Fig. 7.10).

Previous suggestions concerning a function of carotenoids in photoprotection have included a de-excitation of the triplet excited state of chlorophyll (though triplet–triplet energy transfer) as well as de-excitation of singlet oxygen (Fig. 7.10; see Chapter 8). On the basis of the association between zeaxanthin and the decrease of fluorescence yield, Demmig-Adams and co-workers have therefore suggested a third distinct mechanism that removes excess excitation energy closer to the point at which it accumulates upon the absorption of excess light, at the singlet excited state of chlorophyll.

Before discussing several different possibilities that have been raised for a mechanism of an involvement of zeaxanthin in thermal energy dissipation it needs to be stated that the presence of zeaxanthin alone is not sufficient to result in thermal energy dissipation. As has been pointed out earlier, it is possible for zeaxanthin to be present in the thylakoid membrane without resulting in a lowering of the yield of chlorophyll fluorescence, as long as the system is in darkness or in non-excessive light. Given the evidence that was reviewed in Section 7.2.5 it may be concluded that zeaxanthin is involved in the thermal energy dissipation process that causes the lowering of fluorescence yield. To reconcile the above two statements, it has been proposed that there are additional requirements for the induction of the thermal energy dissipation in addition to the presence of zeaxanthin. These additional conditions would then govern the activity of zeaxanthin in the energy dissipation process, as opposed to the formation of zeaxanthin in the xanthophyll cycle as a prerequisite for this type of energy dissipation. However, there are some similarities in the regulatory factors governing zeaxanthin formation from violaxanthin and the postulated activation of the zeaxanthin-associated energy dissipation process. The decreases in the yield of chlorophyll fluorescence that have now been shown to be associated with zeaxanthin had previously been reported to be abolished by the addition of uncouplers and are therefore dependent on the presence of a ΔpH across the thylakoid membrane, or on a low lumen pH (Krause et al. 1982). In addition to the existence of a ΔpH, a second condition is involved. Decreases in fluorescence yield did not develop even in the presence of a ΔpH when antimycin A was present before illumination (Oxborough and Horton 1987; Gilmore and Yamamoto 1990). This condition, induced by, but not identical with, the ΔpH, promotes thermal energy dissipation and has been addressed as the "high energy state" of the thylakoid membrane under excess light. It is assumed in several of the suggestions for the mechanism of an involvement of zeaxanthin in energy dissipation that are discussed below, that such a "high energy state" can alter the energetics/probabilities of de-excitation pathways for excited states of chlorophyll.

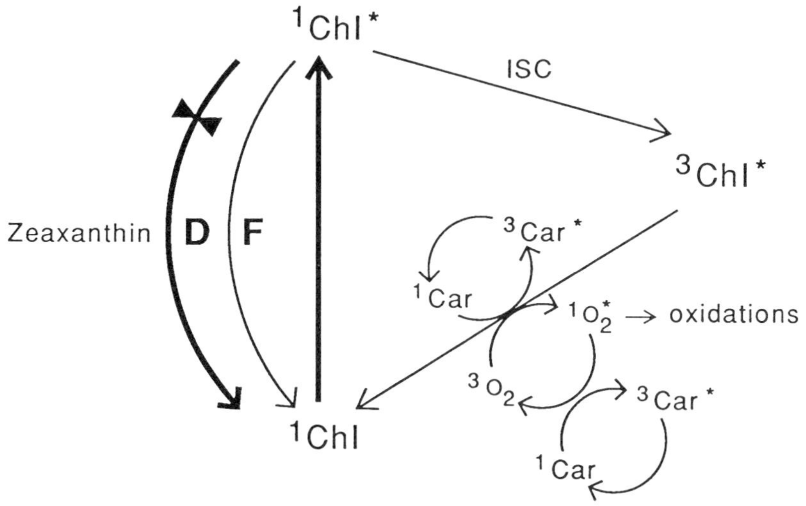

Figure 7.10 Diagram of the possible fates of excessive excitation energy when chlorophyll enters the singlet excited state after receiving energy either from a photon or from a neighbouring chlorophyll (Chl) molecule. When energy is not excessive, it usually leads to photochemistry (not shown) upon transfer to a reaction centre chlorophyll molecule. Energy can be lost through fluorescence (F), through thermal energy dissipation (D) associated with zeaxanthin, or by intersystem crossing (ISC) by which the chlorophyll molecule enters the triplet excited state. The excitation energy can then be transferred to a carotenoid molecule, or to oxygen. The singlet excited state of oxygen can also undergo de-excitation in a reaction with a carotenoid molecule. (Modified after Demmig-Adams 1990.)

The involvement of zeaxanthin in a de-excitation of the singlet excited state of chlorophyll could be either direct, through some unknown mechanism of energy transfer between chlorophyll and zeaxanthin, or indirect, through an alteration of the environment of the chlorophyll molecules. As discussed previously (Demmig-Adams 1990; Siefermann-Harms 1985), a direct singlet–singlet energy transfer from chlorophyll to zeaxanthin is energetically highly improbable, although the precise energy levels of singlet states that could be involved, such as the $^1A_g{}^*$ state (see Chapters 8 and 9; Siefermann-Harms 1985; Demmig-Adams 1990) for zeaxanthin, have not been determined. It is theoretically possible that other mechanisms such as, for example, a rapidly reversible electron transfer reaction could accomplish an energy transfer from chlorophyll to zeaxanthin (cf. Beddard et al. 1977; Liddell et al. 1982). Other mechanisms have also been discussed (see Schreiber and Neubauer 1990). Alternatively, zeaxanthin could cause a change in the environment of the chlorophyll molecule thereby changing the probabilities for internal de-excitation of the singlet excited state as compared to energy transfer or fluorescence. It has been suggested by Horton et al. (1991) that the aggregation state of light-harvesting complex (LHC) II may be changed by protonation, leading to thermal energy dissipation as indicated by a low yield of chlorophyll fluorescence, and that zeaxanthin dramatically alters the pH requirement for the formation of this aggregated state, such that high activities of thermal energy dissipation are possible at moderate pH gradients, i.e. those likely to occur in photosynthesising tissues.

In leaves that are photosynthesising, high levels of thermal energy dissipation occur exclusively in the presence of zeaxanthin whereas levels of energy dissipation are minimal in DTT-treated zeaxanthin-free leaves (Bilger and Björkman 1990; Demmig-Adams et al. 1990). In zeaxanthin-free leaves the thermal energy dissipation activity, as quantified from the rapidly reversible lowering of the fluorescence yield, accounted for only 10% of that in zeaxanthin-containing leaves. In several studies on isolated chloroplasts the maximal levels of thermal energy dissipation observed were less than 50% of those in leaves (Demmig-Adams et al. 1990a-d; Gilmore and Yamamoto 1991d; Horton et al. 1991). At moderately excessive PFD or at relatively small ΔpH, ratios between the levels of thermal energy dissipation in the presence and absence of zeaxanthin were observed that were similar to those for leaves (Gilmore and Yamamoto 1991d; Horton et al. 1991). However, at a large ΔpH that is apparently not encountered in photosynthesising leaves, a lowering of the fluorescence yield indicative of thermal energy dissipation was reported for zeaxanthin-free chloroplasts; it was similar in extent to that in zeaxanthin-containing chloroplasts (Horton et al. 1991), although always less than 50% of that in zeaxanthin-containing leaves.

One related and unresolved point is the question of whether or not there is principally only one mechanism for thermal energy dissipation in the photochemical/light-capturing system that lowers the yield of chlorophyll fluorescence. Horton et al. (1991) have reported that all characteristics are the same for the lowering of fluorescence yield observed in the absence and in the presence of zeaxanthin, including a lowering of the yield of both F_o and F_m, indicative of thermal energy dissipation in the chlorophyll pigment bed. In contrast, Gilmore and Yamamoto

(1991a) and Adams et al. (1990) have reported that in the absence of zeaxanthin no effect on F_o was observed. In leaves that were manipulated so as to have only minimal rates of electron transport (in 2% O_2, 0% CO_2) a pronounced decrease in the yield of F_m but not of F_o could be induced in the absence of zeaxanthin. It cannot be excluded that this type of decrease in fluorescence yield is indeed distinct from that observed in isolated chloroplasts by Horton and co-workers (Horton et al. 1991; Noctor et al. 1991). This point needs to be clarified. Other characteristics of the process that lowers the fluorescence yield in the absence of zeaxanthin were the same in all studies on isolated chloroplasts (Gilmore and Yamamoto 1991a,b,d; Horton et al. 1991; Noctor et al. 1991) as those for the zeaxanthin-associated process, such as its pH dependence and sensitivity to antimycin. It has been reported earlier that antimycin inhibits the decrease in fluorescence yield that is indicative of thermal energy dissipation but not the formation of a ΔpH across the thylakoid membrane (Oxborough and Horton 1987). Antimycin has now been shown to inhibit the lowering of the fluorescence yield both in the absence and in the presence of zeaxanthin (Gilmore and Yamamoto 1991b,d; Horton et al. 1991). It has recently been suggested by Gilmore and Yamamoto (1991d) that the antimycin site is pH sensitive so that when a ΔpH has already been formed, antimycin does not inhibit thermal energy dissipation. It has also been suggested by Horton et al. (1991) that antimycin prevents the formation of the aggregated state of LHCII that is caused by protonation and that is proposed to facilitate thermal energy dissipation.

Considering the close relationship between the occurrence of thermal energy dissipation and a change in the light-scattering signal from thylakoids (indicative of a conformational change) it has been suggested by several authors (Demmig-Adams 1990; Horton et al. 1991) that an altered state of the thylakoid membrane is involved in the activation of thermal energy dissipation. It has been reported that the change in the light-scattering signal that is induced by excess light is greatly reduced in the absence of zeaxanthin (Bilger et al. 1989) as well as in the presence of antimycin (Horton et al. 1991). It has been speculated by Demmig-Adams (1990) that such an altered state of the thylakoid membrane, as indicated by the light-scattering signal, might possibly cause changes that allow an interaction between chlorophyll and zeaxanthin that would not otherwise be possible. As an alternative, it is suggested in the model of Horton et al. (1991) that the altered (aggregated) state of LHCII itself promotes thermal energy dissipation which is greatly enhanced by the conversion of violaxanthin into zeaxanthin.

It may be that the kinetics of the return of this altered state of the thylakoid membrane to the normal state upon return to non-excessive light can differ. It has been proposed that the relaxation kinetics of thermal energy dissipation in the chlorophyll pigment bed are quite variable; similarly the kinetics of the reconversion of zeaxanthin into violaxanthin can vary from minutes to days (Demmig et al. 1987, 1988; Demmig-Adams et al. 1989b). One possible speculation is that the altered state of the thylakoid membrane might affect the accessibility of zeaxanthin to the epoxidase.

At this point it is unknown whether zeaxanthin is the only carotenoid that can function in the thermal dissipation of excessive excitation energy, or if the monoepoxide, antheraxanthin, is also capable of promoting thermal energy dissipation. As mentioned above, a relationship between the epoxidation state of the xanthophyll cycle and the level of thermal energy dissipation has been reported (Bilger and Björkman 1991), but this relationship does not seem to hold under all circumstances (W.W. Adams and B. Demmig-Adams, unpublished data). This relationship may depend on the quantity of antheraxanthin that is actually present. This problem remains unresolved at this point.

The question has also been raised as to whether the epoxide cycle involving diatoxanthin and diadinoxanthin seen in a large number of algae has a similar function to that of the cycle involving violaxanthin, antheraxanthin and zeaxanthin. Such is suggested by the fact that the total pool of diatoxanthin and diadinoxanthin increases with increasing growth PFD in the diatom *Nitzschia palea* and that the ratio of the epoxide-free form diatoxanthin to the monoepoxide diadinoxanthin becomes greater (Willemönas and Monas 1991).

7.3 ECOPHYSIOLOGY OF THE XANTHOPHYLL CYCLE

7.3.1 Diurnal de-epoxidation and epoxidation in plants in the field

Before the xanthophyll cycle in plants in the field had been completely characterised, there were some indications that leaves growing in natural sunlight might rely on zeaxanthin for dissipating rather large amounts of excess light energy which can be absorbed by chlorophyll during peak irradiance. Diurnal measurements of changes in the chlorophyll fluorescence yield in several cactus species growing in full sunlight revealed that, during the course of the day, there were diurnal changes in thermal energy dissipation, with strong increases in energy dissipation which parallelled the increase in incident PFD (Adams 1988; Adams et al. 1987, 1988, 1989). This was not unexpected given that these plants exhibit rather low rates of photosynthesis compared to the majority of plants that grow in full sunlight. However, Björkman and Schäfer (1989) demonstrated that even a well-watered, vigorous crop plant (cotton), with very high rates of photosynthesis, dissipated the majority of the absorbed energy thermally at midday when exposed to full sunlight. They found that approximately 25% of the light which was absorbed was used to fix CO_2 in photosynthesis, 19% was used in photorespiration (the photosynthetic fixation of O_2), and fully 56% of the energy was dissipated thermally. In all these studies the changes in chlorophyll fluorescence yield were of the type associated with the dissipation of excessive energy via zeaxanthin (see Section 7.2.3). However, no measurements of zeaxanthin were made in any of these studies.

High levels of zeaxanthin and massive decreases in chlorophyll fluorescence yield indicative of energy dissipation were first reported for the leaves of the chaparral shrub *Arbutus unedo* during peak exposure to light under natural conditions in Portugal at the end of the summer (Demmig-Adams et al. 1989a). The de-epoxidation of violaxanthin to zeaxanthin in response to increases in PFD under controlled conditions in the laboratory has been described in Section 7.1.2. More recently the complete diurnal changes in the epoxidation state of the xanthophyll cycle that occur in the leaves of a variety of species growing in full sunlight throughout the course of the day have been examined (Demmig-Adams and Adams 1992a; Adams and Demmig-Adams 1992; Adams et al. 1992). In exposed field sites the changes in incident PFD were closely parallelled by changes in the epoxidation state of the xanthophyll cycle. The diurnal changes in incident PFD, parallelled by changes in the ratio of zeaxanthin:(V+A+Z), are shown for two species, *Euonymus kiautschovicus* and *Malva neglecta*, in Figs. 7.11 and 7.12. Similar diurnal patterns of changes in the xanthophyll cycle components as depicted in *E. kiautschovicus* and *M. neglecta* were also found in the crop species sunflower, pumpkin, cucumber (Adams and Demmig-Adams 1992), and in a separate study of cotton (B. Demmig-Adams, W. Bilger, O. Björkman, W.W. Adams, S.S. Thayer, and C. Shih, unpublished data). The above species are all plants with the C_3 pathway of photosynthesis. The carotenoid composition of seven species of C_4 plants growing in natural sunlight has recently been investigated and found to be very similar to that of C_3 species growing in the same habitat (H.Y. Yamamoto and O. Björkman, unpublished data). The xanthophyll cycle pool sizes were in the same range in both C_3 and C_4 species at peak irradiance at noon. The C_4 species had epoxidation states of 0.45–0.55 which are similar to the epoxidation states of C_3 species with high rates of photosynthesis. Thus, major changes in the epoxidation state of the xanthophyll cycle, fine-tuned to the diurnal changes in irradiance, appear to occur daily in plants growing in natural sunlight.

Euonymus kiautschovicus, which had a lower rate of photosynthesis than did *M. neglecta*, formed much more zeaxanthin at midday and experienced a decrease in the violaxanthin content to a much greater extent than did *M. neglecta* at a comparable incident PFD (Figs. 7.11 and 7.12). This same type of response was observed among numerous other species, with those species possessing the higher rates of photosynthesis forming relatively less zeaxanthin at peak irradiance than did those species with the lower rates of photosynthesis (Thayer and Björkman 1990; Demmig-Adams and Adams 1992a,b; Adams and Demmig-Adams 1992). This is as one might expect, since a higher ratio of PFD:photosynthesis would require a higher energy dissipation activity in order to avoid an accumulation of potentially damaging excitation energy. From this correlation between the state of the xanthophyll cycle and the ratio of photosynthesis and PFD, or between EPS and ((photosynthesis:PFD)x(V+A+Z)) as reported by Thayer and Björkman (1990), it may be possible to predict the actual photosynthesis rate in plants in the field based on their carotenoid composition at a particular time.

Another interesting phenomenon, which is not generally observed for well-watered plants cultivated in growth chambers or in a glasshouse, is the presence of a considerable level of zeaxanthin in the leaves of plants growing in the field prior to sunrise after an entire night of darkness (Demmig-Adams et al. 1989a; Adams and Demmig-Adams 1992; Adams et al. 1992; Demmig-Adams and Adams 1992a). This was particularly pronounced, in a pattern similar to that observed for peak zeaxanthin levels at midday, in those species with the lower photosynthesis rates.

This, in turn, can lead to differently shaped light response curves of zeaxanthin formation in plants from field sites when compared to those generated under laboratory conditions. Light response curves of zeaxanthin formation obtained from plants in the laboratory typically exhibit a pronounced biphasic pattern, with no zeaxanthin formation at low PFDs, followed by zeaxanthin formation with increasing levels of excessive PFD (see Figs. 7.3 and 7.4). Plants in the field, however, often do not exhibit a biphasic pattern of zeaxanthin formation. Instead, they contain a certain level of zeaxanthin prior to dawn, and exhibit further increases in zeaxanthin which parallel the increases in PFD up to midday.

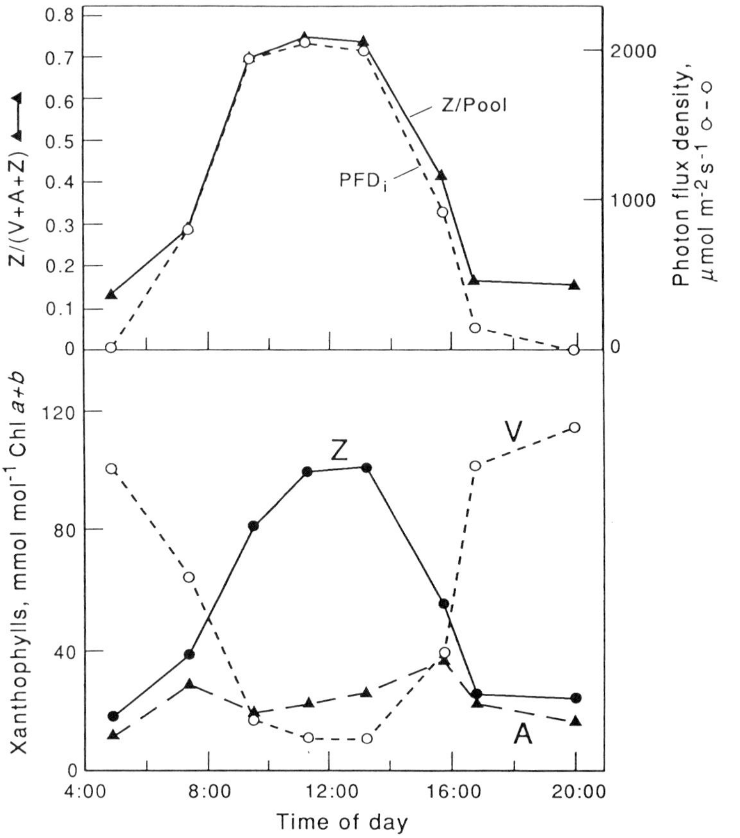

Figure 7.11 Diurnal changes in incident photon flux density (PFD), the ratio of zeaxanthin to the total xanthophyll cycle pool (violaxanthin + antheraxanthin + zeaxanthin, V+A+Z), and in the levels of violaxanthin, antheraxanthin, and zeaxanthin in a *Euonymus kiautschovicus* leaf growing in full sunlight over the course of a clear day. (Data from Adams and Demmig-Adams 1992.)

It has previously been reported that experimental transfer of leaves from limiting to excessive to limiting light resulted in the retention of considerable levels of zeaxanthin due to slow removal of zeaxanthin (Demmig-Adams et al. 1989d; Bilger and Björkman 1990), as well as hysteresis in the changes in fluorescence yield in some cases (Demmig and Winter 1988). In the field one might anticipate that the removal of zeaxanthin could occur concomitantly with the decrease in PFD during the afternoon/evening hours because the changes in PFD over this time period are rather gradual. In some species this was indeed the case, whereas other species, particularly those which reached very high degrees of de-epoxidation at midday, exhibited hysteresis in the zeaxanthin levels versus PFD during the afternoon relative to the morning hours, with more zeaxanthin present during the descending than during the ascending PFD curves (Adams and Demmig-Adams 1992).

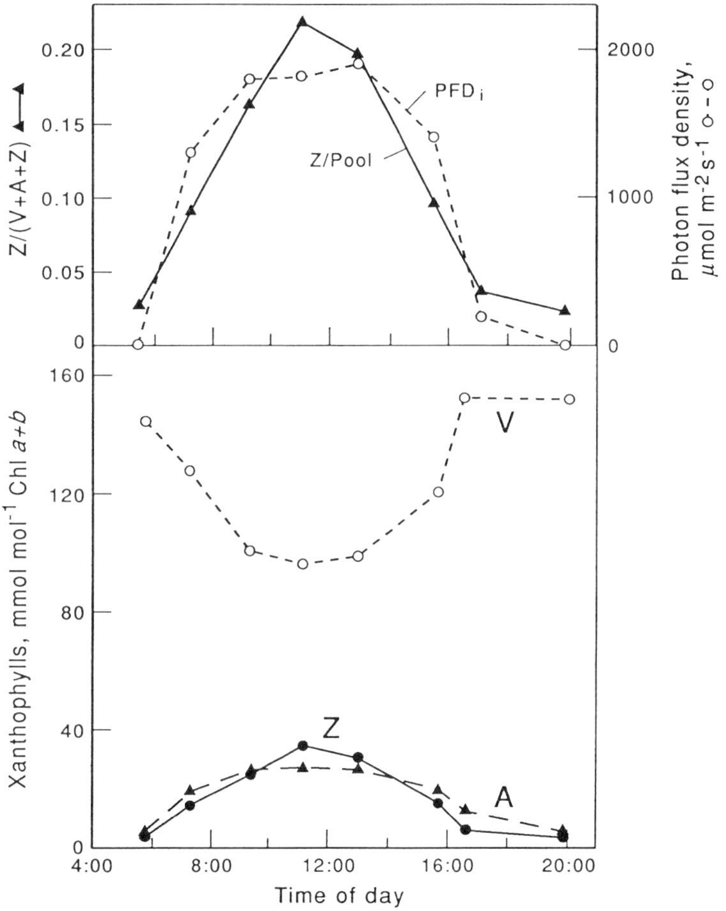

Figure 7.12 Diurnal changes in incident PFD, the ratio of zeaxanthin to the total xanthophyll cycle pool (violaxanthin + antheraxanthin + zeaxanthin), and in the levels of violaxanthin, antheraxanthin, and zeaxanthin in a *Malva neglecta* leaf growing in full sunlight over the course of a clear day. For abbreviations,see Fig. 7.11. (Data from Adams and Demmig-Adams 1992.)

7.3.2 Involvement of the xanthophyll cycle in the response of plants to a combination of high light and other environmental stress factors

Whenever an environmental stress factor (e.g. reduced water availability or low temperatures) causes a decrease in the rate of photosynthesis, the prevailing PFD to which the plants are exposed can become excessive in situations where such PFDs were not excessive before, or the prevailing PFD can become more excessive if they were excessive prior to the increased stress. The evidence indicating that zeaxanthin is involved in dissipating greater amounts of energy when plants are exposed to different environmental stresses in combination with light will be examined.

(a) Water stress

The involvement of zeaxanthin in thermal energy dissipation in leaves of a plant subjected to water stress was first demonstrated in *Nerium oleander* (Demmig et al. 1988). In this investigation *N. oleander* which had developed under high light and had been watered frequently to maintain a high water potential in the leaves was used. Upon cessation of watering, the level of zeaxanthin in the leaves after 12 h of darkness (each night) rose progressively over the subsequent 12 days. The increases in zeaxanthin content were parallelled by proportional increases in thermal energy dissipation which were sustained after the 12 h night. These changes occurred during exposure each day to a constant PFD which, nevertheless, became more and more excessive as photosynthesis declined with each progressive day of water stress.

N. oleander is, however, rather unusual in its tolerance to drought in that it does not exhibit osmotic adjustment in response to decreases in water potential. Instead, metabolic activity decreases and the uptake of carbon from the atmosphere (i.e. net photosynthetic CO_2 fixation) ceases. Thus, a particular propensity for alternative thermal energy dissipation directly within the photosynthetic apparatus might not be unexpected in *N. oleander* under water stress. It is possible, on the other hand, that plants which do exhibit osmotic adjustment in response to water stress, and are thereby able to maintain some level of stomatal opening even during pronounced changes in their water relations, may continue to allow photosynthetic electron transport to CO_2 to occur, as well as to other acceptors such as O_2 (via photorespiration and direct reduction of O_2 in the Mehler reaction).

Further evidence to indicate that thermal energy dissipation, probably via zeaxanthin, becomes increasingly important with water stress was obtained by Björkman and Schäfer (1989), who compared irrigated cotton with cotton which had received no water for up to two months. In the irrigated cotton, with a water potential of -1.1 MPa, photosynthetic CO_2 fixation utilised 25% of the incident PFD, photorespiration (O_2 fixation) 19%, and 56% of the incident PFD was dissipated within the photosynthetic apparatus, as mentioned above. At an intermediate level of water stress (water potential of -1.8 MPa), CO_2 fixation utilised 18% of the incident PFD, photorespiration 16%, and 66% of the energy was dissipated thermally. Under severe water stress (water potential of -2.8 MPa), only 13% of the energy was used for photosynthetic CO_2 fixation, an additional 13% was used in photorespiration, and approximately 73% of the energy was dissipated thermally. Thus thermal dissipation of excessive energy absorbed by leaves in the photochemical system became increasingly important as water availability decreased and plants were subjected to increasing water stress.

There is also an interesting relationship between the biosynthesis of zeaxanthin and abscisic acid (ABA) synthesis. ABA has been implicated in the response of plants to water stress, particularly with regard to the closure of stomata in response to decreasing water potential. An ABA-deficient mutant of *Arabidopsis thaliana* can be grown only in an atmosphere of very high humidity, otherwise the plants wilt and die due to the rapid loss of water associated with stomata which remain totally open (Rock and Zeevaart 1991). This mutant also accumulates large amounts of ß-carotene, and even larger quantities of zeaxanthin, but contains only very small amounts of violaxanthin, presumably due to the fact that it lacks the epoxidase for forming violaxanthin from zeaxanthin. This study also provides strong correlative evidence that the main pathway for ABA synthesis in *A. thaliana* is from xanthophyll epoxides. Thus, the same biosynthetic pathway apparently provides two of the compounds (zeaxanthin and ABA) that afford acclimation to high light/arid habitats.

(b) Salinity

In a study of mangroves (particularly *Rhizophora stylosa*) growing in northern Australia on the eastern coast of Queensland, Björkman et al. (1988) found that such halophytes exhibited high levels of thermal energy dissipation from their leaves in the field. A closely related species, *R. mangle*, was also shown to have the capacity to form large amounts of zeaxanthin under controlled conditions (Demmig-Adams et al. 1989f). To date, however, there have been no studies which have examined both thermal energy dissipation and zeaxanthin content or formation in either halophytes or non-halophytes exposed to saline conditions in the field.

As with water stress, different species of halophytes might rely on different strategies to cope with excessive excitation energy, e.g. thermal energy dissipation within the photosynthetic apparatus involving zeaxanthin versus increases in proportion to electron flow to electron acceptors other than CO_2 .

(c) Temperature

Temperature response of changes in thermal energy dissipation and epoxidation state

Adams et al. (1990) reported that, in spinach leaves, the rate at which thermal energy dissipation developed (upon exposure to high light) increased with increasing temperature. Although thermal energy dissipation activity develops more rapidly at higher temperature, once steady-state conditions are reached, leaves subjected to low temperatures actually dissipate larger amounts of energy than do leaves maintained at more moderate temperature, because photosynthesis is suppressed at low temperatures. Bilger and Björkman (1991) obtained evidence that linked zeaxanthin more directly to this temperature response of thermal energy dissipation in cotton. They found that the rate of violaxanthin de-epoxidation increased continuously from 10 to 40 °C, but that the total extent of de-epoxidation, once steady-state conditions were attained (this may require a considerable amount of time at low temperature, since the enzymic conversion of violaxanthin to zeaxanthin is very slow), actually increased with decreasing temperature.

Effect of season on de-epoxidation rates and zeaxanthin levels in field-grown plants
In plants which are not acclimated to chilling temperatures the de-epoxidation of violaxanthin to zeaxanthin (as an enzymic reaction) proceeds very slowly at low temperature, as does the development of thermal energy dissipation that is estimated from the decrease in the yield of chlorophyll fluorescence (Demmig-Adams et al. 1989e). Furthermore, there were no consistent differences in the rates of violaxanthin de-epoxidation nor in the rates of the decrease in fluorescence yield between warm-grown spinach leaves from a glasshouse and overwintering spinach leaves growing outside in Colorado (B. Demmig-Adams and W.W. Adams, unpublished data), suggesting that there was no acclimation with regard to the rapidity of zeaxanthin formation at low temperatures in spinach. In other overwintering species which were examined in the field, substantial amounts of zeaxanthin could be retained overnight during periods when temperatures fell below 0 °C (W.W. Adams and B. Demmig-Adams, unpublished data). After some of the colder nights there was as much zeaxanthin present in the leaves of these perennial species prior to sunrise as could be found at midday during the previous or subsequent day. The retention of large amounts of zeaxanthin throughout the night, thus apparently negating the necessity to form any zeaxanthin the following day, would be consistent with a lack of acclimation in the capacity to form zeaxanthin more rapidly at chilling temperatures. These pre-dawn levels of zeaxanthin were also accompanied by sustained decreases in the yield of fluorescence which, in part or mostly, may be due to a sustained increase in thermal energy dissipation involving zeaxanthin. Similar sustained reductions in chlorophyll fluorescence in response to chilling temperatures have also been reported in field-grown spinach by Somersalo and Krause (1990).

Some degree, however, of seasonal acclimation concerning the de-epoxidation rates and pool sizes of the xanthophyll cycle was reported by Bilger and Björkman (1991). The xanthophyll cycle pool size and violaxanthin de-epoxidation rates were compared in leaves of *Malva parviflora* removed from the plants in the spring versus those which were removed from the plants in the summer (all in California). Those examined in the spring had larger xanthophyll cycle pools and exhibited rates of violaxanthin de-epoxidation which were higher at both 15 and 30 °C than those examined in the summer. Thus, the plants which developed from the winter into the spring apparently had a larger capacity for the formation of zeaxanthin and could form zeaxanthin more rapidly (at all temperatures) than the plants which developed through the summer.

(d) Nitrogen limitation

Ferrar and Osmond (1986) examined the effects of transferring, from low light to high light, plants which were grown with either adequate or limiting nitrogen. In general, those plants which were nitrogen deficient exhibited considerable photoinhibition of photosynthesis (a decrease in the efficiency of photosynthetic energy conversion), whereas those which contained sufficient nitrogen did not. However, it is not known to what degree these changes were due to an increase in thermal energy dissipation involving zeaxanthin (photoprotection), as opposed to photoinhibitory damage to the photochemical apparatus. Similar results were also obtained with the green alga *Ulva rotundata* (Henley et al. 1991).

The first study to examine the xanthophyll cycle specifically with regard to nitrogen nutrition was that of Khamis et al. (1990). In this study thermal energy dissipation developed to a greater degree, and more zeaxanthin was found to be present upon transfer to high light in those plants which were deficient in nitrogen than in those that were not. Despite the fact that leaves of the nitrogen-deficient plants had lower rates of photosynthesis, they exhibited smaller increases in the reduction state of PSII upon transfer to high light, presumably due to the higher rates of zeaxanthin-associated thermal energy dissipation.

7.3.3 Shade-sun acclimation of the xanthophyll cycle pool size

The carotenoid compositions of a shade leaf and a sun leaf are compared in Fig. 7.13. These differences are typical for those found between shade and sun species, as well as between shade and sun leaves of the same species (Thayer and Björkman 1990; Demmig-Adams and Adams 1992a,b). The major difference between these two groups of leaves from very different light environments is the size of the pool of the xanthophyll cycle components, violaxanthin + antheraxanthin + zeaxanthin. The leaves from the high light habitats contain much more of the xanthophyll cycle components than do the shade leaves. ß-Carotene is also present in greater amounts in the sun leaves than in shade leaves, and indeed the xanthophyll cycle pool and ß-carotene have been reported to vary concomitantly (Demmig-Adams et al. 1989f; Thayer and Björkman 1990). Thus, there seems to be a specific stimulation of ß-carotene synthesis in high light as opposed to synthesis of α-carotene and lutein. No clear trend has been observed for any changes in the other xanthophylls, lutein and neoxanthin, in response to differences in light regime during growth. α-Carotene is, however, found in greater quantities in shade-acclimated plants (cf. Thayer and Björkman 1990; see Chapter 6).

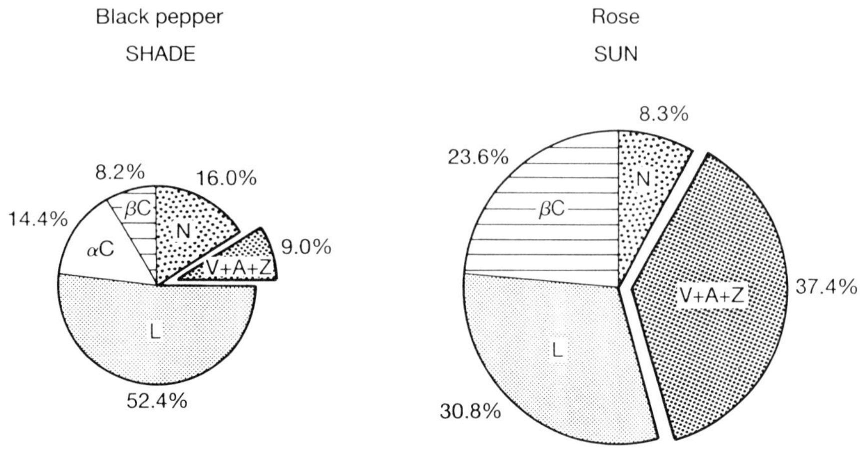

Figure 7.13 Carotenoid composition (expressed as percent of total carotenoid content) of leaves of an understorey rainforest plant (Black pepper) which developed in deep shade (<10 μmol photons m^{-2} s^{-1}) and a high light plant (rose) which developed in full sunlight. The relative size of the two pie diagrams reflects the difference in the total content of the carotenoids (on a chlorophyll *a* + *b* basis) between the two species. (αC, α-carotene; ßC, ß-carotene; L, lutein; N, neoxanthin; V+A+Z, violaxanthin + antheraxanthin + zeaxanthin). (Data from Demmig-Adams and Adams 1992a.)

In the mutant of *Arabidopsis thaliana* which accumulates ß-carotene and zeaxanthin, and possesses extremely low levels of violaxanthin, neoxanthin and ABA, significantly reduced levels of lutein (a derivative of α-carotene) were also reported (Rock and Zeevaart 1991). It is very interesting that the ß-carotene pathway appears to produce several compounds involved in the acclimation of plants to high light: zeaxanthin is required for dissipation of excessive energy absorbed by chlorophyll, and ABA is required for adequate stomatal control, especially in high light habitats which are likely to be subjected to periods of reduced water availability. It is also interesting that the pool of ascorbate which is involved in the xanthophyll cycle (as well as in other high light induced responses of plants) undergoes changes in its size in response to excess light, apparently much like the xanthophyll cycle pool itself (Foyer et al. 1992).

The transfer from low light to high light of several species representing mesophytic and xerophytic plants resulted in a large increase in the xanthophyll cycle pool over a period of a week (Demmig-Adams et al. 1989f). Similar results have been observed in green algae (Henley et al. 1991; Franklin et al. 1992). In two extensive surveys that included 36 different species, representing a variety of different life forms, high light acclimated plants were found to contain much larger xanthophyll cycle pools than low light acclimated plants, similar to those depicted in Fig. 7.13 (Thayer and Björkman 1990; Demmig-Adams and Adams 1992a). The shade/sun acclimation of the xanthophyll cycle pool is thus a very general phenomenon in plants, and is also in contrast to previous reports that the xanthophyll to carotene ratio was higher in shade leaves than in sun leaves (Lichtenthaler and Meier 1984; Lichtenthaler et al. 1982; Anderson 1986).

The size of the xanthophyll cycle pool, and thereby the capacity of the plant to form zeaxanthin (and perform thermal energy dissipation) thus increases with increases in the amount of excessive light experienced by a particular leaf (cf. Thayer and Björkman 1990). Recent results, however, suggest that the degree of excess light experienced by the plant may not be the only factor which determines the size of the xanthophyll cycle pool. In full sunlight, species with higher rates of photosynthesis experience lower degrees of excessive light, and species with lower rates of photosynthesis experience greater degrees of excessive light. Interestingly, species which experienced the lower degrees of excessive light had xanthophyll cycle pools which were of a similar size or exceeded those seen in other species (on both a leaf area and a chlorophyll $a + b$ basis; Demmig-Adams and Adams 1992a). This may suggest that a very large xanthophyll cycle pool, with a considerably larger amount of violaxanthin than is required for conversion into zeaxanthin at midday, is also associated with a high photosynthesis rate, i.e. high metabolic activity and possibly high rates of carotenoid biosynthesis.

At peak irradiance under otherwise favourable conditions, species with high rates of photosynthesis (such as a variety of annual crop species) typically converted only 50% (or less) of the total xanthophyll cycle pool into zeaxanthin + antheraxanthin (cf. also Fig. 7.12). On the other hand, several perennial shrubs and vines with lower photosynthesis rates formed more zeaxanthin (on either a leaf area or chlorophyll $a + b$ basis) at peak irradiance, but did not contain larger xanthophyll cycle pools than did species with higher rates of photosynthesis. Therefore it is not uncommon for species such as the perennial shrubs and vines to convert more than 80% of the total xanthophyll cycle pool into zeaxanthin + antheraxanthin (cf. also Fig. 7.11; as mentioned in Section 7.1, violaxanthin was only 5%–12% of the total xanthophyll cycle pool at midday in these species; Demmig-Adams and Adams 1992a; see also Demmig-Adams et al. 1989a).

The acclimation of the xanthophyll cycle pool size is consistent with a specific function for these xanthophylls in photoprotection. A leaf with a larger quantity of the xanthophyll components has a greater potential for zeaxanthin formation when excessive excitation energy is accumulated by chlorophyll, and thus a greater potential for the thermal dissipation of that excessive energy. Thus, acclimation to high light involves not only a high capacity for photosynthesis (i.e. the utilisation of the absorbed light through photosynthesis; Björkman 1981) but also a high capacity for alternative (thermal) dissipation.

ACKNOWLEDGEMENTS

We thank Olle Björkman, Wolfgang Bilger, Sue Thayer and Harry Yamamoto for making their unpublished data available to us. Some of our unpublished data was supported by the United States Department of Agriculture, Competitive Research Grants Office, award number 90-37130-5422, and by a Faculty Development

Award from the University of Colorado to W.W. Adams III.

REFERENCES

Adams, W.W. III (1988) Photosynthetic acclimation and photoinhibition of terrestrial and epiphytic CAM tissues growing in full sunlight and deep shade. *Aust. J. Plant Physiol.*, **15**:123-134.

Adams, W.W. III and Demmig-Adams, B. (1992) Operation of the xanthophyll cycle in higher plants in response to diurnal changes in incident sunlight. *Planta*, **186**:390-398.

Adams, W.W. III, Smith, S.D. and Osmond, C.B. (1987) Photoinhibition of the CAM succulent *Opuntia basilaris* growing in Death Valley: Evidence from 77K fluorescence and quantum yield. *Oecologia*, **71**:221-228.

Adams, W.W. III, Terashima, I., Brugnoli, E. and Demmig, B. (1988) Comparisons of photosynthesis and photoinhibition in the CAM vine *Hoya australis* and several C_3 vines growing on the coast of eastern Australia. *Plant Cell Environ.*, **11**:173-181.

Adams, W.W. III, Díaz, M. and Winter, K. (1989) Diurnal changes in photochemical efficiency, the reduction state of Q, radiationless energy dissipation, and non-photochemical fluorescence quenching in cacti exposed to natural sunlight in northern Venezuela. *Oecologia*, **80**:553-561.

Adams, W.W. III, Demmig-Adams, B. and Winter, K. (1990) Relative contributions of zeaxanthin-related and zeaxanthin-unrelated types of "high-energy-state" quenching of chlorophyll fluorescence in spinach leaves exposed to various environmental conditions. *Plant Physiol.*, **92**:302-309.

Adams, W.W. III, Volk, M., Hoehn, A. and Demmig-Adams, B. (1992) Leaf orientation and the response of the xanthophyll cycle to incident light. *Oecologia*, **90**:404-410.

Aihara, M.S. and Yamamoto, H.Y. (1968) Occurrence of antheraxanthin in two Rhodophyceae *Acanthophora spicifera* and *Gracilaria lichenoides*. *Phytochemistry*, **7**:497-499.

Anderson, J.M. (1986) Photoregulation of the composition, function, and structure of thylakoid membranes. *Annu. Rev. Plant Physiol.*, **37**: 93-136.

Anderson, J.W., Foyer, C.H. and Walker, D.A. (1983) Light-dependent reduction of dehydroascorbate and uptake of exogenous ascorbate by spinach chloroplasts. *Planta*, **158**:442-450.

Aristarkov, A.I., Nikandrov, V.V. and Krasnovskii, A.A. (1987) Ascorbate permeability of chloroplast thylakoid membrane: Reduction of plastoquinones and cytochrome f. [Translated from *Biokhimiya*, **52**:2051-2060] *Biochemistry*, **52**:1776-1784.

Asada, K. and Takahashi, M. (1987) Production and scavenging of active oxygen in photosynthesis. In: Kyle, D.J., Osmond, C.B. and Arntzen, C.J. (eds.), *Photoinhibition*. Elsevier, Amsterdam, pp. 227-284.

Beck, A., Burkert, A. and Hofmann, M. (1983) Uptake of L-ascorbate by intact spinach chloroplasts. *Plant Physiol.*, **73**:41-45.

Beddard, G.S., Davidson, R.S. and Tretheway, K.R. (1977) Quenching of chlorophyll fluorescence by ß-carotene. *Nature*, **267**: 373-374.

Bilger, W. and Björkman, O. (1990) Role of the xanthophyll cycle in photoprotection elucidated by measurements of light-induced absorbance changes, fluorescence and photosynthesis in leaves of *Hedera canariensis*. *Photosynth. Res.*, **25**:173-185.

Bilger, W. and Björkman, O. (1991) Temperature dependence of violaxanthin de-epoxidation and non-photochemical fluorescence quenching in intact leaves of *Gossypium hirsutum* L. and *Malva parviflora* L. *Planta*, **184**:226-234.

Bilger, W., Björkman, O. and Thayer, S.S. (1989) Light-induced spectral absorbance changes in relation to photosynthesis and the epoxidation state of xanthophyll cycle components in cotton leaves. *Plant Physiol.*, **91**:542-551.

Björkman, O. (1981) Responses to different quantum flux densities. In: Lange, O.L., Nobel,

P.S., Osmond, C.B. and Ziegler, H. (eds.), *Physiological Plant Ecology* I. Encyclopedia of Plant Physiology, Vol. 12A. Springer, Berlin, pp. 57-107.

Björkman, O. and Demmig, B. (1987) Photon yield of O_2 evolution and chlorophyll fluorescence characteristics at 77K among vascular plants of diverse origins. *Planta*, 170:489-504.

Björkman, O. and Schäfer, C. (1989) A gas exchange–fluorescence analysis of photosynthetic performance of a cotton crop under high-irradiance stress. (Extended abstract.) *Phil. Trans. R. Soc. Lond. B*, **323**:309-311.

Björkman, O., Demmig, B. and Andrews, T.J. (1988) Mangrove photosynthesis: Response to high-irradiance stress. *Aust. J. Plant Physiol.* **15**:43-61.

Bjørnland, T. and Aguilar-Martinez, M. (1976) Carotenoids in red algae. *Phytochemistry*, **15**:291-296.

Blass, U., Anderson, J.M. and Calvin, M. (1959) Biosynthesis and possible relations among the carotenoids and between chlorophyll *a* and *b*. *Plant Physiol.*, **34**:329-333.

Braumann, T.H. and Grimme, L.H. (1981) Reversed-phase high performance liquid chromatography of chlorophylls and carotenoids. *Biochim. Biophys. Acta*, **637**:8-17.

Brown, L.M. and McLachlan, J. (1982) Atypical carotenoids for the Rhodophyceae in the genus *Gracilaria* (Gigartinales). *Phycologia*, **21**:9-16.

Costes, C., Burghoffer, C., Joyard, J. *et al.* (1979) Occurrence and biosynthesis of violaxanthin in isolated spinach chloroplast envelope. *FEBS Lett.*, **103**:17-21.

Demmig, B. and Björkman, O. (1987) Comparison of the effect of excessive light on chlorophyll fluorescence (77K) and photon yield of O_2 evolution in leaves of higher plants. *Planta*, **171**:171-184.

Demmig, B. and Winter, K. (1988) Characterisation of three components of non-photochemical fluorescence quenching and their response to photoinhibition. *Aust. J. Plant Physiol.*, **15**:163-177.

Demmig, B., Winter, K., Krüger, A. and Czygan, F.-C. (1987) Photoinhibition and zeaxanthin formation in intact leaves. A possible role of the xanthophyll cycle in the dissipation of excess light energy. *Plant Physiol.*, **84**:218-224.

Demmig, B., Winter, K., Krüger, A. and Czygan, F.-C. (1988) Zeaxanthin and the heat dissipation of excess light energy in *Nerium oleander* exposed to a combination of high light and water stress. *Plant Physiol.*, **87**:17-24.

Demmig-Adams, B. (1990) Carotenoids and photoprotection in plants. A role for the xanthophyll zeaxanthin. *Biochim. Biophys. Acta*, **1020**:1-24.

Demmig-Adams, B. and Adams, W.W. III (1992a) Carotenoid composition in sun and shade leaves of plants with different life forms. *Plant Cell Environ.*, **15**:411-419.

Demmig-Adams, B. and Adams, W.W. III (1992b) Light, photosynthesis, and the xanthophyll cycle. In: Pell, E.J. and Steffen, K.L. (eds.), *Active Oxygen/Oxidative Stress and Plant Metabolism*. American Society of Plant Physiologists, Rockville, pp. 171-179.

Demmig-Adams, B., Adams, W.W. III, Winter, K. *et al.* (1989a) Photochemical efficiency of photosystem II, photon yield of O_2 evolution, photosynthetic capacity, and carotenoid composition during the "midday depression" of net CO_2 uptake in *Arbutus unedo* growing in Portugal. *Planta*, **177**:377-387.

Demmig-Adams, B., Winter, K., Krüger, A. and Czygan, F.-C. (1989b) Light stress and photoprotection related to the carotenoid zeaxanthin in higher plants. In: Briggs, W.R. (ed.) *Photosynthesis*. Alan R. Liss, New York, pp. 375-391.

Demmig-Adams, B., Winter, K., Krüger, A. and Czygan, F.-C. (1989c) Light response of CO_2 assimilation, dissipation of excess excitation energy, and zeaxanthin content of sun and shade leaves. *Plant Physiol.*, **90**:881-886.

Demmig-Adams, B., Winter, K., Krüger, A. and Czygan, F.-C. (1989d) Zeaxanthin and the induction and relaxation kinetics of the dissipation of excess excitation energy in leaves in 2% O_2, 0% CO_2. *Plant Physiol.*, **90**:887-893.

Demmig-Adams, B., Winter, K., Krüger, A. and Czygan, F.-C. (1989e) Zeaxanthin synthesis, energy dissipation, and photoprotection of photosystem II at chilling temperatures. *Plant*

Physiol., **90**:894-898.
Demmig-Adams, B., Winter, K., Winkelmann, E. *et al.* (1989f) Photosynthetic characteristics and the ratios of chlorophyll, ß-carotene, and the components of the xanthophyll cycle upon a sudden increase in growth light regime in several plant species. *Bot. Acta*, **102**:319-325.
Demmig-Adams, B., Adams, W.W. III, Czygan, F.-C. *et al.* (1990a) Differences in the capacity for radiationless energy dissipation in green and blue-green algal lichens associated with differences in carotenoid composition. *Planta*, **180**:582-589.
Demmig-Adams, B., Adams, W.W. III, Green, T.G.A. *et al.* (1990b) Differences in the susceptibility to light stress in two lichens forming a phycosymbiodeme, one partner possessing and one lacking the xanthophyll cycle. *Oecologia*, **84**:451-456.
Demmig-Adams, B., Adams, W.W. III, Heber, U. *et al.* (1990c) Inhibition of zeaxanthin formation and of rapid changes in radiationless energy dissipation by dithiothreitol in spinach leaves and chloroplasts. *Plant Physiol.*, **92**:293-301.
Demmig-Adams, B., Máguas, C., Adams, W.W. III, *et al.* (1990d) Effect of high light on the efficiency of photochemical energy conversion in a variety of lichen species with green and blue-green phycobionts. *Planta*, **180**:400-409.
Duckham, S.C., Linforth, R.S.T and Taylor, I.B. (1991) Abscisic-acid-deficient mutants at the *aba* gene locus of *Arabidopsis thaliana* are impaired in the epoxidation of zeaxanthin. *Plant Cell Environ.*, **14**: 601-606.
Dujardyn, M. and Foyer, C.H. (1989) Limitation of CO_2 assimilation and regulation of Benson–Calvin cycle activity in barley leaves in response to changes in irradiance, photoinhibition, and recovery. *Plant Physiol.*, **91**:1562-1568.
Ferrar, P.J. and Osmond, C.B. (1986) Nitrogen supply as a factor influencing photoinhibition and photosynthetic acclimation after transfer of shade-grown *Solanum dulcamara* to bright light. *Planta*, **168**: 563-570.
Foyer, C.H., Lelandais, M., Edwards, E.A. and Mullineaux, P.M. (1992) The role of ascorbate in plants: Interactions with photosynthesis and regulatory significance. In: Pell, E.J. and Steffen, K.L. (eds.), *Active Oxygen/Oxidative Stress and Plant Metabolism*. American Society of Plant Physiologists, Rockville, pp. 131-144.
Franklin, L.A., Levavasseur, G., Osmond, C.B. *et al.* (1992) Two components of onset and recovery during photoinhibition of *Ulva rotundata*. *Planta*, **186**:399-408.
Gamon, J.A., Field, C.B., Bilger, W. *et al.* (1990) Remote sensing of the xanthophyll cycle and chlorophyll fluorescence in sunflower leaves and canopies. *Oecologia*, **85**:1-7.
Gilmore, A.M. and Yamamoto, H.Y. (1990) Zeaxanthin formation in q_E-inhibited chloroplasts. In: Baltscheffsky, M. (ed.) *Current Research in Photosynthesis*, Vol. 2. Kluwer Academic Publishers, Dordrecht, pp. 495-498.
Gilmore, A.M. and Yamamoto, H.Y. (1991a) Zeaxanthin-dependent and independent nonphotochemical quenching are quantitatively related to ΔpH. *Plant Physiol.* Suppl., **96**:41.
Gilmore, A.M. and Yamamoto, H.Y. (1991b) Zeaxanthin-dependent and independent nonphotochemical quenching are resistant to antimycin in proportion to ΔpH. *Plant Physiol.* Suppl., **96**:119.
Gilmore, A.M. and Yamamoto, H.Y. (1991c) Resolution of lutein and zeaxanthin using a nonendcapped, lightly carbon-loaded C-18 high-performance liquid chromatographic column. *J. Chromatogr.*, **543**:137-145.
Gilmore, A.M. and Yamamoto, H.Y. (1991d) Zeaxanthin formation and energy-dependent fluorescence quenching in pea chloroplasts under artificially-mediated linear and cyclic electron transport. *Plant Physiol.*, **96**:635-643.
Greer, D.H., Berry, J.A. and Björkman, O. (1986) Photoinhibition of photosynthesis in intact bean leaves: Role of light, temperature and requirement for chloroplast–protein synthesis during recovery. *Planta*, **168**:253-260.
Hager, A. (1967) Untersuchungen über die Rückreaktionen im Xanthophyll-Cyclus bei *Chlorella*, *Spinacia* und *Taxus*. *Planta*, **76**:138-148.

Hager, A. (1969) Lichtbedingte pH-Erniedrigung in einem Chloroplasten-Kompartiment als Ursache der enzymatischen Violaxanthin → Zeaxanthin-Umwandlung; Beziehungen zur Photophosphorylierung. *Planta*, *89*:224-243.

Hager, A. (1980) The reversible, light-induced conversions of xanthophylls in the chloroplast. In: Czygan, F.-C. (ed.) *Pigments in Plants*. Fischer, Stuttgart, pp. 57-79.

Hager, A. and Meyer-Bertenrath, T. (1966) Die Isolierung und quantitative Bestimmung der Carotinoide und Chlorophylle von Blättern, Algen und isolierten Chloroplasten mit Hilfe dünnschichtchromatographischer Methoden. *Planta*, **69**:198-217.

Hager, A. and Perz, H. (1970) Veränderung der Lichtabsorption eines Carotinoids im Enzym(De-Epoxidase)-Substrat(Violaxanthin)-Komplex. *Planta*, **93**:314-322.

Hager, A. and Stransky, H. (1970) Das Carotinoidmuster und die Verbreitung des lichtinduzierten Xanthophyll-Cyclus in verschiedenen Algenklassen. V. Einzelne Vertreter der Cryptophyceae, Euglenophyceae, Bacillariophyceae, Chrysophyceae und Phaeophyceae. *Arch. Microbiol.*, **73**:77-89.

Harbinson, J., Genty, B. and Foyer, C.H. (1990) Relationship between photosynthetic electron transport and stromal enzyme activity in pea leaves. *Plant Physiol.*, **94**:545-553.

Henley, W.J., Levavasseur, G., Franklin, L.A. *et al.* (1991) Photoacclimation and photoinhibition in *Ulva rotundata* as influenced by nitrogen availability. *Planta*, **184**:235-243.

Horton, P. (1990) Regulation of light harvesting by metabolic events. In: Baltscheffsky, M. (ed.) *Current Research in Photosynthesis*, Vol. IV. Kluwer Academic Publishers, Dordrecht, pp. 111-118.

Horton, P., Ruban, A.V., Rees, D. *et al.* (1991) Control of the light-harvesting function of chloroplast membranes by the proton concentration in the thylakoid lumen: aggregation states of the LHCII complex and the role of zeaxanthin. *FEBS. Lett.*, **1059**:355-360.

Jones, B.L. and Porter, J.W. (1986) Biosynthesis of carotenes in higher plants. *Crit. Rev. Plant Sci.*, **3**:295-324.

Khamis, S., Lamaze, T., Lemoine, Y. and Foyer, C. (1990) Adaptation of the photosynthetic apparatus in maize leaves as a result of nitrogen limitation. *Plant Physiol.*, **94**:1436-1443.

Kitajima, M. and Butler, W.L. (1975) Quenching of chlorophyll fluorescence and primary photochemistry in chloroplasts by dibromothymoquinone. *Biochim. Biophys. Acta*, **376**:105-115.

Krause, G.H. (1988) Photoinhibition of photosynthesis. An evaluation of damaging and protective mechanisms. *Physiol. Plant.*, **74**:566-574.

Krause, G.H. and Behrend, U. (1986) ΔpH-dependent chlorophyll fluorescence quenching indicating a mechanism of protection against photoinhibition of chloroplasts. *FEBS Lett.*, **200**:298-302.

Krause, G.H., Vernotte, C. and Briantais, J.-M. (1982) Photoinduced quenching of chlorophyll fluorescence in intact chloroplasts and algae. Resolution into two components. *Biochim. Biophys. Acta*, **679**:116-124.

Krinsky, N.I. and Welankiwar, S. (1984) Assay of carotenoids. *Meth. Enzymol.*, **105**:155-162.

Lemoine, Y., Zabulon, G. and Cornu, A. (1987) Chlorophyll–protein complexes; changes associated with chloroplast development in a virescent *Petunia hybrida* mutant. In: Biggins, J. (ed.) *Progress in Photosynthesis Research*, Vol. II. Martinus Nijhoff, Dordrecht, pp. 371-374.

Liaaen-Jensen, S. (1978) Algal carotenoids and chemosystematics. In: Faulkner, D.J. and Fenical, W.H. (eds.), *Marine Natural Products Chemistry*. Plenum, New York, pp. 239-259.

Lichtenthaler, H.K. and Meier, D. (1984) Regulation of chloroplast photomorphogenesis by light intensity and light quality. In: Ellis, R.J. (ed.) *Chloroplast Biogenesis*. Cambridge University Press, Cambridge, pp. 261-281.

Lichtenthaler, H.K., Burgstahler, R., Buschmann, C. *et al.* (1982) Effect of high light and high light stress on composition, function and structure of the photosynthetic apparatus.

In: Marcelle, R. (ed.) *Stress Effects on Photosynthesis*. Dr W. Junk, The Hague, pp. 353-370.

Liddell, P.A., Nemeth, G.A., Lehman, W.R. *et al.* (1982) Mimicry of carotenoid function in photosynthesis: Synthesis and photophysical properties of a carotenopyropheophorbide. *Photochem. Photobiol.*, **36**: 641-645.

Mantoura, R.F.C. and Llewellyn, C.A. (1983) Rapid determination of algal chlorophyll and carotenoid pigments and their breakdown products in natural waters by reverse-phase high-performance liquid chromatography. *Anal. Chim. Acta*, **151**:297-314.

Neubauer, C. and Yamamoto, H.Y. (1991) The Mehler-peroxidase reaction generates the ΔpH that is required for zeaxanthin-related fluorescence quenching. *Plant Physiol.* Suppl., **96**:119.

Noctor, G., Rees, D. and Horton, P. (1989) Uncoupler titrations of energy-dependent quenching of chlorophyll fluorescence in chloroplasts. In: Baltscheffsky, M. (ed.) *Current Research in Photosynthesis*, Vol. 1. Kluwer Academic Publishers, Dordrecht, pp. 627-630.

Noctor, G., Rees, D., Young, A. and Horton, P. (1991) The relationship between zeaxanthin, energy-dependent quenching of chlorophyll fluorescence, and trans-thylakoid pH gradient in isolated chloroplasts. *Biochim. Biophys. Acta*, **1057**:320-330.

Oxborough, K. and Horton, P. (1987) Characterization of the effects of antimycin A upon high energy state quenching of chlorophyll fluorescence (q_E) in spinach and pea chloroplasts. *Photosynth. Res.*, **12**:119-128.

Pfündel, E., Dilley, R.A., Gilmore, A. and Yamamoto, H.Y. (1991) Delocalized ΔH^+ drives the violaxanthin → zeaxanthin conversion much more effectively than localized $\Delta\mu H^+$ having the equivalent capacity for ATP formation. *Plant Physiol.* Suppl., **96**:16.

Prenzel, U. and Lichtenthaler, H.K. (1982) Localization of ß-carotene in chlorophyll a–proteins and changes in its levels during short-term high light exposure of plants. In: Wintermans, J.F.G.M. and Kuiper, P.J.C. (eds.), *Biochemistry and Metabolism of Plant Lipids*. Elsevier, Amsterdam, pp. 565-572.

Rees, D., Young, A., Noctor, G. *et al.* (1989) Enhancement of the ΔpH-dependent dissipation of excitation energy in spinach chloroplasts by light-activation; correlation with the synthesis of zeaxanthin. *FEBS Lett.*, **256**:85-90.

Rees, D., Noctor, G. and Horton, P. (1990) The effect of high-energy-state excitation quenching on maximum and dark level chlorophyll fluorescence yield. *Photosynth. Res.*, **25**:199-211.

Rock, C.D. and Zeevaart, J.A.D. (1991) The *aba* mutant of *Arabidopsis thaliana* is impaired in epoxy-carotenoid biosynthesis. *Proc. Natl. Acad. Sci. USA*, **88**: 7496-7499.

Sapozhnikov, D.I. (1973) Investigation of the violaxanthin cycle. *Pure Appl. Chem.*, **35**:47-61.

Sapozhnikov, D.I., Kyzsovskaya, T.A. and Maevskaya, A.N. (1957) Change in the interrelationship of the basic carotenoids of the plastids of green leaves under the action of light. *Doklady Akademy Nauk SSSR, Botanical Sciences Sections*, **113**:74-76.

Schreiber, U. and Neubauer, C. (1990) O_2-dependent electron flow, membrane energization and the mechanism of non-photochemical quenching of chlorophyll fluorescence. *Photosynth. Res.*, **25**: 279-293.

Schultz, G., Heintze, A., Hoppe, P. *et al.* (1992) Tocopherol and carotenoid synthesis in chloroplasts. Tight linkage to plastidic carbon metabolism in developing chloroplasts. In: Pell, E.J. and Steffen, K.L. (eds.), *Active Oxygen/Oxidative Stress and Plant Metabolism*. American Society of Plant Physiologists, Rockville, pp. 156-170.

Siefermann, D. and Yamamoto, H.Y. (1974a) Light-induced de-epoxidation of violaxanthin in lettuce chloroplasts. III. Reaction kinetics and effect of light intensity on de-epoxidase activity and substrate availability. *Biochim. Biophys. Acta*, **357**:144-150.

Siefermann, D. and Yamamoto, H.Y. (1974b) Light-induced de-epoxidation of violaxanthin in lettuce chloroplasts. V. Dehydroascorbate, a link between photosynthetic electron transport and de-epoxidation. In: Avron, M. (ed.) *Proceedings of 3rd International*

Congress on Photosynthesis, Vol. 3. Elsevier, Amsterdam, pp. 1991-1998.
Siefermann, D. and Yamamoto, H.Y. (1975) Properties of NADPH and oxygen-dependent zeaxanthin epoxidation in isolated chloroplasts. A transmembrane model for the violaxanthin cycle. *Arch. Biochem. Biophys.*, **171**:70-77.
Siefermann, D. and Yamamoto, H.Y. (1976) Light-induced de-epoxidation in lettuce chloroplasts. VI. De-epoxidation in grana and stroma lamellae. *Plant Physiol.*, **57**:939-940.
Siefermann-Harms, D. (1977) The xanthophyll cycle in higher plants. In: Tevini, M. and Lichtenthaler, H.K. (eds.), *Lipids and Lipid Polymers in Higher Plants*. Springer, Berlin, pp. 218-230.
Siefermann-Harms, D. (1985) Carotenoids in photosynthesis. I. Location in photosynthetic membranes and light-harvesting function. *Biochim. Biophys. Acta*, **811**: 325-355.
Siefermann-Harms, D. (1988) High performance liquid chromatography of chloroplast pigments. One-step separation of carotene and xanthophyll isomers, chlorophyll and pheophytin. *J. Chromatogr.*, **448**:411-416.
Siefermann-Harms, D., Joyard, J. and Douce, R. (1978) Light-induced changes of the carotenoid levels in chloroplast envelopes. *Plant Physiol.*, **61**:530-533.
Somersalo, S. and Krause, G.H. (1990) Photoinhibition at chilling temperatures and effects of freezing stress on cold acclimated spinach leaves in the field. A fluorescence study. *Physiol. Plant.*, **79**:617-622.
Stransky, H. and Hager, A. (1970a) Das Carotinoidmuster und die Verbreitung des lichtinduzierten Xanthophyll-Cyclus in verschiedenen Algenklassen. II. Xanthophyceae. *Arch. Mikrobiol.*, **71**:164-190.
Stransky, H. and Hager, A. (1970b) Das Carotinoidmuster und die Verbreitung des lichtinduzierten Xanthophyll-Cyclus in verschiedenen Algenklassen. IV. Cyanophyceae und Rhodophyceae. *Arch. Mikrobiol.*, **72**:84-96.
Thayer, S.S. and Björkman, O. (1990) Leaf xanthophyll content and composition in sun and shade determined by HPLC. *Photosynth. Res.*, **23**:331-343.
Weis, E. and Berry, J.A. (1987) Quantum efficiency of photosystem II in relation to "energy"-dependent quenching of chlorophyll fluorescence. *Biochim. Biophys. Acta*, **894**:198-208.
Willemönas, M. and Monas, E. (1991) Relationship between growth irradiance and the xanthophyll cycle in the diatom *Nitzschia palea*. *Physiol. Plant.*, **83**:449-456.
Wright, S.W. and Shearer, J.D. (1984) Rapid extraction and high-performance liquid chromatography of chlorophylls and carotenoids from marine phytoplankton. *J. Chromatogr.*, **294**:281-295.
Yamamoto, H.Y. (1979) Biochemistry of the violaxanthin cycle in higher plants. *Pure Appl. Chem.*, **51**:639-648.
Yamamoto, H.Y. and Higashi, R.M. (1978) Violaxanthin de-epoxidase; lipid composition and substrate specificity. *Arch. Biochem. Biophys.*, *190*:514-522.
Yamamoto, H.Y. and Kamite, L. (1972) The effects of dithiothreitol on violaxanthin de-epoxidation and absorbance changes in the 500-nm region. *Biochim. Biophys. Acta*, **267**:538-543.
Yamamoto, H.Y., Nakayama, T.O.M. and Chichester, C.O. (1962) Studies on the light and dark interconversions of leaf xanthophylls. *Arch. Biochem.*, **97**:168-173.
Yamamoto, H.Y., Chang, J.L. and Aihara, M.S. (1967) Light-induced interconversions of violaxanthin and zeaxanthin in New Zealand spinach-leaf segments. *Biochim. Biophys. Acta*, **141**:342-347.
Yamamoto, H.Y., Chenchin, E.E. and Yamada, D.K. (1974) Effect of chloroplast lipids on violaxanthin de-epoxidase activity. In: Avron, M. (ed.) *Proceedings of 3rd International Congress on Photosynthesis*. Elsevier, Amsterdam, pp. 1999-2006.

8

The photochemistry and function of carotenoids in photosynthesis

H.A. Frank and R.J. Cogdell

8.1 INTRODUCTION

8.1.1 General overview

In photosynthetic systems, carotenoids act as light-harvesting molecules and provide photoprotection of the plant and bacterial species (Cogdell and Frank 1987; Siefermann-Harms 1985). In many cases, the manner in which a particular carotenoid functions depends on its photochemical properties. Carotenoids participate in an abundance of photochemical reactions including singlet–singlet energy transfer, triplet–triplet energy transfer, oxidation, reduction and isomerisation. Carotenoid molecules are capable of reporting information about the course of these reactions from readily observable changes in many of their molecular spectroscopic properties.

In this chapter we concentrate on the photochemical reactions of carotenoids in photosynthesis. Over the past few years there have been a number of reviews on the function of carotenoids in photosynthesis (Cogdell and Frank 1987; Siefermann-Harms 1985; Koyama 1991). However, progress has been so swift since these were written that many of the problems that were previously highlighted in these reviews are now well on the way to being solved. An abundance of novel experimental and theoretical work has been carried out on carotenoids in recent years, providing a basis from which our understanding of their photochemical properties is being rapidly advanced. We believe that for the reader of this chapter, the conceptual framework is most naturally built by introducing the excellent work on model polyenes that has either preceded or been concurrent with the studies on carotenoids. Our approach is to follow the guidance set out by researchers in the field of polyene spectroscopy, making use of those aspects of polyene models that are applicable to carotenoids in accounting for the photochemical properties of these more extensively conjugated systems.

In general in this chapter, the results of such spectroscopic studies are used to describe and explain the photochemistry of carotenoids in relation to their functioning in photosynthesis. Details of spectroscopic methods themselves and of their application to the study of carotenoids in photosynthesis are given in Chapter 9.

8.1.2 Photochemical reactions

Many of the important functions of carotenoids in photosynthesis come from the fact that the carotenoids are bound into the same pigment–protein complexes as the chlorophylls and bacteriochlorophylls. One such example is as follows: if cells of photosynthetic organisms lacking carotenoids (e.g. the well-known carotenoidless mutant *Rhodobacter sphaeroides* R-26) are illuminated in the presence of oxygen they sensitise their own death (Griffiths et al. 1955). This is an example of photooxidative killing and is the result of triplet excited chlorophyll or bacteriochlorophyll sensitising the formation of singlet molecular oxygen (i.e. the $^1\Delta_g$ state of O_2). Singlet oxygen is a powerful enough oxidant to react with lipids, proteins and nucleic acids (Foote 1968a,b; Wilson and Hastings 1970), so that cells that are exposed to high concentrations of it die rapidly (Griffiths et al. 1955). Carotenoids bound in proximity to the chlorophyll or bacteriochlorophyll prevent this reaction (Krinsky 1966). Thus, the first, and arguably the essential, function of carotenoids in photosynthesis is to prevent harmful photo-oxidative reactions related to the presence of oxygen.

The second function of carotenoids in photosynthesis is to act as accessory light-harvesting pigments, i.e. to absorb incident solar radiation in the ~450–570 nm region where the chlorophylls do not absorb efficiently, and to transfer the energy to the chlorophylls (Cogdell and Frank 1987). If light absorbed by the chlorophylls and bacteriochlorophylls were only to be used to drive photosynthesis, then the efficiency of light usage in this wavelength range would be very poor. Carotenoids, therefore, enhance the wavelength range over which light is able to promote photosynthesis.

The light-harvesting and photoprotection examples of carotenoid functions in photosynthesis involve molecular singlet and triplet–states (Fig. 8.1). The photoprotective function utilises triplet–triplet energy transfer from chlorophyll or bacteriochlorophyll to carotenoids to prevent the chlorophyll-sensitised formation of singlet oxygen (Foote 1976; Borland et al. 1987). For photoprotection to be complete reactions 8.1–8.4

$$Chl \xrightarrow{h\upsilon} {}^1Chl^* \quad (chlorophyll\ \ excitation) \quad (8.1)$$

$$^1Chl^* \rightarrow {}^3Chl^* \quad (triplet-state\ \ formation) \quad (8.2)$$

$$^{3}Chl^{*} + Car \rightarrow Chl + {}^{3}Car^{*} \quad (triplet-triplet \;\; energy \;\; transfer) \quad (8.3)$$

$$^{3}Car^{*} \rightarrow Car + heat \quad (carotenoid \;\; deactivation) \quad (8.4)$$

must compete effectively with reaction 8.5

$$^{3}Chl^{*} + O_2 \rightarrow Chl + {}^{1}\Delta_g O_2^{*} \quad (singlet \;\; oxygen \;\; formation) \quad (8.5)$$

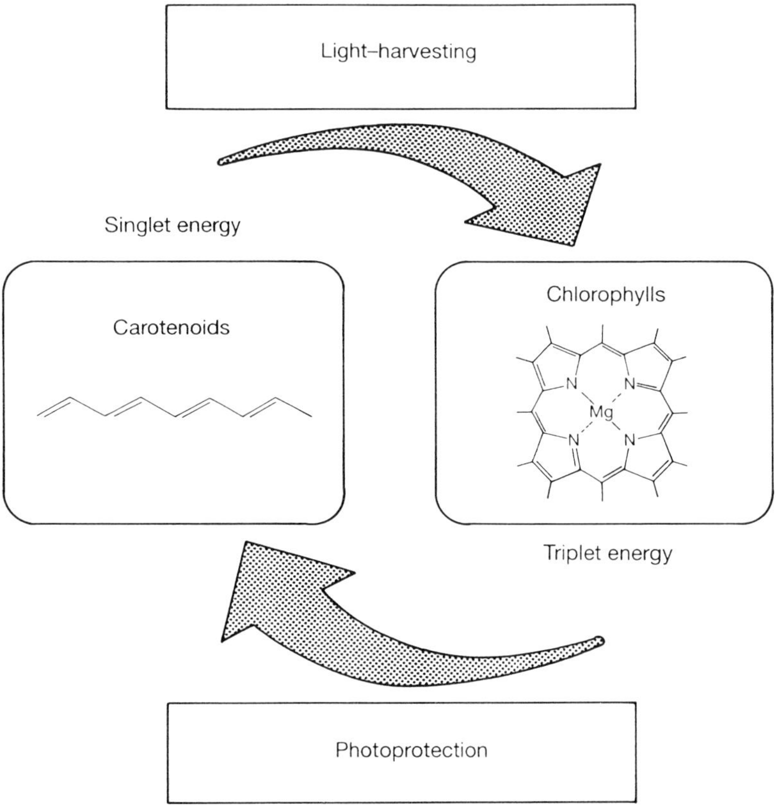

Figure 8.1 Figurative representation of the two roles of carotenoids in photosynthesis: light-harvesting and photoprotection.

It is also thought that carotenoids may scavenge singlet oxygen directly – a diffusion-limited reaction in solution (Krinsky 1968; Foote 1968a); i.e.

$$^{1}\Delta_g O_2^{*} + Car \rightarrow O_2 + {}^{3}Car^{*} \quad (singlet \;\; oxygen \;\; scavenging) \quad (8.6)$$

$$^{3}Car^{*} \rightarrow Car + heat \quad (carotenoid \;\; deactivation) \quad (8.7)$$

The light-harvesting function makes use of singlet–singlet energy transfer between carotenoids and chlorophylls (Cogdell and Frank 1987).

$$Car \xrightarrow{h\upsilon} {}^1Car^* \quad (carotenoid \ \ excitation) \quad (8.8)$$

$$^1Car^* + Chl \rightarrow Car + {}^1Chl^* \quad (singlet-singlet \ \ energy \ \ transfer) \quad (8.9)$$

The captured energy ends up ultimately in the reaction centre of the photosynthetic organism where it aids in the generation of a transmembrane potential via electron transfer between a series of donors and acceptors. This electrical potential is then converted into a chemical potential which serves as the thermodynamic driving force for the photosynthetic process.

8.2 ENERGY LEVELS

The origin of many of the photochemical properties of carotenoids lies in the disposition of their low-lying excited energy states. Here we discuss these states *in vitro* and *in vivo* .

8.2.1 *In vitro*

(a) Singlet-states

Carotenoids and polyenes possess two low-lying electronic excited singlet-states that account for many of their spectroscopic and functional properties (Hudson et al. 1982). These are denoted the 1^1B_u and the 2^1A_g states. Even for non-linear polyenes and carotenoids, it is customary to refer to these excited states by their C_{2h} symmetry state designations. This is only strictly appropriate for the linear polyenes. However, because the excited states of the non-linear molecules retain many of the characteristics of the linear systems (Birge et al. 1978; Hudson et al. 1982), the notation is carried over.

The 1^1B_u state

The pronounced absorption in the near ultraviolet (UV) or visible regions that is characteristic of all polyenes and carotenoids is attributable to an allowed transition from the ground 1^1A_g state to a 1^1B_u state. In general, the $1^1A_g \rightarrow 1^1B_u$ transition is very strong (oscillator strength of ~1) and increases in intensity with the extent of π-electron conjugation. The energy of the 1^1B_u state, which can readily be determined from the absorption spectrum, provided the spectral origin (i.e. the transition from the lowest vibrational level (v = 0) of the ground state to the lowest vibrational level (v′ = 0) of the excited state) is observed, appears to decrease to a constant value at long carbon–carbon double bond chain lengths. (Fig. 8.2 shows the relationship between carbon–carbon double bond chain length and the energy of the 1^1B_u state

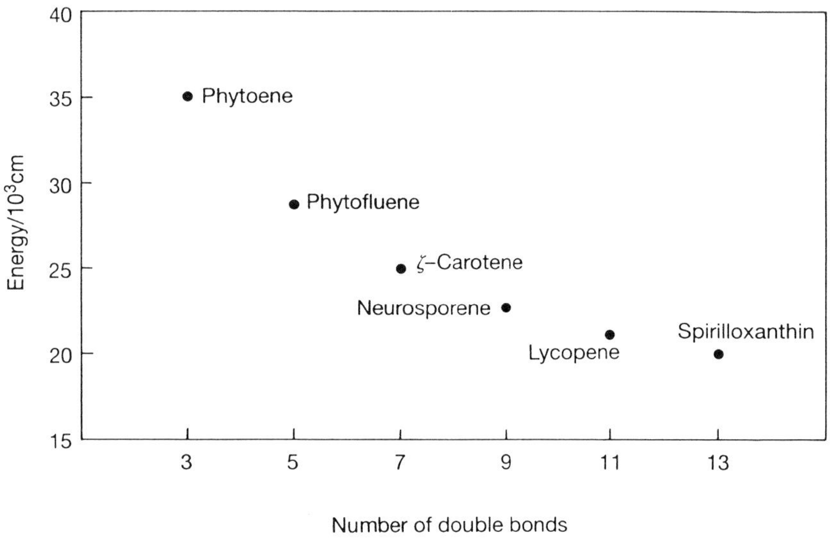

Figure 8.2 The dependence on carbon–carbon double bond chain length of the energy of the 1^1B_u state deduced from the λ_{max} (in 40-60 b.p. petroleum ether) of the absorption bands of various carotenoids. Ideally, the 1^1B_u state energy should be deduced from the spectral origin of the transition. However, in many cases this is not easily discernable.

deduced from the λ_{max} of the absorption bands of various carotenoids.) Molecular orbital theory calculations using semi-empirical configuration interaction approaches yield adequate descriptions of the energy of the 1^1B_u state and the oscillator strength of the $1^1A_g \rightarrow 1^1B_u$ transition (Hudson et al. 1982; Zhang et al. 1989; Connors and Frank, unpublished work). The calculations support the view that the 1^1B_u state is largely ionic in character (Birge 1986; Hudson et al. 1982). Also, owing to dispersion forces, the $1^1A_g \rightarrow 1^1B_u$ transition frequency undergoes a pronounced shift to lower energy when the non-polar polyenes and carotenoids are dissolved in non-polar solvents of increasing refractive index (Hudson et al. 1982; Andersson et al. 1991). The spectral shift depends on the the solvent polarisability rather than on the polarity of the solvent (Hudson et al. 1982). It is found experimentally that the $1^1A_g \rightarrow 1^1B_u$ transition frequency follows a linear dependence with the ratio $(n^2 - 1)/(n^2 + 2)$, where n is the solvent refractive index (Andersson et al. 1991; Andrews and Hudson 1978; Basu 1964; LeRosen and Reid 1952, Hudson et al. 1982), see Fig. 8.3. This ratio is related to the solvent polarisability, α, by the equation

$$\frac{n^2-1}{n^2+2}=\frac{N_A\rho\alpha}{3\varepsilon_0 M} \quad (8.10)$$

where N_A is Avogadro's number, ρ is the density, ε_0 is the vacuum permittivity and M is the molar mass (Barrow 1973). For a direct comparison with theoretical calculations, the transition energies for the carotenoids in solution should be extrapolated to the gas phase by use of solvent shift theory (Longuet-Higgins and Pople 1957; McCrae 1957; Basu 1964; Sklar et al. 1977; Andrews and Hudson 1978; Hudson et al. 1982). Andersson et al. (1991) applied this theory to spheroidene and ß-carotene in a series of different solvents and analysed the data according to the equation

$$\upsilon\ (solv)=\upsilon\ (gas)-\frac{K(n2-2)}{(n2+2)} \quad (8.11)$$

where

$$\mathrm{K}=\frac{2}{\mathrm{hca}^3}\frac{E}{E+\varepsilon}\left[\left|P_{ge}\right|^2+(\alpha_e-\alpha_g)\right] \quad (8.12)$$

and h is Planck's constant, c is the speed of light, a is the solute cavity radius, E is the average excitation energy for the solvent, ε is the average excitation energy for the solute, P_{ge} is the transition dipole moment for the ground-to-excited state electronic transition and α_e -α_g is the difference in polarisability of these two states. The numerical values obtained for K by Andersson et al. (1991) were 11 700 (±500) cm^{-1} and 11 800 (± 500) cm^{-1} for spheroidene and ß-carotene, respectively. These are similar to the values of K obtained for shorter polyenes (Sklar et al. 1977, 1985) The corresponding values of υ (gas) were 24 800 cm^{-1} and 24 900 cm^{-1}. It is important to note that since K depends on the electronic dipole moment of the transition (*viz,* P_{ge}) weaker transitions will also shift with solvent polarisability, albeit less markedly than strongly allowed transitions.

In recent years, weak fluorescence has been observed from the 1^1B_u state of carotenoids and polyenes that have eight or more conjugated carbon–carbon double bonds (Snyder et al. 1985; Watanabe et al. 1987; Bondarev et al. 1989; Gillbro and Cogdell 1989; Cosgrove et al. 1990; DeCoster et al. 1992; Shreve et al. 1991a,b;). Although the fluorescence yield is typically very low (~10^{-4}, Table 8.1), the spectra show good mirror image symmetry with the absorption profiles of the molecules (see Fig. 8.4). However, some variations in the fluorescence band shapes with excitation wavelength have been noted (Gillbro and Cogdell 1989). Carotenoids and polyenes having eight or fewer conjugated carbon–carbon double bonds display fluorescence that is significantly "Stokes-shifted" from the absorption spectrum (Cosgrove et al. 1990; DeCoster et al. 1992). The abrupt crossover from the "1^1B_u" to "Stokes-shifted" emission at a conjugated chain length of eight is best explained by assuming that two competing processes are responsible for deactivation of the 1^1B_u state. These are fluorescence and rapid internal conversion to another state that lies between the 1^1B_u state and the ground state.

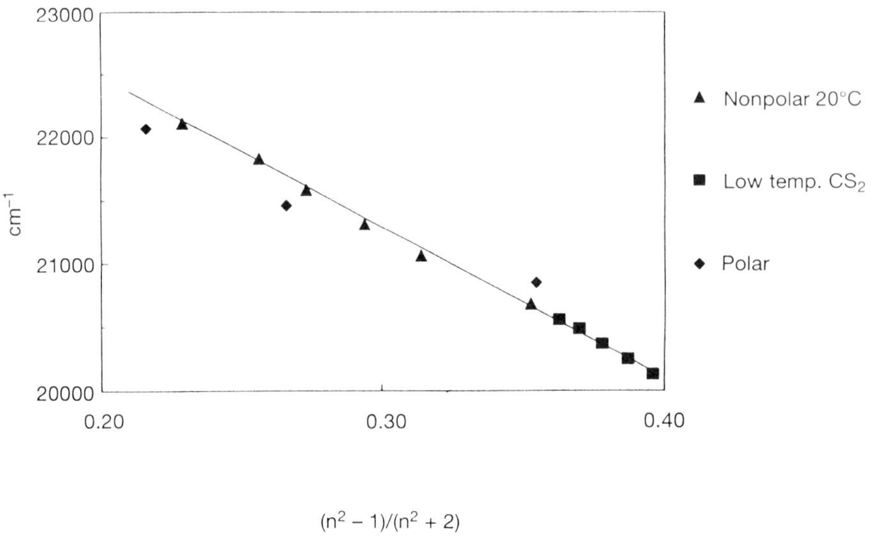

Figure 8.3 The effect on the λ_{max} of spheroidene of the quantity $(n^2 - 1)/(n^2 + 2)$ where n is the index of refraction of the solvent. This quantity is related to the solvent polarizability. (See the discussion in the text).

Table 8.1 The state origin assignments, spectral peaks, yields and extent of conjugation of fluorescent carotenoid molecules

Carotenoid	*No. of conjugated double bonds*	*State origin assigment*	*Spectral peaks (nm)*	*Yield*	*Reference*
ß-Carotene in isopentane	11	S_2	510,545,575	4×10^{-5} (77K) 8×10^{-5} (4.2K)	Bondarev et al. 1989
ß-Carotene in CS_2	11	S_2	555	6×10^{-5}	Gillbro & Cogdell 1989; Mimuro et al. 1992
ß-Carotene in CS_2	11	S_2	575	1.6×10^{-4}	Shreve et al. 1991a
ß-Carotene in ethanol	11	S_2	525	1.9×10^{-4}	Shreve et al. 1991a

Rhodopin in CS_2	11	S_2	550,590	ND	Gillbro & Cogdell 1989
Spheroidenone in CS_2	11	S_2	570,610	3×10^{-5}	Gillbro & Cogdell 1989
ß-Apo-8'-carotenal in CS_2	10	S_1, S_2	535,585,759	1.8×10^{-5}	Mimuro et al. 1991
ß-Apo-8'-carotenal in CCl_4	10	S_1, S_2	559,750	2.0×10^{-5}	Mimuro et al. 1991
ß-Apo-8'-carotenal in *n*-hexane	10	S_1, S_2	550	2.4×10^{-5}	Mimuro et al. 1991
ß-Apo-12'-carotenol in EPA at 77K	7	S_1	600	ND	Cosgrove et al. 1990
ß-Apo-8'-carotenol in EPA at 77K	9	S_2	480,520,540	ND	Cosgrove et al. 1990
ß-Apo-6'-carotenol in EPA at 77K	10	S_2	495,530,570	ND	Cosgrove et al. 1990
Isozeaxanthinin EPA at 77K	11	S_2	510,454,580	2.3×10^{-4}	Cosgrove et al. 1990
Fucoxanthin in ethanol	10*	S_1, S_2	750,540	7.9×10^{-4} (S_1) 6.6×10^{-5} (S_2)	Shreve et al. 1991c
Fucoxanthin in CS_2	10*	S_1	750	5.0×10^{-5}	Katoh et al. 1991; Mimuro et al. 1992
Neoxanthin in CS_2	10*	S_2	550	1.0×10^{-5}	Mimuro et al. 1992
Siphonaxanthin in CS_2	9	S_1	750	3.1×10^{-5}	Mimuro et al. 1992
Peridininin CS_2	9*	S_1	750	ND	Mimuro et al. 1992
ß-Carotene analogue mini-5 in 3-methylpentane	5	S_1	500	7×10^{-3} (in *n*-hexane)	Andersson et al. 1992
ß-Carotene analogue mini-7 in 3-methylpentane	7	S_1	580	8×10^{-4} (in *n*-hexane)	Andersson et al. 1992
ß-Carotene analogue mini-9 in 3-methylpentane	9	S_1, S_2	510,740	1×10^{-4} (S_2 in *n*-hexane)	Andersson et al. 1992

ß-Carotene in 3-methylpentane	11	S_2	500	2 x 10^{-4} (in *n*-hexane)	Andersson et al. 1992
Spheroidene in methanol	10	S_2	495,530	ND	DeCoster et al. 1992
3,4-Dihydro-spheroidene in methanol	9	S_1, S_2	475,510,750	ND	DeCoster et al. 1992
3,4,5,6-Tetrahydro-spheroidene in methanol	8	S_1, S_2	450,480,710	ND	DeCoster et al. 1992
3,4,7,8-Tetrahydro-spheroidene in methanol	7	S_1, S_2	430,460,480,630	ND	DeCoster et al. 1992

Spectral observations were done at room temperature except where otherwise noted
ND not detertmined
* Including allene

The 2^1A_g state

For many years there were several spectral characteristics of polyenes that could not be easily reconciled with molecular orbital theory based on linear combinations of atomic orbitals (LCAOMO) (Hudson et al. 1982). These included: (i) very large energy separations (3000–7000 cm^{-1}) between the spectral origins of absorption and emission; (ii) anomalously long intrinsic fluorescence lifetimes of polyenes; (iii) the fact that changes in the solvent polarisability were found to shift the absorption and emission spectra differently (Andrews and Hudson 1978). The most compelling explanation for these anomalies is that the first singlet excited state is not the 1^1B_u state, but instead is a state having A_g symmetry (see Fig. 8.5). This idea has completely revolutionised the picture of how the excited electronic states of polyenes and carotenoids are ordered. Data in support of the hypothesis of a low-lying 2^1A_g state have come from absorption and fluorescence studies on unsubstituted polyenes (see Bouwman et al. 1990, and references therein), and from investigations of several α,ω-diphenylpolyenes and α,ω-dimethylpolyenes. These various molecules exhibit the behaviour noted above and also show that, in general, the quantum yield for the $2^1A_g \rightarrow 1^1A_g$ fluorescence transition decreases as the chain length increases. This has led to a description of the polyene $2^1A_g \rightarrow 1^1A_g$ symmetry-forbidden transition where its oscillator strength derives from a vibronically-induced borrowing of intensity from the 1^1B_u state (Hudson et al. 1982). Increasing the chain length, which widens the 2^1A_g to 1^1B_u energy gap, would decrease the amount of state mixing and lead to a reduction in the observed fluorescence yield. High resolution, low-temperature (4.2 K), Shpolskii matrix, spectral data from various polyenes (e.g. diphenyloctatetraene in tetradecane) show a series of weak absorption lines at an energy lower than the pronounced $1^1A_g \rightarrow 1^1B_u$ absorptions (D'Amico et al. 1980; Hudson et al. 1982; Snyder et al. 1985). The weak lines are very close to the origins of the emission spectra and are attributable

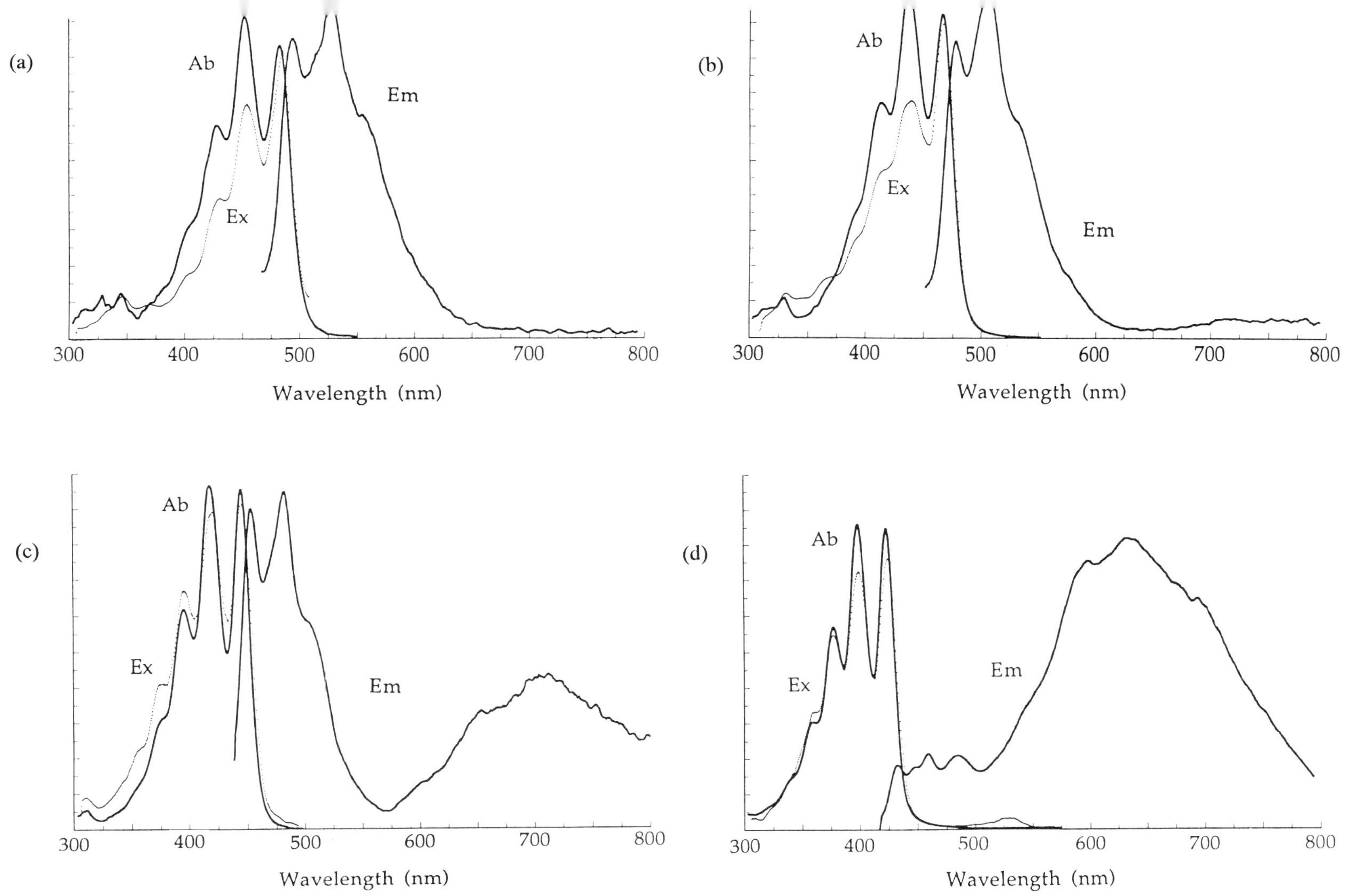
(a)
Ab
Ex
Em
300
400
500
600
700
800
Wavelength (nm)
(b)
Ab
Ex
Em
300
400
500
600
700
800
Wavelength (nm)
(c)
Ab
Ex
Em
300
400
500
600
700
800
Wavelength (nm)
(d)
Ab
Ex
Em
300
400
500
600
700
800
Wavelength (nm)

Figure 8.4. Absorption (Ab, solid line), fluorescence (Em), and fluorescence excitation (Ex, dotted line) spectra of (**a**) spheroidene (10 carbon–carbon double bonds), (**b**) 3,4-dihydrospheroidene, also known as methoxyneurosporene (9 carbon–carbon double bonds); (**c**) 3,4,5,6-tetrahydrospheroidene (8 carbon–carbon double bonds); and (**d**) 3,4,7,8-tetrahydrospheroidene (7 carbon–carbon double bonds) in methanol. Spheroidene has its most prominent emission from the 1^1B_u state. As the π-electron conjugation is shortened, the emission gradually becomes dominated by fluorescence from the 2^1A_g state. The details of the experiments are given in DeCoster et al. (1992).

to the $1^1A_g \rightarrow 2^1A_g$ absorptions. These high resolution studies provide the clearest confirmation of the low-lying 2^1A_g state hypothesis and vibronic coupling between this state and the 1^1B_u state.

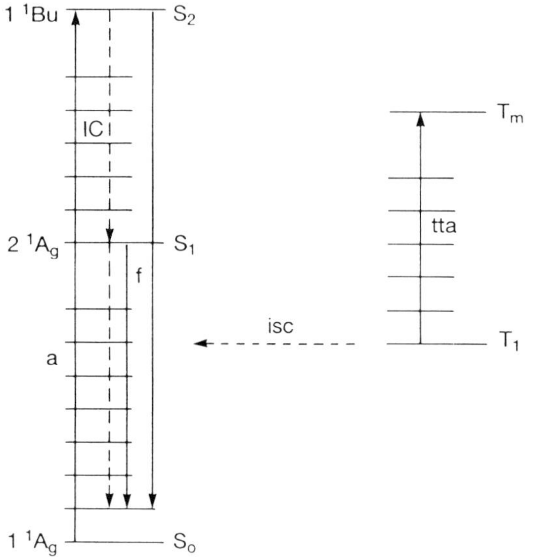

Figure 8.5 Energy level diagram for carotenoid and polyene low-lying singlets and triplets, summarizing the various intramolecular photochemical processes the molecules can undergo. a, absorption; ic, internal conversion; f, fluorescence; isc, intersystem crossing; tta, triplet–triplet absorption.

It is now widely accepted that a low-lying A_g state is a general feature of all polyene and carotenoid molecules. Simple molecular orbital theory calculations are not well suited for a quantitative description of the 2^1A_g states, although in principle the approach should work given enough configuration interaction. Suitable qualitative trends of polyenes may be obtained by use of the semi-empirical molecular orbital theory configuration interaction approach if double (or higher) excitations are included in the configuration interaction. More appropriate are excitonic descriptions (Schulten and Karplus 1972; Schulten et al. 1976; Tavan and Schulten 1979). These calculations support the idea that the 2^1A_g state is largely covalent in character (Hudson et al. 1982). It should be emphasised, however, that the energies of the 2^1A_g

states for the very long-chain carotenoids are not yet well defined experimentally or theoretically. This is because long-chain carotenoids (greater than eight carbon–carbon double bonds) tend to fluoresce from their 1^1B_u states, and thus provide no information about the absolute energy of their 2^1A_g states. From a theoretical standpoint, the extremely large number of excitations required in the basic set (i.e. up to and including quadruple excitations are required to predict the tendency of the $1^1B_u - 2^1A_g$ energy gap to increase with conjugated chain length) (Tavan and Schulten 1979), and hence the inordinate amount of computer time required in calculations on the long-chain carotenoids, preclude a detailed computational analysis of the 2^1A_g state at present. Recently, Kohler (1991) has proposed a simpler way to approach this problem.

Because knowledge of the energies of the 2^1A_g states of carotenoids is important in understanding the mechanism of energy transfer by carotenoids in photosynthesis, a significant effort from numerous groups has gone into attempting to establish values for the 2^1A_g state energies of various carotenoids. One of the first such attempts was made by Thrash et al. (1977) using Raman spectroscopy. Features in the Raman excitation profile spectrum of the υ_1 band of all *trans*-ß-carotene (11 conjugated carbon–carbon double bonds) in the preresonance region were interpreted as vibrational structure of a low-lying excited ^1A_g state with its origin near 17 230 (± 100) cm^{-1}. An energy gap of 3 470 (± 100) cm^{-1} between the 1^1B_u and 2^1A_g states was deduced, the 1^1B_u spectral origin appearing at 20 700 cm^{-1} (i.e. 483 nm).

More recently, a series of ß-apo-carotenols (Cosgrove et al. 1990), spheroidenes (DeCoster et al. 1992) and mini-carotenes (i.e. a series of ß-carotene homologues with decreasing chain length; Andersson et al. 1992) having varying extents of double bond conjugation have been analysed by absorption, fluorescence and fluorescence excitation spectroscopy (see Fig. 8.4 and Table 8.1). Carotenoids with fewer than eight conjugated carbon–carbon double bonds exhibit the "Stokes-shifted" 2^1A_g-state emission that is seen in the shorter polyenes. The 2^1A_g fluorescence has also been reported from an allenic carotenoid, fucoxanthin (Mimuro et al. 1992; Shreve et al. 1991c; Table 8.1). Because the longer chromophore carotenoids have not been observed to fluoresce from their lowest singlet-states it is difficult to determine the energies of the 2^1A_g states. To a first approximation one may extrapolate the energies of the 2^1A_g states of the shorter polyene/carotenoid molecules to the longer conjugated systems (see Fig. 8.6). The trends observed for the state energies of the ß-apo-carotenols, spheroidenes and shorter polyenes indicate that whereas both the 1^1B_u and 2^1A_g state energies decrease with extent of π-electron conjugation, the $1^1B_u - 2^1A_g$ energy gap increases. An extrapolation yields a $1^1B_u - 2^1A_g$ energy gap of approximately 6000–7000 cm^{-1} for molecules such as ß-carotene which has eleven carbon–carbon double bonds. The value (3470 (± 100) cm^{-1}) determined by Thrash et al. (1977) for ß-carotene is significantly out of this range and calls into question the validity of the Raman approach to determining the 2^1A_g state energy. Indeed, recent resonance Raman measurements strongly suggest that this technique is not capable of detecting the weak $1^1A_g \rightarrow 2^1A_g$ transition on top of the strong $1^1A_g \rightarrow 1^1B_u$ absorption (Watanabe et al. 1987). The extrapolation of the fluorescence data would put the 2^1A_g state energy of ß-carotene at approximately 14 000 cm^{-1}.

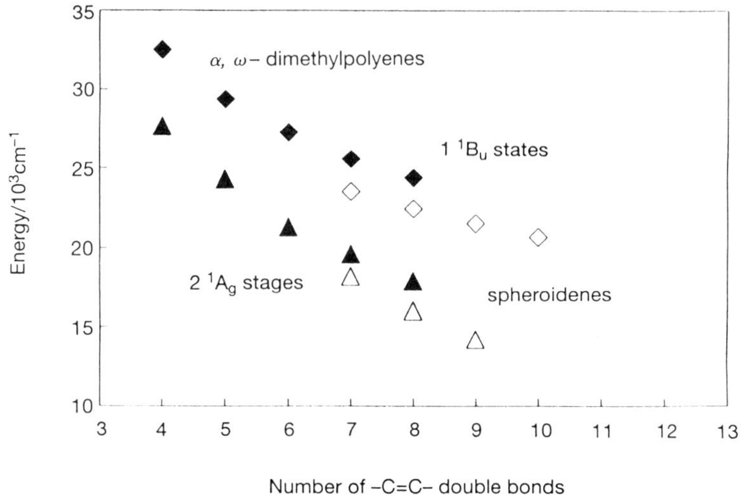

Figure 8.6 The dependence on extent of π-electron conjugation of the 1^1B_u and 2^1A_g state energies of the low-lying singlet-states of a series of α,ω-dimethylpolyenes and spheroidenes.

As mentioned above, the crossover from 1^1B_u to "Stokes-shifted" 2^1A_g emission at a conjugated chain length of approximately eight is best explained by fluorescence and internal conversion to the 2^1A_g state competing for deactivation of the 1^1B_u state. In this model the crossover can be qualitatively understood in the following manner: the energy gap between the 2^1A_g and the 1^1B_u states increases at longer chain lengths, while the absolute energies of both states decrease. This should decrease fluorescence yields from both 1^1B_u and 2^1A_g because of enhanced internal conversion to 1^1A_g. The widening $1^1B_u - 2^1A_g$ energy gap with increasing extent of conjugation diminishes the probability of internal conversion between the 1^1B_u and the 2^1A_g states and allows fluorescence from the 1^1B_u state to compete effectively with radiationless decay for deactivation to the ground state. Hence, all long chromophore carotenoids are observed to be 1^1B_u emitters. Shorter carotenoids will emit from their 2^1A_g states.

A promising approach to determining directly the energies of the A_g states of the carotenoids is two-photon spectroscopy. The $1^1A_g \rightarrow 2^1A_g$ transition is not symmetry forbidden in a two-photon absorption experiment. This is because the transition may proceed by absorption through a virtual B_u state intermediate. However, a favourable selection rule does not mean that a transition will be observed (Birge 1986). Although two-photon spectroscopic techniques have been used for polyenes such as octatetraene (Granville et al. 1979) and retinal (Birge 1986), they have met with little success for isolated carotenoids *in vitro*. This is because fluorescence detection of two-photon excitation is usually the method of choice for determining the forbidden state energies in molecules that emit efficiently. Carotenoids fluoresce only very weakly. In principle, however, with improved detection techniques this experiment should be possible. Other two-photon spectroscopic methods such as the thermal lens or double resonance techniques have been discussed (Birge 1986), but not applied successfully to carotenoids. Apparently, overtones of solvent vibrations interfere with the measurements as one goes farther into the infrared spectral region. Fluorescence detection of two-photon excitation has been used to probe the forbidden states of the protein-bound carotenoids in the light-harvesting pigment–protein complex from *Phaeodactylum tricornutum* (Shreve et al. 1990). In this study, energy transfer from the carotenoids to fluorescing, protein-bound, chlorophyll *a* molecules facilitated the investigation. However, owing to complexity in the spectra, the assignment of the carotenoid A_g state energy from the two-photon profile was not unambiguous. Two-photon spectroscopic techniques, however, continue to hold great promise for understanding the excited state complexion of carotenoids, particularly in native photosynthetic proteins, where use can be made of the natural tendency of the carotenoids to transfer energy to the strongly fluorescing chlorophylls.

(b) Triplet-states

The triplet–states of carotenoid molecules are poorly understood. This is because direct population of triplet–states of isolated carotenoids via singlet–triplet intersystem crossing is not very efficient. The low yields of intersystem crossing make investigations of carotenoid triplets feasible only through the use of energy donors which sensitise the formation of the carotenoid triplet-state (Bensasson et al. 1976). The triplets are then detected by pulse radiolysis (Bensasson et al. 1976), flash photolysis (Mathis and Kleo 1973; Bensasson et al. 1976) or electron paramagnetic resonance (EPR) spectroscopic techniques (Frank et al. 1980). Pulse radiolysis and flash photolysis optical spectroscopic experiments have been used to determine triplet-state energies (Mathis and Kleo 1973; Bensasson et al. 1976), triplet–triplet absorption spectra of carotenoids (Mathis and Kleo 1973) and quantum yields of carotenoid triplet formation. EPR spectroscopy has been used to elucidate the magnetic interactions between the two electrons that comprise the triplet–state (Frank et al. 1980).

Triplet-state energies

Direct information on the absolute energies of triplet–states of polyenes and carotenoids is sparse (Hudson and Kohler 1974). Electron impact spectroscopy has been used on 1,3-butadiene (Mosher et al. 1973) and 1,3,5-hexatriene (Knoop and Oosterhoff 1973) and their lowest triplet-state energies have been determined to be 3.2 eV and 2.6 eV, respectively. Magnetic perturbation by oxygen at high pressures to enhance the singlet–triplet transition in ethylene, butadiene, hexatriene, octatetraene, and several polyene aldehydes including retinal (Evans 1960, 1961; Raubach and Guzzo 1971; Hudson and Kohler 1974) has been employed to determine the triplet energies of these molecules. The data of Evans (1960, 1961) have been extrapolated by Bensasson et al. (1976) to estimate the triplet-state energies of the longer-chain carotenoids (see Fig. 8.7). The carotenoid triplet-state energies have important consequences concerning the ability of carotenoids to scavenge singlet oxygen or quench the triplet-states of various chlorophyll species. As an example, an extrapolation from the shorter polyenes (Fig. 8.7) leads to a value of ~ 6 200 cm^{-1} for the triplet-state energy of ß-carotene. This would suggest that the energy is sufficiently low to quench the triplet-state energy of chlorophyll *a* (10 752 cm^{-1} in diethyl ether (Krasnovskii et al. 1974) and to scavenge the $^1\Delta_g$ state of oxygen (7 867 cm^{-1}, Krinsky 1971; 7 855 cm^{-1}, Bensasson et al. 1976).

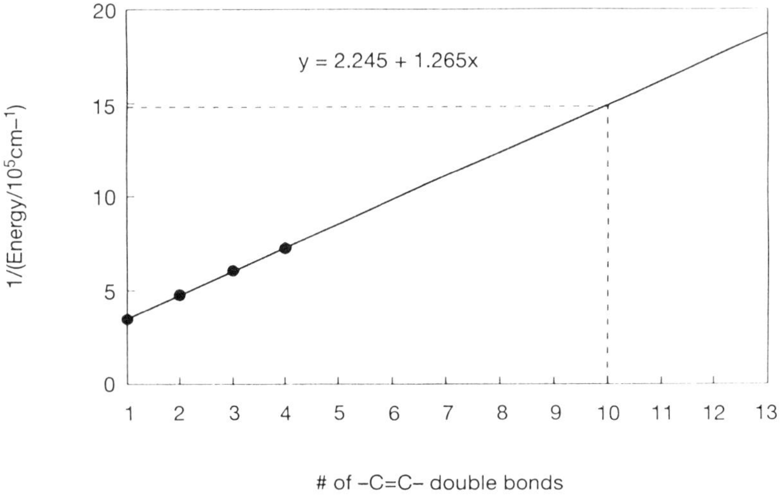

Figure 8.7 Plot of the reciprocal triplet energy (based on the spectral origins of the singlet-triplet transitions) of four polyenes, ethylene, butadiene, hexatriene and octatetraene (Evans (1960, 1961) as a function of chain length. The solid line is an extrapolation to longer carotenoids, e. g. the reciprocal energy of spheroidene (ten carbon–carbon double bonds) is represented by the horizontal dashed line. the coefficients of the equation ($y = 2.245 + 1.265x$, where y is the inverse of the triplet energy and x is the number of carbon–carbon double bonds) were derived from a linear fit to the data. This plot reproduces the original analysis presented by Bensasson et al. (1976) and is adaped from their work by permission of Pergamon Press PLC.

Recently, laser-induced optoacoustic calorimetry has been applied to a few carotenoids in an attempt to obtain the absolute triplet-state energies of these molecules (R.W Redmond, unpublished). The experiment makes use of laser excitation of a triplet energy donor in solution with the molecule of interest (a carotenoid in this case) in a cuvette in contact with an acoustic transducer (Burkey et al. 1986). After excitation of the donor, the triplet energy is transferred to the carotenoid where it is stored in a time domain that is long compared to the transducer response time. By measuring the heat generated with and without the carotenoid present in solution, the stored energy (related to the triplet-state energy of the carotenoid) may be determined (see Burkey et al. 1986, for details of the method). The experiment has, as yet, been applied to only a few carotenoids, but recent data are, at best, more accurate than other determinations and, at worst, consistent with the indirect determinations, as discussed below.

An interesting rule-of-thumb concerning triplet-state energies derives from a valence bond theory description of the 2^1A_g states of polyenes (Hudson et al. 1982). In this model the 2^1A_g state wavefunction is described as the composite of two triplet pairs coupling to give a singlet spin state. This description suggests that the triplet energies of polyenes and carotenoids should be approximated by about one-half the energies of the 2^1A_g states. If one compares the 2^1A_g state energies determined for the spheroidene series (see Table 8.2) with the triplet energies for the same series extrapolated from the data of Evans (1960, 1961) (Fig. 8.7, and Table 8.2) one can see that there is excellent adherence to the $E_T \sim 0.5\,E_{2^1Ag}$ rule. For example, the 2^1A_g state energy of 3,4,5,6-tetrahydrospheroidene is determined by fluorescence spectroscopy to be ~16 700 cm^{-1}. The extrapolated value for its triplet energy (eight carbon–carbon double bonds) is ~8 100 cm^{-1}. It is important to note, however, that due to uncertainties in the location of the 2^1A_g state for longer polyenes, the rule-of-thumb must be applied with caution to the carotenoids.

Table 8.2 Singlet- and triplet-state energies of some carotenoids. The singlet-state energies were deduced from the absorption and fluorescence spectra of four carotenoids: spheroidene (10 carbon–carbon double bonds), 3,4-dihydrospheroidene (9), 3,4,5,6-tetrahydrospheroidene (8) and 3,4,7,8-tetrahydrospheroidene (7). The spectra are shown in Fig. 8.4. The first four carotenoid triplet energies were determined by Evans (1960, 1961). The values given in parentheses are obtained from the linear extrapolation shown in Fig. 8.7. Energies are in cm^{-1}

No. of conjugated carbon-carbon double bonds	*$S_2(1B_u)$ state energies*	*$S_1(1A_g)$ state energies*	*T_1 triplet-state energies*
1			≤ 28 700
2			20 830
3			16 450
4			13 750
5			(11 670)
6			(10 170)
7	23 600	18 400	(9 010)
8	22 500	16 700	(8 090)
9	21 500	15 300	(7 340)
10	20 800	(14 000)	(6 710)
11			(6 190)
12			(5 740)
13			(5 350)

An indirect means of determining the triplet-state energies of carotenoids is

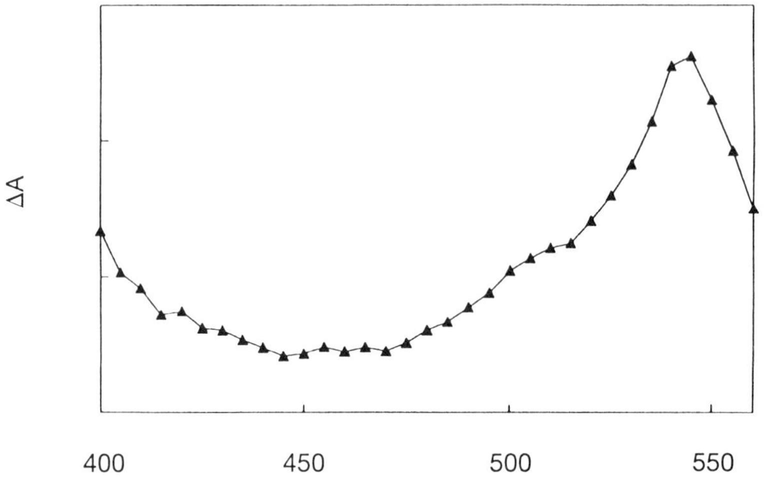

Figure 8.8 Room temperature (295 K) triplet absorption spectrum of spheroidene incorporated into the reaction centres of *Rb. sphaeroides* R-26. The triangles represent the intensities of the transients measured 135 nm after an excitation flash. Data taken from Frank et al. (1986).

through the use of triplet donors of known energy to sensitise the formation of carotenoid triplet-states. If, after selective excitation of the energy donor, carotenoid acceptor triplets are observed, the triplet energy of the carotenoid must lie below that of the donor. If, however, no carotenoid triplets are formed, the negative result is interpreted as indicating that the carotenoid triplet is lying above that of the donor. This approach has been used on spheroidene, spirilloxanthin, ζ-carotene, phytoene (Bensasson et al. 1976), ß-carotene and lycopene (Land et al. 1971), and others (Mathis and Kleo 1973; Rodgers and Bates 1980). The triplet donors included naphthacene (i.e. tetracene, $E_T = 10\,282$ cm^{-1}), pyrene ($E_T = 16\,802$ cm^{-1}), anthracene (E_T = 14 712 cm^{-1}) and chlorophyll *a* (E_T = 10 115 cm^{-1}). A problem with these determinations is that usually only an upper limit can be placed on the carotenoid triplet energies. This upper limit may be considerably higher than the actual carotenoid triplet energy. This is because it is difficult to find donor molecules that form triplets efficiently *and* have low enough energies to allow a precise determination of the triplet energies of the carotenoids. Nevertheless, these data are valuable and are consistent with the energies derived from the extrapolation described above. In related studies, Foote et al. (1970a) found that ß-carotene inhibited reactions involving the $^1\Delta_g$ state of oxygen. This would be the case only if the triplet-state energy of ß-carotene lies below that of the $^1\Delta_g$ state of oxygen. This is also consistent with the energy extrapolations previously mentioned. Chantrell et al. (1977) studied the triplet energy transfer from protoporphyrin IX dimethyl ester (E_T = 12 579 cm^{-1}) to a series of polyenes of differing conjugated chain length. Pariser–Parr–Pople molecular orbital calculations were performed to elucidate the dependence of the triplet-state energy on the π-electron chain length of the polyenes. Surprisingly, Chantrell et al. (1977) found that very long carotenoid homologues (i.e. those having 15 or 19 carbon–carbon double bonds) did not quench the protoporphyrin triplet-state as efficiently as did ß-carotene. This was explained by suggesting that close proximity of triplet donor to acceptor, a feature essential for high quenching efficiencies (Hammond et al. 1964; Gust and Moore, 1991), was not realised with the protoporphyrin and the long-carotenoid acceptor molecules.

Triplet-triplet absorption

The optical properties of the triplet-states of carotenoids are frequently characterised by their triplet–triplet absorption spectra. These spectra are usually recorded after formation of the carotenoid triplet using energy donors in conjunction with flash photolysis or pulse radiolysis techniques. The triplet–triplet absorption spectra of several natural and synthetic carotenoids have been reported (e.g. see Mathis and Kleo 1973; Bensasson et al. 1976). In general, carotenoid triplets display a pronounced absorption feature in the visible region that has a bandwidth in the range 10-50 nm (Mathis and Kleo 1973; Bensasson et al. 1976) (Fig. 8.8). The frequencies corresponding to the λ_{max} values for the triplet–triplet absorptions follow a linear dependence with the frequencies of the lowest singlet–singlet energies (Mathis and Kleo 1973). Also, triplet–triplet transitions in carotenoids have very high molar absorptivities (~2 x 10^5 M^{-1} cm^{-1}, Mathis and Kleo 1973; Cogdell et al. 1983). When these are known for a particular carotenoid (Table 8.3), quantum yields of carotenoid triplet-state formation can be determined from triplet–triplet absorption spectra.

Table 8.3 Singlet and triplet extinction coefficients of some carotenoids. The values were determined at the maximum spectral intensity wavelength positions

Carotenoid	*Triplet extinction coefficients (1(mol cm)$^{-1}$)*	*Reference*
ß-Carotene	1.3 x 10^5	Land et al. 1971
Lycopene	3.9 x 10^5	Land et al. 1971
Lutein	4.0 x 10^5	Mathis 1969
Neurosporene	2.74 x 10^5	Cogdell et al. 1983
Spheroidene	3.09 x 10^5	Cogdell et al. 1983
Spheroidenone	6.06 x 10^4	Cogdell et al. 1982
Spirilloxanthin	9.20 x 10^4	Cogdell et al. 1983
Carotenoid	*Singlet extinction coefficients (1(mol cm)$^{-1}$)*	*Reference*
Spheroidene	1.49 x 10^5	Goodwin et al. 1956
Spheroidenone	1.29 x 10^5	Liaaen-Jensen and Jensen 1971
Neurosporene	1.69 x 10^5	Liaaen-Jensen and Jensen 1971
Chloroxanthin	1.37 x 10^5	Nakayama 1958

EPR spectroscopy

EPR spectroscopic determinations of carotenoid triplet-states have been carried out in two ways: by optical detection at zero magnetic field and by conventional high field EPR spectroscopy. Because these experiments require the use of cryogenic temperatures to observe the triplet-state spectra, it is difficult to bring a triplet donor in proximity to sensitise the formation of carotenoid triplet *in vitro*. Hence, only a few such reports of carotenoid triplet-state EPR spectra have appeared. Before delving into these, it is appropriate to review briefly how triplet-state EPR spectra are characterised.

The triplet-state of a molecule consists of two unpaired electrons with a total spin quantum number of one. For $\pi\pi^*$ triplet-states, the interaction between the two electrons is dominated by dipolar coupling (Carrington and McLachlan 1967). The anisotropic nature of the dipolar interaction leads to a splitting of the triplet spin sub-levels even in the absence of an externally applied magnetic field. This so-called zero-field splitting is characterized by two static, linearly independent, spin Hamiltonian parameters denoted $|D|$ and $|E|$. Both $|D|$ and $|E|$ are spatial averages over the electronic part of the triplet-state wavefunction. $|D|$ assesses the extent of the dipolar interaction, and $|E|$ measures the deviation from axial symmetry. Thus, in principle, it should be possible to distinguish between various stereochemical isomers of carotenoids on the basis of their zero-field splitting parameters. These parameters may be determined at zero-field by directly inducing transitions between the spin sub-levels by means of microwaves and monitoring an optical property of the molecule sensitive to the triplet-state population changes. Absorption, fluorescence, phosphorescence, and triplet–triplet absorption have all been used. This is referred to as optically detected magnetic resonance (ODMR). (See Clarke, 1982; for a review of this subject see also Fig. 8.9). Alternatively, the parameters may be obtained from the observed line splittings in a conventional high field EPR spectroscopic experiment (Hoff 1979). Both approaches may be carried out in steady-state or in time-resolved modes. The latter approach yields the individual, spin-sub-level, intersystem crossing rate constants for the decay of the triplet to the ground state (McGann and Frank 1985a). These kinetic parameters are thought to be sensitive functions of the structure of the molecule (Clarke and Frank 1976).

The triplet-state spectra are characterised not only by their static (zero-field splitting) and kinetic (intersystem crossing rate constant) parameters but also by their spin polarisation patterns (Thurnauer et al. 1975). Because triplet-states have spin sub-level-selective routes of population and decay, and carotenoid triplets have extremely short lifetimes (<10 ms), the triplet-state spin sub-levels decay to the ground state before spin lattice relaxation occurs (McGann and Frank 1985b). This results in a non-Boltzmann population distribution in the spin sublevels of the triplet and leads to “spin-polarised” EPR signals. These are signals observed as either emission or enhanced absorption lines.

The particular pattern of the lines (e.g. *eaa eea,* where *e* denotes a signal in emission and *a* denotes a signal in enhanced absorption for the six different lines in the triplet spectrum) contains information about the mechanism of formation of the triplet (Frank et al. 1980). The pattern *eaa eea* or its complement *aee aae* are indicative that the triplet is being formed via the radical pair mechanism (or having accepted triplet-state energy from a donor born via that mechanism). Any other pattern (e.g. *eae aea*) indicates that the triplet was born via intersystem crossing or has quenched a triplet born via that mechanism. (See Frank 1992 for a review of this topic; see also Fig. 8.10.)

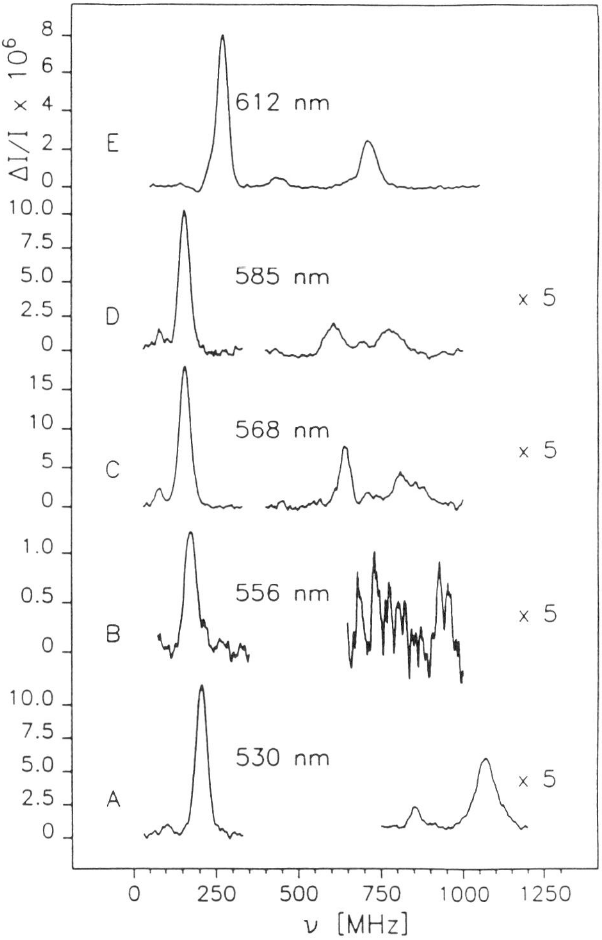

Figure 8.9 ODMR spectra (by detection of microwave-induced absorption changes (ADMR)) of five samples: (A) B880-RC complexes from *Rc. gelatinosus 149*; (B) B880-RC complex from *Rps. acidophila* 7050; (C) B880-RC complex from *Rps. palustris 8252*; (D) B880-RC complex from *Rsp. rubrum S1*; (E) RC complex from *Rsp. rubrum S1*. The wavelengths of detection are given for each spectrum. On the vertical scale the microwave-induced changes in the transmitted light intensity, DI, are normalised by the transmission, I. The figure is from Aust et al. (1991).

Attempts to observe the triplet-state EPR spectrum of ß-carotene or other carotenoids in organic solvents have never succeeded. However, when ß-carotene

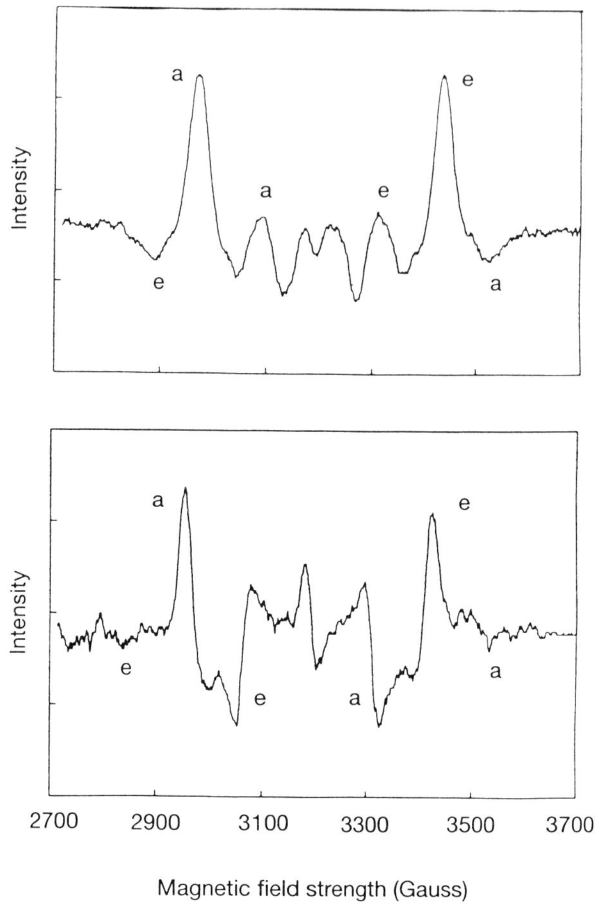

Figure 8.10 High-field triplet-state EPR spectra. The top trace displays the spectrum of spheroidene reconstituted into the reaction centres of *Rb. sphaeroides* R-26. This spectrum shows the radical pair polarization pattern *eaa eea*. The bottom trace displays the spectrum of spheroidene in the B800–850 light-harvesting complex from anaerobically grown *Rb. sphaeroides* wild type strain 2.4.1. This spectrum shows the intersystem crossing mechanism polarization pattern, *eae aea*.

was solubilised in non-ionic detergent micelles, it gave rise to a triplet-state EPR spectrum characterised by zero-field splitting parameters $|D| = 0.0333$ ($\pm$ 0.0010) cm^{-1} and $|E| = 0.0037$ ($\pm$ 0.0010) cm^{-1} and the polarisation pattern *eae aea* (Frank et al. 1980). The same spectrum was observed for ß-carotene dissolved in phospholipid bilayer vesicles (Frank et al. 1980). Recently, the triplet-state of ß-carotene was reported in single crystalline samples (Frick et al. 1990). Time-resolved EPR was used to detect the signals which were characterised by zero-field splitting parameters $|D| = 0.0331$ cm^{-1} and $|E| = 0.0040$ cm^{-1}. These values are in agreement with those from the previous ß-carotene/detergent micelle sample EPR experiments (Frank et al. 1980). The single-crystal studies of Frick et al. (1990) also allowed the orientation of the fine-structure tensor with respect to the molecular axes to be determined. It was found that these axes coincide to within 20°, and the signals that maximise the fine-structure splitting correspond to the direction along the molecular chain-axis.

Other EPR studies on carotenoids *in vitro* have been carried out on a series of synthetic carotenoporphyrin complexes (Frank et al. 1987a). In these complexes carotenoid triplet-state formation was facilitated by the covalent linking of a porphyrin triplet-state donor to a carotene derivative which served as the triplet acceptor. The carotenoid zero-field splitting parameters were $|D| = 0.0356$ ($\pm$ 0.0010) cm^{-1} and $|E| = 0.0036$ ($\pm$ 0.0010) cm^{-1} and the spin polarisation pattern was *eae aea*. Surprisingly, the polarisation pattern that was observed for the carotenoid acceptor was not sensitive to the initial polarisation pattern of the porphyrin triplet donor. Further studies on triplet energy transfer from the standpoint of spin sub-levels would be valuable in attempts to understand this observation.

(c) Singlet-and triplet-state energies of chlorophylls, bacteriochlorophylls and oxygen

In attempting to discover how carotenoids carry out light-harvesting and photoprotection in photosynthesis, it is important to know the energies of the states to or from which the transfer takes place. Although the singlet-state energies of numerous chlorophylls are readily discernable from their absorption spectra, the triplet-state energies are not as easily determined. An important contribution to this area was made by Takiff and Boxer (1988a,b). Using a germanium photodiode detector cooled with liquid nitrogen, this group observed phosphorescence spectra from a series of bacteriochlorophylls and bacteriophaeophytins *in vitro* and from the primary electron donors in reaction centres isolated from *Rhodobacter sphaeroides* and *Rhodopseudomonas viridis*. These data, along with previous determinations of chlorophyll singlet and triplet energies (e. g. Krasnovskii et al. 1974), have helped to clarify the energetic possibilities for transfer to and from carotenoids.

The question of whether the triplet of the reaction centre primary donor can sensitise the formation of singlet oxygen is an interesting one from the perspective of where the photoprotection actually occurs in photosynthetic organisms. The triplet-state energy determinations based on phosphorescence spectra (Takiff and Boxer 1988a,b) suggest that the antenna and primary donor bacteriochlorophyll of *Rps. viridis* lie at too low an energy to populate the $^1\Delta_g$ state of oxygen, whereas the antenna and primary donor bacteriochlorophyll triplets of *Rb. sphaeroides* lie either above or at least within kT of the energy of singlet oxygen. Boucher et al. (1977) found that the reaction centres from *Rhodospirillum rubrum* sensitised the formation of the $^1\Delta_g$ state of oxygen. More experiments in this area would be helpful in understanding the precise mechanism of $^1\Delta_g$ formation in the reaction centre.

8.2.2 *In vivo*

(a) Antenna complexes

The antenna complexes which can be isolated from the photosynthetic membranes of purple phototrophic bacteria can be divided into two main classes, the "core" antenna complexes and the "variable" or "peripheral" antenna complexes (Cogdell and Thornber 1980). These complexes are usually known by the approximate NIR (near-infrared) absorption maxima of bacteriochlorophyll, i.e. B880 (a "core" complex) or B800–850 (a "variable" complex). The "core" complexes are found as a stoichiometric conjugate with the reaction centres, while the "variable" antenna complexes are arranged more peripherally and, as their name suggests, are present in variable amounts, relative to the reaction centres. All these antenna complexes appear to be constructed on the same modular principle. They consist of oligomers of two types of small, hydrophobic apoprotein (α and ß) and a small number of pigments per $\alpha\beta$ pair, usually three or four (Thornber et al. 1983). The ratio of carotenoids to bacteriochlorophyll in these complexes depends upon the specific complex being considered, but usually is either 1:1, 1:2 or 1:3 (Thornber et al. 1983; Evans et al. 1988).

The "core" antenna complexes so far studied may be subdivided into two types, the B890 and the B875 (Thornber et al. 1983). The B890 type has a bacteriochlorophyll:carotenoid ratio of 2:1. This type of complex is found in *R. rubrum* and *Chr. vinosum,* for example (Thornber et al. 1983). The B875 type has a bacteriochlorophyll:carotenoid ratio of 1:1. This type of "core" complex appears to be more widespread and it is found typically in *Rb. sphaeroides* and *Rhodobacter capsulatus* (Thornber et al. 1983). The "variable" antenna complexes form quite a diverse group, based upon their absorption spectra. However, they all contain carotenoids, usually one per two or three bacteriochlorophylls. This group of antenna complexes can also have multiple "α" and "ß" apoproteins within a single complex, e.g. the B800–820 complex from *Rhodopseudomonas acidophila* probably contains two types of α and two types of ß apoproteins (Thornber et al. 1983). These different antenna complexes contain different carotenoid types. The efficiency of singlet–singlet energy transfer from the carotenoid to the bacteriochlorophyll can vary among complexes (Angerhofer et al. 1986; Noguchi et al. 1990), and there are marked differences between the circular dichroism (CD) spectra of the different carotenoids, depending upon the complex (Cogdell and Scheer 1985).

Absorption

The absorption spectra of carotenoid molecules embedded in pigment–protein complexes are shifted compared to the spectra in various non-polar solvents. For example, the λ_{max} of spheroidene in pentane is 447 nm whereas in the B800–850 complex prepared from *Rb. sphaeroides* wild type strain 2.4.1 the λ_{max} is 480 nm (Frank et al. 1986). This shift could arise from the dispersion interactions as discussed above for carotenoids *in vitro* but, in principle, may also come from exciton interactions and/or interactions between the transition moment and nearby ionic charges. Andersson et al. (1992) proposed that, given the remarkable agreement between the absorption spectrum of spheroidene in the B800–850 complex and the same molecule dissolved in the aromatic solvent, quinoline, dispersive interactions between the carotenoid and aromatic amino acid residues are responsible for the red-shifted spectra of carotenoids *in vivo*. Because a crystal structure of this complex is not yet available, it is too early to rule out interactions with point charges or other groups.

An important avenue of investigation of the properties of carotenoids in pigment–protein complexes has come from the ability to incorporate these molecules into the proteins prepared from the various carotenoidless mutant strains. These carotenoid reconstitution studies are extremely valuable because they offer the means of systematically analysing different carotenoids bound in the **same** protein environment. Various carotenoids have been reconstituted into several different light-harvesting complexes (Davidson and Cogdell 1981; Hayashi et al. 1989; Noguchi et al. 1990; H. Frank and R. Farhoosh, unpublished data). By reconstituting a systematic series of carotenoids into the same protein, the effect of chain length, energy levels, carotenoid functional groups, etc., on carotenoid function can be assessed. By incorporating the same carotenoid (e.g. spheroidene) into different proteins, the effect of the protein environment can be studied. A discussion of the energy transfer behaviour of these reconstituted complexes is given below.

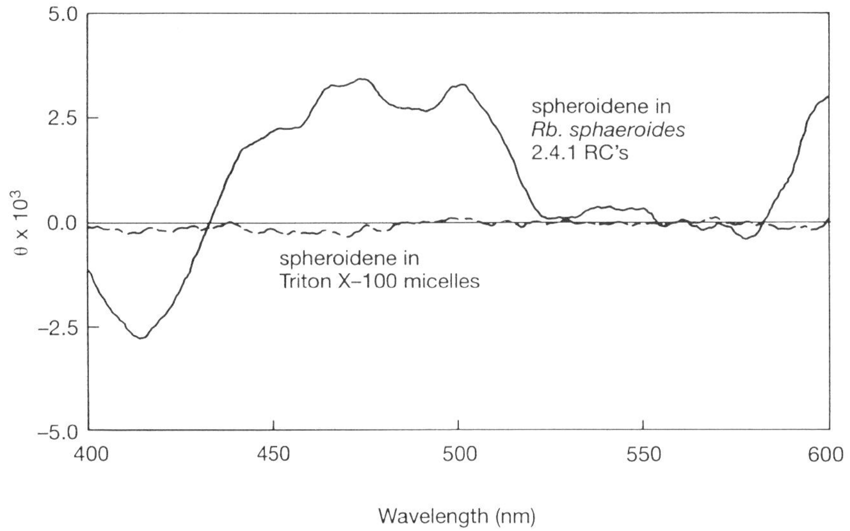

Figure 8.11 Circular dichroism (CD) spectra of reaction centres from the carotenoidless strain *Rb. sphaeroides* R-26 and from the carotenoid-containing *Rb. sphaeroides* wild type strain 2.4.1. Also shown is the fact that spheroidene in Triton-X-100 detergent micelles does not display a CD spectrum. Data from Frank and Violette (1989).

Circular dichroism

CD spectra in the region of the main carotenoid absorption bands of a series of B800–850 and other antenna complexes have been used to probe the details of carotenoid binding (Cogdell and Scheer 1985). It is interesting to note that achiral carotenoids in solution do not display CD spectra (Boucher et al. 1977; Frank and Violette 1989; see Fig. 8.11) but when bound in the protein they have significant optical activity which gives rise to pronounced circular dichroism (Fig. 8.11). This may arise from an asymmetry in the binding environment that induces optical activity. Alternatively, it has been proposed that dimeric exciton interactions are responsible for the CD in the B880 complex (Lozano et al. 1990; Zurdo et al. 1991).

Electrochromism

Reversible red shifts in the absorption spectra of carotenoids, induced either by light or oxygenation, were first observed in chromatophores from *R. rubrum* (Chance and Smith 1955) and *Rb. sphaeroides* (Chance 1958). A typical difference spectrum of this band shift is shown in Fig. 8.12. Similar carotenoid band shifts have also been seen in chloroplasts, e.g. the so-called "515 nm shift" (Duysens 1954). It has been shown that these carotenoid band shifts are induced by the membrane potential, which is generated across the chromatophore or chloroplast membrane as a result of "coupled" electron transport (see Wraight et al. 1978, for reviews of this topic). Indeed, this electrochromic response of the carotenoids has been used to monitor the size of the membrane potential (Jackson and Crofts 1969).

This effect of membrane potential inducing a band shift in a pigment is an example of electrochromism or a Stark effect (Liptay 1969). Electrochromism will occur owing to a differential interaction between the membrane field and the ground and excited states of the molecule. The change in the energies of the states is manifested as a change in the frequency of the absorption such that

$$\Delta \upsilon = \frac{1}{h}\left[(\mu_g^{\circ} - \mu_g^{\circ}) + 1/2(\alpha_e - \alpha_g)E\right]E \quad (8.13)$$

where $\Delta\upsilon$ is the frequency shift, h is Planck's constant, the subscripts e and g refer to the excited and ground states, respectively, E is the electric field strength, μ° is the dipole moment in the absence of the electric field and α is the polarisability. For an absorption band with an absorption maximum at wavelength λ, the wavelength shift, $\Delta\lambda$, is given by

$$\Delta\lambda = \frac{\lambda^2}{hc}(\Delta\mu^{\circ} + 1/2\Delta\alpha E)E \quad (8.14)$$

where c is the speed of light. The direction and magnitude of the shift will depend upon the field strength, the change in dipole moment ($\Delta\mu^{\circ}$), the change in polarisability ($\Delta\alpha$), and the orientation of these changes with respect to the electric field vector. Explicitly writing the dependence on orientation one obtains;

$$\Delta \upsilon = \frac{1}{h}\left[\left|\Delta \vec{\mu}^{\circ}\right| \cos\theta + 1/2\left|\Delta \vec{\alpha}\right|\left|\vec{E}\right|^{2} \cos^{2}\phi\right] \quad (8.15)$$

where θ is the angle between the electric field and the change in permanent dipole moment, and ϕ is the angle between the electric field and the change in polarisability. Carotenoids in native biological membranes are uniaxially distributed owing to their unique and fixed orientations relative to the membrane field (Gottfried et al. 1991a). In this case, two terms survive the orientational ensemble averaging, one being linear with respect to the electric field and one quadratic. Thus, depending on whether it is the change in dipole moment or that in polarisability that dominates the interaction, the observed band shift is either a linear or quadratic function of the electric field (Reich and Schmidt 1972; Kakitani et al. 1982). It is interesting to note that for an isotropic or biaxial distribution, the Stark effect will always be quadratic with respect to the external field regardless of the origin (dipole moment or polarisability change) of the interaction (Gottfried et al. 1991a). This will be the case as long as the Stark splitting is less than the inhomogeneous linewidth of the transition. This is nearly always true (Gottfried et al. 1991a).

Carotenoids, to a first approximation, can be considered as having a centre of symmetry in their conjugated double bond system (e.g. ß-carotene). They lack, therefore, a significant permanent dipole moment in the ground state, and this can probably also be said of the molecule in its low-lying excited states. Electrochromism would, thus, be expected to occur only via a change in polarisability and lead to a quadratic external field dependence. This expectation has been borne out in several studies *in vitro* (Reich and Sewe 1977; Sewe and Reich 1977). Also, Gottfried et al. (1991a,b) have reported essentially first-derivative Stark effect spectra of spheroidene and spheroidenone *in vitro*. These are interpreted by invoking changes in the polarisability of the molecules upon excitation.

Not all the carotenoids show a band shift induced by membrane potential (Scolnik et al. 1980). In *Rb. capsulatus,* for example, only carotenoids which are associated with the B800–850–antenna complex are sensitive to the applied electric field (Scolnik et al. 1980). The reaction centre carotenoid also shows a band shift (Cogdell et al. 1977). However, in this case it is not affected by the transmembrane potential, but by the local electric field set up when the reaction centre bacteriochlorophyll "special pair" is oxidised (Cogdell et al. 1977).

The carotenoid electrochromism has also been used to monitor the kinetics of light-induced electric potential changes across the photosynthetic membrane (Jackson and Dutton 1973; Scolnik et al. 1980; Kakitani et al. 1982). The mechanism of these carotenoid band shifts has been investigated by Gottfried et al. (1991a,b) using B800–850 complexes. These authors found unusually large, essentially second derivative-shaped, Stark spectra for spheroidene and spheroidenone in the B800–

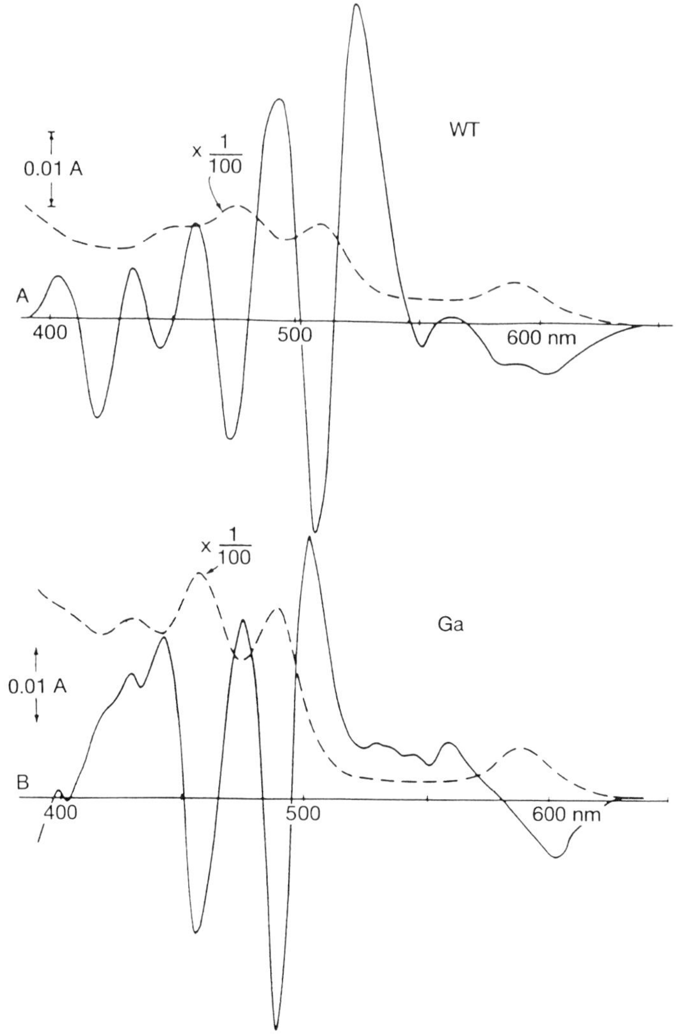

Figure 8.12 Light-induced difference spectra in chromatophores of *Rhodobacter sphaeroides* wild type (A) and green mutant Ga (B). The absorption spectra are also shown, reduced 100-fold. Figure taken from Wraight et al. 1978.

850 complexes from *Rb. sphaeroides*. The results were interpreted in terms of a protein-induced perturbation of the electronic structure of the carotenoids that leads to a large difference in the permanent dipole moment between the molecules in their ground and excited 1^1B_u states *in vivo*. The molecular features that may lead to such a large difference in the permanent dipole moment include interactions with charged (Kakitani et al. 1982), polar (Symons and Swysen 1983) or polarisable residues or helix dipoles (as suggested by Gottfried et al. 1991b). Whatever the case, the data indicate that the large difference dipole moment ($\Delta\mu°$) for the carotenoids bound in the B800–850 complexes is attributable to a significant protein matrix

field. Also, a change in the dipole moment of the carotenoid upon excitation predicts a linear shift of the electronic transition frequency with transmembrane potential (in a uniaxially oriented sample) as discussed above. This has been observed (Reich and Sewe 1977; Sewe and Reich 1977).

Triplet-states

There are no direct determinations of carotenoid triplet-state energies in bacterial or higher plant antenna complexes of which we are aware, but certainly an upper limit can be placed on these energies from the fact that carotenoid triplets are formed with high efficiency *in vivo* by triplet energy transfer from chlorophylls that have known triplet energies. Owing to their small oscillator strength, the triplet energies of chlorophylls and carotenoids would not be expected to shift very much with changes in their environment.

(b)Reaction centres

Virtually all reaction centres that have been isolated from wild type bacterial species contain one bound carotenoid molecule per reaction centre protein. For example, the reaction centre complex from anaerobically grown *Rb. sphaeroides* wild type strain 2.4.1 contains spheroidene (Fig. 8.13) in a 1:1 stoichiometry with the primary donor. Under aerobic (or semi-anaerobic) growth conditions, the reaction centre still has a ratio of carotenoid to primary donor of 1:1, but it binds a higher percentage of spheroidenone relative to spheroidene than exists in the whole cells (R.J. Cogdell, unpublished data).

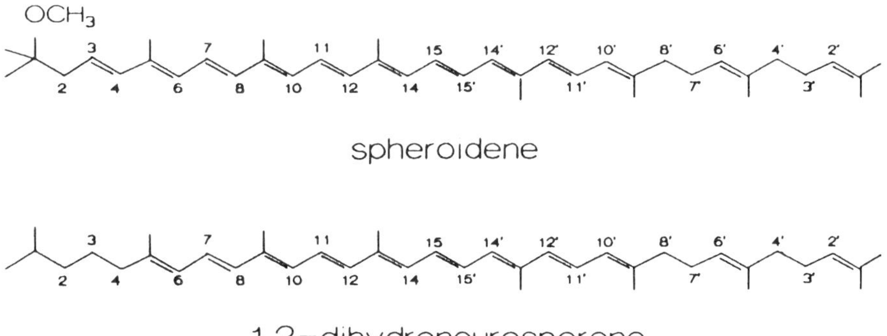

Figure 8.13 The structures of spheroidene and 1,2-dihydroneurosporene. These carotenoids are found in the reaction centres of *Rb. sphaeroides* wild type strain 2.4.1 and *Rps. viridis*, respectively.

An exception to the rule that reaction centres of all wild type phototrophic bacteria have bound carotenoids is the reaction centre from *Chloroflexus aurantiacus*. This bacterium has large quantities of carotenoids in its whole cells, yet has no bound carotenoid in the purified reaction centre preparation (Blankenship et al. 1988). It is not clear whether the carotenoid is present in the reaction centre *in vivo* and is lost during the reaction centre isolation procedure or whether the carotenoid is not present *in vivo* at all.

The D1/D2/Cyt *b*559 photosystem (PS) II reaction centre preparation from higher plant chloroplasts has been shown to contain two ß-carotene molecules per reaction centre complex (van Dorssen et al. 1987; Newell et al. 1991). PS I has not been isolated in a form where it is possible to distinguish between reaction centre-bound and antenna-bound carotenoids.

Singlet-states

Carotenoids bound in the reaction centre have absorption spectra that are red-shifted from their spectra in non-polar organic solvents. This is the same behaviour as is observed for carotenoids in the light-harvesting complexes discussed above. Although it is not possible to rule out exciton interactions or interactions between carotenoids and charged amino acid residues in the light-harvesting complexes as being responsible for the shifts in the spectra, these factors can probably be dismissed in the reaction centre. The crystal structures of the reaction centre proteins from *Rb. sphaeroides* 2.4.1 and *Rps. viridis* have been determined to atomic resolution by X-ray diffraction techniques. In both of these complexes the environment of the carotenoid is clearly defined as largely hydrophobic, consisting of several residues possessing aromatic sidechains. The residues constrain the carotenoid to a specific stereochemistry which, along with dispersive interactions, most likely accounts for the shift in the absorption spectrum. The reaction centre structures show no charged residues within 10 Å of the carotenoid. Also, although the bound carotenoids display pronounced CD spectra, there is no evidence that this CD and chirality arise from exciton interactions between the carotenoid and other cofactors.

In a manner similar to that mentioned above for the light-harvesting complexes, it is possible to incorporate carotenoids into reaction centres isolated from the carotenoidless mutants of two phototrophic bacteria, *R. rubrum* G-9 and *Rb. sphaeroides* R-26 (Boucher et al. 1977; Agalidis et al. 1980; Chadwick and Frank 1986). Carotenoids can be incorporated into these reaction centres in such a way that all of their various characteristics (i.e. absorption, CD, EPR, and resonance Raman spectra, singlet and triplet energy transfer properties) appear to be identical with those in the wild type reaction centre. The reconstituted carotenoids, however, do not show spectra analogous (albeit red-shifted) to their spectra observed in solution. For example, one study compared the spectral properties of spheroidene and methoxyneurosporene reconstituted into the reaction centre of *Rb. sphaeroides* R-26 (Frank et al. 1986). These two carotenoids are structurally identical except for the fact that spheroidene has an additional double bond at the C-3,4 position in its π-electron conjugated system giving it a total chain length of 10 carbon–carbon double bonds. Methoxyneurosporene (also known as 3,4-dihydrospheroidene) has nine carbon–carbon double bonds. Surprisingly, the absorption spectra of these two carotenoids, which are distinctly different *in vitro,* shifted to precisely the same position upon reconstitution into the *Rb. sphaeroides* R-26 reaction centre protein.

Furthermore, the triplet–triplet absorption and EPR spectra of these two different pigment-protein complexes reconstituted with carotenoid, were found to be the same. These data have been explained in terms of a combined effect of protein-induced band shifts and carotenoid structural changes that cause the molecules to have identical spectroscopic properties (Frank et al. 1986).

It is interesting to ask: where are the 1^1B_u and 2^1A_g states of carotenoids in the reaction centres? The 1^1B_u state energy is easily determined from the spectral origin of the carotenoid absorption spectrum. For spheroidene in reaction centres of *Rb. sphaeroides* this is at 501 nm or ~20 000 cm^{-1}. No experiments have directly addressed the 2^1A_g energy of the reaction centre-bound carotenoid. If one takes as a first approximation the 1^1B_u - 2^1A_g energy gap of ~6 800 cm^{-1} suggested for spheroidene from the fluorescence experiments discussed above, one arrives at an energy of 13 200 cm^{-1} for the 2^1A_g state of the reaction centre-bound spheroidene. This energy is probably slightly too low because the dispersive interactions with the protein are likely to stabilise the 1^1B_u state more than they are the 2^1A_g state (Hudson et al. 1982). A better approximation might be to assume that owing to its extremely small transition dipole, the 2^1A_g transition energy is essentially unaffected by binding to the protein, whereas the 1^1B_u energy shifts from ~20 700 cm^{-1} to ~20 000 cm^{-1}. This would lead to a 1^1B_u - 2^1A_g energy gap of ~6 100 cm^{-1} and place the 2^1A_g state energy at ~13 900 cm^{-1}. This is sufficient for singlet energy transfer from the 2^1A_g state to the reaction centre-bound bacteriochlorophyll and is discussed below.

Triplet-states

The energies of the reaction centre-bound carotenoid triplet-states are important in understanding photoprotection. It is generally agreed that photoprotection in the reaction centre occurs because the carotenoid quenches the primary donor triplet-state before this can sensitise the formation of the $^1\Delta_g$ state of oxygen. For this process to occur, the triplet energies of carotenoids must lie lower than the energy of the primary donor (7 590 cm^{-1} for the bacteriochlorophyll *a*-containing reaction centres; Takiff and Boxer 1988a), but the precise carotenoid energies have never been determined. Several authors have examined the triplet-state quenching properties of various carotenoids in native and carotenoid-reconstituted reaction centres by using transient optical (Cogdell et al. 1975; Frank et al. 1983, 1986; Schenck et al. 1984) and EPR spectroscopy (Chadwick and Frank 1986). The authors concur that carotenoids possessing nine or more conjugated double bonds are capable of quenching the primary donor triplet in the bacteriochlorophyll *a*-containing reaction centres. An interesting anomaly is provided by the bacteriochlorophyll *b*-containing reaction centre of *Rps. viridis,* the carotenoid of which, 1,2-dihydroneurosporene (nine conjugated double bonds), does not quench the primary donor triplet-state at any temperature (Holten et al. 1978; Frank et al. 1980). There are three possible explanations for this: (i) the triplet-state energy of the carotenoid lies at a higher energy than the bacteriochlorophyll *b* triplet energy donor; (ii) the carotenoid is not properly positioned to accept the triplet energy (an unlikely explanation because the crystal structures have shown that the 1,2-dihydroneurosporene is in the same location in *Rps. viridis* as spheroidene is in *Rb. sphaeroides*); or (iii) the triplet energy transfer to the carotenoid involves an intermediate bacteriochlorophyll molecule whose triplet energy relative to the primary donor ($\Delta E \sim 1000$ cm^{-1}) is too high to be populated at any temperature (as suggested by Takiff and Boxer 1988a). The factors controlling triplet energy transfer are discussed below.

8.3 DETERMINATIONS OF CAROTENOID STRUCTURES

Determination of the structure of a carotenoid can simply be the identification of the molecular formula (e.g. by mass spectrometry), but may also involve the geometry of the carotenoid relative to other pigments, or the elucidation of its stereochemistry. In this section we deal with all of these as they pertain to the ability of carotenoids to function in the photosynthetic apparatus.

8.3.1 In vitro

(a)X-ray diffraction

In principle, the most precise information on the stereochemistry of carotenoids *in vitro* should come from single-crystal X-ray diffraction studies. Indeed, the absolute three-dimensional structures of crystalline ß-carotene, canthaxanthin and their 15,15′-didehydro derivatives have been determined (Sly 1964; Sterling 1964; Bart and MacGillavry 1968a, b). It is important to emphasise, however, that there is good evidence that the structures deduced for these long-chain carotenoids from the X-ray diffraction studies are perturbed by the crystalline fields in the lattice. The value of these studies in attempting to make correlations with other means of structure determination (e.g. spectroscopic studies in solution) is therefore somewhat depreciated. Nevertheless, the studies do show some features that are probably also present in solution. These include the facts that the polyene chains of ß-carotene and canthaxanthin are essentially planar but show in-plane bending into an "S-shape" owing to steric interactions between hydrogens and the side chain methyl groups (see Fig. 8.14.). More precise details, such as the degree of bond length variation with position in the polyene chain, are ambiguous.

(b) Absorption spectroscopy

Absorption spectroscopy provides a means of distinguishing different geometrical isomers of carotenoids. For a detailed description of the application of absorption spectroscopy to carotenoids, the reader is referred to Chapter 9. Here, we discuss only the two most prominent absorption spectral features which are used in this manner. These are the high oscillator strength transition in the visible region associated with the $1^1A_g \rightarrow 1^1B_u$ absorption (usually accompanied by some vibronic structure) and the so-called "*cis* peak" which occurs at shorter wavelengths and is

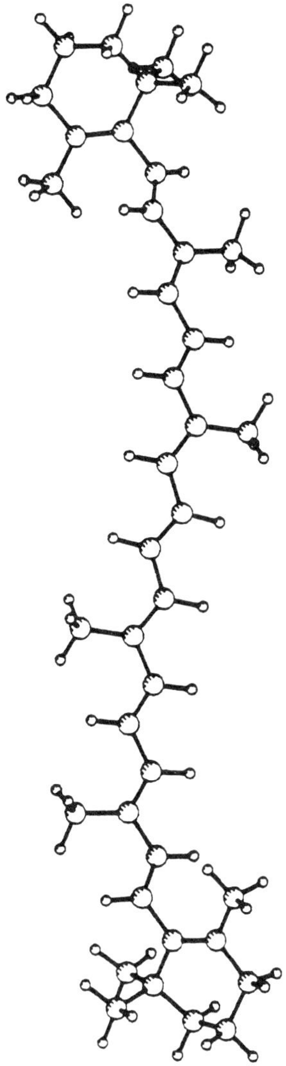

Figure 8.14 Structure of ß-carotene derived from an energy minimization using an MMX force field

indicative of a change in the linear all-*trans* structure of the carotenoid to a *cis* configuration. For example, the *cis* peak of the 15,15′-*cis* configuration occurs at approximately 140 nm below the longest wavelength band of the strong, visible-absorption feature (see Fig. 8.15). In general, the *cis* peak arises (at the expense of some intensity in the main absorption feature) because of a relaxation in the singlet absorption selection rules when the symmetry of the molecule is lowered. The intensity of the *cis* peak is highest for *cis*-isomers in which the *cis*-double bond is near to the centre of the chromophore.

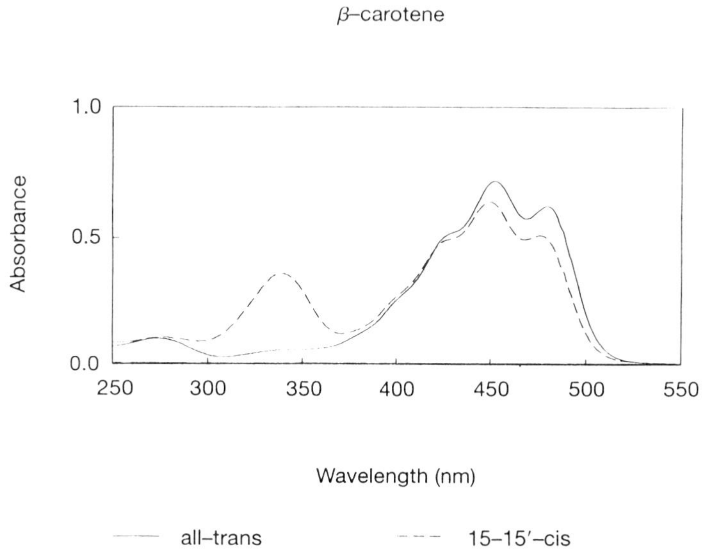

Figure 8.15 The absorption spectra of the all-*trans* and 15,15′-*cis* configuration of ß-carotene. The *cis* peak occurs at approximately 140 nm below the longest wavelength band of the strong, visible-absorption feature.

(c) Resonance Raman spectroscopy

Resonance Raman spectroscopy has been employed by many investigators to study the ground and excited state structures of carotenoids (Lutz 1984; Lutz et al. 1978, 1982, 1987; Dallinger et al. 1979, 1981a; Agalidis et al. 1980; Jensen et al. 1980, 1982; Wilbrandt and Jensen 1981; Koyama et al. 1982, 1983; Kito et al. 1983). This topic is dealt with in detail in Chapter 9. It is important to note, however, that there is still much disagreement about the interpretation of the resonance Raman data pertaining to structural assignments. For example, according to Dallinger et al. (1981a) carotenoids *in vitro* undergo no changes in either the structural configuration of the chain or in the delocalised π-electron densities upon being excited into their triplet-states. However, Jensen et al. (1980) interpreted their resonance Raman data on all-*trans*-ß-carotene in terms of changes in chain configuration and/or conformation. The problem with many of these investigations is that it is difficult to decide whether a spectral difference between the ground state and an excited state is related to configurational changes or to changes in electronic structure.

8.3.2 In vivo

(a) X-ray diffraction studies of carotenoids bound to the reaction centre

The crystallisation and X-ray analysis of the phototrophic bacterial reaction centres of *Rps. viridis* and the carotenoid-containing strains of *Rb. sphaeroides* (2.4.1 and Y1) have been central to understanding how carotenoids carry out photoprotection in photosynthesis (Michel 1982; Deisenhofer et al. 1984,1985; Michel et al. 1986; Frank et al. 1987b; Allen et al. 1988a,b; Deisenhofer and Michel 1988, 1989a, b; Feher et al. 1989; Komiya et al. 1988; Yeates et al. 1988; Arnoux et al. 1989). In this section we deal only with the structural determinations and defer to the next section the discussion of the relationships between these structures and carotenoid function. The X-ray structure determinations of the reaction centre have revealed that there is an approximate two-fold rotation symmetry that relates the bound bacteriochlorophylls, bacteriophaeophytins and quinones (see Fig. 8.16).

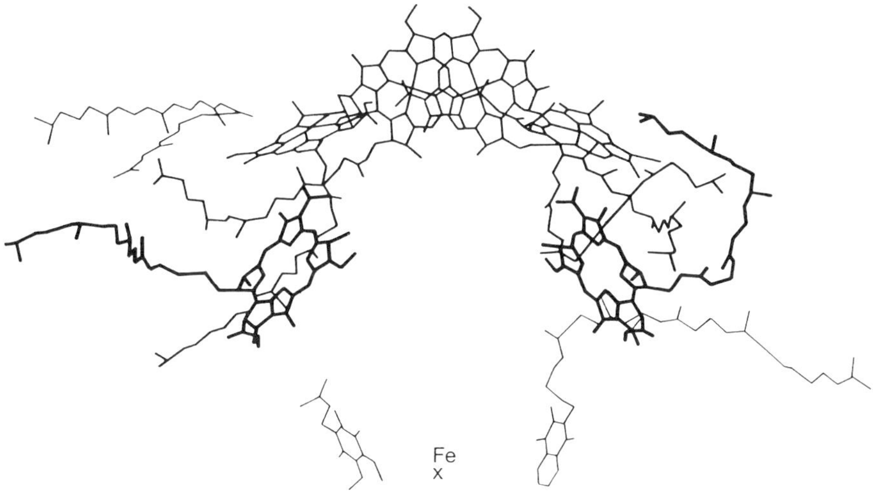

Figure 8.16 The structure of the reaction centre from *Rps. viridis*

The carotenoid molecule is the only chromophore in the reaction centre that does not conform to the approximate two-fold symmetry. The X-ray diffraction data on all the reaction centre proteins clearly indicate that the carotenoid is located near the monomeric accessory bacteriochlorophyll (B_B) that lies between the carotenoid and the primary donor. The carotenoid lies in a plane perpendicular to the two-fold symmetry axis of the reaction centre, and between the B and C helices of the M subunit. However, the exact stereochemistry of the bound carotenoid derived from the X-ray determinations remains uncertain. This is due primarily to the fact that the electron density map contained only part of the carotenoid structure. The part that was observed included only the π-electron conjugated chain of the carotenoid. The fits to the electron density maps for *Rb. sphaeroides* 2.4.1 and *Rps. viridis* are best represented by molecules that are bent out of plane into a quasi-*cis* configuration near C-13. The uncertainty in these determinations precluded a unique positioning of the *cis* bond, however. A 15,15´-*cis* isomeric structure for the carotenoids, as identified by resonance Raman (Koyama et al. 1983) and NMR spectroscopy (DeGroot et al. 1990; Gebhard et al. 1991), is also consistent with the electron density. This was shown to be the case by Arnoux et al. (1989) from X-ray diffraction data taken from crystals of *Rb. sphaeroides* strain Y1. It is not yet known if the uncertainty in the structure of the reaction centre-bound carotenoid stems from inherent structural disordering that arises from multiple configurations and/or conformations of the carotenoid in its binding site. Higher quality crystals have become available, and soon the data will be available to address this issue.

(b)Structure determinations by resonance Raman spectroscopy

Reaction centres

Resonance Raman spectroscopy has been used extensively to elucidate the structure of reaction centre-bound carotenoid. Lutz et al. (1978) initially proposed that carotenoids bound to reaction centres of phototrophic bacteria adopt strained, di-*cis* configurations. Koyama et al. (1983) carried out studies on several different isomers of ß-carotene and concluded that the properties of the 15,15´-*cis* configuration were closest to those of the carotenoid in the reaction centre, suggesting that this also assumes a planar 15,15´-*cis* configuration rather than a di-*cis* isomeric structure. Subsequently, Lutz et al. (1987) have revised their structural preference based on their own proton NMR investigation of carotenoids extracted rapidly from the reaction centre. These authors now conclude that the overall carotenoid stereochemistry is 15,15´-*cis*, but that the molecule is twisted in the C-7 to C-11 regions, rather than planar. A twisted carotenoid structure was previously proposed by Chadwick and Frank (1986) from EPR spectra of the photoexcited triplet-state of the reaction centre-bound carotenoid (see below). It may be that this twisting is responsible for the lack of resolution in the electron density map generated from the X-ray diffraction experiments. The resolution for the carotenoid is poorest at the ends of the molecules where there is likely to be conformational flexibility (rotations about single bonds).

An interesting question is whether carotenoids isomerise when they enter their triplet-states. Several carotenoids are thought to exhibit this behaviour *in vitro* (Claes 1960; Foote et al. 1970a; Jensen et al. 1982; Hashimoto et al. 1989). The situation in native complexes is not as clear. Boucher and Gingras (1984) monitored the absorption spectral changes of spirilloxanthin bound to reaction centres of *R. rubrum* S1 and the carotenoidless mutant *R. rubrum* G9 reconstituted with spirilloxanthin, and interpreted their data in terms of 15% of the bound carotenoid isomerising from *cis* to all-*trans*. Lutz et al. (1982) employed time-resolved resonance Raman spectroscopy to reaction centres of *Rb. sphaeroides* strains 2.4.1 and Ga and found marked differences between the spectra and those of triplet, all-*trans* carotenoids studied *in vitro*. They concluded that no isomerisation of the carotenoid from *cis* to all-*trans* occurs upon $S_0 \rightarrow T_1$ excitation and that the carotenoid retains its inherent *cis* configuration.

Resonance Raman spectroscopy has also been carried out on ß-carotene bound to the PS II reaction centre of chloroplasts. Using the D1/D2/Cyt *b*559 reaction centre preparation, Fujiwara et al. (1987) and Ghanotakis et al. (1989) found that the bound carotenoid is in an all-*trans* configuration.

Light-harvesting complexes

There have been several resonance Raman spectroscopic studies of the structure of the light-harvesting complex from photosynthetic systems. In contrast to those in reaction centres, the carotenoids in the antenna complexes appear largely to be all-*trans* isomers (Lutz 1984; Iwata et al. 1985; Robert and Frank 1988a,b; Hayashi et al. 1989; Noguchi et al. 1990). Possible reasons for this high degree of natural selection have been addressed (Koyama et al. 1990). Some authors have grouped the light-harvesting carotenoids into two types of all-*trans* structures; *viz* planar and twisted (Iwata et al. 1985; Hayashi et al. 1989; Noguchi et al. 1990). This distinction is based primarily on the fact that in the latter type a Raman line around 950 cm^{-1} (C–H out-of-plane wag) has higher intensity, suggesting twisting around the C-11,12 and/or the C-7,8 double bond positions. There also appear to be differences in the energy transfer characteristics of these different types of carotenoid, as discussed below.

(c)Structure determinations by absorption and CD spectroscopy

Reaction centres

On the basis of their electronic absorption and CD data, Boucher et al. (1977) proposed that carotenoids reconstituted into the reaction centre of *R. rubrum* G9 adopt a twisted, protohelical-shaped, 15,15′-*cis* configuration. This was deduced largely on the basis of an apparently high intensity *cis* peak in the 350–400 nm region of the absorption spectrum of the bound carotenoid. In contrast to this, Agalidis et al. (1980) found a very small oscillator strength in the same absorption spectral region for spheroidene bound in *Rb. sphaeroides* 2.4.1 or reconstituted into *Rb. sphaeroides* R-26 reaction centres. The problem with attempting to deduce the structure of the reaction centre-bound carotenoid from the *cis* peak intensity is that the *cis* peak is likely to be under the maxima of the very intense Soret bands from the reaction centre bacteriochlorin pigments (see Fig. 8.17). This unfortunate circumstance makes it virtually impossible to distinguish between an actual *cis* peak of the carotenoid and a perturbation of the Soret bands induced by carotenoid binding. Lutz et al. (1987) investigated this by comparing the absorption spectrum of spheroidene in *Rb. sphaeroides* 2.4.1 with that of methoxyneurosporene in *Rb. sphaeroides* GA. Whereas the strong $1^1A_g \rightarrow 1^1B_u$ transition of methoxyneurosporene was shifted by approximately 20 nm to the blue versus that of spheroidene, no pronounced spectral shift was observed 140 nm below the $1^1A_g \rightarrow 1^1B_u$ spectral origins where the *cis* peaks should lie. On the contrary, the spectral feature at 367 nm suspected to be the carotenoid *cis* peak, remained unshifted. This was taken as confirmation that this band arises from carotenoid-induced hyperchromic and bathochromic effects on the Soret bands of the bound bacteriochlorophyll (and bacteriophaeophytin) pigments and not from the carotenoid *cis* peak. The intensity of the actual carotenoid *cis* peak must be very low.

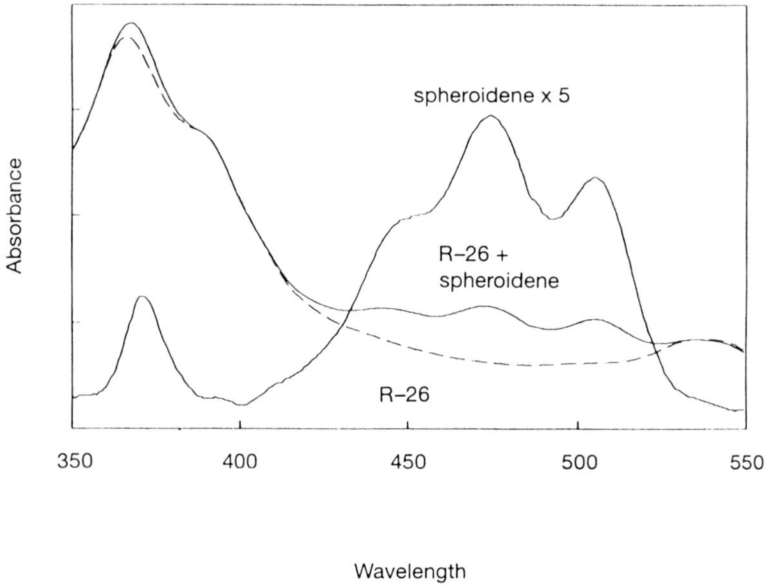

Figure 8.17 The absorption spectra of *Rb. sphaeroides* R-26 reaction centres with (——) and without (- - - -) incorporated spheroidene. The difference spectrum (scaled up by a factor of 5) is also shown. The feature at ~370 nm is not a "*cis* peak" but, rather due to a shift of the bacteriochlorophyll/ bacteriophaeophytin Soret bands upon binding of the carotenoid to the reaction centre. No peaks attributable to spheroidene were detected below 350 nm. These spectra reproduce the original experimental results of Agalidis et al. (1980) and are adapted from their work by permission of Elsevier Science Publishers.

If the carotenoid bound in the reaction centre is indeed adopting a *cis* conformation, then why does it lack a *cis* peak? Lutz et al. (1987) suggested that out-of-plane twisting of the carotenoid may result in a drastic weakening of the oscillator strength observed for the *cis* peak. Semi-empirical molecular orbital calculations carried out by R.E. Connors and H. Frank (unpublished data) have suggested that twisting of the carotenoid polyene chain can lead to separation of the *cis* peak into multiple transitions in this region (Fig. 8.18). The resulting *cis* peak lineshape in the 350–400 nm region could then be quite broad and difficult to observe. Additional broadening of the absorption spectral features of carotenoids is seen for carotenoids bound in the protein matrix versus their spectra *in vitro*. All of these factors may conspire to obscure the *cis* peak.

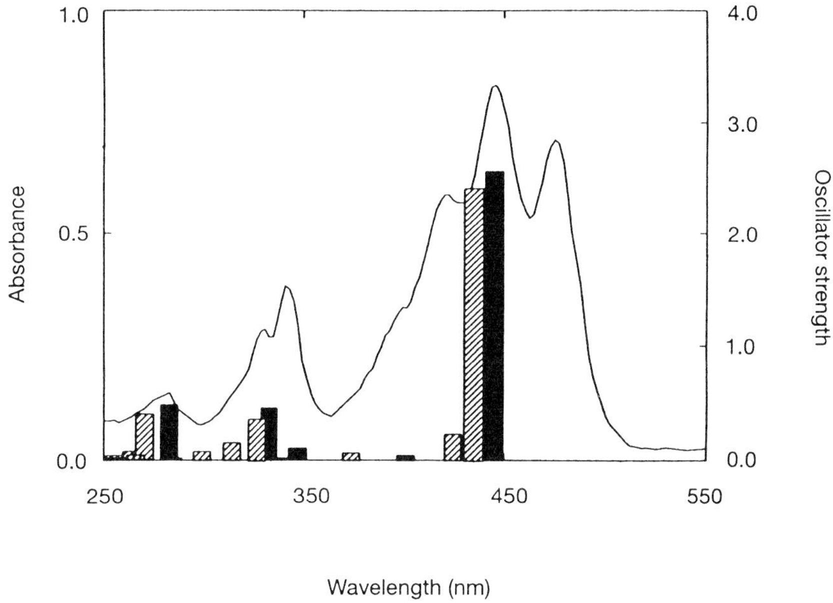

Figure 8.18 Comparison of the *in vitro* 15,15′-*cis*-spheroidene (in methanol) spectrum with the predicted positions of the allowed transitions from semiempirical molecular orbital theory calculations. The solid bars represents the results of a calculation on the undistorted 15,15′-*cis*-spheroidene using the geometry obtained by computing the fully optimised structure of the carotenoid with an AM1 Hamiltonian and followed by an INDO calculation of the excitation energies and relative oscillator strengths. The shaded bars represent the results of an INDO calculation carried out on the 15,15′-*cis*-spheroidene twisted by 90° about the C4-C5 single bond. (See Fig. 8.2). The "*cis* peak" region around 350 nm is seen to break up into multiple transitions upon distortion of the carotenoid away from planarity.

Finally, the CD spectra of carotenoids bound to the reaction centres are very intense, positive in sign, and correspond to molar ellipticities in the visible region in the order of 10^5 deg. M^{-1} cm^{-1} (Cogdell et al. 1976; Boucher et al. 1977; Frank and Violette 1989). The acyclic bacterial carotenoids, in general, show no optical activity when dispersed in solution. This is suggestive of the protein inducing the CD from interactions between the carotenoid transition moment and asymmetric amino acid residues in the vicinity of the chromophore. Because of this, it is difficult to deduce anything about the configuration of the carotenoid *in situ* from its CD spectrum. Evidently, the CD offers more information about the nature of the binding site than it does about the carotenoid stereochemistry.

Light-harvesting complexes

The carotenoid structures in the light-harvesting pigment–protein complexes are thought to be reasonably well understood on the basis of their absorption spectra. Longer carotenoids absorb more to the red than the shorter molecules, as is expected from their greater extent of π-electron conjugation. Also, there is essentially no evidence for *cis* peaks in the 350–400 nm region of the absorption spectra. This is consistent with the resonance Raman structural assignments discussed above that indicate that the light-harvesting protein-bound carotenoids are largely in all-*trans* configurations. Strong CD spectra from numerous antenna complexes have been reported (Cogdell and Scheer 1985). These tend to appear close to the same wavelengths as the corresponding absorption spectra. Like the reaction centre-bound carotenoid CD spectra, the spectra from the light-harvesting complexes may also be environmentally induced (Bolt et al. 1981), or alternatively may be due to exciton interactions (Lozano et al. 1990; Zurdo et al. 1991).

(d) Structure determinations by EPR spectroscopy

Reaction centres

EPR of the photoexcited triplet-states of carotenoids has been used extensively to probe the structures, geometries and dynamics of carotenoids in whole cells, chromatophores, and isolated pigment–protein complexes of various strains of phototrophic bacteria (Frank et al. 1982, 1983, 1984; McGann and Frank 1985a,b). Chadwick and Frank (1986) incorporated a selected series of carotenoids into reaction centres of *Rb. sphaeroides* R-26 in an attempt to reveal the effects of variations in the conjugated π-electron chromophore of carotenoids on their zero-field splitting parameters. Table 9.12 summarises the zero-field splitting parameters of many carotenoid triplets. Surprisingly, the zero-field splitting parameters of two different reaction centre-bound carotenoids, methoxyneurosporene and spheroidene, which differ by one double bond in their conjugated chain (see discussion above on absorption spectra), were indistinguishable. The zero-field splitting parameters of two other carotenoids, 3,4-dihydroanhydrorhodovibrin and spirilloxanthin, having 11 and 13 double bonds, respectively, were also found to be similar to each other, but smaller than those of methoxyneurosporene and spheroidene. The results suggested that the carotenoids were twisted in such a way that the extents of their π-electron conjugation were interrupted. It was suggested that the protein binding induces twists of the carotenoids near the ends of the π-electron chains. From subsequent resonance Raman spectroscopic studies, Lutz et al. (1987) also postulated that the reaction centre carotenoids adopt twisted structures.

Owing to the anisotropic nature of the dipolar interaction that dominates the coupling between the two π-electron spins in a $\pi\pi^*$ triplet-state of a carotenoid, the use of an oriented sample (e.g. chromatophores dried onto mylar strips) can reveal the orientations of the principal magnetic axes of the triplet relative to the normal to the membrane plane and relative to other magnetic axes. This analysis was carried out for spheroidene in the reaction centre of *Rb. sphaeroides* 2.4.1 (Frank et al. 1984). The projections of the magnetic axes with respect to the membrane normal and with respect to the primary donor triplet-state magnetic axes were determined.

Another approach to the use of triplet-states to deduce the geometrical arrangements of carotenoids employs magnetophotoselection, a technique that combines the optical photoselection and triplet-state EPR experiments. The method is carried out by exciting a random sample with polarised light and measuring the light-induced EPR spectrum with the light polarisation parallel or perpendicular to the static EPR field direction. Spheroidene in the reaction centre of *Rb. sphaeroides* 2.4.1 displays pronounced magnetophotoselection (McGann and Frank 1985b). The data were used to calculate a value of 75 (± 8)° for the angle between the primary donor Q_Y and the carotenoid transition moments. The angles between the individual triplet principal magnetic axes were also determined.

One of the problems with using triplet-state magnetic axes orientations to deduce the geometries of carotenoids is that the relationship between the magnetic axes of the carotenoid and its molecular axes is not well defined. The first step in making this connection more concrete was recently carried out by Frick et al. (1990). As alluded to in Section 8.2.1 (b), time-resolved EPR was used to detect the triplet-state signals from single crystals of ß-carotene (Frick et al. 1990). These studies allowed the orientation of the fine-structure tensor with respect to the molecular axes to be determined. It was found that these axes coincide to within 20°, and the signals that maximise the fine-structure splitting correspond to the direction along the molecular chain axis. How the directions of the magnetic axes are affected by changes in the stereochemistry of carotenoids is, however, still unknown.

It is generally believed that the individual triplet-state spin sublevel rate constants for intersystem crossing are sensitive functions of molecular symmetry (Clarke and Frank 1976). Thus, if these kinetic parameters can be measured for carotenoids, information about the structure of the molecule in its photoexcited triplet-state can be obtained. An investigation of the individual triplet-state spin sublevel dynamics of the carotenoid, spheroidene, in the reaction centre of *Rb. sphaeroides* 2.4.1 was carried out by McGann and Frank (1985a). The parameters were measured by direct-detection, rapid-transient, high field EPR spectroscopy on chemically reduced chromatophores prepared from the bacterium. The transient signals were analysed as a function of microwave power to obtain the rate constants, $k_1 = (2.5 \pm 0.2) \times 10^5\ s^{-1}$, $k_2 = (9.0 \pm 2.2) \times 10^4\ s^{-1}$ and $k_1 = (1.2 \pm 0.4) \times 10^5\ s^{-1}$. The inverse of the average of these values yields a value of 4.2 (± 0.3) ms and is in good agreement with the triplet-state lifetime of spheroidene in the reaction centre measured by optical spectroscopic techniques (Parson and Monger 1976). The fact that these individual spin-sublevel rate constants are all reasonably similar in magnitude is suggestive of a spin system of rather low symmetry. This would be consistent with the current hypothesis that spheroidene, when bound to the reaction centre, adopts a twisted *cis* configuration. Additional theoretical and experimental studies of the excited triplet-states of carotenoids and other linear conjugated polyenes are needed, however, before one can make detailed assignments of structure based on triplet-state dynamics data.

Absorption-detected ODMR of reaction centres from different strains of *Rb. sphaeroides* has been reported (Ullrich et al. 1987, 1989). The zero-field splitting parameters for the carotenoid triplets are, for the most part, in good agreement with those determined by high-field EPR. In reaction centres from *Rb. sphaeroides* 2.4.1 two carotenoid triplet-states which differ in their zero-field splitting $|D|$ values were found at low temperatures ($T < 40$ K). These probably correspond to different conformational isomers of the spheroidene molecule that are frozen in at the low temperature. This would be consistent with the current sentiment that twisting of the carotenoid at the ends of the molecule is likely. An observation of two different spectral forms of spheroidene was also reported by Kolaczkowski (1989) using high field EPR techniques.

Light-harvesting complexes

EPR has been used to study the triplet-states in B800–850 light-harvesting complexes isolated from the phototrophic bacteria, *Rb. sphaeroides* GA, aerobically and anaerobically grown *Rb. sphaeroides* 2.4.1 (Fig. 8.10) and *Rps. acidophila* 7750 (Frank et al. 1987a). These spectra show a remarkable similarity to those observed for the carotenoporphyrin complexes discussed above. The polarisation pattern for all the carotenoid triplets was observed to be *eae aea,* consistent with an intersystem crossing mechanism of formation. It is likely that the carotenoid triplet was formed via triplet–triplet energy transfer from bacteriochlorophyll which underwent intersystem crossing. This would suggest that the carotenoid is in very close spatial proximity to the bacteriochlorophyll, as required for triplet energy transfer (see below). Computer simulations of the observed EPR spectral lineshapes yielded values for the zero-field splitting parameters which could be understood in terms of the varying extents of π-electron conjugation in the carotenoids. The rate constants for population and decay of the observed triplet-states were also obtained from the computer simulations.

Carotenoid triplet-states in various light-harvesting pigment–protein complexes have been studied by ODMR also (Angerhofer et al. 1986; Ullrich et al. 1989). The data are in reasonable agreement with the high field measurements.

(e) Solid-state magic-angle sample-spinning (MASS) NMR

A novel approach to determining the structure of the carotenoid, spheroidene, was recently undertaken by using solid-state magic-angle sample-spinning (MASS) nuclear magnetic resonance (NMR) (DeGroot et al. 1990; Gebhard et al. 1991). The methodology proceeded in several stages as follows: First, the complete synthesis of the carotenoid spheroidene, containing ^{13}C isotopes at various chromophoric carbon positions was undertaken. Second, the carotenoids were reconstituted into the *Rb. sphaeroides* R-26 reaction centre. Third, the solid-state MASS NMR spectra of the complexes were taken. Finally, the spectra were interpreted to elucidate the structure of the spheroidene molecule in the reaction centre. The ^{13}C label was incorporated at the 14′ and 15′ positions of spheroidene. The presence of ^{13}C at these positions makes it possible to distinguish between the 15-*cis* and all-*trans* isomers of spheroidene. The reconstitution of the ^{13}C-labelled spheroidene into the *Rb. sphaeroides* R-26 reaction centres was carried out according to previously published procedures (Chadwick and Frank 1986). The measurement of the solid-state MASS NMR spectrum of the labelled spheroidenes reconstituted into the *Rb. sphaeroides* R-26 reaction centres showed several peaks, the most interesting of which were in the region from 100 p.p.m. to 150 p.p.m. where one finds the aromatic and conjugated carbons. A subtraction of the spectrum of the unreconstituted complex from that of the [14′-^{13}C]-spheroidene-reconstituted complex revealed a peak at 125.3 p.p.m. (see Fig. 8.19). This is most likely the 14′-^{13}C signal arising from the labelled spheroidene.

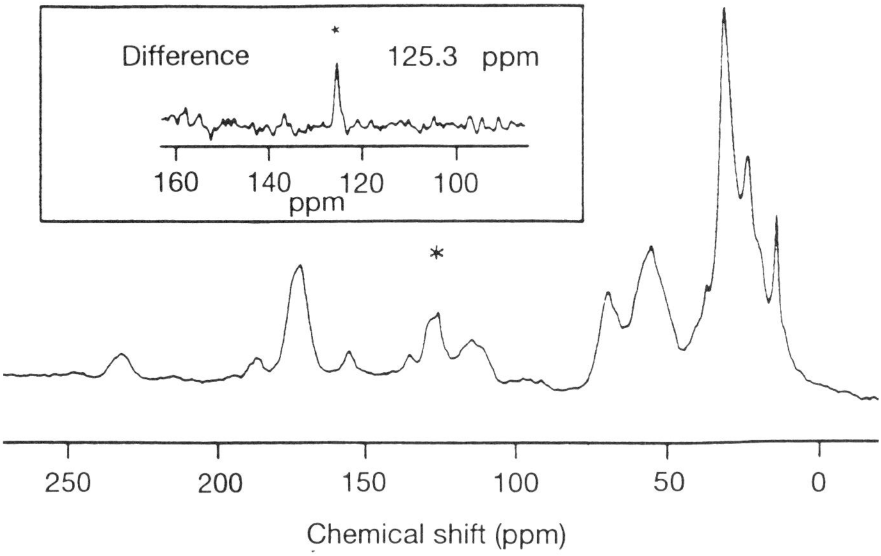

Figure 8.19 Cross polarization magic angle spinning (CP/MAS) ^{13}C NMR spectrum of *Rb. sphaeroides* R-26 reaction centres reconstituted with [14′-^{13}C]-spheroidene. Figure from Gebhard et al. (1991).

The elucidation of the precise structure of spheroidene in the reaction centre required an analysis of the chemical shift changes expected upon isomerisation. Thus, a solid-state natural abundance ^{13}C MASS NMR analysis of all-*trans*- and 15,15′-*cis*-ß-carotene was undertaken (DeGroot et al. 1990; Gebhard et al. 1991). The spectra indicated that when this carotenoid underwent a configurational change from all-*trans* to 15-*cis*, the ^{13}C resonance assigned to the 14-position (ß-carotene is symmetric) shifts from 132 p.p.m. to 127 p.p.m., which is very close to where the spheroidene 14′-^{13}C resonance was observed. This suggests that the spheroidene is also in a 15-*cis* configuration. Of course, the elucidation of the entire structure of spheroidene requires selective isotopic labelling at numerous positions in the molecule as well as solid-state ^{13}C MASS NMR spectra of the various labelled compounds in solution. These experiments are in progress.

(f)Linear dichroism and photoselection of carotenoids

Linear dichroism (LD) is the differential absorption of polarised light by an oriented sample. The technique has been applied extensively to several reaction centre and antenna complexes in an attempt to decipher the relative orientations of the chromophores in these systems (Breton and Vermeglio 1982). Photoselection uses the preferential absorption (proportional to $\cos^2\theta$) of polarised light to generate a non-random distribution of excited systems which show either photochemistry or emission. The orientation dependence of the photochemistry or emission can be used to reveal the relative orientations of transition moments. Here we focus on the elucidation of the orientations of the $1^1A_g \rightarrow 1^1B_u$ transition moments of carotenoids in these complexes.

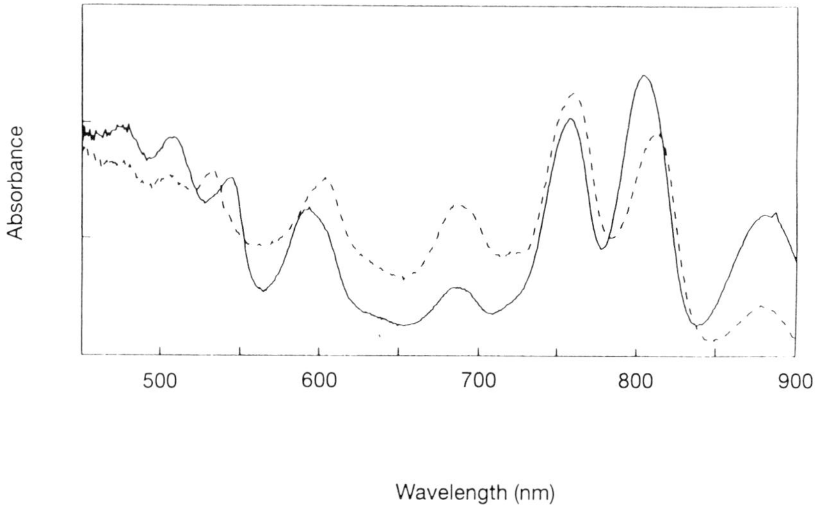

Figure 8.20 Low (85K) temperature polarised absorption of a single crystal of *Rb. sphaeroides* wild type 2.4.1 reaction centres. The dashed line represents the polarised absorption when the polariser is aligned parallel to the long axis of the crystal, and the solid line represents the polarised absorption when the polariser is perpendicular to this axis or parallel to the short axis of the crystal.

Reaction centres

Linear dichroism carried out on the carotenoids in reaction centres of *Rb. sphaeroides* 2.4.1 and *Rps. viridis* oriented in polyacrylamide gels shows the rather similar result that the carotenoid transition moment axis is oriented perpendicular to the C-2 symmetry axis of the reaction centre (Breton 1985). This was also found to be the case for the carotenoid in reaction centres from *Rb. sphaeroides* G1C oriented in an electric field (Gagliano et al. 1986). From photoselection (Vermeglio et al. 1978) and magnetophotoselection (McGann and Frank 1985b) experiments the angle between the carotenoid transition moment and the primary donor Q_Y transition was deduced to be approximately 75°. Linear dichroism studies on the carotenoids in crystalline reaction centres were being carried out by Frank et al. (1989a and 1991a) in an attempt to correlate the spectroscopic transition moment orientations with the atomic coordinates derived from X-ray diffraction studies. Fig. 8.20 presents a typical spectral trace obtained from such experiments.

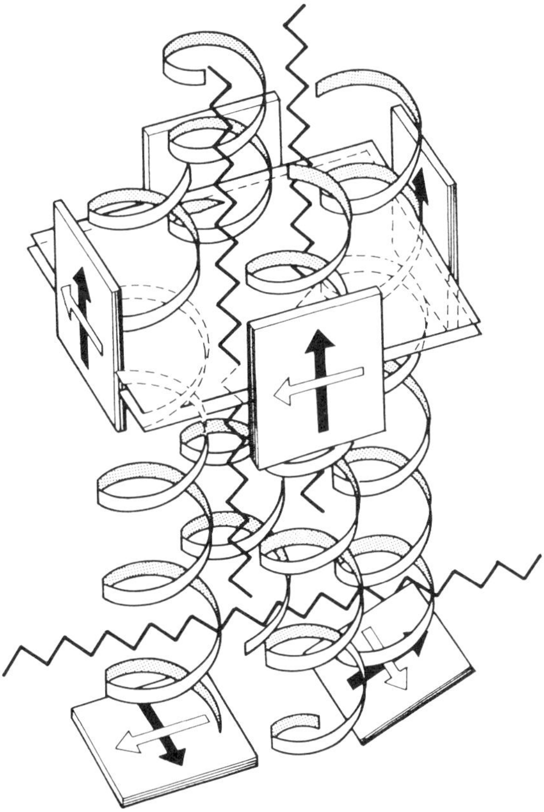

Figure 8.21 Model of the B800–850 light-harvesting complex. Six bacteriochlorophylls and three carotenoids are displayed. The carotenoids are postulated to exist in two orientations, parallel or perpendicular to the membrane plane. See the text for discussion of this point. This Figure was redrawn by Breton and Nabedryk (1987) from the original by Kramer (1984).

Light-harvesting complexes
Kramer et al. (1984) measured the LD and fluorescence polarisation spectra of the B800–850 complex from *Rb. sphaeroides* strain 2.4.1. They investigated the complex solubilised either in LDAO, where a native absorption spectrum was retained, or in LDS, where the 800 nm absorption band was diminished. The LD in the region where the carotenoids absorb was found to be negative. In this case the alignment of the complex was achieved by squeezing polyacrylamide slices containing the antenna complex. On the basis of these studies, Kramer et al. (1984) proposed a model for the organisation of the pigments within this antenna complex. This is illustrated in Fig. 8.21. This model proposes two pools of carotenoids, one of which is oriented perpendicular to the membrane, and the other, which is proposed to absorb at slightly longer wavelengths, being oriented parallel to the membrane. However, Breton and Nabedryk (1987) have argued that if this were the case, then at cryogenic temperatures, when the carotenoid absorption bands are markedly sharpened, the LD spectrum in the region of the carotenoid absorption should take on a derivative shape. This was not observed. Indeed, the LD spectrum remained negative throughout the region in which the carotenoid absorbed. Breton and Nabedryk (1987) concluded that there was, in fact, no evidence for two pools of carotenoid. Instead they advanced a model in which a single geometric arrangement of the carotenoids exists. This apparent conflict in interpretation illustrates the need for a high-resolution structure of a bacterial antenna complex.

LD spectra have also been recorded for a variety of pigment–protein complexes isolated from various algae and higher plants (see Breton and Nabedryk 1987, for a review). However, owing to the large number of chromophores present in typical preparations, the structural information obtained from these experiments yields only an average picture about the organisation of the carotenoids.

8.4 THE DYNAMICS OF CAROTENOID EXCITED STATES AND THE EFFICIENCY OF ENERGY TRANSFER

8.4.1 *In vitro*

The excited state lifetimes of carotenoids are essential pieces of information when possible mechanisms of carotenoid–chlorophyll energy transfer are being considered. In particular, they provide data that can be used to determine which of the two carotenoid singlet-state states, the 1^1B_u or 2^1A_g state, is the most likely energy donor. The data can also be used to place limits on the distances over which this energy can be transferred. A variety of methods, both indirect and direct, have been employed for measuring these dynamics. These include ground state depletion (Dallinger et al. 1981b; Wylie and Koningstein 1984), lifetime broadening of resonance Raman spectral lines (Haley and Koningstein 1983), time-resolved resonance Raman spectroscopy (Noguchi et al. 1989; Hashimoto and Koyama 1990; Kuki et al. 1990), emission yield determinations (Gillbro and Cogdell 1989; Cosgrove et al. 1990;

Andersson et al. 1991; Katoh et al. 1991; Mimuro et al. 1991; Shreve et al. 1991a) and picosecond (Wasielewski and Kispert 1986; Gillbro and Cogdell 1989) and femtosecond (Shreve et al. 1991b,c) transient absorption measurements.

(a) The 2^1A_g state

The indirect methods of ground state depletion and lifetime broadening yielded values for the lifetime of the 1^1A_g state of ß-carotene of between 0.25 and 1.5 ps. However, direct picosecond transient absorption spectroscopy produced a value of 8.4 ps for ß-carotene, with similar results for canthaxanthin and ß-8´-apocarotenal (Wasielewski and Kispert 1986). The discrepancy between these results has been discussed by the authors and explained in terms of the inherent technical difficulties of using indirect methods to deduce dynamics. The lifetime of the 2^1A_g state of a range of carotenoids has now been investigated by several other workers and it is generally accepted that the longer lifetimes determined by Wasielewski and Kispert (1986) are correct. Gillbro and Cogdell (1989) used picosecond ground state depletion/recovery experiments and measured the lifetime of the 2^1A_g state of spheroidenone in CS_2 to be 15 ps. Shreve et al. (1991a,b) used femtosecond ground state depletion/recovery techniques and recorded lifetimes of the 2^1A_g state of ß-carotene in CS_2 and of spheroidene in cyclohexane of 11 ps and 9.1 ps, respectively. Noguchi et al. (1989) Hashimoto and Koyama (1990) and Kuki et al., (1990) used picosecond time-resolved resonance Raman spectroscopy to look at the 2^1A_g state of ß-carotene and spheroidene. The time resolution of these experiments was in the order of 10 ps. Because the 2^1A_g state was detected, these authors argued that the lifetime of the 2^1A_g state must be about 10 ps or longer. If it were in the sub-picosecond time domain, it would not have been detected.

Recently, fluorescence emission from the 2^1A_g state of several carotenoids has been detected. (Cosgrove et al. 1991; Katoh et al. 1991; Mimuro et al. 1991; Andersson et al. 1992; DeCoster, et al. 1992) (see Fig. 8.4). The time-resolved fluorescence decay profile of fucoxanthin ($n = 8$) in CS_2 gave a major lifetime of 41 ps, while that of ß-8´-apocarotenal in the same solvent was 16 ps. Andersson et al. (1992) measured emission from the 2^1A_g state of a range of carotenoid analogues with 5, 7, 9 and 11 double bonds. The lifetime of the emission from the analogue with 5 double bonds, in *n*-hexane, was 2.0 ns. The yield of this emission was $7 \pm 3 \times 10^3$, which corresponds to a natural radiative lifetime of 300 nsec for the 2^1A_g state. If a similar natural radiative lifetime is assumed for the analogue with 7 double bonds and ß-carotene ($n = 11$), then based upon fluorescence yield measurements the lifetime of the 2^1A_g state can be estimated to be ~240 ps and 3 - 10 ps respectively. The lifetime of the 2^1A_g state of carotenoids is clearly in the picosecond time range, although it varies somewhat depending upon the number of conjugated double bonds and on the solvent.

(b)The 1^1B_u state

The lifetime of the 1^1B_u state of a range of carotenoids has been estimated from emission yield determinations. However, as far as we know, its lifetime has been determined only twice by direct kinetic experiments. Shreve et al. (1991a) used femtosecond transient absorption measurements and determined that the 1^1B_u state of ß-carotene lived for 200–250 fs at room temperature. The 1^1B_u state of spheroidene lived for 340 fs (Shreve et al. 1991b). The lifetime varied slightly depending upon the solvent. Interestingly, it was found that the 1^1B_u state decayed in this time into a second excited state which had a lifetime of 11 ps (ß-carotene) or 9.1 ps (spheroidene) (i.e. it apparently decayed into the 2^1A_g state). Table 8.1 gives a list of emission yields for a range of carotenoids. The emission yield can be expressed as

$$\Phi_f = \frac{\tau}{\tau_r} \quad (8.16)$$

where τ is the lifetime and τ_r is the natural radiative lifetime of the state. The natural radiative lifetime can be estimated by integrating the absorption profile of the emitting state, and has been estimated to be about 10^{-9} s for carotenoids such as ß-carotene (Gillbro and Cogdell 1989; Shreve et al. 1991a). When this value is used, the lifetimes of ß-carotene, spheroidene, neurosporene and okenone can be estimated to be 200 fs, 200 fs and 70 fs, respectively. These values agree rather well with those given by the direct kinetic measurements of Shreve et al. (1991a,b) in view of the errors inherent in quantifying these low fluorescence yields. For the ensuing discussions of carotenoid-to-chlorophyll energy transfer, it is important to remember that the 1^1B_u state decays into the 2^1A_g state in a few hundred femtoseconds. The 2^1A_g state then lives for a few picoseconds, with the exact lifetime depending on such factors as the number of conjugated double bonds present in the molecule.

(c) The triplet-state

As described above, carotenoids are not readily directly excited into their triplet-states. Carotenoid triplets have been studied *in vitro* following their formation by sensitisation via other triplet donors. In general, the triplet-state of most carotenoids of biological interest decays in the microsecond time scale. There is a small effect of degree of conjugation upon the decay time – the greater the degree of conjugation the faster the decay rate. In organic solvents most carotenoids react with triplet sensitisers at or near diffusion limited rates.

8.4.2 Model systems

Gust and co-workers have reported the synthesis of a variety of model compounds consisting of carotenoids linked to a range of porphyrins and pyrophaeophorbides (for a recent review see Gust and Moore 1991). The model compounds have been used to determine the structural requirements for the photosynthetic functions of carotenoids. Both the photoprotective and the antenna functions have been

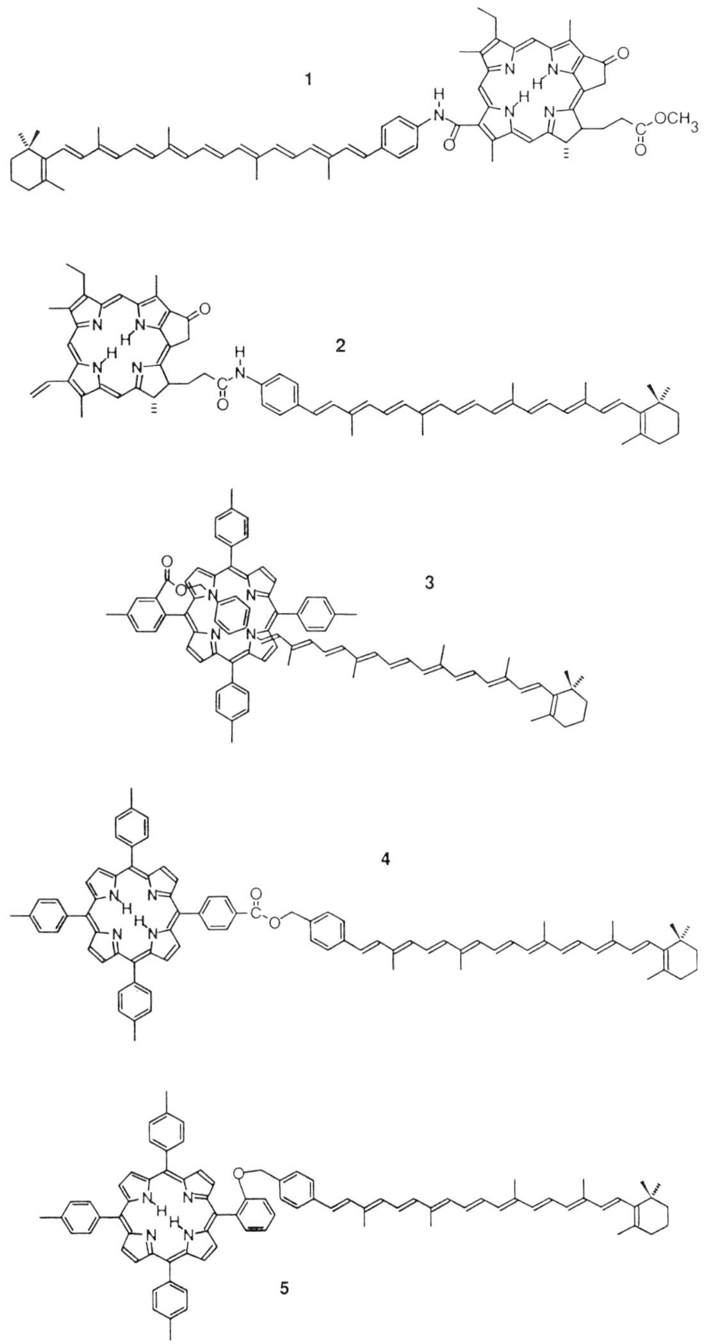

Figure 8.22 A series of carotenoporphyrin molecules synthesised and studied by Gust and Moore and coworkers. Figure taken from Moore et al. (1990).

successfully reproduced in these models, but only when the π-electron systems of the carotenoid and the macrocycles can come into close proximity. Mimicry of the photoprotective function has been assayed both by measuring the generation of the carotenoid triplet upon excitation of the porphyrin or the pyrophaeophorbide and by inhibition of the production of singlet oxygen (Gust and Moore 1991). Figure 8.22 shows a range of typical model compounds. The triplet energy transfer rate in compound 1 is very fast ($7 \times 10^9\ s^{-1}$). In compound 2, the rate of triplet exchange is dependent upon the solvent conditions. In a rigid solution the rate is $<10^4\ s^{-1}$ but in fluid solution this rate rises to $\sim 10^7\ s^{-1}$. In model 3, this triplet energy transfer rate is $\sim 10^9\ s^{-1}$, but in compound 4, where the molecular structure does not allow intramolecular motion to bring the two π-systems into contact, the rate drops again to $\sim 10^6\ s^{-1}$. The conclusion from these experiments is that the dominant factor which controls the rate of this triplet–triplet reaction is exchange coupling, the magnitude of which depends directly on the extent of the overlap between the π-electron systems of the donor and the acceptor. Mimicry of the antenna function of the carotenoid has a different distance dependence from that of the triplet exchange reaction. This is not surprising because the donor triplet-state lives longer by between six and nine orders of magnitude than the donor singlet-state of the carotenoid. The efficiency of singlet–singlet energy transfer in model 1 is ~55% and the rate constant for this reaction is $7 \times 10^{10}\ s^{-1}$ (Wasielewski et al. 1986a). Increasing the intra-chromophore distance from ~2 Å to ~5 Å as in model 2, reduces the efficiency of energy transfer to less than 5% and the rate to $<3 \times 10^9\ s^{-1}$. NMR analysis has shown that the π-electron systems of the two chromophores interact directly in model 1 (Chachety et al. 1984). The general conclusion from all these model studies is that very strong direct electronic coupling is necessary between the π-electron system of the carotenoid donor and the chlorophyll-derivative acceptor in order to achieve efficient singlet–singlet energy transfer. In other words, these workers interpret their results to mean that this singlet–singlet energy transfer occurs by an electron exchange mechanism (Gust and Moore 1991).

8.4.3 *In vivo*

The two different roles that carotenoids play in photosynthesis, namely light-harvesting and photoprotection, occur via singlet and triplet energy transfer, respectively. The molecular details of how these processes occur *in vivo* have been, and are being, studied, in a wide range of native and carotenoid-reconstituted reaction centre and light-harvesting complexes (Cogdell and Frank 1987). In reaction centres, the energy transfer takes place to and from the primary electron donor, whereas in the light-harvesting complexes carotenoids transfer their energy to and from chlorophylls whose environments are as yet not well defined. The controlling features are being addressed primarily by steady state fluorescence excitation experiments and by picosecond and femtosecond transient absorption methods. In this section, we survey the current status of results in this rapidly expanding area.

(a)Reaction centres

Singlet energy transfer

The efficiency of singlet energy transfer from carotenoid-to-primary donor bacteriochlorophyll (P865) was measured in reaction centres from *Rb. sphaeroides* 2.4.1. by fluorescence excitation spectroscopy (Cogdell et al. 1976). The reaction centres were reduced chemically and P865 fluorescence was monitored from the emission above 900 nm. The excitation spectrum was measured throughout the carotenoid and bacteriochlorophyll *a* Q_x absorption regions. The results were then compared to the absorption spectrum in these same regions and an efficiency of about 80% for this singlet–singlet energy transfer was deduced. However, the light-harvesting ability of the reaction centre carotenoid probably has little physiological significance because the reaction centre-bound carotenoids represent only a small percentage of the total carotenoid absorption of the intact photosynthetic membranes.

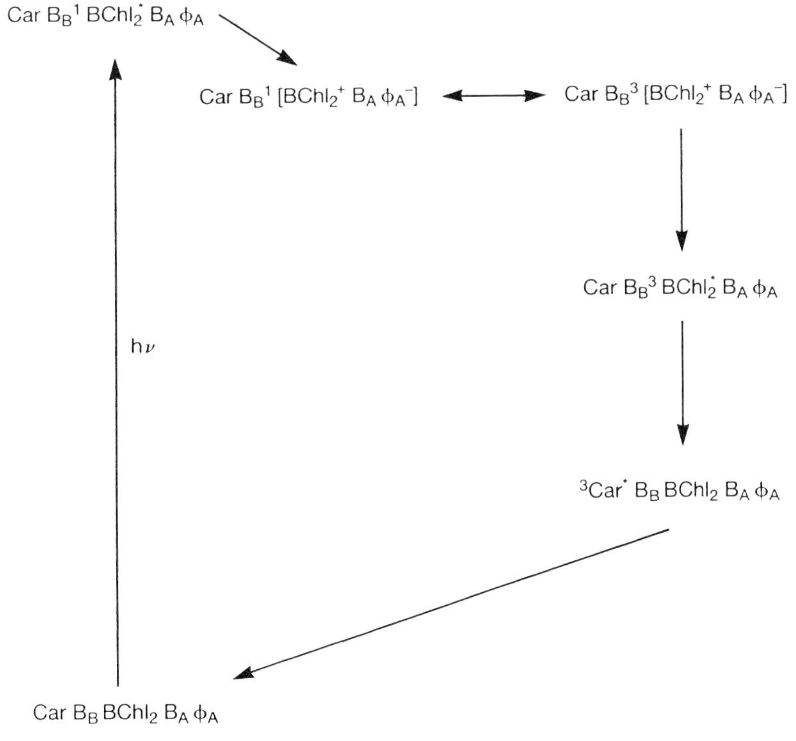

Figure 8.23 Schematic representation of the formation of a carotenoid triplet-state $^3Car^*$ in reaction centres of the photosynthetic bacteria under "blocked" conditions; i.e. where the forward electron transfer beyond the bacteriophaeophytin, f_A is not able to occur owing to the removal of chemical reduction of the subsequent quinone electron acceptors.

Triplet energy transfer

When the photochemical charge separation in the reaction centre is blocked (e.g. by reduction of the quinone acceptor in the purple bacterial reaction centres) illumination induces the formation of the triplet-state of the primary donor (Cogdell et al. 1975). Depending upon the conditions this triplet can be produced in high yields. This now creates a potentially damaging situation because, in principle, the triplet primary donor could react with molecular oxygen to produce the very harmful singlet oxygen. In most reaction centres the triplet-state of the primary donor is rapidly quenched by a carotenoid triplet, at a rate which exceeds the rate of the interaction with oxygen by a factor of between 10^2 and 10^3. The question of the mechanism of this process, however, remains unanswered.

It is well established that the reaction centres of phototrophic bacteria contain a primary electron donor consisting of a bacteriochlorophyll dimer ($BChl_2$), an initial electron acceptor known to be a bacteriophaeophytin molecule (ϕ_A), and subsequent quinone electron acceptors (Q_A and Q_B) which interact with a non-haem iron (Fe) atom (Feher et al. 1989). After absorption of light, or after excitation transfer from the antenna pigments, the primary donor is excited to an upper singlet electronic state which reduces the ϕ_A acceptor in ~4 ps (Kirmaier and Holten 1987). Under conditions where a chemical reductant has been added (~ 300 mV) or in a reaction centre that lacks Q_A, the forward photochemistry is blocked, and the charge-separated state of $BChl_2^+\phi_A^-$ undergoes a back reaction in ~10 ns. However, not all the reaction centres which back react in this manner return directly to the ground state. Many form a triplet-state on the $BChl_2$ dimer (see Fig. 8.23). In the carotenoidless mutant *Rb. sphaeroides* R-26, the $BChl_2$ triplet decays to the ground state in ~50 ms (Parson and Monger 1976). In virtually all carotenoid-containing reaction centres, the carotenoid reduces the lifetime of the $BChl_2$ triplet by quenching it in ~30 ns (Cogdell et al. 1975). An exception to this rule is *Rps. viridis*, the main carotenoid of which, 1,2-dihydroneurosporene (Fig. 8.13), does not quench the triplet-state of the primary donor (Holten et al. 1978). In the systems that carry out the quenching reaction, the process takes place with approximately 100% efficiency at room temperature (Cogdell et al. 1975) and approximately 50% efficiency at ~35–45 K (Cogdell et al. 1975; Parson and Monger 1976; Frank et al. 1983; Schenck et al. 1984). At ~10 K the reaction is virtually entirely inhibited (Parson and Monger 1976).

There have been many different suggestions concerning the origin of the temperature dependence of the primary donor-to-carotenoid triplet-state quenching process in the reaction centre. (For a recent discussion of this topic see Frank 1993.) The issue is briefly reviewed here. The first to characterise the quenching process were Parson and Monger (1976) who proposed that the triplet excitation is shared in an equilibrium between the triplet-states of the primary donor and the carotenoid. They suggested that the reduced carotenoid triplet yield at low temperatures could be due to a shifting of the equilibrium toward the primary donor triplet. This could be brought on by an inversion of the relative triplet-state energies of the donor and acceptor species. Schenck et al. (1984) proposed a different kind of equilibrium to account for the behaviour of the quenching process. They stated that temperature dependence of the triplet transfer had its origin in a thermal equilibrium between the primary donor triplet and another donor state, ^3D, more energetic by ~0.025 eV (~200 cm^{-1}), out of which the transfer was made. Alternatively, Frank et al. (1983) proposed, from EPR studies, that the donor state was actually a specific vibrational mode that promoted the transfer. From experiments that used linear-dichroic magneto optical difference spectroscopy (LD-MODS), Lous (1988) also suggested that the triplet transfer involved higher vibrational states. He was not able to rule out, however, the possibility that a higher energy triplet exciton (electronic) state of the primary donor was involved as an intermediate in the transfer. Kolaczkowski (1989) tested the hypothesis that vibrational states are involved as intermediates in the triplet energy transfer. Using normal or per-deuterated spheroidene incorporated into native or deuterated reaction centres of *Rb. sphaeroides* R-26 and flash transient optical spectroscopy, he found two different rates of triplet energy transfer to the carotenoid. One of these was fast and was suggested to arise from a vibrationally "hot" primary donor. The other was slower and was interpreted as originating from a thermally equilibrated primary donor triplet-state. The slower transfer rate was proposed to dominate the high temperature (>200 K) regime. In this model the inhibition of the transfer at low temperatures could be accounted for by assuming that at low temperatures the transfer becomes activated by thermal excitation of the vibrationally relaxed primary donor triplet to a specific (130 cm^{-1}) transition state (Kolaczkowski 1989). Needless to say, the models have become more complex as experiments are brought to bear on the problem.

As discussed above, Takiff and Boxer (1988a, b) measured the phosphorescence spectra of the primary donors from reaction centres of the bacteriochlorophyll *a*-containing *Rb. sphaeroides* and the bacteriochlorophyll *b*-containing *Rps. viridis* and the phosphorescence spectra of monomeric bacteriochlorophylls *a* and *b in vitro*. These determinations place constraints on the absolute triplet energy of the carotenoid relative to the primary donor and monomeric bacteriochlorophyll species in the reaction centre. Further, they suggested a mechanism for the triplet energy transfer that accounts for the temperature dependence of the quenching process. If the triplet energy transfer proceeds from the primary donor to the carotenoid via the structurally bridging monomeric B_B molecule, the difference in triplet energies between the primary donor and the B_B molecule would render the energy transfer in *Rb. sphaeroides* reaction centres active at low temperatures. In contrast to this, owing to a larger primary donor–B_B triplet-state energy gap, the transfer in *Rps. viridis* reaction centres would be energetically unfavourable at all temperatures up to room temperature. The X-ray structure of the reaction centre makes it tempting to propose that the B_B molecule is involved in the transfer. The structure shows that the B_B molecule is positioned on the path from the primary donor to the carotenoid, ~3 Å edge-to-edge distance from both of these cofactors (Yeates et al. 1988; Deisenhofer and Michel 1989a).

A verification of the involvement of the B_B molecule in the triplet energy transfer process was recently presented by Frank and Violette (1989). Using *Rb. sphaeroides* R-26 reaction centres treated with sodium borohydride (sodium borohydride is thought to remove or dislocate the B_B molecule from the reaction centre. Ditson et al. 1984; Maroti et al. 1985), Frank and Violette demonstrated that the primary donor–carotenoid triplet energy transfer reaction was inhibited, despite the fact that the treated reaction centres were reconstituted with the carotenoid spheroidene to the same extent as native (untreated) *Rb. sphaeroides* R-26 reaction centres. Several questions remain to be answered about triplet energy transfer in the reaction centre. (i) Should the triplet energy transfer reaction be described as a "hopping" of the excitation between distinct molecular sites or in a "super-exchange" model where a quantum mechanical mixing of states contributed from the primary donor, the B_B molecule and the carotenoid give rise to wavefunctions appropriate to describe and predict the temperature dependence of the reaction? (ii) What specific electronic and/or vibrational states are involved in the process? (iii) Can the temperature dependence be altered by incorporating different bridging B_B molecules? These questions are sure to be answered as work proceeds on several different reaction centre preparations.

Carotenoid triplet-states have been reported from green plant preparations in which the primary electron acceptors were oxidised (Frank et al. 1979). However, no time-resolved studies have been reported to date, probably because of the problem of interfering changes from the antenna pigments that are present in these preparations. Interestingly, in the fundamental D1/D2/Cyt *b*559 reaction centre preparations, carotenoid triplet formation no longer occurs (Frank et al. 1989b).

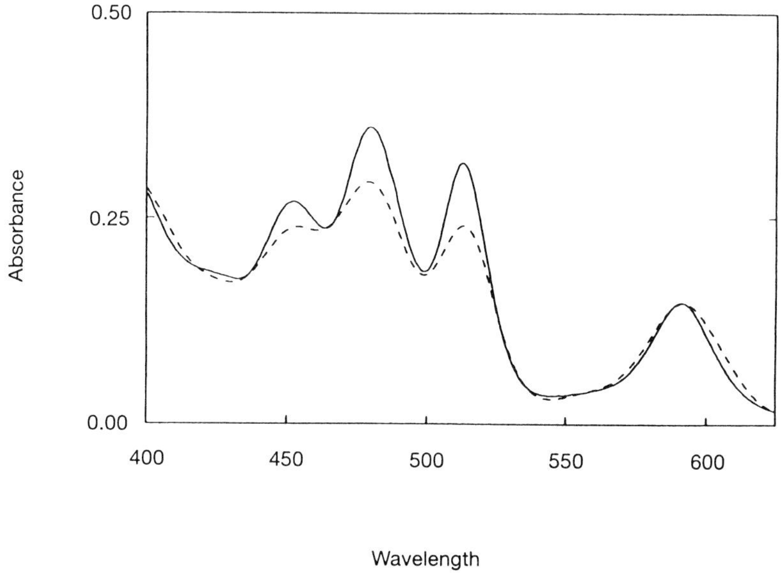

Figure 8.24. Absorption (——) and fluorescence excitation (- - - - -) spectra indicating carotenoid–bacteriochlorophyll singlet–singlet energy transfer in the B800–850 light-harvesting complex from *Rb. sphaeroides* wild type strain 2.4.1. The bacteriochlorophyll fluorescence was monitored at 850 nm. The traces were normalised to the absorption at 590 nm.

b)Light-harvesting complexes

Singlet energy transfer

Singlet-singlet energy transfer from carotenoids to chlorophylls has been demonstrated clearly for many years (see Cogdell and Frank 1987 for a review). This is illustrated in Fig. 8.24 by a comparison of the absorption and bacteriochlorophyll fluorescence excitation profiles for the B800–850 complex from *Rb. sphaeroides* 2.4.1. The remaining problem in this area of research is the need to explain the exact molecular mechanism(s) by which singlet–singlet energy transfer occurs. Three issues need to be settled: (i) What is the electronic excited state from which energy transfer takes place? (ii) What is the energy transfer mechanism? (iii) What is the nature of the acceptor state? The two possible donor states of the carotenoids are the 1^1B_u and 2^1A_g states. The mechanistic extremes of energy transfer are either a dipole–dipole (Förster) mechanism or an electron exchange mechanism (Dexter) (Förster 1948, 1965, 1968; Dexter, 1953; see also Fig. 8.25). Before going into the details of the light-harvesting process it is worth while pointing out the essential elements in these prospective energy transfer mechanisms and highlighting how they can be differentiated.

The fundamental equation describing the dependence of the rate of energy transfer on distance in this mechanism is given by

$$K_{Förster} = \frac{9000(\ln 10)\kappa^2 \phi_d}{128\pi^5 n^4 N_o \tau_d r^6} J_{Förster} \quad (8.17)$$

where

$$J_{Förster} = \frac{\int_{-\infty}^{\infty} F_d(v)\varepsilon_a(v)v^{-4}dv}{\int_{-\infty}^{\infty} F_d(v)dv} \quad (8.18)$$

and κ describes the relative orientations of the dipoles, ϕ_d is the donor fluorescence quantum yield, $J_{Förster}$ is the spectral overlap integral, n is the refractive index of the solvent, N_o is Avogadro's number, r is the donor-acceptor distance, τ_d is the donor fluorescence lifetime, $F_d(\upsilon)$ is the fluorescence emission intensity at wavenumber υ, and $\varepsilon_a(\upsilon)$ is the absorption coefficient of the energy acceptor at wavenumber υ.

It is now generally accepted that singlet–singlet energy transfer between chlorophylls in photosynthesis occurs by the Förster mechanism (van Grondelle 1985; Sundström and van Grondelle 1991). Typical chlorophyll-chlorophyll separations are thought to be of the order of 10–20 Å and their overlap integrals are generally assumed to be large. The energy transfer by this mechanism can be efficient even though its rate decreases with distance by $1/r^6$ because the lifetime of the excited singlet-state of the donor is quite long (usually hundreds of picoseconds to a few nanoseconds). Obviously, if a Förster mechanism is to be invoked for the singlet energy transfer from carotenoid to chlorophyll, then the distance between the molecules must be very close to accommodate the short lifetime of the 2^1A_g and 1^1B_u donor states. Indeed, the distance would probably have to be so close, i.e. 2–4 Å, that it is debatable whether the weak interaction hypothesis central to the Förster mechanism would still apply.

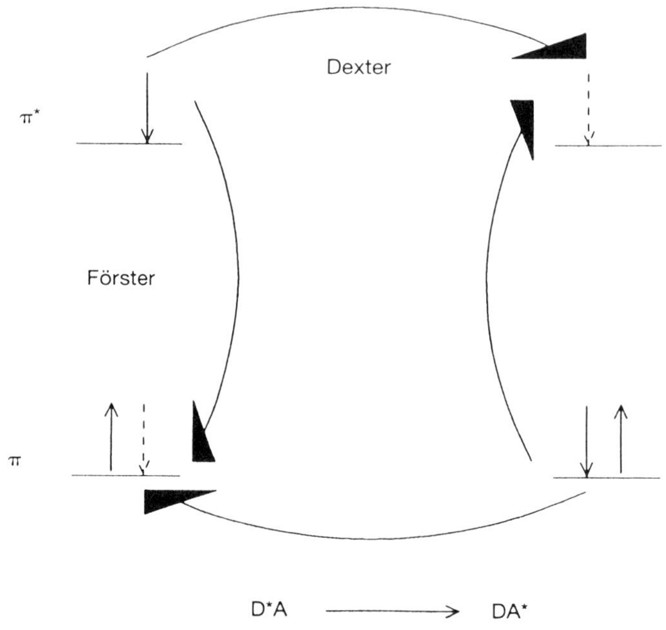

Figure 8.25 Schematic representation of the energy states involved in Förster (dipole-dipole) and Dexter (exchange) mechanisms of energy transfer.

The fundamental equation describing the dependence of the rate of energy transfer on distance in the exchange (Dexter) mechanism is given by

$$k_{Dexter} = \frac{4\pi}{h} Z^2 J_{Dexter} \quad (8.19)$$

where Z is a function of the donor–acceptor distance, r, as

$$Z^2 = K^2 \exp\left(\frac{-2r}{L}\right) \quad (8.20)$$

where K is related to the nature of the orbitals involved and L represents the van der Waals' radius (Tuua 1978). The overlap integral in this mechanism is given by

$$J_{Dexter} = \frac{\int_{-\infty}^{\infty} F_d(v)\varepsilon_a(v)v^{-4}dv}{\int_{-\infty}^{\infty} F_d(v)dv \int_{-\infty}^{\infty} \varepsilon_a(v)dv} \quad (8.21)$$

Note that, in principle, the mechanism (Förster or Dexter) which operates can be determined by a consideration of the effect of donor–acceptor separation and spectral overlap on the rate constant for energy transfer. The efficiency of energy transfer, ε_{ET}, is related to the rate constant for energy transfer, k_{ET}, by the expression

$$\varepsilon_{ET} = \frac{k_{ET}}{k_{ET} + k_c} \quad (8.22)$$

where k_c represents the sum of the rate constants for all the processes that compete with energy transfer.

The efficiencies of energy transfer can be measured by steady state absorption and fluorescence excitation techniques. Table 8.4 presents a comprehensive list of energy transfer efficiencies for carotenoid–chlorophyll singlet–singlet transfer from a wide range of antenna complexes. The data are most reliable for the purple bacterial complexes and lower algal complexes, where the spectral separation of the carotenoid and chlorophyll absorption bands is well defined. This spectral separation makes the data more reliable than those from higher plant antenna complexes, where extensive spectral deconvolution is required to separate the relative contributions of the chlorophylls and the carotenoids. We therefore concentrate our discussion on the bacterial and lower algal complexes.

Table 8.4 Carotenoid–chlorophyll singlet energy transfer efficiences in photosynthetic bacterial and algal light-harvesting complexes. These efficiences were determined by steady-state, carotenoid excitation-induced, bacteriochlorphyll fluorescence

Carotenoid	*Bacterium or alga*	*Light-harvesting*	*Efficiency (%)*	*Reference*
Spheroidene, spheroidenone (92%, 8%)	*Rb. sphaeroides* 2.4.1 **N**	B800–850 (anaerobic)	100, 87, >90	Cogdell et al. 1981 Noguchi et al. 1990 van Grondelle et al. 1982
Spheroidene, spheroidenone (70%, 30%)	*Rb. sphaeroides* 2.4.1 **N**	B800–850 (anaeobic–aerated)	95	Cogdell et al. 1981

Spheroidene, spheroidenone (2%, 98%)	*Rb. sphaeroides* 2.4.1 **N**	B800–850 (semi-aerobic)	75	Cogdell et al. 1981
Spheroidene, spheroidenone	*Rb. capsulatus* **N**	B800–850	82, >90	Noguchi et al. 1990, van Grondelle et al. 1982
Spheroidene, spheroidenone	*Rc. gelatinosus* **N**	B800–850	84	Noguchi et al. 1990
Neurosporene	*Rb. sphaeroides* G1C **N**	B800–850	100, >90	Cogdell et al. 1981, van Grondelle et al. 1982
Neurosporene, methoxy-neurosporene, chloroxanthin (60%,14%,26%)	*Rb. sphaeroides* GA **N**	B800–850	94	Cogdell et al. 1981, Nogouchi et al. 1990
Spheroidene	*Rb. sphaeroides* LH-R-26 **R**	B850	60–70	Davidson and Cogdell 1981
Spheroidene	*Rb. sphaeroides* LH-R-26 **R**	B870	70	Noguchi et al. 1990
Spheroidene	*Rb. sphaeroides* **N**	B875	70	
Spheroidene	*Rb. sphaeroides* LH-R-26 **R**	B850	56 ± 2	Frank et al. unpublished data
3,4-Dihydro-spheroidene	*Rb. sphaeroides* LH-R-26 **R**	B850	71 ± 3	Frank et al. unpublished data
3,4,5,6-Tetrahydro-spheroidene	*Rb. sphaeroides* LH-R-26 **R**	B850	65 ± 3	Frank et al. unpublished data
3,4,7,8-Tetrahydro-spheroidene	*Rb. sphaeroides* LH-R-26 **R**	B850	53 ± 5	Frank et al. unpublished data
Neurosporene	*Rb. sphaeroides* LH-R-26 **R**	B850	60–70	Davidson and Cogdell 1981
Neurosporene	*Rb. sphaeroides* LH-R-26 **R**	B870	72	Noguchi et al. 1990

Methoxy-neurosporene	*Rb. sphaeroides* LH-R-26 **R**	B870	70	Noguchi et al. 1990
Chloroxanthin	*Chr. vinosum* **N**	B800–850 B800–820[1]	50	Hayashi et al. 1989
Spheroidene	*Chr. vinosum* **R**	B800–850 B800–820[2]	40, 37	Hayashi et al. 1989 Noguchi et al. 1990
Neurosporene	*Chr. vinosum* **R**	B800–850 B800–820[2]	56	Noguchi et al. 1990
Methoxy-neurosporene	*Chr. vinosum* **R**	B800–850 B800–820[2]	44	Noguchi et al. 1990
Spheroidene	*Rc. gelatinosus* **N**	B875	65	
Spheroidene, spheroidenone	*Rc. gelatinosus* **N**	B870	72	Nogouchi et al. 1990
Spheroidene, spheroidenone	*Rc. gelatinosus* **N**	B800–850[3]	70–75	Kramer et al. 1984
Rhodopin	*Chr. vinosum* **R**	B800–850 B800–820[2]	29	Nogouchi et al. 1990
Spirilloxanthin	*Chr. vinosum* **R**	B800–850 B800–820[2]	27	Nogouchi et al. 1990
Spirilloxanthin, rhodopin, lycopene	*Rps. acidophila* **N**	B880	25	Angerhofer et al. 1986
Rhodopin	*Rps. acidophila* **N**	B875	25	
Spirilloxanthin, rhodopin, lycopene	*Rps. acidophila* **N**	B800–850 type I	50–55, 51	Angerhofer et al. 1986 Chadwick et al. 1987
Spirilloxanthin, rhodopin, lycopene	*Rps. acidophila* **N**	B800–850 type I[3]	33	Chadwick et al. 1987
Spirilloxanthin, rhodopin, lycopene	*Rps. acidophila* **N**	B800–850 type II	70–75	Angerhofer et al. 1986
Spirilloxanthin, rhodopin,	*Rps. acidophila* **N**	B800–820	70–75	Angerhofer et al. 1986

lycopene				
Rhodopinal	*Rps. acidophila* **N**	B800–820	75	
Spirilloxanthin, rhodopin, lycopene	*Chr. vinosum* **N**	B800–820 B800–850[4]	32	Noguchi et al. 1990
Spirilloxanthin, rhodopin, lycopene	*Rp. palustris* **N**	B800–850	50	Angerhofer et al. 1986
Spirilloxanthin	*R. rubrum* **N**	B890	30	
Rhodopin	*Rb. sphaeroides* LH-R-26 **R**	B870	43	Noguchi et al. 1990
Spirilloxanthin	*Rp. palustris* **N**	B875	35	
Okenone	*Chr. purpuratum* **N**	B830	65	
Peridinin		Peridinin/Chl *a*	100	
Fucoxanthin		Fucoxanthin/ Chl *a*	100	

[1] DPA cultured >5 days
[2] DPA cultured 2–3 days
[3] LDS treated
[4] No DPA
N = Native, **R** = Reconstituted

The singlet energy transfer efficiencies range from 25% to 100%, and it is not yet clear what controls this efficiency. Ideally, several complexes would be desirable in determining the controlling features. These include: (i) well-defined pigment–protein complexes where a range of different carotenoids can be incorporated into a single controlled binding site; (ii) a variety of antenna complexes whose structures are known and which have the same carotenoid molecule bound in a uniform site. In the latter systems the distances between donor and acceptor molecules can be correlated with the rates of the energy transfer. Unfortunately, no three-dimensional structures of antenna complexes that contain chlorophylls and carotenoids are yet available. H. Frank and R. Farhoosh (unpublished data) have studied the energy transfer properties of the B850 antenna complex from *Rb. sphaeroides* R-26 reconstituted with a range of carotenoids of different extents of conjugation. A multifaceted experimental approach involving structural (X-ray diffraction, NMR, EPR, etc.) and spectroscopic (static and dynamic) tools needs to be brought to bear on this problem to elucidate the molecular details of the energy transfer process.

One of the earliest picosecond time-resolved measurements of the carotenoid-to-bacteriochlorophyll singlet–singlet energy transfer process was carried out by Wasielewski et al. (1986b). Using the B800–850 complex from *Rps. acidophila* strain 7750, these workers excited at 510 nm, into the absorption band of the carotenoid (in this case mainly rhodopin), with a 4 ps pulse. The ground state bleaching of the carotenoid was observed to occur instantaneously and then to decay with a time constant of 5.6 (± 0.9) ps. The rise time of the bleaching at 860 nm, which occurs as the energy arrives at the bacteriochlorophyll, was measured to be 6.1 (± 0.9) ps, i.e. within the experimental error of the decay time of the bleaching of the carotenoid's absorption band. Similar experiments on the picosecond time scale were also carried out by Gillbro et al. (1988) working with the B800–850 complex from *Rps. acidophila* strain 7050 (major carotenoid rhodopinal). The energy transfer process took place in 3 (± 1) ps. So far only two systems have been investigated on the femtosecond time scale. In the first, Trautman et al. (1990a) studied the B800–850 complex isolated from *Rb. sphaeroides* strain 2.4.1. The complex was excited at 510 nm with a 240 fs excitation pulse. In a more recent paper, Shreve et al. (1991b) have extended and re-interpreted the original data. Their current interpretation of the multiphasic, complex kinetics is that the carotenoid transfers its energy to both the 800 nm- and 850 nm-absorbing bacteriochlorophylls. In the new model the majority of the carotenoid molecules present in the complex transfer their energy to the 850 nm bacteriochlorophylls in ~300–400 fs. The remainder transfer their energy via the 800 nm bacteriochlorophylls to the 850 nm-absorbing molecules. Interestingly, these workers also suggest that energy transfer can originate from both the 1^1B_u and the 2^1A_g states of the carotenoid.

Similar experiments were also carried out by the same group (Trautman et al. 1990b) on thylakoid membrane preparations from the diatom *Phaeodactylum tricornutum* and from the eustigmatophyte *Nannochloropsis* sp. Excitation with pulses characterised by 240 fs pump-probe cross-correlations allowed the time for the carotenoid–chlorophyll energy transfer in the eustigmatophyte to be measured at 240 (± 40) fs. The transfer in the diatom was found to be bi-exponential, with transfer times 500 (± 100) fs and 2.0 (± 0.5) ps and relative amplitudes 1.7 (± 0.7):1. The time-resolved studies summarised above show quite clearly that carotenoid-to-chlorophyll singlet–singlet energy transfer is indeed very fast. However, it is still an open question as to what controls the partitioning of the energy transfer between the 2^1A_g state route and the 1^1B_u state route. The answer at this point might very well be that it will depend precisely upon which antenna complex is being considered. In two cases, the B800–850 complex from *Rb. sphaeroides* containing either spheroidene (Shreve et al. 1991b) or neurosporene (Cogdell et al. 1992), the fluorescence yield of the 1^1B_u emission has been determined for the carotenoid both *in vitro* and *in vivo*. For spheroidene the yields are 3.3 (± 0.6) x 10^{-4} and 1.9 (± 0.4) x 10^{-4} respectively, while for neurosporene they were estimated to be 2.2 x 10^{-4} both *in vitro* and *in vivo*. If a large fraction of the energy transfer were occurring from the 1^1B_u state, the yield of the emission from this state might have been expected to be dramatically reduced *in vivo* compared with that *in vitro*. This is not observed, a fact which serves to reinforce the view that, in those cases where there are no other mitigating factors, most of the transfer should originate from the 2^1A_g state.

Triplet energy transfer

Carotenoid triplet-states in antenna systems were first described in higher plants (Mathis 1969; Witt 1971). It was shown that, upon flash excitation of chlorophyll, especially under conditions where the energy in the flash was above that required to saturate photosynthesis, carotenoid triplets were formed. These workers proposed that the carotenoid triplets were formed in a triplet–triplet transfer reaction from chlorophyll triplets. This type of mechanism was then shown to occur *in vitro* between chlorophyll *a* and carotenoids in organic solvents (Mathis 1969). Earlier, Parson (1967) had described what has now been shown to be antenna and reaction centre carotenoid triplet-states in chromatophores from *R. rubrum* strain S1. When the primary electron acceptor was chemically reduced, flash excitation caused the formation of a transient state with an absorption band near 420 nm. This transient was initially interpreted as a second photosystem capable of operating at very low redox potential. However, subsequent work in Parson's laboratory (Cogdell et al. 1976; Monger et al. 1976) demonstrated clearly that this 420 nm transient indicated the formation of carotenoid triplets. Monger et al. (1976) studied the formation of carotenoid triplets in chromatophores from *R. rubrum* and a variety of strains of *Rb. sphaeroides*. In the carotenoidless *Rb. sphaeroides* R-26 membranes, flash excitation into the long wavelength bacteriochlorophyll absorption bands induced the formation of bacteriochlorophyll triplets. In the carotenoid-containing *Rb. sphaeroides* strains only carotenoid triplets were detected. These triplets were formed in less than 10 ns and decayed in a few microseconds at room temperature. The position of the new absorption due to carotenoid triplet-state formation, just to the red of the ground state absorption bands, was shifted in wavelength in the carotenoid-containing strains exactly as expected from the change in number of conjugated double bonds ($n = 13$ for spirilloxanthin, $n = 10$ for spheroidene and $n = 9$ for neurosporene and its derivatives). This confirmed that the transient species was a carotenoid triplet. Carotenoid triplet formation also occurs in isolated purple bacterial antenna complexes (Cogdell et al. 1981) with kinetics similar to those that have been measured in the membrane.

It is now generally accepted that antenna carotenoid triplets are formed in a triplet–triplet transfer reaction with triplet excited chlorophyll. Moreover, because the $S_0 \rightarrow T_1$ transition is spin forbidden, it is thought that the process takes place by an exchange mechanism. Triplet bacteriochlorophyll in the antenna complex from the carotenoidless strain of *Rb. sphaeroides* R-26.1 is formed in ~10 ns following laser-flash excitation. In carotenoid-containing complexes, bacteriochlorophyll triplet formation is not seen. Instead, carotenoid triplets are seen with a rise time indistinguishable from that of the carotenoidless case. This suggests that even though a nanosecond formation time is slow compared with the time over which singlet–singlet energy transfer occurs, the rate of the triplet transfer reaction may be limited only by the time required for intersystem crossing to occur from singlet- to triplet-excited bacteriochlorophyll.

8.5 MISCELLANEOUS PROPERTIES OF CAROTENOIDS

8.5.1 Carotenoid radicals

Carotenoid radical cations have recently been studied by Grant et al. (1988). Using a combination of cyclic voltammetry, absorption spectroscopy and EPR these authors have detected radical cations of ß-carotene, ß-apo-8′-carotenal and canthaxanthin in various solvents. These data are important in understanding whether carotenoids participate in electron transfer reactions in photosynthetic systems. This participation is certainly the case in numerous model systems (Gust and Moore 1991).

8.5.2 Oxidation-reduction reactions of carotenoids

Radical ions have been generated for a range of polyenes, including carotenoids (Mathis and Vermeglio 1972). These radicals have been generated both by pulse radiolysis (Lafferty et al. 1978) and by light-driven electron transfer in the presence of suitable dyes (McVie et al. 1979; Chauvet et al. 1983). In general, the radical anions and cations have similar λ_{max} values though the cations usually absorb at longer wavelengths than do the anions. However, both radical ions always absorb to the red of the parent molecule. The functional significance of carotenoid radicals in biology is not clear; carotenoid cations have been shown to occur in PS II preparations (Telfer et al. 1991) but only under conditions where the primary reactions are partially blocked on the donor side.

8.5.3 Carotenoids as fluorescence quenchers

In organic solvents, carotenoids such as ß-carotene have been shown to quench chlorophyll fluorescence (Beddard et al. 1977). It was suggested from these experiments that the quenching mechanism could involve electron transfer between the chlorophyll and the carotenoid, with a rapid, subsequent back reaction, i.e.

$$^{1}Chl\,a^{*} + Car \rightarrow Chl\,a^{-} \rightarrow Chl\,a + Car \quad (8.21)$$

Similar fluorescence quenching of chlorophyll is seen in mixtures of chlorophyll and quinones (Parson and Cogdell 1975). One-electron redox reactions have been demonstrated *in vitro* between mixtures of chlorophylls and carotenoids, and between porphyrins and carotenoids (Chauvet et al. 1983). The rate of reaction between chlorophyll *a* and either ß-carotene or lycopene has been shown to occur at near the diffusion limit.

Until a few years ago, these observations may have been viewed simply as curiosities in terms of any biological relevance; however, quite recently in plants there has been work which shows a strong correlation between zeaxanthin content and the amount of non-photochemical fluorescence quenching (Demmig-Adams et al 1989; Demmig-Adams 1990; see Chapter 7). In chloroplasts there is an enzyme system which reversibly converts violaxanthin into zeaxanthin (the "violaxanthin" or "xanthophyll cycle", see Yamamoto 1979; Fig. 8.26). It has been suggested that the xanthophyll cycle could be part of a photoprotective mechanism, but until recently there was no direct evidence for this idea.

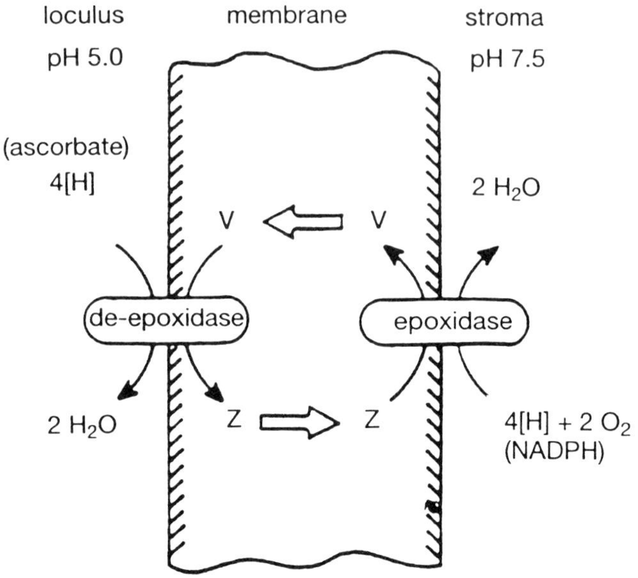

Figure 8.26 A schematic representation of the xanthophyll cycle - an enzyme system in chloroplasts which reversibly converts violaxanthin into zeaxanthin (Krinsky 1978). The conversion utilises epoxidation and deepoxidation reactions which have been found in both plants and algae.

Demmig-Adams et al. (1989) have shown that there is a close correlation between the zeaxanthin content and one type of photosynthetic radiationless deactivation. When plants are over excited, PS II is damaged in a process called photoinhibition. It has been suggested that a fluorescence quenching induced by zeaxanthin would be protective since it would reduce the effective number of photons reaching the reaction centre and therefore reduce the magnitude of the photoinhibitory response. It must be pointed out though that, as yet, there is no direct evidence that zeaxanthin, and not violaxanthin, does physically quench chlorophyll fluorescence.

Triplet-excited carotenoids can also act as fluorescence quenchers of both chlorophyll and bacteriochlorophyll (Breton and Geacintov 1979; Kingma 1983). Within the photosynthetic apparatus the light-absorbing pigments are effectively organised into "domains". Within these domains, excited states, both singlets and triplets, are mobile. If two excited states meet and collide then they have a certain probability of annihilating (McGann and Frank 1983). One example of this is singlet heterofission to form a triplet pair; e.g.

$$Car+{}^{1}Chl^{*} \rightarrow \left[{}^{3}Car \bullet {}^{3}Chl^{*}\right] \quad (8.22)$$

This could then result in a reduction in the concentration of excited singlet-states and hence produce fluorescence quenching.

8.6 CONCLUSIONS

Because of the intense coloration of carotenoids, they have proved readily amenable to isolation and characterisation over the years. However, only recently has a somewhat more detailed knowledge of their photochemistry emerged. This has happened because of significant technical advances in optical and magnetic resonance spectroscopy which have extended the sensitivity and dynamic range of the methodology. These advances have generated a renewed interest in the biological functions of carotenoids and, as a consequence, this field has become a very active and competitive research area. This review has attempted to deal with what is currently known about carotenoid photochemistry, concentrating on recent results which are important for understanding how carotenoids function in photosynthesis. The activity in this area is progressing at such a rapid pace that one may expect many of the details which are missing to be supplied, in the very near future.

ACKNOWLEDGEMENTS

H.A.F. and R.J.C. wish to express their sincere gratitude to the following individuals who have in many different ways contributed to the completion of this review: Dr Per-Ola Andersson, Professor Ronald Christensen, Professor Robert E. Connors, Ms Beverly DeCoster, Professor Tomas Gillbro, Dr Huub deGroot, Dr Ronald Gebhard, Professor Johan Lugtenburg, Professor Tom Owens, and Dr A. P. Shreve and Professor T. George Truscott; and the members of the Storrs research group, Ms Mila Aldema, Dr Barry Chadwick, Ms Agnes Cua, Ms Roya Farhoosh, Dr Jennifer Innes, Dr William McGann, Mr John Machnicki, Dr Shahriar Taremi and Dr Carol Violette. Support for the research in the group of H.A.F. was provided by the National Institutes of Health (GM-30353), the Competitive Research Grants Office of the U.S. Department of Agriculture (88-37130-3938), the University of Connecticut Research Foundation, and a NATO grant for International Collaboration in Research (0107/88) awarded to H.A.F. and R.J.C. Support for the research in the group of RJC was provided by an EEC grant and a British Council grant.

REFERENCES

Agalidis, I., Lutz, M. and Reiss-Husson, F. (1980) Binding of carotenoids on reaction centres from *Rhodopseudomonas sphaeroides* R-26. *Biochim. Biophys. Acta*, **589**:264-274.

Allen, J.P., Feher, G., Yeates, T.O. *et al.* (1988a) Structure of the reaction centre from *Rhodobacter sphaeroides* R-26 and 2.4.1. In: Breton, J. and Vermeglio, A. (eds.), *The Photosynthetic Bacterial Reaction Centre*. NATO ASI Series A: Life Sciences. Plenum, New York, pp.149:5-11,

Allen, J. P., Feher, G., Yeates, T. O. *et al.* (1988b) Structure of the reaction centre from *Rhodopseudomonas sphaeroides* R-26: Protein-cofactor (quinones and Fe^{2+}) interactions. *Proc. Natl. Acad. Sci. USA*, **85**:8487-8491.

Andersson, P.O., Gillbro, T., Ferguson, L. and Cogdell, R.J.(1991) Absorption spectral shifts of carotenoids related to medium polarizability. *Photochem. Photobiol.*, **54**:353-360.

Andersson, P.O., Gillbro, T., Asato, A.E. and Liu, R.S.H. (1992) Dual singlet-state emission in a series of mini-carotenes. *J. Lumin.*, **51**:11-20.

Andrews, J. R. and Hudson, B.S. (1978) Environment effects on radiative rate constants with application to linear polyenes. *J. Chem. Phys.*, **68**:4587-4594

Angerhofer, A., Cogdell, R.J. and Hipkins, M.F. (1986) A spectral characterization of the light-harvesting pigment-protein complexes from *Rhodopseudomonas acidophila*. *Biochim. Biophys. Acta*, **848**:333-341.

Arnoux, B., Ducruix, A., Reiss-Husson, F., *et al.* (1989) Structure of spheroidene in the photosynthetic reaction centre from *Rhodobacter sphaeroides*. *FEBS Lett.*, **258**:47-50.

Aust, V., Angerhofer, A., Ullrich, J. *et al.* (1991) ADMR of carotenoid triplet states in bacterial photosynthetic antenna and reaction centre complexes. *Chem. Phys. Lett.* **181**:213-221.

Barrow, G.M. (1973) *Physical Chemistry*, 3rd Ed. McGraw Hill, New York.

Bart, J.C.J. and MacGillavry, C.H. (1968a) The crystal and molecular structure of canthaxanthin. *Acta Crystallogr.*, **B24**:1587-1606.

Bart, J.C.J. and MacGillavry, C.H. (1968b) The crystal structure of 15,15′-dehydrocanthaxanthin. *Acta Crystallogr.*, **B24**:1569-1586.

Basu, S. (1964) Theory of solvent effects on molecular electronic spectra. *Adv. Quantum Chem.*, **1**:145-169.

Beddard, G.S., Davidson, R.S. and Thethewey, K.R. (1977) Quenching of chlorophyll fluorescence by ß-carotene. *Nature*, **267**:373-374.

Bensasson, R., Land, E.J. and Maudinas, B. (1976) Triplet states of carotenoids from photosynthetic bacteria studied by nanosecond ultraviolet and electron pulse irradiation. *Photochem. Photobiol.*, **23**:189-193.

Birge, R.R. (1986) Two photon spectroscopy of protein-bound chromophores. *Acc. Chem. Res.*, **19**:138-146.

Birge, R., Bennett, J.A., Fang, H.L.-B. and Leroi, G.E. (1978) The two photon spectroscopy of all-*trans* retinal and related polyenes. In: Zewail, A. H., (ed.), *Springer Series in Chem. Phys.*, Vol. 3. Springer, Berlin, pp. 347-354.

Blankenship, R.E., Trost, J.T. and Mancino, L.J. (1988) In Breton, J. and Vermeglio, A., (eds.)*The Photosynthetic Bacterial Reaction Centre: Structure and Dynamics*. New York: Plenum Press, pp. 119-127.

Bolt, J.D., Hunter, C.N., Niederman, R.A. and Sauer, K. (1981) Linear and circular dichroism and fluorescence polarization of the B875 light-harvesting bacteriochlorophyll-protein complex from *Rhodopseudomonas sphaeroides*. *Photochem. Photobiol.*, **34**:653-656.

Bondarev, S. L., Bachilo, S. M., Dvornikov, S. S. and Tikhomirov, S. A. (1989) $S_2 \rightarrow S_0$ fluorescence and transient $S_n \leftarrow S_1$ absorption of all-*trans*-ß-carotene in solid and liquid solutions. *J. Photochem. Photobiol.*, **46A**:315-322.

Borland, C.F., McGarvey, D.J., Truscott, T.G. *et al.* (1987) Photophysical studies of bacteriochlorophyll *a* and bacteriopheophytin *a*–singlet oxygen generation. *J. Photochem. Photobiol.*, **1B**:93-101.

Boucher, F. and Gingras, G. (1984) Spectral evidence for photo-induced isomerisation of carotenoids in bacterial photoreaction centre. *Photochem. Photobiol.*, **40**:277-281.

Boucher, F., Rest, M. van der and Gingras, G. (1977) Structure and function of carotenoids in the photoreaction centre from *Rhodospirillum rubrum*. *Biochim. Biophys. Acta*,

461:339-357.

Bouwman, W. G., Jones, A. C., Phillips, D. *et al.* (1990) Fluorescence of gaseous tetraenes and pentaenes. *J. Phys. Chem.*, **94**:7429-7434.

Breton, J. (1985) Orientation of the chromophores in the reaction centre of *Rhodopseudomonas viridis*. Comparison of low temperature linear dichroism spectra with a model derived from X-ray crystallography. *Biochim. Biophys. Acta*, **810**:235-245.

Breton, J. and Geacintov, N.E. (1979) Chlorophyll orientation and energy migration in the p/s membrane. *Ciba Found. Symp.*, **61**:217-236.

Breton, J. and Nabedryk, E. (1987) Pigment and protein organization in reaction centre and antenna complexes. In: Barber, J. (ed.), *The Light Reactions*. Elsevier Science Publishers, Amsterdam, pp. 159-195.

Breton, J. and Verméglio, A. (1982) Orientation of photosynthetic pigments *in vivo*. In: Govindjee, (ed.), *Photosynthesis. Energy Conversion by Plants and Bacteria*. Vol 1. Academic Press, New York, pp. 153-194.

Burkey, T.J., Majewski, M. and Griller, D. (1986) Heats of formation of radicals and molecules by a photoacoustic technique. *J. Am. Chem. Soc.*, **108**:2218-2221.

Carrington, A. and McLachlan, A.D. (1967) *Introduction to Magnetic Resonance*. Harper and Row, New York.

Chachety, C., Gust, D., Moore, T.A. *et al.* (1984) NMR spectra of carotenoporphyrins. Computer assisted conformational analysis. *Org. Mag. Res.*, **22**: 39-45.

Chadwick, B.W. and Frank, H.A. (1986) Electron-spin resonance studies of carotenoids incorporated into reaction centres of *Rhodobacter sphaeroides* R-26.1. *Biochim. Biophys. Acta*, **851**:257- 266.

Chadwick, B.W., Zhang, C., Cogdell, R.J. and Frank, H.A. (1987) The effects of lithium dodecyl sulfate and sodium borohydride on the absorption spectrum of the B800–850 light harvesting complex from *Rhodopseudomonas acidophila* 7750. *Biochim. Biophys. Acta*, **893**:444-457.

Chance, B. (1938) Oxygen-linked absorbancy changes in photosynthetic cells. *Brookhaven Symp. Biol.*, **11**: 74-86.

Chance, B. and Smith, L. (1955) Respiratory pigments of *Rhodospirillum rubrum*. *Nature*, **175**:803-806.

Chantrell, S.J., McAuliffe, C.A., Munn, R.W. *et al.* (1977) Excited states of protoporphyrin IX dimethyl ester: Reaction of the triplet with carotenoids. *J. Chem. Soc. Farad. Trans. I*, **73**:858-865.

Chauvet, J.-P., Viovy, R., Land, E.J. *et al.* (1979) One-electron oxidation of carotene and electron transfers involving carotene cations and chlorophyll pigments in micelles, *J. Phys. Chem.*, **87**:592-601.

Claes, H. (1960) Interaction between chlorophyll and carotenes with different chromophoric groups. *Biochem. Biophys. Res. Commun.*, **3**:585-590.

Clarke, R.H. (1982) *Triplet State ODMR Spectroscopy*. John Wiley and Sons, New York.

Clarke, R.H. and Frank, H.A. (1976) Triplet state radiationless transitions in polycyclic hydrocarbons. *J. Chem. Phys.*, **65**:39-47.

Cogdell, R.J. and Frank, H.A. (1987) How carotenoids function in photosynthetic bacteria. *Biochim. Biophys. Acta*, **895**:63-79.

Cogdell, R.J. and Scheer, H. (1985) Circular dichroism of light-harvesting complexes from purple photosynthetic bacteria. *Photochem. Photobiol.*, **42**:669-678.

Cogdell, R.J. and Thornber, J.P. (1980) Light-harvesting pigment–protein complexes of purple photosynthetic bacteria. *FEBS Lett.*, **122**:1-8.

Cogdell, R.J., Monger, T.G., and Parson, W.W. (1975) Carotenoid triplet states in reaction centres from *Rhodopseudomonas sphaeroides* and *Rhodospirillum rubrum*. *Biochim. Biophys. Acta*, **408**:189-199.

Cogdell, R.J., Parson, W.W., and Kerr, M.A. (1976) The type, amount, location, and energy transfer properties of the carotenoid in reaction centres from *Rhodopseudomonas sphaeroides*. *Biochim. Biophys. Acta*, **430**:83-93.

Cogdell, R.J., Celis, S., Celis, H. and Crofts, A.R. (1977) Reaction centre carotenoid band shifts. *FEBS Lett.*, **80**:190-193.

Cogdell, R.J., Hipkins, M.F., MacDonald, W. and Truscott, T.G. (1981) Energy transfer between the carotenoid and bacteriochlorophyll within the B800–850 light-harvesting pigment–protein complex of *Rps. sphaeroides*. *Biochim. Biophys. Acta*, **634**:191-202.

Cogdell, R.J., Land, E.J. and Truscott, T.G. (1983) The triplet extinction coefficients of some bacterial carotenoids. *Photochem. Photobiol.*, **38**: 723-725.

Cogdell, R.J., Andersson, P.O. and Gillbro, T. (1992) Carotenoid singlet-states and their involvement in photosynthetic light-harvesting. *J. Photochem. Photobiol.* **B15**:105-112.

Cosgrove, S.A., Guite, M.A., Burnell, T.B. and Christensen, R.L. (1990) Electron relaxation in long polyenes. *J. Phys. Chem.*, **94**:8118-8124.

D'Amico, K.L., Manos, C. and Christensen, R.L. (1980) Electronic energy levels in a homologous series of unsubstituted linear polyenes. *J. Am. Chem. Soc.*, **102**:1777-1782.

Dallinger, R.F., Guanci, J.J., Woodruff, W.H. and Rodgers, M.A. (1979) Vibrational spectroscopy of the electronically excited state: pulse radiolysis/time-resolved resonance Raman study of triplet ß-carotene. *J. Am. Chem. Soc.*, **101**:1355-1357.

Dallinger, R.F., Farquharson, S., Woodruff, W.H. and Rodgers, M.A. (1981a) Vibrational spectroscopy of the electronically excited state. 4. Nanosecond and picosecond time-resolved resonance Raman spectroscopy of carotenoid excited states. *J. Am. Chem. Soc.*, **103**:7433-7440.

Dallinger, R.F., Woodruff, W.H. and Rodgers, M.A. (1981b) The lifetime of the excited singlet-state of ß-carotene: Consequences to photosynthetic light harvesting. *Photochem. Photobiol.*, **33**:275-277.

Davidson, E. and Cogdell, R.J. (1981) Reconstitution of carotenoids into the light-harvesting pigment–protein complex from the carotenoidless mutant of *Rhodopseudomonas sphaeroides* R-26. *Biochim. Biophys. Acta*, **635**:295-303.

DeCoster, B., Christensen, R. L., Gebhard, R. *et al.* (1992) Low-lying electronic states of carotenoids. *Biochim. Biophys. Acta*, **1102**:107-114.

DeGroot, H., Gebhard, R., van der Hoef, K. *et al.* (1990) The structure of the carotenoid in reaction centres of photosynthetic bacteria: A solid-state magic-angle sample spinning NMR investigation of spheroidene reconstituted into *Rhodobacter sphaeroides* R-2. *Biophys. J.*, **57**:567a.

Deisenhofer, J. and Michel, H. (1988) The crystal structure of the photosynthetic reaction centre from *Rhodopseudomonas viridis*. In: J. Breton and Vermeglio, A. (eds.), *The Photosynthetic Bacterial Reaction Centre*, NATO ASI Series A: Life Sciences. Plenum, New York, 149:1-3.

Deisenhofer, J. and Michel, H. (1989a) The photosynthetic reaction centre from the purple bacterium *Rhodopseudomonas viridis*. *Science*, **245**:1463-1473.

Deisenhofer, J. and Michel H. (1989b) The photosynthetic reaction centre from the purple bacterium *Rhodopseudomonas sphaeroides* and *Rhodospirillum rubrum*. *Chemica Scripta*, **29**:205-220.

Deisenhofer, J., Epp, O., Miki, K. *et al.* (1984) X-Ray structure analysis of a membrane protein complex. Electron density map at 3 angstrom resolution and a model of the chromophores of the photosynthetic reaction centre from *Rhodopseudomonas viridis*. *J. Mol. Biol.*, **180**:385-398.

Deisenhofer, J., Epp, O., Miki, K. *et al.* (1985) Structure of the protein subunits in the photosynthetic reaction centre of *Rhodopseudomonas viridis* at 3 angstrom. resolution. *Nature*, **318**:618-624.

Demmig-Adams, B. (1990) Carotenoids and photoprotection in plants: A role for the xanthophyll zeaxanthin. *Biochim. Biophys. Acta*, **1020**:1-24.

Demmig-Adams, B., Winter, K., Kruger, A. and Czygan, F.-C. (1989) Zeaxanthin and the heat dissipation of excess light energy in *Nerium oleander* exposed to a combination of high light and water stress. *Plant Physiol.*, **87**:17-24.

Dexter, D.L. (1953) A theory of sensitised luminescence in solids. *J. Chem. Phys.*, **21**:836-

860.

Ditson, S.L., Davis, R.C. and Pearlstein, R.M. (1984) Relative enrichment of P-870 in photosynthetic reaction centres treated with sodium borohydride. *Biochim. Biophys. Acta*, **766**:623-629.

Dorssen, R.J. van, Breton, J., Plijter, J.J. *et al.* (1987) Spectroscopic properties of the reaction centre and of the 47kDa chlorophyll protein of photosystem II. *Biochim. Biophys. Acta*, **893**:267-274.

Duysens, L.N.M. (1954) Reversible changes in the absorption spectrum of *Chlorella* upon irradiation. *Science*, **120**:353-354.

Evans, D.F. (1960) Magnetic perturbation of single-triplet transitions. Part IV. Unsaturated compounds. *J. Chem. Soc.*, **1960**:1735-1745.

Evans, D.F. (1961) Magnetic perturbation of singlet-triplet transitions. Part VI. Octa-1,3,5,7-tetraene. *J. Chem. Soc.*, **1961**:2566-2569.

Evans, M.B., Cogdell, R.J., and Britton, G. (1988) Determination of the bacteriochlorophyll: carotenoid ratios of the B890 antenna complex of *Rhodospirillum rubrum* and the B800–850 complex of *Rhodobacter sphaeroides*. *Biochim. Biophys. Acta*, **935**:292-298.

Feher, G., Allen, J.P., Okamura, M.Y. and Rees, D.C. (1989) Structure and function of bacterial photosynthetic reaction centres. *Nature*, **339**:111-116.

Foote, C.S. (1968a) Photosensitized oxygenations and the role of singlet O_2. *Acc. Chem. Res.*, **1**:104-110.

Foote, C.S. (1968b) Mechanisms of photosensitized oxidation. *Science*, **162**:963-970.

Foote, C.S. (1976) In: Pryor, W.A. (ed.), *Free Radicals and Biological Systems*. Academic Press, New York, pp. 85-133.

Foote, C.S., Chang, Y.C. and Denny, R.W. (1970a) Chemistry of singlet oxygen. XI. *cis-trans* isomerisation of carotenoids by singlet oxygen and a probable quenching mechanism. *J. Am. Chem. Soc.*, **92**:5218-5219.

Foote, C. S., Chang, Y. C. and Denny, R. W. (1970b) Chemistry of singlet oxygen. X. Carotenoid quenching parallels biological protection. *J. Am. Chem. Soc.*, **92**:5216-5218.

Förster, Th. (1948) Intermolecular energy transfer and fluorescence. *Ann. Phys.*, **2**:55-75.

Förster, Th. (1965) In: Sinanoglu, O. (ed.), *Modern Quantum Chemistry*, Part III: Action of light and organic crystals. Academic Press, New York, pp. 93-137.

Förster, Th. (1968) Intermolecular energy transfer and fluorescence. *Ann. Physik. Leipzig*, **2**:55-75.

Frank, H.A. (1992) Electron paramagnetic resonance studies of carotenoids. *Methods Enzymol.*, **213**:305-312.

Frank, H.A. (1993) Carotenoids in photosynthetic bacterial reaction centres: Structure, spectroscopy and photochemistry. In: Norris, J. R. and Deisenhofer, J. (eds.), *The Photosynthetic Reaction Centre*. (In press).

Frank, H.A. and Violette, C.A. (1989) Monomeric bacteriochlorophyll is required for triplet energy transfer between the primary donor and the carotenoid in photosynthetic bacterial reaction centres. *Biochim. Biophys. Acta*, **976**:222- 232.

Frank, H.A., McLean, M.B. and Sauer, K. (1979) Triplet states in photosystem I of spinach chloroplasts and subchloroplast particles. *Proc. Natl. Acad. Sci. USA*, **76**:5124-5128.

Frank, H.A., Bolt, J.D., De B. Costa, S.M. and Sauer, K. (1980) Electron paramagnetic resonance detection of carotenoid triplet states. *J. Am. Chem. Soc.*, **102**:4893-4898.

Frank, H.A., Machnicki, J. and Felber, M. (1982) Carotenoid triplet states in photosynthetic bacteria. *Photochem. Photobiol.*, **35**:713-718.

Frank, H.A., Machnicki, J. and Friesner, R. (1983) Energy transfer between the primary donor bacterio-chlorophyll and carotenoids in *Rhodopseudomonas sphaeroides*. *Photochem. Photobiol.*, **38**:451-456.

Frank, H.A., Machnicki, J. and Toppo, P. (1984) The orientation of the principal magnetic axes of the carotenoid in *Rhodopseudomonas sphaeroides* wild type. *Photochem. Photobiol.*, **39**:429-432.

Frank, H.A., Chadwick, B.W., Taremi, S. *et al.* (1986) Singlet and triplet absorption spectra of carotenoids bound in the reaction centres of *Rhodopseudomonas sphaeroides* R-26. *FEBS Lett.*, **203**:157-163.

Frank, H.A., Chadwick, B.W., Oh, J.J. *et al.* (1987a) Triplet-triplet energy transfer in B800–850 light-harvesting complexes of photosynthetic bacteria and synthetic carotenoporphyrin molecules investigated by electron spin resonance. *Biochim. Biophys. Acta*, **892**:253-263.

Frank, H.A., Taremi, S.S. and Knox, J.R. (1987b) Crystallization and preliminary X-ray and optical spectroscopic characterization of the photochemical reaction centre from *Rhodobacter sphaeroides* strain 2.4.1. J. *Mol. Biol.*, **198**:139-141.

Frank, H.A., Violette, C.A., Taremi, S.S. and Budil, D. (1989a) Linear dichroism of single crystals of the reaction centre from *Rhodobacter sphaeroides* wild type strain 2.4.1. *Photosynth. Res.*, **21**:107-116.

Frank, H.A., Hansson, Ö. and Mathis, P. (1989b) Electron paramagnetic resonance and optical changes of the photosystem II reaction centre complex produced by low temperature illumination. *Photosynth. Res.*, **20**:279-289.

Frank, H.A., Aldema, M.A., Violette, C.A. and Parot, P.H. (1991a) Low temperature polarised absorption microspectroscopy of single crystals of the reaction centre from *Rhodobacter sphaeroides* wild type strain 2.4.1. *Photochem. Photobiol.*, **54**:151-155.

Frick, J., Von Schütz, J.U., Wolf, H.C. and Kothe, G. (1990) First detection of the (nonphosphorescent) triplet state in single crystals of ß-carotene. *Mol. Cryst. Liq. Cryst.*, **183**:269-272.

Fujiwara, M., Hayashi, H., Tasumi, M. *et al.* (1987) Structural studies on a Photosystem II reaction centre complex consisting of D-1 and D-2 polypeptides and cytochrome *b*-559 by resonance Raman spectroscopy and high-performance liquid chromatography. *Chem Letts.* **1987**:2005-2008.

Gagliano, A.G., Breton, J. and Geacintov, N.E. (1986) Electric linear dichroism of chromatophores, pigment-protein and reaction centre complexes derived from *Rhodopseudomonas sphaeroides*. *Photobiochem. Photobiophys.*, **10**:213-221.

Gebhard, R., Hoef, K. van der, Violette, C.A. *et al.* (1991) ^{13}C MAS NMR evidence for a 15,15′-*Z* configuration of the spheroidene chromophore in the *Rhodobacter sphaeroides* reaction centre. *Pure Appl. Chem.*, **63**:115-122.

Ghanotakis, D. F., dePaula, J. C., Demetriou, D. M. *et al.* (1989) Isolation and characterization of the 47 kDa protein and the D_1-D_2 cytochrome *b*-559 complex. *Biochim. Biophys. Acta*, **974**:44-53.

Gillbro, T. and Cogdell, R. J. (1989) Carotenoid fluorescence. *Chem. Phys. Lett.*, **158**:312-316.

Gillbro, T., Cogdell, R. and Sundström, V. (1988) Energy transfer from carotenoid to bacteriochlorophyll *a* in the B800-820 antenna complexes from *Rhodopseudomonas acidophila* strain 7050. *FEBS Lett.*, **235**:169-172.

Goodwin, T. W., Land, D. G. and Sissins, M. E. (1956) Studies in carotenogenesis. 23. The nature of the carotenoids in the photosynthetic bacterium *Rhodopseudomonas sphaeroides* (Athiorhodaceae). *Biochem. J.*, **64**:486-492.

Gottfried, D.S., Steffen, M.A. and Boxer, S.G. (1991a) Stark effect spectroscopy of carotenoids in photosynthetic antenna and reaction centre complexes. *Biochim. Biophys. Acta*, **1059**:76-90.

Gottfried, D.S., Steffen, M.A. and Boxer, S.G. (1991b) Large protein-induced dipoles for a symmetric carotenoid in a photosynthetic antenna complex. *Science*, **251**:662-665.

Grant, J.L., Kramer, V.J., Ding, R. and Kispert, L.D. (1988) Carotenoid cation radicals: Electrochemical, optical and EPR study. *J. Am. Chem. Soc.*, **110**:2151-2157.

Granville, M.F., Holtom, G.A. and Kohler, B.E. (1979) Experimental confirmation of the dipole forbidden character of the lowest excited singlet-state in 1,3,5,7-octatetraene. *J. Chem. Phys.*, **70**:593-594.

Griffiths, M., Sistrom, W.R., Cohen-Bazire, G. and Stanier, R.Y. (1955) Functions of carotenoids in photosynthesis. *Nature*, **176**:1211-1214.

Grondelle, R. van (1985) Excitation energy transfer, trapping and annihilation in photosynthetic systems. *Biochim. Biophys. Acta*, **811**:147-195.
Grondelle, R. van, Kramer, H.J.M. and Rijgersberg, C.P. (1982) Energy transfer in the B800–850 carotenoid light-harvesting complex of various mutants of *Rhodopseudomonas sphaeroides* and of *Rhodopseudomonas capsulata*. *Biochim. Biophys. Acta*, **682**:208-215.
Gust, D. and Moore, T.A. (1991) Mimicking photosynthetic electron and energy transfer. In: Volman, D., Hammond, G., and Neckers, D. (eds.), *Advances in Photochemistry*, Vol. 16. John Wiley & Sons, New York, pp. 1-63.
Haley, L.V. and Koningstein, J.A. (1983) Space and time-resolved resonance-enhanced vibrational Raman spectroscopy from femtosecond-lived singlet excited state of ß-carotene. *Chem. Phys.*, **77**:1-9.
Hammond, G.S., Saltiel, J., Lamola, A.A. *et al.* (1964) Mechanisms of photochemical reactions in solution. XXII. Photochemical *cis-trans* isomerisation. *J. Am. Chem. Soc.*, **86**:3197-3217.
Hashimoto, H. and Koyama, Y. (1990) The 2-Ag–state of a carotenoid bound to spinach chloroplasts as revealed by picosecond transient Raman spectroscopy. *Biochem. Biophys. Acta*, **1017**:181-186.
Hashimoto, H., Koyama, Y., Ichimura, K. and Kobayashi, T. (1989) Time-resolved absorption spectroscopy of the triplet state produced from the all-*trans*, 7-*cis*, 9-*cis*, 13-*cis*, and 15-*cis* isomers of ß-carotene. *Chem. Phys. Lett.*, **162**:517-522.
Hayashi, H., Noguchi, T. and Tasumi, M. (1989) Studies on the interrelationship among the intensity of a Raman marker band of carotenoids, polyene chain structure, and efficiency of the energy transfer from carotenoids to bacteriochlorophyll in photosynthetic bacteria. *Photochem. Photobiol.*, **49**:337-343.
Hoff, A. J. (1979) Application of ESR in photosynthesis. Phys. Rep., 54:75-200.
Holten, D., Windsor, M.W., Parson, W.W. and Thornber, J.P. (1978) Primary photochemical processes in isolated reaction centres of *Rhodopseudomonas viridis*. *Biochim. Biophys. Acta*, **501**:112-126.
Hudson, B.S. and Kohler, B.E. (1974) Linear polyene electronic structure and spectroscopy. *Annu. Rev. Phys. Chem.*, **25**:437-460.
Hudson, B.S., Kohler, B.E. and Shulten, K. (1982) Linear polyene electronic structure and potential surfaces. In: Lim, E. C., (ed.), *Excited States*, Vol. 6. Academic Press, New York, pp. 22-95.
Iwata, K., Hayashi, H. and Tasumi, M. (1985) Resonance Raman studies of the conformations of all-*trans* carotenoids in light-harvesting systems of photosynthetic bacteria. *Biochim. Biophys. Acta*, **810**:269-273.
Jackson, J.B. and Crofts, A.R. (1969) The high energy state in chromatophores from *Rps. sphaeroides*. *FEBS Lett.*, **4**:185-189.
Jackson, J.B. and Dutton, P.L. (1973) The kinetic and redox potentiometric resolution of the carotenoid shifts in *Rps. sphaeroides* chromatophores: their relationship to electric field alterations in electron transport. *Biochim. Biophys. Acta*, **325**:102-113.
Jensen, N.-H., Wilbrandt, R., Pagsberg, P.B. *et al.* (1980) Time resolved resonance Raman spectroscopy: The excited triplet state of all-*trans*-beta-carotene. *J. Am. Chem. Soc.*, **102**:7441-7444.
Jensen, N.-H., Nielsen, A.B. and Wilbrandt, R. (1982) Chlorophyll *a*-sensitized *trans-cis* photoisomerisation of all-*trans*-ß-carotene. *J. Am. Chem. Soc.*, **104**:6117-6119.
Kakitani, T., Honig, B. and Crofts, A.R. (1982) Theoretical studies of the electrochemic response of carotenoids in photosynthetic membranes. *Biophys. J.*, **39**:57-63.
Katoh, T., Nagashima, U. and Mimuro, M. (1991) Fluorescence properties of the allenic carotenoid fucoxanthin: Implication for energy transfer in photosynthetic systems. *Photosyn. Res.*, **27**:221-226.
Kingma, H. (1983) Redox states of the reaction centre in relation to energy transfer and mechanism of carotenoid triplet formation in photosynthetic bacteria. Ph.D. Thesis,

University of Leiden.
Kirmaier, C. and Holten, D. (1987) Primary photochemistry of reaction centres from the photosynthetic purple bacteria. *Photosynth. Res.*, **13**:225-260.
Kito, M., Yamashita, J. and Koyama, Y. (1983) Configurations of the carotenoid in carotenoproteins from *Rhodospirillum rubrum*. A resonance Raman study. *Photobiochem. Photobiophys.*, **5**:209-217.
Knoop, F.W.E. and Oosterhoff, L.J. (1973) Low-energy electron impact excitation of 1,3,5-*trans*-hexatriene. *Chem. Phys. Lett.*, **22**:247-248.
Kohler, B.E. (1991) Electronic properties of linear polyenes. In: Bredas, J. L. and Silbey, R. (eds.), *Conjugated polymers: The Novel Science and Technology of Conducting and Non-linear Optically Active Materials*. Kluwer Academic Publishers, Dodrecht.
Kolaczkowski, S.V. (1989) On the mechanism of triplet energy transfer from the primary donor to spheroidene in photosynthetic reaction centres from *Rhodobacter sphaeroides* 2.4.1. Ph.D. Thesis, Brown University.
Komiya, H., Yeates, T.O., Rees, D.C. *et al.* (1988) Structure of the reaction centre from *Rhodopseudomonas sphaeroides* R-26 and 2.4.1: Symmetry relations and sequence comparisons between different species. *Proc. Natl. Acad. Sci. USA.*, **85**:9012-9016.
Koyama, Y. (1991) Structures and functions of carotenoids in photosynthetic systems. *J. Photochem. Photobiol.*, **9B**:265-280.
Koyama, Y., Kito, M, Takii, T. *et al.* (1982) Configuration of the carotenoid in the reaction centres of photosynthetic bacteria. Comparison of the resonance Raman spectrum of the reaction centre of *Rhodopseudomonas sphaeroides* G1C with those of *cis-trans* isomers of beta-carotene. *Biochim. Biophys. Acta*, **680**:109-118.
Koyama, Y., Takii, T., Saiki, K. and Tsukida, K. (1983) Configuration of the carotenoid in the reaction centres of photosynthetic bacteria-2. Comparison of the resonance Raman lines of the reaction centres with those of 14 different *cis-trans* isomers of ß-carotene. *Photobiochem. Photobiophys.*, **5**:139-150.
Koyama, Y., Takatsuka, I., Kanaji, M. *et al.* (1990) Configurations of carotenoids in the reaction centre and the light-harvesting complex of *Rhodospirillum rubrum*: Natural selection of carotenoid configurations by pigment–protein complexes. *Photochem. Photobiol.*, **51**:119-128.
Kramer, H.J.M., Grondelle, R. van , Hunter, C.N. *et al.* (1984) Pigment organization of the B800–850 antenna complex of *Rhodopseudomonas sphaeroides. Biochim. Biophys. Acta*, **765**:156-165.
Krasnovskii, A.A., Lebedev, N.N. and Litvin, F.F. (1974) Spectral characteristics of phosphorescence of chlorophylls and pheophytins *a* and *b*. *Dokl. Akad. Nauk SSSR*, **216**:1406-1409.
Krinsky, N.I. (1966) In: Goodwin, T.W. (ed.), *Biochemistry of Chloroplasts*. Academic Press, New York, pp. 423-430.
Krinsky, N.I. (1968) The protective function of carotenoid pigments. In: Giese, A.C. (ed.), *Photophysiology*, Vol. III. Academic Press, New York, pp. 123-195.
Krinsky, N.I. (1971) Function. In: Isler, O. (ed.), *Carotenoids*. Birkhäuser Verlag, Basel, pp. 669-716.
Krinsky, N.I. (1978) Non-photosynthetic function of carotenoids. *Phil. Trans. Roy. Soc.*, **B284**:581-590.
Kuki, M., Hashimoto, H and Koyama, Y. (1990) The $2^1A_g^-$-state of a carotenoid bound to the chromatophore membrane of *Rhodobacter sphaeroides* 2.4.1 as revealed by transient resonance Raman spectroscopy. *Chem. Phys. Lett.*, **165**:417-422.
Lafferty, J., Land, E.J. and Truscott, T.G. (1978) Electron-transfer reactions involving chlorophyll *a* and carotenoids. *J. Chem. Soc. Faraday I*, **74**:2760-2762.
Land, E.J., Sykes, A. and Truscott, T.G. (1971) The *in vitro* photochemistry of biological molecules-II. The triplet states of ß-carotene and lycopene excited by pulse radiolysis. *Photochem. Photobiol.*, **13**:311-320.
LeRosen, A.L. and Reid, C.E. (1952) An investigation of certain solvent effects in absorption

spectra. *J. Chem. Phys.*, **20**:233-236.

Liaaen-Jensen, S. and Jensen, A. (1971) Quantitative determination of carotenoids in photosynthetic tissues. *Meth. Enzymol.*, **23**:586-602.

Liptay, W. (1969) Elektrochromie-solvatochromie. *Angew. Chem.*, **81**:195-232 (Int. Ed. Engl. **8**:177-188).

Longuet-Higgins, H.C. and Pople, J.A. (1957) Electronic spectral shifts of non-polar molecules in non-polar solvents. *J. Chem. Phys.*, **27**:192-194.

Lous, E.K. (1988) Interactions between pigments in photosynthetic protein complexes. An optically-detected magnetic resonance and magnetic field effect study. Ph.D. Thesis. University of Leiden.

Lozano, R.M., Fernández-Cabrera, C. and Ramírez, J.M. (1990) The contribution of the carotenoid to the visible circular dichroism of the light-harvesting antenna of *Rhodospirillum rubrum. Biochem. J.*, **270**:469-472.

Lutz, M. (1984) Resonance Raman studies in photosynthesis. In: Clark, R. J. H. and Hester, R. E. (eds.), *Advances in Infrared and Raman Spectroscopy.* Vol. 11, Wiley Heyden, Amsterdam, pp. 211-300.

Lutz, M., Agalidis, I., Hervo, G. *et al.* (1978) On the state of carotenoids bound to reaction centres of photosynthetic bacteria: A resonance Raman study. *Biochim. Biophys. Acta,* **503**:287-303.

Lutz, M., Chinsky, L. and Turpin, P. (1982) Triplet states of carotenoids bound to reaction centres of photosynthetic bacteria: Time-resolved resonance Raman spectroscopy. *Photochem. Photobiol.*, **36**:503-515.

Lutz, M., Szponarski, W., Berger, G. *et al.* (1987) The stereoisomerisation of bacterial, reaction-centre-bound carotenoids revisited: an electronic absorption, resonance Raman and -NMR study. Biochim. Biophys. Acta, 894:423- 433.

Maroti, P., Kirmaier, C., Wraight, C. *et al.* (1985) Photochemistry and electron transfer in borohydride-treated photosynthetic reaction centres. *Biochim. Biophys. Acta,* **810**:132-139.

Mathis, P. (1969) Triplet-triplet energy transfer from chlorophyll *a* to carotenoids in solution and in chloroplasts In: Metzner, H. (ed.), *Progress in Photosynthesis Research,* Vol. 2. Tübingen, Germany, pp. 881-822.

Mathis, P. and Kleo, J. (1973) The triplet state of ß-carotene and of analog polyenes of different length. *Photochem. Photobiol.,* **18**:343-346.

Mathis, P. and Verméglio, A. (1972) Formes transitoires des caroténoides; état triplet et radical-cation. *Photochem. Photobiol.*, **15**:157-164.

McGann, W.J. and Frank, H.A. (1983) Magnetic field effects on the fluorescence of mutant strains of *Rhodopseudomonas capsulata. Biochim. Biophys. Acta*, **725**:178-189.

McGann, W.J. and Frank, H.A. (1985a) Transient electron spin resonance spectroscopy of the carotenoid triplet state in *Rhodopseudomonas sphaeroides* wild type. *Chem. Phys. Lett.*, **121**:253-261.

McGann, W.J. and Frank, H.A. (1985b) Magnetophotoselection of *Rhodopseudomonas sphaeroides* wild type reaction centres. *Biochim. Biophys. Acta*, **807**:101-109.

McRae, E.G. (1957) Theory of solvent effects on molecular electronic spectra. Frequency shifts. *J. Phys. Chem.*, **61**:562-572.

McVie, J., Sinclair, R.S., Tait, D. *et al.* (1979) Electron transfer reactions involving porphyrins and carotenoids. *J. Chem. Soc. Faraday I*, **75**:2869-2872.

Michel, H. (1982) Three-dimensional crystals of a membrane protein complex. The photosynthetic reaction centre from *Rhodopseudomonas viridis. J. Molec Biol.*, **158**:567-572.

Michel, H., Epp, O. and Deisenhofer, J. (1986) Pigment-protein interactions in the photosynthetic reaction centre from *Rhodopseudomonas viridis. EMBO J.*, **5**:2445-2451.

Mimuro, M., Nishimura, Y., Yamazaki, I. *et al.* (1991) Fluorescence properties of the allenic carotenoid fucoxanthin: Analysis of the effect of keto carbonyl group by using a model compound, all-*trans*-beta-apo-8´-carotenal. *J. Luminesc.*, **50**:1-10.

Mimuro, M., Nagashima, U., Takaichi, S., *et al.* (1992) A special molecular structure in carotenoids is necessary for the *in vivo* energy transfer function in the algal photosynthetic pigment system. *Biochim. Biophys. Acta,* **1098**:271-274

Monger, T.G., Cogdell, R.J. and Parson, W.W. (1976) Triplet states of bacteriochlorophyll and carotenoids in chromatophores of photosynthetic bacteria. *Biochim. Biophys. Acta,* **449**:136- 153.

Moore, T.A., Gust, D. and Moore, A.L. (1990) The function of carotenoid pigments in photosynthesis and their possible role in the evolution of higher plants In: Krinsky, N., Mathews-Roth, M.M. and Taylor, R.F. (eds.), *Carotenoids: Chemistry and Biology.* Plenum Press, New York, pp. 223-228.

Mosher, O.A., Flicker, W.M. and Kuppermann, A. (1973) Triplet states in 1,3-butadiene. *Chem. Phys. Lett.*, **19**:332-333.

Nakayama, T.O.M. (1958) The carotenoids of *Rhodopseudomonas*. II. A comparative study of mutants and the wild type. *Arch. Biochem. Biophys.*, **75**:356-360.

Newell, W.R., Amerongen, H. van, Barber, J. and Van Grondelle, R. (1991) Spectroscopic characterisation of the reaction centre of Photosystem II using polarised light: Evidence for ß-carotene excitons in PS II reaction centres. *Biochim. Biophys. Acta,* **1057**:232-238.

Noguchi, T., Kolaczkowski, S., Arbour, C. *et al.* (1989) Resonance Raman spectrum of the excited 2^1A_g state of ß-carotene. *Photochem. Photobiol.*, **50**:603-609.

Noguchi, T., Hayashi, H, and Tasumi, T. (1990) Factors controlling the efficiency of energy transfer from carotenoids to bacteriochlorophyll in purple photosynthetic bacteria. *Biochim. Biophys. Acta,* **1017**:280-290.

Parson, W.W. (1967) Flash-induced absorbance changes in *Rhodospirillum rubrum* chromatophores. *Biochim. Biophys. Acta,* **131**: 154-172.

Parson, W.W. and Cogdell, R.J. (1975) The primary photochemical reaction of bacterial photosynthesis. *Biochim. Biophys. Acta,* **416**:265-278.

Parson, W.W. and Monger, T.G. (1976) Interrelationships among excited states in bacterial reaction centres. *Brookhaven Symp. Biol.*, **28**:195-212.

Raubach, R.A. and Guzzo, A.V. (1971) Singlet-triplet absorption spectrum of all-*trans* retinal. *J. Phys. Chem.*, **75**:983-984.

Reich, R. and Schmidt, S. (1972) Uber den Einblub elektrischer Felder auf das Absorptionsspektrum von Farbstoffmolekülen in Lipidschichten. I. Theorie. *Ber. Bunsen. Ges. Phys. Chem.*, **76**:589-598.

Reich, R. and Sewe, K.-U. (1977) The effect of molecular polarization on the electrochromism of carotenoids. I. The influence of a carboxylic group. *Photochem. Photobiol.*, **26**:11-17.

Robert, B. and Frank, H.A. (1988a) Effect of lithium dodecyl sulfate on B800–850 antenna complexes from *Rhodopseudomonas acidophila* 7750 : A resonance Raman study. In Scheer, H. and Schneider, S. (eds.), *Photosynthetic Light-Harvesting Systems*, Vol. I. Walter de Gruyter & Co., pp.349-353

Robert, B. and Frank, H.A. (1988b) A resonance Raman investigation of the effect of lithium dodecyl sulfate on the B800–850 light-harvesting protein of *Rhodopseudomonas acidophila* 7750. *Biochim. Biophys. Acta,* **934**:401-405.

Rodgers, M.A.J. and Bates, A.L. (1980) Kinetic and spectroscopic features of some carotenoid triplet states: Sensitisation by singlet oxygen. *Photochem. Photobiol.*, **31**:533-537

Schenck, C.C., Mathis, P. and Lutz, M. (1984) Triplet formation and triplet decay in reaction centres from the photosynthetic bacterium *Rhodopseudomonas sphaeroides*. *Photochem. Photobiol.*, **39**:407-417.

Schulten, K. and Karplus, M. (1972) On the origin of a low-lying forbidden transition in polyenes and related molecules. *Chem. Phys. Lett.*, **15**:305-309.

Schulten, K., Ohmine, I. and Karplus, M. (1976) Correlation effects in the spectra of polyenes. *J. Chem. Phys.*, **64**:4422-4441.

Scolnik, P.A., Zannoni, D. and Marrs, B.L. (1980) Spectral and functional comparisons between the carotenoids of the two antenna complexes of *Rhodopseudomonas capsulata*. *Biochim. Biophys. Acta*, **593**:230-240.

Sewe, K.-U. and Reich, R (1977) The effect of molecular polarization on the electrochromism of carotenoids. II. Lutein–chlorophyll complexes: The origin of the field-indicating absorption change at 520 nm in the membranes of photosynthesis. *Z. Naturforsch.*, **32C**:161-177.

Shreve, A.P., Trautman, J.K., Owens, T.G. and Albrecht, A.C. (1990) Two-photon excitation spectroscopy of thylakoid membranes from *Phaeodactylum tricornutum*: Evidence for an *in vivo* two-photon-allowed carotenoid state. *Chem. Phys. Lett.*, **170**:51-56.

Shreve, A.P, Trautman, J.K., Owens, T.G. and Albrecht, A.C. (1991a) Determination of the S_2 lifetime of ß-carotene. *Chem. Phys. Lett.*, **178**:89-96.

Shreve, A.P., Trautman, J.K., Frank, H.A. *et al.* (1991b) Femtosecond energy-transfer processes in the B800–850 light-harvesting complex of *Rhodobacter sphaeroides* 2.4.1. *Biochim. Biophys. Acta*, **1058**:280-288.

Shreve, A.P., Trautman, J.K., Owens, T.G. and Albrecht, A.C. (1991c) A femtosecond study of *in vivo* and *in vitro* electronic state dynamics of fucoxanthin and implications for photosynthetic carotenoid-to-chlorophyll energy transfer mechanisms. *Chem. Phys.*, **154**:171-178.

Siefermann-Harms, D. (1985) Carotenoids in photosynthesis. I. Location in photosynthetic membranes and light-harvesting function. *Biochim. Biophys. Acta*, **811**:325-355.

Sklar, L.A., Hudson, B.S., Petersen, M. and Diamond, J. (1977) Conjugated polyene fatty acids on fluorescent probes: spectroscopic characterisation. *Biochemistry*, **16**: 813-819.

Sly, W.G. (1964) The crystal structure of 15,15′-dehydro-ß-carotene. *Acta Crystallogr.*, **17**:511-528.

Snyder, R., Arvidson, E., Foote, C. *et al.* (1985) Electronic energy levels in long polyenes: $S_2 \rightarrow S_0$ emission in all-*trans*-1,2,5,7,9,11,13-tetradecaheptaene. *J. Am. Chem. Soc.*, **107**:4117-4122.

Sterling, C. (1964) Crystal structure analysis of ß-carotene. *Acta Crystallogr.*, **17**:1224-1228.

Sundström, V. and Grondelle, R. van (1991) The dynamics of excitation energy transfer in photosynthetic bacteria In: Scheer, H. (ed.), *Chlorophylls*. CRC Press, Boca Raton, pp. 1092-1124.

Symons, M. and Swysen, C. (1983) On the location of the carotenoids in the light-harvesting pigment–protein complexes of the photosynthetic bacterium *Rhodopseudomonas capsulata*. *Biochim. Biophys. Acta*, **723**:454.

Takiff, L. and Boxer, S.G. (1988a) Phosphorescence from the primary electron donor in *Rhodobacter sphaeroides* and *Rhodopseudomonas viridis* reaction centres. *Biochim. Biophys. Acta*, **932**:325-334.

Takiff, L. and Boxer, S.G. (1988b) Phosphorescence spectra of bacteriochlorophylls. *J. Am. Chem. Soc.*, **110**:4425-4426.

Tavan, P., and Schulten, K. (1979) The $2^1A_g - 1^1B_u$ energy gap in the polyenes: An extended configuration interaction study. *J. Chem. Phys.*, **70**:5407-5413.

Telfer, A., De Las Rivas, J. and Barber, J. (1991) ß-Carotene within the isolated photosystem II reaction centre: photooxidation and irreversible bleaching of this chromophore by oxidized P680. *Biochim. Biophys. Acta*, **1060**:106-114.

Thornber, J.P., Cogdell, R.J., Pierson, B.K. and Seftor, R.E.B. (1983) Pigment–protein complexes of purple photosynthetic bacteria: An overview. *J. Cell. Biochem.*, **23**:159-169.

Thrash, R.J., Fang, H.L.-B. and Leroi, G.E. (1977) The Raman excitation profile spectrum of ß-carotene in the preresonance region: Evidence for a low-lying singlet-state. *J. Chem. Phys.*, **67**:5930-5933.

Thurnauer, M.C., Katz, J.J. and Norris, J.R. (1975) The triplet state in bacterial photosynthesis: Possible mechanisms of the primary photo-act. *Proc. Natl. Acad. Sci. USA.*, **72**:3270-3274.

Trautman, J.K., Shreve, A.P., Violette, C.A. *et al.* (1990a) Femtosecond dynamics of energy transfer in B800–850 light-harvesting complexes of *Rhodobacter sphaeroides*. *Proc. Natl. Acad. Sci. USA.*, **87**:215-219.

Trautman, J.K., Shreve, A.P., Owens, T.G. and Albrecht, A.A. (1990b) Femtosecond dynamics of carotenoid-to-chlorophyll energy transfer in thylakoid membrane preparations from *Phaeodactylum tricornutum* and *Nannochloropsis sp*. *Chem. Phys. Letts.*, **166**, 369-376.

Tuua, N.J. (1978) *Modern Molecular Photochemistry*. Benjamin Cummins, Menlo-Park, CA.

Ullrich, J., Angerhofer, A., Schütz, J.U.von and Wolf, H.C. (1987) Zero-field absorption ODMR of reaction centres of *Rhodobacter sphaeroides* at temperatures between 4.2 and 75 K. *Chem. Phys. Lett.*, **140**:416-420.

Ullrich, J., Speer, R., Greis, J. *et al.* (1989) Carotenoid triplet states in pigment-protein complexes from photosynthetic bacteria: Absorption-detected magnetic resonance from 4 to 225 K. *Chem. Phys. Lett.*, **155**:363-370.

Vermeglio, A., Breton, J., Paillotin, G. and Cogdell, R.J. (1978) Orientation of chromophores in reaction centres of *Rhodopseudomonas sphaeroides*: A photoselection study. *Biochim. Biophys. Acta*, **501**:514-530.

Wasielewski, M.R. and Kispert, L.D. (1986) Direct measurement of the lowest excited singlet-state lifetime of all-*trans*-ß-carotene and related carotenoids. *Chem. Phys. Letts.*, **128**: 238-243.

Wasielewski, M.R., Liddell, P.A., Barrett, D. *et al.* (1986a) Ultrafast carotenoid to pheophorbide energy transfer in a biomimetic model for antenna function in photosynthesis. *Nature*, **322**:570-572.

Wasielewski, M.R., Tiede, D.M. and Frank, H.A. (1986b) Ultrafast electron and energy transfer in reaction centre and antenna proteins from photosynthetic bacteria. In: Fleming, G. R. and Siegman, A. E. (eds.), *Ultrafast Phenomena*, Vol. V. Springer-Verlag, Berlin, pp. 388-392.

Watanabe, J., Kinoshita, S. and Kushida, T. (1987) Effects of nonzero correlation time of system-reservoir interaction on the excitation profiles of second-order optical processes in ß-carotene. *J. Chem. Phys.*, **87**:4471-4477.

Wilbrandt, R. and Jensen, N.-H. (1981) Time resolved resonance Raman spectroscopy: The triplet state of all-*trans*-retinal. *J. Am. Chem. Soc.*, **103**:1036-1041.

Wilson, T. and Hastings, J.W. (1970) In: Giese, A.C. (ed.), *Photophysiology*, Vol. 5. Academic Press, New York. p. 49.

Witt, H. T. (1971) Coupling of quanta, electrons, fields, ions, and phosphorylation in the functional membrane of photosynthesis. *Q. Rev. Biophys.*, **4**:365-477.

Wraight, C.A., Cogdell, R.J., and Chance, B. (1978) Ion transport on electrochemical gradients in photosynthetic bacteria. In: Clayton, R. K. and Sistrom, W. R. (eds.), *The Photosynthetic Bacteria*. Plenum Press, New York, pp. 471-511.

Wylie, I.W. and Koningstein, J.A. (1984) Photoisomerisation and time-resolved Raman studies of 15,15′-*cis*-ß-carotene and 15,15′-*trans*-ß-carotene. *J. Phys. Chem.*, **88**:2950-2953.

Yamamoto, H.Y. (1979) Biochemistry of the violaxanthin cycle in higher plants. *Pure Appl. Chem.*, **51**:639-648.

Yeates, T.O., Komiya, H., Chirino, A. *et al.* (1988) Structure of the reaction centre from *Rhodopseudomonas sphaeroides* R-26 and 2.4.1: Protein-cofactor (bacteriochlorophyll, bacteriopheophytin, and carotenoid) interactions. *Proc. Natl. Acad. Sci. U.S.A.*, **85**:7993-7997.

Zhang, C.-F., Violette, C.A., Frank, H.A. and Birge, R.R. (1989) Electronic states of isolated and reaction centre-bound carotenoids. *Biophys. J.*, **55**:223a.

Zurdo, J., Lozano, R.M., Fernández-Cabrera, C. and Ramírez, J.M. (1991) Dimeric carotenoid interaction in the light-harvesting antenna of purple phototrophic bacteria. *Biochem. J.*, **274**:881-884.

9

Spectroscopic studies of carotenoids in photosynthetic systems

Y. Koyama and H. Hashimoto

9.1 INTRODUCTION

This chapter is designed to be complementary to Chapter 8. It is intended to be an introduction to a few very powerful but still developing spectroscopic techniques, i.e. electronic, Raman and magnetic resonance spectroscopies, for biologists, chemists and physicists who are interested in "carotenoids in photosynthesis". It is not a comprehensive catalogue of the results of spectroscopic techniques. The main emphasis of this chapter is on spectroscopic studies of structures and excited-state properties of carotenoids both free in solution and bound to the reaction centre (RC) or the light-harvesting complex (LHC) of phototrophic bacteria; ß-carotene, however, is also discussed as a prototype or model carotenoid. Each section of this chapter is divided into two parts: the first is written for a newcomer to the particular field of spectroscopy, is entitled "Fundamentals", and provides the reader with a brief description of spectroscopic background that is brief but sufficient for one to understand and appreciate the results obtained when the technique is applied. The second part, which is entitled "Examples of Application", is intended to serve as a handbook or a manual for a carotenoid spectroscopist and consists of a list of classical and up-to-date examples. Here, fundamental data which may be necessary or useful for day-to-day research work are included; some unpublished data which are expected to be important in the future are also given. Because of shortage of space all the experimental details have not been described; the original papers should be consulted. The results from different spectroscopic methods have not been correlated in this chapter. Such correlations form the basis of Chapter 8, and can also be found in recent reviews by Frank et al. (1991), Koyama (1991) and Mimuro and Katoh (1991).

9.2 ELECTRONIC SPECTROSCOPY

9.2.1 $S_n \leftarrow S_0$ Absorption: S_0 state of carotenoids

(a)Fundamentals

Intermolecular interaction of a carotenoid with the surroundings

In carotenoids, it has recently been established (see below and Chapter 8) that the optically allowed transition is actually the S_2 ($^1B_u^+$) $\leftarrow$ S_0 ($^1A_g^-$) transition and that the S_1 ($2^1A_g^-$) level is "hidden" below the S_2 ($^1B_u^+$) level. The S_2 ($^1B_u^+$) level is expected to be strongly dependent on intermolecular interactions between the carotenoid and surrounding molecules, which include interactions with (i) aromatic groups (dispersion interaction), (ii) local dipoles and (iii) point charges. These interactions can cause a red shift of the $S_2 \leftarrow S_0$ absorption. Exciton coupling of parallel transition dipoles can also give rise to red shift of the absorption.

For an apolar solute in apolar solvents, the dispersive interaction is predominant, and a quantum mechanical perturbation theory predicts the relation (Shulten et al. 1980)

$$\tilde{\upsilon}(gas) = \tilde{\upsilon}(solv) = \frac{K(n^2-1)}{(n^2+2)} \quad (9.1)$$

where, $\tilde{\upsilon} = \upsilon/c$; υ is the frequency, c is the velocity of light and n is the refractive index. An approximate expression for K is given by (Basu 1964)

$$K = 7.048 \times 10^{15} \times a^{-3}\left[\left|P_{ge}\right|^2 + E(\alpha_g - \alpha_e)\right] \quad (9.2)$$

where $|P_{ge}|$ is the difference in the dipole moment between the ground and excited states to be calculated from the oscillator strength, the radius of the cavity (a), average excitation energy of the solute (E), and polarisability in the ground and excited states (α_g and α_e).

Electrochromism

When a carotenoid bound to a pigment–protein complex is embedded in the photosynthetic membrane, a red or blue shift of the S_2 ($^1B_u^+$) $\leftarrow$ S_0 ($^1A_g^-$) absorption can take place, reflecting the transmembrane electric potential (Wraight et al. 1978).

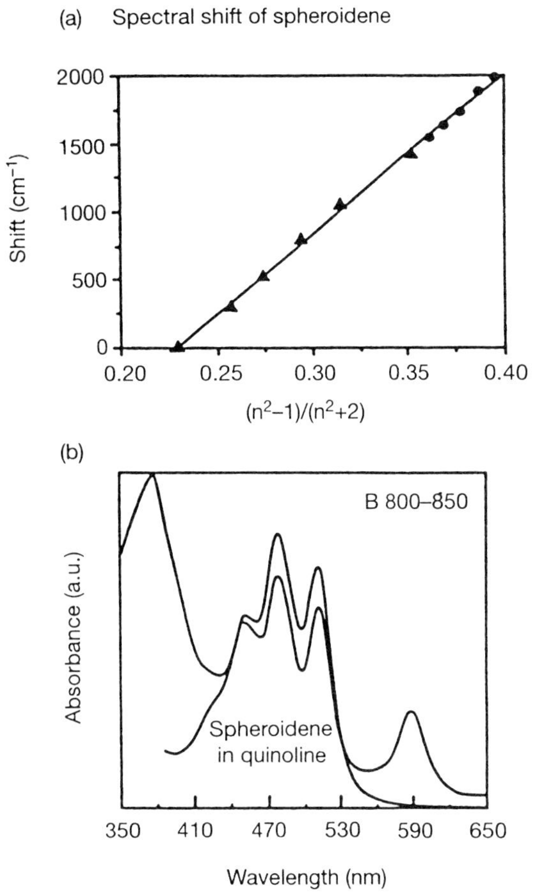

Figure 9.1 (a) Red-shift of the $S_2 \leftarrow S_0$ absorption (with reference to the absorption in *n*-hexane) as a function of the polarisabilities of various solvents (Δ) and of CS_2 at various temperatures (•) (b) Absorption spectra of spheroidene bound to the B800-850 LHC of *Rb. sphaeroides* 2.4.1 and of spheroidene dissolved in quinoline. (Reproduced from Andersson et al. 1990)

This electrochromism of the carotenoid can be substantiated by measurement of Stark effects when an external electric field is applied to the sample, which is embedded homogeneously in a film. Further, Stark effect spectroscopy can provide direct information on changes in the dipole moment (Δμ) and the polarisability (Δα) when the molecule in the S_0 state is excited to the S_2 state ($|P_{ge}| = |\Delta\mu|$). For an immobilised isotropic sample, changes in the dipole moment (Δμ) lead to band-broadening (second-derivative shaped change of the original absorption spectrum), and changes in the polarisability (Δα) lead to band shift (first-derivative shaped change, see Fig. 9.2-(a) C and D for the first- and second-derivative shapes). Since all orientations of $\Delta\mu$ are present relative to the external applied field, shifts to both higher and lower energies can occur, and the Stark effect becomes quadratic with the field (as in the case when $\Delta\alpha$ is predominant). If an apolar, essentially symmetrical carotenoid shows (i) the first-derivative effect when unbound and (ii) the second-derivative effect when bound to the pigment–protein complex, the changes in the dipole moment for the bound carotenoid are ascribed to a local field which is generated by the apoprotein and is much larger than the external field applied.

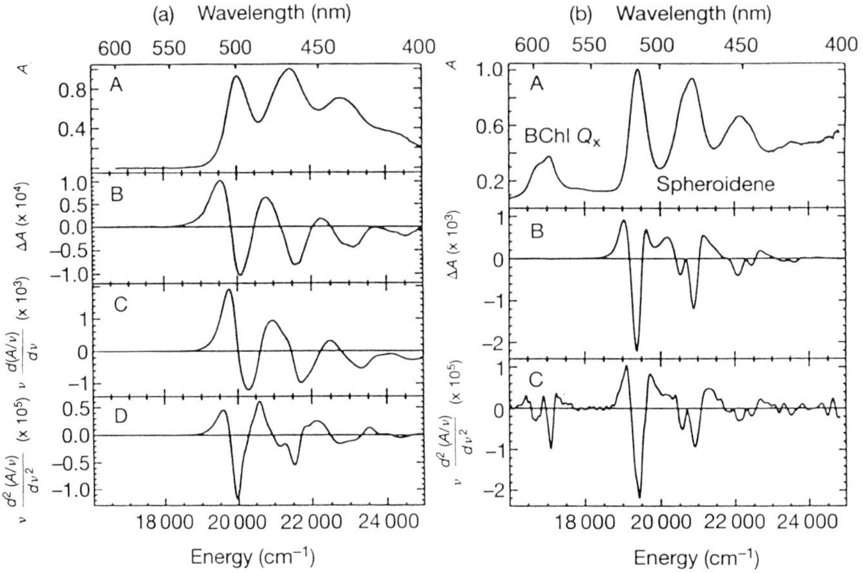

Figure 9.2 (a) Unbound spheroidene in a poly(methyl methacrylate) film. A, absorption spectrum; B, Stark effect spectrum; C, first-derivative and D, second-derivative of the absorption spectrum. (b) Spheroidene bound to the B800-850 LHC of *Rb. sphaeroides* in the film. A, absorption spectrum; B, Stark effect spectrum; C, second-derivative of the absorption spectrum. The external field applied to measure the Stark effect spectrum was 1 x 10^5 V/cm. (Reproduced from Gottfried et al. 1991a)

Configuration

The transition moment of the strongly optically-allowed S_2 ($^1B_u^+$) $\leftarrow$ S_0 ($^1A_g^-$) transition is along the long axis of the conjugated chain of a carotenoid, and the S_3 ($^1A_g^+$) $\leftarrow$ S_0 ($^1A_g^-$) transition is perpendicular to it. Therefore, the latter transition is forbidden for the all-*trans* isomer, and becomes weakly and strongly allowed for peripheral-*cis* and central-*cis* isomers, respectively. On the other hand, the former transition becomes weaker when the position of the *cis* bend moves toward the centre of the conjugated chain. Thus, the intensity ratio of the latter transition versus the former transition can be used to predict the configuration of the carotenoid. Further, the **energy** of the $S_2 \leftarrow S_0$ transition is also sensitive to the *cis–trans* configuration.

Optical activity

A molecule for which the chromophore has no plane nor centre of symmetry, nor the four-fold rotation-reflection symmetry axis exhibits optical rotating power (optical activity). The optical activity can be measured as either differential absorption of right and left circularly-polarised light (circular dichroism, CD) or wavelength dependence of the rotating power of plane-polarised light (optical rotatory dispersion, ORD). The physical bases of CD and ORD are the same, and the CD and ORD curves can be correlated with each other by the Kronig–Kramers relation.

Carotenoids lacking chiral groups show no optical activity when they are free in organic solution. However, when they are twisted as a result of binding to the apo-complex, optical activity may be observed. In particular, when two equivalent transition dipoles which are not in the same plane interact with each other (called "exciton coupling"), a pair of CD bands with the same magnitude but with the opposite direction appear on both sides of the electronic absorption which is expected to appear when no exciton coupling is present. The pair is called a "CD couplet".

(b)Examples of Application

Red shift of the $S_2 \leftarrow S_0$ absorption upon binding of spheroidene to the B800–850 light-harvesting complex

The red shift of the S_2 ($^1B_u^+$) ← S_0 ($^1A_g^-$) absorption of spheroidene (**1**) upon binding was interpreted in terms of dispersion interaction, which depends not on the polar character but on the polarisability of the environment of the carotenoid (Andersson et al. 1990). In order to evaluate the effect of polarisability, apolar solvents with different refractive indices, i.e. *n*-hexane, cyclohexane, CCl_4, mixtures of *n*-hexane/CS_2, and CS_2 in the temperature range from -90 to 20 °C, were used. Figure 9.1(a) shows the relationship between the red shift (with reference to the absorption band in *n*-hexane) and the polarisability $(n^2-1)/(n^2+2)$, where n is the refractive index. The linear relation was expressed by the equation,

$$\Delta\tilde{\upsilon} = -2680 + 11700 \ \frac{(n^2-1)}{(n^2+2)} \quad (9.3)$$

based on which, values for $\tilde{\upsilon}(gas) = 24\,800\ cm^{-1}$ and $K = 11\,700\ cm^{-1}$ were obtained.

By the use of the approximate expression for *K* (see equation 9.2) and an assumed value of a = 10 Å, changes in the dipole moment (P_{ge}) and the polarisability ($\Delta\alpha = \alpha_g - \alpha_e$) upon the $S_2 \leftarrow S_0$ excitation were estimated to be: P_{ge} = 12 D and $\Delta\alpha$ = 370 $Å^3$.

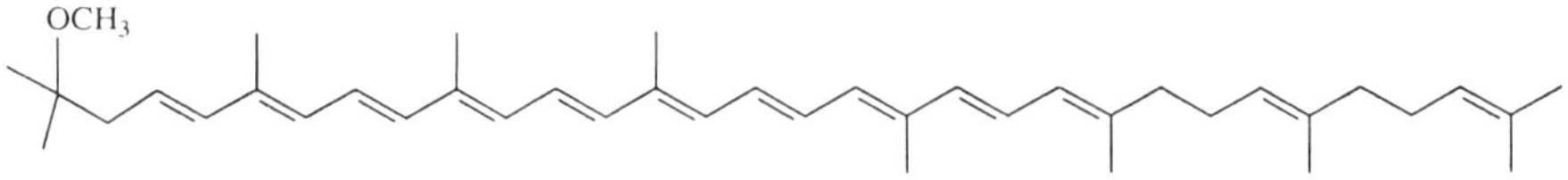

1

The perfect overlap of the absorption spectrum of spheroidene dissolved in quinoline with that bound to the LHC (Fig. 9.1b) supports the idea that the large red shift of the $S_2 \leftarrow S_0$ absorption originates from dispersion interaction of the carotenoid with an environment having high polarisability, and that the particular carotenoid is surrounded by several aromatic amino acids. Further experiments in polar solvents showed only a small extra contribution to the red shift owing to other forms of interaction (Andersson et al. 1991).

Stark effect on the $S_2 \leftarrow S_0$ absorption for free and bound spheroidene

Shifts in the carotenoid absorption (termed the "carotenoid shift") have been widely used to measure the transmembrane potential and the electrogenicity of charge separation steps (Wraight et al. 1978). Stark effect spectroscopy of the unbound and bound carotenoids was used in an attempt to reveal the underlying mechanism (Gottfried et al. 1991a). Figure 9.2(a) shows that the Stark effect spectrum of spheroidene in a poly(methyl methacrylate) film (i.e. an **unbound** carotenoid) has nearly the same shape as the first derivative of the absorption spectrum except that it is shifted by ≈250 cm^{-1} to lower energy. All the vibrational structures could be fitted well by a sum of the first-derivative and the second-derivative contributions, which are ascribable to $\Delta\alpha$ and $\Delta\mu$, respectively, of the molecule upon excitation.

An extraordinarily large electric-field effect is seen on comparison of the ordinate scales and the factors for the Stark effect spectra between the the absorption spectrum (A) and the Stark effect spectrum (B) of spheroidene **bound** to the B800–850 LHC of *Rhodobacter sphaeroides* as shown in Fig. 9.2b. The close agreement of the Stark effect spectrum (C) with the second-derivative of the absorption spectrum indicates that the electrochromic effect is essentially due to a change in the dipole moment between the ground and excited states, the value of which was determined to be:

$$|\Delta\mu| = 15.3\ (\pm\ 0.3)\ \mathrm{D/f, where\ f} = 1 \approx 1.1.$$

Although the first derivative-shaped contribution due to $|\Delta\alpha|$ should be comparable in the Stark effect spectra of spheroidene in the protein (Fig. 9.2b) and in an apolar glass matrix (Fig. 9.2a), the above value of $|\Delta\mu|$ induced by the protein environment is so large that the former contribution is not detectable. The results demonstrate the presence of a large perturbation to the electronic structure of this nearly symmetrical carotenoid due to the organised environment in the protein. (It is also to be noted that the Stark effect spectrum of the **unbound asymmetrical** carotenoid, spheroidenone (**2**), embedded in the film, resembled the second derivative of its absorption spectrum, although the extent of the Stark effect was smaller (Gottfried et al. 1991b).

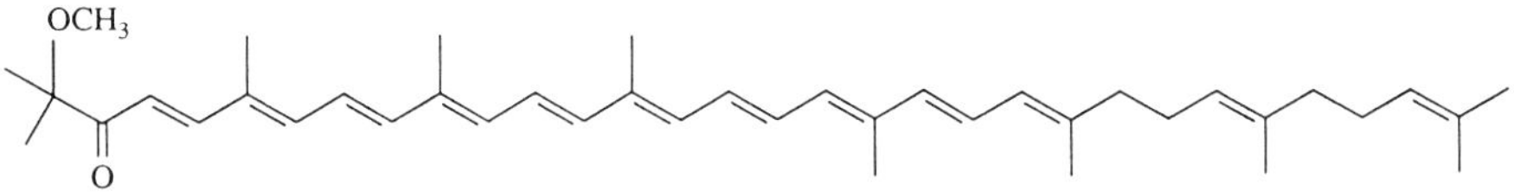

2

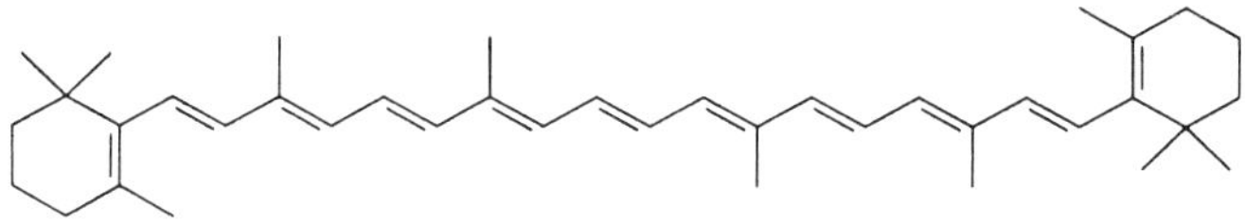

3

Prediction of configuration of carotenoids in the reaction centres

Figure 9.3 shows the $S_n \leftarrow S_0$ absorptions of the geometrical isomers of ß-carotene (**3**; Koyama et al. 1982; Tsukida et al. 1982). Table 9.1 lists the wavelengths of the ${}^1A_g^+ \leftarrow {}^1A_g^-$ and the ${}^1B_u^+ \leftarrow {}^1A_g^-$ absorptions ("isomerisation shifts" in parentheses) and the intensity ratio of the ${}^1A_g^+ \leftarrow {}^1A_g^-$ versus ${}^1B_u^+ \leftarrow {}^1A_g^-$ transition ("relative intensity of the *cis* peak"). It can be seen that the relative intensity of the *cis* peak is in the order, (all-*trans*) < 7-*cis* < 9-*cis* < 13-*cis* < 15-*cis*, and that the isomerisation shift is in the order, 15-*cis* < 7-*cis* < 9-*cis* < 13-*cis*.

The isomerisation shift and the relative intensity of the *cis* peak were used, together with Raman spectral patterns (see Section 9.3.1) and the isomerisation shift of olefinic ${}^1H$ NMR signals (see Section 9.4.1 (b)), to deduce the configuration of neurosporene (**4**) and spirilloxanthin (**5**) extracted from the reaction centres (RC) of *Rb. sphaeroides* G1C and *Rhodospirillum rubrum* S1, respectively, as 15-*cis* (15-Z) (Koyama et al. 1988a, 1990). Comparison of the electronic absorption isomerisation shifts and the relative intensities of the *cis* peak of the carotenoids listed in Table 9.2, supports the above configurational assignment.

Figure 9.4 shows the difference spectra ((after reconstitution) minus (before reconstitution)) of a number of carotenoids which were incorporated into the RC of *R. rubrum* G9 (a carotenoidless mutant). The strong *cis* peaks suggest that the configuration of the RC-bound carotenoids is 15-*cis*, although hyperchromism of bacteriochlorophyll (BChl) absorptions is also seen (Boucher et al. 1977).

Table 9.1 Electronic absorption spectra of geometrical isomers of ß-carotene, neurosporene and spirilloxanthin in *n*-hexane at room tempertaure.

	Wavelength (nm)						*Intensity ratio*
		$(1A_g^+ \leftarrow 1A_g^-)$		$(1B_u^+ \leftarrow 1A_g^-)$			$(1A_g^+ \leftarrow 1A_g^-)$
		$(1 \leftarrow 0)$	$(0 \leftarrow 0)$	$(2 \leftarrow 0)$	$(1 \leftarrow 0)$	$(0 \leftarrow 0)$	$(1B_u^+ \leftarrow 1A_g^-)$
ß-Carotene							
All-*trans*		337		426	451	478	0.06
7-*cis*		336		423	446	474(4)[a]	0.06
9-*cis*		337		421	445	472(6)	0.09
13-*cis*		336		418	442	467(11)	0.36
15-*cis*		335		425	448	476(2)	0.52
Neurosporene							
All-*trans*		319	332	414	438	468	0.05
15-*cis*		322	332	411	435	465(3)	0.33
Spirilloxanthin							
All-*trans*	367		386	463	492	526	0.10
13-*cis*	368		386	460	488	520(6)	0.55
15-*cis*	367		385	462	490	524(2)	0.65

[a] Shift relative to the all-*trans* isomer

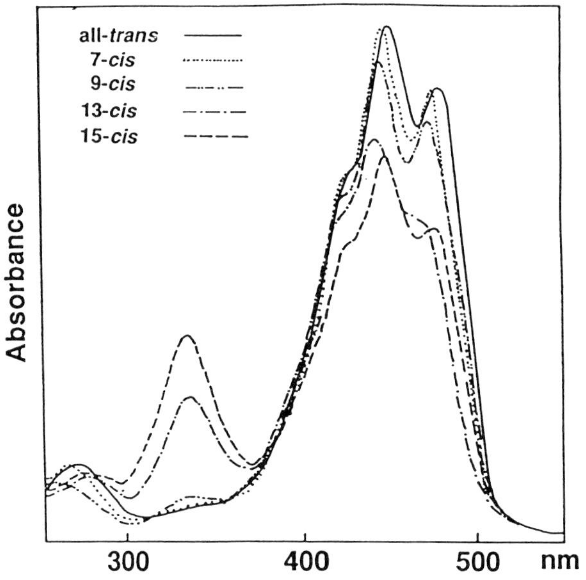

Figure 9.3 The $S_n \leftarrow S_1$ absorption spectra of geometrical isomers of ß-carotene.

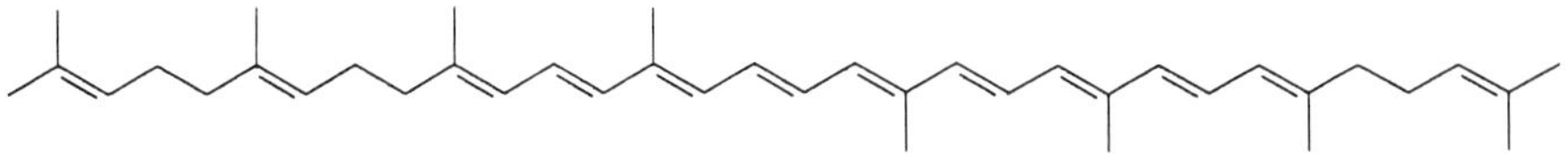

4

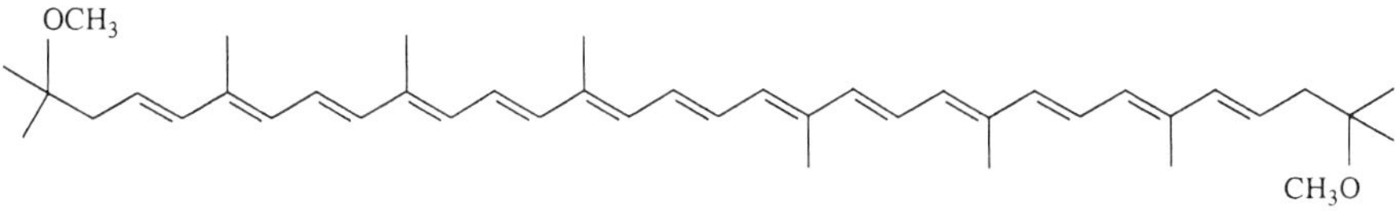

5

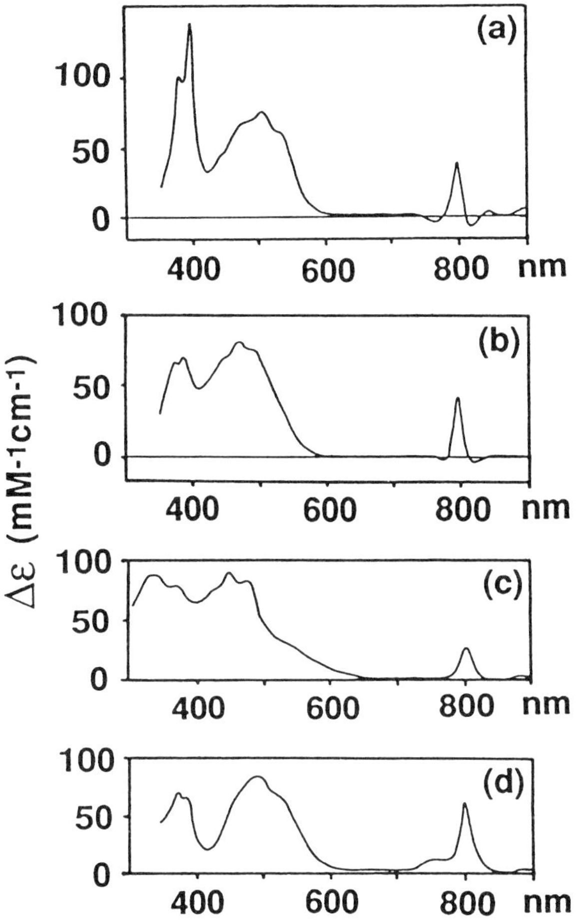

Figure 9.4 Difference spectra, [(after reconstitution) minus (before reconstitution)] of a carotenoid in the RC of *R. rubrum* G9 (carotenoidless mutant); (a) spirilloxanthin, (b) spheroidene, (c) chloroxanthin and (d) spheroidenone. (Revised from Boucher et al. 1977)

Twisting of carotenoids upon binding to the reaction centres and light harvesting complexes detected by CD spectroscopy

The optical activity detected for a carotenoid bound to the RC or light-harvesting complex (LHC) provided a unique piece of information on the twisting of the conjugated backbone (Cogdell et al. 1976). For example, spheroidene in the RC of the wild type *Rb. sphaeroides* shows three strong CD peaks at 503, 472 and 446 nm whereas these bands are absent for the carotenoidless mutant, R-26 (Fig. 9.5). Further, the carotenoid extracted in organic solvents exhibited no optical activity. Therefore, the strong CD bands must result from asymmetry of the carotenoid binding site. However, spheroidene in the chromatophores (LHCs) showed a CD spectrum different from that of the RC in that the long wavelength band was very intense relative to the two bands at shorter wavelengths. The above results indicate some twisting of the conjugated backbone for both RC- and LHC-bound spheroidene.

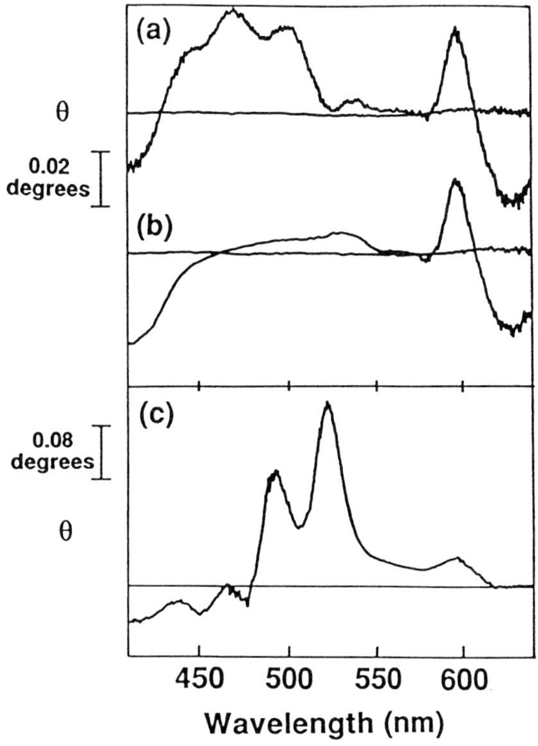

Figure 9.5 CD spectra of (a) the RC of *Rb. sphaeroides* 2.4.1, (b) the RC of *Rb. sphaeroides* R26, and (c) the chromatophores of *Rb. sphaeroides* 2.4.1 (Revised from Cogdell et al. 1976)

Purified carotenoids were attached to the RC of *R. rubrum* G9 (a carotenoidless mutant), and their configurations and conformations were discussed on the basis of electronic absorption and CD spectroscopy (Boucher et al. 1977). When bound to the G9 RC, spheroidene exhibits a positive CD band in the region of the $^1B_u^+ \leftarrow {}^1A_g^-$ transition and a negative CD band in the region of the $^1A_g^+ \leftarrow {}^1A_g^-$ transition (*cis* peak) (Fig. 9.6). The CD spectrum suggests that the 15-*cis* configuration, which was predicted based on the strong *cis* peak as described above, is twisted in a protohelical shape. Since the $^1B_u^+ \leftarrow {}^1A_g^-$ transition dipole is parallel to the long axis and the $^1A_g^+$

$\leftarrow {}^1A_g^-$ transition dipole is perpendicular to it, the CD bands are expected to be of opposite signs in the above twisted conformation. The CD spectrum shows that this is the case.

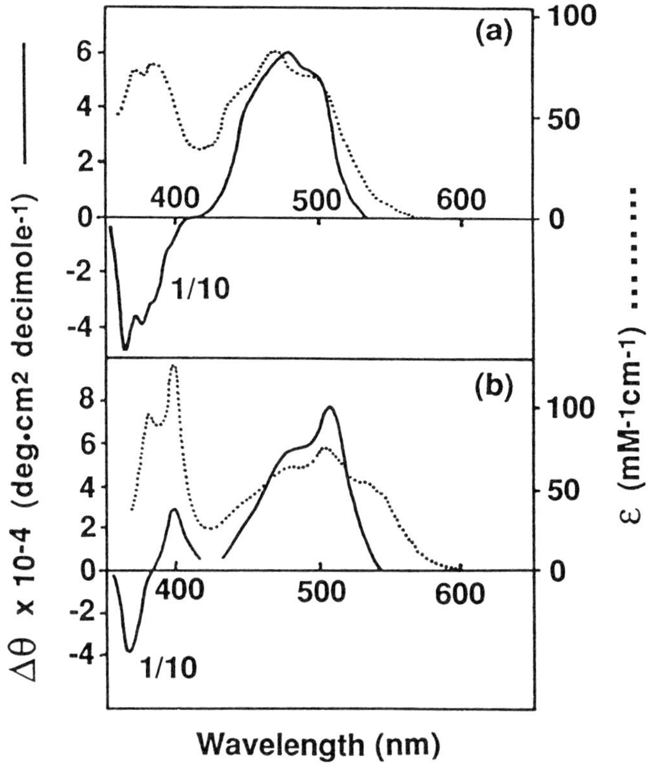

Figure 9.6 Difference CD (———) and difference absorption (•••••••••••••) spectra of [(reconstituted) minus (carotenoidless)] preparations of the RC of *R. rubrum* G9; (a) spheroidene and (b) spirilloxanthin. (Revised from Boucher et al. 1977)

Spirilloxanthin shows similar CD and absorption spectra. However, the CD band corresponding to the ${}^1A_g^+ \leftarrow {}^1A_g^-$ transition is split into two components with opposite signs but with similar magnitudes. This is ascribed to a CD couplet due to exciton coupling, because the 15-*cis*-spirilloxanthin molecule can generate a pair of identical transition dipoles when twisted around the C15,15′ double bond (due to symmetry of the chemical structure of spirilloxanthin).

9.2.2 $T_n \leftarrow T_1$ Absorption: T_1 state of carotenoids

(a)Fundamentals

$T_n \leftarrow T_1$ Absorption, triplet structure and triplet energy transfer of carotenoids
The efficiency of intersystem crossing is very low in carotenoids (see Section 9.2.3 (b)) and, therefore, the T_1 states of carotenoids are usually generated by pulse radiolysis or flash photolysis using a sensitiser (naphthalene, anthracene, chlorophyll, fluorene, etc.). Since the optical parameters of the $T_n \leftarrow T_1$ absorption (the spectral pattern and the absorption coefficient, ε) and the T_1 lifetime (or the decay constant, k_d) are dependent on the length and the configuration of the conjugated chain, these parameters can be used to identify or differentiate the structure of carotenoids. Typical T_1 lifetimes of carotenoids are of the order of 10^0–10^1 μs, and most of the $T_n \leftarrow T_1$ absorption spectra have been recorded by means of transient absorption spectroscopy using ≈10 ns pumping and probing pulses.

For carotenoids free in solution, the rate constant of energy transfer (k_T) from a sensitiser to a carotenoid,

$$^3S^* + {}^1Car \rightarrow {}^1S + {}^3Car^* \quad (9.4)$$

can be determined by using the dependence of the rate of decay of the sensitiser on the concentration of the carotenoid,

$$\text{rate} = k_0 + k_T[Car] \quad (9.5)$$

On the other hand, the rate constant of quenching (k_Q) of T_1 carotenoid by oxygen,

$$^3Car^* + {}^3O_2 \rightarrow {}^1Car + {}^1O_2^* \quad (9.6)$$

can be determined by the dependence of the decay rate of the T_1 carotenoid on the concentration of oxygen,

$$\text{rate} = k_d + k_Q[O_2] \quad (9.7)$$

For carotenoids bound to the LHC or the RC, triplet energy transfer processes between BChl and carotenoid,

$$^3BChl^* + {}^1Car \rightarrow {}^1BChl + {}^3Car^* \quad (9.8)$$

$$BChl^{+\bullet}\text{-}BChl^{-\bullet} + {}^1Car \rightarrow BChl\text{-}BChl + {}^3Car^* \quad (9.9)$$

can be established when the following relation is shown for a triplet donor (D) and a triplet acceptor (A),

$$^3D^* + {}^1A \rightarrow {}^1D + {}^3A^* \quad (9.10)$$

$$k_{decay}(^3D^*) = k_{rise}(^3A^*) \quad (9.11)$$

These energy transfer reactions underline the mechanisms of generation (or decay) of T_1 carotenoids, and reveal the basis of photoprotective roles of carotenoids in the photosynthetic systems.

(b)Examples of Application

ß-Carotene homologues and bacterial carotenoids with different lengths of the conjugated chain

The T_1 state properties of ß-carotene and its strict homologues containing a ß-ring at both ends and differing only by the length of the conjugated chain, i.e. ß-carotene (C_{40}, *n* = 11 conjugated double bonds), and C_{30} (*n* = 7), C_{35} (*n* = 9), C_{50} (*n* = 15) and C_{60} (*n* = 19) homologues, were compared by flash photolysis with chlorophyll *a* as the sensitiser (Mathis and Kleo 1973).

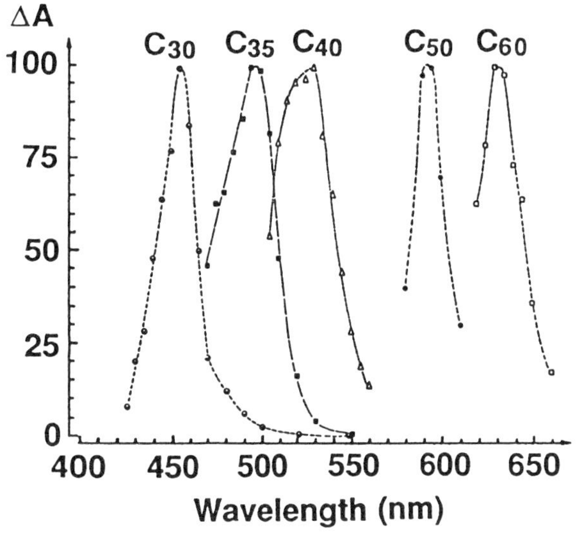

Figure 9.7 $T_n \leftarrow T_1$ absorption bands of ß-carotene (C_{40}) and its homologues (C_{30}, C_{35}, C_{50} and C_{60}). (Revised from Mathis and Kleo 1973)

Figure 9.7 shows a regular displacement of the $T_n \leftarrow T_1$ absorption band depending on the chain length. Figure 9.8a shows a linear relationship between the energies of the $T_n \leftarrow T_1$ ($\tilde{\upsilon}_T$) and $S_2 \leftarrow S_0$ ($\tilde{\upsilon}_S$) transitions which is expressed as, $\upsilon_T = 0.75\ \upsilon_S + 2\,600$ (cm^{-1}); a similar relation was found for a variety of polyenes (Truscott et al. 1973). In Fig. 9.8b, k_d is plotted on a logarithmic scale versus the estimated T_1 energy (E_T). Table 9.2 lists the triplet data for the ß-carotene homologues; the rate constants of (i) first-order decay (k_d), (ii) energy transfer from chlorophyll *a* (k_T), and (iii) quenching by oxygen (k_Q) are shown together with the number of the conjugated double bonds and the wavelengths of the $S_2 \leftarrow S_0$ and $T_n \leftarrow T_1$ absorptions. The decay rate constant (k_d) increases smoothly with increasing number of conjugated double bonds. The energy transfer rate constant (k_T) varies slightly within the series, but it is always much smaller than the rate of a diffusion-controlled reaction (k_{diff} = 10.8 x 10^9 M^{-1} s^{-1}), indicating a low probability of energy transfer during the collisional process. The quenching rate constant (k_Q) is much higher and reaches the same order of magnitude as k_{diff}; k_Q should be related to the energy of $^1O_2$ ($^1\Delta_g$), which is above the E_T of C_{35} to C_{60} homologues and below the E_T of the C_{30} homologue. For quenching by chlorophyll *a*, k_Q = 2.2 x 10^9 M^{-1} s^{-1} was obtained.

The T_1 state properties of bacterial carotenoids were also examined by anthracene-sensitised flash photolysis (Bensasson et al. 1976). Similar dependencies of both the $T_n \leftarrow T_1$ absorption and the first order decay constant (k_d) on the length of the conjugated chain are also seen for the bacterial carotenoids listed in Table 9.3. The rates of energy transfer from anthracene (k_T) for these carotenoids are comparable to k_{diff} = 10.8 x 10^9 $M^{-1}s^{-1}$, indicating that their E_T values are lower than that of anthracene (E_T = 123 kJ mol^{-1}). The ε values for the $T_n \leftarrow T_1$ absorptions were determined by the energy transfer techniques, and the low ϕ_{353} values (quantum yield of intersystem crossing at 353 nm excitation) show that the T_1 state is not populated intramolecularly and that the light energy absorbed by bacterial carotenoids is not wasted in crossing over to the T_1 state.

Table 9.2 Triplet data for ß-carotene homologues in benzene at 23°C

Polyene	*No. of conjugated double bonds*	$S_2 \leftarrow S_0$ *absorption (nm)*	$T_n \leftarrow T_1$ *absorption (nm)*	$10^4 k_d$ [a] (s^{-1})	$10^{-9} k_T$ [b] $(M^{-1}s^{-1})$	$10^{-9} k_Q$ [c] $(M^{-1}s^{-1})$
C_{30} homologue	7	385	457	5	0.9	2.8
C_{35} homologue	9	427	495	10	0.8	3.2
ß-Carotene (C_{40})	11	465	526	17	1.2	3.6
C_{50} homologue	15	518	595	31	1.8	7.0
C_{60} homologue	19	552	620	40	1.8	5.6

[a] Rate constant of first order decay
[b] Rate constant of energy transfer from chlorophyll *a*.
[c] Rate constant of quenching by oxygen.

T_1 species generated from geometrical isomers of ß-carotene.

The T_1 species which were generated by anthracene-sensitised flash photolysis from several geometrical isomers of ß-carotene were compared by transient absorption spectroscopy (Hashimoto et al. 1989). Figure 9.9 shows the transient absorption spectra of T_1 species which were generated from geometrical isomers of ß-carotene. The all-*trans*, 7-*cis*, 9-*cis* and 13-*cis* isomers each show a unique spectral pattern, which indicates that each isomer generates its own T_1 species (the "all-*trans*", "7-*cis*", "9-*cis*" and "13-*cis*" T_1). However, the T_1 species generated from the 15-*cis* isomer shows the same spectral pattern as that of the "all-*trans*" T_1.

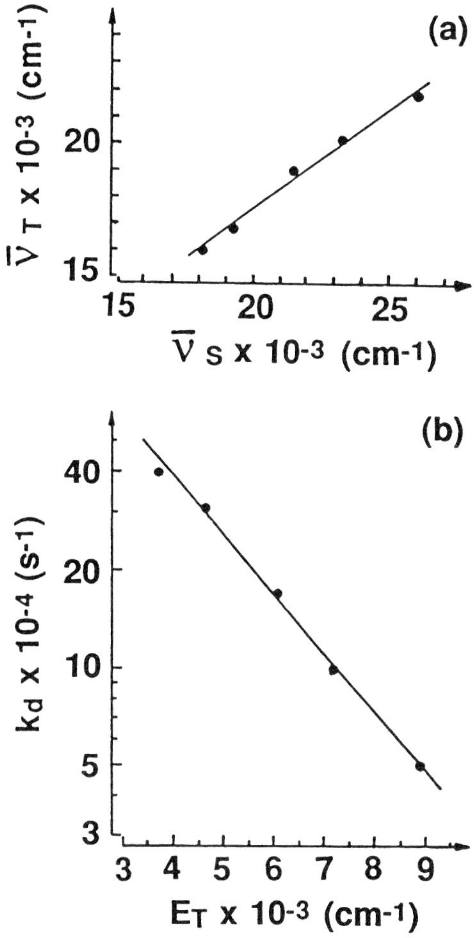

Figure 9.8 (a) Relation between the energies of the $T_n \leftarrow T_1$ ($\tilde{\upsilon}_T$) and $S_2 \leftarrow S_1$ ($\tilde{\upsilon}_S$) absorptions, and (b) relation between the first order decay constant (k_d) and the estimated T_1 energy for ß-carotene and its homologues (revised after Mathis and Cleo 1973)

Since the 15-*cis* isomer isomerises very efficiently upon excitation to the triplet-state (see Section 9.3.3 (b)), the observation was ascribed to extremely efficient isomerisation in the T_1 state, i.e. it was interpreted that the "15-*cis*" T_1 is too short-lived to be detected, and that the "all-*trans*"T_1, the result of isomerisation in the T_1 state, is detected. The results are consistent with those of time-resolved Raman spectroscopy (see Section 9.3.3 (b)). The wavelength of the $T_n \leftarrow T_1$

absorption (nm) and the decay rate constant (k_d x 10^{-4} s^{-1}) for each isomer are as follows: all-*trans*, 520 and 21; 7-*cis*, 510 and 16; 9-*cis*, 520 and 27; 13-*cis*, 522 and 28; and 15-*cis*, 520 and 21.

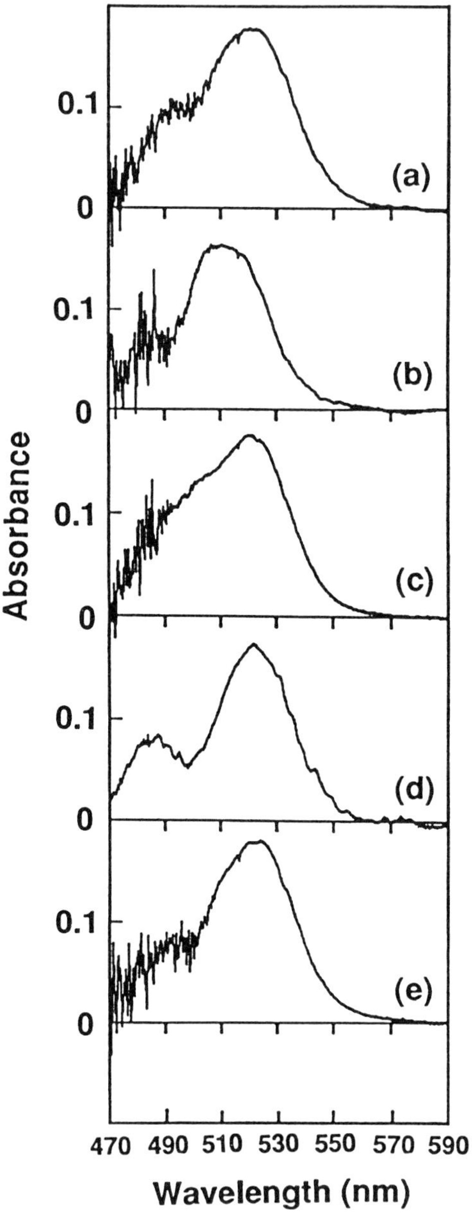

Figure 9.9 $T_n \leftarrow T_1$ absorption spectra of geometrical isomers of ß-carotene; (a) all-*trans*, (b) 7-*cis*, (c) 9-*cis*, (d) 13-*cis* and (e) 15-*cis*. (1.8 µs after excitation)

Table 9.3 Triplet data for bacterial carotenoids in cyclohexane

Carotenoids	*Number of conjugated double bonds*	$T_n \leftarrow T_1$ *absorption (nm)*	$10^{-4}k_d$ (s^{-1})	$10^{-9}k_T{}^a$ $(M^{-1}s^{-1})$	$\varepsilon(T_n \leftarrow T_1)$ $(M^{-1}cm^{-1})$	$\phi_{353}{}^b$
ζ-Carotene	7	445	4.4	14	81000	<0.002
Spheroidene	10	510	8.2	10	29000	<0.002
Spirilloxanthin	13	560	16	20	43000	0.028

a Rate constant of energy transfer from naphthalene.
b Quantum yield of intersystem crossing upon direct excitation at 353 nm.

T_1 carotenoids bound to the reaction centres

The RCs of carotenoidless and carotenoid-containing strains of *Rb. sphaeroides* and *R. rubrum* were reduced with $Na_2S_2O_4$ to block their electron transfer reactions. They were then excited with 834 nm, 20 ns pulses to investigate the photoprotective role of carotenoids in the RCs (Cogdell et al. 1975). In the case of carotenoidless RCs (from *Rb. sphaeroides* R-26 and *R. rubrum* G9), a transient state (P^F) was immediately formed, and the decay of P^F yielded a longer-lived state (P^R) which was ascribed to the T_1 state of BChl. The long lifetime of the P^F state (10 ns) ruled out the possibility that it is the S_1 state. In the case of carotenoid-containing RCs (from *Rb. sphaeroides* 2.4.1 and Ga, and *R. rubrum* S1), the P^R state was not detected, and the decay of P^F resulted in the generation of the T_1 state carotenoids; the result supported the view that the P^F state is a biradical which is triplet in character. Table 9.4 lists the kinetic data which support the above conclusions. Note that the rate constant of the decay of the P^F state agrees with that of the rise of the T_1 state carotenoid (Car^T) for both *Rb. sphaeroides* 2.4.1 and Ga. It is also noted that the rate constant of decay of Car^T increases when the length of the conjugated chain increases, i.e. chloroxanthin (**6**) ($n = 9$) < spheroidene ($n = 10$) < spirilloxanthin ($n = 13$), as in the case of carotenoids free in solution. The $T_n \leftarrow T_1$ absorption shifts to the longer wavelengths with increasing length of the conjugated chain of the RC-bound carotenoids (Fig. 9.10), as for free all-*trans* carotenoids), and the second absorption appears for spirilloxanthin in the shorter wavelength region.

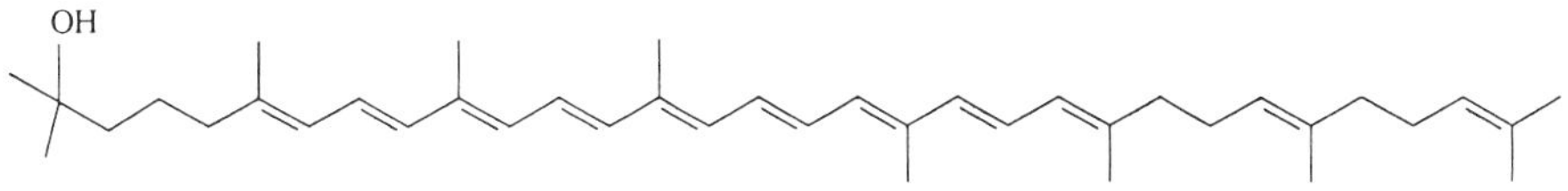

6

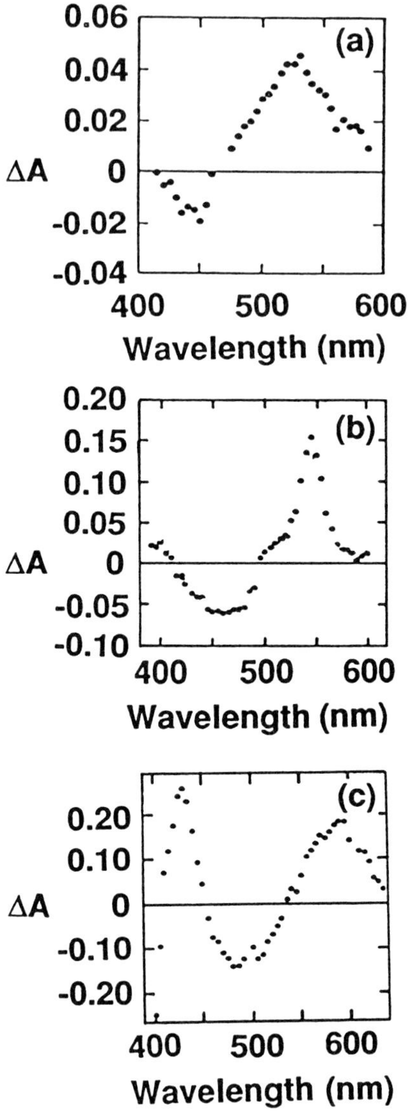

Figure 9.10 $T_n \leftarrow T_1$ absorption spectra of native carotenoids bound to the RC (a) chloroxanthin (*Rb. sphaeroides* Ga), (b) spheroidene (*Rb. sphaeroides* 2.4.1) and (c) spirilloxanthin (*R. rubrum* S1). (Revised from Cogdell et al. 1975)

When the RCs from any of the types used were excited with supersaturating flashes at redox potentials that were positive enough to hold the primary electron acceptor in the oxidised form prior to flash, only absorbance changes due to the formation of P^+ were seen. There was no evidence for the generation of T_1 carotenoids under these conditions.

Table 9.4 Triplet rate constants (s^{-1}) for bacteriochlorophyll *a* and carotenoids bound to the reaction centres and the light-harvesting complexes of phototrophic bacteria

Phototrophic bacteria (carotenoids)	*Reaction centre* [a]				*Light-harvesting complex*		
	P^F decay	*P^R decay*	*Car^T rise*	*Car^T decay*	*P^R decay*	*Car^T rise*	*Car^T decay*
	$10^{-7}k$	*$10^{-4}k$*	*$10^{-7}k$*	*$10^{-5}k$*	*$10^{-4}k$*	*$10^{-7}k$*	*$10^{-5}k$*
Rb. spheaeroides							
R-26	10	~20	—	—	1.4	—	—
Ga (chloroxanthin)	5-10	—	5-10	~2	—	~3	8
2.4.1 (spheroidene)	~10	—	~10	~3	—	~3	14
Rs. rubrum							
G9	~5	2	—	—	1.4	—	—
S1 (spirilloxanthin)	ND	—	~2	~4	—	~2	25

a) Under reducing conditions.
ND not determined.

T_1 carotenoids bound to the chromatophores

Chromatophores from *Rb. sphaeroides* and *R. rubrum* were excited with ≈15 ns, 694 or 834 nm pulses to investigate the photoprotective role of carotenoids in the LHCs (Monger et al. 1976). In such complexes, the triplet states of pigments are generated (i) if the primary electron transfer reaction is blocked by prior chemical oxidation of P870 or reduction of the primary electron acceptor, (ii) if the photochemical apparatus is oversaturated with light, or (iii) if the bacteria are genetically devoid of RCs. (In the case of the RCs, as described above, only the reduction of the electron acceptor led to the formation of the triplet-states.) Table 9.4 lists the kinetic data for triplet BChl and carotenoids in the LHCs. In the antenna BChl, the T_1 (P^R) state arises from the S_1 state by intersystem crossing, and the P^R state decays with time constant of 70 μs ($k_d \approx 1.4 \times 10^4$ s). The T_1 state of the antenna carotenoids is probably formed by energy transfer from the T_1 BChl. The energy transfer is very efficient, and the rise time of Car^T, at low redox potentials, is 30 ns ($k \approx 3 \times 10^7 s^{-1}$) for chloroxanthin and spheroidene (*Rb. sphaeroides* Ga and 2.4.1) and 45 ns ($k \approx 2 \times 10^7 s^{-1}$) for spirilloxanthin (*R. rubrum*). Here also, the T_1 decay rate depends on the length of the conjugated chain, increasing as the length of the conjugated chain increases. The chromatophores of the RC-less mutant of *Rb. sphaeroides* PM-8 dpl show essentially the same $T_n \leftarrow T_1$ absorption spectrum as that of the RC-containing mutant (nine conjugated double bonds in both carotenoids). It should also be noted that the $T_n \leftarrow T_1$ absorption shifts to the longer wavelengths as the length of the conjugated chain increases.

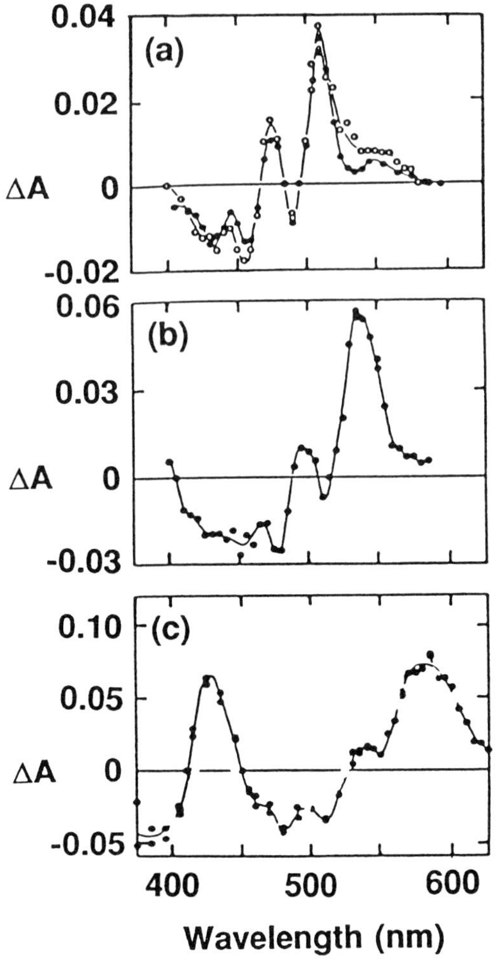

Figure 9.11 $T_n \leftarrow T_1$ absorption spectra of native carotenoids in the chromatophores; (a) (•) neurosporene (*Rb. sphaeroides* PM-8 dpl) and (○) chloroxanthin (Ga), (b) spheroidene (*Rb. sphaeroides* 2.4.1), and (c) spirilloxanthin (*R. rubrum* S1). All spectra were recorded at low redox potentials. (Revised from Monger et al. 1976)

Structural constraint applied to carotenoids bound to the reaction centre

The $S_2 \leftarrow S_0$ and the $T_n \leftarrow T_1$ absorption spectra were recorded for two carotenoids, methoxyneurosporene (**7**) ($n = 9$) and spheroidene ($n = 10$), which were incorporated into the RC of a carotenoidless mutant of *Rb. sphaeroides,* R26 (Frank et al. 1986). The T_1 state of the RC-bound carotenoid was observed only when electron transfer beyond the initial electron acceptor was blocked. In quinone-containing RCs this can be accomplished by adding sodium ascorbate and flashing the RC with an excitation light (to pump electrons from the $BChl_2$ donor to secondary acceptor quinones); the $BChl_2$ donor is subsequently reduced by ascorbate while the quinones remain reduced. Table 9.5 lists the wavelengths of the $S_2 \leftarrow S_0$ and $T_n \leftarrow T_1$ absorptions of the two carotenoids both free in solution and bound to the RC.

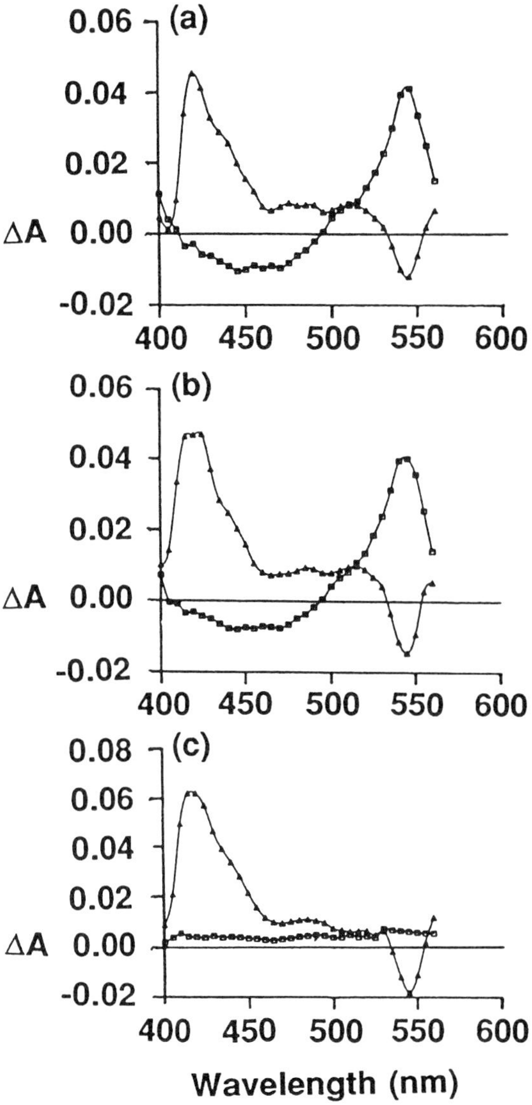

Figure 9.12 Transient absorption spectra of carotenoids incorporated into the RC of *Rb. sphaeroides* R26 (carotenoidless mutant); (a) spheroidene and (b) methoxyneurosporene (Δ) during laser excitation and (□) 135 ns after excitation. (c) Transient absorption for the carotenoidless RC is also shown for comparison. (Revised from Frank et al. 1986)

Spheroidene ($n = 10$) shows the $S_2 \leftarrow S_0$ absorption peaks in a longer wavelength region than methoxyneurosporene ($n = 9$) when free in solution, whereas they show essentially the same absorption peaks when they are bound to the RC. The transient absorptions of spheroidene-bound RC and methoxyneurosporene-bound RC for the longer delay time (135 ns after excitation) shown in Fig. 9.12 are almost identical, and they are definitely ascribed to the $T_n \leftarrow T_1$ transition (both at 545 nm).

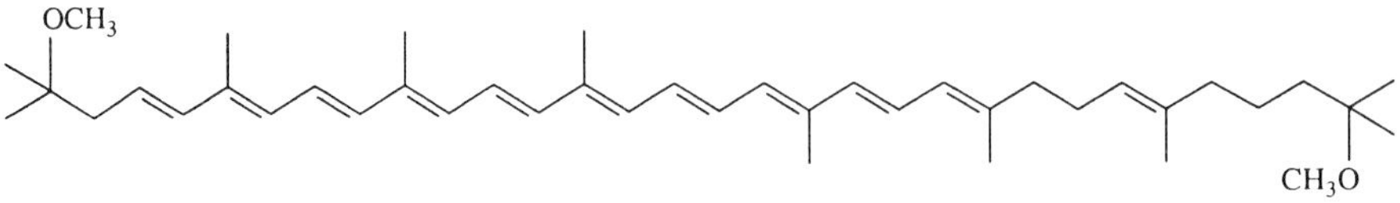

7

Table 9.5 $S_2 \leftarrow S_0$ and $T_n \leftarrow T_1$ absorptions of spheroidene and methoxyneurosporene (nm)

	Free in solution		*Bound to reaction centre*	
	Spheroidene	*Methoxy-neurosporene*	*Spheroidene*	*Methoxy-neurosporene*
	475	463	498	500
$S_2 \leftarrow S_0$	447	434	470	471
absorption	421	411	437	438
$T_n \leftarrow T_1$ absorption	510	489	545	545

The results suggest that there exists a single, well-defined binding site for these carotenoids in the protein, and that strong binding of the carotenoids to the RC is accompanied by a twisting of the carotenoids in a manner that interrupts their π-electron conjugation into the same effective length.

9.2.3 $S_n \leftarrow S_1$ Absorption: S_1 State of Carotenoids

(a)Fundamentals

The S_1 and S_2 states of carotenoids

The experimental results described below establish that the S_1 and S_2 states of carotenoids are the $2A_g$ and B_u states. (Only for ß-carotene which has the symmetry C_{2h} does the notation of $2^1A_g^-$ and $^1B_u^+$ hold rigorously, but the notation of $2A_g$ and B_u is still used for carotenoids that have a "symmetrical" polyene chromophore).

The S_2 (B_u) state is reached directly by photoexcitation from the S_0 (A_g) state, but the S_1 ($2A_g$) state is reached only by internal conversion via the S_2 (B_u) state. The lifetimes of the S_1 and S_2 states are in the order of 10 and 0.1 ps, respectively. Therefore, picosecond transient absorption spectroscopy can be used to probe the S_1 state, whereas femtosecond transient absorption spectroscopy (described in the next section) is needed to probe the S_2 state. The S_1 lifetime can be determined either by bleaching of the $S_2 \leftarrow S_0$ ($S_n \leftarrow S_0$) absorption or by the $S_n \leftarrow S_1$ absorption. Further, the $S_n \leftarrow S_1$ absorption should provide information concerning the S_1 state structure, because it should depend on the length and configuration of the conjugated chain, as in the case of the $T_n \leftarrow T_1$ absorption. It also provides information on energy transfer. Carotenoid–BChl energy transfer can be established when the decay of S_1 carotenoid and the rise of S_1 BChl show the same value for their rate constant. The short S_1 lifetime calls for a mechanism of efficient energy transfer which most probably takes advantage of close contact between the carotenoid and BChl molecules.

(b)Examples of application

*The S_1 state of all-*trans*-ß-carotene*

The $S_n \leftarrow S_1$ absorption and the bleaching of the $S_2 \leftarrow S_0$ absorption for all-*trans*-ß-carotene and its completely deuterated derivative ($C_{40}D_{56}$) were recorded by using 450 nm, 4 ps pulses for excitation (Wasielewski et al. 1989). Figure 9.13 shows the entire transient spectra, which indicate both bleaching of the $S_2 \leftarrow S_0$ absorption with vibronic structures in the 420–510 nm region and the appearance of a new absorption band above 510 nm. Figure 9.14 shows transient absorption changes at 505 nm and 560 nm. They recover with similar kinetics, and deconvolution of these data with the instrument response yielded the S_1 state lifetime: 8.1 (± 0.5) ps for all-*trans*-ß-carotene and 10.5 (± 0.6) ps for all-*trans*-ß-carotene-D_{56} both in 3-methylpentane at 21 °C. These recovery times increase by about a factor of 2 in glassy 3-methylpentane, and are nearly independent of temperature from -173 to -253 °C.

The short S_1 ($2^1A_g^-$) lifetime indicates efficient non-radiative decay which presumably occurs through vibronic coupling (see Section 9.3.3 (b)). The lifetimes of the S_1 state of ß-carotene and fully deuterated ß-carotene, determined as a function of temperature, suggest that both the C–H stretching (bending) modes and large-amplitude skeletal motions are **not** primarily responsible for the rapid non-radiative decay. Instead, the most likely candidate for the normal mode that contributes significantly to the non-radiative decay is a small-amplitude carbon–carbon stretching motion along the conjugated backbone. Molecular orbital calculations support the idea that changes in the carbon–carbon bondlengths are necessary to provide a large Frank–Condon factor, which facilitates rapid non-radiative relaxation from the S_1 ($2^1A_g^-$) to the S_0 ($^1A_g^-$) state.

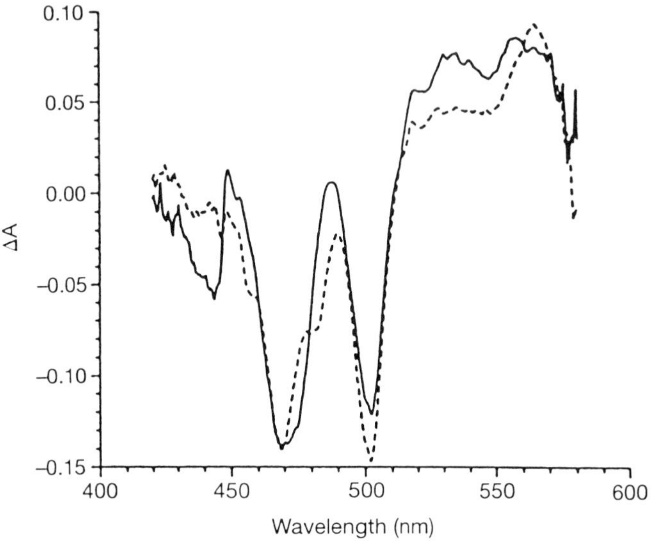

Figure 9.13 Transient absorption spectra of all-*trans*-ß-carotene (———) and fully deuterated all-*trans*-ß-carotene (D_{56}) (•••••••••••••). (Reproduced from Wasielewski et al. 1989)

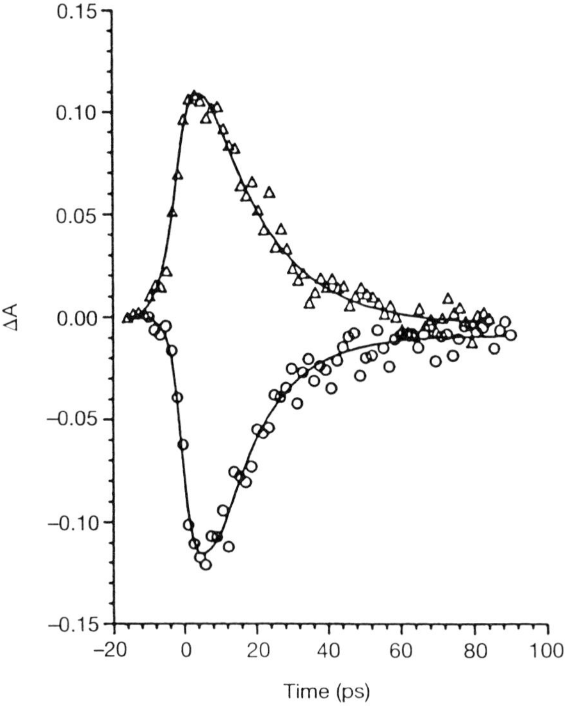

Figure 9.14 Transient absorption changes at 505 nm (O) and 560 nm (Δ) for all-*trans*-ß-carotene. (Reproduced from Wasielewski et al. 1989)

The S_1 state of geometrical isomers of β-carotene

Time-resolved absorption spectra of geometrical isomers of β-carotene were recorded by using 20–25 ps, 355 nm pump and white-continuum interrogating pulses to examine the possibility of isomerisation in the S_1 state and of intersystem crossing to the T_1 state (Hashimoto et al. 1991). Figure 9.15 shows a set of time-resolved absorption spectra for the all-*trans*, 9-*cis*, 13-*cis* and 15-*cis* isomers. Each isomer shows a strong absorption around 560 nm immediately after excitation. The absorption maximum differs from one isomer to another; presumably, the "all-*trans*" and each "*cis*" S_1 species are detected. No time-dependent shift of each $S_n \leftarrow S_1$ absorption was observed, a fact which strongly suggests that no isomerisation takes place in the S_1 state. The lifetimes which were estimated approximately from the time-resolved spectra were similar for all isomers; the lifetimes for the all-*trans* and 15-*cis* isomers were determined precisely, by using 0.5 ps pump and probing pulses, to be 12.4 (± 0.5) and 14.0 (± 0.5) ps, respectively.

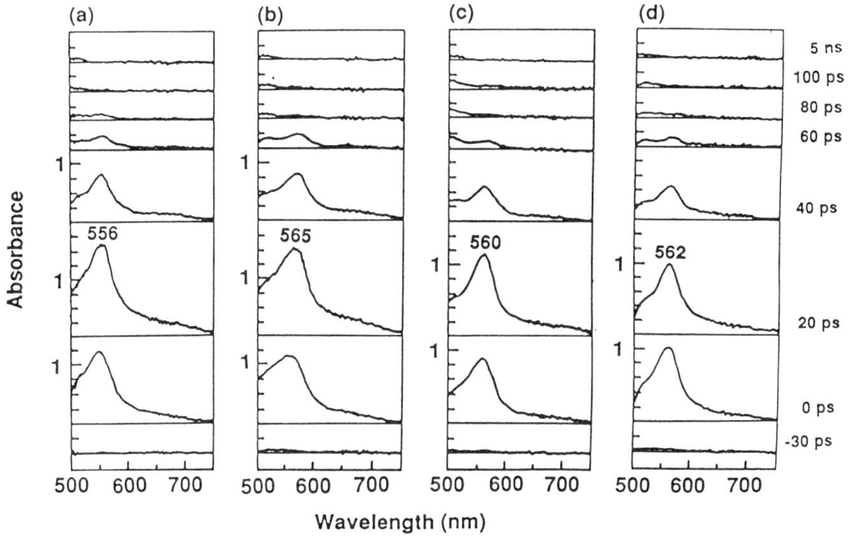

Figure 9.15 Time-resolved absorption spectra of geometrical isomers of β-carotene (a) all-*trans*, (b) 9-*cis*, (c) 13-*cis* and (d) 15-*cis*. (Revised from Hashimoto et al. 1991)

The set of spectra show also a very weak absorption profile near 520 nm in the 5 ns spectrum of each isomer; this becomes evident at about 60 ps after excitation. On the basis of their wavelengths and longer lifetimes, the absorptions were ascribed to the $T_n \leftarrow T_1$ transition. The result indicates that intersystem crossing does take place in β-carotene; the quantum yield was estimated to be of the order of 1 x 10^{-3}.

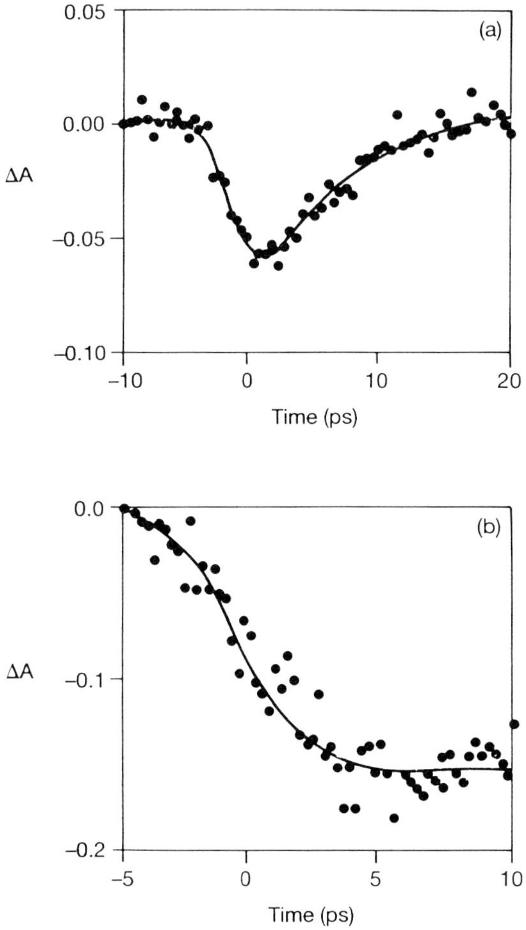

Figure 9.16 Carotenoid–BChl energy transfer in the B800-850 LHC from *Rps. acidophila* 7750. (a) Bleaching and recovery of the $S_2 \leftarrow S_0$ absorption of carotenoids at 480 nm and (b) bleaching of the B850 BChl at 860 nm. The time constant of the recovery of the carotenoid absorption and the rise of BChl bleaching were determined to be 5.6 ± 0.9 ps and 6.1 ± 0.9 ps, respectively. (Revised from Wasielewski et al. 1986)

Carotenoid–Bchl energy transfer in the light-harvesting complex

Singlet energy transfer from carotenoid to BChl in the B800–850 LHC from *Rhodopseudomonas acidophila* 7750 was measured directly by selective excitation of the carotenoid with 515 nm pulses (Wasielewski et al. 1986). The total instrument response function was 4 ps, and a 20 μJ pulse was used to excite the sample with, at most, one photon per protein molecule. The time constant of the recovery of the $S_2 \leftarrow S_0$ absorption of carotenoids (rhodopin, **8**; rhodopin glucoside, **9**; and lycopene, **10**) and that of the rise of the BChl bleaching have been determined to be 5.6 (± 0.9) ps and 6.1 (± 0.9) ps, respectively (Fig. 9.16). These time constants agree with each other, within the limit of experimental error, and establish the carotenoid–BChl energy transfer. No bleaching of the B800 BChl was observed.

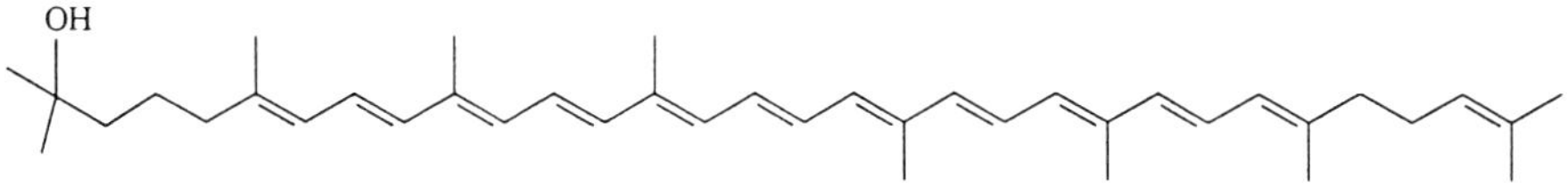

8

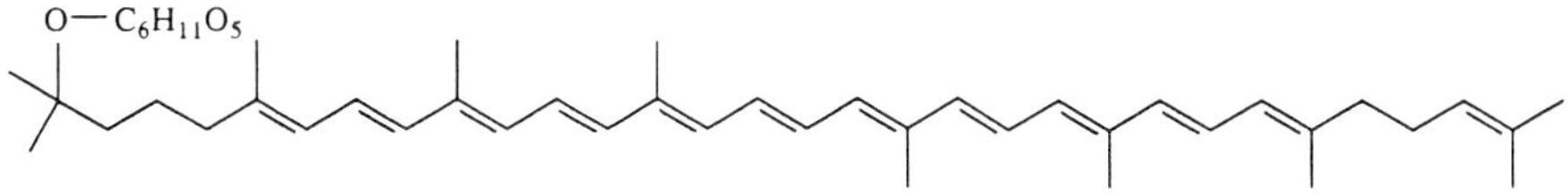

9

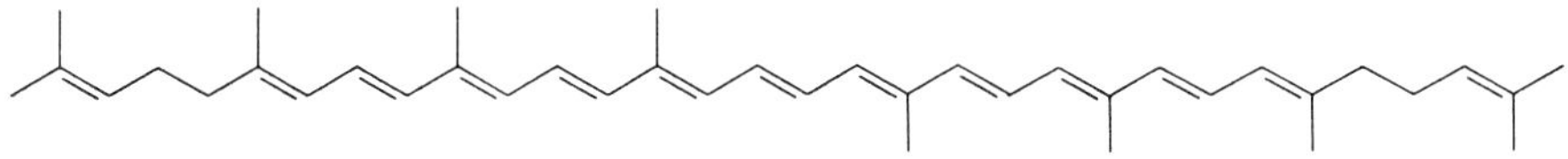

10

The efficiency of singlet energy transfer, as determined from its fluorescence excitation spectrum, is 50%. The above transient absorption result shows that the sum of the rate constants for energy transfer and for the intrinsic decay ($S_1 \rightarrow S_0$) is 1.7 x 10^{11} s^{-1}. Thus, the rate constant for energy transfer is estimated to be 8.5 x 10^{10} s^{-1}. This rate is comparable to that for a model system with a donor–acceptor distance of only 2 Å. Therefore, the efficient energy transfer implies that the electronic interaction between the carotenoid donor and the chlorophyll acceptor must be very strong.

9.2.4. Bleaching of $S_2 \leftarrow S_0$ Absorption and $S_2 \rightarrow S_0$ Fluorescence: S_2 State of Carotenoids

(a)Fundamentals

Determination of the S_2 state lifetime

The S_2 lifetimes can be determined either directly by femtosecond transient absorption spectroscopy or indirectly by fluorescence spectroscopy. The former utilises recovery of the bleaching of the $S_2 \leftarrow S_0$ absorption; no $S_n \leftarrow S_2$ transient absorption has been reported. It takes advantage of the fact that the S_2 lifetime (10^{-1} ps) is two orders of magnitude shorter than the S_1 lifetime (10^1 ps). By selecting the wavelength of the isosbestic point between the bleaching of the $S_2 \leftarrow S_0$ absorption and the appearance of the $S_n \leftarrow S_1$ absorption, the S_2 lifetime can be correctly determined from the transient data. Fluorescence spectroscopy takes advantage of the fact that the S_2 (B_u) $\rightarrow$ S_0 (A_g) transition is optically-allowed. The quantum yield of fluorescence (ϕ_f) is given, in terms of the radiative and non-radiative rate constants (k_r and k_{nr}) or lifetimes ($\tau_r = k_r^{-1}$ and $\tau_{nr} = k_{nr}^{-1}$),

$$\phi_f = \frac{k_r}{k_r + k_{nr}} = \frac{\tau}{\tau_r} \quad (9.12)$$

and therefore

$$\tau = \phi_f \times \tau_r \quad (9.13)$$

When $\phi_f << 1$ ($k_r << k_{nr}$), an approximation holds as

$$\tau_{nr} = \phi_f \times \tau_r = \tau \quad (9.14)$$

On the other hand, the radiative rate constant (k_r) is obtained from the absorption coefficient (ε_{max}) by using the relation,

$$k_r = 10^4\, \varepsilon_{max} \quad (9.15)$$

A typical observed value of ε_{max}, $\approx 10^5$ M^{-1} cm^{-1} leads to a value for $\tau_r \approx 10^{-9}$ s. When it is combined with another typical value of $\phi_f \approx 10^{-4}$, $\tau (= \tau_{nr}) \approx 10^{-13}$ s is obtained.

(b) Examples of application

The S_2 state of all-trans-ß-carotene

The S_2 lifetime of all-*trans*-ß-carotene was determined by both femtosecond transient absorption and fluorescence spectroscopy to be 0.25 ps* in ethanol and 0.2 ps* in CS_2 at room temperature (Shreve et al. 1991). (* The values cited in the original report are interchanged.) Figure 9.17a shows induced $S_n \leftarrow S_1$ absorption at 540 nm following excitation at 480 nm in ethanol; the fit of the decay yielded the S_1 lifetime of 9.5 ps. Figure 9.17b shows the primary transient features, which indicate the presence of an additional excited state with a lifetime of about 0.25 ps; it decays into a second state, whose subsequent relaxation matches the above S_1 lifetime of 9.5 ps. In particular, the isosbestic point for the S_0 and S_1 species is 490 nm in ethanol; the recording at 498 nm shows clearly an initial bleach and decay to the null absorption. The result establishes the presence of the extremely short-lived S_2 ($^1B_u^+$) state. Transient data and fits, as shown in the figures, were obtained also for CS_2 solutions (time constants, 0.2 ps and 11 ps). Figure 9.17c shows the fluorescence spectrum for ß-carotene in ethanol solution excited at 440 nm. The integrated emission yields (ϕ_f) are 1.9 x 10^{-4} in ethanol and 1.6 x 10^{-4} in CS_2. Further, if the entire absorption features from about 20 000 cm^{-1} to 27 000 cm^{-1} are taken from

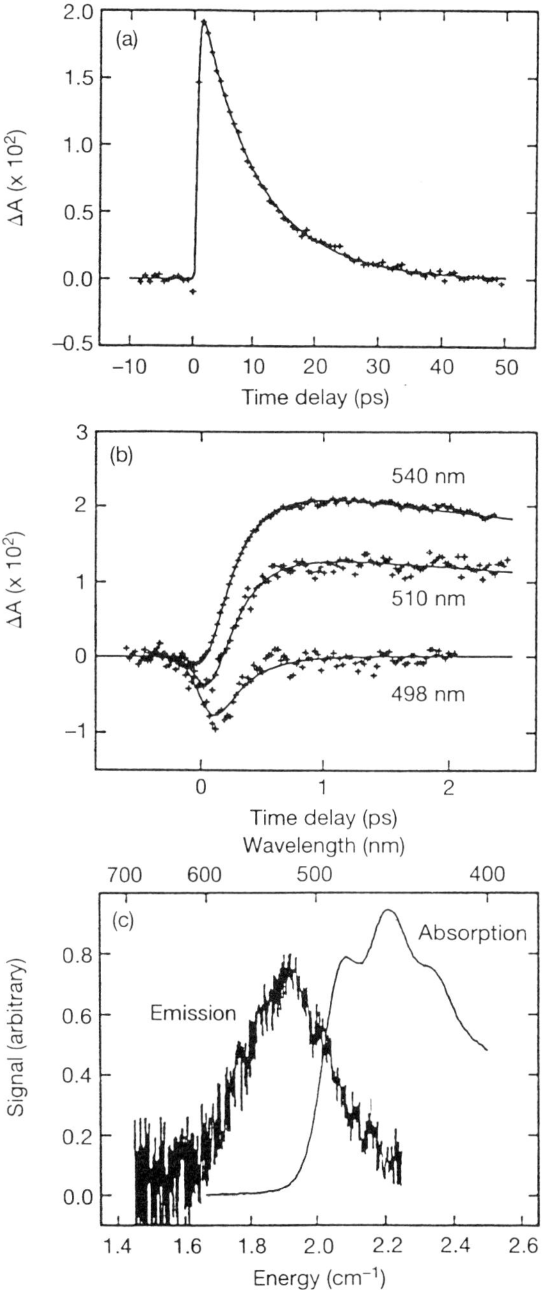

Figure 9.17 (a) Transient absorption and fluorescence of all-*trans*-ß-carotene in ethanol at room temperature. (a) Induced absorption and decay at 540 nm; fit of the decay yielded an S_1 lifetime of 9.5 ps. (b) Primary transient features at various wavelengths; fit of the initial decay yielded the S_2 lifetime of 0.25 ps. (c) Fluorescence spectrum at 440 nm excitation; the absorption spectrum is also shown for comparison. (Reproduced from Shreve et al. 1991).

one electronic state, and the emission spectrum is taken from the particular state, one obtains $\tau_r \approx 1 \times 10^{-9}$ s for the ethanol solution and $\approx 8 \times 10^{-10}$ s for the CS_2 solution. By using the above values of ϕ_f and τ_r, the S_2 state lifetime is estimated to be 0.19 and 0.13 ps in ethanol and CS_2, respectively. These are in rough agreement with that determined directly by transient absorption spectroscopy. The relative insensitivity of the S_2 lifetime to a change of solvents from ethanol to CS_2 suggests that the internal state dynamics of carotenoids *in vitro* would be relatively unperturbed *in vivo*.

Carotenoid–BChl energy transfer in the light-harvesting complex

The electronic state dynamics of spheroidene both free in cyclohexane solution and bound to the B800–850 LHC of *Rb. sphaeroides* 2.4.1 has been examined at room temperature by femtosecond transient absorption (pump-probe cross correlations, 200–300 fs) and fluorescence spectroscopies, and evidence for direct energy transfer from the S_2 state of the carotenoid to the BChl has been presented (Shreve et al. 1992). Figure 9.18a shows the recovery of induced absorption of free spheroidene at 540 nm following excitation with a 480 nm pulse. The exponential fit of the data showed a ground state recovery time of 9.1 ps. Figure 9.18b shows the time-resolved difference absorption of free spheroidene at 498, 510 and 540 nm immediately after excitation at 480 nm. Fitting of the data to a three-state model

$$S_2 \xrightarrow{k_1} S_1 \xrightarrow{k_2} S_0$$

which, with a value of $k_2^{-1} = 9.1$ ps (fixed), yielded a value of $k_1^{-1} = 0.34$ ps. Determination of the quantum yield of fluorescence by integration of the emission spectrum (data not shown) resulted in $\phi_f = 3.3\ (\pm\ 0.6) \times 10^{-4}$. Assuming a radiative lifetime of $\tau_r = 10^{-9}$ ps (see Section 9.2.4 (a)), the S_2 lifetime ($\tau_r\phi_f = k_1^{-1}$) of 0.33 ps is obtained, in agreement with the above value.

The time-resolved difference absorption of the B800–850 complex with a probe wavelength of 800 nm following excitation with an 800 nm pump pulse (data not shown) showed a decay time of 0.7 ps for the B800 excited state. Figure 9.18c shows the time-resolved difference absorption probed at 800 nm following excitation of the carotenoid with a 510 nm pump pulse (the data have been corrected for the contribution of the B850 excited state). Figure 9.18d shows a kinetic model used for fitting of the data when the carotenoid is excited to the optically active S_2 (B_u) state (Fig. 9.18c). In this fitting the following were considered: (i) two energy transfer channels from both the S_2 and S_1 states, (ii) the B800-to-B850 transfer time (the lifetime of the B800 excited state) was fixed at 0.7 ps, and (iii) the rate constants for the internal conversion dynamics of the carotenoids were fixed at the *in vitro* values. The fitting resulted in the transfer times of 1.7 ps and 3.8 ps for the carotenoid S_2–B800 and S_1–B800, respectively. Thus, energy transfer from both the S_2 and S_1 states of the carotenoid to BChl have been established.

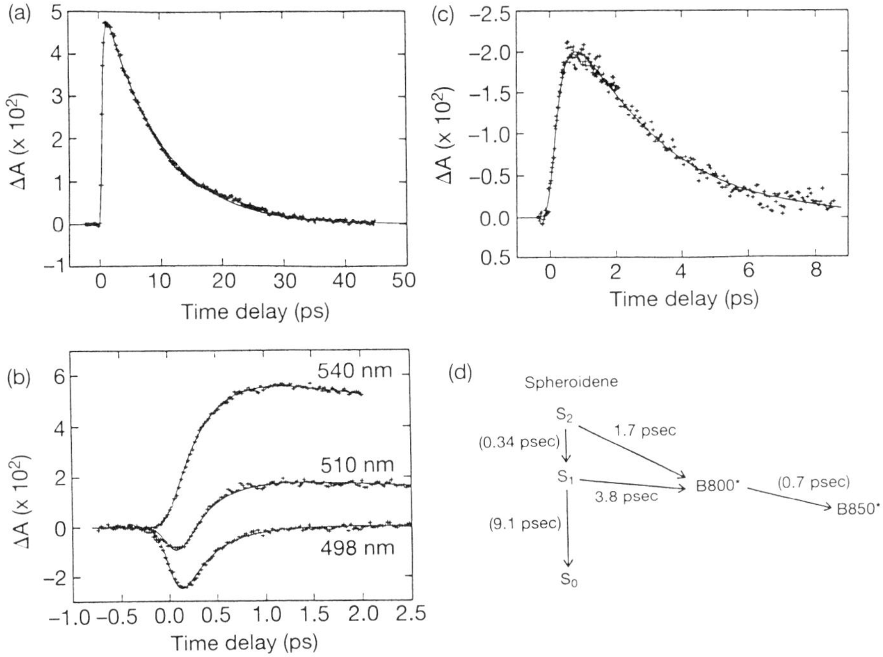

Figure 9.18 Carotenoid–BChl energy transfer in the B800–850 LHC of *Rb. sphaeroides* 2.4.1 as revealed by femtosecond absorption spectroscopy. (a) Recovery of induced absorption at 540 nm following excitation of free spheroidene (in cyclohexane) with a 480 nm pulse. (b) Time-resolved difference absorption of free spheroidene at 498, 510 and 540 nm immediately after excitation at 480 nm. (c) Time-resolved difference absorption probed at 800 nm following excitation of spheroidene bound to the LHC with a 510 nm pump pulse (the data have been corrected for the contribution of the B850 excited state). (d) A kinetic model and the result of fitting of the data shown in Fig. 9.18c; the time constants in parentheses are fixed.

Fluorescence from linear polyenes and carotenoids

The $S_2 \rightarrow S_0$ and $S_1 \rightarrow S_0$ fluorescence for all-*trans*-tetradeca-1,3,5,7,9,11,13-heptaene (**11**) were recorded at -196 °C in EPA (Cosgrove et al. 1990), and are shown in Fig. 9.19. In this polyene, both the $S_2 \rightarrow S_0$ and $S_1 \rightarrow S_0$ emissions are clearly seen. The presence or absence of these emissions depends, however, on the length of the conjugated chain: for simple unsubstituted (or methyl substituted) polyenes, the $S_2 \rightarrow S_0$ emission is observed in molecules with more than six conjugated double bonds. This trend can be attributed, in part, to a larger S_2–S_1 energy gap and the resultant decrease in $S_2 \rightarrow S_1$ radiationless decay rate in longer polyenes. For example the series, octatetraene, decapentaene, dodecahexaene, and tetradecaheptaene in *n*-hexane at room temperature shows a systematic increase in the S_2–S_1 energy gap from 4 900 to 6 300 cm^{-1}. Therefore, it is reasonable that ß-carotene shows only the $S_2 \rightarrow S_0$ emission. However, the trend predicts a 5 500–6 500 cm^{-1} S_2–S_1 energy gap for polyenes such as ß-carotene (n = 11); this is much larger than the value of 3 500 cm^{-1} reported by Thrash et al. (1977).

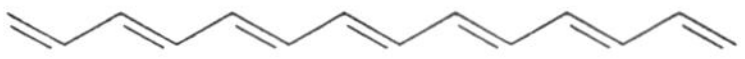

11

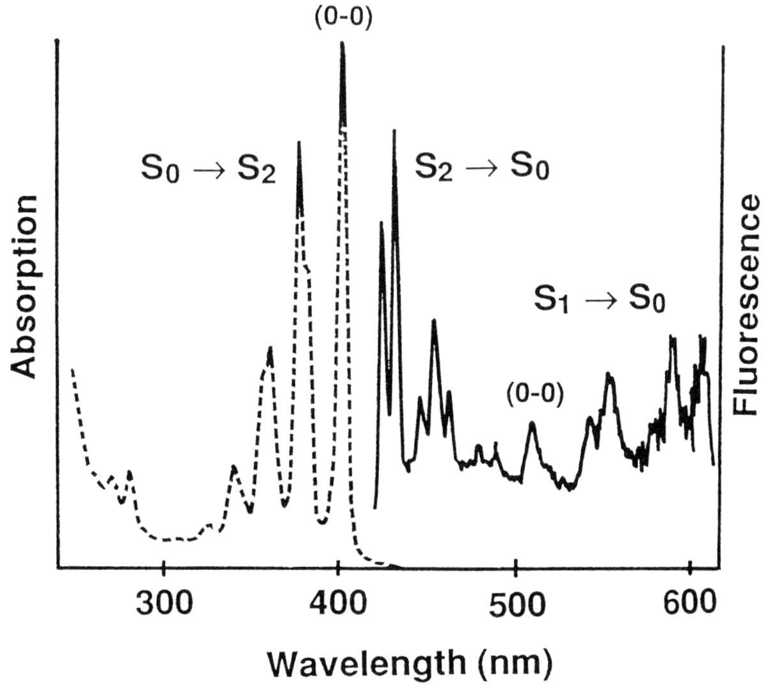

Figure 9.19 $S_2 \rightarrow S_0$ and $S_1 \rightarrow S_0$ fluorescence for all-*trans*-tetradeca-1,3,5,7,9,11,13-heptaene recorded at 405.4 nm excitation in EPA at -196 °C. (Reproduced from Cosgrove et al. 1990).

It has been reported recently that some carotenoids do show $S_1 \rightarrow S_0$ emissions. From Fig. 9.20 it can be seen that (i) spheroidene (n = 10) and neurosporene (n = 9) show only the $S_2 \rightarrow S_0$ emission, (ii) 3,4,5,6-tetrahydrospheroidene (**12**, n = 8) shows both the $S_2 \rightarrow S_0$ and $S_1 \rightarrow S_0$ emissions, and (iii) 3,4,7,8-tetrahydrospheroidene (**13**, n = 7) shows minor $S_2 \rightarrow S_0$ emission and major $S_1 \rightarrow S_0$ emission. This observation (DeCoster et al. 1992) parallels that of the linear polyenes discussed above. Mimuro et al. (1992) have shown that carotenoids having an asymmetrical polyene chain (e.g. fucoxanthin, **14**; siphonaxanthin, **15**; peridinin, **16**; and ß-apo-8´-carotenal, **17**), to which a carbonyl group is attached, show the $S_1 \rightarrow S_0$ emission, and that carotenoids having a symmetric polyene chain (e.g. neoxanthin, **18**; and ß-carotene) show the $S_2 \rightarrow S_0$ emission (Fig. 9.21). Most probably this is due to the symmetry of the S_1 state.

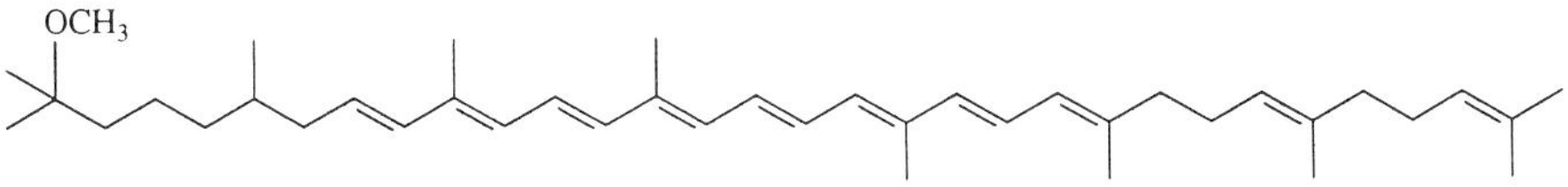

12

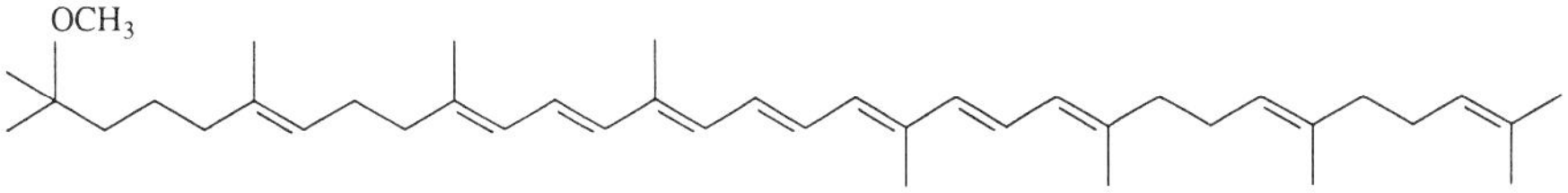

13

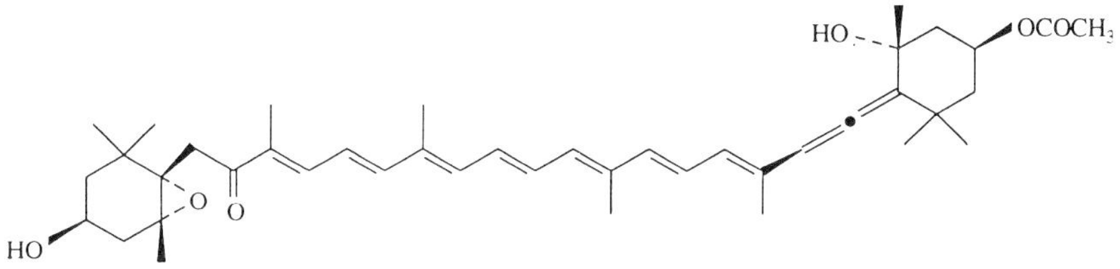

14

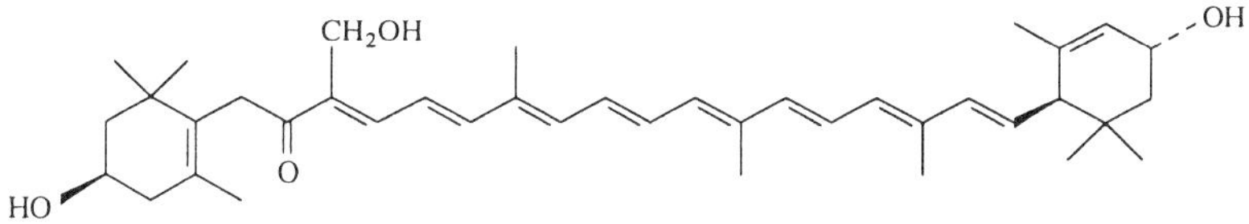

15

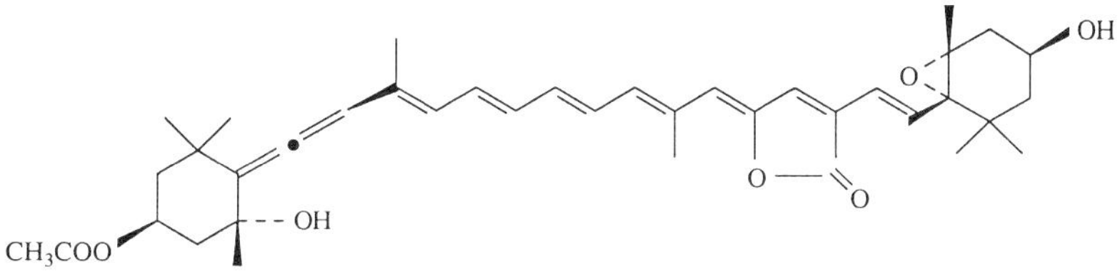

16

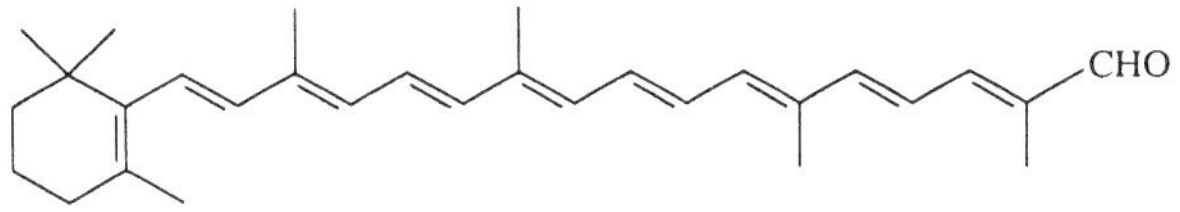

17

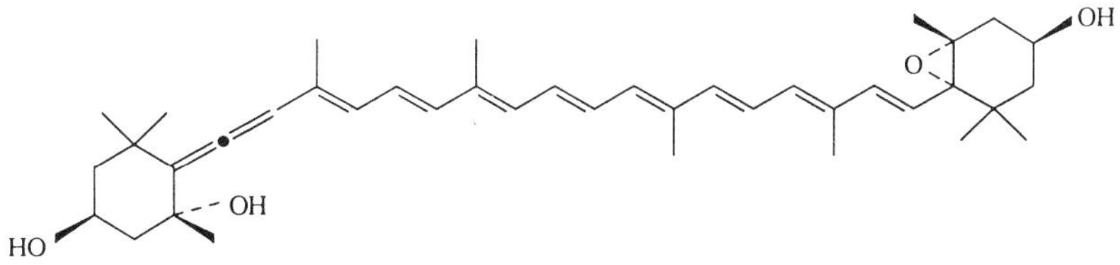

18

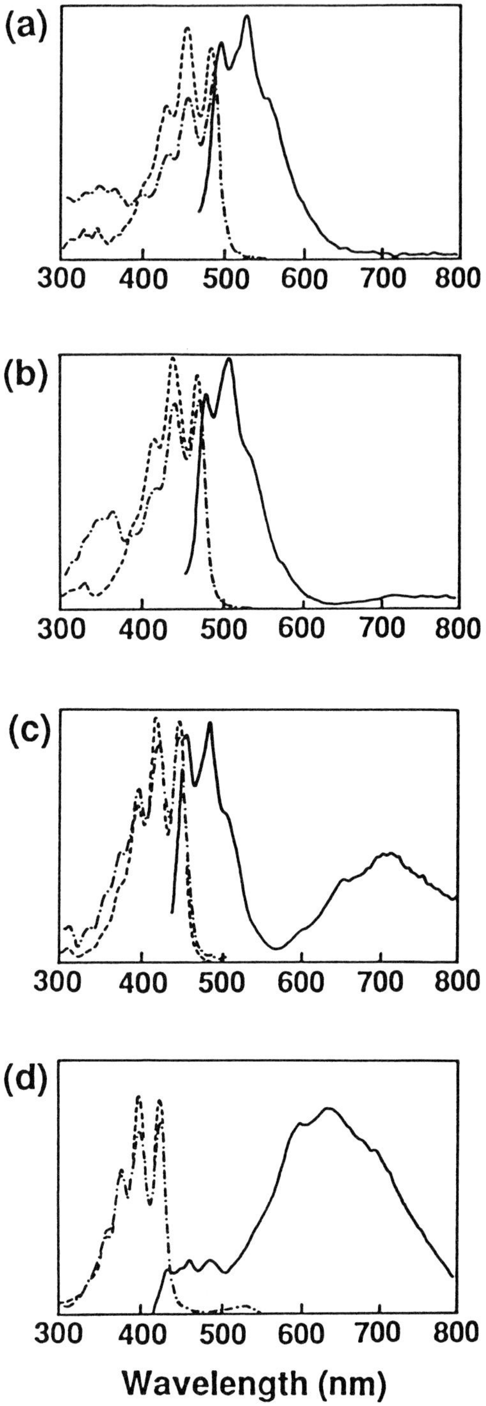

Figure 9.20 Fluorescence from (a) spheroidene, (b) methoxyneurosporene, (c) 3,4,5,6-

tetrahydrospheroidene and (d) 3,4,7,8-tetrahydrospheroidene. Emission spectra (——)were recorded at (a) 450, (b) 437, (c) 417 and (d) 400 nm excitation. Excitation spectra (–•–•–•) were recorded at (a) 560, (b) 530, (c) 710 and (d) 700 nm emission. Absorption spectra (- - - - - -) are also shown for comparison. (Revised from DeCoster et al. 1992)

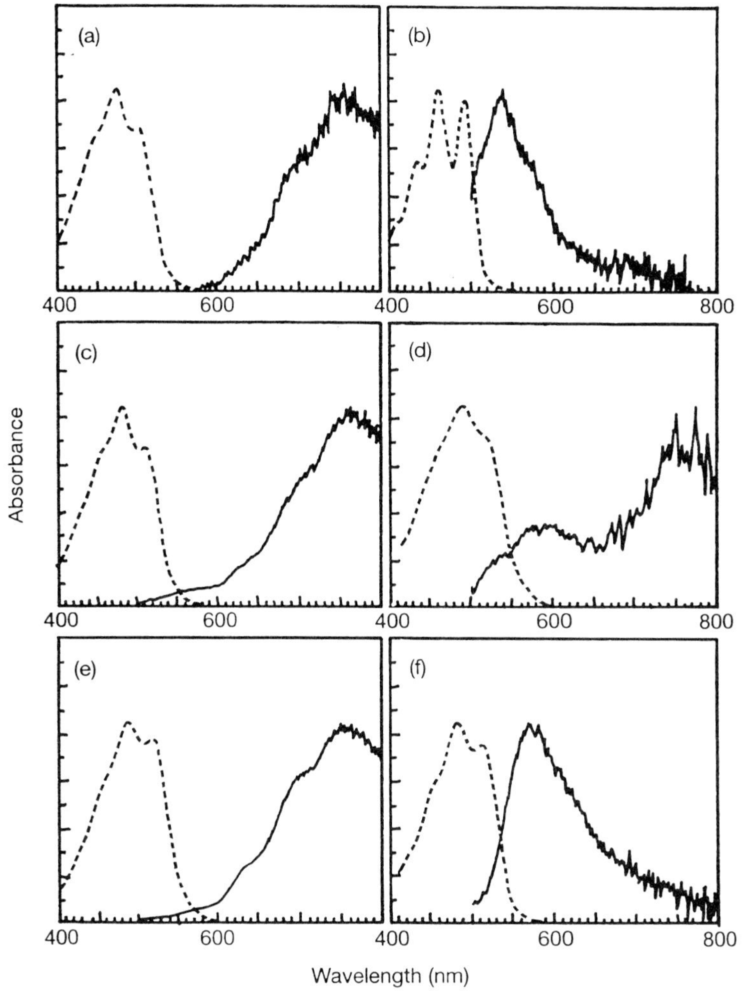

Figure 9.21 Absorption spectra and fluorescence from (a) fucoxanthin, (b) neoxanthin, (c) siphonaxanthin, (d) ß-apo-8′-carotenal, (e) peridinin, and (f) ß-carotene. (Revised from Mimuro et al. 1992)

All the above results indicate that the energy of the S_1 ($2^1A_g^-$) state is much lower than that proposed by Thrash et al. (1977). An energy diagram for ß-carotene and BChl is shown in Fig. 9.22. However, the S_1 energy level of ß-carotene needs to be established; the possibility of singlet energy transfer from the S_1 ($2A_g$) state of a carotenoid to the Q_x level of BChl is now completely excluded.

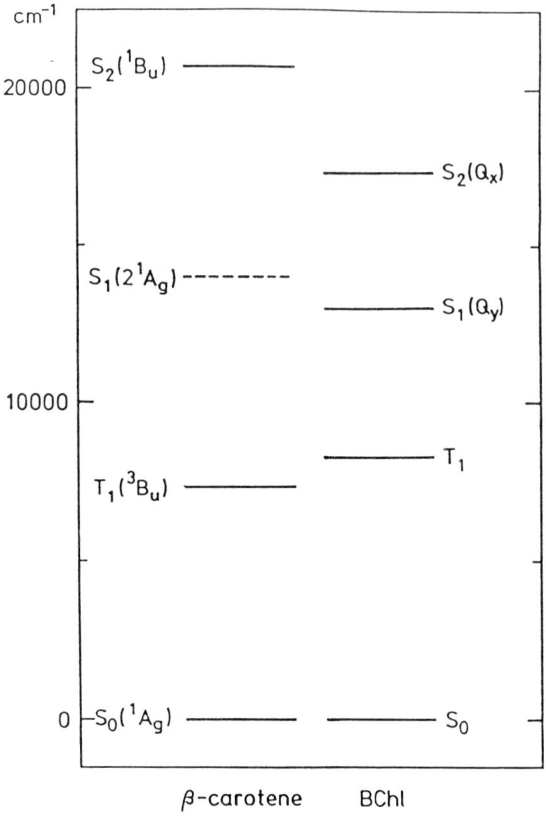

Figure 9.22 Energy diagram of ß-carotene and bacteriochlorophyll *a* (BChl). (Revised from Koyama 1991)

9.3 RAMAN SPECTROSCOPY

9.3.1 Resonance Raman: S_0 State Carotenoids

(a)Fundamentals

Normal modes of β-carotene

The characteristics of the structure of ß-carotene lead to the following classifications of the normal vibrational modes:

(i) "Skeletal modes" and "peripheral modes". The "skeletal modes" include the vibrations of the conjugated backbone and of the methyl groups attached to it, whilst the "peripheral modes" include the vibrations of the ß-rings.

(ii) "In-plane (ip) modes" and "out-of-plane (op) modes". Assuming a planar, conjugated polyene backbone in the *cis* or *trans* configuration, we can classify the skeletal modes into "ip modes" and "op modes".

(iii) Classification by symmetry. If the conjugated backbone is completely planar, the all-*trans* isomer has C_{2h} symmetry and its normal modes are classified into Raman-active A_g and B_g species and infrared-active A_u and B_u species. If there is twisting of the conjugated backbone, its symmetry can be C_i and the normal modes are classified into Raman-active A_g species and infrared-active A_u species. In the case of 7-*cis* and 9-*cis* isomers, their normal modes in the central, symmetrical part of the chain (pseudo-C_i symmetry) can be classified into "A_g-like" and "A_u-like" modes. The completely planar 15-*cis* configuration has C_{2v} symmetry, and its normal modes are classified into Raman and infrared-active A_1, B_1 and B_2 species and Raman-active A_2 species. If there is a twist around the C15,15′ double bond, the symmetry is degraded to C_2, and the normal modes are classified into A and B species (both Raman and infrared-active).

(iv) "Local modes" and "extended modes". Complicated mixing of the same type of vibrations, and coupling between two different types of vibrations can take place in the polyene backbone. In an extended all-*trans* chain, a normal vibration which is approximately localised within the group of atoms between two methyl groups is termed a "local mode", whereas a normal vibration which extends beyond the two methyl groups is termed an "extended mode". When a vibration takes place around a *cis* or *trans* double bond, it can be classified as "*cis*-mode" or "*trans*-mode". The former is always a "local mode", but the latter can be either a "local mode" or an "extended mode".

The normal modes of vibrations can be calculated by Wilson's GF matrix method. In order to reduce the vibrational degree of freedom, $3m$-6 (where m is the number of atoms), "a ß-carotene model", in which the ß-rings were removed, was used for the calculations. The normal modes can be expressed in terms of displacements of atoms (the L_x matrix).

Resonance Raman process

The uniqueness of **resonance** Raman spectroscopy is that it detects only the Raman lines of a chromophore, for which the energy gap of the transition matches the energy of the probe light. (A complete description of the resonance Raman process is given in the next section.) Under this rigorous resonance condition, the intensity of the Raman lines is increased by a factor of $\approx 10^5$ when compared to a completely off-resonance condition. The Raman intensity is strongly enhanced even under a pre- (or post-) resonance condition, when the energy of the probe light is lower (or higher) than that of the transition. Thus, by tuning the wavelength of light, one can selectively probe the Raman spectrum of the carotenoid (or BChl) molecule *in situ* bound to the RC or the LHC. When the probe light is in resonance with the π–π^* transition, only the vibrational modes that take place within the conjugated backbone ("skeletal modes") are detected.

In the resonance Raman process enhancement is seen of those vibrational modes in which the displacements of atoms are in accordance with the distortion of the molecule when the transition from the ground state to the resonant excited–state takes place. In the case of polyene molecules, change in the bond lengths of the C=C and C–C bonds, torsion around the C=C bond, and deformation of the C–H bond(s) are expected to take place upon excitation. Therefore, these vibrational modes in carotenoids ("skeletal modes") should cause high resonance Raman intensities. In particular, totally symmetrical modes should be strongly enhanced. This selective enhancement of the particular modes makes the Raman spectra of carotenoids much simpler than their infrared spectra, and makes them similar for different carotenoids.

Raman lines sensitive to the structures of the conjugated backbone

Raman lines which are expected to be sensitive to changes in *cis-trans* configurations are as follows: (i) the C=C stretchings which are coupled in-phase to give rise to strong Raman lines, (ii) "local modes" characteristic of the *cis* or *trans* -HC=CH-group in which the C=C stretching and the C–H bendings are coupled to one another, and (iii) "local or extended modes" of a *trans* zigzag chain, in which the C–H bendings and the C–C stretchings are coupled to one another. The ways of mixing and coupling of vibrations as well as force constants themselves may be altered when the *cis* bend is introduced at various positions of the conjugated chain.

A Raman line which is expected to be sensitive to a twisting of the conjugated backbone is that due to the C–H op wagging coupled with the C=C torsion. The particular mode is strongly resonance-enhanced when there is a twisting around the C=C bond.

(b) Examples of Application

Prediction of the configuration of carotenoids in the reaction centre; spectral comparison with geometrical isomers of ß-carotene

The Raman lines of carotenoids bound to the RCs *in situ* were compared with those of ß-carotene isomers, and the configuration of the RC bound carotenoids was determined to be 15-*cis* (Koyama et al. 1983b). Figure 9.23 compares the Raman lines of spheroidene and neurosporene bound to the RCs of *Rb. sphaeroides* 2.4.1 and G1C with those of 14 different *cis-trans* isomers of ß-carotene in both the 1 570–1 520 cm^{-1} (C=C stretching) and the 1 300–1 100 cm^{-1} (C–H ip bending and C–C stretching) regions. The Raman lines are classified schematically into strong, medium and weak. For the former region (1 570–1 520 cm^{-1}), the shifts of the C=C stretching Raman lines of the RC-bound carotenoids, with reference to the Raman lines of their all-*trans* isomers (shown by dotted lines), are 10 cm^{-1}. 15-*cis*-ß-

Carotene shows exactly the same degree of shift; additional isomers giving a shift within the limit of 10 (± 5) cm^{-1} are 9-*cis*, 13-*cis*, 9,13-di-*cis*, 9,15-di-*cis* and the uncharacterised isomers no. 5 and no. 16. For the latter region (1 300–1 100 cm^{-1}), the RC-bound carotenoids exhibit a characteristic, medium Raman line around 1240 cm^{-1}; the 15-*cis* and 9,15-di-*cis* isomers of ß-carotene show a corresponding Raman line. Further, the RC-bound carotenoids show a strong Raman line around 1 160 cm^{-1}, but they show no strong or medium Raman line below 1 150 cm^{-1}. The latter criterion excludes the possibility of the 9,15-di-*cis* configuration. The above comparisons lead to the conclusion that the RC-bound carotenoids have a 15-*cis* configuration.

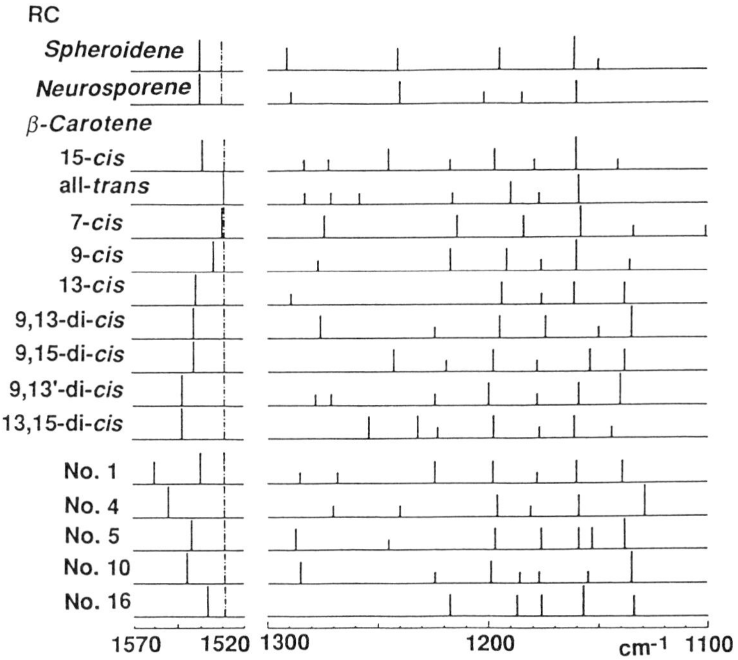

Figure 9.23 Comparison of the Raman lines of spheroidene and neurosporene bound to the RCs of *Rb. sphaeroides* 2.4.1 and G1C with those of 14 different *cis-trans* isomers of ß-carotene. Raman lines are classified into strong, medium, and weak; dotted broken lines indicate the Raman lines of the all-*trans* isomers for reference.(Revised from Koyama et al. 1983b)

*Key Raman lines of the all-*trans *and various* cis *configurations, and their vibrational modes*

Raman lines characteristic of each *cis* or *trans* configuration were identified experimentally, and their normal modes were analysed by means of calculations of normal vibrations (Koyama et al. 1988b). Figure 9.24 compares the spectral patterns for some geometrical isomers of ß-carotene.

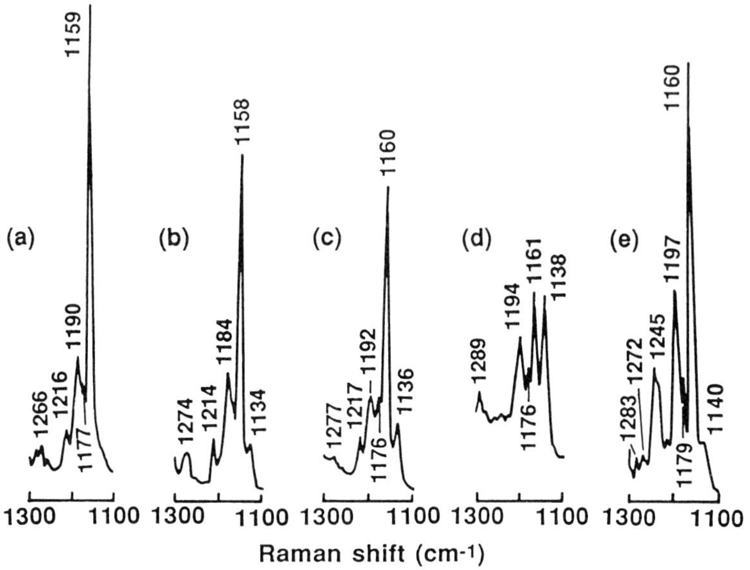

Figure 9.24 Comparison of the Raman spectral patterns of geometrical isomers of ß-carotene in the most configuration-sensitive region; (a) all-*trans*, (b) 7-*cis*, (c) 9-*cis*, (d) 13-*cis* and (e) 15-*cis*. (Revised from Koyama et al. 1983b)

There are several important features: (i) The Raman line at 1 274 cm^{-1} is regarded as a key Raman line of the 7-*cis* configuration. Figure 9.25a shows the normal mode which is responsible for this Raman line. It is a coupled mode of the C7–H and C8–H ip bendings, and is characterised as an A_1-type "*cis*" ("local") mode around that particular double-bond, C-7,8 which has local symmetry C_{2h}. (ii) The Raman line at 1 245 cm^{-1} is a key Raman line of the 15-*cis* configuration. Figure 9.25b shows two vibrational modes responsible for this Raman line, namely coupled modes of the C15–H and C15′–H ip bendings and the C15=C15′ stretching. These are also "*cis*" ("local") modes. (iii) The Raman line around 1 160 cm^{-1} decreases in intensity in the order, all-*trans* > 7-*cis* > 9-*cis* > 13-*cis*. Figure 9.25c shows the vibrational modes responsible for this Raman line. They consist of the C14–C15 and C14′–C15′ stretchings coupled with the C15–H and C15′–H ip bendings. The electronic and vibrational transition moments are parallel to each other in the case of the all-*trans* isomer, but in the other isomers they make larger angles in the order all-*trans* < 7-*cis* < 9-*cis* < 13-*cis*. This may be a reason for the change in intensity. (iv) The Raman line around 1140 cm^{-1} increases in intensity in the order (all-*trans*) < 7-*cis* < 9-*cis* < 13-*cis*. Its intensity is comparable to that of the 1 160 cm^{-1} line in the case of the 13-*cis* isomer. Figure 9.26 shows the vibrational modes responsible for this Raman line, namely a pair of vibrational modes of the C10–C11 stretching coupled with the C10–H, C11–H and C12–H ip bendings, and of the C10′–C11′ stretching coupled with the C10′–H, C11′–H and C12′–H ip bendings. (These are typical "extended modes".) These modes are coupled to form the A_g (left) and A_u (right) modes in the all-*trans* isomer. The same types of mode occur on the same part of the molecule in the 7-*cis* and 9-*cis* isomers around the centres of the different all-*trans* chain section (to the left and right) in the 13-*cis* isomer. The trend may be reflected in the Raman intensities. Thus, two sets of key Raman lines which (i) identify the unmethylated 7-*cis* and 15-*cis* configurations, and (ii) distinguish the stretched (all-*trans*), terminal-bent (7-*cis* and 9-*cis*) and central-bent (13-*cis*) configurations, were found. Thus, the configurational prediction of the RC-bound carotenoids, described above, is now rationalised.

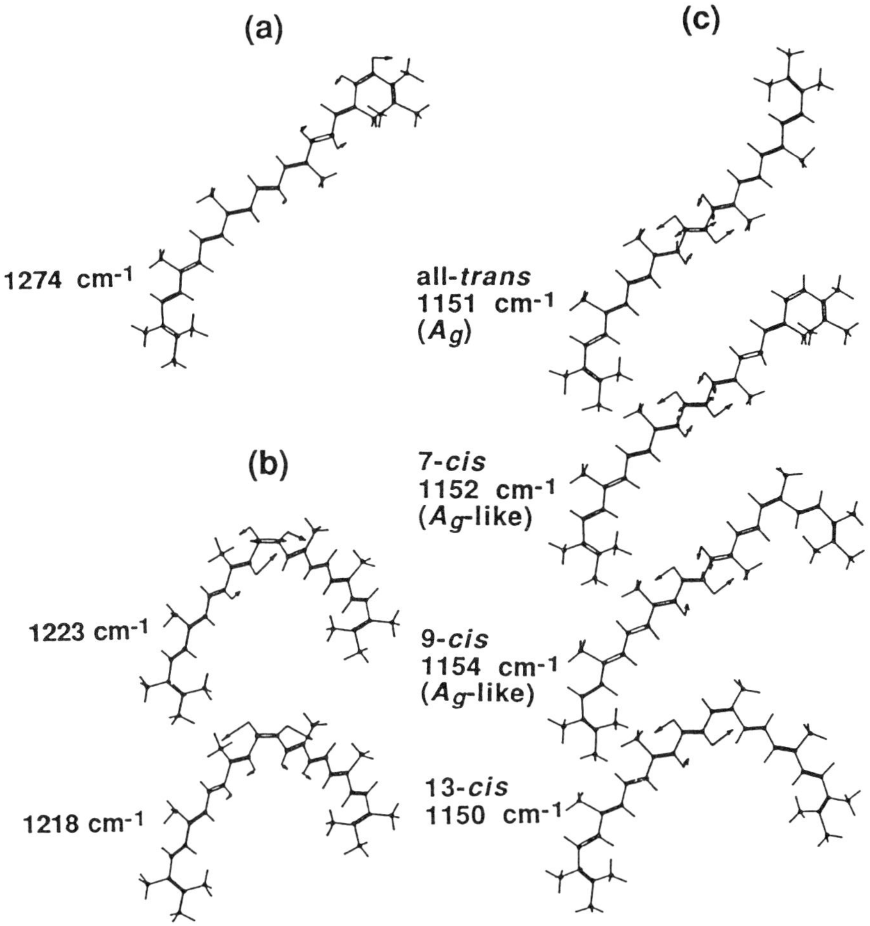

Figure 9.25 Normal modes responsible for the key Raman lines of the (a) 7-*cis* configuration at 1 274 cm^{-1}, (b) 15-*cis* configuration at 1 245 cm^{-1}, and (c) all-*trans*, 7-*cis*, 9-*cis* and 13-*cis* configurations at ~1 160 cm^{-1}. Calculated frequencies are given in the figure. (Revised from Koyama et al. 1988b)

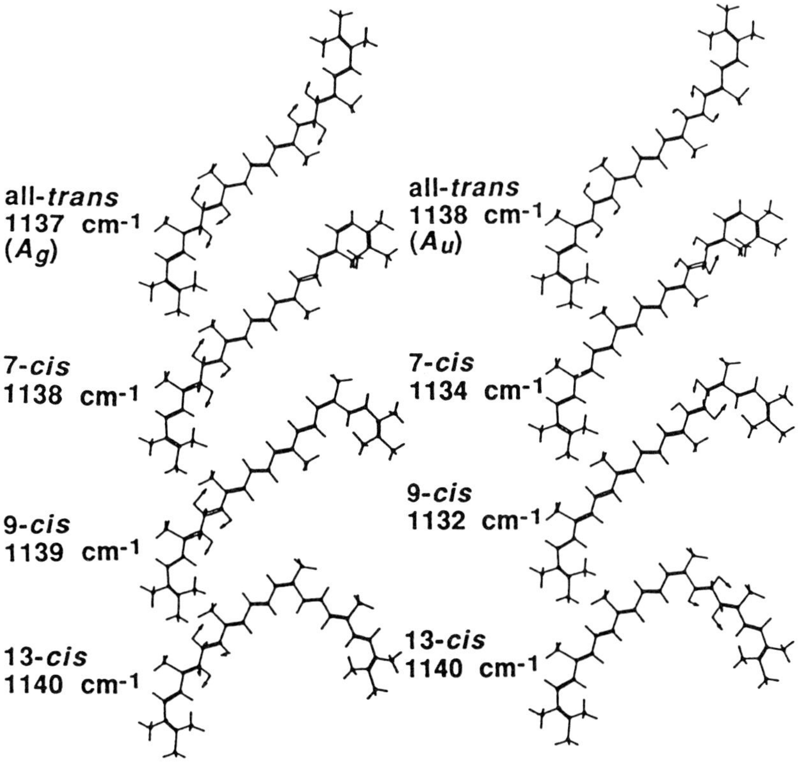

Figure 9.26 Normal modes responsible for the key Raman line around 1140 cm^{-1} of the all-*trans*, 7-*cis*, 9-*cis* and 13-*cis* configurations. (Revised from Koyama et al. 1988b)

*Distortion of the backbone of 15-*cis*-neurosporene bound to the reaction centre*
The configuration of neurosporene which was extracted from the RC and the LHC was determined to be 15-*cis* and all-*trans*, respectively, by means of resonance Raman, electronic absorption (see Section 9.2.1 (b)) and ^{1}H NMR spectroscopies (see Section 9.4.1 (b)). In particular, Raman spectroscopy was used to prove that the configuration of the carotenoid was not altered during the process of extraction and purification, and to detect some distortion of the backbone of the 15-*cis* carotenoid upon binding to the apo-complex (Koyama et al. 1988a). The spectral patterns for the bound and free all-*trans*-neurosporene shown in Fig. 9.27 are nearly identical as are the spectral patterns for the bound and free 15-*cis*-neurosporene. These observations prove that no isomerisation has taken place during the extraction procedure. Close comparison of the bound and free 15-*cis*-neurosporene, however, reveals that the relative intensity of the Raman line at 1 240 cm^{-1} (see above for the normal mode responsible for this Raman line) and that at 960 cm^{-1} (most probably the C15–H and C15′–H op waggings coupled with the C15=C15′ torsion) are enhanced upon binding. The enhancement of the resonance Raman intensities strongly suggests that a distortion (elongation and twisting) around the *cis* C-15,15′

double bond takes place in the bound, 15-*cis* carotenoid.

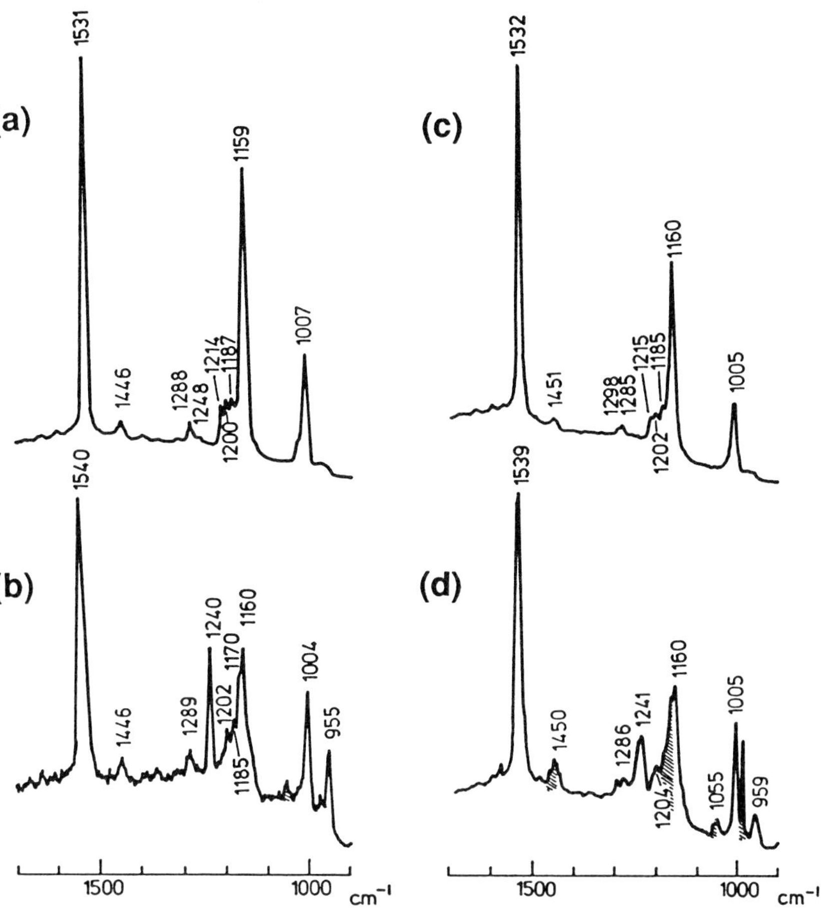

Figure 9.27 Comparison of the Raman spectra of neurosporene isomers bound to the pigment–protein complexes of *Rb. sphaeroides* G1C and those free in *n*-hexane solution. (a) All-*trans*-neurosporene bound to the LHC and (b) 15-*cis*-neurosporene bound to the RC; (c) all-*trans*-neurosporene and (d) 15-*cis*-neurosporene in *n*-hexane (shaded lines are those due to the solvent). (Revised from Koyama et al. 1988a)

Twisting of the conjugated backbone and efficiency of carotenoid–BChl energy transfer in the light-harvesting complexes

The relationship between the intensity of the 960 cm^{-1} Raman line and the energy transfer efficiency was examined for all-*trans* carotenoids bound to the B800–850 LHCs (chromatophores) of a variety of phototrophic bacteria (Hayashi et al. 1989; Noguchi et al. 1990a). Figure 9.28 shows two typical examples. The intensity of the fluorescence excitation spectrum relative to that of the absorption spectrum, when the absorption and emission of the Q_x band of BChl around 590 nm are normalised, is high for *Rb. sphaeroides* (a) and low for *Chromatium vinosum* (b). Thus, energy transfer efficiency is high for the former, and low for the latter. On the other hand, the intensity of the 960 cm^{-1} Raman line, relative to the 1 010 cm^{-1} line (the methyl ip rockings which are not structure-sensitive), is low for *Rb. sphaeroides* (c) and high for *Chr. vinosum* (d). Therefore, the conjugated all-*trans* backbone is expected to be planar in the former and twisted in the latter. Table 9.6 lists the relative intensity of the 960 cm^{-1} Raman line and the efficiency of energy transfer for various phototrophic bacteria.

Table 9. 6 Intensity of the 960 cm^{-1} Raman line and energy transfer efficiency

	Carotenoid[a]	*Number of conjugated double-bonds*	*960cm*$^{-1}$ *Raman line*[b]	*Energy transfer*[c] *efficiency (%)*[c]
Rb. sphaeroides 2.4.1	spheroidene	10	W	94(87)
Ga	chloroxanthin	9	W	90(94)
Rb. capsulatus	spheroidene	10	W	81(82)
Rc. gelatinosus	hydroxyspheroidene	10	W	75(84)
Chr. vinosum	rhodopin	11	S	28(32)
	spirilloxanthin	13		
Rps. palustris	spirilloxanthin	13	M	49(50)
R. rubrum	spirilloxanthin	13	M	32

[a] Major component(s).
[b]Relative intensity; W, weak; M, medium; and S, strong.
[c] Chromatophore membrane (B800–850 complex).

They are classified into two groups: (i) spheroidene-series carotenoids (number of conjugated double bonds, $n = 9$ to ≈ 10) bound to *Rb. sphaeroides, Rb. capsulatus, Rc. gelatinosus* adopt planar structures and show high energy transfer efficiencies (>70 %); (ii) spirilloxanthin series carotenoids (n = 11 to ≈ 13) bound to *Chr. vinosum, Rps. palustris* and *R. rubrum* adopt twisted structures and show lower energy transfer efficiencies (<50 %). When the LHC of *Chr. vinosum* was denatured with sodium dodecyl sulphate (SDS), the intensity of the 960 cm^{-1} Raman line decreased, indicating the relaxation of the twisted form into a planar form (data not shown). Further, when the major carotenoids of *Rb. sphaeroides* (spheroidene and spheroidenone) were incorporated into the LHC of *Chr. vinosum,* they exhibited a much stronger 960 cm^{-1} Raman line than when they are bound to their intrinsic LHC (data not shown). Both results support the idea that the carotenoids are distorted through strong intermolecular interaction in the LHC of *Chr. vinosum*. A variety of carotenoids with different lengths of the conjugated chain ($n = 9$ to ≈ 13) have been incorporated into the LHC of *Chr. vinosum*, where they exhibited strong or medium relative intensities of the 960 cm^{-1} line as well as low energy transfer efficiencies (Table 9.7).

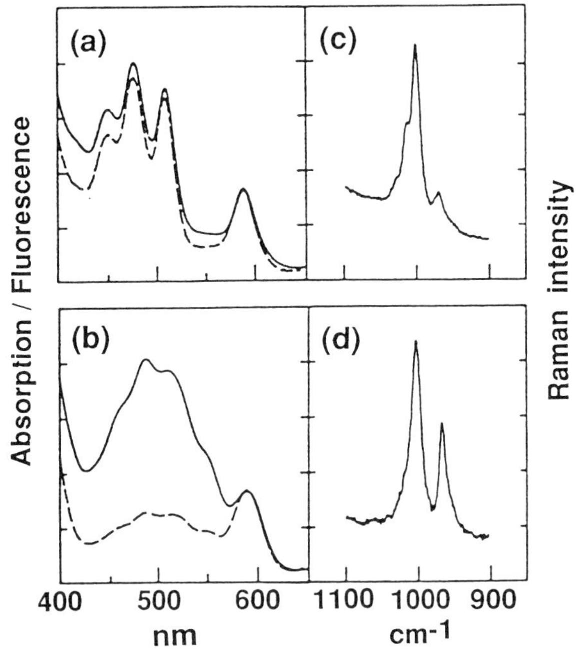

Figure 9.28 Absorption (———) and fluorescence excitation (••••••••••) spectra of the chromatophores from (a) *Rb. sphaeroides* and (b) *Chr. vinosum* (the absorption and emission spectra are normalised at the Q_x band of BChl around 590 nm) and Raman lines of the chromatophores from (c) *Rb. sphaeroides* and (d) *Chr. vinosum*. (Revised from Noguchi et al. 1990a)

Table 9.7 Intensity of the 960 cm^{-1} Raman line and energy transfer efficiency for carotenoids incorporated into B800–850 and B800–820 complexes (mixture) of *Chromatium vinosum*.

Carotenoids	*Number of conjugated double bonds*	*960cm⁻¹ Raman line*	*Energy transfer efficiency*
Neurosporene	9	M	56
Methoxyneurosporene	9	M	44
Spheroidene	10	M	37
Rhodopin	11	S	29
Spirilloxanthin	13	S	27

9.3.2 Resonance Raman excitation profile: S_2 and S_0 states of carotenoids

(a) Fundamentals

Resonance Raman excitation profile as a function of structural change upon excitation

The intensity (I_{mn}) of a resonance Raman process from the initial state (m) to the final state (n) is given by

$$I_{mn} \propto I_0 (\upsilon_0 - \upsilon_{mn})^4 \sum_{\rho\sigma} \left|(\alpha_{\rho\sigma})_{mn}\right|^2 \quad (9.16)$$

where I_0 and υ_0 are the intensity and frequency of the incident light, and $\upsilon_{mn} = (E_n - E_m)/h$, the Raman shift. The scattering tensor, $\alpha_{\rho\sigma}$, is given by the Kramers–Heisenberg dispersion formula,

$$\sum_r \left[\frac{<m|R_\sigma|r><r|R_\rho|n}{\upsilon_{rm} - \upsilon_0} + \frac{<m|R\rho|r<>r|R_\sigma|n>}{\upsilon_{rn} - \upsilon_0} \right] \quad (9.17)$$

where r is an intermediate state, R_σ is the electronic dipole moment operator, and $\upsilon_{rm} = (E_r - E_m)/h$. Under a rigorous resonance condition, $\upsilon_{rm} \approx \upsilon_0$ and the second term is negligible when compared to the first term (the second term will therefore be abbreviated hereafter). Adiabatic approximation leads to the expressions,

$$\begin{aligned} |m> &= |g> \; |i> \\ |n> &= |g> \; |j> \quad (9.18) \\ |r> &= |e> \; |v> \end{aligned}$$

where $|g>$ and $|e>$ are the electronic wave functions of the ground and excited states, and $|i>$, $|j>$ and $|v>$, the vibrational wave functions. The Herzberg–Teller expansion is a formal expansion of an electronic wave function in a Taylor's series of displacement of the nuclear coordinates from the equilibrium position in the ground state

$$|e> = |e^\circ> + \sum_a \sum_{s \neq e}{}' \frac{(h_a)^\circ_{es} \bullet \Delta Q_a}{E^\circ_e - E^\circ_s} \left| s^\circ> \right. \quad \left| g = \right| g^\circ> + \sum_a \sum_{t \neq g}{}' \frac{(h_a)^\circ_{gt} \bullet \Delta Q_a}{E^\circ_g - E^\circ_t} |t^\circ> \quad (9.$$

(9.19) where $h^\circ_a = (H/\ Q_a)^\circ$ is the vibronic coupling operator for the normal mode a. Introducing Eqs. 9.18 and 9.19 into Eq. 9.17 leads to the Albrecht *A* and *B* terms (Tang and Albrecht 1970). (The *C* term is abbreviated, because it can be shown to be much smaller.)

$$(\alpha_{\rho\sigma})_{gi,gj} = A + B \quad (9.20)$$

$$A = \sum_{e \neq g}{}' \sum_{v} \left[\frac{(g^\circ|R_\sigma|e^\circ)(e^\circ|R_\rho|g^\circ)}{E_{ev} - E_{gi} - E_o} \right] < i|v >< v|j > \quad (9.21)$$

$$B = \sum_{e \neq g}{}' \sum_{v} \sum_{s \neq e}{}' \sum_{a} \left[\frac{(g^\circ|R_\sigma|e^\circ)(g^\circ|h_a|s^\circ)(s^\circ|R_\rho|g^\circ)}{E_{ev} - E_{gi} - E_o} \right] \times \frac{< i|v >< v|Q_a|j >}{E^\circ_e - E^\circ_s}$$

$$+ \left[\frac{(g^\circ|R_\sigma|s^\circ)(s^\circ|h_a|e^\circ)(e^\circ|R_\sigma|g^\circ)}{E_{ev} - E_{gi} - E_o} \right] \times \frac{< i|Q_a|v >< v|j >}{E^\circ_e - E^\circ_s} \quad (9.22)$$

As for the particular case of the totally symmetrical vibrations of the C=C stretching (ν_1) and the C–C stretching (ν_2) of a carotenoid (ß-carotene) which are in resonance with the $S_2\,(^1B_u^+) \leftarrow S_1\,(^1A_g^-)$ electronic transition, the *A* term was found to be mainly responsible for the resonance Raman intensities (Inagaki et al. 1974). Under these conditions, only the R_z component and only one excited state in resonance need to be taken into account in the *A* term. Thus, the equations are simplified as

$$(\alpha_{zz})_{gi,gj} \approx A \quad (9.23)$$

$$A = \left[\frac{(g^\circ|R_z|e^\circ)(e^\circ|R_z|g^\circ)}{h} \right] f_A \quad (9.24$$

$$f_A = \sum_{v} \frac{< i|v >< v|j}{\upsilon_{ev,gi} - \upsilon_o + i\gamma_e'} \quad (9.25)$$

where $\upsilon_{ev,gi} = (E_{ev} - E_{gi})/h$, $\upsilon_0 = E_0/h$, and γ_e' is a damping constant phenomenologically introduced. Thus, Eq. 9.25 gives the dependence of resonance Raman intensity on υ_0, in other words a resonance Raman excitation profile. The next step is to evaluate the numerator of Eq. 9.25 for various intermediate vibrational states *v*

$$f_A' = < i|v >< |j > \quad (9.26)$$

This can be easily calculated if one assumes (i) exactly the same harmonic potential functions for the ground ($|i>$ and $|j>$) and the excited ($|v>$) states, (ii) the dimensionless coordinates in common, which are related to the normal coordinates by

$$\xi_a = 2\pi(\upsilon_a/h)^{1/2} \quad (9.27)$$

and (iii) the vibrational wave functions of a particular normal coordinate a for the g and e states are given as

$$\Psi_{i,a^g} = N_i H_i(\xi_a)\exp\left[\frac{-\xi_a^{\ 2}}{2}\right]$$

$$\Psi_{v,a^e} = N_v H_v(\xi_a - \Delta_a)\exp\left[\frac{-(\xi_a - \Delta_a)^2}{2}\right] \quad (9.28)$$

where $H_i(H_v)$ represent the ith (vth) Hermite polynomial, N_i (N_v), the normalisation constant, and, most importantly, Δa, the displacement of the potential minimum of the resonant excited state with reference to that of the ground state.

The resonance Raman excitation profile for the fundamental Stokes Raman line ($i = 0$ and $j = 1$) for normal mode a is characterised as follows: the denominator of Eq. 9.25 indicates that it can give a peak every time when h υ_0 matches the energy of one of the vibronic levels, i.e. E_{ev} - E_{g0}. However, the intensity of each peak is determined by the overlap integrals in the numerator, which is a function of Δa defined above. In other words, the resonance Raman excitation profile is a function of structural change upon excitation to the resonant excited state. The displacement in the dimensionless coordinate can be translated into that in the normal coordinate and then that in the internal coordinate(s), e.g. a change in the bond length in the case of a pure stretching coordinate. It is also noted that, in general, Δa (and thus, the excitation profile) differs from one normal coordinate to another.

(b) Examples of Application

Changes in the C=C and C–C bondlengths of ß-carotene upon $S_2 \leftarrow S_0$ excitation
Resonance Raman excitation profiles of all-*trans*-ß-carotene in the region of the $S_2 \leftarrow S_0$ absorption were recorded for υ_1 (the C=C stretching at 1 525 cm^{-1}) and υ_2 (the C–C stretching at 1 155 cm^{-1}) Raman lines, and they were analysed in terms of the Albrecht theory to determine the displacements of the equilibrium positions upon $S_2 \leftarrow S_0$ excitation along the C=C or C–C stretching coordinates (Inagaki et al. 1974). Nine different laser lines (441.6, 457.9, 465.8, 472.7, 476.5, 488.0, 496.5, 501.7 and 514.5 nm) were used to record the excitation profiles. Resonance Raman intensities were determined at -150 °C by using a Raman line of the solvent isopentane as the internal standard, and they were expressed as

$$S = [I(\lambda)/I^{\circ}(\lambda^{\circ})] \times [\lambda/\lambda^{\circ}]^4 \quad (9.29)$$

where λ and $I(\lambda)$ represent, respectively, the wavelength and intensity of each

Raman line of ß-carotene, and $\lambda°$ and $I°(\lambda°)$ those of an internal standard line. By taking the intensity ratio, all the factors due to the effects of self-absorption are cancelled out. Figure 9.29 shows the excitation profiles for the υ_1 and υ_2 Raman lines. The υ_1 excitation profile shows peaks at the 0–0 and 1–0 transitions, while the υ_2 excitation profile shows peaks at the 0–0 and 2–0 transitions. Figure 9.30 shows the results of simulation of the excitation profile for the (a) υ_1 and (b) υ_2 Raman lines, which were based on the assumptions described above. In this simulation, the intermediate vibrational states $v = 0$ up to $v = 4$ were taken into account. The ordinate scale of calculated $|f_A|^2$ is adjusted to fit the observed S value of v_1 at 488.0 nm. The best fit was obtained for $|\Delta_1| = 1.1$ and $|\Delta_2| = 1.4$. Assuming that all the unit vibrations of the C=C (C–C) stretching, with exactly the same magnitude and phase, form the normal vibration of υ_1 (υ_2), the absolute values for the displacement of the equilibrium position along the C=C (C–C) bond were determined tobe 0.02 (0.03) Å.

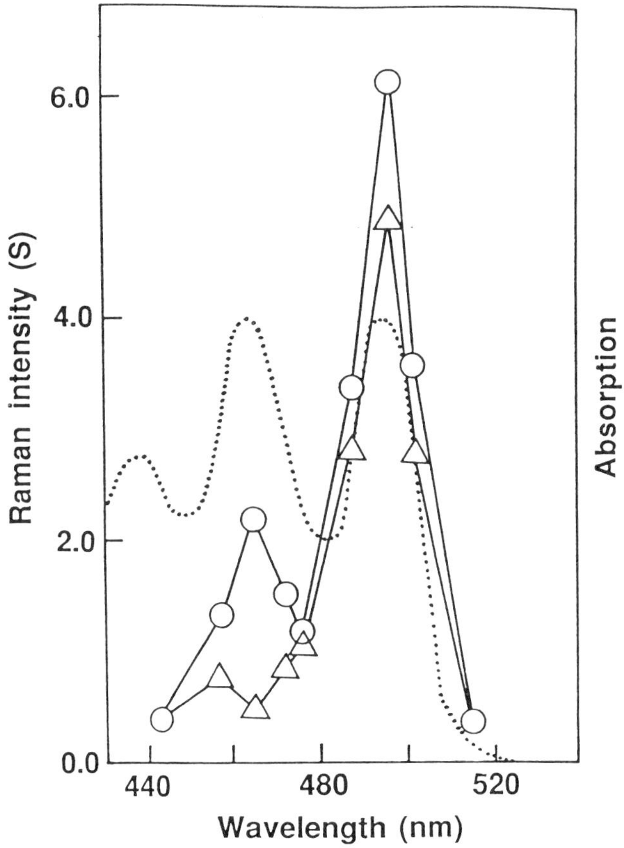

Figure 9.29 Excitation profiles for the υ_1, C=C stretching Raman line at 1525 cm^{-1} (O) and the υ_2, C–C stretching Raman line at 1155 cm^{-1} (Δ). See Eq. 9.20 for the definition of S. The dotted line indicates the $S_2 \leftarrow S_0$ absorption spectrum. (Revised from Inagaki et al. 1974)

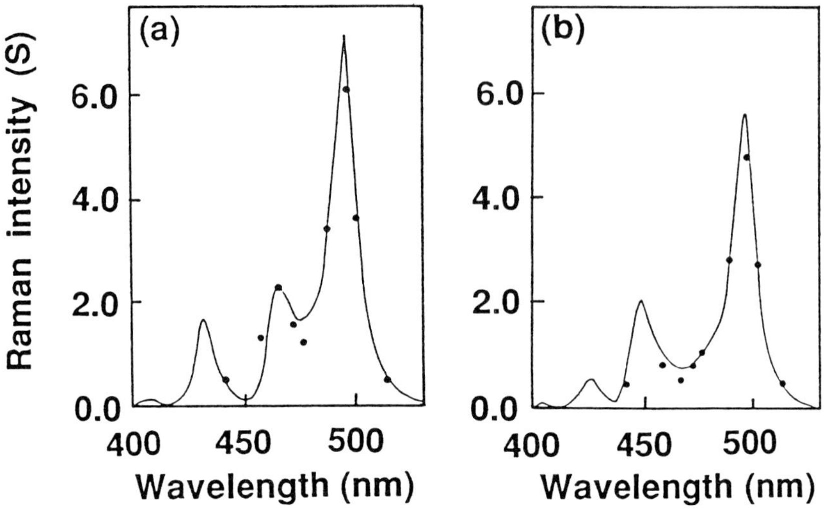

Figure 9.30 Simulation of the excitation profiles for the (a) υ_1 and (b) υ_2 Raman lines. The ordinate scale of the simulation has been adjusted to fit that of the observed intensity of υ_1 at 488.0 nm. (Revised from Inagaki et al. 1974)

The intensity ratio of the υ_1/υ_2 Raman lines of a carotenoid in the chromatophores used to probe the membrane potential

The shift of the $S_2 \leftarrow S_0$ absorption of carotenoids ("carotenoid shift") is widely used to probe a change in the membrane potential as described above. By taking advantage of the difference in the excitation profiles of the υ_1 and υ_2 Raman lines of neurosporene in the chromatophores of *Rb. sphaeroides* G1C, it was shown that the relative intensity of the υ_1 and υ_2 Raman lines could be used as a sensitive probe for the membrane potential, when an appropriate probe wavelength was selected (Koyama et al. 1979). Figure 9.31 (left) shows the excitation profiles for the υ_1 and ν_2 Raman lines; the Raman line of water around 3 300 cm^{-1} was used as an internal standard. The idea behind this work was that, if the 472.7 nm line of an Ar^+-laser is used for probing, the intensity ratio of the υ_1/υ_2 Raman lines should be very sensitive to small shifts of the entire envelope, because the slope of the excitation profiles for ν_1 and ν_2 are of opposite sign in this region. In order to test this idea, the effect of diffusion potential on the relative intensity was examined. It has been shown by electronic absorption spectroscopy that a valinomycin-induced K^+-efflux from the chromatophores produces a blue shift of the $S_2 \leftarrow S_0$ absorption due to an inside-negative diffusion potential. Figure 9.31 (right) shows the resonance Raman result. The decrease in the υ_1/υ_2 intensity ratio from 1.51 to 1.25 in experiment (b) indicates a blue shift and thus reflects the perturbation due to an inside-negative membrane potential. In the presence of 50 mM KCl (experiment c) the ratio remained at 1.50. The resonance Raman detection of the membrane potential was also tested by the effects of oligomycin, FCCP (carbonylcyanide-*p*-trifluoromethylphenyl hydrazone) and sodium cholate (Koyama et al. 1979).

Opposite changes in the υ_1/υ_2 intensity ratio were detected by probing at 457.9 nm (Koyama et al. 1983a), as expected from the excitation profiles shown in the Fig. 9.31. The resonance Raman probe for the membrane potential is very useful for strongly scattering samples such as intact cells.

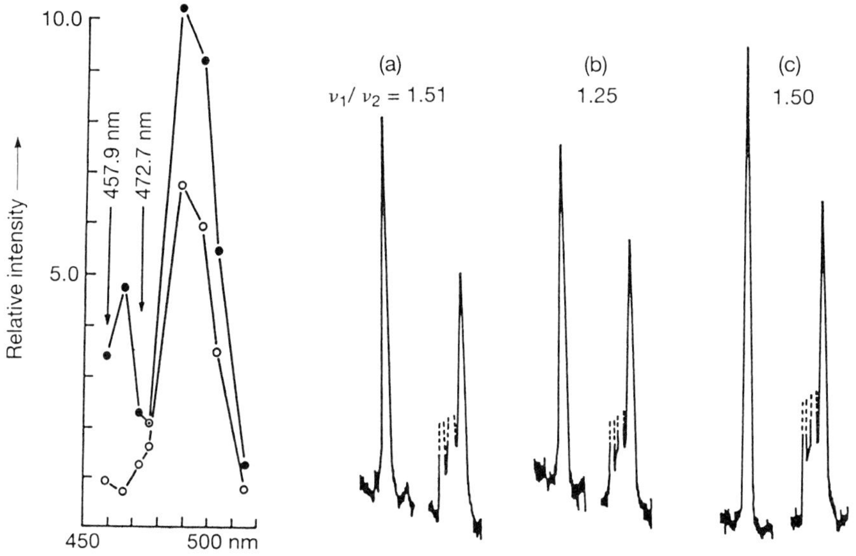

Figure 9.31 Excitation profiles of the υ_1 (•) and υ_2 (○) Raman lines of neurosporene bound to the chromatophores of *Rb. sphaeroides* G1C. Arrows indicate the wavelengths used to probe the "carotenoid shift" by the relative Raman intensity of υ_1/υ_2. The K^+-loaded (100mM) chromatophores were suspended in a K^+-free buffer, and the solution was mixed, through a T-junction, with an equal volume of (a) the same K^+-free buffer (control), (b) the K^+-free buffer + valinomycin (high inside-negative membrane potential) and, (c) a buffer containing 50 mM KCl + valinomycin (low inside-negative membrane potential). See text for a detailed description. (Revised from Koyama et al. 1979)

9.3.3 Transient resonance Raman: T_1 and S_1 states of carotenoids

(a) Fundamentals

Excited states of carotenoids probed by time-resolved (transient) resonance Raman spectroscopy

The T_1 state of a carotenoid is generated by (i) energy transfer from a triplet sensitiser such as BChl, (ii) singlet homo-fission between a pair of carotenoid molecules bound side-by-side (when high photon-density is applied), or (iii) intersystem crossing from the S_1 state (quantum yields, $\approx 10^{-3}$). The S_2 and S_1 states are generated by direct photoexcitation and by subsequent relaxation, respectively. The lifetimes of the T_1, S_1 and S_2 states are of the order, 10^{-6}–10^{-5} s, 10^{-11} s and 10^{-13} s, respectively.

In order to obtain a set of time-resolved Raman spectra, the duration of the pump and probe pulses should be smaller than the lifetime of the excited state in question. However, a transient Raman spectrum (the Raman spectrum of the transient species to be detected) can be obtained even when the pulse duration is one order of magnitude larger than the lifetime of the excited state .

To obtain time-resolved spectra, the interval between the pump (probe) pulses should be larger than the lifetime. This is not a necessary condition to obtain a transient spectrum.

Time-resolved Raman spectra of free and bound T_1 carotenoids are readily obtained by two-colour pump-and-probe experiments (the first colour for pumping the sensitiser or BChl, and the second for probing the resultant T_1 carotenoid) using 10^{-9}–10^{-8} s pulses with repetition 10^1 Hz. Transient Raman spectra of T_1 carotenoids bound to the LHC can also be obtained by one-colour experiments using a picosecond pulse train having high peak power (10^1–10^2 kW) and a high repetition rate (76 MHz); T_1 carotenoids are generated by singlet homo-fission. Another type of transient Raman spectrum of a T_1 carotenoid free in solution can be obtained by one-colour or two-colour experiments using the above picosecond pulse train(s); a small amount of the T_1 carotenoid generated by intersystem crossing is probed by a large number of pulses during the T_1 lifetime.

Transient Raman spectra of free and bound S_1 carotenoids have been obtained by one-colour and two-colour experiments using 30 - 100 ps pulses. (The first S_1 Raman spectrum of a carotenoid was reported as late as 1989 (Hashimoto and Koyama 1989a), whereas the first T_1 Raman spectrum was reported as early as in 1979 (Dallinger et al. 1979)). No transient Raman spectra of S_1 carotenoids have been reported. The uncertainty principle

$$\Delta E \bullet \Delta t \geq \frac{h}{2\pi} \quad (\Delta\tilde{\upsilon} \bullet \Delta t \geq \frac{1}{4\pi c}) \quad (9.30)$$

sets the limit for the shortest pulses that can be used in Raman spectroscopy; if one uses 10^{-13} s pulses to record a Raman spectrum, the uncertainty of $\Delta\,\upsilon$ becomes 10–100 cm^{-1} and no practical Raman spectra are obtained. It may be possible in the future to obtain transient Raman spectrum with 10^{-12} s pulses.

The $T_n \leftarrow T_1$ and $S_n \leftarrow S_1$ transient absorptions appear on the longer wavelength side of the $S_2 \leftarrow S_0$ absorption at $\approx$450 nm, in the case of ß-carotene for example, at $\approx$520 and $\approx$560 nm, respectively. Here also, **resonance** Raman spectroscopy is a powerful method to probe selectively the particular excited state (the population of which is generally low) of a single carotenoid in a mixture of pigments. Therefore, time-resolved (transient) Raman spectroscopy is the only tool now available with which to investigate the excited state structures of carotenoids in real photosynthetic systems. As in the case of ground state resonance Raman spectroscopy, it provides unique pieces of information about the configurations (conformations) and the bond orders of the C=C and C–C bonds in the excited states.

(b) Examples of application

T_1 species generated from geometrical isomers of ß-carotene

A comparison has been made of time-resolved resonance Raman spectra (probed at 532 nm) of T_1 species, which were generated by anthracene-sensitised flash photolysis (pump at 337 nm) of ß-carotene isomers, and extremely rapid "15-*cis*" to "all-*trans*" isomerisation in the T_1 state was found (Hashimoto and Koyama 1988). A set of time-resolved spectra showed excitation from and subsequent relaxation to the original isomer in the case of the all-*trans*, 7-*cis* and 9-*cis* isomers, and excitation from the original isomer and subsequent isomerisation into the all-*trans* isomer in the case of the 13-*cis* and 15-*cis* isomers. Figure 9.32 compares the T_1 Raman spectra at maximum triplet population; spectral differences are seen in the 1 200–1 150 cm^{-1} region. T_1 species generated from the all-*trans*, 7-*cis*, 9-*cis* and 13-*cis* isomers (the "all-*trans*", "7-*cis*", "9-*cis*" and "13-*cis*" T_1) each show their own unique spectrum, but the T_1 species generated from the 15-*cis* isomer shows exactly the same spectrum as that generated from the all-*trans* isomer. The quantum yields of triplet-sensitised isomerisation (defined as the decrease in the starting isomer per T_1 species generated) were as follows (Kuki et al. 1991): all-*trans*, 0.044; 7-*cis*, 0.12; 9-*cis*, 0.15; 13-*cis*, 0.87; and 15-*cis*, 0.98. Figure 9.33 compares the quantum yield of triplet-sensitised isomerisation defined for each isomerisation pathway. The results establish extremely efficient one-way isomerisation from 15-*cis* to all-*trans* via the T_1 state. Thus, it is concluded that the "15-*cis*" T_1 is too short lived to be detected, and that the "all-*trans*" T_1, the result of the extremely efficient isomerisation in the T_1 state, is detected by time-resolved Raman spectroscopy. (Compare this with the results of transient absorption spectroscopy, Section 9.2.2 (b)). Figure 9.34 shows the result of PPP-SD-CI calculations of the π-bond orders for a model for ß-carotene (docosaundecaene, **19**) in the S_0 and T_1 states (Kuki et al. 1991). A large decrease in the π-bond order in the central part of the conjugated chain upon triplet excitation most probably facilitates the efficient isomerisation from the "15-*cis*" T_1 to the "all-*trans*" T_1.

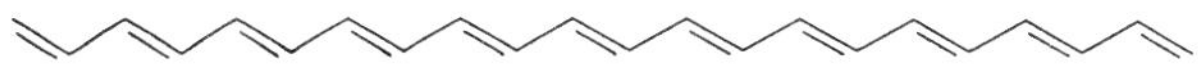

19

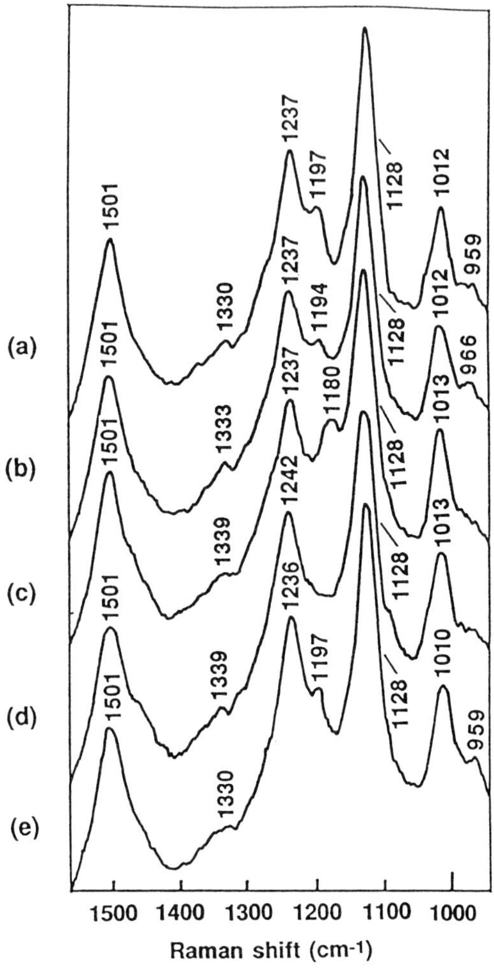

Figure 9.32 Raman spectra of T_1 species generated from the geometrical isomers of ß-carotene (a) all-*trans*, (b) 7-*cis*, (c) 9-*cis*, (d) 13-*cis* and (e) 15-*cis*. (Revised from Hashimoto and Koyama 1988)

T_1 carotenoids bound to the reaction centres

The T_1 Raman spectrum of spheroidene in the RC of *Rb. sphaeroides* 2.4.1 was recorded and its unique T_1 structure was found (Lutz et al. 1982). The RC was poised at low redox potential at -193 °C; 600 nm pulses were used to pump BChl through the Q_x band, and 528.7 nm CW light was used to probe (with a synchronised gate) the T_1 carotenoid (Fig. 9.35). Upon triplet excitation, the C=C stretching Raman line shifts from 1 530 to 1 504 cm^{-1}, whereas the C–C stretching Raman line at 1 160 cm^{-1} splits into three components, at 1 186, 1 155 and 1 134 cm^{-1}. The spectral pattern for the bound carotenoid in the 1 300–1 100 cm^{-1} region, which is expected to be most structure-sensitive, is completely different from those of free all-*trans* carotenoids in solution. Lutz et al. (1982) claimed that the RC-bound carotenoid still takes a *cis* configuration in the T_1 state. However, the spectral pattern is completely different from those of *cis*-isomers of ß-carotene free in solution, and the 934 cm^{-1} Raman line that can be assigned to the C–H out-of-plane wagging is strongly enhanced. These results strongly suggest that the T_1 carotenoid in the RC takes a twisted conformation.

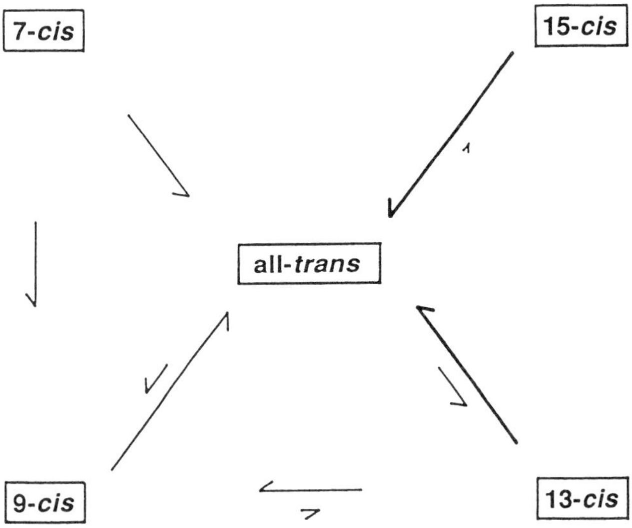

Figure 9.33 Scheme of triplet-sensitised isomerisation among the all-*trans* and mono-*cis* isomers of ß-carotene. The length of each arrow is proportional to the quantum yield; the length of a thicker arrow should be multiplied by 10 to compare with a thinner arrow. (Revised from Kuki et al. 1991)

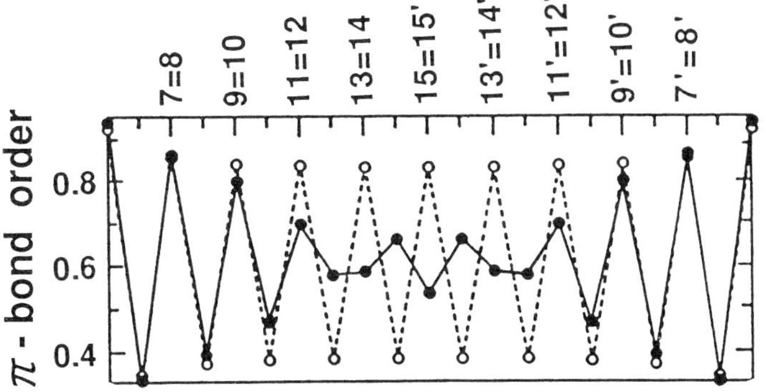

Figure 9.34 PPP-SD-CI calculations of the π-bond order in the S_0 (○) and T_1 (•) states for docosaundecaene (a model for ß-carotene). Each double bond is equivalent to that found in ß-carotene.

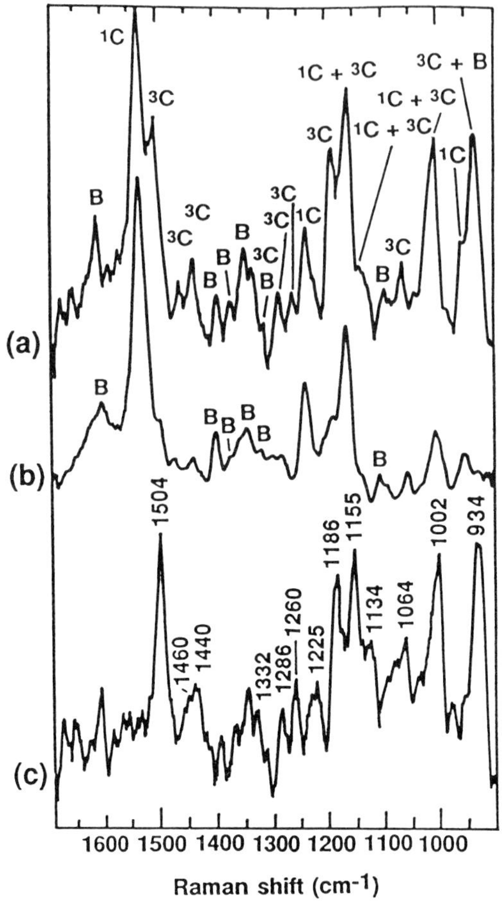

Figure 9.35 Raman spectrum of T_1 spheroidene in the RC of *Rb. sphaeroides* 2.4.1. Spectra recorded (a) with and (b) without pumping; the difference spectrum (c) was obtained by cancelling the S_0 1 535 cm^{-1} line. B, ^1C and ^3C indicate Raman lines due to BChl(S_0), Car(S_0) and Car(T_1), respectively. (Revised from Lutz et al. 1982)

Triplet excitation of precursors of spirilloxanthin in the chromatophores of R. rubrum *S1*

Triplet excitation through singlet homo-fission of intermediates in carotenogenesis in the chromatophores (LH) was detected by transient Raman spectroscopy, a result which indicates that these intermediates are also tightly bound to the LHC and involved in energy transfer (Naruse et al. 1991). A 6 day culture of *R. rubrum* S1 contained only spirilloxanthin, whereas a 3 day culture contained spirilloxanthin, 3,4-dihydrospirilloxanthin (**20**), rhodovibrin (**21**), and anhydrorhodovibrin (**22**) in the proportions of 6:1.5:2:1 (spirilloxanthin:total precursors = 4:3). All these precursors have exactly the same conjugated chain, with 12 double bonds (therefore rhodovibrin was used as a representative), whereas spirilloxanthin has a conjugated chain with 13 double bonds. Spectral comparison of T_1 species of chromatophores and free carotenoids reveals that the transient Raman spectrum of the chromatophores from 3 day cultures can be explained in terms of overlap of the T_1 Raman spectra of the precursors and spirilloxanthin (Fig. 9.36). First, the spectrum of the chromatophores (3 day culture) is characterised by two split peaks at 1 274 and 1 255 cm^{-1} and also by the presence of a weak line at 972 cm^{-1} (vertical arrows). These characteristics are also seen in the spectrum of T_1 rhodovibrin, i.e. a broad profile at 1 265 cm^{-1} and a weak line at 968 cm^{-1}. Secondly, the spectrum of the chromatophores (3-day culture) is characterised by the higher frequencies of the Raman lines at 1 478, 1 175, 1 120 and 1 009 cm^{-1} in comparison with the corresponding Raman lines in the spectrum of the chromatophores from the 6-day culture (horizontal arrows). This characteristic is explained in terms of the contribution of T_1 rhodovibrin in the spectrum of T_1 spirilloxanthin. In the recording of the spectra of T_1 carotenoids bound to the chromatophores, only the carotenoids were excited at 532 nm. Since the quantum yield of intersystem crossing is very low, the T_1 carotenoids must be generated through singlet homo-fission. The results indicate that two carotenoid molecules, which may be either spirilloxanthin or the precursors, are tightly bound together in the LHC.

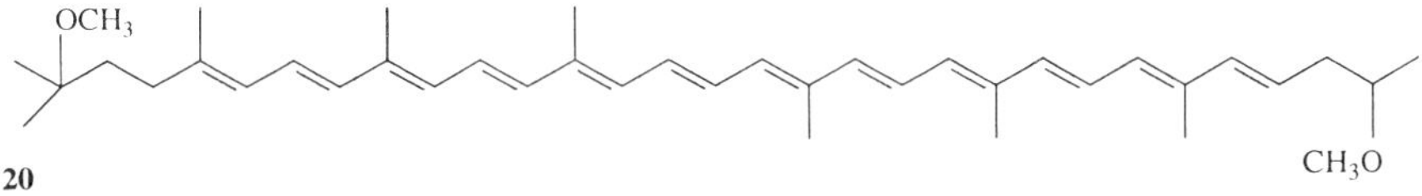

20

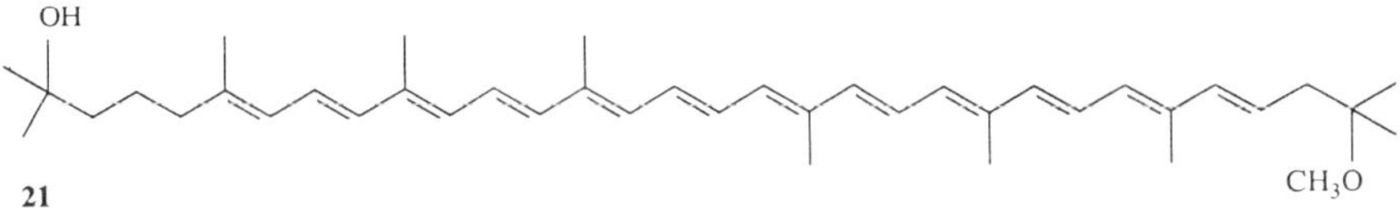

21

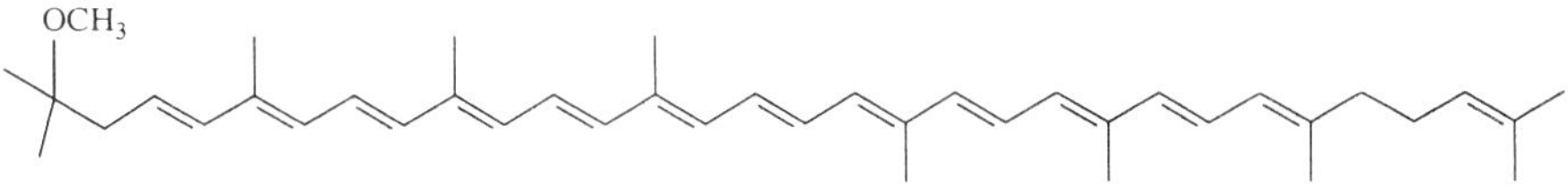

22

Vibronically coupled $2^1A_g^-$ state as the S_1 state of ß-carotene

An S_1, C=C stretching Raman line of all-*trans*-ß-carotene with an extraordinarily high frequency was detected by using ≈100 ps, 355 nm pump and 532 nm probe, mode-locked and Q-switched pulse trains. The S_1 Raman line indicated that the S_1 state of ß-carotene is actually the $2^1A_g^-$ state, which is vibronically coupled with the S_0 ($^1A_g^-$) state (Hashimoto and Koyama 1989a). The entire Raman spectra with higher signal:noise ratios were reported later (Hashimoto and Koyama 1989b; Noguchi et al. 1989). The S_1 Raman spectrum of all-*trans*-ß-carotene shows a Raman line at as high as 1 793 cm^{-1} (in THF, Fig. 9.37). A plausible explanation for the abnormally high frequency is vibronic coupling between this state and the ground (^1A_g) state through this totally symmetrical (a_g-type) vibration, which indicates that the S_1 state is actually the $2^1A_g^-$ state. Through this vibronic coupling, the S_1, C=C stretching Raman line is supposed to shift to the higher frequencies and the S_0, C=C stretching Raman line to the lower frequencies. The presence of this type of vibronic coupling was then established for shorter polyenes; it is now established for longer polyenes both experimentally (Simpson et al. 1987; Kohler et al. 1988) and theoretically (Negri et al. 1989).

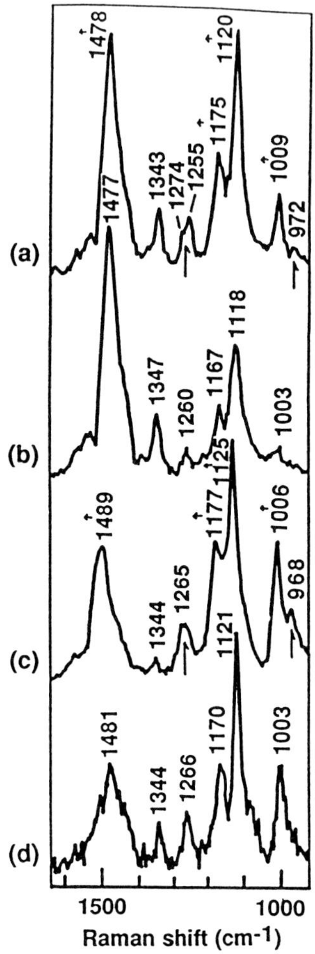

Figure 9.36 Transient Raman spectra of the chromatophores from cells of (a) 3 day culture and (b) 6 day culture of *R. rubrum* S1. T_1 Raman spectra of (c) rhodovibrin and (d) spirilloxanthin in THF solutions are included for comparison. (Revised from Naruse et al. 1991)

The implications of the identification of the vibronically coupled $2^1A_g^-$ state as the S_1 state in the energy-transfer and energy-dissipation functions of carotenoids in photosynthetic systems are considered to be three-fold. (i) It provides a route for carotenoid–BChl singlet energy transfer, in which the electron exchange mechanism should play an important role (the S_1 ($2^1A_g^-$) → S_0 ($^1A_g^-$) transition is symmetrically forbidden). (ii) Its lifetime is much longer than that of the S_2 state because of the forbidden nature of the transition; it may provide an energy pool for efficient energy transfer. (iii) The a_g, C=C stretching vibrational motion, which is responsible for the vibronic interaction, can provide a pathway of energy dissipation by enhancing the $S_1 \rightarrow S_0$ deactivation.

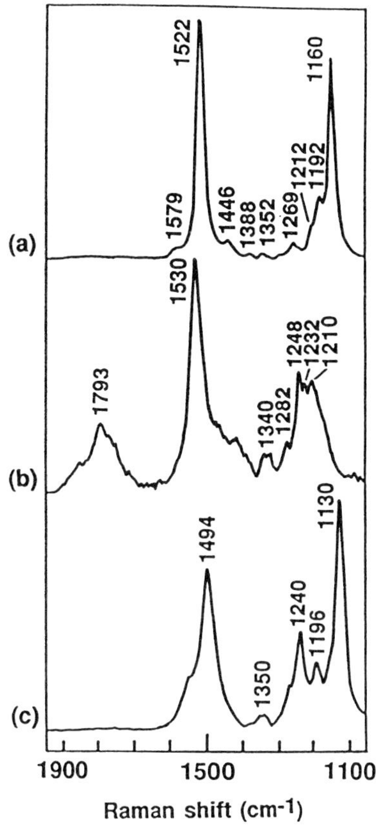

Figure 9.37 Raman spectra of all-*trans*-ß-carotene in the (a) S_0, (b) S_1 and (c) T_1 states, probed at 532 nm. (Reproduced from Hashimoto and Koyama 1989b)

All the Raman lines shown in Fig. 9.37 should provide very important pieces of information on the C=C and C–C bond orders in the S_1 and T_1 states. However, in order to extract this information as a set of force constants, a combination of PPP-CI type theoretical calculations and normal coordinate analyses of isotope-substituted derivatives are necessary.

The S_1 state of spheroidene bound to the chromatophores

The transient Raman spectrum of the S_1 and T_1 states of carotenoids bound to the chromatophore membranes of *Rb. sphaeroides* 2.4.1, which was recorded by using a 532 nm, ≈100 ps, mode-locked and Q-switched pulse train as a difference spectrum of high power minus low power, has been compared with the S_1 and T_1 Raman spectra of spheroidene and the T_1 Raman spectrum of spheroidenone free in solution. The vibronically coupled C=C stretching Raman line of bound spheroidene at 1 766 cm^{-1} identified the $2A_g$ state as the S_1 state, and its frequency, which was lowered by 51 cm^{-1} from that of free spheroidene, indicated weakened vibronic coupling for the bound carotenoid (Fig. 9.38, Kuki et al. 1990).

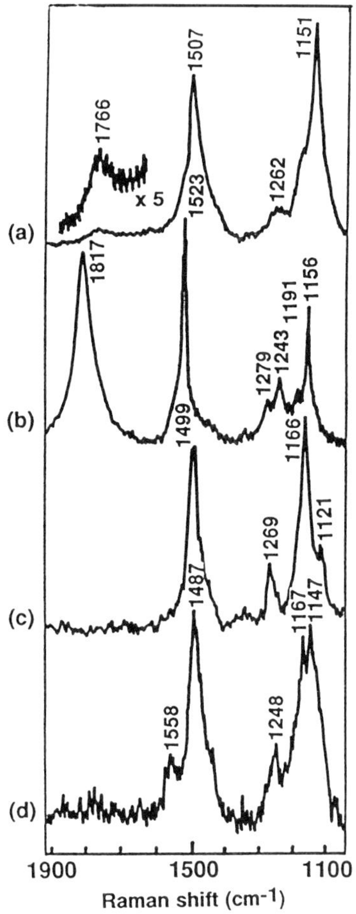

Figure 9.38 Raman spectra of bound and free carotenoids in the S_1 and T_1 states. (a) S_1 and T_1 carotenoids bound to the chromatophore of *Rb. sphaeroides* 2.4.1, (b) S_1 spheroidene in *n*-hexane, (c) T_1 spheroidene in *n*-hexane, and (d) T_1 spheroidenone in THF. (Reproduced from Kuki et al. 1990)

The chromatophores contained spheroidene and spheroidenone in the ratio of 84:16. The $S_n \leftarrow S_1$ ($T_n \leftarrow T_1$) absorptions are estimated to appear at 575 (532) and 605 (560) nm for spheroidene and spheroidenone, respectively. Therefore, when probed at 532 nm, mainly S_1 spheroidene (off resonance) and T_1 spheroidene (rigorous resonance) bound to the chromatophores are expected to be detected. Thus, the strong 1 507 and 1 151 cm^{-1} Raman lines and the weak 1 262 cm^{-1} Raman line are ascribed to T_1 spheroidene, while a weak but distinct feature at 1 766 cm^{-1} is ascribed to S_1 spheroidene.

The strength of the vibronic coupling is proportional to

$$< \Psi_1 |(\partial H/\partial Q)| \Psi_0 > /(E_1 - E_0) \quad (9.31)$$

where H is the Hamiltonian operator; Q, the stretching normal coordinate; and E_1 and E_0 (ψ_1 and ψ_0) are the energies (wavefunctions) of the S_1 and S_0 states. Therefore, the weakened vibronic coupling is ascribable either to a change in the wavefunction (electronic distribution) or to a change in the difference between the S_1 and S_0 state energies.

Vibronic coupling between the S_2 (B_u) and S_1 ($2A_g$) states for asymmetrical carotenoids

The frequencies of the S_1 and S_0, vibronically coupled C=C stretching Raman lines were examined for a variety of carotenoids, and the strength of the vibronic coupling was found to be almost constant for carotenoids with 9 to ≈13 conjugated double bonds; exceptions were two asymmetric carotenoids (Noguchi et al. 1990b). Figure 9.39 shows a plot of the S_1 and S_0, C=C stretching frequencies against the S_2–S_0 energy gap ($1/\lambda_{max}$) for a variety of carotenoids. The parallel linear relations for the S_1 and S_0, C=C stretching Raman lines indicate that the vibronic coupling between the S_1 and S_0 states is almost constant for most of the carotenoids. However, the S_1, C=C stretching frequencies of ß-apo-8´-carotenal and ethyl ß-apo-8´-carotenoate (**23**) are much lower, which strongly suggests an additional vibronic coupling between the S_2 and S_1 states.

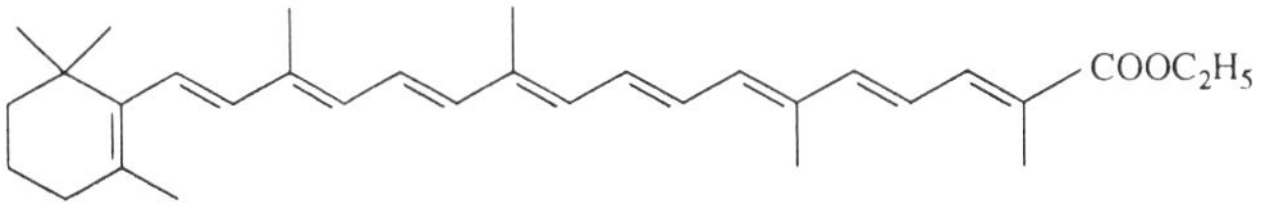

23

The frequencies of the S_1 and S_0, C=C stretching Raman lines of the carotenoids shown in Fig. 9.39 were examined in a variety of solvents with various degrees of polarisability, $(n^2\text{-}1)/(n^2\text{+}2)$, in order to identify the S_2–S_1 vibronic coupling (Noguchi et al. 1991). As for polyenes, the solvent-induced shifts of the absorption and emission spectra were studied as a function of the polarisability of a variety of solvents; it has been shown that the S_2 state energy is dependent on the polarizability, but that the S_1 state energy is not. Therefore, the same type of the S_2 and S_1 solvent effects were assumed for these carotenoids. The C=C stretching frequencies of a number of carotenoids in various solvents plotted as a function of the S_2–S_0 energy gap are shown in Fig. 9.40. There are three low-lying energy levels of carotenoids, i.e. S_2, S_1 and S_0. If the carotenoid molecule has C_{2h} symmetry, the S_2 (B_u) state couples neither to the S_1 ($2A_g$) state nor to the S_0 (A_g) state through the a_g C=C stretching vibration. However, if the C_{2h} symmetry is broken (as in the case of the particular asymmetric carotenoids), the S_2–S_1 and S_2–S_0 vibronic coupling can take place. Further, as mentioned above, the strength of vibronic coupling is inversely proportional to the energy gap of the states involved. The S_0, C=C stretching frequencies are found to be independent of the S_2–S_0 energy gap, a fact which excludes the possibility of S_2–S_0 vibronic coupling. However, the S_1, C=C stretching frequency for each of the three carotenoids decreases when the S_2–S_0 energy gap, and thus, the S_2–S_1 energy gap decreases (the S_1–S_0 energy gap is assumed to be constant), a fact which supports the idea of the S_2–S_1 vibronic coupling. The trend is more pronounced for the asymmetrical carotenoids.

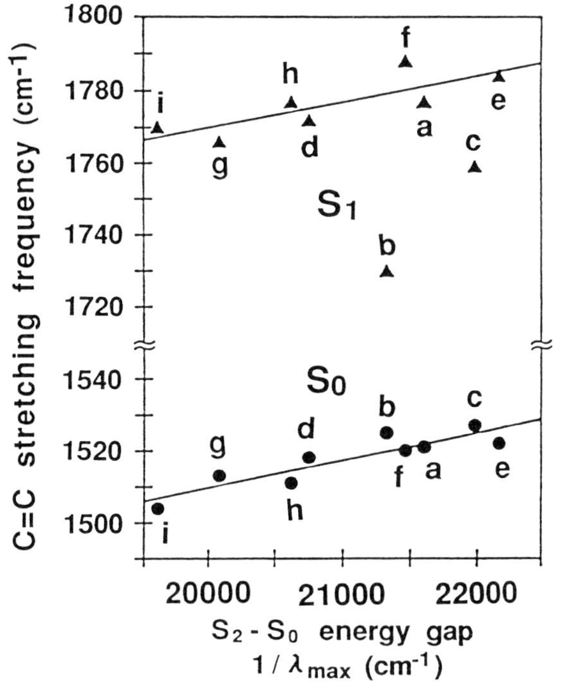

Figure 9.39 Plot of the S_1 and S_0, C=C stretching frequencies against the S_2 - S_0 energy gap for symmetrical and unsymmetrical carotenoids; (a) ß-carotene, (b) ß-apo-8´-carotenal, (c) ethyl ß-apo-8´-carotenoate, (d) canthaxanthin, (e) neurosporene, (f) spheroidene, (g) spheroidenone, (h) rhodopin and (i) spirilloxanthin. (Reproduced from Noguchi et al. 1990b)

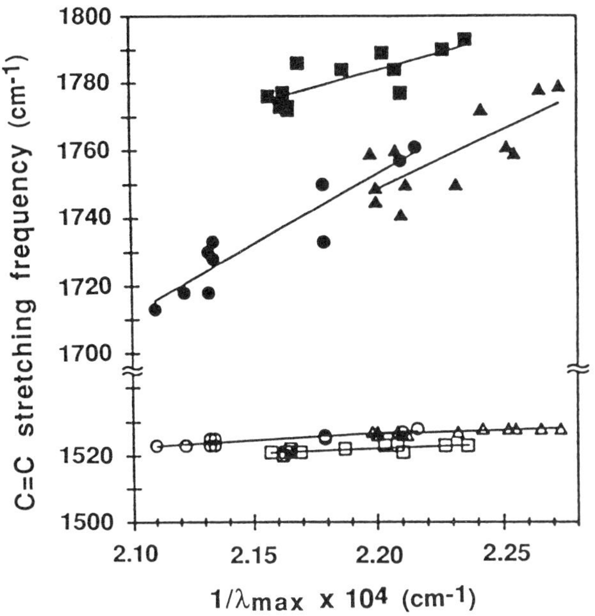

Figure 9.40 Vibronically-coupled C=C stretching frequencies of ß-carotene (S_0; S_1), ß-apo-8´-carotenal (O, S_0; O, S_1), and ethyl ß-apo-8´-carotenoate (Δ, S_0; Δ, S_1) in various solvents plotted as a function of the S_2- S_0 energy gaps. (Reproduced from Noguchi et al. 1991)

9.4 MAGNETIC RESONANCE SPECTROSCOPY

9.4.1 Nuclear magnetic resonance (NMR): S_0 state of carotenoids

(a)Fundamentals

Chemical shifts

When a magnetic field (H_0) is applied along the *z*-axis, a magnetic moment (μ) which is generated by a spinning nucleus has energy given by

$$E = -\mu_z H_0 = -\gamma(h/2\pi)I_z H_0 \quad (9.32)$$

where γ is the magnetogyric ratio and I_z the nuclear spin along the *z*-axis. In the case of ^1H and ^{13}C, there are two different quantum states, $I_z = +1/2$ and $I_z = -1/2$, which give rise to a transition

$$v_0 = \gamma H_0 / 2\pi \quad (9.33)$$

In a molecule, the nuclear spin is also under the influence of the local magnetic field (H') which is generated by electronic motion in the molecule in response to the external magnetic field (H_0); the former has the opposite direction and the amount is proportional to the latter,

$$H = H_0 + H' = H_0(1 - \sigma) \quad (9.34)$$

$$v = v_0(1 - \sigma) \quad (9.35)$$

The shielding constant, σ, is expressed in terms of the "chemical shift" (δ, in p.p.m.) with reference to that (σ_r) of a standard (tetramethyl silane),

$$\delta = \sigma_r - \sigma = \frac{v - v_r}{v_0} = \frac{\Delta v}{v_0} \quad (9.36)$$

The van der Waals interaction is one of very many factors which determine the value of chemical shift. When a methine and a methyl group attached to the conjugated chain are brought into close contact, the shielding effect is reduced for ^1H, and increased for ^{13}C, due to a change in electronic distribution around the nuclei, and a low field shift (lfs) of the ^1H signal (a high field shift (hfs) of the ^{13}C signal) takes place. Note that a "high field shift" is by reference to the external field, H_0, that has to be applied to cause the spin transition; it corresponds to a larger shielding constant, σ, and thus, to a smaller chemical shift, δ.

Changes in the values of chemical shifts are a sensitive probe for a *cis* configuration of the carotenoid backbone; hfs or lfs of the ^1H signals take place on the convex or concave side of a *cis* bend, repectively. A combination of the hfs and lfs of the olefinic ^1H signals for each *cis* configuration is summarised as "isomerisation shifts" (a set of changes in the values of chemical shifts relative to those of the all-*trans* isomer) (Englert 1982), and therefore the determination of the configuration of a carotenoid is straightforward once a set of assignments is available for the olefinic ^1H signals. Shifts of the ^{13}C signals in the opposite direction can also be used to identify a *cis* bend.

Coupling constants

When a pair of spins interact with each other through valence electrons, splitting of the signal of each spin takes place. Let us consider the two protons of adjacent CH groups in the polyene backbone connected by the C=C or C–C bond. This constitutes an AB spin system. The energy of spin A splits into two, depending on the state of spin B, +1/2 or -1/2, and gives rise to two split resonance lines. The strength of the interaction is expressed in terms of a vicinal coupling constant, *J*, which is determined by factors that include (i) the substituent groups at both carbon atoms, (ii) the dihedral angle, (iii) a pair of bond angles of C=C–H (or C–C–H), and (iv) the C=C (or C–C) bond lengths. In the case of the polyene backbone of a carotenoid, the factors (i)–(iii) are similar for each and every C=C or C–C bond, and thus the vicinal coupling constant can be used to probe the carbon–carbon bondlength (i.e. the bond order) by using factor (iv).

Solid state NMR

Recently, a new solid state NMR technique called cross polarisation (CP) magic angle spinning (MAS) spectroscopy has been developed. This is applied to a carotenoid bound to a pigment–protein complex in order to obtain its high resolution ^{13}C NMR spectrum. The technique takes advantage of (i) high power decoupling to cancel the effect of local field induced by ^{1}H, (ii) cancellation of the through-space interaction of magnetic dipoles having the ($3\cos^2 q$-1) dependence by spinning the sample along the magic angle $q = 54°44'$, and (iii) cross-polarisation to enhance ^{13}C spin-lattice relaxation by energy transfer to ^{1}H. When a ^{13}C atom in the polyene chain of the bound carotenoid is enriched, the technique is powerful for determining the carotenoid configuration *in situ*.

Two-dimensional NMR techniques

Various kinds of two-dimesional NMR technique, which take advantage of (i) coupling of spins through chemical bonds (COSY, TOCSY etc.) and (ii) coupling of spins through space, i.e. nuclear Overhauser effect, (NOESY, ROESY etc.), have been developed to facilitate the assignment of the ^{1}H and ^{13}C signals. By combining these techniques, assignment of the ^{1}H and ^{13}C signals and of the configurations can be done without any empirical rule. However, the principles of the two-dimensional NMR techniques and their application are beyond the scope of this chapter, and the reader is referred to the textbooks that describe the techniques and also to excellent reviews by Englert (1982, 1985, 1991) describing application to carotenoids.

(b)Examples of Application

^{1}H NMR of ß-carotene isomers

^{1}H NMR spectra of ß-carotene isomers have been recorded in benzene and $CDCl_3$ solutions (Koyama et al. 1989) and are shown in Fig. 9.41. Table 9.8 lists the values of chemical shifts; the isomerisation shifts for each isomer are shown in Table 9.9. The all-*trans* configuration of ß-carotene is characterised by the 11H (11´H) and 15H (15´H) signals in the low-field region. Each *cis* configuration is characterised by the following isomerisation shifts (see Table 9.9): 7-*cis*, the hfs of the 7H and 8H

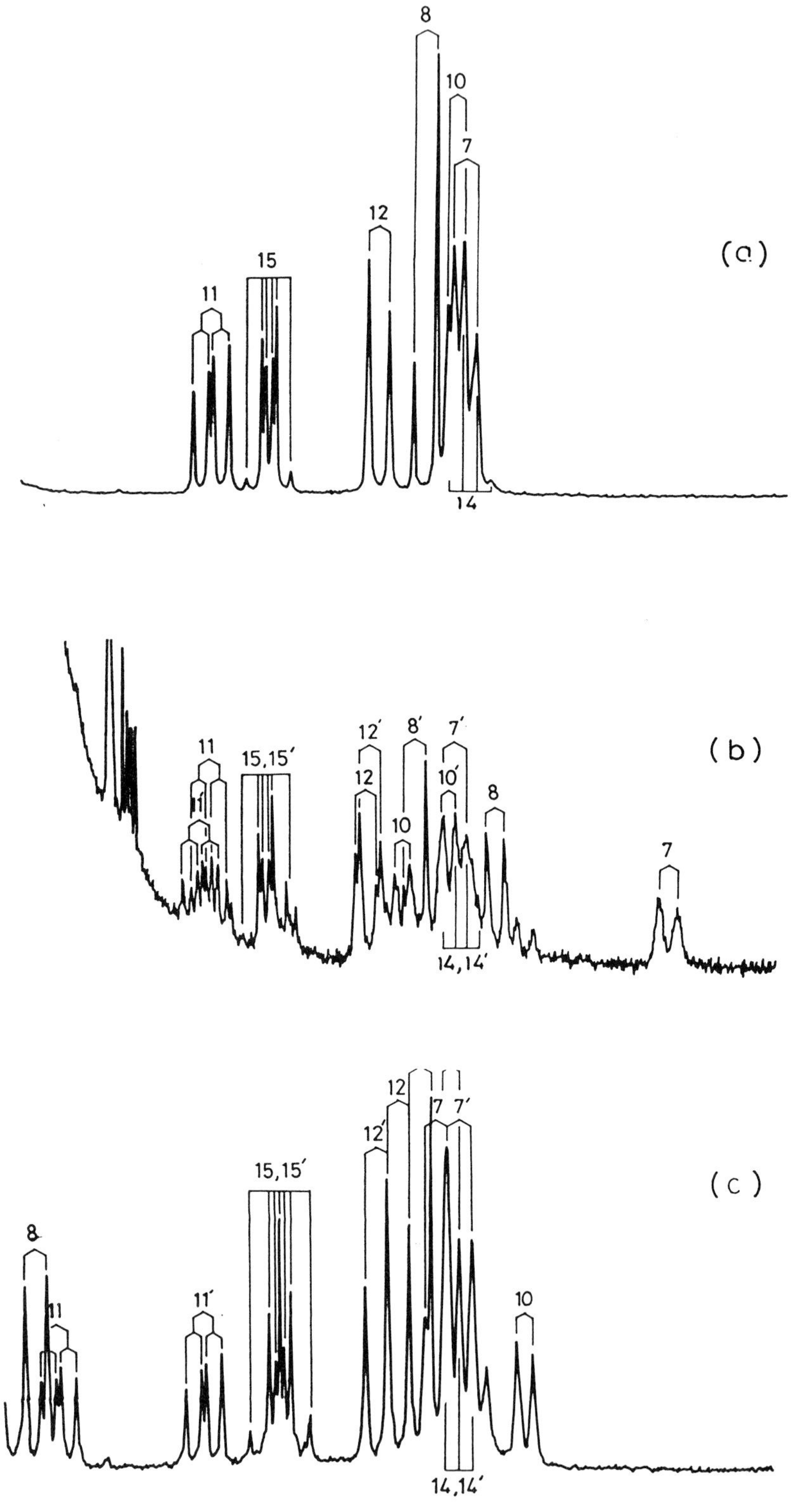
(a)
8
10
7
12
15
11
14
(b)
12′
8′
7′
11
15,15′
12
10′
8
11′
10
7
14,14′
(c)
12
7
7′
12′
15,15′
8
11
11′
10
14,14′

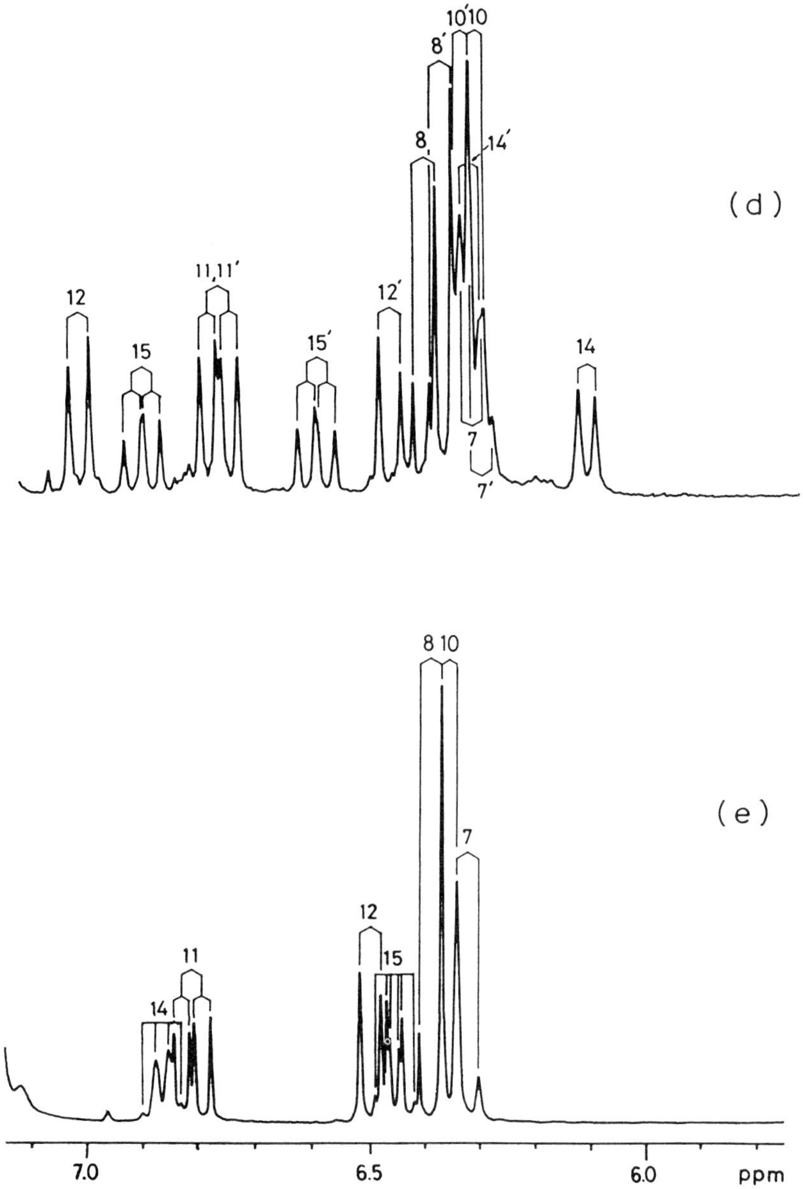

Figure 9.41 ^{1}H NMR spectra of the olefinic proton region of the geometrical isomers of ß-carotene in benzene solution; (a) all-*trans*, (b) 7-*cis*, (c) 9-*cis*, (d) 13-*cis* and (e) 15-*cis*. (Reproduced from Koyama et al. 1989)

signals and the lfs of the 10H signal; 9-*cis*, the hfs of the 10H signal and the lfs of the 8H and 11H signals; 13-*cis*, the hfs of the 14H signal and the lfs of the 12H and 15H signals; and 15-*cis*, the hfs of the 15H (15´H) signal and the lfs of the 14H (14´H) signal. The spectral pattern and the chemical shifts (isomerisation shifts) for each isomer in benzene solution can be used as a standard to determine the *cis-trans* configurations of other carotenoids. (The polar solvent, chloroform, should be avoided, whenever possible, because central-*cis* isomers tend to isomerise into the all-*trans* isomer in this solvent even at low temperatures.)

Table 9.8 ^{1}H-NMR Chemical shifts (p.p.m) for geometrical isomers of ß-carotene at 27 °C

Benzene solution						*Chloroform solution*				
	all-*trans*	7-*cis*	9-*cis*	13-*cis*	15-*cis*	all-*trans*	7-*cis*	9-*cis*	13-*cis*	15-*cis*
7 H	6.33	5.94	6.37	6.34	6.32	6.17	5.83	6.17	6.17	6.18
7'H		6.33	6.32	6.32			6.17			
8 H	6.40	6.26	7.09	6.43	6.39	6.12	6.09	6.66	6.13	6.13
8'H		6.40	6.40	6.40			6.12	6.11		
10 H	6.35	6.42	6.21	6.33	6.36	6.14	6.21	6.04	6.15	6.15
10'H		6.34	6.34	6.35			6.14	6.14		
11 H	6.79	6.78	7.05	6.79	6.81	6.64	6.56	6.74	6.63	6.68
11'H		6.79	6.78				6.63	6.64	6.64	
12 H	6.48	6.49	6.44	7.08	6.50	6.34	6.32	6.28	6.87	6.41
12'H		6.48	6.47	6.49			6.34	6.34	6.34	
14 H	6.32	6.32	6.28	6.13	6.86	6.24	6.23	6.23	6.09	6.65
14'H			6.31	6.34				6.22		
15 H	6.68	6.67	6.62	6.93	6.46	6.62	6.61	6.61	6.79	6.38
15'H			6.65	6.61					6.55	

Vicinal coupling constants provide unique information about carbon–carbon bondlengths (bond orders) (Koyama et al. 1989). In the case of a linear polyene, a PPP-SD-CI calculation (see Section 9.3.3 (b)) for docosaundecaene, a model of ß-carotene without the methyl groups, predicted respective decreases and increases in the π-bond order of the C=C and C–C bonds from both ends toward the centre of the molecule, i.e. from 0.922 to 0.833 and from 0.347 to 0.384 (Kuki et al. 1991). The prediction is supported by the coupling constants, J (see Table 9.10), since the vicinal coupling constant decreases or increases when the bond order decreases or increases. For the all-*trans* isomer, the J(CH=CH) value decreases toward the centre of the molecule, i.e. J(7H=8H) = 16, J(11H=12H) = 15 and J(15H=15´H) = 14 Hz, whereas the J(CH-CH) value increases, i.e. J(10H-11H) = 11 and J(14H-15H) = 12 Hz; this observation was first reported by Wernly and Lauterwein (1985). The same trends are seen for the all-*trans* parts of the *cis* isomers. The J(CH=CH) value for the *cis* bond also decreases as this bond is nearer to the centre; J(15H=15´H) of the 15-*cis* isomer (11 Hz) is smaller than J(7H=8H) of the 7-*cis* isomer (13 Hz).

Table 9. 9 Chemical shifts for the all-*trans* isomer and isomerization shiftsa for *cis* isomers of ß-carotene, canthaxanthin, neurosporene and spirilloxanthin in benzene

	ß-Carotene[c]					Canthaxanthin[c]				Neurosporene[d]				Spirilloxanthin[d]		
	all-*trans*	7-*cis*	9-*cis*	13-*cis*	15-*cis*	all-*trans*	9-*cis*	13-*cis*	15-*cis*	all-*trans*	9-*cis*	13'-*cis*	15-*cis*	all-*trans*	13-*cis*	15-*cis*
7 H	6.33	-0.39	+0.04			6.15	+0.03			6.66				6.79	-0.08	-0.04
7'H																
8 H	6.40	-0.14	+0.69	+0.03		6.38	+0.75	+0.03	-0.03	6.42	+0.66			6.53		
8'H															-0.05	
10 H	6.35	+0.07	-0.14			6.29	-0.07			6.34	-0.18		+0.03	6.39	-0.06	
10'H										6.14						
11 H	6.79		+0.26			6.73	+0.20			6.75	+0.26		+0.04	6.78	-0.04[b]	
11'H										6.66					-0.04[b]	
12 H	6.48		-0.04	+0.60		6.50	-0.08	+0.55		6.48	-0.06			6.54	+0.52	
12'H										6.42		+0.56	+0.05		-0.03	
14 H	6.32		-0.04	-0.19	+0.54	6.36	-0.03	-0.20	+0.54	6.33[b]			+0.57[b]	6.38	-0.25	+0.54
14'H								+0.04				-0.23				
15 H	6.68		-0.06	+0.25	-0.22	6.70	-0.04	+0.24	-0.23	6.69[b]		-0.10	-0.22[b]	6.73	+0.21	-0.26
15'H			-0.03	-0.07				-0.07				+0.26			-0.12	

[a] Isomerisation shifts of which absolute values are larger than 0.03 are shown; major isomerisation shifts are underlined.
[b] Approximate values; [c] At 27 °C; [d] At 7 °C

¹H NMR of carotenoids extracted from the reaction centre and the light harvesting complex

Figure 9.42 shows ¹H–¹H chemical shift-correlated (COSY) spectra of neurosporene isolated from the LHC (a) and RC (b) of *Rb. sphaeroides* G1C (Koyama et al. 1988a). The details of the assignment of the olefinic ¹H signals are given in the original paper. It is important to note here that the COSY spectra have been helpful for identifying the 14H (14′H) or 15H (15′H) signals, which are overlapped with other ¹H signals, and are not clearly seen in the one-dimensional spectrum. These signals played a key role in the configurational determinations. The values of chemical shifts and isomerisation shifts lead to straightforward configuration assignments, i.e. (a) all-*trans* and (b) 15-*cis*. The results, together with those of the carotenoids from *R. rubrum* S1 (Koyama et al. 1990), established the natural selection of the carotenoid configurations, i.e. of all-*trans* by the LHCs and of 15-*cis* by the RCs.

Table 9.9 lists the chemical shifts of the all-*trans* isomer and the isomerisation shifts of mono-*cis* isomers also for canthaxanthin (**24**), neurosporene and spirilloxanthin in benzene. Note that the chemical shifts of the all-*trans* isomer are in the order, 11H (11′H) > 15H (15′H) > 12H (12′H) > 8H (8′H) > 10H (10′H) ≈ 14H (14′H). The pattern of isomerisation shifts for each *cis* isomer (larger values are underlined) is similar when one carotenoid is compared with another, showing the validity of this pattern in configurational determinations.

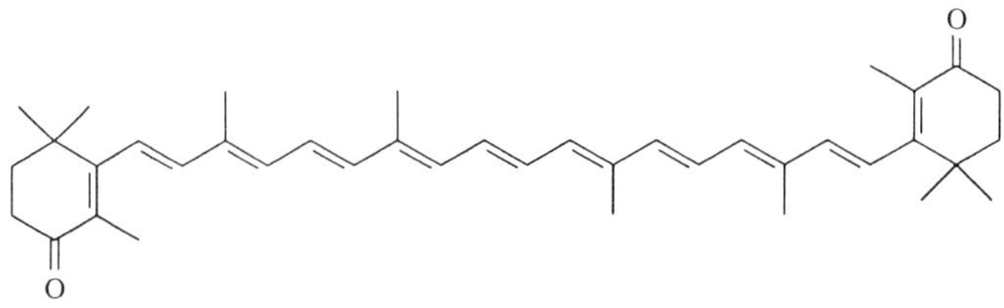

24

¹³C-NMR of a carotenoid bound to the reaction centre

Figure 9.43 shows the CP/MAS ¹³C-NMR spectrum of the RC of *Rb. sphaeroides* R-26 reconstituted with [14′-¹³C] spheroidene (see also Chapter 8). After subtraction of the spectrum of the R-26 RC before reconstitution (the natural abundance ¹³C-NMR spectrum) from the above spectrum, the spectrum of the label, [14′-¹³C] spheroidene, appears (inset). The MAS NMR spectrum of solid all-*trans* [14′-¹³C]spheroidene shows a chemical shift of 133.9 p.p.m. The value of the chemical shift for the bound spheroidene, 125.3 p.p.m. (corresponding to a Ifs of 8.6 p.p.m.), supports the conclusion that the RC-bound carotenoid has a 15-*cis* configuration (Gebhard et al. 1991).

Figure 9.42 ¹H-¹H COSY spectra of neurosporene isolated from (a) LH all-*trans*- and (b) RC 15-*cis* neurosporene of *Rb. sphaeroides* G1C. in benzene solution. (Reproduced from Koyama et al. 1988a)

(a)

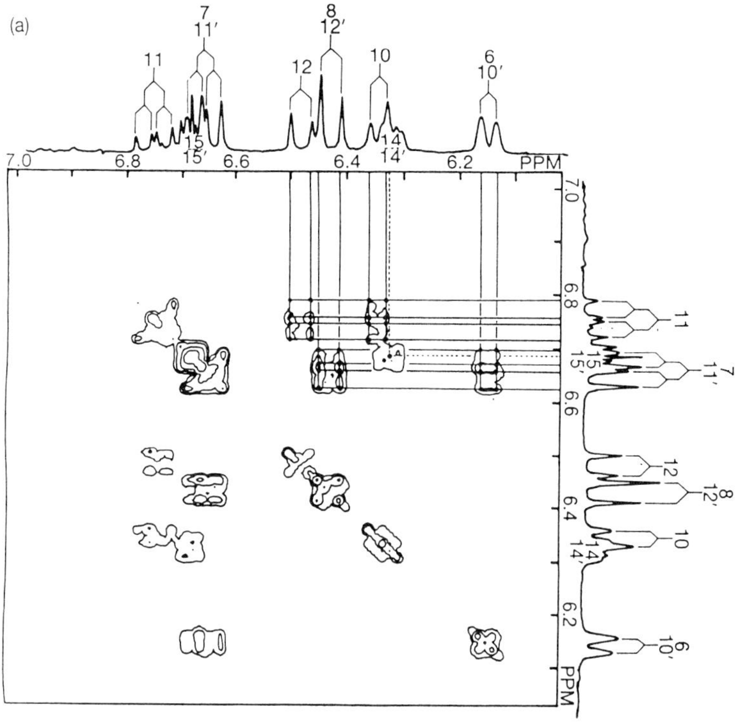

7
11′
8
12′
11
12
10
6
10′
15
15′
14
14′
7.0
6.8
6.6
6.4
6.2
PPM

(b)

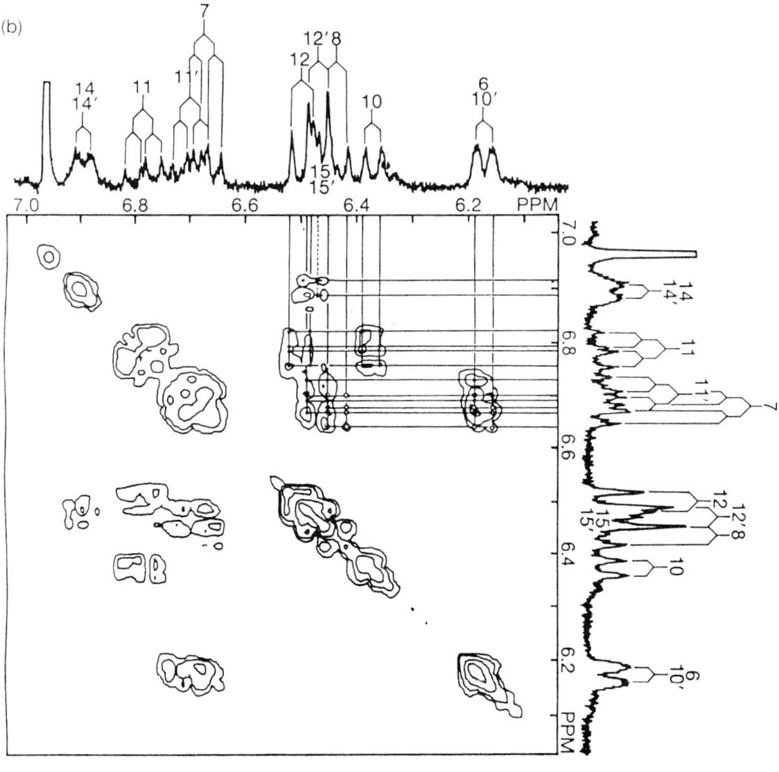

7
12′8
11′
12
14
14′
11
10
6
10′
15
15′
7.0
6.8
6.6
6.4
6.2
PPM

Table 9.10 ^{1}H NMR Coupling constants, J (Hz) for geometrical isomers of ß-carotene. Data for both benzene and chloroform solutions were combined or averaged

	all-trans	*7*-cis	*9*-cis	*13*-cis	*15*-cis
7H=8H 7'H=8'H	16	13[a] 16	16 17	16 16	16
10H-11H 10'H-11'H	11	11 11	12 10	12 11	11
11H=12H 11'H=12'H	15	15 15	15 15	15 15	15
14H-15H 14'H-15'H	12	12	12	12 12	13
15H=15'H	14	14	14	14	11[a]
14H•••15'H 14'H•••15H	-1	-1	-1		-1

[a] Coupling constants for H substituents about a *cis* double bond. All the other values are for the *trans* double or single bonds

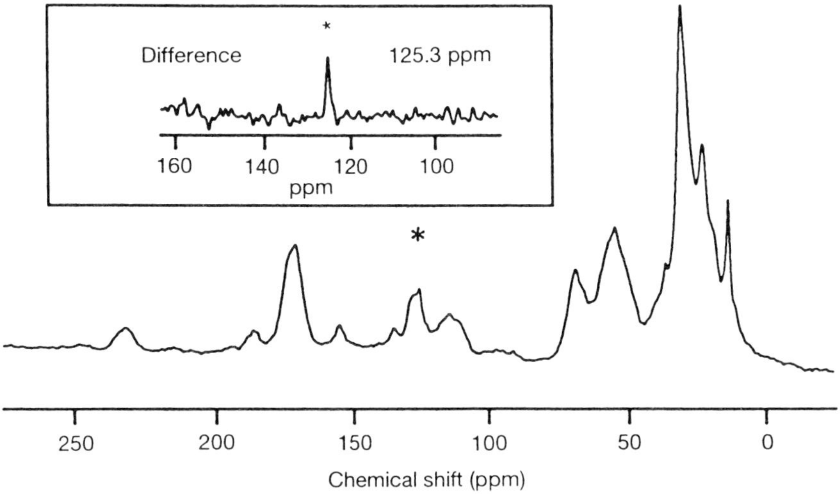

Figure 9.43 CP/MAS ^{13}C-NMR spectrum of the RC of *Rb. sphaeroides* reconstituted with [14´-^{13}C] spheroidene. After subtraction of the spectrum of the R-26 RC before reconstitution (the natural abundance ^{13}C-NMR spectrum), the spectrum of the label, introduced from [14´-^{13}C] spheroidene appears (inset). (Reproduced from Gebhard et al. 1991)

9.4.2 Electron paramagnetic resonance (EPR): T_1 state of carotenoids

(a) Fundamentals

Zero-field splitting

The spin Hamiltonian of the triplet state is given in a principal axis system by

$$H_s = \mathbf{S} \times \mathbf{D} \times \mathbf{S} = D_x S_x^2 + D_y S_y^2 + D_z S_z^2 \quad (9.37)$$

where $\mathbf{S}$ is the spin tensor and $\mathbf{D}$ the fine-structure tensor. The principal value, D_x, for example, is described as:

$$D_x = (1/2)(g_e \beta)^2 \iint \Psi^*(\mathbf{r}_1, \mathbf{r}_2)\{(r_{12}^2 - 3X_{12}^5\}\Psi(\mathbf{r}_1, \mathbf{r}_2)d\mathbf{r}_1 d\mathbf{r}_2 \quad (9.38)$$

where g_e is the g value for the free electron, ß the Bohr magneton, ψ the orbital wave function, $\mathbf{r}_1$ and $\mathbf{r}_2$ the coordinates of the unpaired electrons, and X_{12} the X component of $\mathbf{r}_1 - \mathbf{r}_2$.

Since

$$D_x + D_y + D_z = 0 \quad (9.39)$$

H_s can be described by using two independent parameters, D and E:

$$H_s = D[S_z^2 - (1/3)S(S+1)] + E(S_x^2 - S_y^2) \quad (9.40)$$

where

$$D_x = -D/3 + E, \ D_y = -D/3 - E, \ D_z = 2D/3 \quad (9.41)$$

D and E are called "zero-field splitting parameters".

The energy values (eigen values of H_s) are given by

$$E_x = -D_x, \ E_y = -D_y, \ E_z = -D_z \quad (9.42)$$

Splitting by external field

In the presence of external field H, the Hamiltonian is given by

$$H = \mathbb{S} \times \mathbb{D} \times \mathbb{S} + \beta \mathbb{S} \times \mathbb{g} \times \mathbb{H} \quad (9.43)$$

When $\mathbb{H}$ // Z, the eigen values are given as follows (for $D_x < D_y$):

$$W_{+1} = -D_x + g_z \beta H \times \tan \theta_z$$
$$W_0 = -D_z$$
$$W_{-1} = -D_y + g_z \beta H \times \tan \theta_z \quad (9.44)$$

where,

$$\tan \theta_z = -2 g_z \beta H / (D_x - D_y) \quad (9.45)$$

For $D_x > D_y$, W_{+1} and W_{-1} should be interchanged.

Similar expressions are obtained for $\mathbb{H}$ // X and $\mathbb{H}$ // Y. An energy diagram and a spectral pattern are shown in Fig. 9.44.

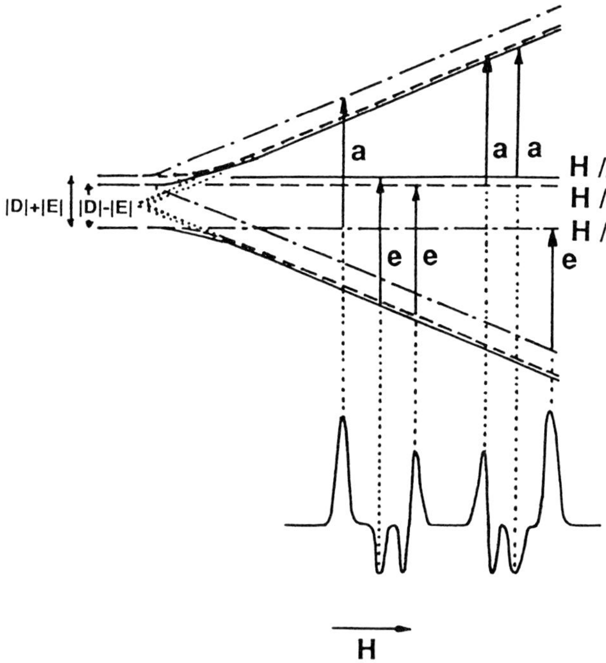

Figure 9.44 Composite triplet energy levels as a function of magnetic field, transitions among them, and an example of the polarization pattern.

Zero-field splitting parameters and polarisation patterns

The zero-field splitting parameter D, which is directly read from the EPR spectrum, is related to the average distance r_{12} between the two unpaired electrons (see Eq. 9.38), and is thus linked to the spatial extension of the π-electrons. The other zero-field splitting parameter E gives additionally a measure of the deviation from axial symmetry (Eq. 9.40).

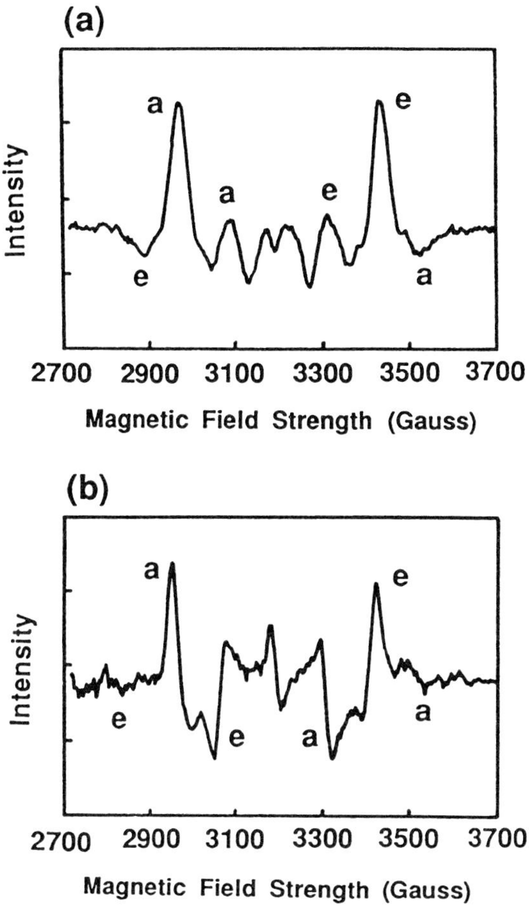

Figure 9.45 Polarisation patterns of T_1 carotenoid generated through energy transfer originally from the BChl dimer in the RC and antenna BChl in the LHC. (a) T_1 state EPR signal ("*eaa eea*") from spheroidene reconstituted into the RC of *Rb. sphaeroides* R26, and (b) that ("*eae aea*") from spheroidene in the B800-850 LHC of *Rb. sphaeroides* 2.4.1. (Revised from Frank 1992)

The EPR spectral pattern, i.e. absorption ("*a*") and emission ("*e*") in each transition (as shown in Fig. 9.44), is determined by the relative steady-state population of each sub-level. When this deviates from Boltzmann distribution it is referred to as "spin polarisation". In particular, when the triplet-state of a carotenoid is generated by energy transfer from the special pair BChls in the RC, whose triplet-state has been generated through the radical pair mechanism, the polarisation pattern *eaa eea* (Fig. 9.45a) emerges for the carotenoid. This is because the T_0 level of the donor is densely populated via S–T_0 mixing in the radical pair (Hoff 1982). When the triplet state of a carotenoid is generated by energy transfer from a donor whose triplet state has been generated through intersystem crossing, the relative population of each sublevel gives rise to the polarization pattern of *eae aea* for the carotenoid (Fig. 9.45b). Therefore, the polarisation pattern reflects the mechanism by which the triplet state of the donor BChls has been generated.

(b)Examples of application

Polarisation patterns and the mechanism of triplet generation
EPR spectra of carotenoids were first detected by Frank et al. (1980). Triplet spheroidene in the RC of *Rb. sphaeroides* 2.4.1 gave the zero-field splitting parameters, $|D| = 0.0290 (\pm 0.0005)$ cm^{-1} and $|E| = 0.044 (\pm 0.0006)$ cm^{-1}, which were different from those of the triplet BChl dimer ($|D|$ = 0.0189 (± 0.0004) cm^{-1} and $|E|$ = 0.0032 (± 0.0004) cm^{-1}) in the RC of *Rb. sphaeroides* R-26 (a carotenoidless mutant).

Table 9.11 lists the polarisation patterns and the zero-field splitting parameters of spheroidene bound to the RC, LHC, reduced cells and untreated cells of *Rb. sphaeroides* 2.4.1. Data for spirilloxanthin bound to the cells of *R. rubrum* are also shown for comparison (Frank et al. 1980). The polarisation pattern of the triplet carotenoid in the RC is *eaa eea*, indicating that it has been transferred from the primary donor whose triplet was formed via the radical pair mechanism. The polarisation pattern of the triplet BChl dimer is *aee aae*. The difference in the polarisation pattern indicates either (i) that the *D* values of the donor and the acceptor have opposite signs and energy transfer occurs via the T_0 levels of the donor and the acceptor, or (ii) that the *D* values have the same signs and energy transfer takes place from the T_0 level of the donor to the $T_{\pm 1}$ levels of the acceptor.
The polarisation pattern of the triplet carotenoid in the LH is *eae aea*, which indicates that the radical pair mechanism is not involved in the triplet formation of the donor. Presumably, the triplet BChls in the LHC have been formed though intersystem crossing.

Interestingly, the triplets of spheroidene and spirilloxanthin generated in reduced cells show the polarisation pattern *eaa eea*, a fact which indicates that their triplet state is generated through energy transfer originally from the BChl dimer in the RCs. On the other hand, the triplet carotenoids generated in untreated cells show the polarisation pattern *eae aea*, indicating that the triplet state is generated through energy transfer from BChls in the LHC.

The zero-field splitting parameter $|D|$ of the RC-bound carotenoid is significantly smaller than that of the LHC-bound carotenoid, suggesting the importance of structural (and environmental) effects. It should also be noted that the $|D|$ value of the reduced or untreated cells is similar to those of the RC (LHC), supporting the idea that the RC-bound or LHC-bound triplet carotenoid can be detected in the cells.

Table 9.11 Polarisation patterns and zero-field splitting parameters

Location of T_1 carotenoid	*Polarisation*	$\|D\|$	$\|E\|$
Rb. sphaeroides 2.4.1 (spheroidene)			
Reaction centre	*eaa eea*	0.0290 ± 0.0005	0.0044 ± 0.0006
Light-harvesting complex	*eae aea*	0.0326 ± 0.0007	0.0036 ± 0.0007
Reduced cells	*eaa eea*	0.0289 ± 0.0010	0.0044 ± 0.0010
Untreated cells	*eae aea*	0.0323 ± 0.0010	0.0033 ± 0.0010
R. rubrum S1 (spirilloxanthin)			
Reduced cells	*eaa eea*	0.0180 ± 0.0004	0.0040 ± 0.0004
Untreated cells	*eae aea*	0.0233 ± 0.0007	0.0026 ± 0.0007

See Eq. 9.31 (Section 9.4.2) for the definition of the zero-field splitting parameters

Zero-field splitting parameters of carotenoids in reconstituted reaction centres
Table 9.12 lists the $|D|$ and $|E|$ values for carotenoids incorporated into the RC of *Rb. sphaeroides* R26.1 (Chadwick and Frank 1986). Comparison between the number of conjugated double bonds and the $|D|$ value reveals the trend that the $|D|$ value decreases as the extent of the π-electron delocalisation increases. The magnitude of $|E|$ for these carotenoids appears to follow the same trend. This observation suggests that these RC-bound carotenoids have the same symmetry and that the observed variation in the zero-field splitting parameters is a result of change in the extent of π-electron delocalisation.

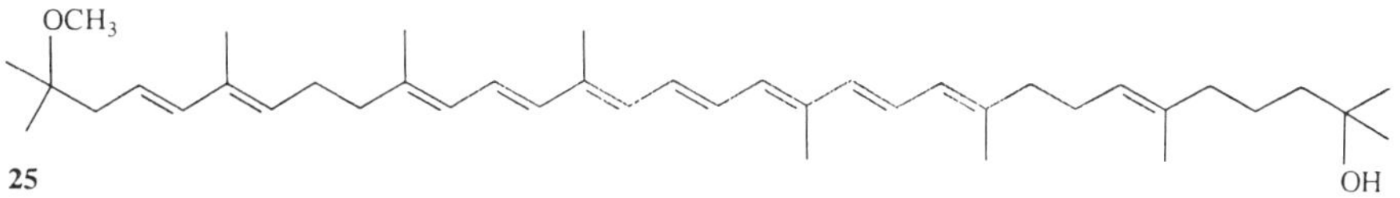

25

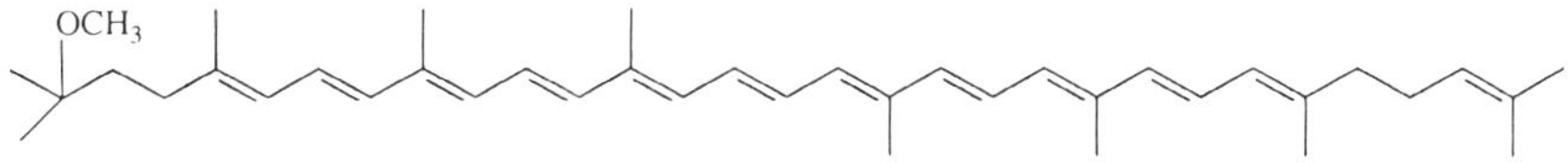

26

Detailed comparison reveals that the $|D|$ and $|E|$ values do not vary systematically with the number of conjugated double bonds; methoxyneurosporene (n = 9), spheroidene (n = 10), hydroxyspheroidene (**25**, n = 10), and hydroxyneurosporene (n=9) give a similar set of $|D|$ and $|E|$ values. Further, 3,4-dihydroanhydrorhodovibrin (**26**, n = 11) and spirilloxanthin (n = 13) give the same $|D|$ and $|E|$ values, which are smaller than the above. In order for the same zero-field splitting parameters for carotenoids having different numbers of conjugated double bonds to be observed, the effective length of the conjugated chain, when bound to the apo-complex, should be the same. Thus, a model for the bound carotenoids, in which the extent of π electron conjugation is restricted by out-of-plane twist at both ends, has been proposed (Chadwick and Frank 1986).

Table 9.12 Zero-field splitting parameters for carotenoids in pigment–protein complexes

Complex/ carotenoids	*Number of conjugated double bonds*	$\lvert D \rvert$	$\lvert E \rvert$
Reaction centre (reconstituted)			
Rb. sphaeroides R-26			
Hydroxyneurosporene	9	0.0284 ± 0.0006	0.0041 ± 0.0003
Methoxyneurosporene	9	0.0286 ± 0.0006	0.0044 ± 0.0003
Spheroidene	10	0.0286 ± 0.0006	0.0044 ± 0.0003
Hydroxyspheroidene	10	0.0286 ± 0.0006	0.0044 ± 0.0003
Spheroidenone	11	0.0271 ± 0.0005	0.0042 ± 0.0003
3,4-Dihydro-anhydrorhodovibrin	11	0.0201 ± 0.0003	0.0037 ± 0.0002
Spirilloxanthin	13	0.0201 ± 0.0003	0.0037 ± 0.0002
B800–850 Complex			
Rb. sphaeroides Ga			
neurosporene	9	0.0365 ± 0.0002	0.0035 ± 0.0002
Rb. sphaeroides 2.4.1			
spheroidene (anaerobic)	10	0.0324 ± 0.0002	0.0036 ± 0.0002
Rb. sphaeroides 2.4.1			
spheroidenone (aerobic)	11	0.0318 ± 0.0002	0.0032 ± 0.0002
Rps. acidophila 7750			
rhodopin	11	0.0279 ± 0.0003	0.0029 ± 0.0003

Zero-field splitting parameters of carotenoids bound to the light-harvesting complexes

Table 9.12 also lists the $\lvert D \rvert$ and $\lvert E \rvert$ values of carotenoids bound to the B800–850 LHCs (Frank et al. 1987). The zero-field splitting parameters do vary systematically with the number of conjugated double bonds. The $\lvert D \rvert$ and $\lvert E \rvert$ values follow the order neurosporene > spheroidene > rhodopin, in which the number of conjugated double bonds is 9, 10 and 11 , respectively. The results indicate that, within the B800–850 class, the apo-complex exerts little or a similar degree of influence on the extent of π-electron delocalisation. This observation for the LHC-bound carotenoids forms a striking contrast to that in the RC-bound carotenoids.

ACKNOWLEDGEMENTS

The authors thank Professor Harry A. Frank and Dr. Mamoru Mimuro for providing their unpublished fluorescence data; information from their most recent and important contributions (DeCoster et al. as well as Mimuro et al.) are included in this chapter. The authors are indebted to Professor Frank also for critical reading of the EPR section and for providing his recent article entitled "Electron paramagnetic resonance studies of carotenoids" (Frank, 1992).

The compiling of the references by Mr Yousuke Miki is gratefully acknowledged.

REFERENCES

Andersson, P.O., Gillbro, T., Ferguson, L. and Cogdell, R.J. (1990) Spectral shift of purple bacterial carotenoids related to solvent and protein polarizability. In: Baltscheffsky, M. (ed.), *Current Research in Photosynthesis*, Vol. II, Kluwer Academic Publishers, Dordrecht, pp. 117-120.

Andersson, P.O., Gillbro, T., Ferguson, L. and Cogdell, R.J. (1991) Absorption spectral shifts of carotenoids related to medium polarizability. *Photochem. Photobiol.*, **54**:353-360.

Basu, S. (1964) Theory of solvent effects on molecular electronic spectra. *Adv. Quantum Chem.*, **1**:145-169.

Bensasson, R., Land, E.J. and Maudinas, B. (1976) Triplet states of carotenoids from photosynthetic bacteria studied by nanosecond ultraviolet and electron pulse irradiation. *Photochem. Photobiol.*, **23**:189-193.

Boucher, F., Rest, M. van der and Gingras, G. (1977) Structure and function of carotenoids in the photoreaction center from *Rhodospirillum rubrum*. *Biochim. Biophys. Acta*, **461**:339-357.

Chadwick, B.W. and Frank, H.A. (1986) Electron-spin resonance studies of carotenoids incorporated into reaction centers of *Rhodobacter sphaeroides* R-26.1. *Biochim. Biophys. Acta*, **851**:257-266.

Cogdell, R.J., Monger, T.G. and Parson, W.W. (1975) Carotenoid triplet states in reaction centers from *Rhodopseudomonas sphaeroides* and *Rhodospirillum rubrum*. *Biochim. Biophys. Acta*, **408**:189-199.

Cogdell, R.J., Parson, W.W. and Kerr, M.A. (1976) The type, amount, location, and energy transfer properties of the carotenoid in reaction centers from *Rhodopseudomonas sphaeroides*. *Biochim. Biophys. Acta*, **430**:83-93.

Cosgrove, S.A., Guite, M.A., Burnell, T.B. and Christensen, R.L. (1990) Electronic relaxation in long polyenes. *J. Phys. Chem.*, **94**:8118-8124.

Dallinger, R.F., Guanci, Jr., J.J., Woodruff, W.H. and Rodgers, M.A.J. (1979) Vibrational spectroscopy of the electronically excited state: pulse radiolysis/time-resolved resonance Raman study of triplet ß-carotene. *J. Am. Chem. Soc.*, **101**:1355-1357.

De Coster, B., Christensen, R.L., Gebhard, R. *et al.* (1992) Low-lying electronic states of carotenoids. *Biochim. Biophys. Acta*, **1102**:107-114.

Englert, G. (1982) N.M.R. of carotenoids. In: Britton, G. and Goodwin, T.W. (eds.), *Carotenoid Chemistry and Biochemistry*. Pergamon , Oxford, pp. 107-134.

Englert, G. (1985) NMR of carotenoids: New experimental techniques. *Pure Appl. Chem.*, **57**:801-821.

Englert, G. (1991) NMR of carotenoids: Novel experimental techniques. *Pure Appl. Chem.*, **63**:59-70.

Frank, H.A. (1992) Electron paramagnetic resonance studies of carotenoids. *Meth. Enzymol.*, **213**:305-312.

Frank, H.A., Bolt, J.D., B. de Costa, S.M. and Sauer, K. (1980) Electron paramagnetic resonance detection of carotenoid triplet states. *J. Am. Chem. Soc.*, **102**:4893-4898.

Frank, H.A., Chadwick, B.W., Taremi, S., *et al.* (1986) Singlet and triplet absorption spectra of carotenoids bound in the reaction centers of *Rhodopseudomonas sphaeroides* R-26. *FEBS Lett.*, **203**:157-163.

Frank, H.A., Chadwick, B.W., Oh, J.J. *et al.* (1987) Triplet-triplet energy transfer in B800-850 light-harvesting complexes of photosynthetic bacteria and synthetic carotenoporphyrin molecules investigated by electron spin resonance. *Biochim. Biophys. Acta*, **892**:253-263.

Frank, H.A., Violette, C.A., Trautman, J.K. *et al.* (1991) Carotenoids in photosynthesis: structure and photochemistry. *Pure Appl. Chem.*, **63**:109-114.

Gebhard, R., van der Hoef, K., Violette, C.A. *et al.* (1991) ^{13}C Magic angle spinning NMR evidence for a 15,15′-Z configuration of the spheroidene chromophore in the *Rhodobacter sphaeroides* reaction center: synthesis of ^{13}C- and ^{2}H-labeled spheroidenes. *Pure Appl. Chem.*, **63**: 115-122.

Gottfried, D.S., Steffen, M.A. and Boxer, S.G. (1991a) Large protein-induced dipoles for a symmetric carotenoid in a photosynthetic antenna complex. *Science*, **251**: 662-665.

Gottfried, D.S., Steffen, M.A. and Boxer, S.G. (1991b) Stark effect spectroscopy of carotenoids in photosynthetic antenna and reaction center complexes. *Biochim. Biophys. Acta*, **1059**: 76-90.

Hashimoto, H. and Koyama, Y. (1988) Time-resolved resonance Raman spectroscopy of triplet ß-carotene produced from all-*trans*, 7-*cis*, 9-*cis*, 13-*cis*, and 15-*cis* isomers and high-pressure liquid chromatography analyses of photoisomerisation via the triplet state. *J. Phys. Chem.*, **92:** 2101-2108.

Hashimoto, H. and Koyama, Y. (1989a) The C=C stretching Raman lines of ß-carotene isomers in the S_1 state as detected by pump-probe resonance Raman spectroscopy. *Chem. Phys. Lett.*, **154**: 321-325.

Hashimoto, H. and Koyama, Y. (1989b) Raman spectra of all-*trans*-ß-carotene in the S_1 and T_1 states produced by direct photoexcitation. *Chem. Phys. Lett.*, **163**: 251-256.

Hashimoto, H., Koyama, Y., Ichimura, K. and Kobayashi, T. (1989) Time-resolved absorption spectroscopy of the triplet state produced from the all-*trans*, 7-*cis*, 9-*cis*, 13-*cis*, and 15-*cis* isomers of ß-carotene. *Chem. Phys. Lett.*, **162**: 517-522.

Hashimoto, H., Koyama, Y., Hirata, Y. and Mataga, N. (1991) S_1 and T_1 species of ß-carotene generated by direct photoexcitation from the all-*trans*, 9-*cis*, 13-*cis*, and 15-*cis* isomers as revealed by picosecond transient absorption and transient Raman spectroscopies. *J. Phys. Chem.*, **95**: 3072-3076.

Hayashi, H., Noguchi, T. and Tasumi, M. (1989) Studies on the interrelationship among the intensity of a Raman marker band of carotenoids, polyene chain structure, and efficiency of the energy transfer from carotenoids to bacteriochlorophyll in photosynthetic bacteria. *Photochem. Photobiol.*, **49**: 337-343.

Hoff, A.J. (1982) ODMR spectroscopy in photosynthesis II: the reaction center triplet in bacterial photosynthesis. In: Clarke, R.J.H. (ed.), *Triplet State ODMR Spectroscopy: Techniques and Applications to Biophysical Systems*, A Wiley-Interscience Publication, New York, pp. 367-425.

Inagaki, F., Tasumi, M. and Miyazawa, T. (1974) Excitation profile of the resonance Raman effect of ß-carotene. *J. Mol. Spectrosc.*, **50**: 286-303.

Kohler, B.E., Spangler, C. and Westerfield, C. (1988) The 2^1A_g state in the linear polyene 2,4,6,8,10,12,14,16-octadecaoctaene. *J. Chem. Phys.*, **89**: 5422-5428.

Koyama, Y. (1991) Structures and functions of carotenoids in photosynthetic systems. *J. Photochem. Photobiol. B: Biol.*, **9**: 265-280.

Koyama, Y., Long, R.A., Martin, W.G. and Carey, P.R. (1979) The resonance Raman spectrum of carotenoids as an intrinsic probe for membrane potential: oscillatory changes in the spectrum of neurosporene in the chromatophores of *Rhodopseudomonas sphaeroides*. *Biochim. Biophys. Acta*, **548**: 153-160.

Koyama, Y., Kito, M., Takii, T. *et al.* (1982) Configuration of the carotenoid in the reaction centers of photosynthetic bacteria: comparison of the resonance Raman spectrum of the reaction center of *Rhodopseudomonas sphaeroides* G1C with those of *cis-trans* isomers of ß-carotene. *Biochim. Biophys. Acta*, **680**: 109-118.

Koyama, Y., Tahara, M. and Segawa, S. (1983a) A slow phase of the carotenoid shift induced by pH jump or by continuous illumination: an electron absorption and resonance Raman study on chromatophores of *Rhodopseudomonas sphaeroides* G1C. *Photobiochem. Photobiophys.*, **6**: 329-338.

Koyama, Y., Takii, T., Saiki, K. and Tsukida, K. (1983b) Configuration of the carotenoid in the reaction centers of photosynthetic bacteria (2): comparison of the resonance Raman lines of the reaction centers with those of the 14 different *cis-trans* isomers of ß-carotene. *Photobiochem. Photobiophys.*, **5**: 139-150.

Koyama, Y., Kanaji, M. and Shimamura, T. (1988a) Configurations of neurosporene isomers isolated from the reaction center and the light-harvesting complex of *Rhodobacter sphaeroides* G1C: a resonance Raman, electronic absorption, and ^{1}H NMR study. *Photochem. Photobiol.*, **48**: 107-114.

Koyama, Y., Takatsuka, I., Nakata, M. and Tasumi, M. (1988b) Raman and infrared spectra of the all-*trans*, 7-*cis*, 9-*cis*, 13-*cis* and 15-*cis* isomers of ß-carotene: key bands distinguishing stretched or terminal-bent configurations from central-bent configurations. *J. Raman Spectrosc.*, **19**: 37-49.

Koyama, Y., Hosomi, M., Hashimoto, H. and Shimamura, T. (1989) ^{1}H NMR spectra of the all-*trans*, 7-*cis*, 9-*cis*, 13-*cis* and 15-*cis* isomers of ß-carotene: elongation of the double bond and shortening of the single bond toward the center of the conjugated chain as revealed by vicinal coupling constants. *J. Mol. Struct.*, **193**: 185-201.

Koyama, Y., Takatsuka, I., Kanaji, M. *et al.* (1990) Configurations of carotenoids in the reaction center and the light-harvesting complex of *Rhodospirillum rubrum* : natural selection of carotenoid configurations by pigment protein complexes. *Photochem. Photobiol.*, **51**: 119-128.

Kuki, M., Hashimoto, H. and Koyama, Y. (1990) The $2^1A_g^-$ state of a carotenoid bound to the chromatophore membrane of *Rhodobacter sphaeroides* 2.4.1 as revealed by transient resonance Raman spectroscopy. *Chem. Phys. Lett.*, **165**: 417-422.

Kuki, M., Koyama, Y. and Nagae, H. (1991) Triplet-sensitized and thermal isomerisation of all-*trans*, 7-*cis*, 9-*cis*, 13-*cis* and 15-*cis* isomers of ß-carotene: configurational dependence of the quantum yield of isomerisation via the T_1 state. *J. Phys. Chem.*, **95**:7171-7180.

Lutz, M., Chinsky, L. and Turpin, P.Y. (1982) Triplet states of carotenoids bound to reaction centers of photosynthetic bacteria: time-resolved resonance Raman spectroscopy. *Photochem. Photobiol.*, **36**: 503-515.

Mathis, P. and Kleo, J. (1973) The triplet state of ß-carotene and of analog polyenes of different length. *Photochem. Photobiol.*, **18**: 343-346.

Mimuro, M. and Katoh, T. (1991) Carotenoids in photosynthesis: absorption, transfer and dissipation of light energy. *Pure Appl. Chem.*, **63**: 123-130.

Mimuro, M., Nagashima, U., Takaichi, S. *et al.* (1992) Molecular structure and optical properties of carotenoids for in vivo energy transfer function in the algal photosynthetic pigment system. *Biochim. Biophys. Acta*, **1098**:271-274.

Monger, T.G., Cogdell, R.J. and Parson, W.W. (1976) Triplet states of bacteriochlorophyll and carotenoids in chromatophores of photosynthetic bacteria. *Biochim. Biophys. Acta*, **449**: 136-153.

Naruse, M., Hashimoto, H., Kuki, M. and Koyama, Y. (1991) Triplet excitation of precursors of spirilloxanthin bound to the chromatophores of *Rhodospirillum rubrum* as detected by transient Raman spectroscopy. *J. Mol. Struct.*, **242**: 15-26.

Negri, F., Orlandi, G., Zerbetto, F. and Zgierski, M.Z. (1989) Theoretical study of the force field of the lowest singlet electronic states of long polyenes. *J. Chem. Phys.*, **91**: 6215-6224.

Noguchi, T., Kolaczkowski, S., Arbour, C. *et al.* (1989) Resonance Raman spectrum of the excited 2^1A_g state of ß-carotene. *Photochem. Photobiol.*, **50**: 603-609.

Noguchi, T., Hayashi, H. and Tasumi, M. (1990a) Factors controlling the efficiency of energy transfer from carotenoids to bacteriochlorophyll in purple photosynthetic bacteria. *Biochim. Biophys. Acta,* **1017**: 280-290.

Noguchi, T., Hayashi, H., Tasumi, M. and Atkinson, G.H. (1990b) Frequencies of the Franck-Condon active a_g C=C stretching mode in the $2^1A_g^-$ exited state of carotenoids. *Chem. Phys. Lett.*, **175**: 163-169.

Noguchi, T., Hayashi, H., Tasumi, M. and Atkinson, G.H. (1991) Solvent effects on the a_g C=C stretching mode in the $2^1A_g^-$ excited state of ß-carotene and two derivatives: picosecond time-resolved resonance Raman spectroscopy. *J. Phys. Chem.*, **95**: 3167-3172.

Shreve, A.P., Trautman, J.K., Frank, H.A. *et al.* (1991) Femtosecond energy-transfer processes in the B800-850 light-harvesting complex of *Rhodobacter sphaeroides* 2.4.1. *Biochim. Biophys. Acta,* **1058**: 280-288.

Shreve, A.P., Trautman, J.K., Owens, T.G. and Albrecht, A.C. (1991) Determination of the S_2 lifetime of ß-carotene. *Chem. Phys. Lett.*, **178**: 89-96.

Shulten, K., Dinur, U. and Honig, B.J. (1980) The spectra of carbonium ions, cyanine dyes, and protonated Schiff base polyenes. *J. Chem. Phys.*, **73**: 3927-3935.

Simpson, J.H., McLaughlin, L., Scott-Smith, D. and Christensen, R.L. (1987) Vibronic coupling in polyenes: high resolution optical spectroscopy of all-*trans*-2,4,6,8,10,12,14-hexadecaheptaene. *J. Chem. Phys.*, **87**: 3360-3365.

Tang, J. and Albrecht, A.C. (1970) Developments in the theories of vibrational Raman intensities. In: Szymanski, H.A. (ed.), *Raman Spectroscopy: Theory and Practice*, Vol. 2. Plenum Press, New York, pp. 33-68.

Thrash, R.J., Fang, H.L.B. and Leroi, G.E. (1977) The Raman excitation profile spectrum of ß-carotene in the preresonance region: evidence for a low-lying singlet state. *J. Chem. Phys.*, **67**: 5930-5933.

Truscott, T.G., Land, E.J. and Sykes, A. (1973) The *in vitro* photochemistry of biological molecules III: absorption spectra, lifetimes and rates of oxygen quenching of the triplet states of ß-carotene, retinal and related polyenes. *Photochem. Photobiol.*, **17**: 43-51.

Tsukida, K., Saiki, K., Takii, T. and Koyama, Y. (1982) Separation and determination of *cis/trans*-ß-carotenes by high-performance liquid chromatography. *J. Chromatogr.*, **245**: 359-364.

Wasielewski, M.R., Tiede, D.M. and Frank, H.A. (1986) Ultrafast electron and energy transfer in reaction center and antenna proteins from photosynthetic bacteria. In: Fleming, G.R. and Siegman, A.E. (eds.), *Ultrafast Phenomena* V. Springer, Berlin (Springer Series in Chemical Physics 46) pp. 388-392.

Wasielewski, M.R., Johnson, D.G., Bradford, E.G. and Kispert, L.D. (1989) Temperature dependence of the lowest excited singlet-state lifetime of all-*trans*-ß-carotene and fully deuterated all-*trans*-ß-carotene. *J. Chem. Phys.*, **91**: 6691-6697.

Wernly, J. and Lauterwein, J. (1985) Two-dimensional NMR studies of polyene systems. *Mag. Res. Chem.*, **23**: 170-176.

Wraight, C.A., Cogdell, R.J. and Chance, B. (1978) Ion transport and electrochemical gradients in photosynthetic bacteria. In: Clayton, R.K. and Sistrom, W.R. (eds.), *The Photosynthetic Bacteria.* Plenum, New York, pp. 471-511.

10

Methods for the isolation and analysis of carotenoids

G. Britton and A.J. Young

10.1 INTRODUCTION

Carotenoid analysis is undertaken for many different purposes and the analytical strategy that is adopted is determined largely by the amount of information that is required. Thus, in many cases, the only objective is a routine screening of the leaf or other carotenoid-containing material to monitor possible and often subtle quantitative changes in pigment content and composition. In other cases, identification of all carotenoids found in a range of photosynthetic organisms may be required. Chromatographic properties and ultraviolet/visible (UV/Vis) light absorption spectra provide a valuable first indication of the compounds present, but other data, especially mass spectra and preferably nuclear magnetic resonance (NMR) spectra, are essential to confirm identifications. If compounds are detected that have not previously been identified, i.e. possible new structures, their characterisation is an exercise in investigative organic chemistry and the non-chemist should seek specialist help. Some of the most complicated and difficult, yet most important, analytical problems involve the separation and characterisation of carotenoid geometrical (*Z*/*E*) isomers.

The need for detailed chemical characterisation varies from one study to the next. At every level, however, the main criterion is that the procedure should be sufficiently rigorous to ensure that no mistakes of identification are possible and that major losses of carotenoid that would result in serious errors in quantitative analysis are avoided.

In all carotenoid work, the same general principles, strategies and procedures apply, but practical details need to be varied to meet different requirements and the variety of problems posed by different biological sources of material, e.g. plants, algae or bacteria. In particular, the conjugated polyene chromophore that is characteristic of carotenoids renders these compounds very sensitive to oxygen, light and heat. Many structural features of carotenoids may also be modified by exposure to acid or alkali. The susceptibilty of carotenoids to these conditions entails the use of much more stringent precautions than are necessary for many other natural products if modifications or loss of carotenoids in natural extracts are to be avoided. Careful and rapid work is therefore essential to prevent the loss of the sample or the appearance of artifacts.

In this chapter, recommended procedures are described for the extraction, isolation, analysis and characterisation of carotenoids. The minimum identification criteria for carotenoids are: (i) UV/Vis absorption spectra in at least two different solvents, (ii) chromatography (R_f, T_r) against standards in at least two different systems, and (iii) mass spectra (molecular ion and characteristic fragment ions).

No attempt is made to describe the historical development of methods nor to give a comprehensive, critical survey of available procedures. The treatment is selective, and the recommended methods are largely those that are used in the authors' laboratories. Further details of the general methods used in the handling, extraction and analysis of carotenoids from biological systems are available (see Britton 1985, 1991; Britton and Goodwin 1971; Davies 1976; Goodwin and Britton 1988; Liaaen-Jensen and Jensen 1971; Lichtenthaler 1987).

10.2 EXPERIMENTAL METHODS

Although they cover a wide range of polarities, from the apolar hydrocarbon carotenes to the polar xanthophylls, the carotenoids are lipid substances, with typical lipid solubility properties; even the carotenoid glycosides have only minimal solubility in water. Many of the experimental procedures used for other classes of lipid can therefore be used for the extraction, purification and analysis of carotenoids.

The special and characteristic chemical and spectroscopic properties of the carotenoids are a consequence of the system of conjugated double bonds, which not only constitutes the light-absorbing chromophore, but also renders the carotenoids sensitive to oxygen, light, heat, acids and, in some cases, alkali, and especially to combinations of these factors. Stringent precautions must be observed in order to avoid or minimise losses of material or unwanted structural changes. It is important to work quickly; all procedures that introduce risks of oxidation, isomerisation, etc., should be carried out as quickly as possible.

The sensitivity of modern high-performance liquid chromatography (HPLC) and other analytical methods is so great that misleading results can be obtained by the detection of artifacts that may be produced during extraction and purification. The strong colour of carotenoids can mask the presence of substantial amounts of colourless contaminants which are not detected during the normal spectrophotometric assay of carotenoids but will render useless any mass spectrometry (MS), NMR analysis, etc.

10.2.1 Protection against oxidation

The susceptibilty of carotenoids to oxidation cannot be over emphasised. Oxygen, especially in combination with light and/or heat, is the most destructive factor. The presence of even traces of oxygen in stored samples, of peroxides in solvents (especially diethyl ether), or of any oxidising agents even in crude samples containing carotenoids can rapidly lead to bleaching or to the formation of artifacts such as epoxides or apocarotenals. The acyclic carotenes, particularly phytofluene and ζ-carotene, seem to be most labile, but ß-carotene is also readily destroyed or epoxidised; xanthophylls in general are more stable. Carotenoid samples or extracts should always be stored in an inert atmosphere (Ar or N_2) or *in vacuo* in the complete absence of oxygen. Solutions should be sparged with N_2 or Ar for a few minutes prior to storage.

It is often assumed that carotenoid molecules are stabilised by protein, and this is frequently the case. When green plant tissues are disrupted, however, especially during the isolation or incubation of chloroplasts or thylakoids or their pigment–protein complexes, the carotenoids, especially ß-carotene, can undergo rapid photochemical or enzymic oxidation.

10.2.2 Protection against light and heat

Exposure of carotenoids to light or heat should be avoided if possible, although direct photoisomerisation of most carotenoids during most manipulations occurs comparatively slowly and usually not to any great extent. With extracts that contain chlorophylls, or any other potential sensitisers, photoisomerisation, proceeding via the carotenoid triplet-state, can occur very rapidly so that appreciable amounts of carotenoid *Z* isomers (*cis* somers) can be produced during even brief exposure of the extract to light. Even in the absence of a sensitiser, direct sunlight or UV light may cause some geometrical isomerisation and must be avoided. Low intensity diffuse daylight or subdued artifical light is acceptable for most carotenoid manipulations, but precautions must be taken to exclude light during chromatography. Thus, glass chromatography columns should be wrapped in aluminium foil or black paper, and developing tanks for thin layer chromatography (TLC) covered with a black cloth or kept in a dark cupboard.

Carotenoids may also undergo isomerisation and structural modification if heated, either as solids or in solution. They should never be subjected to excessive heat. Solvents with low boiling points should be used whenever possible, as these can subsequently be removed at low temperature. Solvents should be evaporated on a rotary evaporator at a temperature not exceeding 40 °C.

10.2.3 Avoidance of acid or alakli

Almost all carotenoids are susceptible to decomposition, dehydration or isomerisation if subjected to acid conditions. Carotenoid 5,6-epoxides undergo particularly facile isomerisation to the corresponding 5,8-epoxides if exposed to even traces of acid during extraction and purification. Many plant tissues are sufficiently acid to bring about this isomerisation, but it can usually be prevented by the inclusion of a neutralising agent such as $NaHCO_3$ during extraction. Acidic adsorbents, especially silica gel, silicic acid and acid alumina, can cause isomerisation during chromatography, and should be avoided, as should acidic solvents, notably chloroform, which often contains traces of HCl. Acidic reagents and strong acids should not be used in rooms where carotenoids are being handled.

Most carotenoids are stable to alkali. Some, however, notably fucoxanthin and related compounds, carotenoids containing the 3-hydroxy-4-oxo-ß-ring as in astaxanthin, and, of course, carotenoid esters, are altered by treatment with even weak alkali. Treatment with alkali, especially saponification, must be avoided if it is suspected that any such compounds may be present.

10.2.4 Purity of solvents, adsorbents and reagents

Pure solvents and reagents should be used for all work with carotenoids. Solvents should be dried and redistilled, and peroxides must be removed from diethyl ether by distillation from, for example, reduced iron powder or calcium hydride. Solvents should be stored in airtight dark glass bottles. Because of its toxicity, benzene is not recommended; the less toxic toluene is usually a satisfactory substitute.

When small samples (e.g. a few micrograms) of a carotenoid are being prepared for analysis by MS or NMR, impurities can be introduced during the purification procedure. Rigorous purification of solvents by double distillation and filtration through an activated material such as alumina are recommended. TLC plates should be pre-washed with a solvent at least as strong in polarity as that which is to be used for elution of carotenoids.

Small amounts of plasticisers, especially phthalates, are readily dissolved by organic solvents, and can cause a major contamination problem. All contact of samples, solvents, etc. with plastic materials must be avoided. Plastic or polythene wash bottles, pipettes, stoppers etc. should **never** be used in carotenoid work. Solvents, even water, should never be stored in plastic bottles.

10.3 EXTRACTION

10.3.1 Higher plants

Undamaged and preferably fresh biological material should be used, and carotenoids extracted as soon as possible in order to minimise oxidative or enzymic degradation. Carotenoids are usually extracted with a water-miscible solvent such as acetone, ethanol or methanol (chloroform-methanol is not recommended because of the possibility of traces of HCl) and is aided by mechanical disruption of the tissues. For leaf material this can be carried out efficiently by means of a mortar and pestle in the presence of clean sand, whereas other tissues may require the use of an electric blender. The addition of small pieces of solid CO_2 helps to minimise exposure to oxygen. The slight acidity of many plant tissues can be sufficient to cause isomerisation of the 5,6-epoxides to their furanoid, 5,8-epoxide, derivatives; this can be prevented by the addition of a small quantity of $NaHCO_3$ during the extraction procedure. In leaves, violaxanthin and neoxanthin are particularly sensitive to furanoid rearrangement, forming auroxanthin and neochrome, respectively. Exposure to such acidic conditions makes any quantitative assessment of pigment analyses particularly difficult. Following filtration of the extract, the solid cell debris usually needs to be re-extracted with solvent until no more colour is recovered (usually two to three times).

Small-scale extracts (i.e. from less than ~0.5 g of biological material) which are to be analysed by HPLC may be evaporated under a steady stream of N_2, then redissolved in ether or HPLC solvent for analysis. For larger samples, the carotenoid-containing lipid extract should be transferred to a separating funnel and an approximately equal volume of diethyl ether added. The solution is then thoroughly mixed and approximately half the total volume of water or saturated sodium chloride solution added. The funnel is then swirled vigorously to ensure efficient extraction and partition, and the two phases are allowed to separate. A smooth swirling action achieves efficient mixing and the risk that emulsions may form is much less than with the more conventional vigorous shaking. The lower aqueous phase is run off and the upper ethereal layer, which contains the pigments, is washed a further two to three times with water. In quantitative work, the combined aqueous washings are re-extracted with fresh diethyl ether, to ensure complete recovery of pigment.

[Extraction with petrol or hexane instead of diethyl ether is sometimes reported, but this cannot be recommended when quantitative extraction of xanthophylls is required; losses of polar components such as violaxanthin and neoxanthin can be quite substantial. Diethyl ether/petrol mixtures (~1:1) are usually satisfactory, if used with care. Similarly the practice of adding first water and then diethyl ether also renders complete recovery difficult and is not recommended.]

The ethereal pigment extracts are evaporated to dryness on a rotary evaporator at a temperature no greater than 40°C. Once dried, the pigments can be redissolved in a small volume of diethyl ether and transferred to a snap-top vial, blown to dryness under a steady stream of N_2 and stored (see Section 10.2.4). The removal of residual water is facilitated by the addition of a small volume of redistilled ethanol.

10.3.2 Algae

Similar procedures are applicable to algae, though variations in details are necessary to take account of the variety of different forms, ranging from microscopic unicellular species to the large thalloid seaweeds, and also because of the range of different carotenoid structures that may be present (see Chapter 2). In particular, many of the abundant algal carotenoids, notably fucoxanthin and peridinin, are alkali-labile so saponification and the use of alumina for chromatography must be avoided.

The macroalgae (seaweeds) are best extracted fresh by mechanical homogenization with acetone or acetone-methanol (7:3 (v/v)). For bulk extraction of a major component, such as fucoxanthin for use as a standard, the alga may first be air-dried for ~48 h.

With many unicellular algae (e.g. *Scenedesmus* spp.) the complete extraction of all pigments from the cells may be very difficult to achieve by using cold solvents. In this case much more efficient extraction may be obtained with hot acetone or methanol; the rapid extraction with the hot solvent should cause less damage to the sample than an otherwise lengthy extraction procedure. Extraction may also be facilitated by first breaking the cells by methods such as ultrasonic disintegration or the French press.

For full details of the isolation and analysis of carotenoids from any particular class of algae, the reader is recommended to consult the original publications from Liaaen-Jensen and co-workers.

10.3.3 Phototrophic bacteria

Essentially the same strategy is used to isolate pigments of the bacteria as that for higher plants and algae, but with some variation to take into account differences in the nature of the biological material and the properties of the carotenoids present.

Carotenoids are extracted from harvested cells with a water-miscible organic solvent, usually acetone, ethanol or methanol, or a mixture of these (usually acetone/methanol, 2:1 (v/v)). In some cases, the carotenoids can be extracted efficiently simply by soaking the cell pellet in solvent at room temperature under N_2 for a few hours. When very polar carotenoids, especially glycosides, are present or when the carotenoids are more tenaciously held in the cell wall or membrane, cells may need to be disrupted by, for example, French press, ultrasonic disintegration, shaking or grinding with glass beads or other abrasive material, treatment with enzymes (especially lysozyme) followed by osmotic shock, or even direct saponification. With most phototrophic bacteria, however, efficient extraction may be achieved, without cell disruption, by the use of hot solvent. As in the algae, the brief exposure to heat which this entails is thought to be less harmful than the effects of prolonged and inefficient extraction with cold solvent, or of cell disruption.

Solvent extraction should be repeated. In some cases, a first extraction with methanol serves mainly to remove water, or to extract bacteriochlorophyll, but liberates little carotenoid. The bulk of the carotenoid is then removed by a second extraction with acetone. When carotenoid glycosides are likely to be present, extraction should be more efficient with methanol or ethanol alone. The carotenoids are then transferred to diethyl ether and evaporated, as described in Section 10.3.1. If any long-chromophore acyclic carotenoids such as spirilloxanthin are present, it is usually advantagous to include dichloromethane or toluene (about 10% of total volume) in any solution containing the carotenoids during the extraction/isolation procedure in order to prevent substantial losses from crystallisation.

Complete extraction of pigments from many bacteria, as with some thick cell-walled algae, can only be achieved by using a probe sonicator or a French press to disrupt the cells. Conventional solvent extraction with acetone, methanol or ethanol can then be used to extract the carotenoids from the disrupted cells.

Nelis and de Leenheer (1989) recommended pre-treatment with KOH (60% (w/v)) or, preferably, with liquified phenol to improve extraction of carotenoids with methanol from freeze-dried bacterial cells. Attention was drawn to possible problems with the stability of carotenoids in phenol.

A more recent method for extracting natural products from biological material is supercritical fluid extraction (SFE). The use of this technique in extracting carotenoids has been limited, although some studies (e.g. Cygnarowicz and Seider 1990; Degnan et al. 1991) have been reported. Although there are a number of problems associated with this technique (e.g. the poor solubility of some carotenoids in CO_2), initial results indicate that SFE may have a role in analytical and preparative scale extractions.

10.3.4 Saponification

The saponification (alkaline hydrolysis) of total extracts prior to the isolation or analysis of carotenoids is a common, and often necessary, practice. It may be used to remove large amounts of neutral lipid which may interfere with chromatographic procedures, to remove the chlorophylls which may mask the presence of carotenoids or for the hydrolysis of carotenoid acyl esters. Great care should be taken to ensure that saponification has no detrimental effect on the carotenoids. Small losses of carotenoid are inevitable during saponification, though this may be limited if care is taken to remove all oxygen from the reaction vessel. With this in mind it is particularly useful to perform a preliminary chromatographic or spectroscopic analysis on a sample prior to saponification, so that an overall picture of the pigment composition can be obtained. Examples of artifact production as a result of saponification are given in Section 10.3.5.

The general procedure is for the carotenoid-containing extract to be saponified in 6% (v/v) ethanolic KOH at room temperature, in the dark, under N_2. A period of 1-2h is sufficient to destroy chlorophylls completely, although the saponification can be left for a longer period (e.g. overnight) for convenience, without damage to the carotenoids. An equal volume of diethyl ether or diethyl ether–petrol (1:1) is then added, followed by water or NaCl solution until two layers form. The lower aqueous phase is re-extracted with the organic solvent and the combined organic extracts are washed with water until they are free of alkali. Usually three to four washings are required, and more diethyl ether or diethyl ether–petrol can be added should this prove necessary. These extracts can then be evaporated and stored in the usual manner.

All traces of acetone must be removed before saponification to avoid the possibility of producing artifacts by facile aldol condensation between apocarotenals and acetone, and of introducing contaminants by the polymerisation of acetone.

In the case of many green algae in which astaxanthin esters are accumulated as major secondary products (e.g. *Haematococcus* spp., see Chapter 6), saponification of the complete pigment extract is desirable for complete analysis of the geometrical and optical isomers of astaxanthin. Astaxanthin is, however, particularly unstable and will readily undergo rapid and irreversible oxidation to its 2,3-dihydro-derivative, astacene, precluding the use of traditional saponification procedures. This transformation can be prevented by carrying out the saponification under completely anaerobic conditions in specially designed apparatus, as described by Müller et al. (1980). Similarly, samples containing fucoxanthin or other compounds containing the 3-hydroxy-4-oxo ß-ring must never be saponified in the normal manner. Jacobs et al. (1982) have described a technique for using cholesterol esterase to hydrolyse small amounts of esters of astaxanthin and fucoxanthin.

Miki (1990) has reported the successful use of microwaves in the saponification of carotenoids. Not only are the reaction times greatly reduced, but also the breakdown of carotenoids through microwave-aided saponification is much less than in conventional conditions.

10.3.5 Production of artifacts

The problems encountered in the analysis of carotenoids from natural sources, especially the production of artifacts, have recently been highlighted by Liaaen-Jensen (1990). Some of the artifacts may be produced as a result of improper handling of the biological material prior to extraction (e.g. enzymic reactions). A useful indicator of this in photosynthetic tissues is the production of breakdown products of chlorophyll. Most commonly, artifacts arise during the extraction and subsequent purification and analysis of carotenoids. The procedures that can be used in order to minimise this are described in the relevant sections in this chapter. Some common examples of the production of artifacts in photosynthetic tissues are given below.

(a) Pre-extraction artifacts

The formation of artifacts through the mishandling (e.g. poor storage conditions) of biological tissues is rarely reported, although it is always best to extract fresh material whenever possible. Many tissues may degrade and liberate acids on prolonged storage, particularly at room temperature. Liaaen-Jensen (1990) has described the production of an artifact, derived from 19´-hexanoyloxyfucoxanthin by the action of acid, produced as a natural consequence of the aging of algal cells. More commonly, acids liberated from plant tissues may bring about the furanoid rearrangement of epoxide groups, although this may be more commonly seen during the isolation of carotenoids. The effect of acid is also thought to be responsible for the production of isofucoxanthin from fucoxanthin (Jensen 1966). The occurrence of manixanthin (9,9´-di*cis*-alloxanthin) in the cryptomonad alga *Chroomonas salina* has now been considered to be artifactual and to be brought about by the 9,9´-*trans* to 9,9´-di*cis* isomerisation of alloxanthin (Liaaen-Jensen

1990).

(b) Isolation artifacts

Geometrical isomers of carotenoids are probably the most common of all artifacts. In many cases *cis–trans* transformations may go undetected, because subsequent work does not involve the separation and analysis of these isomers. Stereochemical transformation of carotenoids may be brought about through the action of heat, light, acids or active agents (e.g. alumina) and means of avoiding these are given elsewhere in this chapter.

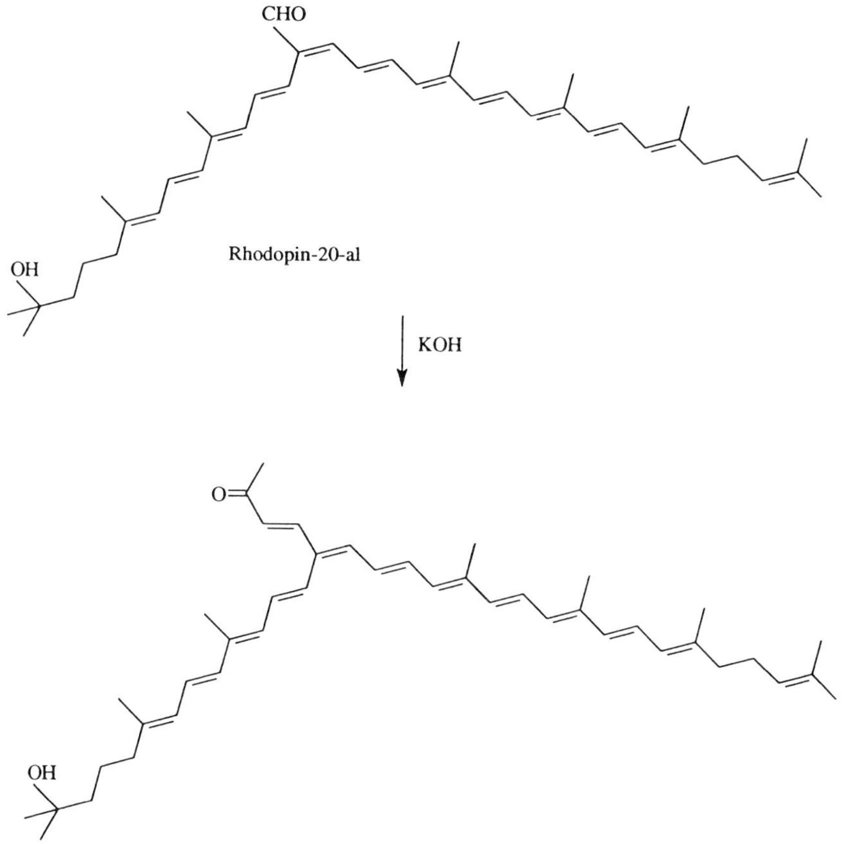

Figure 10.1 Scheme showing the production of the C_{43} artifact of lycopene-20-al under alkaline conditions in the presence of acetone (Liaaen-Jensen 1990).

With many isolated carotenoids, though, it is particularly difficult to prevent *cis–trans* isomerisation from taking place. Photosynthetic tissues naturally contain examples of both all-*trans* (e.g. ß-carotene) and *cis* isomers (e.g. 9′-*cis*-neoxanthin). Improper handling, especially in the presence of a sensitiser such as chlorophyll or bacteriochlorophyll, may result in alterations to the isomeric composition of the carotenoids.

The furanoid rearangement of epoxide groups in natural extracts that have been exposed to acidic conditions is a common feature and has been discussed in earlier sections. It is particuarly important to ensure that any natural acids present in tissues are neutralised and that the carotenoids are not exposed to acidic or certain active surfaces (e.g. alumina or silicic acid). The details of the furanoid rearrangement have been elucidated and were discussed by Liaaen-Jensen (1990) and by Eugster (1982, 1985).

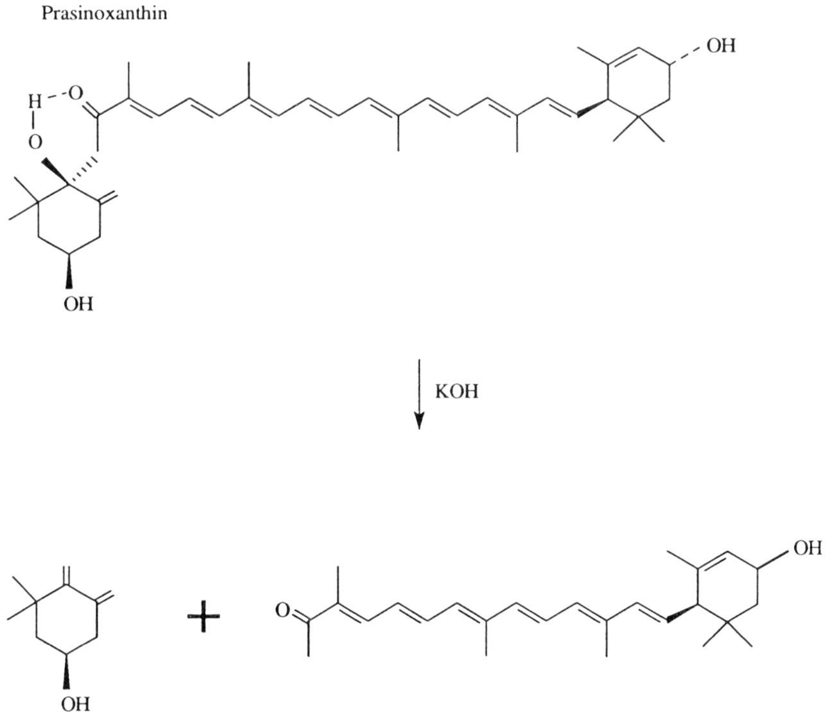

Figure 10.2 Scheme showing the formation of the carotenoid artifact micronone from the common algal carotenoid prasinoxanthin during saponification (Liaaen-Jensen 1990).

Not only may exposure to acids cause the formation of artifacts, care must also be taken when carrying out saponification procedures, during which samples are treated with alkali (see Section 10.2.3). Carotenals rapidly undergo aldol condensation when KOH is used in the presence of acetone, resulting in the formation of ß-hydroxy ketones followed by dehydration to conjugated methyl ketones. As an example, Liaaen-Jensen has reported the detection of high levels of C_{43} artifacts of the carotenoids from *Rhodopseudomonas acidophila*, e.g. lycopene-20-al (Fig. 10.1).

Alkaline conditions may also result in the dehydration of ß-hydroxy ketones to form cross-conjugated diketones. The presence of such ketones in natural extracts which have been saponified should be carefully examined (Liaaen-Jensen 1990). Similar conditions may also result in the *retro*-aldol cleavage of ß-hydroxyketones, for example, the formation of micronone from prasinoxanthin (Fig. 10.2; Foss et al. 1984). Fucoxanthin and peridinin are particularly sensitive to alkali, and recoveries are generally very low. A number of artifacts can be found, including an unstable product from fucoxanthin which is thought to be a hemiketal, and peridininol.

Another common alkali-catalysed reaction is the conversion of α-ketols and their esters into enolised α-diketones or diosphenols. These reactions are typified by the instability of astaxanthin and astaxanthin esters in natural extracts. Improper handling will readily lead to the formation of astacene, which is considered to be an artifact. Great care must therefore be taken to exclude oxygen from any saponification procedure in which astaxanthin is present (see Section 10.2.3).

Certain carotenoids will, even under neutral conditions, react rapidly in the presence of oxygen, for example rhodopin-20-al can be formed from rhodopin-20-ol in bacterial extracts.

10.4 SEPARATION AND PURIFICATION

10.4.1 General comments

Extracts of photosynthetic organisms or tissues normally contain a mixture of different carotenoids. For some purposes, particularly for the routine quantitative determination of carotenoid compositions, all that is necessary is to separate the individual carotenoids from each other; the presence of colourless, non-carotenoid contaminants will not interfere with the subsequent spectrophotometric assay. In other kinds of work, however, especially when characterisation by MS, NMR or infrared spectroscopy is required or in work that involves isotopic labelling, all contaminants must be removed and the purification of carotenoids must be rigorous.

10.4.2 Non-chromatographic methods

Although they are used much less extensively than they have been in earlier studies, some non-chromatographic procedures can sometimes still be employed to advantage, particularly in large-scale preparative work. The most useful of such methods are precipitation procedures to remove major contaminants such as sterols and partition between immiscible solvents to separate polar from non-polar carotenoids. Britton (1985) has described the use of these non-chromatographic procedures for the separation and purification of carotenoids.

Two methods have been used frequently to remove large amounts of sterols from plant and algal extracts, to facilitate the subsequent purification of carotenoids. A substantial amount of sterol can be precipitated from petrol solutions that are left overnight at -10 °C, but a more efficient procedure involves precipitation of the sterols as their digitonides.

Phase separation involves the partitioning of an extract between two immiscible solvent phases of different polarity. The most commonly used solvents are petrol or hexane and aqueous 90% methanol; non-polar carotenes, carotene epoxides and esters are recovered in the petrol epiphase, polar xanthophylls from the lower or hyperphase. More specific separations can sometimes be achieved by modifying the solvent compositions.

Large-scale isolation techniques based on liquid–liquid partitioning have been successfully employed in the purification of two major algal carotenoids, fucoxanthin and peridinin (Haugan et al. 1992). These and other techniques are also described in more detail by Davies (1976).

10.4.3 Chromatography

(a)General considerations

The technique of chromatography was originally developed to separate chlorophylls and carotenoids in plant extracts on columns of starch or sugar. Column chromatography remains extremely valuable, especially for preliminary separations and large-scale work. For smaller samples and delicate separations TLC is widely used. Because carotenoids are so easy to detect by their absorption of visible light, HPLC has now become the method of choice for carotenoid analysis. Gas–liquid chromatography (GLC or GC) is not suitable for work with carotenoids because of the general instability of these compounds at elevated temperatures. The relatively new technique of supercritical fluid chromatography (SFC) shows considerable potential and merits further exploration. Chromatography on paper impregnated with aluminium oxide or Kieselguhr can be a valuable analytical technique especially for the detection of geometrical (*Z/E*) isomers.

(b) Strategy

As almost all the common chromatographic methods can be used with carotenoids, a combination of these techniques can be employed for the separation and purification of these pigments. Except for work on a very large scale or a very small scale, a recommended strategy is first to use column chromatography to achieve separation into fractions that contain common groups of carotenoids of approximately similar polarity and then to use TLC to separate the individual pigments in each fraction and bring them to a satisfactory state of purity. For very large scale separation and purification it may be more practical to use only column chromatography, and for small-scale work to use only TLC or HPLC.

A preliminary examination of a small sample of the extract by TLC or HPLC will allow a strategy to be devised for the subsequent purification.

(c)Column and thin-layer chromatography

Throughout all the procedures described the solvents used must be of the highest quality. All solvents should be dried and redistilled before use and peroxides removed from diethyl ether by distillation from reduced iron powder, calcium hydride or $LiAlH_4$. This is particularly important when samples are being prepared for detailed spectroscopic analysis.

A combination of column chromatography and TLC is generally used for the isolation and purification of carotenoids from total lipid extracts. Column chromatography is usually used first to fractionate the sample on the basis of the polarity of the pigments. This is followed by separation of components with similar polarity by TLC. A selection of different adsorbents can be used effectively to separate the individual carotenoids. Britton (1991), for example, described a general strategy of successive TLC on silica gel and MgO-Kieselguhr G.

With an unfamiliar extract a small amount of sample should first be subjected to preliminary investigation by TLC. This will provide an idea of what carotenoid types are present in the extract and of what separation strategy to adopt. This may best be achieved by applying a single spot of sample on to silica gel G and developing successively in a series of solvents of increasing polarity, as shown in Table 10.1.

Table 10.1 Solvents suitable for the chromatography of carotenoids on silica gel G TLC or on silica gel or alumina (Brockmann grade III) columns.

Solvent	*Carotenoids groups eluted*
Petrol	Carotene hydrocarbons
5% Diethyl ether/petrol	Carotene epoxides
10% Diethyl ether/petrol	Xanthophyll esters
20%–30% Diethyl ether/ petrol	Monooxocarotenoids, monomethoxy- and dimethoxycarotenoids
50%–60% Diethyl ether/petrol	Monohydroxy- and dioxocarotenoids
Diethyl ether or 5% ethanol/diethyl ether	Di- and trihydroxy carotenoids, xanthophyll epoxides
20% Ethanol/ether or ethyl acetate/propan-2-ol/water	Tetrahydroxycarotenoids and carotenoid glycosides

The most commonly used adsorbents for column chromatography and TLC of carotenoids are silica and alumina. The general instability of some carotenoids can cause problems even with these adsorbents. Neutral adsorbents should be used and acid-washed alumina or acidic materials such as silica gel or silicic acid avoided as these may result in the furanoid rearangement of 5,6-epoxide groups (silicic acid can be neutralised to pH 7.0 before use). Both of these adsorbents separate carotenoids on the basis of the polarity of the carotenoid molecule and are therefore most suitable for the initial fractionation of an extract. The carotenes are very weakly adsorbed and are generally eluted with a low polarity solvent such as petrol. Elution of the more polar compounds is generally achieved by using more polar solvents such as diethyl ether and ethanol (Table 10.1). With alumina columns, the sample is usually applied in petrol and the sample eluted with increasing concentrations of diethyl ether in petrol.

In the case of alumina the activation strength of the adsorbent may be selected by altering the water content prior to chromatography. In general, Brockmann grade III (deactivated by the addition of 6% (v/w) water) is suitable for most purposes, as given in Table 10.1. Whilst activated alumina (grade 0 or I) has particular uses for the separation of carotenes and their geometrical isomers (see Table 10.2), its indiscriminate use may lead to large sample losses and isomerisation of the carotenoids present. Irreversible losses of certain carotenoids (e.g. astaxanthin and some of the seco-carotenoids) can occur on alumina. A small-scale test should therefore always be made before committing the whole sample.

Table10. 2 Separation of carotene *cis-trans* isomers on activated (Brockmann grade I) alumina columns

Compound	*Eluting solvent*
15-*cis*-Phytoene	2% Diethyl ether/petrol
all-*trans*-Phytoene	4% Diethyl ether/petrol
15-*cis*-Phytofluene	10% Diethyl ether/petrol
cis-α- and ß-Carotene	15%–20% Diethyl ether/petrol
15-*cis*-ζ-Carotene	25%–30% Diethyl ether/petrol
all-*trans*-α- and ß-Carotene	30%–40% Diethyl ether/petrol
all-*trans*-ß-Zeacarotene	50% Diethyl ether/petrol
all-*trans*-ζ-Carotene	60%–100% Diethyl ether/petrol
all-*trans*-Neurosporene	0%–5% Acetone/diethyl ether
all-*trans*-Lycopene	10%–20% Acetone/diethyl ether

Sample purification following column chromatography is best achieved by successive TLC on silica and/or MgO. As with silica column chromatography, diethyl ether or acetone in petrol mixtures are used to separate the individual carotenoids. A second TLC step, with alternative solvents such as methanol/toluene, propan-2-ol/petrol or ethyl acetate/carbon tetrachloride can be effective for resolving and purifying individual pigments.

Magnesium oxide has been used as an adsorbent for carotenoid separations where silica does not perform very satisfactorily (e.g. resolution of ß- and ε-ring carotenoid isomers). Separation on MgO depends on the number and arrangement of double bonds in the molecule; thus carotenoids with the most extensive conjugated polyene systems are most strongly adsorbed, e.g. lycopene > neurosporene > ζ-carotene (see Table 10.3). $Ca(OH)_2$ will separate geometrical isomers. Acyclic carotenoids are much more strongly adsorbed than those with cyclic end-groups containing the same number of double bonds: e.g. lycopene > γ-carotene > β-carotene. Spirilloxanthin and other acyclic carotenoids with a very long chromophore are so strongly adsorbed that they are difficult to elute, and losses during chromatography can be substantial. Compounds with β-rings are held more strongly than the related ε-ring compounds: e.g. β-carotene > α-carotene and zeaxanthin > lutein. Carotenoids with 5,6-epoxide groups are much less strongly adsorbed than 5,8-epoxides, but other polar groups such as hydroxy groups have little influence on the adsorption of the molecule. Chromatography on MgO is particularly useful for removing colourless non-carotenoid impurities which are less strongly adsorbed and run towards the top of the TLC plate.

The practice of using columns of MgO or $Ca(OH)_2$ that incorporate a filter aid such as celite, is common in some laboratories but cannot be generally recommended for preliminary fractionation of a mixture of carotenoids because of the somewhat unpredictable and apparently illogical adsorption affinities and order of elution. These adsorbents can, however, be useful for subsequent specialised separation of different mixtures of geometrical isomers on a preparative scale.

Table 10.3 Solvents suitable for TLC of carotenoids on MgO:Kieselguhr (1:1, w/w).

Carotenoids	*Solvent*
Phytoene	Petrol
Phytofluene, α- and β-carotene	2%–6% Acetone/petrol or 10% toluene/petrol
ζ-Carotene	6%–8% Acetone/petrol or 10%–15% toluene/petrol
α- and β-Zeacarotene	10%–15% Acetone/petrol or
δ- and γ-Carotene	30% toluene/petrol
Lycopene and spirilloxanthin	Acetone/petrol/toluene (2:1:2; by vol.)
Violaxanthin, lutein and lutein epoxide	20% Acetone/petrol
Zeaxanthin	25% Acetone/petrol
Neoxanthin	30% Acetone/petrol
Spheroidene, spheroidenone and hydroxyspheroidene	Acetone/petrol/toluene (1:1:1; by vol.)

Many other adsorbents (cellulose, starch, $CaCO_3$, $Ca_3(PO)_4$, CaO, $MgCO_3$, $ZnCO_3$, microcell C) have been successfully applied for the isolation and separation of carotenoid mixtures (Britton, 1985).

Procedure

Column chromatography The adsorbent that is recommended for column chromatography of carotenoids, whenever possible, is neutral alumina. The alumina is first deactivated to the required grade by the addition of the appropriate amount of water (e.g. 6% (w/w) for grade III), with stirring to ensure that the water is evenly distributed. For small scale work (up to ~200 mg of lipid) approximately 10 g of alumina is satisfactory, although large–scale preparations can employ a higher ratio of lipid:alumina. Ideally the column length should be 10–20 times the width. Deactivated alumina may be kept as a slurry in redistilled petrol for subsequent use.

The total lipid extract is dissolved in a small volume of petrol and applied to the top of the column (it may be necessary to add no more than 10% (v/v) toluene or dichloromethane if an extract contains large amounts of acyclic carotenes that possess a long chromophore or up to 5% (v/v) diethyl ether for samples containing polar xanthophylls which do not readily dissolve in petrol alone). Note: The first fraction collected when 5% diethyl ether is used will contain carotenes, but it may also have carotene epoxides, esters and non-carotenoid impurities. This fraction can, if necessary, be evaporated to dryness and rechromatographed, in petrol, on a second alumina column.

Elution is continued with petrol until all the carotenes have been collected. When this first eluate becomes colourless, elution is continued with selected solvents that contain increasing amounts of diethyl ether, and suitable fractions are collected (Table 10.1). When a certain solvent mixture is used to elute fractions that contain only colourless substances that might otherwise contaminate subsequent carotenoid fractions, 100 ml of eluate should be collected for each 10 g of alumina in the column. Otherwise, elution is continued with each solvent until any moving coloured band has been collected.

During the chromatography, the column must be protected from light by enclosing it in a sheath of aluminium foil, black paper, or a black cloth which can be easily opened or removed to allow inspection of the column.

For general semi-preperative work, when the sample is especially concentrated, or if there is a likelihood that the carotenoid may crystallise out then a small layer (~1 cm) of clean sand applied to the top of the column will help to prevent or overcome any blockages. This sand layer can be agitated without disturbing the column itself.

Thin-layer chromatography Introduced in the late 1950s, TLC soon became the most versatile and effective method for purifying carotenoids, and it remains valuable even in laboratories that have good HPLC facilities. TLC is widely employed for the purification of carotenoids for spectroscopic analysis, and for partial identification by comparison with authentic standards. A particular advantage is that, apart from the biosynthetic intermediates phytoene and phytofluene, the carotenoids are coloured and can be seen on TLC plates with great sensitivity. Phytoene and phytofluene and their derivatives are located by examining the chromatogram under UV light. Phytofluene fluoresces intensely greenish-white. The weak violet fluorescence of phytoene is not so easily seen, but phytoene is readily detected by its quenching of the fluorescence of silica gel GF_{254} plates or after spraying with a fluorescent dye such as Rhodamine 6G.

The lability of carotenoids can be a disadvantage and general TLC practice may need to be modified to avoid excessive losses. Carotenoids on TLC are particularly susceptible to oxidation so samples should be applied to the TLC plate and removed as rapidly as possible. All chromatographic steps should be carried out in the dark (e.g by covering the developing tank with a black cloth), and, if necessary, the tank can be flushed with N_2. For preliminary work a small piece (2 x 5 cm) cut from aluminium foil-backed silica gel plates may be used or silica gel-coated microscope slides. These can be developed in 2–3 min in a small jar or beaker with only 2–5 ml of solvent.

TLC on silica gel Silica gel is the most widely used adsorbent in all TLC, including that of the carotenoids. The separations that are obtained depend on polarity, the most polar compounds being the most strongly adsorbed. When it is suspected that colourless compounds such as phytoene or its derivatives may be present silica that incorporates a fluorescent dye should be used (e.g GF_{254}).

Silica gel is often acidic enough to cause isomerisation of 5,6-epoxide groups, and it may therefore be necessary to prepare the silica gel slurry with dilute KOH (~ 0.05 M) or pH 7.0 buffer in place of water. TLC of astaxanthin and related ketocarotenoids usually benefits from the inclusion of citric acid (~ pH 4.0). The chromatography of acyclic carotenes can be hampered by their crystallisation on the TLC plate, causing streaking and poor resolution. This may be overcome by including a small quantity of dichloromethane (<10%) in the developing solvent. Diethyl ether–petrol mixtures are very suitable for most carotenoid separations, and the solvent composition that was used to elute a fraction from a preliminary alumina column usually gives effective separations on TLC (Table 10.1). Good resolution is also obtained with acetone–petrol mixtures (1% acetone is used in place of each 5% of ether in the solvent mixtures listed in Table 10.1). For carotenoid glycosides ethyl acetate/propan-2-ol/water (60:40:2, by vol.) is recommended.

After separation, the individual carotenoid bands are recovered from the developing chromatogram by elution with diethyl ether or acetone, a little ethanol being added if necessary for more polar carotenoids. The carotenoid zone is rapidly scraped from the plate and eluted with a solvent more polar than that used for development. The adsorbent is removed by filtration or centrifugation. Filtration through a sintered glass funnel is most satisfactory, but if such funnels are not available a simple but effective alternative is a small glass funnel with a small plug of absorbent (defatted) cotton wool in the stem. The eluting solvent (ether or ethanol) is added dropwise until the surface of the adsorbent is colourless, and elution is then continued with ether until all the carotenoid has been recovered (usually 2–5 ml). The solvent is then removed under a gentle stream of N_2.

TLC on magnesium oxide Magnesium oxide does not adhere well, if at all, to glass plates. When used in a 1:1 mixture with Kieselguhr G, however, MgO does give stable layers which are extremely valuable for carotenoid TLC, and which will separate mixtures of carotenoids that co-chromatograph on silica, e.g. α- and β-carotenes, lutein and zeaxanthin, chloroxanthin and rhodopin. TLC on MgO does have some disadvantages, notably that carotenoids tend to run as broad, tailing bands especially when the chromatogram is overloaded (which happens very easily). Also carotenoids may be very difficult to remove from MgO unless they are eluted as rapidly as possible, preferably before the solvent has fully evaporated. This can be a particular problem with the long-chromophore acyclic carotenoids such as spirilloxanthin which are adsorbed extremely strongly.

The most effective solvents for TLC on MgO-Kieselguhr G are mixtures of petrol with acetone and/or toluene. The adsorptive strength of MgO is somewhat variable, so the solvent compositions need to be optimised for each batch. As a guideline, the following solvent mixtures should be approximately correct: 4% acetone in petrol for α- and ß-carotene, 25% acetone in petrol for zeaxanthin, acetone/toluene/petrol (1:1:4, by vol.) or (1:1:1) for lycopene and spirilloxanthin, respectively (see Table 10.3). Elution from MgO is best achieved with acetone, plus some toluene or ethanol for the strongly adsorbed carotenoids. Should solvent mixtures containing acetone be used directly with basic adsorbents such as MgO (in an undiluted state and not mixed with Kieselguhr or an inert filter aid such as celite) polymerisation of the acetone may occur leading to further contamination of the carotenoid samples. The conversion of fucoxanthin into isofucoxanthin on MgO has been reported (Jensen 1966). Following TLC with MgO it is recommended that carotenoids are rechromatographed on silica TLC.

As well as separating mixtures of carotenoids, TLC on MgO also provides a very useful and efficient means of removing colourless, non-carotenoid impurities. A developing solvent is used such that the carotenoid leaves the origin but runs with a low Rf value. Colourless impurities usually run with the solvent front and are thus easily removed.

Other TLC systems The complex TLC system ($CaCO_3$/MgO/$Ca(OH)_2$, 30:60:4) developed by Hager and Stransky (1970) gives good separations of a number of carotenoids and has many applications.

Stauber and Jeffrey (1988) have described a TLC procedure to resolve the pigments commonly found in diatom species. A two-dimensional washed-cellulose TLC system (1st solvent: *n*-propanol/ petrol (60–80) 2.5:97.5 (v/v); 2nd solvent: chloroform/petrol (60–80):acetone 30:70:0.5; Jeffrey 1981) was used initially to separate chlorophylls *a* and *c*, ß-carotene and the main xanthophylls.

Reversed-phase TLC systems can also be employed for the separation of carotenoids. Wright and Jeffrey (1987) have used RP-8 plates with methanol/water (9:1 (v/v)) for the separation of fucoxanthin, 19´-butanoyloxyfucoxanthin and 19´-hexanoyloxyfucoxanthin. Bjørnland et al. (1988, 1989) have also described the use of TLC, and HPTLC, procedures for the separation of fucoxanthin and its 19´-hexanoyloxy and 19´-butanoyloxy derivatives.

Partition chromatography of carotenoids by reversed-phase TLC on Kieselguhr, silica gel G, or cellulose layers that are impregnated with liquid paraffin or vegetable oil has been used to separate polar xanthophylls and xanthophyll esters.

Paper chromatography Paper chromatography is now used mainly for analysing isomeric sets that are obtained by I_2-catalysed photoisomerisation of pure carotenoids. Specially prepared impregnated papers are normally used, the most effective ones being those filled with Kieselguhr or alumina (Schleicher and Schull Nos. 287 and 288, respectively).

The chromatograms are developed for a short time with acetone, to concentrate the sample into a narrow line to give optimum resolution. The acetone is evaporated rapidly under a steady stream of N_2 and the chromatogram is then developed in the required solvent. The usual solvents employed are mixtures of acetone (2%–25%) in petrol. Details are given by Jensen and Liaaen-Jensen (1959). Acetone is normally used for the elution of carotenoids from the paper.

(d)Purification of samples for MS, NMR

The chromatographic procedures described above will usually give a carotenoid sample that is free from other contaminating carotenoids and is suitable for UV/Vis absorption spectroscopic and spectrophotometric analysis. However, substantial amounts of impurities that originate from the chromatographic adsorbents can be introduced into the sample. When samples are to be submitted for MS or NMR analysis, the solvents used for their preparation must be purified rigorously, and TLC plates must be prewashed by running in solvent that is at least as polar as that which is to be used for the chromatography. If possible, samples should be eluted from TLC adsorbents with ether; acetone or ethanol are more likely to extract contaminants, so if they are required only the minimum amounts should be used. Carotenoid samples from TLC should be given an additional purification immediately before being subjected to MS or NMR analysis. A satisfactory procedure is to filter the sample through a small column of neutral alumina (3–4 cm in a Pasteur pipette is usually sufficient), the activity of which has been adjusted by addition of water so that the carotenoid will be eluted with diethyl ether or diethyl ether–petrol mixtures. The sample is applied to the column in petrol, and elution is begun with petrol or a solvent mixture that is not sufficiently polar to remove the carotenoid. This eluate is discarded, and the carotenoid is then eluted with a more polar solvent mixture, evaporated and, if intended for NMR analysis, dried under vacuum to remove traces of solvent. A small column of silica may be used in a similar way for those carotenoids that are not stable on alumina.

(e) High performance liquid chromatography

HPLC has now largely replaced the use of thin-layer and open column chromatography for the analysis of carotenoids. Recently the use of advanced detectors such as photodiode-array systems has allowed UV/Vis spectra to be obtained on-line for a wide wavelength range. Computing facilities permit the storage of a number of chromatograms and spectra from a single chromatographic run. These detectors have proved particularly useful for the analysis of carotenoids and chlorophylls, although their detection limits (sensitivity) are much lower than those of conventional detectors.

To date, the relatively new techniques of supercritical fluid extraction (SFE) and chromatography (SFC) have not been applied systematically to the study of carotenoids. SFC has been used in our laboratories in preliminary studies with leaf extracts and conventional C_{18}-bonded HPLC columns. Baseline separation of lutein and zeaxanthin has been achieved. The high pressures involved may, however, restrict the use of this technique to certain detectors. In theory, an integrated system combining the techniques of SFE and SFC should reduce the possible formation of artifacts in samples, allowing, for example, a more accurate assessment of the geometrical isomers present in natural samples. Schmitz et al. (1989) have reported the use of capillary SFC for the analysis of carotenoids in extracts from vegetables.

In addition to improvements in analytical equipment, advances have been made in recent years in the techniques used for the separation of geometrical and optical isomers of carotenoids. The use of HPLC in carotenoid analyses has been reviewed by Rüedi (1985); Goodwin and Britton (1988) and Khachik et al. (1991).

Higher plants

Reversed-phase chromatography A large number of chromatographic procedures have been reported that are suitable for the separation of the carotenoids commonly found in the leaves of higher plants. Reversed-phase chromatography is the preferred separation method in most cases. Unlike normal-phase (adsorption) chromatography (see below) this affords little risk of decomposition or structural modification of the carotenoids during analysis.

The stationary phases most commonly employed are those with C_{18}-bonded chains (ODS); the use of C_8-bonded columns has been less common. Resolution of carotenoids is highly dependent on the degree of end-capping and the extent of carbon loading of the stationary phase. With an increasing number of column types becoming available the scope for separations is increasing (see Craft 1992). Lauren and McNaughton (1986) and Lauren et al. (1986) have examined the choice of stationary phase and the use of modifiers, respectively, in non-aqueous reversed-phase HPLC of carotenoids. Most commercial columns available today give good separations of the main carotenoids found in the photosynthetic tissues of higher plants.

Reversed-phase HPLC columns can be classified broadly into two types on the basis of the method used in the manufacture of the bonded phases, namely monomeric (in which the silica is reacted with monofunctional silanes) or polymeric (a reaction between silica and di- or tri-functional silanes in the presence of water). Some columns can be regarded as intermediate between the two. Of the solvents commonly employed, acetonitrile or methanol are generally used for carotenoids. Care must be taken with aqueous systems that incorporate acetonitrile, as these are particularly prone to problems with de-gassing. Nelis and de Leenheer (1983) have used non-aqueous chromatographic systems for the analysis of carotenoids although most solvent separation systems used for the separation and analysis of carotenoids contain some water.

The chromatographic separation used routinely in our laboratories for carotenoids and chlorophylls from higher plant and some algal sources is shown in Fig. 10.3. This system uses a Spherisorb ODS2 (25.0 cm x 0.46 cm) column, usually with a pre-filter and 2 cm pre-column (same packing material). The solvent gradient employed is given in the legend to Fig. 10.3. This chromatographic system is particularly versatile, and provides good resolution of the polar xanthophylls, chlorophylls *a* and *b* and of the carotenes. It can therefore be used for an initial screening of an unknown sample, and the gradient can readily be modified to provide improved specific separations.

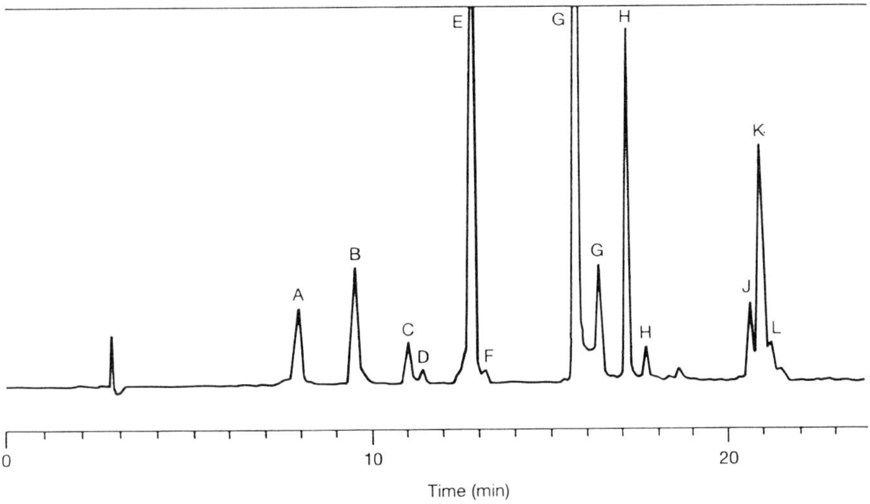

Figure 10.3 Separation of carotenoids and chlorophylls from a leaf extract by reversed-phase HPLC, monitored at 447 nm. Column: Spherisorb-ODS 2 5 μm (25 cm x 4.6 mm). Mobile phase: solvent A, ethyl acetate; solvent B, acetonitrile/water (9:1 (v/v)). 0–16 min, 0%–60% A; 16–30 min, 60% A; 30–35 min 100% A. Flow rate, 1.0 ml min^{-1}. Peak identifications: (A) neoxanthin, (B) violaxanthin, (C) lutein-5,6-epoxide, (D) antheraxanthin, (E) lutein, (F) zeaxanthin, (G) chlorophyll *b*, (H) chlorophyll *a*, (I) ß-carotene-5,6-epoxide, (J) α-carotene, (K) ß-carotene, (L) *cis* isomers of ß-carotene.

The chromatographic resolution of carotenoids on HPLC is particularly sensitive to the nature of the solvent that is used to dissolve the sample for injection on to the HPLC column. Care must be taken to ensure that all polar components are dissolved. On the system described above the best resolution is obtained when the sample is dissolved and injected in a small volume of a 1:1 mixture of acetonitrile–water (9:1)/ ethyl acetate and not in ethyl acetate alone. Noticeable improvements in resolution of the xanthophylls and their geometrical isomers are seen when this mixture is used.

Internal standards are used widely in the quantitative analysis of carotenoids by

HPLC. Among the most common are the synthetic compounds decapreno-ß-carotene and ethyl-8´-apo-ß-caroten-8´-oate (Khachik and Beecher 1985). Some naturally occurring campounds, especially some of the xanthophylls, may prove to be useful standards for particular applications.

In the past few years the interest in the functioning of the xanthophyll cycle has resulted in improved HPLC separation of lutein and zeaxanthin (Thayer and Björkman 1990; Gilmore and Yamamoto 1991; Rock and Zeevaart 1991). Separation of these isomers is difficult and care must be taken that some *cis* isomers of lutein are not co-eluted with zeaxanthin. The chromatographic system shown above provides near-baseline separation of lutein and zeaxanthin and has been shown to be accurate quantitatively down to very low levels of zeaxanthin. Other chromatographic systems that may prove useful in the analysis of carotenoids from higher plants are shown in Table 10.4.

Table 10. 4 HPLC columns used in the analysis of carotenoids found in higher plants.

Compounds	*Stationary phase*	*Reference*
Carotenes and xanthophylls	Microsorb RP18	Khachik et al. 1986.
	Lichrosorb RP8	Braumann and Grimme 1981.
	Lichrosorb RP18	Braumann and Grimme 1981.
	Nova-Pak RP18	Siefermann-Harms 1988.
	Spherisorb ODS	Hoque 1988; Gilmore and Yamamoto 1991; Young et al. 1989.
	Ultrasphere ODS	Hoque 1988.
	Vydac 201SC	Hoque 1988.
	Zorbax ODS	Thayer and Björkman 1990; Young et al. 1989
Xanthophyll acyl esters	Spherisorb ODS	Barry et al. 1992; Khachik and Beecher 1988; Khachik et al 1988.
	Waters RP18	Philip et al. 1988; Philip and Chen 1988.
	Waters Silica	Philip et al. 1988; Philip and Chen 1988.

The natural occurrence of xanthophyll esters in plant and algal tissues is becoming of great interest (see Chapter 7). The best resolution of these compounds is obtained by reversed-phase chromatography with, for example, the separation system shown in Fig. 10.4. The majority of samples contain one or more carotenoids esterified with a range of different fatty acids. The chain length and degree of unsaturation of the esterifying fatty acids greatly affect the retention time of the ester, so that xanthophylls esterified with shorter chain saturated acids (e.g. lutein bis-palmitate) are eluted before those with longer chain saturated acids (e.g lutein bis-stearate) and esters with unsaturated fatty acids are eluted before esters with saturated fatty acids of the same chain length. The absorption spectra of xanthophylls are not altered by esterification, so that a provisional identification of the carotenoids present is possible. The ester fraction can also be collected from the column, saponified in KOH as described earlier (usually for a few hours or overnight) and re-chromatographed to provide a more accurate determination of the parent xanthophyll and to allow any apolar carotenoids previously hidden by the xanthophyll esters to be identified. Both carotenoid and esterifying fatty acid can be identified by mass spectrometry (Khachik et al. 1988; Barry et al. 1992).

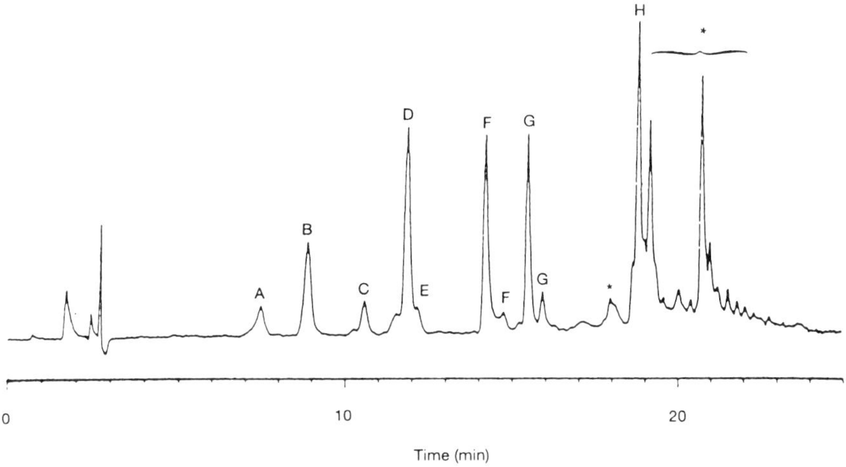

Figure 10.4 Separation of pigments from a senescing leaf, containing high levels of xanthophyll acyl esters, monitored at 447 nm. The chromatographic conditions are similar to those used above, except that a linear gradient of 0%–100% ethyl acetate over 25 minutes was used. Peak identifications: (A) neoxanthin, (B) violaxanthin, (C) antheraxanthin, (D) lutein, (E) zeaxanthin, (F) chlorophyll *b*, (G) chlorophyll *a*, (H) ß-carotene, (*) complex mixture of xanthophyll acyl esters, the main component of which has been shown to be lutein bis-linolenate (Barry et al. 1992).

As with most chromatographic analyses, the maintenance of the stationary phase is very important in determining the resolution of carotenoids. In general use, a reversed-phase column may last 6–12 months, without the need for regeneration. Difficult separations such as lutein/zeaxanthin may require a high quality, relatively new, column.

Normal phase (adsorption) chromatography The extensive use of silica TLC has lead logically to the wide use of silica in HPLC separations. Most solvent systems employ hexane to which increasing amounts of a more polar solvent (e.g. propan-2-ol) are added (either isocratically or, more commonly, as a stepwise gradient) to elute carotenoids of increasing polarity. However, substantial degradation of carotenoids may occur on silica. Problems may also arise because some silica preparations may be slightly acidic and cause furanoid rearrangement of 5,6-epoxides. Quantitative analysis by HPLC on silica, or indeed any other adsorption column, should be underatken with caution.

In our laboratories we have employed a normal phase chromatographic system for the separation of the most common xanthophylls (lutein, violaxanthin, zeaxanthin and neoxanthin) and some of their geometrical isomers (see Section 10.4.3) found in leaf tissues (Britton 1991). The separation of these carotenoids and chlorophylls in a leaf extract is shown in Fig. 10.5. A major disadvantage of this and other adsorption chromatography systems is, however, the poor retention of carotenes and other apolar carotenoids (for example, the rapid elution of ß-carotene seen in Fig. 10.5). This can be partially overcome by altering the eluting solvent composition, e.g. by the use of mixtures of hexane with ethyl acetate or methyl-*t*-butyl ether. The resolution of the more polar xanthophylls on silica is, however, particularly good.

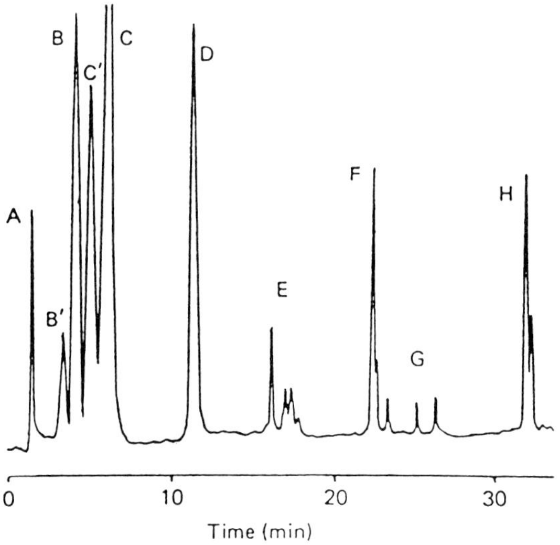

Figure 10.5 Separation of carotenoids and chlorophylls from a leaf extract using normal-phase (adsorption) HPLC, monitored at 447 nm. Column: Spherisorb Silica 3 µm (25 cm x 4.6 mm) Mobile phase: solvent A, hexane/propan-2-ol (4:1 (v/v)); solvent B, hexane. 0–8 min, 10% A; 8–20 min, 10%–25% A; 20–30 min, 40% A; 30–40 min, 70% A. Flow rate 2.0 ml min^{-1}. Peak identifications: (A) ß-carotene, (B) chlorophyll *a*, (C) chlorophyll *b*, (D) all-*trans* lutein, (E) *cis* isomers of lutein, (F) all-*trans* violaxanthin, (G) *cis* isomers of violaxanthin, (H) 9-*cis*-neoxanthin, (I) all-*trans*-neoxanthin.

Bonded nitrile phases can be used as an alternative to silica for the HPLC separation of xanthophylls. Rüedi (1985) has discussed the use of nitrile phases to separate carotenoids that are structurally closely related. They can also be used to achieve efficient separations of geometrical isomers (see Section 10.4.3). The best resolution is generally obtained with samples that have been previously fractionated into groups based on their polarity. The solvent systems can be either gradient or isocratic, although the best resolution is usually obtained with the latter. Rüedi (1985) has described a number of solvent systems which can be employed to resolve carotenoids from different polarity groups on a nitrile column.

Algae and cyanobacteria
The chromatographic systems described above for the analysis and identification of the carotenes and xanthophylls found in higher plants are generally suitable for use, often with only minor modifications, for the analysis of algal carotenoids. Problems may, however, be encountered due to the particularly large range of structural features found in the algal carotenoids, and more specialised chromatographic systems may be needed. It is suggested that readers consult recent publications from the laboratory of Liaaen-Jensen, which usually provide a detailed description of the isolation, purification (including HPLC) and analysis of a wide range of algal carotenoids. Table 10.5 lists a number of recent publications in which details of the chromatographic separations of algal carotenoids are given.

Table 10. 5 HPLC columns used in the study of algal carotenoids.

Algal class	*Stationary phase*	*Reference*
Bacillariophyceae	Spherisorb RP18	Pennington et al. 1988
	RSil RP18	Brown 1988
	Econosphere C8	Brown 1988
Chlorophyceae*	RP18	Humbeck et al. 1990
	Zorbax ODS	Young and Britton 1988
(secondary carotenoids)	Spherisorb CN	Grung et al. 1989
(secondary carotenoids)	Spherisorb Silica	Grung et al. 1989
Cryptophyceae	Zorbax ODS	Lichtlé et al. 1987
Cyanophyceae	Zorbax ODS	Berland et al. 1989
Eustigmatophyceae	Porosil C18	Owens et al. 1987
	Econosphere C18	Brown 1987
Prymnesiophyceae	Spherisorb CN	Bjerkeng et al. 1990; Bjørnland et al. 1988
	Econosphere C18	Fawley 1987
Raphidophyceae	Whatman ODS	Kohata and Wantanabe 1988

* The majority of the HPLC systems used for the analysis of the carotenoids found in the photosynthetic tissues of higher plants (see Table 10.4) are suitable for use in the analysis of carotenoids found in theChlorophycean algae.

The chromatography of pigments from chlorophyll *c*-containing algae can be greatly improved by the addition of modifiers such as 1% (v/v) triethylamine to the mobile phase. A number of studies have reported the use of ion-pairing reagents (e.g. tetrabutylammonium hydroxide and ammonium acetate, neutralised with acetic acid; Kohata and Wantanabe 1988) for improved resolution of chlorophyllides and phaeophorbides during the HPLC of pigment extracts from algae.

The reversed-phase chromatographic system described in Fig. 10.3 is based on a procedure that was originally developed for the separation of algal pigments (Wright and Shearer 1983). This method can be used routinely as it resolves the carotenes and many of the common algal xanthophylls including fucoxanthin (including resolution of the all-*trans* and some *cis*-isomers), siphonein, siphonaxanthin, diatoxanthin and diadinoxanthin. Minor adjustments to the solvent gradient (e.g. the inclusion of isocratic flow) can greatly improve the resolution of some of these carotenoids.

Brown (1987) described another reversed-phase system (Econosphere C18, methanol:water:acetone (8:1:1) to methanol:acetone (4:1) in 10 min) which separated ß-carotene and a number of xanthophylls found in the eustigmatophycean alga *Nannochloropsis salina*. The separation of chlorophyll *a*, neoxanthin, violaxanthin, vaucheriaxanthin and its esters (three peaks) was good and this system may prove to be useful with some of the other algal classes.

Much interest has recently been focused on fucoxanthin, its isomeric composition and its 19´-acyloxy derivatives in marine algae. The separation of fucoxanthin, 19´-hexanoyloxyfucoxanthin and 19´-butanoyloxyfucoxanthin by reversed-phase HPLC has been reported and is shown in Fig. 10.6 (Bjørnland et al. 1988, 1989).

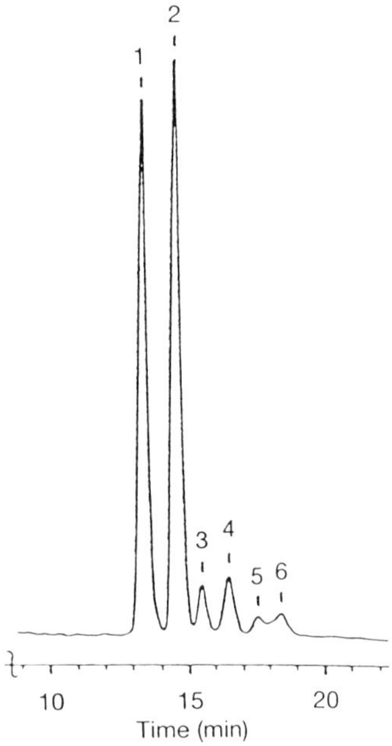

Figure 10.6 Separation of the main carotenoids from the prymnesiophyte alga *Phaeocystis* sp. clone 677-3, monitored at 445 nm. Column: Brownlee CN 5 µm (25 cm x 4.6 mm). Mobile phase: *n*-hexane/propan-2-y acetate/acetone (76:17:7 by vol.). Flow rate 1.6 ml min^{-1}. Peak identifications: (1) fucoxanthin (T_r = 13.3 min), (2) 19´-hexanoyloxyfucoxanthin (T_r = 15.5 min), (3) *neo U*-fucoxanthin (T_r = 15.5 min), (4) 19´-butanoyloxyfucoxanthin (T_r = 16.6 min), (5) *neo V*-fucoxanthin (T_r = 17.6 min), (6) *neo U*-19´-hexanoyloxyfucoxanthin (T_r = 18.4

min). (Bjørnland et al. 1988).

In comparison to the number of chromatographic systems that have been developed for the analysis of the carotenoids that are commonly found in the photosynthetic tissues of higher plants and in many algae, few HPLC analyses have been described of the carotenoids found in the cyanobacteria. The method illustrated in Fig. 10.3 gives acceptable results in many cases but the wide range of polarity of the carotenoids present, and especially the presence of glycosides, can make it particularly difficult to achieve satisfactory resolution of all components. A reversed-phase HPLC system has been described by Berland et al. (1989) which is effective in separating up to ten carotenoids in extracts from *Aponotheca* sp. Further development is needed to establish chromatographic systems that are suitable for these compounds and extracts.

Phototrophic bacteria

As with the cyanobacteria, few HPLC systems that are suitable for the acyclic carotenoids commonly found in the phototrophic bacteria have been published. A procedure used in our laboratories is generally suitable and some details of this have been published (Connor and Britton 1990). This system has been used to resolve the carotenoids and bacteriochlorophylls found in a variety of bacteria, including *Rhodocylus gelatinosus, Rhodobacter sphaeroides, Rhodopseudomonas capsulatus, Rps. viridis, Rps. acidophila, Rhodospirillum rubrum, Rhodomicrobium vannielii, Chromatium minutissimum and C. purpuratum.* A reversed-phase column is used, but the solvent gradient differs from that used with higher plants and algae. The separation of pigments from two species, *Rhodocyclus gelatinosus* and *Rmb. vannielii* is illustrated in Fig. 10.7. The solvent gradient can be modified to give optimum resolution of the range of carotenoids that are present in any particular organism. This procedure will resolve the hydrocarbons and carotenoids containing combinations of functional groups, as well as bacteriochlorophylls and bacteriophaeophytins esterified with different isoprenoid alcohols.

Separation of the carotenoids is influenced by three main structural features. First, the nature of the functional groups present is a major factor, with the more polar compounds generally being eluted first, e.g. compounds with the same carbon skeleton and chromophore are eluted in the order dihydroxy > hydroxymethoxy > monohydroxy > dimethoxy > monomethoxy > hydrocarbon.

Secondly, if a series of compounds is present with the same functional groups but differing in the length of the polyene chromophore, these are well resolved, and eluted in order of decreasing chromophore length, e.g. lycopene (11 conjugated double bonds) – neurosporene (9) – tetrahydrolycopene (7) – phytofluene (5) – phytoene (3).

Thirdly, if two compounds have the same chromophore and the same two substituents, a compound in which the two substituents are located in the same half of the molecule, so that the other half-molecule remains as an unsubstituted hydrocarbon end-group, will have a longer retention time than will a compound in which the two groups are located at the two ends of the molecule.

Nelis and de Leenheer (1989) and de Leenheer and Nelis (1992) have described quantitative methods for the extraction and determination of bacterial carotenoids by non-aqueous HPLC on a Zorbax ODS column. This system provided good, rapid (<10 min), separation of the main bacterial carotenoids, spheroidene and spheroidenone and some of their geometrical isomers. Pigments were eluted with an isocratic mixture of acetonitrile-methanol-dichloromethane (40:50:10 (w/w)) or (70:15:15 (w/w)).

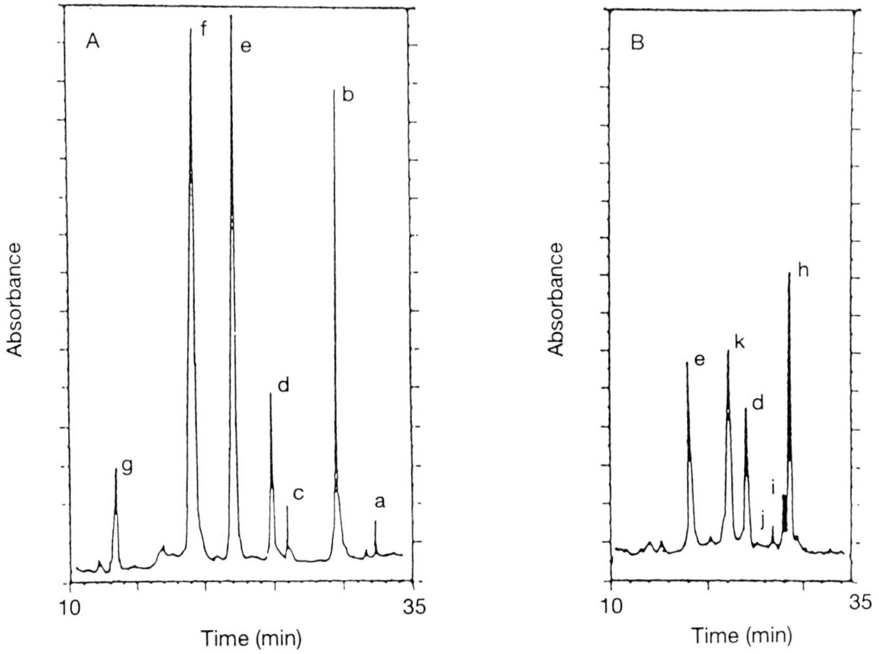

Figure 10.7 Separation of pigments from *Rhodocyclus gelatinosus* 151 (A) and *Rhodomicrobium vanielli* RB4 (B). Column: Spherisorb ODS2, 5 µm (25 cm x 4.6 mm). Mobile phase: solvent A, ethyl acetate, solvent B, acetonitrile/water (9:1) containing 0.1% triethylamine. 0–15 min, 23% A; 15–30 min, 23%–70% A; 30–40 min, 70%–100% A; 40–45 min, 100% A. Flow rate 1.0 ml min^{-1}. Peak identifications: (a) neurosporene, (b) spheroidene, (c) spheroidenone, (d) spirilloxanthin, (e) bacteriochlorophyll, (f) hydroxyspheroidene, (g) hydroxyspheroidenone, (h) lycopene, (i) bacteriophaeophytin, (j) anhydrorhodovibrin, (k) rhodopin (Connor and Britton 1990).

In general, these reversed-phase chromatographic systems do not permit the resolution, with any degree of reliability, of geometrical (*Z/E* or *cis/trans*) isomers. As with higher plants, adsorption chromatography on silica or bonded nitrile stationary phases is most suitable for the resolution of these compounds (G. Britton and A. Connor, unpublished data). Hexane/ethyl acetate solvent mixtures are very effective, but the optimum percentage of ethyl acetate has to be determined for each compound. As an example, isocratic elution with 10% ethyl acetate in hexane resolves up to nine geometrical isomers of okenone with retention times ranging between 14 and 30 min.

Resolution of carotenoid isomers by HPLC

Resolution of geometrical isomers of ß-carotene A number of HPLC systems have been used to separate the *cis–trans* isomers of ß-carotene from leaves, algae, fruit and vegetables. These chromatographic systems rely mainly on adsorption chromatography, for example Vecchi et al. (1981) have used Al_2O_3 and Tsukida et al. (1982) $Ca(OH)_2$ to resolve the geometrical isomers of ß-carotene. Ben Amotz et al. (1982) have described the use of Alox-T for the analysis of geometrical isomers of ß-carotene. Reversed-phase chromatography is traditionally used to separate carotenoids with a wide range of polarities but is less good for the separation of isomers. Two *cis* isomer peaks eluting after the main all-*trans* ß-carotene peak are generally observed in leaf extracts, with the chromatographic system described in Fig. 10.3. Similar observations have been made by Khachik and Beecher (1985). Reversed-phase chromatography has been used successfully in a few studies for the specific separation of geometrical isomers, Jensen et al. (1982) using a Nucleosil C18 column and Bushway (1986) using a Vydac C18 column. O'Neill et al. (1991) have recently performed a comparative study of the range of chromatographic systems available for the separation of geometrical isomers of ß-carotene. They evaluated the performance of the different columns and of a range of solvent systems for the resolution of the *cis–trans* isomers of ß-carotene found in plant tissues. They concluded that whereas the Vydac C18 column (201 TP) was particularly efficient at separating the all-*trans*, 9-*cis* and 13-*cis* isomers, only the $Ca(OH)_2$ column developed by Tsukida et al. (1982) could further resolve the 15-*cis* from the 13-*cis* isomer (Fig. 10.8). The Vydac systems are now well refined, but the separation of the geometrical isomers with these stationary phases is highly temperature dependent and the use of a temperature-regulated column jacket is recommended (G. Beecher and F. Khachik, personal communication).

Alumina can also be used for the separation of the geometrical isomers of ß-carotene (Vecchi et al. 1981), although the high degree of sensitivity of this stationary phase to the water content of the solvents (and of the atmosphere) and to temperature makes its routine use difficult unless a rigorous moisture control system is employed (Rüedi 1985).

Recent developments in SFC which uses CO_2, usually in conjunction with a polar modifier (e.g. methanol), have looked promising for the analysis of carotenoids. Aubert et al. (1991), for example, have described the resolution of *cis*- and *trans*-α- and ß-carotenes by SFC, on a conventional HPLC column.

A combined SFE and SFC system may offer several advantages over more conventional extraction and HPLC procedures in the analysis of natural geometrical isomers (e.g. from reaction centres). In particular, sample manipulation is minimal and this, in combination with low temperatures, should minimise *cis–trans* isomerisation of the carotenoids present.

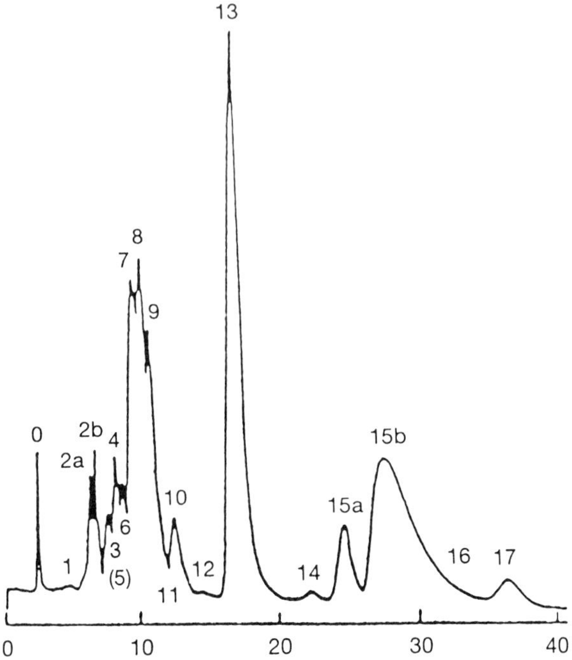

Figure 10.8 Elution profile for a mixture of geometrical isomers of ß-carotene, obtained by thermal isomerisation of the all-*trans* isome,r on a self-packed (400 kg cm^{-2}) Ca$(OH)_2$ (Nakarai) column. The eluting solvent was 0.1% acetone in *n*-hexane at a flow rate of 0.5 ml min^{-1}. Peak identifications: (2) 13,15-di-*cis*, (3) 15-*cis*, (4) 13,13′-di-*cis*, (6) 9,13-di-*cis* (?), (7) 13-*cis*, (8) 9,15-di-*cis*, (9) 9,13-*cis*, (10) 9,13′-di-*cis*, (11) 9,9′,13′-tri-*cis* (?), (12) 7,13′-di-*cis*, (13) all-*trans*, (14) 7,9-di-*cis*, (15a) 9,9′-di-*cis*, (15b) 9-*cis*, (16) 9,9′-di-*cis* (?), (17) 7-*cis*. The remaining peaks are unidentified. Reproduced from Tsukida et al. (1982), additional peak identifications from Koyama et al. (1988).

Resolution of *cis–trans* isomers of other carotenoids Chemically-bonded nitrile phases are probably the most satisfactory phases for the separation of geometrical isomers of the xanthophylls found in higher plants. In most cases the best results are obtained when the individual carotenoid is isolated prior to this analysis, usually by TLC or preferably by semi-preparative reversed-phase HPLC. Reversed-phase chromatography is less efficient for separating geometrical isomers of carotenoids, although Märki-Fischer et al. (1982) have described the resolution of the all-*E* and 5*Z* isomers of lycopene on a C18 column. Geometrical isomers of other carotenes (e.g ζ-carotene, phytofluene and phytoene) can be seen in the chromatographic system described in Fig. 10.3 in extracts from tissues treated with inhibitors of carotenoid biosynthesis (Barry and Pallett 1988, 1990; Young et al. 1989). Other bonded phases (e.g. bonded nitrile and amine phases) can give good resolution of geometrical isomers of carotenoids (see Rüedi 1985). Stationary phases with bonded amino groups have proved to be useful for separating *Z/E* isomers using hexane as the mobile phase (containing 0.1% *N*-ethyldiisopropylamine).

Rüedi (1985) also recommended the use of a reversed-phase chromatographic system (Spherisorb ODS 5 µm; acetonitrile-tetrahydrofuran or acetonitrile-propan-2-ol) for the separation of geometrical isomers of α-carotene, ß-carotene, γ-carotene and δ-carotene. This reversed-phase system also gives good separation of carotenoids in which the terminal double bond of an acyclic end-group is in the 5*Z* configuration, e.g. (5*Z*) and (5*Z*,5′*Z*)-lycopene. The all-*E* isomers are eluted first (see Märki-Fischer et al. 1982).

In most cases, the best resolution of geometrical isomers is achieved by using a pre-fractionated sample (e.g. from semi-preparative reversed-phase HPLC, or TLC). Columns of $Ca(OH)_2$ have been used to resolve the geometrical isomers of canthaxanthin (Hashimoto et al. 1988), spirilloxanthin (Koyama et al. 1990) and neurosporene (Katayama et al. 1990). The use of $Ca(OH)_2$ columns is limited to relatively apolar compounds (Y. Koyama, personal communication) and therefore other chromatographic systems must be employed for the geometrical isomers of carotenoids more polar than canthaxanthin.

Berset and Pfander (1985) have described a chromatographic procedure which uses both silica and bonded nitrile phases to separate six geometrical isomers of lutein (all-*E*, 9*Z*, 9′*Z*, 13*Z*, 13′*Z* and 15*Z*). Märki-Fischer et al. (1982) described the resolution of geometrical isomers of antheraxanthin.

The resolution of geometrical isomers by HPLC can be complicated by a high degree of variability between columns. The choice of stationary phase may also be important in determining extents of loss of the carotenoids and in the relative ease and reliability of separation. For example, HPLC on alumina or $Ca(OH)_2$ requires very precise control of the water content of the solvents and even the atmospheric water content. Slight changes in the water content may cause considerable alterations in the retention times and in the degree of resolution obtained. It is also frequently not appreciated that the resolution and retention times on a reversed-phase column can be highly dependent on temperature.

Resolution of optical isomers HPLC separation of carotenoid optical isomers usually involves the preparation of diastereoisomeric esters through reaction with optically active (-)-camphanoyl chloride (e.g. of astaxanthin, Vecchi and Müller 1979). This method is only effective with carotenoids that possess a 3-hydroxy-4-oxo-ß-ring (e.g. astaxanthin). The optical isomers as the camphanate esters are readily resolved by HPLC on silica. Derivatisation with (*S*)-(+)-α-(1-naphthyl)ethyl isocyanate will permit the resolution of the optical isomers of many hydroxycarotenoids as their diastereoisomeric carbamates (e.g. lutein, zeaxanthin, Rüttimann et al. 1983).

Direct separation of enantiomers (e.g. of ε,ε-carotene-3,3′-diol, Ikuno et al. 1985) on optically active columns has been achieved, permitting a more direct approach to the study of optical isomers in natural systems. Maoka and Matsuno (1989) have desribed the use of a chiral resolution column (Sumipax OA-2000) to resolve the benzoates of the optical isomers of carotenoids that have a 2-hydroxy-ß-end group (e.g. ß,ß-caroten-2-ol and ß,ß-carotene-2,2′-diol) or a 2-hydroxy-4-oxo-ß-end group (e.g. 2-hydroxyechinenone).

10.5 UV/VISIBLE LIGHT ABSORPTION SPECTROSCOPY

Because of their long conjugated double bond systems, the carotenoids absorb in the visible or, in some cases, in the UV, region of the spectrum. Absorption spectra are determined routinely as an important criterion for the identification of a carotenoid and for quantitative analysis. Valuable information can be obtained by examining the position of the absorption maxima (λ_{max}) and the shape or fine structure of the spectrum.

10.5.1 Position of the absorption maxima

Most carotenoids have three absorption maxima (λ_{max}), the positions of which are characteristic of the carotenoid chromophore, although the solvent does have some influence over this. Absorption spectra are usually determined in petrol or hexane for the carotenes and in ethanol for the more polar xanthophylls. The values recorded in these solvents are almost identical, although other solvents may greatly influence the λ_{max} of the carotenoid (see Table 10.6). Thus, λ_{max} values recorded in acetone are greater by ~4 nm, those recorded in chloroform or benzene greater by 10–12 nm, and those in carbon disulphide by as much as 35–40 nm. When spectra are determined on-line by HPLC, it must be remembered that the values for λ_{max} in the eluting solvent frequently do not correspond to the published values that were recorded in a different pure solvent.

Table 10.6 Effect of solvent on the absorption maxima of lycopene.

Solvent	λ_{max} *(nm)*
Petrol	444,470,502
Hexane	447,472,504
Diethyl ether	446,471,503
Ethanol	446,472,503
Acetone	448,474,503
Chloroform	458,484,518
Toluene	457,484,519
Benzene	455,487,522
Pyridine	462,490,525
Carbon Disulphide	477,508,548

The λ_{max} of a carotenoid are largely a function of the length of the conjugated double bond chromophore, and the λ_{max} increase as the number of conjugated double bonds increases. This is clearly seen in the acyclic series of carotenoids as the chromophore increases from three conjugated double bonds in phytoene to 13 in spirilloxanthin (see Table 10.7). Double bonds that are not in conjugation do not contribute to the chromophore.

Extension of the conjugated double bond system into a ring (particularly the C-5,6 double bond of the ß-ring) does extend the chromophore but, because the ring double bond is not coplanar with the main polyene chain, the λ_{max} occur at shorter wavelengths than those of the acyclic carotenoid with the same number of conjugated double bonds. Thus, although they are all conjugated undecaenes, the acyclic, monocyclic and dicyclic isomers lycopene, γ-carotene and ß-carotene have λ_{max} at 444, 470 and 502 nm, at 437, 462, 494 nm and at 425, 450, 478 nm, respectively. Isolated double bonds of the γ- and ε- end-groups do not contribute to the chromophore. ε,ε-Carotene therefore is essentially a conjugated nonaene and has an absorption spectrum identical with that of the acyclic conjugated nonaene, neurosporene.

Table 10.7 Effect of chromophore length (number of conjugated double bonds) and rings on the absorption maxima of carotenoids.

Carotenoid	*Number of conjugated bonds*	λ_{max} *(nm) in petrol*
Phytoene	3	276,286,297
Phytofluene	5	331,348,367
ζ-Carotene	7	378,400,425
Neurosporene	9	414,439,467
Lycopene	11	444,470,502
Spirilloxanthin	13	465,492,525
3,4,3',4'-Tetrahydrolycopene	15	480,510,540

Carotenoid	*Number of rings*	*Number of conjugated double bonds*		λ_{max} *(nm) petrol*
		Polyene chain	*Ring*	
Lycopene	0	11	0	444,470,502
γ-Carotene	1	10	1	437,462,494
δ-Carotene	1	10	0	431,456,489
ß-Carotene	2	9	2	425,449,476
α-Carotene	2	9	1	422,444,473
ε-Carotene	2	9	0	416,440,470

Carbonyl groups, in conjugation with the polyene system, also extend the chromophore. A conjugated keto-group in a ring increases the λ_{max} by 10-20 nm, so that echinenone and canthaxanthin, the 4-oxo and 4,4′-dioxo derivatives of ß-carotene, have λ_{max} at 461 and 478 nm, respectively. The addition of a conjugated keto-group to an acyclic chromophore increases λ_{max} by a much greater amount, ~30 nm; spheroidene and spheroidenone have λ_{max} at 455 nm and 487 nm, respectively.

The presence of hydroxy- and methoxy groups and non-conjugated oxo-groups generally does not affect the chromophore or the λ_{max} of the carotenoid, so that all such substituted carotenoids possess λ_{max} that are virtually identical with those of their parent hydrocarbon molecule. Thus, ß-carotene and its hydroxy-derivatives ß-cryptoxanthin, zeaxanthin, isocryptoxanthin, isozeaxanthin, caloxanthin and nostoxanthin all have virtually identical spectra, with λ_{max} at 425, 450 and 478 nm (in ethanol).

To ensure accuracy of λ_{max} values, the spectrophotometer should be calibrated with the appropriate bands from a holmium oxide filter.

The λ_{max} values in several solvents for the main carotenoids found in plants, algae and phototrophic bacteria are given, along with other physical data, in the Appendix.

10.5.2 Spectral fine structure

The overall shape or degree of fine structure of the absorption spectrum of a carotenoid is also diagnostic and is generally dependent on the extent of planarity that the chromophore can achieve. Thus, acyclic carotenoids, in which the conjugated double bond system can adopt an almost planar conformation (e.g lycopene), are characterised by having an absorption spectrum which has sharp maxima and minima (= persistence). The degree of fine structure decreases a little when the chromophore exceeds nine double bonds. Cyclic carotenoids in which the conjugation does not extend into the rings behave as linear polyenes and have similar well-defined spectral fine structure.

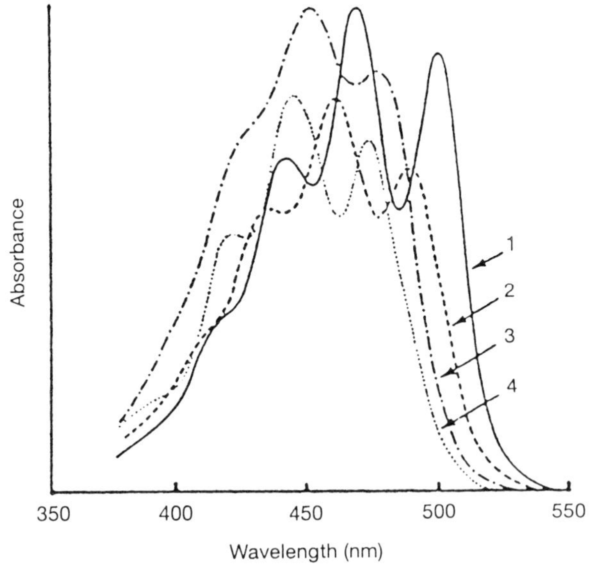

Figure 10.9 Light absorption spectra of some carotenes: (1) Lycopene, (2) γ-Carotene, (3) ß-Carotene, (4) α-Carotene.

When the conjugation extends into a ß-ring, steric strain or hindrance, and the resulting twisting of the molecule to adopt a conformation in which the ring double bond is not coplanar with the π-electron system of the polyene chain, give rise to spectra that show a lower degree of fine structure. The effect is even more pronounced in carotenoids that have carbonyl groups in conjugation with the polyene chain. Spectral fine structure therefore decreases, in the order lycopene (ψ,ψ–acyclic) > γ-carotene (ß,ψ–monocyclic, one ring double bond conjugated) ≈ α-carotene (ß,ε–dicyclic, one ring double bond conjuagted) > ß-carotene (ß,ß–dicylic, two ring double bonds in conjugation) (Fig. 10.9). The spectrum of ketocarotenoids such as canthaxanthin in ethanol have only a single, rounded, almost symmetrical absorption peak, but a slight degree of fine structure remains if the spectrum is determined in a non-polar solvent such as petrol or hexane.

In general, the solvent used to determine the absorption spectra of carotenoids can have an influence on the degree of fine structure observed, as well as λ_{max}, and therefore at least two solvents should normally be used for the determination of spectra.

10.5.3 Geometrical isomers

The absorption spectra of carotenoids that contain one or more *cis* double bonds in the chromophore show several characteristic differences from the spectrum of the all-*trans* compound (Zechmeister 1962; Vetter et al. 1971). For the *cis* isomers, the λ_{max} are generally 1–5 nm lower, the spectral fine structure is decreased and a new absorption peak, commonly referred to as the "*cis*-peak" appears at a characteristic wavelength in the UV region 142 (± 2) nm below the longest wavelength peak in the main visible absorption region (see Chapter 9). These effects, especially the intensity of the the "*cis*-peak", are greatest when the *cis*-double bond is located at or near the centre of the chromophore. This is illustrated in the series of *cis* isomers of ß-carotene shown in Fig. 9.3. With the acyclic carotenoids that have a very long chromophore, such as spirilloxanthin, an overtone peak is usually present in the "*cis*-peak" region of the spectrum of the all-*trans* isomer, though this peak is usually of lower intensity than in the spectra of the *cis* isomers.

10.5.4 Simple diagnostic spectroscopic tests

A rapid and simple test for the presence of 5,6-epoxide groups in a carotenoid can be carried out in a cuvette. The spectrum of the carotenoid is determined in ethanol and redetermined following the addition of 1 drop of dilute acid (usually 0.1 M HCl). A spectral shift is observed of about 20 nm for the mono-epoxide and 40 nm for the di-epoxide due to the furanoid rearrangement to 5,8-epoxide groups. Little overall change in the fine structure of the spectrum is seen (Fig. 10.10).

Carotenoids containing conjugated carbonyl groups (e.g. astaxanthin, canthaxanthin) usually exhibit a spectrum (in ethanol) with little or no fine structure. Again, the presence of such groups can be rapidly determined in a spectrophotometer cuvette. The carotenoid is dissolved in ethanol and the absorption spectrum recorded. A small amount (1-2 mg) of $NaBH_4$ is added and thoroughly mixed and the spectrum redetermined after 30 s, 5 min and 30 min. This will show whether aldehyde groups (these are usually reduced within 30 s—1 min) or keto-groups (5–30 min) are present in conjugation with the main polyene chain; reduction of the carbonyl groups to alcohol groups results in an increase in spectral fine structure and a corresponding reduction in the λ_{max}.

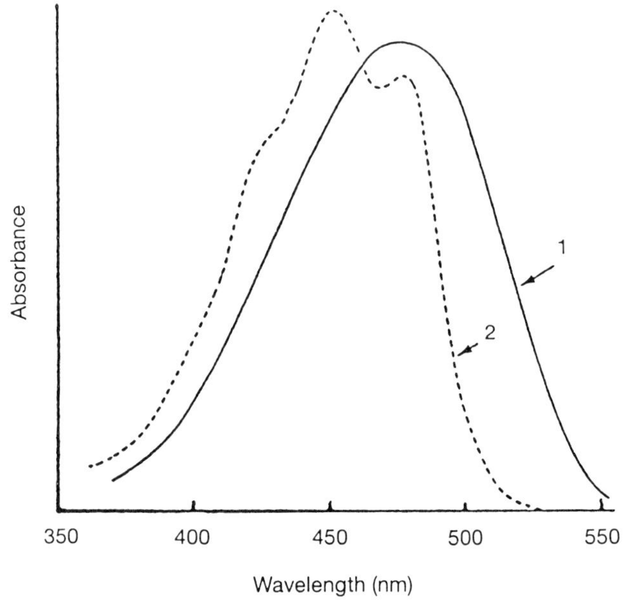

Figure 10.10 Light absorption spectra of (1) canthaxanthin and (2) its $NaBH_4$ reduction product isozeaxanthin.

10.6 QUANTITATIVE DETERMINATION OF CAROTENOIDS

10.6.1 Spectrophotometry

Quantitative determination of carotenoids is generally carried out by simple spectrophotometric techniques. The absorption is determined in the appropriate solvent (usually hexane or ethanol for carotenes and xanthophylls, respectively) at the λ_{max} of the carotenoid. The specific absorption coefficient $A_{1cm}^{1\%}$ (= specific extinction coefficient $E_{1cm}^{1\%}$) is commonly used to quantify carotenoids, although the molar absorption or extinction coefficent ε_{mol} (the absorbance of a 1 M solution) is sometimes used. The $A_{1cm}^{1\%}$ is the absorbance of a 1% (w/v) solution in a 1cm path cuvette at a defined wavelength. An arbitrary value of $A_{1cm}^{1\%} = 2500$ is usually used when no specific values are available for an individual carotenoid or for a mixture of pigments.

The amount of carotenoid present (x g) in y ml of solvent is determined as;

$$x = \frac{A_y}{\left(A_{1cm}^{1\%} \times 100\right)}$$

Tabulated values for $A_{1cm}^{1\%}$ of carotenoids have been published (Davies, 1976; Bauernfeind, 1981; De Ritter, 1981; Britton, 1985, 1991). When available those for the most common carotenoids found in photosynthetic tissues are given in the Appendix.

10.6.2 Quantitative determination by HPLC

HPLC provides the most sensitive, accurate and reproducible method for quantitative analysis of carotenoids, particularly when the instrumentation includes automatic integration facilities for measuring peak areas. The relative amounts of each component in the chromatogram can be determined, provided the peak area can be calculated for each component at its λ_{max} by use of a multi-wavelength detector. If monitoring is possible at only a single wavelength, corrections must be made for the difference between the absorbance at that wavelength and at λ_{max} for each pigment. This procedure allows the relative proportions of each component in a mixture to be determined but, to estimate absolute amounts or concentrations, calibration is necessary. This can be achieved by injecting known amounts (determined spectrophotometrically) of pure carotenoid, determining peak areas, and creating a calibration graph (this is usually linear over the range 10–2000 pmol) which allows the amount of each carotenoid to be estimated with great precision, at the nanogram level. An alternative strategy is to use an internal standard. The usual procedure is to add to the sample an accurately determined known amount of a pure standard compound. The amount of each component in the chromatogram is then estimated by comparing its peak area with that of the standard, again correcting for the difference between absorbance at λ_{max} and that at the monitoring wavelength, if necessary. The internal standard may be a purely synthetic compound or a pure natural compound. The main criteria for its selection are that it should not be present in the sample under investigation, that it should clearly separate from all components in that extract, its stability and λ_{max} should be similar to those of the natural components, and that its $A_{1cm}^{1\%}$ should be known accurately.

An alternative strategy is to use as an internal standard the substance or one of the substances under investigation. The experimental sample is first analysed by HPLC and the peak areas are determined in the normal way. Then a known amount of pure standard compound is added to the sample and a second HPLC analysis performed. The increment in the peak area for the standard compound is then related to the amount that was added, to give a calibration that can be used to determine the amounts of all substances that are present in the extract.

If automatic integration is not available, compounds may be separated by HPLC and then collected and the concentration of each pigment determined in a spectrophotometer. The virtually quantitative recovery of carotenoids from HPLC allows greater accuracy than is possible when other separation methods (such as TLC) are used.

10.7 OTHER SPECTROSCOPIC AND PHYSICAL METHODS

Chromatography, especially HPLC, and UV/Vis absorption spectroscopy provide very valuable first identification criteria for carotenoids. This, however, does not constitute rigorous characterisation, for which other sophisticated methods, particularly MS and NMR must be used, together with circular dichroism (CD) if the chirality of a carotenoid is to be determined. Infrared (IR) spectroscopy has also been widely used, but the information given about carotenoid structures is usually limited. An outline of the main features and applications of these techniques is given here. The interpretation of the large amount of data which these methods generate requires a great deal of experience, and the characterisation of novel or unusual carotenoids is best left to experts.

The brief description in this chapter is restricted to the application of the methods to the identification and characterisation of carotenoids that are isolated from natural extracts. In Chapter 9, Koyama and Hashimoto describe the application of several of these techniques to study carotenoid structures and interactions *in situ*.

10.7.1 Mass spectrometry

The technique of MS is of considerable importance in the elucidation of the structure of carotenoids and, whenever possible, all identifications of carotenoids should be supported by MS data. The information presented here is meant only as a guide to the confirmation of the structure of carotenoids which may be found in photosynthetic systems. More detailed reviews of mass spectrometry of carotenoids have been given by Enzell et al. (1969), Vetter et al. (1971), Johannes et al. (1974, 1979), Moss and Weedon (1976) and Enzell and Wahlberg (1979).

Preparation of samples for MS, as for all other spectroscopic techniques, must be rigorous. Particularly important is the quality of the solvents used during sample preparation. These should be dried and redistilled and, when necessary, peroxides removed. It is also advisable to pass the solvents that are to be used for MS (or NMR) sample preparation through a small column of activated alumina or silica. Distillation through a fractionating column is also recommended. If these procedures are not followed then contamination of the carotenoid with solvent-derived impurity will probably result. Glassware should be acid-washed, although care must be taken to remove all traces of acid before use. The use of plastic or polythene bottles etc. should be avoided at all times as these contain plasticisers which are readily leached out by solvents and will interfere greatly with the subsequent analysis.

The carotenoids lend themselves well to analysis by ionisation by electron impact (EI). Chemical ionisation (CI) has also been used, especially in the study of xanthophyll acyl esters (Khachik et al. 1988; Barry et al. 1992). Ammonia and isobutane are particularly suited to the analysis of carotenoids by desorption chemical ionisation (DCI), and will generally provide a good molecular ion. The results from other techniques such as fast atom bombardment (FAB) are much less satisfactory. Carotenoids have very low volatility, and samples are usually inserted by means of a direct probe, which is heated to 200–220 °C. Sample requirements are small: a good mass spectrum can be obtained from only 1–2 μg of pure carotenoid, though a realistic goal is to obtain 10–20 μg of the compound. An outline of some of the most important diagnostic features of the MS of carotenoids is given here. Further details of fragmentations and their mechanisms are given by Vetter et al. (1971), Moss and Weedon (1976) and Enzell and Broman (1993).

(a) Molecular mass and formula

Almost all carotenoids give good molecular ions, from which the molecular mass can be obtained. Unambiguous molecular formulae can be determined with a high resolution instrument.

(b) Fragmentations

Many fragmentations have been identified which are diagnostic of particular structural features. Some of these provide information primarily about the carbon skeleton of the molecule, others identify the presence of particular functional groups or end-groups.

Polyene chain fragmentations

All carotenoids undergo reactions in which the polyene chain is folded and portions of the chain are then excised. The most intensively studied of these reactions are the losses of toluene (92 mass units (m.u.)) and *m*-xylene (106 m.u.) which occur almost universally with carotenoids. Similar losses of 79 and 158 m.u. frequently occur, but the fragment ions thus produced are of lower intensity. The abundance ratio of the [*M*-92] and [*M*-106] fragment ions can give a good indication of the carbon skeleton of a carotenoid, i.e. for acyclic, monocyclic and dicyclic carotenoids the abundance ratio of $[M\text{-}92]^+$: $[M\text{-}106]^+$ is in the range 0.02–0.03:1, 0.6–1.0:1, and 2–10:1, respectively. Carotenoids with a higher degree of saturation, e.g. phytoene, phytofluene and ζ-carotene show analogous losses of 94 m.u., whereas those in which one of the methyl substituents at C-9 or C-13 is replaced by CH_2OH or CHO lose the corresponding toluene or xylene derivative, e.g. benzaldehyde (106 m.u.) and *m*-tolualdehyde (120 m.u.) for rhodopin-20-al.

Functional groups and end-groups

As with other classes of compound, the presence of functional groups in a carotenoid is indicated by characteristic fragmentations, e.g. hydroxy- and methoxy-groups, especially if in an allylic position, give rise to strong losses of water (18 m.u.) or methanol (32 m.u.), repectively. Acetylation or silylation of a hydroxyl group increase the molecular mass by 42 or 73 m.u., respectively. The so-called "bis-allylic" fragmentations are especially useful. Those single bonds in the carotenoid molecule which are in a position allylic to the main polyene chain and also to an isolated double bond in an end-group undergo particularly facile cleavage (e.g. C-3,4 of lycopene, C-7,8 of ζ-carotene and C-11,12 of phytoene). This allows the length and position of the conjugated chromophore in any acyclic carotenoid to be determined.

It is usually possible to detect the presence of an ε-ring in a carotenoid by losses of 56 m.u. (by a *retro*-Diels–Alder fragmentation to lose C-1, C-2 and the C-1 geminal methyl groups) and of 123 m.u. due to cleavage of the C-6,7 bond (138 m.u. for a 3-hydroxy-ε-ring).

Carotenoid epoxides are readily identified. Both 5,6- and 5,8-epoxides show strong loss of 80 m.u. (two successive losses for di-epoxides) and also abundant ions at *m/z* 165 and 205 (181 and 221 if a 3-hydroxy-group is also present).

Characteristic fragment ions for many of the carotenoids found in photosynthetic systems are given, along with other data, in the Appendix.

10.7.2 Nuclear Magnetic Resonance

NMR is undoubtedly the most powerful technique for investigating carotenoid structures, and its use is essential for full characterisation and structure elucidation. Some bassic principles of NMR theory are discussed in chapter 9. Both proton (^{1}H) NMR and ^{13}C NMR are used routinely by chemists in such work. Moss and Weedon (1976) discussed the general features of NMR spectra of carotenoids, and presented a considerable amount of tabulated data. A survey of progress since then was given by Goodwin and Britton (1988) and further tabulated values for ^{1}H and ^{13}C NMR assignments of a range of carotenoid end-groups were presented. More detailed progress reports by Englert (1982, 1985, 1991) are also available, and the same author has published a most authoritative review of NMR spectroscopy of carotenoids, which includes comprehensive tables of ^{1}H and ^{13}C NMR assignments of carotenoid end-groups (Englert 1993). The interested reader is recommended to consult these articles, and to survey the primary literature for full details of the NMR investigations and assignments for individual carotenoids. Selected assignments for the carotenoid end-groups are given in Tables 10.8 and 10.9. These can be cross-referenced with the end-groups given in the Appendix for individual compounds.

In a high-field ^{1}H or ^{13}C NMR spectrum, e.g. at 400 MHz or 100 MHz, respectively, signals due to the olefinic region of the carotenoid molecule are well resolved and can be assigned and their coupling relationships identified, and the geometrical configuration of the carotenoid determined. Again, details are given by Englert (1993). It must be stressed, however, that the use of restricted information by inexperienced workers can easily lead to erroneous conclusions. The interpretation

Table 10.8 ^{1}H-NMR chemical shift assignments for carotenoid end-groups (solvent $CDC\ell_3$)

		16	*17*	*18*	*19*	*20*	*2*	*2*	*3*	*3*	*4*	*4*	*6*	*7*	*8*	*10*	*11*	*12*	*14*	*15*	
1	ß	1.03	1.03	1.72	1.98	1.98	1.46	1.46	1.62	1.62	2.02	2.02	—	6.16	6.14	6.16	6.65	6.35	6.25	6.63	
2	3,4,7,8-Tetradehydro-ß	1.10	1.10	1.98	2.03	1.96	2.12		5.82		5.90										
3	5,6-Epoxy-5,6-dihydro-ß	1.10	0.94	1.15	1.93	1.97	1.08	1.48	1.43	1.43	1.75	1.91	—	5.87	6.29	6.19	6.62	6.37	6.26	6.62	
4	5,8-Epoxy-5,8-dihydro-ß	1.11	1.18	1.47	1.81	1.96							—	5.24	5.08	6.20	6.51	6.33	6.23	6.62	
5	2-Hydroxy-ß	1.09	1.04	1.71	1.97		3.55^{ax}	1.44^{HO}	1.75^{β}	1.85^{α}		~2.15		—	6.11	6.12	6.16				
6	3-Hydroxy-ß	1.07	1.07	1.74	1.97	1.97	1.48^{a}	1.77^{c}	4.00^{a}	1.34^{HO}	2.04^{a}	2.39^{c}	—	6.10	6.16	6.16	6.63	6.35	6.24	6.62	
7	3-Hydroxy-7,8-didehydro-ß	1.15	1.20	1.92	2.01		1.45^{β}	1.84^{α}	3.99^{ax}	1.41^{HO}	2.07^{β}	2.43^{α}	—	—	—	6.45	6.54				
8	3-Acetoxy-7,8-dihydro-ß	1.20	1.18	1.92	2.01		$1.57^{(\beta)}$	$1.83^{(\alpha)}$	$5.04^{(ax)}$		2.13^{β}	2.49^{α}									OAC 2.05
9	3-Hydroxy-5,6-epoxy-5,6-dihydro-ß	0.98	1.15	1.19	1.93	1.97	1.25^{β}	1.63^{α}	3.91^{α}	1.25^{HO}	1.63^{β}	2.39^{α}	—	5.87	6.29	6.19	6.50	6.37	6.25	6.63	
10	2,3-Dihydroxy-ß	1.00	1.13	1.72	1.97		3.33^{ax}		3.83^{ax}		2.49		—	6.06	6.12	6.17					
11	3,5-Dihydroxy-6,7-didehydro-5,6-dihydro-ß		1.34	1.07	1.35	1.81	1.96	1.34^{β}	1.95^{α}		4.32^{ax}	1.41^{β}	2.27^{α}	—	—	6.08	6.12	6.56	6.35		
12	3-Acetoxy-5-hydroxy-6,7-didehydro- -5,6-dihydro-ß [≡ 14-acetate]	1.39	1.07	1.35	1.81	1.99	1.41^{β}	1.99^{α}	5.37^{ax}	—	1.51^{β}	2.28^{α}	—	—	6.06	6.13	6.60	6.35	6.27	6.76	AC 2.0
13	3,19-Dihydroxy-ß	1.08	1.08	1.75	4.55		1.48^{β}	1.77^{α}	1.25^{OH}	4.03^{ax}	2.39^{α}	2.08^{β}	—	6.35	6.05	6.24	6.71				
14	3,19-Dihydroxy-5,6-epoxy-5,6-dihydro-ß	0.98	1.15	1.18	4.51																
15	3,19-Dihydroxy-8-oxo-7,8-dihydro-ß	0.95	1.00	1.50	4.49	2.00							—	3.50							
16	3,5,6-Trihydroxy-5,6-dihydro-ß	0.88	1.19	1.26	1.98	1.98			4.02												
17	4-Oxo-ß	1.20	1.20	1.88	2.00	2.00	1.86		2.51		—	—	—	6.24	6.37	6.28	6.63	6.41	6.26	6.65	
18	3-Hydroxy-4-oxo-ß	1.21	1.32	1.95	2.00	2.00	1.81^{β}	2.16^{α}	4.32	3.68	—	—	—	6.22	6.43	6.30	6.66	6.45	6.30	6.67	
19	3-Hydroxy-4-oxo-7,8-didehydro-ß	1.36	1.31	2.03	2.05	1.99	1.80^{β}	2.22^{α}	3.62^{HO}	4.33^{Hax}	—	—	—		—	6.65	6.54	6.46	6.34	6.69	
20	3-Hydroxy-8-oxo-5,6-epoxy- -5,6,7,8-tetrahydro-ß	1.04	0.96	1.23	1.95	1.99	1.36^{β}	1.50^{α}	3.82^{ax}		2.33^{α}	1.78^{β}		2.60, 3.66		—	7.15	6.58	6.67	6.41	6.64
21	3,6-Epoxy-4-oxo-5,6-dihydro-ß	0.97	1.13	1.01	1.97	1.97			4.33					5.67							
22	3-Hydroxy-7,8-dihydro-ß-19,11-butenolide	0.92	1.02	1.77	—	2.21	1.80	1.40	4.20		5.49		na	1.5, 1.7	2.4, 2.5	7.01		5.68	6.50	6.65	
23	3-Hydroxy-5,6-epoxy-5,6-dihydro-ß- -19,11-butenolide	0.98	1.21	1.20	—	2.23	1.64^{α}	1.26^{β}	3.91^{ax}		2.40^{α}	1.64^{β}	—	7.19	6.38	7.03	—	5.74	6.46	6.64	

Table 10.8 (cont.)

		16	*17*	*18*	*19*	*20*	*2*	*2*	*3*	*3*	*4*	*4*	*6*	*7*	*8*	*10*	*11*	*12*	*14*	*15*	
24	5,6-Dioxo-5,6-seco-ß	1.17	1.17	2.10	1.98	1.99	~1.54		~1.49		2.40		—	6.54	7.38	6.51	6.65	6.52	6.35	6.70	
25	3-Oxo-retro-ß (6,7-*trans*)	1.39	1.39	2.16	2.03	2.00	2.40				5.94		—	6.90	6.80	6.46	6.79	6.26	6.48	6.43	
26	3-Oxo-retro-ß (6,7-*cis*)	1.25	1.25	2.31									—	6.67	6.55						
27	3,6-Dihydroxy-8-oxo-7,8-dihydro-γ	0.92	1.08	4.46, 4.86	1.94	1.99			3.76		[–2.0—2.2–]		5.47	2.68, 3.41	7.20						
28	ε	0.82	0.92	1.59	1.91	1.95	1.18	1.43	2.00	2.00	5.41	—	2.18	5.53	6.11	6.12	6.60	6.33	6.23	6.62	
29	3-Hydroxy-ε	0.85	1.00	1.63	1.91	1.96	1.37	1.85	4.35	1.30	5.55	—	2.40	5.43	6.14	6.14	6.61	6.35	6.25	6.63	
30	ϕ (phi)	2.23	2.26+	2.27+	2.08		—	—	6.96	—	6.96	—	—	6.60	6.28	6.22					
31	χ (chi)	2.30+	2.21	2.29+	2.06		—	—	—	—	6.99 (H-5)	7.27	—	6.86	6.70	6.31					
32	ψ	1.62	1.69	1.82	1.97	1.97	5.11	—	2.11	2.11	2.11	2.11	5.95	6.49	6.25	6.19	6.64	6.35	6.23	6.63	
33	7,8-Dihydro-ψ	1.63	1.70	1.63	1.83	1.96	5.12	—	2.13	2.13	2.09	2.09	5.12	2.13	2.13	5.97	6.51	6.25	6.25	6.63	
34	1,2,7,8-Tetrahydro-ψ	0.89	0.89	1.61	1.83	1.97	[------1.1—1.4--------]				[--1.8—2.3--]		5.2	2.13	2.13						
35	7,8,11,12-Tetrahydro-ψ (15*Z*)	1.62	1.68	1.60	1.60	1.76	5.09		2.10	2.10	2.10	2.10	5.09	2.10	2.10	5.09	2.10	2.10	6.30	6.10	
	(15*E*)																		5.88	6.32	
36	1-Hydroxy-1,2-dihydro-ψ	1.22	1.22	1.81	1.97		[------------1.4—1.5------------]				2.11		5.96	6.49	6.25						1-HO 1.18
37	1-Hydroxy-1,2,7,8-tetrahydro-ψ	1.21	1.21	1.61	1.81	1.97	[--2.0—2.2--]		2.05		2.05		5.12	2.05	2.12	5.95	6.50	6.23	6.20	6.62	
38	1-Hydroxy-3,4-didehydro-1,2-dihydro-ψ	1.24	1.24	1.93	1.98		2.31		5.76		6.21		6.14	6.60	6.36						HO 1.44
39	1-Methoxy-3,4-didehydro-1,2-dihydro-ψ	1.16	1.16	1.93	1.98	1.95	2.32	2.32	5.72	—	6.17	—	6.11	6.60	6.35	6.22	6.63	6.37	6.26	6.61	OCH3 3.26
40	1-Methoxy-3,4-didehydro-1,2-dihydro--2-oxo-ψ	1.35	1.35	1.98	1.98		—	—	6.76	—	7.47	—	~6.61	~6.61							OCH3 3.22
41	1-Methoxy-1,2-dihydro-4-oxo-ψ	1.20	1.20	1.95	2.00	2.00	1.84		2.77		—	—	[-----------------------na-----------------------]								OCH3 3.28

A blank indicates that no specific assignment has been made for that position in the molecule.
∝, ß, a (axial) are used when specific assignments have been made to non-equivalent protons at that position.
\+ is used when assignments can be interchanged.
na, not available

Table 10.9 ^{13}C-NMR chemical shift assignments for carotenoid end-groups (solvent CDCℓ_3)

		16	*17*	*18*	*19*	*20*	*1*	*2*	*3*	*4*	*5*	*6*	*7*	*8*	*9*	*10*	*11*	*12*	*13*	*14*	*15*	
1	ß	29.0	29.0	21.7	12.8	12.9	34.2	39.8	19.4	33.2	129.4	137.8	126.7	137.6	136.0	130.8	125.3	137.3	136.4	132.4	130.0	
3	5,6-Epoxy-5,6-dihydro-ß	25.9	25.9	21.1	12.9	12.9	33.8	35.9	17.2	30.2	65.3	71.2	124.0	137.2	134.5	132.1	124.8					
6	3-Hydroxy-ß	30.3	28.8	21.6	12.8	12.8	37.2	48.5	65.0	42.5	126.2	137.9	125.7	138.4	135.8	131.3	125.0	137.6	136.5	132.6	130.0	
7	3-Hydroxy-7,8-didehydro-ß	30.5	28.8	22.5	18.1	12.8	36.6	46.7	64.9	41.5	137.3	124.3	89.1	96.6	119.1	135.1	124.3	138.0	136.4	133.4	130.3	
8	3-Acetoxy-7,8-didehydro-ß	28.7	30.2	22.4	18.1		36.1	42.3	67.9	37.5	137.3	124.3	90.1	98.5	121.0	134.6						CH_3 CO 21.4 170.4
9	3-Hydroxy-5,6-epoxy-5,6-dihydro-ß	25.1	29.6	20.3	12.9		35.4	47.9	63.0	41.8	67.0	70.3	125.2	137.2	134.5	132.5	125.2	138.2	136.4	132.9	130.2	
11	3,5-Dihydroxy-6,7-didehydro- -5,6-dihydro-ß	32.2	29.4	31.4	14.0	12.9	35.8	49.5	64.2	49.0	73.0	117.2	202.3	103.2	132.2	128.4	125.2	137.2	137.0	132.4	129.8	
12	3-Acetoxy-5-hydroxy-6,7-didehydro- -5,6-dihydro-ß	32.1	29.2	31.3	14.0	12.9	35.8	45.5	68.0	45.3	72.7	117.5	202.4	103.4	132.5	128.6	125.7	137.2	138.1	132.2	132.5	CH_3 CO 21.4 170.4
17	4-Oxo-ß	27.7	27.7	13.8	12.6	12.8	35.9	37.6	34.3	198.9	130.0	161.0	124.3	141.2	134.8	134.4	124.8	139.3	136.6	133.6	130.5	
18	3-Hydroxy-4-oxo-ß	30.7	26.2	14.0	12.6	12.8	36.8	45.5	69.3	200.4	127.0	162.1	123.3	142.4	134.3	135.2	124.4	139.8	136.7	133.8	130.9	
19	3-Hydroxy-4-oxo-7,8-didehydro-ß	31.0	26.2	14.4	17.6	12.8	36.5	44.3	69.3	199.5	133.5	147.7	88.1	111.0	117.6	140.5	123.9	139.0	136.7	134.8	131.0	
20	3-Hydroxy-8-oxo-5,6-epoxy- -5,6,7,8-tetrahydro-ß	28.2	25.1	21.2	11.8	12.8	35.8	47.1	64.4	41.7	66.2	67.1	40.8	197.9	134.6	139.1	123.4	145.0	135.6	136.6	129.4	
23	3-Hydroxy-5,6-epoxy-5,6-dihydro- -ß-19,11-butenolide	24.9	29.5	19.9	168.7	15.4	35.3	47.1	64.2	40.9	67.5	70.5	133.7	121.8	124.9	136.3	146.9	119.1	134.4	137.8	129.8	
24	5,6-Dioxo-5,6-seco-ß	24.5	24.5	29.7	12.9		46.5	39.4	19.3	44.0	208.4	203.7	119.9	147.8	134.0							
25	6,7-*Trans*-3-oxo-retro-ß	29.9	29.9	22.3	12.4	12.9	38.6	54.4	199.0	126.1	154.7	142.7	128.2	128.3	141.3	138.0	126.9	132.6	137.5	137.7	129.9	
26	6,7-*Cis*-3-oxo-retro-ß	28.3	28.3	25.4	12.4	12.9	41.3	52.5	199.0	128.8	155.5	143.8	125.5	128.8	139.6	137.5	126.4					
28	ε	27.1	27.6	23.0	13.1	12.8	32.6	31.8	23.1	120.7	134.5	55.0	130.3	136.5	135.5	131.0	125.0	137.2	136.4	132.3	130.0	
29	3-Hydroxy-ε	24.4	29.5	22.8	13.1		34.1	44.7	65.9	124.6	137.9	55.2	128.8	137.7	135.0	130.8	124.5	137.6	136.5	132.6	130.0	
30	φ	17.0	20.4⁺	20.9⁺	12.7				127.5⁺	126.5⁺		126.5	126.5	139.0	135.5	132.2	124.8					
31	χ	15.9	15.9	20.9	13.1		134.1	135.3	136.0	127.5	123.1	134.6	128.4	134.1	139.4	130.9						
32	ψ	25.7	17.7	17.0	12.9	12.8	131.6	124.1	26.8	40.3	139.3	125.9	124.9	135.5	136.1	131.6	125.2	137.5	136.6	132.7	130.2	
33	7,8-Dihydro-ψ	25.7	17.7	16.0	16.9	12.7	131.2	124.0	26.6	39.7	134.9	123.9	26.7	40.2	139.7	125.8	125.1	135.3	136.0	131.6	130.1	
35	7,8,11,12-Tetrahydro-ψ(15*Z*)	25.7	17.7	16.0	16.0	16.9	131.2	124.0	26.8	39.8	134.9	124.3	26.8	39.8	135.3	124.4	26.8	40.5	139.4	120.3	123.4	
	(15*E*)																		138.3	125.7	127.3	
36	1-Hydroxy-1,2-dihydro-ψ	29.5	29.5	17.0	13.0		71.2	43.6	22.8	40.7	139.5	126.2	124.9	135.7								
39	1-Methoxy-3,4-didehydro- -1,2-dihydro-ψ	24.9	24.9	13.1	12.9	12.8	75.1	43.7	125.4	137.5	135.4	130.6	124.7	137.5	135.9	132.6	124.9	138.0	136.1	133.0	129.4	OCH_3 49.3
40	1-Methoxy-2-oxo-3,4-didehydro- -1,2-dihydro-ψ	23.0	23.0	12.7	12.8		81.7	203.0	118.1	148.5	133.8	142.4	123.8	141.3								

A blank indicates that no specific assignment has been made for that position in the molecule of NMR spectra of carotenoids is such a sophisticated and specialised field that the help and advice of experts should always be sought.

With a modern, high resolution PFT NMR instrument, good ^1H spectra can be obtained from about 100–200 μg of sample, whereas ^{13}C spectra require 2–5 mg.

It is essential that carotenoid samples for NMR analysis are rigorously purified and that all traces of residual solvent are removed under vacuum.

10.7.3 Circular dichroism

Full characterisation of a carotenoid includes determination of its chirality. This is usually achieved by a combination of NMR spectroscopy and circular dichroism, together with chemical correlations. Since the CD spectra of carotenoids are critically affected by factors such as temperature, solvent and geometrical configuration, erroneous conclusions are easily drawn from CD data. It is strongly recommended that CD study and interpretation should be left to experts. A brief general summary of the use of CD in the carotenoid field was given by Goodwin and Britton (1988) and expert reviews of the subject have been presented (e.g. Noack 1982). The topic is reviewed in detail, and extensive tabulated data presented in a major article by Noack and Buchecker (1993).

10.7.4 Infrared and Raman spectroscopy

Infrared spectroscopy does not play a major role in carotenoid analysis, though it can assist in the identification of carbonyl groups and special structural features such as acetylene and allene groups.

Resonance Raman spectroscopy has made great advances in recent years. No extensive systematic study of carotenoids has been reported, but carotenoids can be detected *in situ* by resonance Raman spectroscopy of intact tissues or isolated photosynthetic membranes (see Chapter 9).

REFERENCES

Aubert, M.-C., Lee, C.R., Krstulovic, A.M. *et al.* (1991) Separation of *trans/cis* α- and ß-carotenes by supercritical fluid chromatography. I. Effects of temperature, pressure and organic modifiers on the retention of carotenes. *J. Chromatogr.*, **557**:47-58.

Barry, P. and Pallett, K.E. (1988) Herbicidal inhibition of carotenogenesis detected by HPLC. In: *Factors Affecting Herbicide Activity and Selectivity*, Proceedings of the European Weed Research Society Symposium, pp. 51-58.

Barry, P. and Pallett, K.E. (1990) The effect of diflufenican on pigment levels in carotenogenic systems. *Z. Naturforsch.*, **45c**:492-497.

Barry, P., Evershed, R., Young, A.J. *et al.* (1992) Characterisation of carotenoid acyl esters produced in drought-stressed barley seedlings. *Phytochemistry*, **31**:3163-3168.

Bauernfeind, J.C. (1981) *Carotenoids as Colorants and Vitamin A Precursors*. Academic Press, New York.

Ben-Amotz, A., Katz, A. and Avron, M. (1982) Accumulation of ß-carotene in halotolerant algae: Purification and characterization of ß-carotene-rich globules from *Dunaliella*

bardawil (Chlorophyceae). *J. Phycol.*, **18**:529-537.
Berland, B., Le Champion, T. and Campos Baeta Nevis, M.H. (1989) Interaction de la salinité et de la temperature sur la morphologie, la croissance et la composition cellulaire d'une Cyanobacterie halotolerante (*Aponotheca* sp.) *Bot. Mar.*, **32**:317-329.
Berset, D. and Pfander, H. (1985) Untersuchungen zur carotinoidzusammensetzung in blüten der gartenhybride *Narcissus* cvv "Golden Harvest". *Helv. Chim. Acta*, **68**:1149-1154.
Bjerkeng, B., Vernett, M., Nielsen, M.V. and Liaaen-Jensen, S. (1990) Carotenoids of *Chrysochromulina polyepis* (Prymnesiophyceae). *Biochem. Syst. Ecol.*, **18**:303-306.
Bjørnland, T., Guillard, R.R.L. and Liaaen-Jensen, S. (1988) *Phaeocystis* sp. clone 677-3 – a tropical marine planktonic Prymnesiophyte with fucoxanthin and 19′-acyloxyfucoxanthins as chemosystematic carotenoid markers. *Biochem. Syst. Ecol.*, **16**:445-452.
Bjørnland, T., Liaaen-Jensen, S. and Thorndsen, J. (1989) Carotenoids of the marine chrysophyte *Pelagococcus subviridis* from the Norwegian Sea. *Phytochemistry*, **28**:3347-3353.
Braumann, T. and Grimme, H. (1981) Reversed-phase high performance liquid chromatography of chlorophylls and carotenoids. *Biochim. Biophys. Acta*, **637**:8-17.
Britton, G. (1985) General carotenoid methods. *Methods Enzymol.*, **111**:113-149.
Britton, G. (1991) Carotenoids. In: Banthorpe, D.V. and Charlwood, B.V. (eds.), *Methods in Plant Biochemistry*, Vol. 7. Academic Press, London, pp. 473-518.
Britton, G. and Goodwin, T.W. (1971) Biosynthesis of carotenoids. *Methods Enzymol.*, **18C**:654-701.
Brown, J.S. (1987) Functional organisation of chlorophyll *a* and the carotenoids in the alga, *Nannochloropsis salina*. *Plant Physiol.*, **83**:434-437.
Brown, J.S. (1988) Photosynthetic pigment organisation in diatoms (Bacillariophyceae). *J. Phycol.*, **24**:96-102.
Budzikiewicz, H. (1974) Biochemical applications of mass spectrometry. *Adv. Mass Spectrom.*, **6**:163-170.
Budzikiewicz, H., Brzezinka, H. and Johannes, B. (1970) Photosynthesis in green plants. II. Mass spectrometric investigation of carotenoids. *Monatsh.*, **101**:579-609.
Bushway, R.J. (1986) Determination of α- and ß-carotene in some raw fruits and vegetables by high performance liquid chromatography. *J. Agric. Food Chem.*, **34**:409-412.
Connor, A.E. and Britton, G. (1990) High performance liquid chromatography analysis of the pigments of purple photosynthetic bacteria. In: Baltscheffsky, M. (ed.), *Current Research in Photosynthesis*, Vol. II. Kluwer, Dordrecht, pp. 53-56.
Craft, N.E. (1992) Carotenoid reversed-phase high-performancce liquid chromatography methods: Reference compendium. *Methods Enzymol.*, **213**:185-189.
Cygnarowicz, M.L. and Seider, W.D. (1990) Design and control of a process to extract ß-carotene with supercritical carbon dioxide. *Biotechnol. Progr.*, **6**:82-91.
Davies, B.H. (1976) Carotenoids. In: Goodwin, T.W. (ed.), *Chemistry and Biochemistry of Plant Pigments*, 2nd. Ed., Vol.2. Academic Press, London, pp.38-165.
Degnan, A.J., Elbe, J.H. von and Hartel, R.W. (1991) Extraction of annatto seed pigment by supercritical carbon dioxide. *J. Food Science*, **56**:1655-1659.
Englert, G. (1982) N.M.R. of carotenoids. In: Britton, G. and Goodwin, T.W. (eds.), *Carotenoid Chemistry and Biochemistry*. Pergamon Press, Oxford, pp. 107-134.
Englert, G. (1985) NMR of carotenoids – new experimental techniques. *Pure Appl. Chem.*, **57**:801-821.
Englert, G. (1991) NMR of carotenoids: Novel experimental techniques. *Pure Appl. Chem.*,

63: 59-70.
Englert, G. (1993) NMR of carotenoids. In: Britton, G., Liaaen-Jensen, S. and Pfander, H. *Carotenoids, Isolation and Analysis,* Vol. 1. Birkhaüser, Basel.(In press)
Enzell, C.R. and Broman, S. (1993) Mass spectrometry. In: Britton, G., Liaaen-Jensen, S. and Pfander, H. *Carotenoids, Isolation and Analysis,* Vol. 1b,. Birkhaüser, Basel.(In press)
Enzell, C.R. and Wahlberg, I. (1979) Carotenoids, In: Waller, G.R. and Dermer, O.C. (eds.), *Biochemical Applications of Mass Spectrometry.* Wiley and Sons, London, pp. 407-438.
Enzell, C.R., Francis, G.W. and Liaaen-Jensen, S. (1969) Mass spectrometric studies of carotenoids. 2. A survey of fragmentation reactions. *Acta Chem. Scand.*, **23**:727-750.
Eugster, C.H. (1982) New carotenoid structures and stereochemistry. In: Britton, G. and Goodwin, T.W. (eds.), *Carotenoid Chemistry and Biochemistry.* Pergamon Press, Oxford, pp.1-26.
Eugster, C.H. (1985) Carotenoid structures, old and new problems. *Pure Appl. Chem.*, **57**:639-647.
Fawley, M.W. (1987) Separation of chlorophylls *c*1 and *c*2 from pigment extracts of *Pavlova gyrans* by reversed-phase high performance liquid chromatography. *Plant Physiol.*, **86**:76-78.
Foss, P., Guillard, R.R.L. and Liaaen-Jensen, S. (1984) Prasinoxanthin. A chemosystematic marker for algae. *Phytochemistry,* **25**:118-124.
Gilmore, A.M. and Yamamoto, H.Y. (1991) Resolution of lutein and zeaxanthin using a nonendcaped, lightly carbon-loaded C-18 high-performance liquid chromatographic column. *J. Chromatogr.*, **543**:137-145.
Goodwin, T.W. and Britton, G. (1988) Distribution and analysis of carotenoids. In: Goodwin, T.W. (ed.), *Plant Pigments.* Academic Press, London, pp. 61-132.
Grung, M., Metzger, P. and Liaaen-Jensen, S. (1989) Primary and secondary carotenoids in two races of the green alga *Botryococcus braunii. Biochem. Syst. Ecol.*, **17**:263-269.
Hager, A. and Stransky, H. (1970) Das carotinoidmuster und die verbreiting des lichtinduzierten xanthophyllcyclus in verschiedenon algenklassen. III. Grünalgen. *Arch. Mikrobiol.*, **72**:68-83.
Hashimoto, H., Koyama, Y. and Shimamura, T. (1988) Isolation of *cis–trans* isomers of canthaxanthin by high-performance liquid chromatography using a calcium hydroxide column and identification of their configurations by ^{1}H NMR spectroscopy. *J. Chromatogr.*, **448**:182-187.
Haugan, J.A., Aackermann, T. and Liaaen-Jensen, S. (1992) Isolation of fucoxanthin and peridinin. *Methods Enzymol.*, **213**:231-235.
Hoque, E. (1988) Isocratic reversed-phase high-performance liquid chromatographic analysis of pigments in Norway spruce. *J. Chromatogr.*, **448**:417-423.
Humbeck, K., Römer, S. and Senger, H. (1990) The role of carotenoids in light-induced assembly of photosystem II in mutant C-6D of *Scenedesmus obliquus*. In: Baltscheffsky, M. (ed.), *Current Research in Photosynthesis* Vol.1. Kluwer Academic Publishers, Dordrecht, pp. 655-658.
Ikuno, Y., Maoka, T., Shimizu, M. *et al.* (1985) Direct diastereoisomeric resolution of carotenoids. II. All ten stereoisomers of tunaxanthin (ε,ε-carotene-3,3´-diol). *J. Chromatogr.*, **328:**387-391.
Jacobs, P.B., Le Boeuf, R.D., McCommas, S.A. and Tauber, J.D. (1982) The cleavage of carotenoid esters by cholesterol esterase. *Comp. Biochem. Physiol.*, **B72**:157-160.
Jeffrey, S.W. (1981) An improved thin-layer chromatographic technique for marine phytoplankton pigments. *Limnol. Oceanogr.*, **26**:191-197.
Jeffrey, S.W. and Wright, S.W. (1987) A new spectrally distinct component in preparations of chlorophylls from the microalga *Emiliana huxleyi* (Prymnesiophyceae). *Biochim. Biophys. Acta,* **894**: 180-188.
Jensen, A. (1966) Algal carotenoids. V. Isofucoxanthin – a rearrangement product of fucoxanthin. *Acta Chem. Scand.*, **20**:1728-1730.

Jensen, A. and Liaaen-Jensen, S. (1959) Quantitative paper chromatography of carotenoids. *Acta Chem. Scand.*, **13**:1863-1872.

Jensen, N-H, Nielsen, A.B. and Wilbrandt, R. (1982) Chlorophyll *a* sensitized *trans-cis* photoisomerization of all-*trans*-ß-carotene. *J. Am. Chem. Soc.*, **104**:6117-6119.

Johannes, B., Brzezinka, H. and Budzikiewicz, H. (1974) Massenspektroskopische fragmentierungsreaktionen. VII. Abbau der polyenkette bei carotinoiden. *Org. Mass Spectrom.*, **9**:1095-1113.

Johannes, B., Brzezinka, H. and Budzikiewicz, H. (1979) Mass spectroscopic fragmentation reactions. XXI. Factors influencing the loss of xylene and toluene from the polyene chain of carotenoids. *Z. Naturforsch.*, **34B**:300-305.

Katayama, N., Hashimoto, H., Koyama, Y. and Shimamura, T. (1990) High-performance liquid chromatography of *cis–trans* isomers of neurosporene: discrimination of *cis* and *trans* configurations at the end of an open conjugated chain. *J. Chromatogr.*, **519**:221-227.

Khachik, F. and Beecher, G.R. (1985) Decapreno-ß-carotene as an internal standard for the quantification of the hydrocarbon carotenoids by high performance liquid chromatography. *J. Chromatogr.*, **346**:237-246.

Khachik, F. and Beecher, G.R. (1988) Separation and identification of carotenoids and carotenol fatty acid esters in some squash products by liquid chromatography. 1. Quantification of carotenoids and related esters by HPLC. *J. Agric. Food Chem.*, **36**:929-937.

Khachik, F., Beecher, G.R. and Whittaker, N.F. (1986) Separation, identification, and quantification of the major carotenoid and chlorophyll constituents in extracts of several green vegetables by liquid chromatography. *J. Agric. Food Chem.*, **34**:603-616.

Khachik, F., Beecher, G.R. and Lusby, W.R. (1988) Separation and identification of carotenoids and carotenol fatty acid esters in some squash products by liquid chromatography. 2. Isolation and characterisation of carotenoids and related esters. *J. Agric. Food Chem.*, **36**:938-946.

Khachik, F., Beecher, G.R. and Goli, M.B. (1991) Separation, identification and quantification of carotenoids in fruits, vegetables and human plasma by high performance liquid chromatography. *Pure Appl. Chem.*, **63**:71-80.

Kohata, K. and Wantanabe, M. (1988) Diel changes in the composition of photosynthetic pigments and cellular carbon and nitrogen in *Chattonella antiqua* (Raphidophyceae). *J. Phycol.*, **24**, 58-66.

Koyama, Y., Hosomi, M., Miyata, A. *et al.* (1988) Supplementary and revised assignment of the peaks of the 7,9-, 9,9′-, 13,13′-, 9,13′-di-*cis* and 9,9′,13-tri-*cis* isomers of ß-carotene in high performance liquid chromatography using a column of calcium hydroxide. *J. Chromatogr.*, **439**:417-422.

Koyama, Y., Takatsuka, I., Kanaji, M. *et al.* (1990) Configurations of carotenoids in the reaction centre and the light-harvesting complex of *Rhodospirillum rubrum*. Natural selection of carotenoid configurations in pigment–protein complexes. *Photochem. Photobiol.*, **51**:119-128.

Lauren, R.L. and McNaughton, D.E. (1986) Comparison of column packings for isocratic HPLC analysis of carotenoids using non-aqueous reversed-phase. *J. Liq. Chromatogr.*, **9**:2013-2033.

Lauren, R.L., Agnew, M.P. and McNaughton, D.E. (1986) The use of decanol for improving chromatographic stability in isocratic non-aqueous reversed-phase analysis of carotenoids by HPLC. *J. Liq. Chromatogr.*, **9**:1997-2012.

Leenheer, A.P. de and Nelis, H.J. (1992) Profiling and quantitation of carotenoids by high-performance liquid chromatography and photodiode array detection. *Methods Enzymol.*, **213**:251-256.

Liaaen-Jensen, S. (1990) Artifacts of natural carotenoids - unintended carotenoid synthesis. In: Krinsky, N.I., Mathews-Roth, M.M. and Taylor, R.F. (eds.), *Carotenoids: Chemistry and Biology*. Plenum Press, New York, pp 149-165.

Liaaen-Jensen, S. and Jensen, A. (1971) Quantitative determination of carotenoids in photosynthetic tissues. *Meth. Enzymol.*, **23**:586-602.

Lichtenthaler, H.K. (1987) Chlorophylls and carotenoids: pigments of photosynthetic biomembranes. *Methods Enzymol.*, **148**:350-382.

Lichtlé, C., Duval, J-C. and Lemoine, Y. (1987) Comparative biochemical, functional and ultrastructural studies of photosystem particles from a Cryptophyceae: *Cryptomonas rufescens*; isolation of an active phycoerythrin particle. *Biochim. Biophys. Acta*, **894**:79-90.

Maoka, T. and Matsuno, T. (1989) Diastereomeric resolution of carotenoids. III ß,ß-Caroten-2-ol, ß,ß-carotene-2,2´-diol and 2-hydroxyechinenone. *J. Chromatogr.*, **478**:379-386.

Märki-Fischer, E., Buchecker, R., Eugster, C.H. *et al.* (1982) Absolute konfiguration von antheraxanthin << *cis*-antheraxanthin >> und der diastereomeren mutatoxanthine. *Helv. Chim. Acta*, **65**:2198-2211.

Moss, G.P. and Weedon, B.C.L. (1976) Chemistry of carotenoids. In: Goodwin, T.W. (ed.), *Chemistry and Biochemistry of Plant Pigments*, 2nd. Ed., Vol.1. Academic Press, London, pp.149-224.

Müller, R.K., Bernhard, K., Meyer, H. *et al.* (1980) Beitrag zur analytik und synthese von 3-hydroxy-4-oxocarotinoiden. *Helv. Chim. Acta*, **63**:1654-1674.

Nelis, H.J. and de Leenheer, A.P. (1983) Isocratic nonaqueous reversed-phase liquid chromatography of carotenoids. *Anal. Chem.*, **55**:270-275.

Nelis, H.J. and de Leenheer, A.P. (1989) Profiling and quantitation of bacterial carotenoids by liquid chromatography and photodiode array detection. *App. Env. Microbiol.*, **55**:3065-3071.

Noack, K. (1982) Circular dichroism of carotenoids and its use in investigations of their structures, configurations and conformations. In: Britton, G. and Goodwin, T.W. (eds.), *Carotenoid Chemistry and Biochemistry*. Pergamon, Oxford, pp. 135-153.

Noack, K. and Buchecker, R. (1993) Circular dichroism. In: Britton, G., Liaaen-Jensen, S. and Pfander, H. (eds.), *Carotenoids, Isolation and Analysis*, Vol. 1b,. Birkhäuser, Basel, (In press).

O'Neil, C.A., Schwartz, S.J. and Catignani, G.L. (1991) Comparison of liquid chromatographic methods for determination of *cis–trans* isomers of ß-carotene. *J. Assoc. Off. Anal. Chem.*, **74**:36-42.

Owens, T.G., Gallagher, J.C. and Alberte, R.S. (1987) Photosynthetic light-harvesting function of violaxanthin in *Nannochloropsis* spp. (Eustigmatophyceae). *J. Phycol.*, **23**:79-85.

Pennington, F.C., Guillard, R.R.L. and Liaaen-Jensen, S. (1988) Carotenoid distribution patterns in Bacillariophyceae (diatoms). *Biochem. Syst. Ecol.*, **16**: 589-592.

Pfander, H. (ed.) (1987) *Key to Carotenoids*. (O. Straub). 2nd Ed., Birkhäuser, Basel.

Philip, T. and Chen, T-S. (1988) Separation and quantitative analysis of some carotenoid fatty acid esters of fruits by liquid chromatography. *J. Chromatogr.*, **435**:113-126.

Philip, T., Chen, T.-S, and Nelson, D.B. (1988) Liquid chromatographic profiles of major carotenoid esters in commercially processed California navel and Valencia orange juice concentrates. *J. Chromatogr.*, **442**:249-265.

De Ritter (1981) Appendix: Absorption maxima and $E_{1cm}^{1\%}$ values for carotenoids In: Bauernfeind, J.C. (ed.), *Carotenoids as Colorants and Vitamin A Precursors*. Academic Press, New York, pp. 883-923.

Rock, C.D. and Zeevaart, J.A.D. (1991) The *aba* mutant of *Arabidopsis thaliana* is impaired in epoxy-carotenoid biosynthesis. *Proc. Natl. Acad. Sci. USA*, **88**: 7496-7499.

Rüedi, P. (1985) HPLC - a powerful tool in carotenoid research. *Pure Appl. Chem.*, **57**:793-800.

Rüttiman, A., Schiedt, K. and Vecchi, M. (1983) Separation of (3*R*,3´*R*)-, (3*R*,3´*S*; meso)-, (3*S*,3´*S*)-zeaxanthin, (3*R*,3´*R*,6´*R*)-, (3*R*,3´*S*,6´*S*)- and (3*S*,3´*S*,6´*S*)-lutein via the dicarbamates of (S)-(+)-α-(1-naphthyl)ethyl isocyanate. *J. High Resol. Chromatogr., Chromatogr. Commun.*, **6**:612-616.

Schmitz, H.H., Artz, W.E., Poor, C.L. *et al.* (1989) High-performance liquid chromatography and capillary supercritical fluid chromatography separation of vegetable carotenoids and carotenoid isomers. *J. Chromatogr.*, **479**:261-268.

Siefermann-Harms, D. (1988) High-performance liquid chromatography of chloroplast pigments. One-step separation of carotene and xanthophyll isomers, chlorophylls and phaeophytins. *J. Chromatogr.*, **448**:411-416.

Stauber, J.L. and Jeffrey, S.W. (1988) Photosynthetic pigments in fifty-one species of marine diatoms. *J. Phycol.*, **24**:158-172.

Sturzenegger, V., Buchecker, R. and Wagnière, G. (1980) Classification of the CD spectra of carotenoids. *Helv. Chim. Acta*, **63**:1074-1092.

Thayer, S.S. and Björkman, O. (1990) Leaf xanthophyll content and composition in sun and shade determined by HPLC. *Photosyn. Res.*, 23: 331-343.

Tsukida, K., Saiki, K., Takii, T. and Koyama, Y. (1982) Separation and determination of *cis/trans* ß-carotenes by high-performance liquid chromatography. *J. Chromatogr.*, **245**:359-364.

Vecchi, M. and Müller, R.K. (1979) Separation of (3*S*,3´*S*)-, (3*R*,3´*R*)- and (3*S*,3´*R*)-astaxanthin via (-)-camphanic acid esters. *J. High Resol. Chromatogr., Chromatogr. Commun.*, **2**:195-196.

Vecchi, M., Englert, G., Maurer, R. and Meduna, V. (1981) Trennung und charakterisierung von ß-carotin-isomeren. *Helv. Chim. Acta*, **64**:2746-2758.

Vetter, W., Englert, G., Rigassi, N. and Schwieter, U. (1971) Spectroscopic methods. In: Isler, O. (ed.), *Carotenoids*, Birkhäuser, Basel, pp.189-266.

Wright, S.W. and Jeffrey, S.W. (1987) Fucoxanthin pigment markers of marine phytoplankton analyzed by HPLC and HPTLC. *Mar. Ecol. Prog. Ser.*, **38**:259-266.

Wright, S.W. and Shearer, J.D. (1984) Rapid extraction and high-performance liquid chromatography of chlorophylls and carotenoids from marine phytoplankton. *J. Chromatogr.*, **294**:281-295.

Young, A.J. and Britton, G. (1988) Photobleaching in the unicellular green alga *Dunaliella parva* 19/9. *Photosynth. Res.*, **25**:129-136.

Young, A.J., Musker, D. and Britton, G. (1989) A rapid method for the analysis of the mode of action of bleaching herbicides. *Pestic. Biochem. Physiol.*, **35**:244-250.

Zechmeister, L. (1962) *Cis-trans Isomeric Carotenoids*, Vitamins A and Arylpolyenes. Springer-Verlag, Vienna.

Appendix

A.J. Young and G. Britton

In this appendix, physical data are presented for a variety of carotenoids commonly found in photosynthetic organisms. In these, the information for an individual compound using a particular technique has been obtained from the authors' own studies or from major review articles: nuclear magnetic resonance (NMR) Moss and Weedon (1976), Englert (1982, 1985, 1991, 1993): mass spectrometry (MS). Enzell et al. (1969), Budzikiewicz et al. (1970), Budzikiewicz (1974), Enzell and Broman (1993): circular dichroism (CD). Sturzenegger et al. (1980), Noack and Buchecker (1993): ultraviolet/visible spectroscopy (UV/Vis), Davies (1976). The full references for these articles are given in Chapter 10.

In the following list, NMR data are not given in full for each compound. Instead, NMR chemical shift (δ) values are given in tabulated form in Chapter 10 (Tables 10.8 and 10.9 for ^{1}H and ^{13}C, respectively) for carotenoid end-groups.

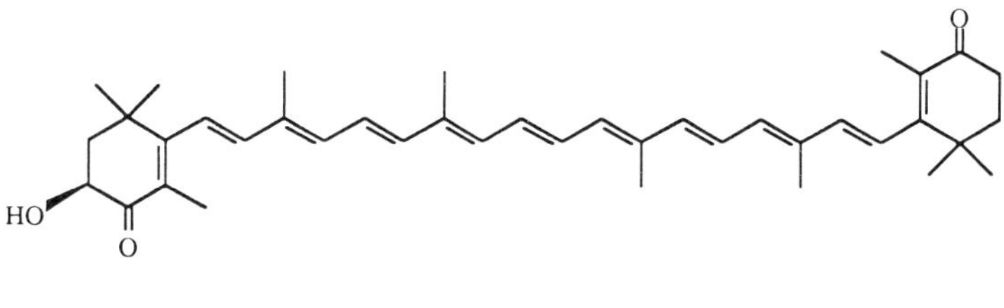

Adonirubin ((3*S*)-3-hydroxy-ß,ß-caroten-4,4'-one)
$C_{40}H_{52}O_3$. m.p. 214–216°C. UV/Vis λ_{max} nm: 294, 468 (hexane); 483 (benzene); 250, 296, 477 (ethanol). MS *m/z* (rel.int): 580 [M]$^+$ (35), 488 [M-92]$^+$ (4), 474 [M-106]$^+$ (3), 203 (50), 91 (100). NMR end-groups: 18-17.

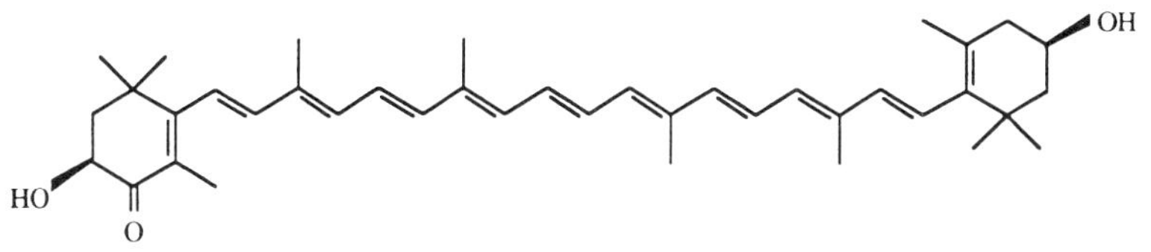

Adonixanthin (3*S*,3'*R*)-3,3-dihydroxy-ß,ß-caroten-4-one)
$C_{40}H_{54}O_3$. m.p. 180–181°C. UV/Vis λ_{max} nm: 296,460 (hexane); 304,481 ($CHCl_3$); 472 (benzene); 299, 465 (ethanol). MS *m/z* (rel. int.): 582 $[M]^+$ (41), 580 $[M-2]^+$ (6), 564 $[M-18]^+$ (5), 490 $[M-92]^+$ (5), 203 (74), 91 (100). CD nm ($\Delta\varepsilon$): 366 (+2.4), 307 (-18.95), 292 Sh. (-11.64), 264 (+11.12), 238 (-9.85). NMR end-groups: 18-6.

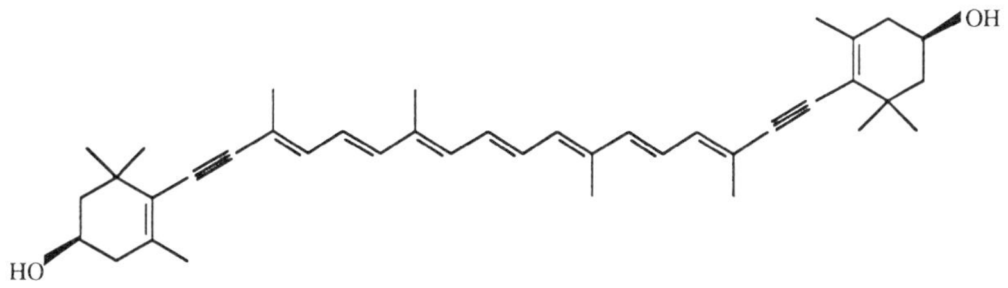

Alloxanthin ((3*R*,3'*R*)-7,8,7',8'-tetradehydro-ß,ß-carotene-3,3'-diol) (9Z,9'*Z* = manixanthin).
$C_{40}H_{52}O_2$. Red needles m.p. 189–190°C. UV/Vis λ_{max} nm: 434, 464, 495 (benzene); 454, 483 (ethanol); 451, 480 (hexane); 454, 488, 518 (CS_2); (436), 460, 489 ($CHCl_3$); 458, 488 (petrol). IR υ_{max} cm^{-1} ($CHCl_3$): 2167*w* (acetylene). MS *m/z* (rel. int.): 564.398 $[M]^+$ (50), 546 $[M-18]^+$ (1), 472 [M-92]+ (<1), 41 (100). NMR end-groups: 7-7.

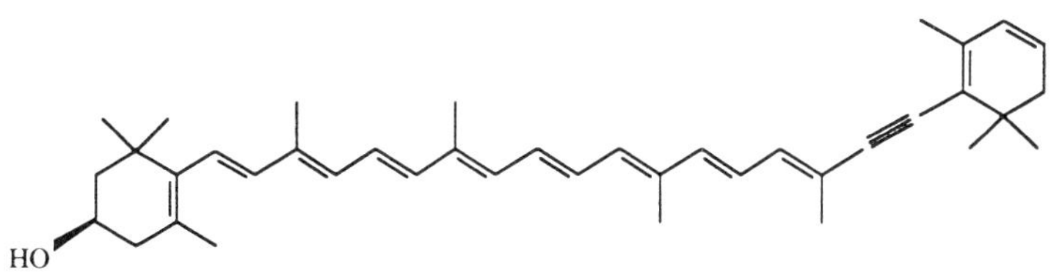

3',4'-Anhydrodiatoxanthin ((3*R*)-3',4',7',8'-tetradehydro-ß,ß-caroten-3-ol)
$C_{40}H_{56}O$. UV/Vis λ_{max} nm: (447), 461, 488 (ethanol). MS *m/z* (rel. int.): 548 $[M]^+$ (100), 546 $[M-2]^+$ (3), 533 $[M-15]^+$ (5), 530 $[M-18]^+$ (1), 456 $[M-92]^+$ (3). For CD, see Fiksdahl et al. 1984. NMR end-groups: 6-2.

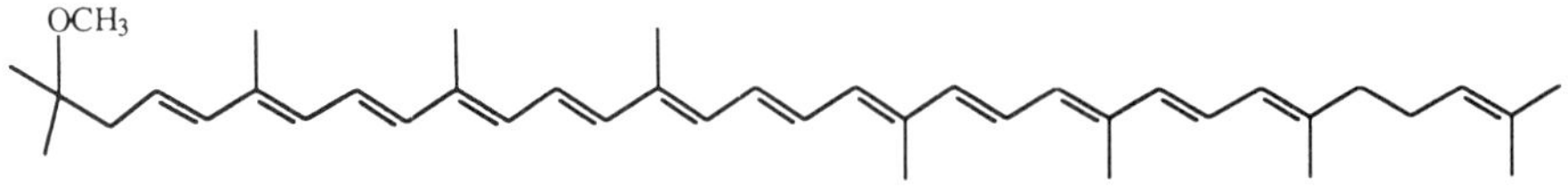

Anhydrorhodovibrin (1-methoxy-3,4-didehydro-1,2-dihydro-ψ,ψ-carotene)
$C_{41}H_{58}O$. UV/Vis λ_{max} nm: 454, 482, 515 (petrol, $A^{1\%}_{1cm}$ 2700); 459, 488, 522 (acetone); 471, 499, 533 ($CHCl_3$). MS *m/z* (rel. int.): 566 $[M]^+$ (100), 534 $[M-32]^+$ (2), 497 $[M-69]^+$ (1), 493 $[M-73]^+$ (2), 474 $[M-92]^+$ (1), 460 $[M-32-106]^+$ (1), 73 (100). NMR end-groups: 39-32.

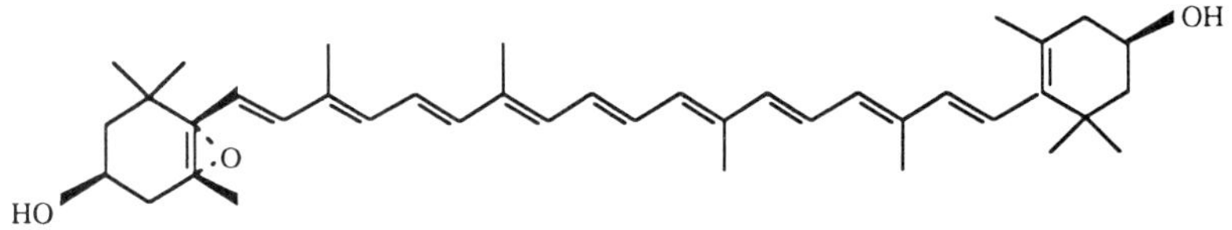

Antheraxanthin ((3*S*,5*R*,6*S*,3'*R*)-5,6-epoxy-5,6-dihydro-ß,ß-carotene-3,3'-diol) $C_{40}H_{56}O_3$. m.p. 205 °C. UV/Vis λ_{max} nm: 422,445,472 (petrol; ethanol); 430,456,484 ($CHCl_3$). CD (EPA, RT) nm (Δε): 216 (-13.2), 239 (+10.3), 275 (-21.1), 335 (+5.7). CD (EPA, -180 °C) nm (Δε): 217 (-38.1), 242 (+36.4), 280 (-55.0), 330 (+6.2), 345 (+10.6), 408 (+2.3), 435 (+2.8), 462 (+3.8), 493 (+5.2). MS *m/z* (rel.int.): 584 $[M]^+$ (60), 504 $[M-80]^+$ (30), 492 $[M-92]^+$ (6), 221 (30), 181, 43 (100). NMR end-groups:9-6.

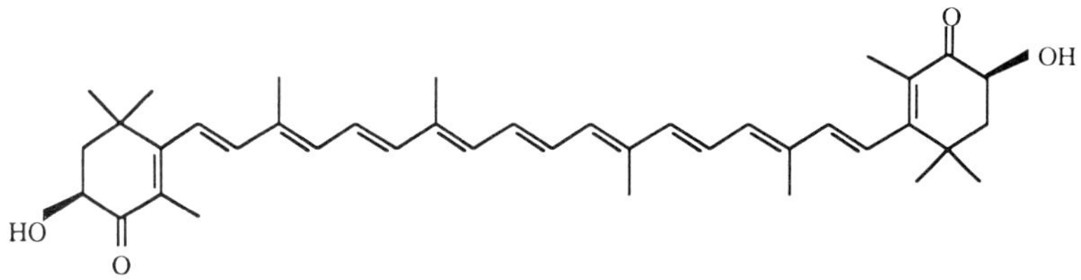

Astaxanthin (3*S*,3'*S*)-3,3'-dihydroxy-ß,ß-carotene-4,4'-dione)
$C_{40}H_{52}O_4$. m.p. 217–219°C (with decomp.). UV/Vis λ_{max} nm: 468 (petrol), 478 (ethanol), 480 (acetone), 483 ($CHCl_3$); 485 (benzene); 503 (CS_2). CD (CH_2Cl_2) nm (Δe): 224 (+12.8), 249 (-14.4), 280 (+12.5), 323 (-23.1), 384 (+6.7), 521 (-3.2). MS *m/z* (rel. int.): 596 $[M]^+$ (10), 578 $[M-18]^+$ (2), 504 $[M-92]^+$ (2), 490 $[M-106]^+$ (3),

203 (10), 91 (100). NMR end groups 18-18.

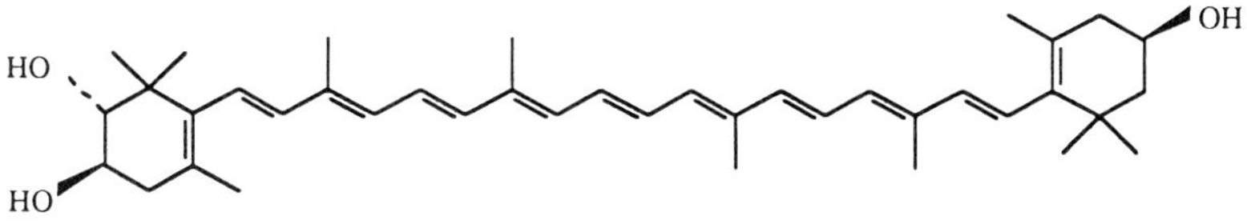

Caloxanthin ((2*R*,3*R*,3'*R*)-ß,ß-carotene-2,3,3'-triol)
$C_{40}H_{56}O_3$. UV/Vis as zeaxanthin. MS *m/z* (rel. int.): 584 $[M]^+$ (100), 566 $[M-18]^+$ (22), 548 $[M-18-18]^+$ (2), 492 $[M-92]^+$ (15), 478 $[M-106]^+$ (3), 474 $[M-18-92]^+$ (4). NMR end-groups: 10-6.

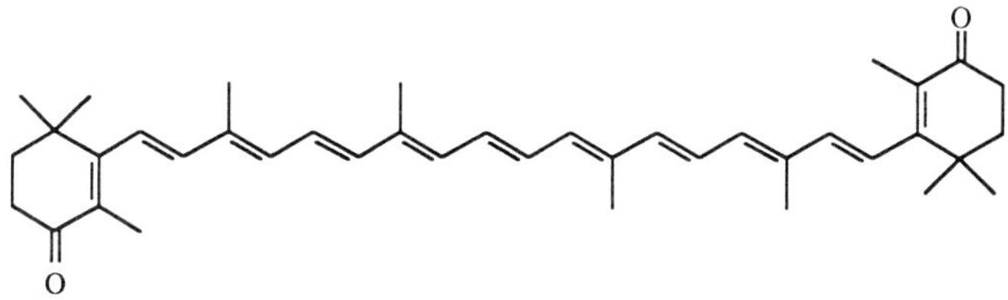

Canthaxanthin (ß,ß-carotene-4,4'-dione)
$C_{40}H_{52}O_2$. UV/Vis λ_{max} nm: 462 (hexane); 466 (petrol $A^{1\%}_{1cm}$ 2200), 474 (ethanol, on reduction with $NaBH_4$ 428, 450, 478); 482 ($CHCl_3$), 500 (CS_2). MS *m/z* (rel.int.): 564 $[M]^+$ (41), 472 $[M-92]^+$ (7), 458 $[M-106]^+$ (2.5), 83 (100). NMR end-groups 17-17.

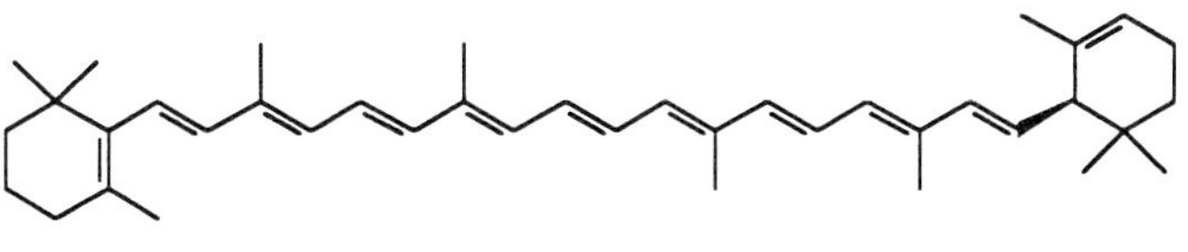

α-Carotene ((6'*R*)-ß,e-carotene)
$C_{40}H_{56}$. Violet prisms from benzene/methanol or petrol. m.p. 187–188°C. UV/Vis λ_{max} nm: 422, 444, 473 (ethanol, petrol, $A^{1\%}_{1cm}$ 2800); 424, 448, 476 (acetone); 433, 457, 484 ($CHCl_3$); 450, 477, 509 (CS_2). CD (dioxane) nm (Δε): 243 (+5.3), 254

(+4.25), 334 (+2.4), 394 (+1.3), 410-490 (+0.9). MS *m/z* (rel. int.): 536 [M]$^+$ (100), 480 [M-56]$^+$ (10), 444 [M-92]$^+$ (25), 430 [M-106]$^+$ (3), 413 [M-123]$^+$ (4). NMR end-groups: 1-28.

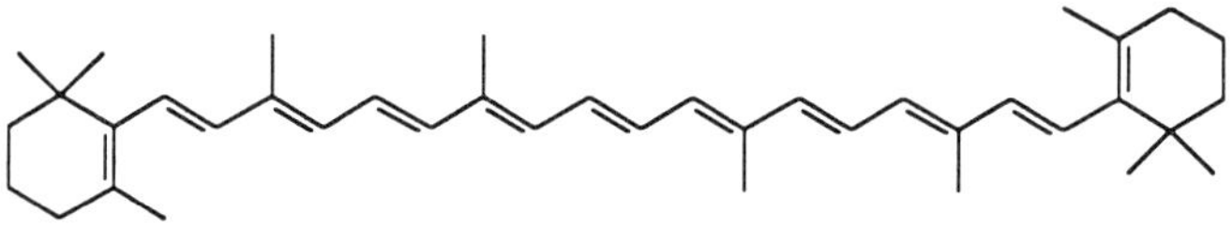

ß-Carotene (ß,ß-carotene)
$C_{40}H_{56}$. m.p. 181–182°C. UV/Vis λ_{max} nm: 425, 450, 476 (petrol, ethanol, $A^{1\%}_{1cm}$ 2592); 429, 452, 478 (acetone); 435, 461, 485 ($CHCl_3$); 435, 462, 487 (benzene); 450, 485, 520 (CS_2). MS *m/z* (rel. int.): 536 [M]$^+$ (100), 444 [M-92]$^+$ (15), 430 [M-106]$^+$. NMR end groups 1-1. Data for the (9*Z*)-, (13*Z*)-, (15*Z*)- and several di-(*Z*)-isomers have been given by Vecchi et al. 1981.

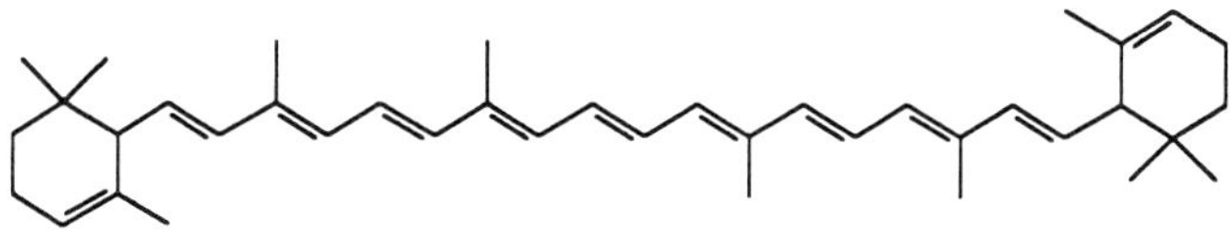

ε-Carotene ((6*R*, 6'*R*)-ε,ε-carotene
$C_{40}H_{56}$. UV/Vis λ_{max} nm: 416, 440, 470 (petrol, ethanol $A^{1\%}_{1cm}$ 3120); 425, 451, 482 (benzene). CD (hexane) nm (Δε): 220 (+), 264 (+19.5), 312 (+2.5), 325 (+3.2), 392 (+4.5), 411 (+7.5), 436 (10.5), 465 (+9.6). MS *m/z* (rel.int.): 536 [M]$^+$ (100), 480 [M-56]$^+$ (6), 444 [M-92]$^+$ (10), 430 [M-106]$^+$ (2), 413 [M-123]$^+$ (2), 123 (65). NMR end-groups: 28-28.

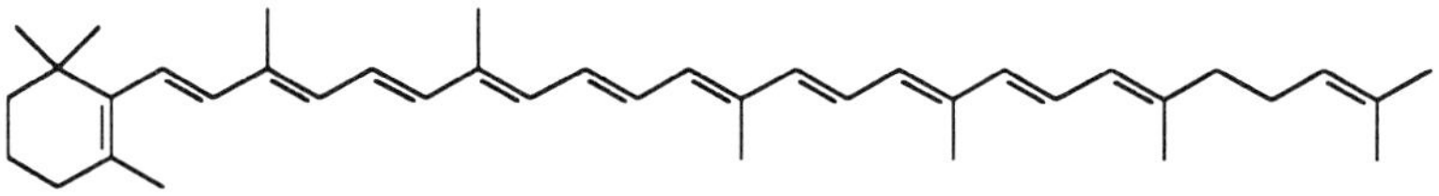

γ-Carotene (ß,ψ-carotene)
$C_{40}H_{56}$. m.p. 178°C (from benzene/methanol). UV/Vis λ_{max} nm: 437, 462, 494 (petrol, $A^{1\%}_{1cm}$ 3100); 439, 461, 491 (acetone); 446, 475, 509 ($CHCl_3$); 463, 496, 533 (CS_2). MS *m/z* (rel.int.): 536 $[M]^+$ (100), 467 $[M\text{-}69]^+$ (5), 444 $[M\text{-}92]^+$ (8), 430 $[M\text{-}106]^+$ (2.5). NMR end-groups: 1-32.

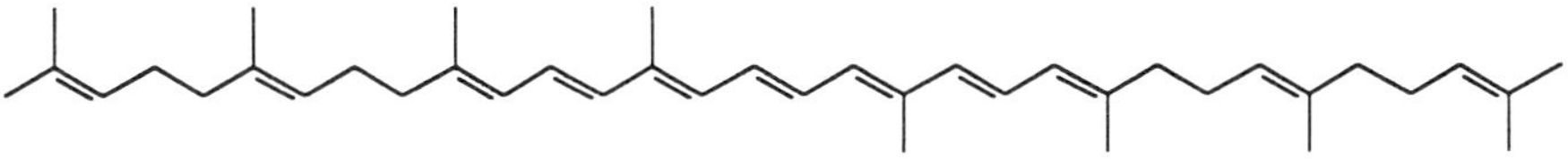

ζ-Carotene (7,8,7',8'-tetrahydro-ψ,ψ-carotene)
$C_{40}H_{60}$. Pale yellow oil. UV/Vis λ_{max} nm: 378, 400, 425 (petrol, hexane $A^{1\%}_{1cm}$ 2555); 384, 405, 530 ($CHCl_3$). MS *m/z* (rel.int.): 540 $[M]^+$ (25), 438 $[M\text{-}92]^+$ (3), 436 $[M\text{-}94]^+$ (2), 403 $[M\text{-}137]^+$ (12), 69 (100). NMR end groups 33-33.

"Unsymmetrical ζ-carotene" 7,8,11,12-tetrahydrolycopene (7,8,11,12-tetrahydro-ψ,ψ-carotene).
$C_{40}H_{60}$. UV/Vis λ_{max} nm: 373, 395, 420 (petrol, hexane); 380, 400, 525 ($CHCl_3$). MS *m/z* (rel. int.): 540 $[M]^+$ (25), 471 $[M\text{-}69]^+$ (4), 438 $[M\text{-}92]^+$ (3), 436 $[M\text{-}94]^+$ (2), 335 $[M\text{-}205]^+$ (20), 69 (100). NMR end-groups: 32-35.

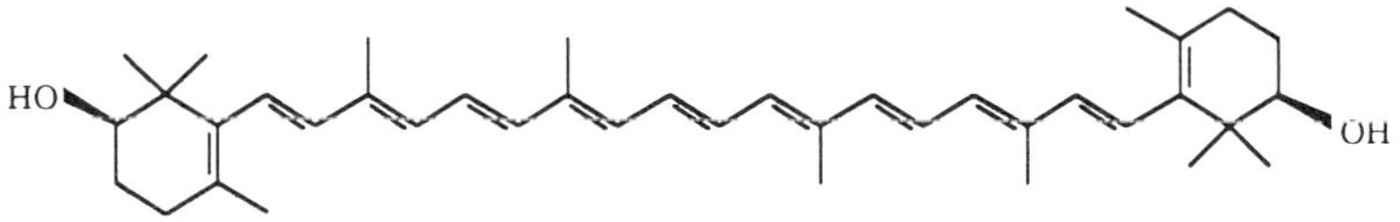

ß,ß-Carotene-2,2'-diol
$C_{40}H_{56}O_2$. UV/Vis λ_{max} as zeaxanthin. MS *m/z*: 568 $[M]^+$, 550 $[M\text{-}18]^+$, 532 $[M\text{-}18\text{-}18]^+$, 476 $[M\text{-}92]^+$, 462 $[M\text{-}106]^+$. CD (Δε) nm: 224 (-1.8), 236 (0), 245 (+2.0), 284 (-3.0), 325 (0), 350 (+0.5). NMR end-groups: 5-5.

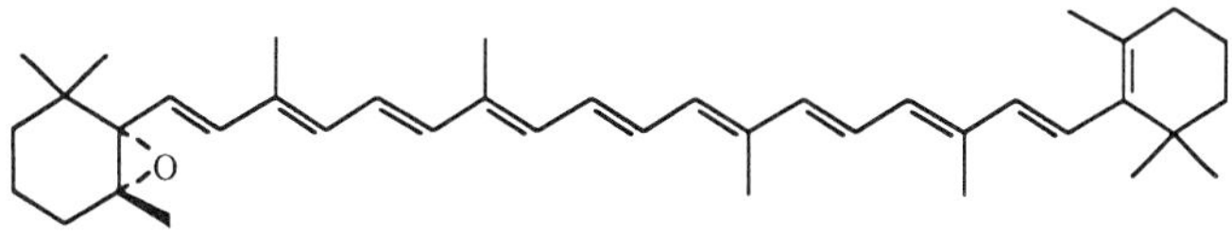

ß-Carotene-5,6-epoxide ((5R,6S)-5,6-epoxy-5,6-dihydro-ß,ß-carotene) $C_{40}H_{56}O$. m.p. 156°C. UV/Vis λ_{max} nm: 422, 445, 472 (hexane); 459, 492 ($CHCl_3$); 479, 511 (CS_2). MS *m/z* (rel. int.): 552 $[M]^+$ (54), 536 $[M-16]^+$ (7), 472 $[M-80]^+$ (39), 460 $[M-92]^+$ (11), 205 (10), 165 (61). Data for the 5*S*,6*R*-isomer are given by Eschenmoser and Eugster (1978). NMR end-groups: 3-1.

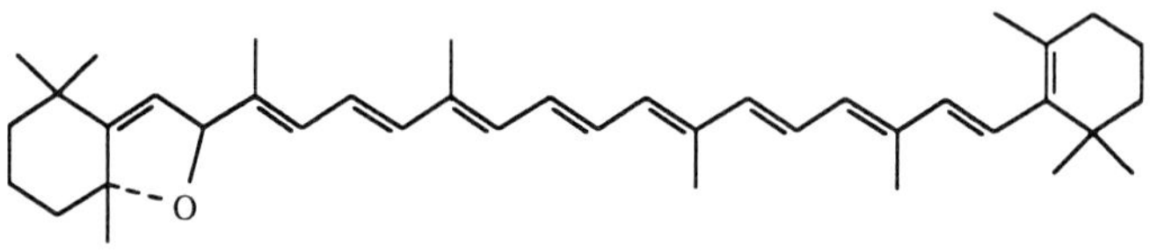

ß-Carotene-5,8-epoxide (= mutatochrome, 5,8-epoxy-5,8-dihydro-ß,ß-carotene)

$C_{40}H_{56}O$. Yellow-orange leaflets from benzene/methanol. m.p. 163–164°C. UV/Vis λ_{max} nm: 409, 428, 452 (petrol); 438, 469 ($CHCl_3$); 416, 437, 463 (benzene); 459, 490 (CS_2). MS as ß-carotene-5,6-epoxide. NMR end-groups: 4-1.

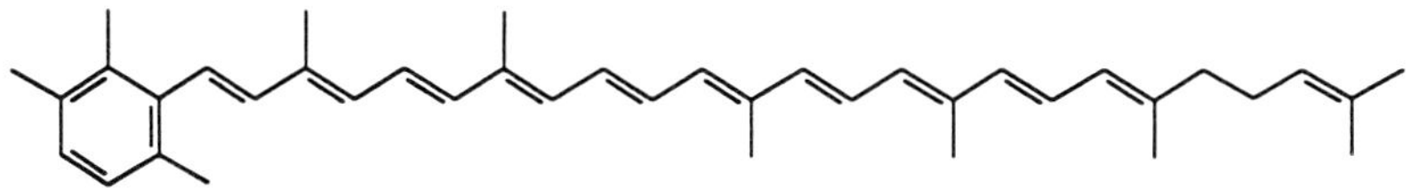

Chlorobactene (ϕ,ψ-carotene)

$C_{40}H_{52}$. Red needles m.p. 145–146°C. UV/Vis λ_{max} nm: 450, 476, 508 (benzene); 435, 460, 491 (petrol, $A^{1\%}_{1cm}$ 3040); (470), 496, 530 (CS_2). MS *m/z* (rel. int.): 532 $[M]^+$ (15), 463 $[M-69]^+$ (2), 440 $[M-92]^+$ (4), 426 $[M-106]^+$ (5), 399 $[M-133]^+$(1), 133 (100), 69 (85). NMR end-groups: 30-32.

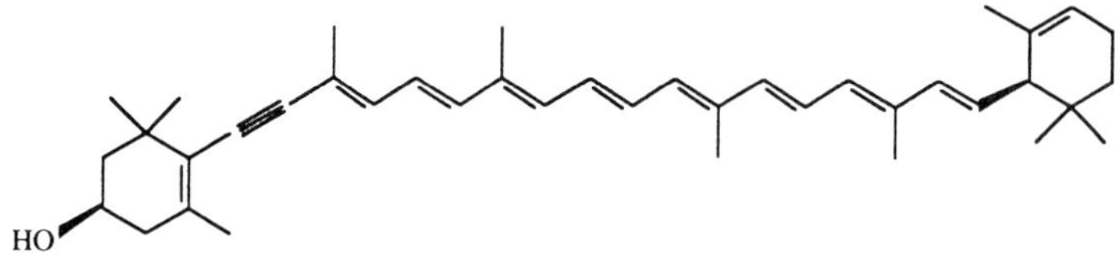

Crocoxanthin ((3*R*,6'*R*)-7,8-didehydro-ß,ε-caroten-3-ol)
$C_{40}H_{56}O$. Red needles, m.p. 163-165°C. UV/Vis λ_{max} nm: 432, 458, 488 (benzene); 447, 477 (ethanol); 422, 445, 475 (hexane); (430), 454, 482 ($CHCl_3$). IR υ_{max} cm^{-1} (CCl_4): 2169*vw* (acetylene). MS *m/z* (rel. int.): 550.421 $[M]^+$ (100), 535 $[M-15]^+$ (2), 532 $[M-18]^+$ (1), 517 $[M-15-18]^+$ (<1), 494 $[M-56]^+$ (<1), 458 $[M-92]^+$(2), 444 $[M-106]^+$(<1), 119 (17), 105 (21), 91 (26). NMR end-groups: 7-28.

A.

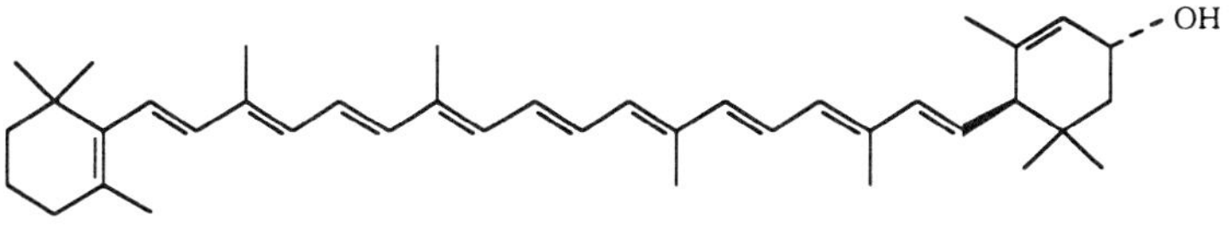

B.

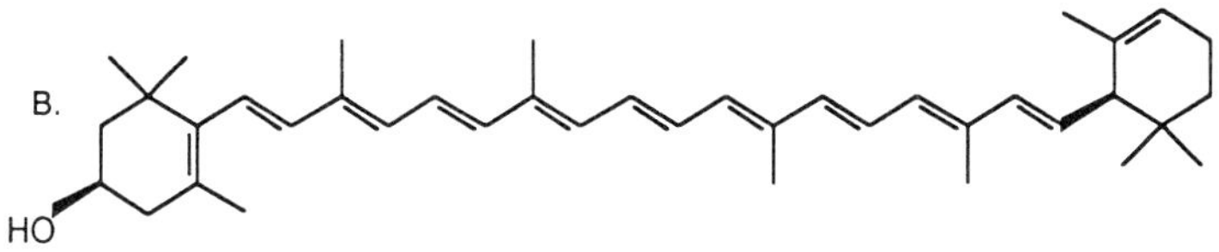

α-Cryptoxanthin (3*R*,6'*R*)-ß,ε-caroten-3-ol or zeinoxanthin (ß,ε-caroten-3'-ol).
$C_{40}H_{56}O$. UV/Vis λ_{max} nm: 421, 444, 472 (petrol); 433, 459, 488 (benzene). Compound A: CD (EPA) nm (Δε): 230 (+6.0), 260 (+4.0), 312 (+1.2). MS *m/z* (rel. int.): 552 $[M]^+$ (85), 534 $[M-18]^+$ (100), 460 $[M-92]^+$ (2), 442 $[M-18-92]^+$ (3). NMR end-groups: 6-28. Compound B: MS *m/z* (rel. int.): 552 $[M]^+$ (100), 534 $[M-18]^+$ (20), 460 $[M-92]^+$ (8), 442 $[M-18-92]^+$ (2). NMR end-groups: 1-29.

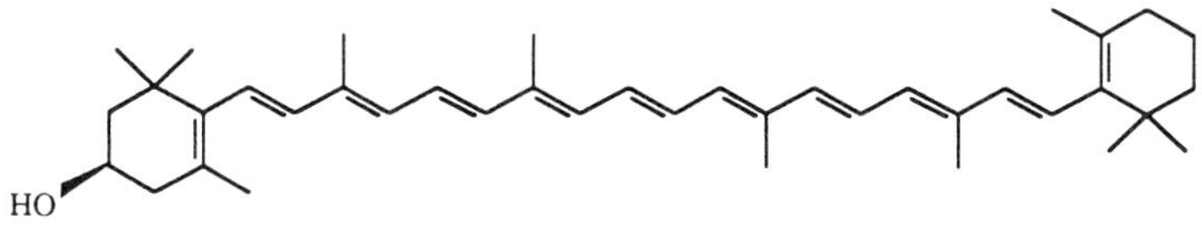

ß-Cryptoxanthin ((3*R*)-ß,ß-caroten-3-ol)
$C_{40}H_{56}O$. Prisms from benzene/methanol, m.p. 169°C. UV/Vis λ_{max} nm: 425, 449, 476 (petrol, $A^{1\%}_{1cm}$ = 2386); 435, 459, 485 ($CHCl_3$); 453, 483, 518 (CS_2). CD (EPA) nm (Δε): 224 (-0.6), 236 (0), 245 (+6.0), 260 (0), 280 (-10.0), 350 (+3.0). MS *m/z* (rel. int.): 552 $[M]^+$ (100), 534 $[M-18]^+$ (25), 460 $[M-92]^+$ (5), 446 $[M-106]^+$ (2). NMR end-groups: 6-1.

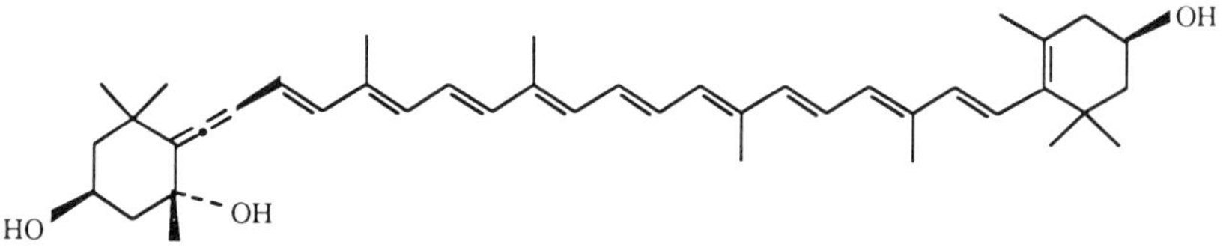

Deepoxyneoxanthin ((3*S*,5*R*,6*R*,3'*R*)-6,7-didehydro-5,6-dihydro-ß,ß-carotene-3,5,3'-triol)
$C_{40}H_{56}O_3$. m.p. 140-143°C. UV/Vis λ_{max} nm: 264, 326, 420, 442, 469 (methanol). MS *m/z* (rel. int.): 584 $[M]^+$ (47.4), 566 $[M-18]^+$ (100), 548 $[M-36]^+$ (33.9), 533 $[M-51]^+$, 530 $[M-54]^+$, 492 $[M-92]^+$, 474 $[M-110]^+$· 456 $[M-128]^+$, 441 $[M-143]^+$ CD (EPA)nm ($\Delta\varepsilon$): 355 (0), 330 (-1.1), 281 (-1.6), 236 (0), 230 (-1.9), 219 (0). NMR end-groups: 11-6.

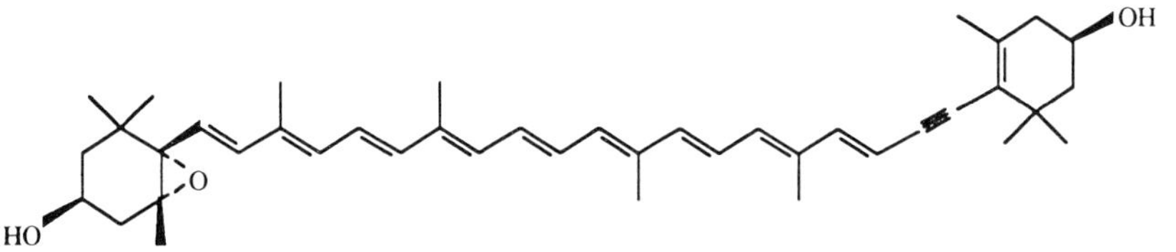

Diadinoxanthin ((3*S*,5*R*,6*S*,3'*R*)-5,6-epoxy-7',8'-didehydro-5,6-dihydro-ß,ß-carotene-3,3'-diol)
$C_{40}H_{54}O_3$. UV/Vis λ_{max} nm: (428), 449, 479 (acetone); 422, 446, 476 (petrol); 449, 474, 506 (CS_2); 432, 455, 482 ($CHCl_3$); 421, 445, 475 (hexane); 424, 445, 474 (ethanol, $A^{1\%}_{1cm}$ = 2250). MS *m/z* (rel. int.): 582 $[M]^+$ (12), 564 $[M-18]^+$(12), 502 $[M-80]^+$ (7), 221 (100), 181 (62). CD (EPA) nm ($\Delta\varepsilon$): 225 (0), 239 (4.6), 249 (0), 280 (-8.5), 297 (6), 237 (4.7) (cryst.). IR υ_{max} cm^{-1}: 2170 *vw* (acetylene). NMR end-groups: 9-7.

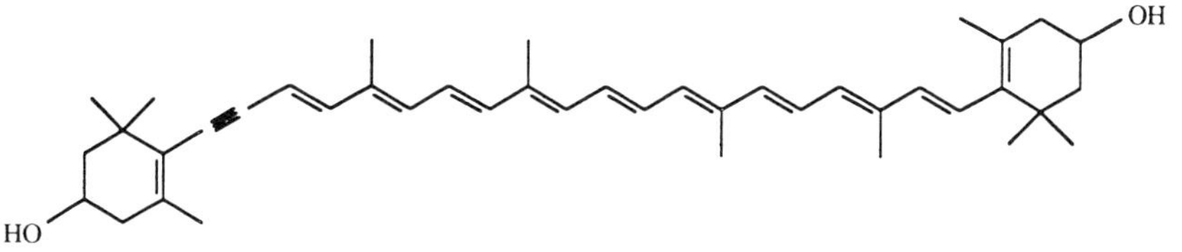

Diatoxanthin ((3*R*,3'*R*)-7,8-didehydro-ß,ß-carotene-3,3'-diol)
$C_{40}H_{52}O_2$. Reddish needles, mp. 201°C. UV/Vis λ_{max} nm: 428, 450, 479 (petrol); (428), 449, 477 (hexane); 425, 449, 475 (ethanol);(434), 454, 482 (acetone); (433), 458, 486 ($CHCl_3$). MS *m/z* (rel. int.): 566 [M]$^+$ (67), 548 [M-18]$^+$ (98), 530 [M-18-18]$^+$ (100), 474 [M-92]$^+$ (12), 283. CD (EPA) nm (Δε): 220 (-3.7), 232 (0), 247 (4.7), 261 (0), 285 (-9.0), 323 (0), 345 (3.3) (cryst.).IR υ_{max} cm^{-1}: 2169 *vw* (acetylene). NMR end-groups: 7-6.

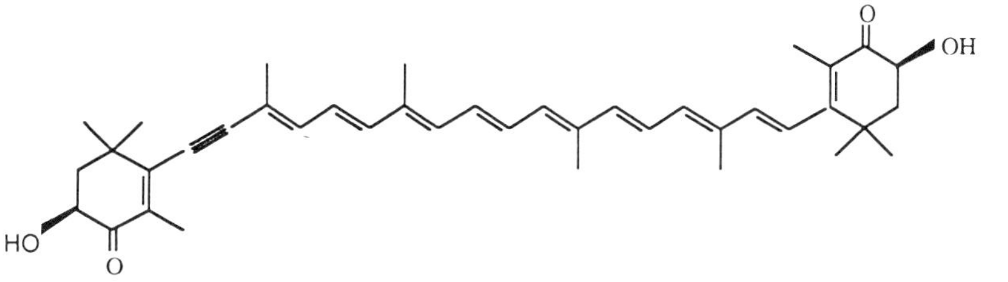

7,8-Didehydroastaxanthin ((3S,3'S)-3,3'-dihydroxy-7,8-didehydro-ß,ß-carotene-4,4'-dione)
$C_{40}H_{50}O_4$. **9-*cis***: m.p. 89–91°C. UV/Vis λ_{max} nm: (390), 487, (515) (benzene). IR υ_{max} cm^{-1}: 2166*m* (acetylene), 1668*s* (C=O, conj. keto). MS m/z (rel. int.): 594 [M]$^+$ (85), 592 [M-2]$^+$ (27), 223 (13), 203 (32), 145 (56), 119 (59), 105 (69), 91 (92), 43 (100). CD nm (Δε): 372 (-10.26), 312 (+7.93), 276 (-21.91), 243 (+5.88).
all-*trans*: m.p. 211–212 °C. IR υ_{max} cm-1: 2160*m* (acetylene), 1664*s* (C=O, conj. keto). CD nm (Δε): 372 (+4.02), 317 (-14.13), 300 (-11.31), 275 (+3.78), 247 (-3.35), 224 (+4.10). MS as 9-*cis*. NMR end-groups:19-18.

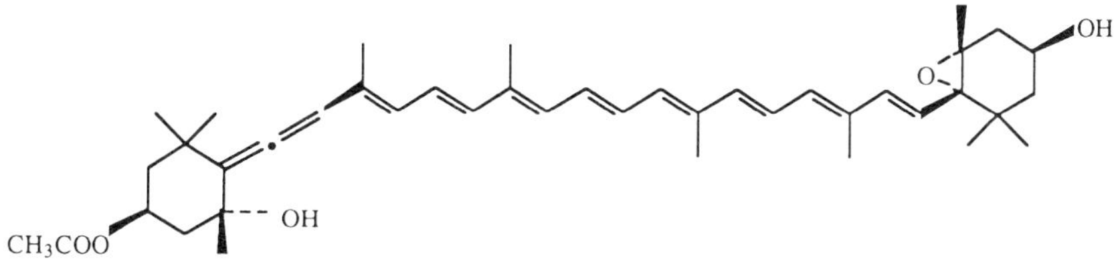

Dinoxanthin ((3*S*,5*R*,6*R*,3'*S*,5'*R*,6'*S*)-5',6'-epoxy-6,7-didehydro-5,6,5',6'-tetrahydro-ß,ß-carotene-3,5,3'-triol 3-acetate) (= neoxanthin 3-acetate])
$C_{42}H_{58}O_5$. UV/Vis as neoxanthin. IR (KBr) υ_{max} cm$^{-1:}$ 1930 (allene). MS *m/z* (rel.

int.): 642 [M]$^+$ (3), 624 [M-18]$^+$ (11), 606 [M-18-18]$^+$(7), 221 (100), 181 (64). NMR end-groups: 12-9.

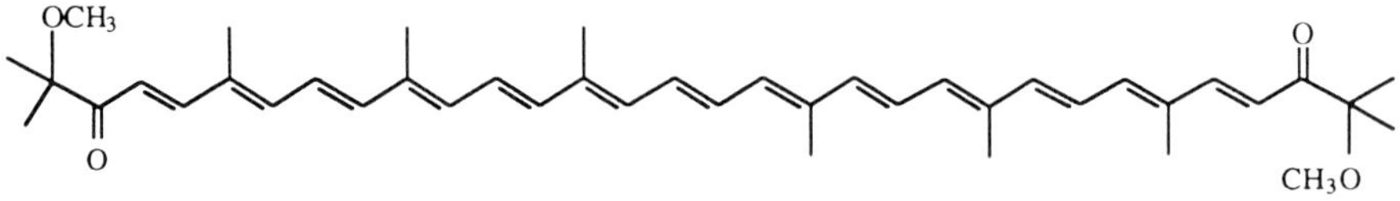

2,2'-Dioxospirilloxanthin (1,1'-dimethoxy-3,4,3',4'-tetradehydro-1,2,1',2'-tetrahydro-ψ,ψ-carotene-2,2'-dione)
$C_{42}H_{56}O_4$ m.p. 225-227°C. UV/Vis λ_{max} nm: 360, 513 ($A^{1\%}_{1cm}$ 2210); 530, 561, 608 (CS_2). MS *m/z*: [M]$^+$, [M-32]$^+$, [M-106]$^+$, [M-106-106]$^+$, 73. NMR end-groups: 40-40.

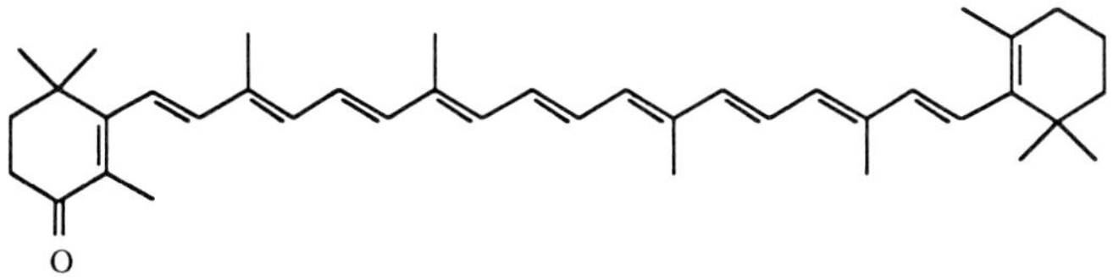

Echinenone (ß,ß-caroten-4-one)
$C_{40}H_{54}O$ UV/Vis λ_{max} nm: 453 (hexane); 456, 482 (petrol; $A^{1\%}_{1cm}$ 2158); 461 (ethanol, shifting to 425, 450, 477 on reduction with $NaBH_4$); 460 (acetone), 471 ($CHCl_3$); 472 (benzene). MS *m/z*: 550 [M]$^+$ (100), 458 [M-92]$^+$ (42). NMR end-groups: 17-1.

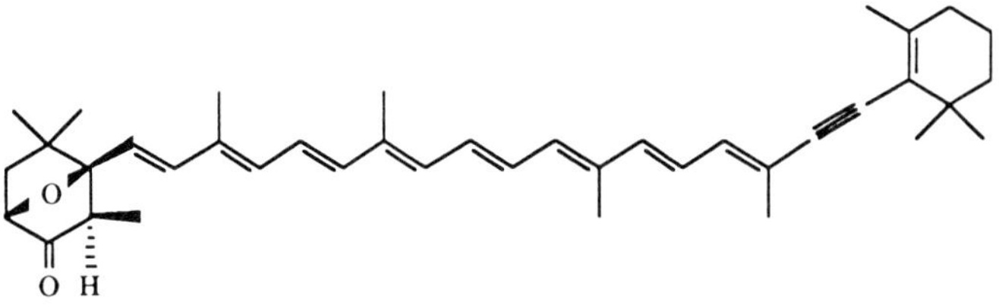

Eutreptiellanone ((3*S*,5*R*,6*S*)-3,6-epoxy-3',4',7',8'-tetradehydro-5,6-dihydro-ß,ß-caroten-4-one)
$C_{40}H_{52}O_2$ UV/Vis λ_{max} nm: (434), 457, 483 (ethanol). MS *m/z* (rel. int.): 562 $[M]^+$ (100), 547 $[M-15]^+$ (6), 506 $[M-56]^+$ (2), 491 $[M-15-56]^+$ (2), 470 $[M-92]^+$ (2), 455 $[M-15-92]^+$ (2), 261 (12). For CD, see Fiksdahl et al. 1984. NMR end-groups: 7-21.

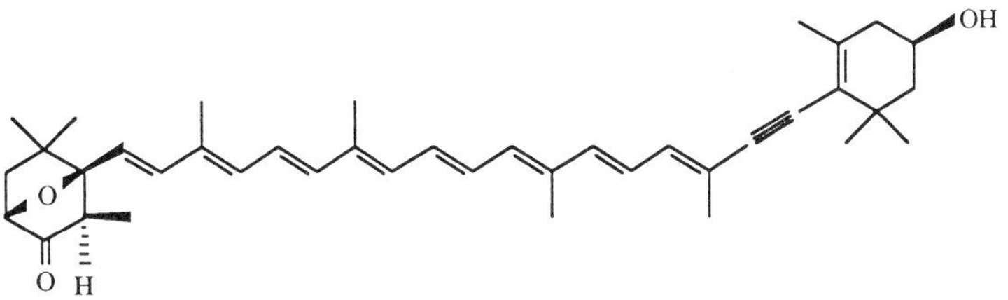

Similar compounds: **ß-Cryptoeutreptiellanone** ((3*S*,5*R*,6*S*,3'*R*)-3'-hydroxy-3,6-epoxy-7',8'-didehydro-5,6-dihydro-ß,ß-caroten-4-one)
$C_{40}H_{56}O_3$. UV/Vis λ_{max} as eutreptiellanone. MS *m/z* (rel. int.): 580 $[M]^+$ (23). CD (EPA) nm ($\Delta\varepsilon$): <246 (neg.), ~246 (0), 273 (-0.9), 300 (0), 328 (1.3), 350 (0). NMR end-groups: 7-21.

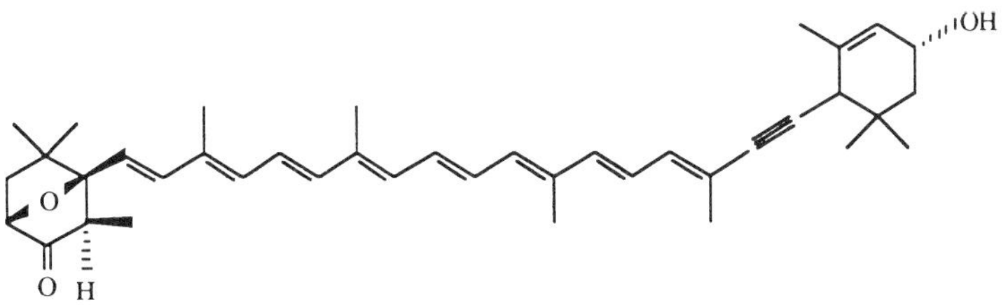

α-Cryptoeutreptiellanone ((3*S*,5*R*,6*S*,3'*R*,6'*R*)-3'-hydroxy-3,6-epoxy-5,6-dihydro-ß,ε-caroten-4-one)
$C_{40}H_{56}O_3$. UV/Vis λ_{max} nm: 414, 438, 468 (petrol); 417, 441, 470 (Me_2CO). MS *m/z* (rel. int.): 582 $[M]^+$ (5), 567 $[M-15]^+$ (1), 564 $[M-18]^+$ (15), 43 (100). CD (MeOH) nm ($\Delta\varepsilon$): 228 (8), 265 (8.5), 302 (0), 322 (-0.8), 337 (0).

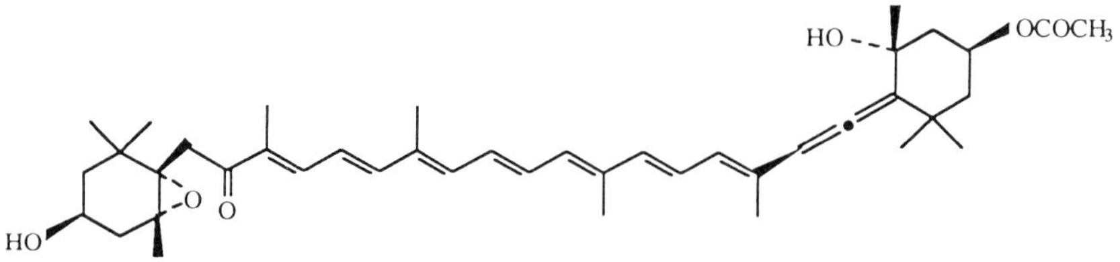

Fucoxanthin ((3*S*,5*R*,6*S*,3'*S*,5'*R*,6'*R*)-5,6-epoxy-3,3',5'-trihydroxy-6',7'-didehydro-5,6,7,8,5',6'-hexahydro-ß,ß-caroten-8-one 3'-acetate)
$C_{42}H_{58}O_6$. Reddish needles from ethanol, m.p. 168-169°C. UV/Vis λ_{max} nm: (420), 447, 469(acetone, $A^{1\%}_{1cm}$ 1650); 460, 478 ($CHCl_3$); 420, 446, 470 (diethyl ether); 426, 449, 475 (ethanol). MS *m/z* (rel. int.): 658 $[M]^+$ (48), 640 $[M-18]^+$ (44), 624 $[M-16-18]^+$ (12), 622 $[M-18-18]^+$ (13), 580 $[M-18-60]^+$ (22), 562 $[M-18-80]^+$ (5), 221 (76), 211 (37), 197 (100). NMR end-groups: 20-12.

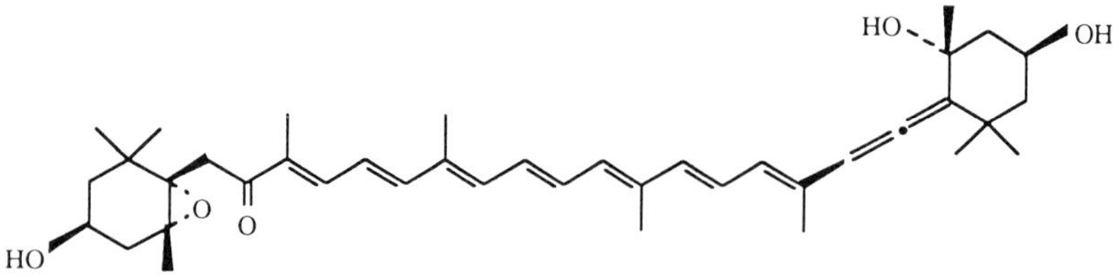

Fucoxanthinol (5,6-epoxy-3,3',5'-trihydroxy-6',7'-didehydro-5,6,7,8,5',6'-hexahydro-ß,ß-caroten-8-one)
$C_{40}H_{56}O_5$. UV/Vis λ_{max} nm: (423), 445, (470). MS *m/z* (rel. int.): 616 $[M]^+$ (5), 598 $[M-18]^+$ (13), 580 $[M-18-18]^+$ (10), 228 (100), 221 (46), 185 (66). NMR end-groups: 20-11.

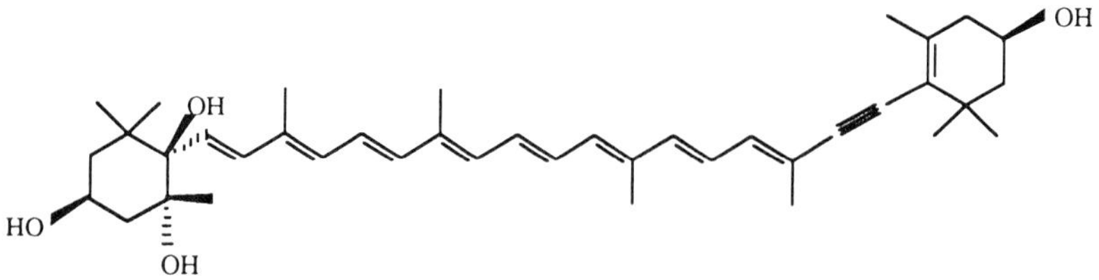

Heteroxanthin ((3*S*,5*S*,6*S*,3'*S*)-7',8'-didehydro-ß,ß-carotene-3,5,6,3'-tetrol) $C_{40}H_{50}O_4$ UV/Vis λ_{max} nm: (418), 442, 470 (ethanol). MS *m/z* (rel. int.): 600 $[M]^+$ (35), 582 $[M-18]^+$ (19), 564 $[M-18-18]^+$ (15), 508 $[M-92]^+$ (17), 502 $[M-18-80]^+$ (8), 256 (100), 239 (65), 221 (49), 181 (75). NMR end-groups: 16-7.

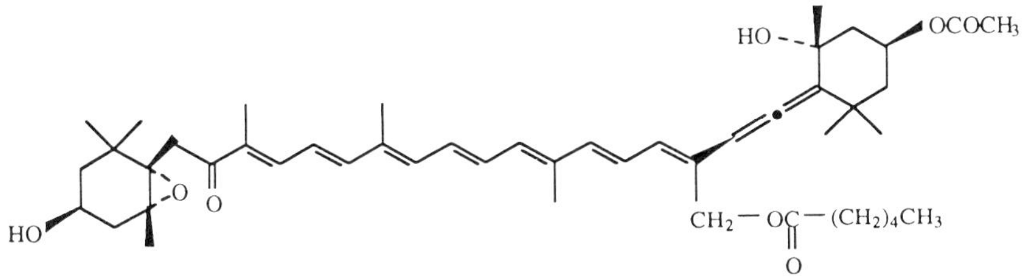

19'-Hexanoyloxyfucoxanthin ((3*S*,5*R*,6*S*,3'*S*,5'*R*,6'*S*)-5,6-epoxy-3,3'5',19'-tetrahydroxy-6',7'-didehydro-5,6,7,8,5',6'-hexahydro-ß,ß-caroten-8-one 3'-acetate 19'-hexanoate)
$C_{40}H_{68}O_8$. UV/Vis λ_{max} nm: (423),444,471 (acetone, $A^{1\%}_{1cm}$ 1500). (423), 445, 474 (hexane); 423, 444, 471. MS *m/z* (rel. int.): 772 $[M]^+$ (1), 770 $[M-2]^+$ (0.3), 754 $[M-18]^+$ (11), 752 $[M-2-18]^+$ (2), 738 $[M-16-18]^+$(4), 736 $[M-18-18]^+$ (0.7), 694 $[M-18-60]^+$ (2), 676 $[M-18-18-60]^+$ (2), 656 $[M-116]^+$ (3), 638 $[M-18-116]^+$ (5), 620 $[M-18-18-116]^+$ (1), 386 (5), 326 (24), 311 (23), 221 (45), 195 (100). IR (KBr) υ_{max} cm^{-1}: 1930*w* (allene). CD (EPA 5:5:2) nm ($\Delta\varepsilon$): 21 (0), 234 (+0.70), 252 (0), 273 (-0.48), 279 (-0.40), 297 (0).

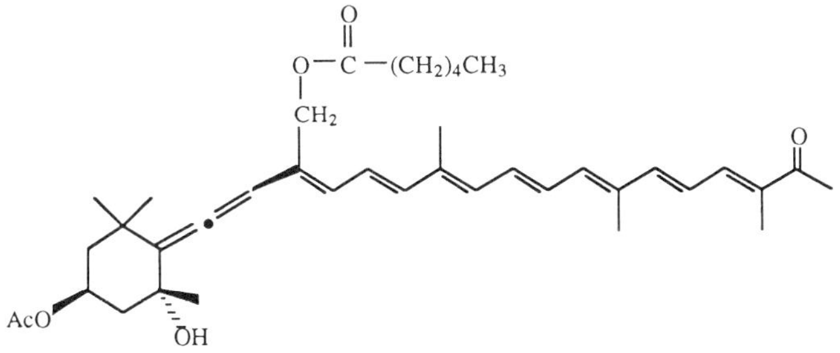

19-Hexanoyloxyparacentrone 3-acetate (3,5,19-trihydroxy-6,7-didehydro-5,6,7',8'-tetrahydro-7'-apo-ß-caroten-8'-one 3-acetate 19-hexanoate)
UV/Vis λ_{max} nm: (427), 447, 474 (hexane); 446, 470 (Me_2CO). MS *m/z* (rel. int.): 618 $[M]^+$ (28), 600 $[M-18]^+$ (51), 502 $[M-116]^+$ (24),326 (33), 284 (54), 232 (100). *Cis* - UV/Vis λ_{max} nm: 327, 442, 463 (hexane). MS *m/z* (rel. int.): 618 $[M]^+$ (14), 600

[M-18]⁺ (36), 586 [M-32]⁺ (1), 558 [M-60]⁺ (4), 540 [M-18-60]⁺ (3), 502 [M-18-60]⁺ (6), 326 (18), 284, 232 (88), 195 (100).

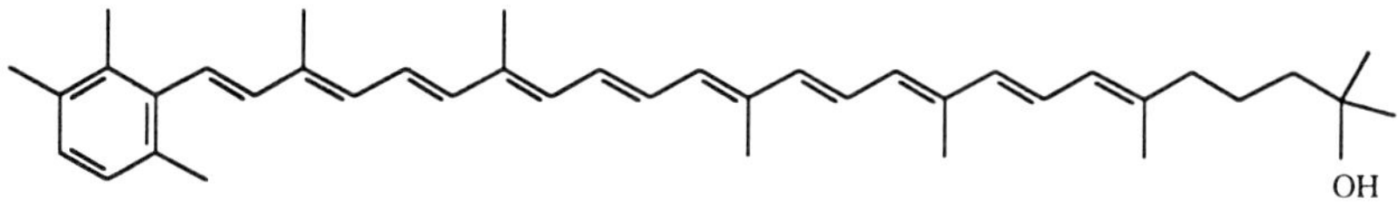

Hydroxychlorobactene (1',2'-dihydro-ϕ,ψ-caroten-1'-ol)
$C_{40}H_{54}O$. m.p. 131–132°C. UV/Vis λ_{max} nm: *trans* - (440), 464, 494 (acetone); (435), 461, 491 (petrol); *neo A* - (436), 460, 489 (acetone); (446), 467, 495 (benzene). NMR end-groups: 30-36.

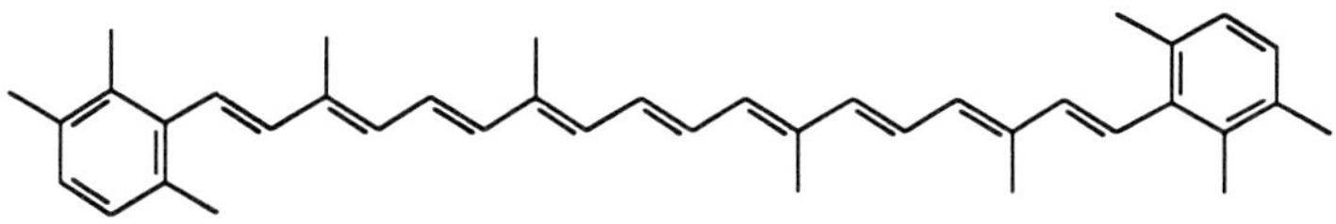

Isorenieratene (= leprotene; ϕ,ϕ-carotene)
$C_{40}H_{48}$. m.p. 191–192°C. UV/Vis λ_{max} nm: (426), 448, 475 (petrol, $A^{1\%}_{1cm}$ 2080); 428, 460, 495 ($CHCl_3$). MS *m/z* (rel. int.): 528 [M]⁺, 436 [M-92]⁺, 422 [M-106]⁺, 133 (100), 105, 91. NMR end-groups: 30-30.

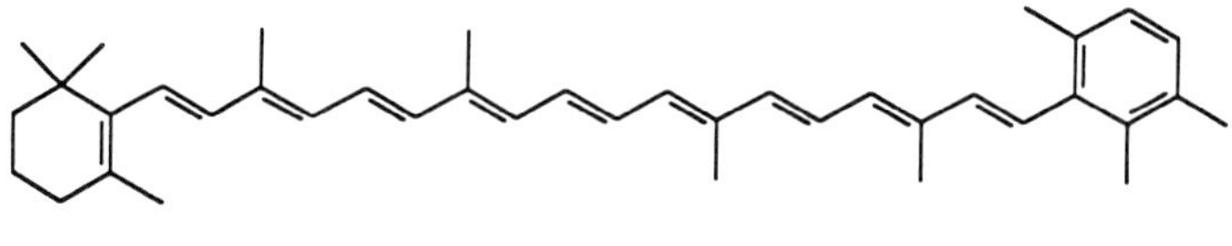

β-Isorenieratene (β,ϕ-carotene)
$C_{40}H_{52}$ Red crystals, m.p. 128°C. UV/Vis λ_{max} nm: 456, 487, 508 (CS_2). MS *m/z* (rel. int.): 532 [M]⁺ (45), 440 [M-92]⁺ (12), 426 [M-106]⁺ (8), 399 [M-133]⁺ (1), 133 (100). NMR end-groups: 1-30.

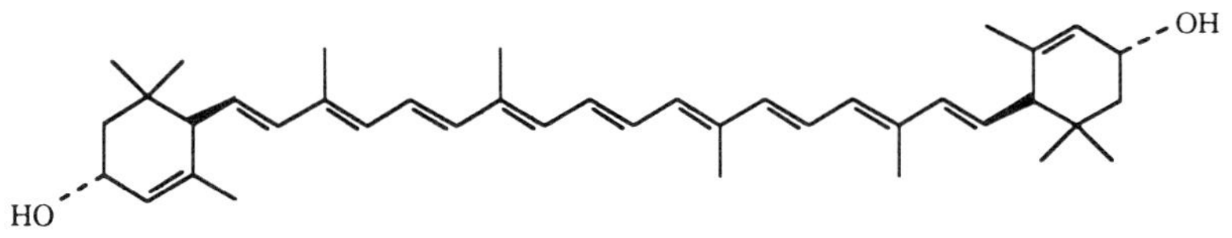

Lactucaxanthin ((3*R*,6*R*,3'*R*,6'*R*)-ε,ε-carotene-3,3'-diol)
$C_{40}H_{56}O_2$. UV/Vis λ_{max} nm: 420, 440, 471 (acetone); 424, 438, 468 (ethanol). CD (EPA) nm (Δε): 215 (+4), 230 (+7), 264 (+17.5), 290 (+1.5), 325 (+2.5), 390 (0). MS *m/z* (rel.int.): 568 $[M]^+$ (25), 550 $[M-18]^+$ (100), 532 $[M-18-18]^+$ (60), 476 $[M-92]^+$ (5), 462 $[M-106]^+$ (1), 458 $[M-18-92]^+$ (32). NMR end-groups: 29-29.

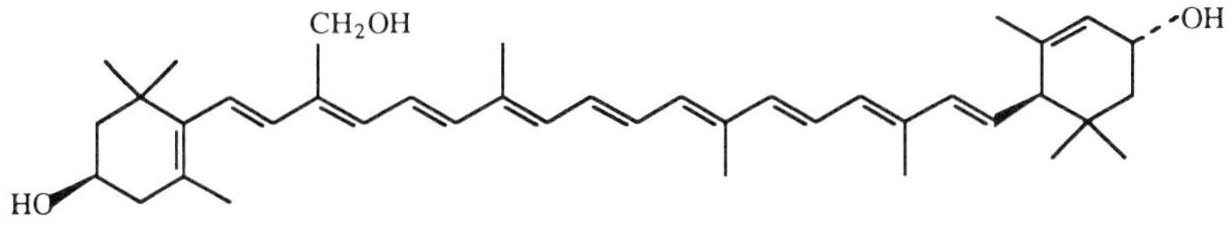

Loroxanthin (ß,ε-carotene-3,19,3'-triol)
$C_{40}H_{56}O_3$. UV/Vis as lutein. CD as lutein. MS *m/z* (rel. int.): 584 $[M]^+$ (40), 566 $[M-18]^+$ (20), 548 $[M-18-18]^+$, 528 $[M-56]^+$, 492 $[M-92]^+$, 478 $[M-106]^+$, 476 $[M-108]^+$, 462 $[M-122]^+$, 446 $[M-138]^+$ 43 (100). Triacetate: 710 $[M]^+$ NMR end-groups: 13-29.

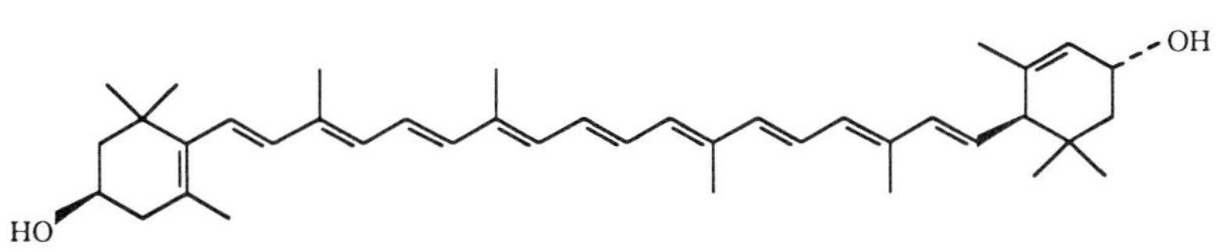

Lutein ((3*R*,3'*R*,6'*R*)-ß,ε-carotene-3,3'-diol)
$C_{40}H_{56}O_2$. Violet prisms from methanol, m.p. 193°C. UV/Vis λ_{max} nm: 421, 445, 474 (petrol); 422, 445, 474 (diethyl ether, ethanol, $A^{1\%}_{1cm}$ 2550); 435, 458, 485 ($CHCl_3$); 433, 458, 487 (benzene, $A^{1\%}_{1cm}$ 2236); 445, 475, 505 (CS_2). CD (hexane) nm (Δε): 212 (+4.2), 242 (+6.8), 280 (-2.8), 327 (+1.1), 392 (-1.0), 447 (+0.8), 468 (+0.5). MS *m/z* (rel. int.) 568 $[M]^+$ (60), 550 $[M-18]^+$ (100), 532 $[M-18-18]^+$ (50), 476 [M-

92]$^+$ (10), 462 [M-106]$^+$ (5). NMR end-groups: 6-29.

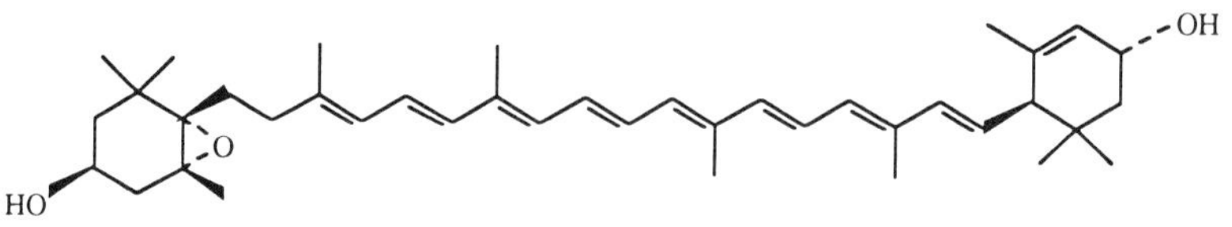

Lutein-5,6-epoxide (= taraxanthin, ((3*S*,5*R*,6*S*,3'*R*,6'*R*)-5,6-epoxy-5,6-dihydro-ß,ε-carotene-3,3'-diol)
$C_{40}H_{56}O_3$. m.p. 192°C (from benzene/methanol). UV/vis λ_{max} nm: 419, 443, 472 (petrol); 420, 442, 471 (ethanol, $A^{1\%}_{1cm}$ 2800); 433, 453, 483 ($CHCl_3$); 429, 455, 485 (benzene, $A^{1\%}_{1cm}$ 2372); 441, 469, 501 (CS_2). CD (dioxane) nm (Δε): 234 (+4.1), 275 (-0.2), 330 (+1.35), 352 (+0.45), 417 (+1.35), 470 (+1.5). MS *m/z* (rel. int.): 584 [M]$^+$ (35), 566 [M-18]$^+$ (12), 548 [M-18-18]$^+$ (2), 504 [M-80]$^+$ (19), 492 [M-92]$^+$ (14), 486 [M-18-80]$^+$ (5), 478 [M-106]$^+$ (3), 474 [M-18-92]$^+$(3), 446 [M-138]$^+$(2), 221 (39), 181 (21), 91 (100). Data for the (all-*E*)-, (9*Z*)-, (13*Z*)-, (9'Z)- and (13'Z)-isomers are given by Deli et al. 1988. NMR end-groups: 9-29.

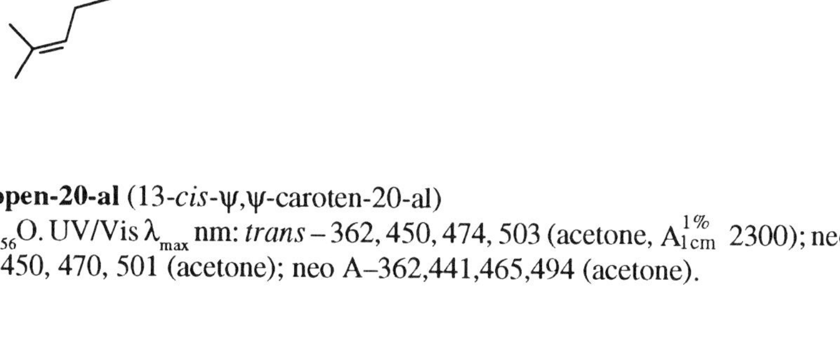

Lycopen-20-al (13-*cis*-ψ,ψ-caroten-20-al)
$C_{40}H_{56}O$. UV/Vis λ_{max} nm: *trans* – 362, 450, 474, 503 (acetone, $A^{1\%}_{1cm}$ 2300); neo B–362, 450, 470, 501 (acetone); neo A–362,441,465,494 (acetone).

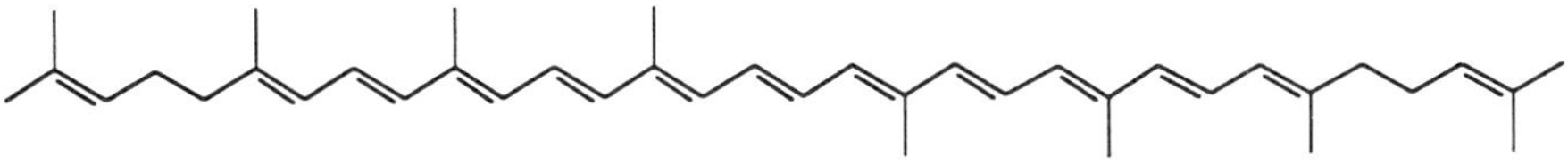

Lycopene (ψ,ψ-carotene)
$C_{40}H_{56}$. Red-violet needles or prisms from petrol or hexane/diethyl ether. m.p. 170°C. UV/Vis λ_{max} nm: 448, 473, 504 (hexane); 446, 472, 505 (petrol, $A^{1\%}_{1cm}$ 3450); 446, 472, 503 (ethanol); 448, 474, 505 (acetone); 458, 484, 518 ($CHCl_3$); 455, 487, 522 (benzene, $A^{1\%}_{1cm}$ 3370); 477, 508, 548 (CS_2). No optical activity. MS *m/z* (rel. int.): 536 $[M]^+$ (70), 467 $[M\text{-}69]^+$ (22), 444 $[M\text{-}92]^+$ (12), 430 $[M\text{-}106]^+$ (28), 69 (100). NMR end-groups: 32-32.

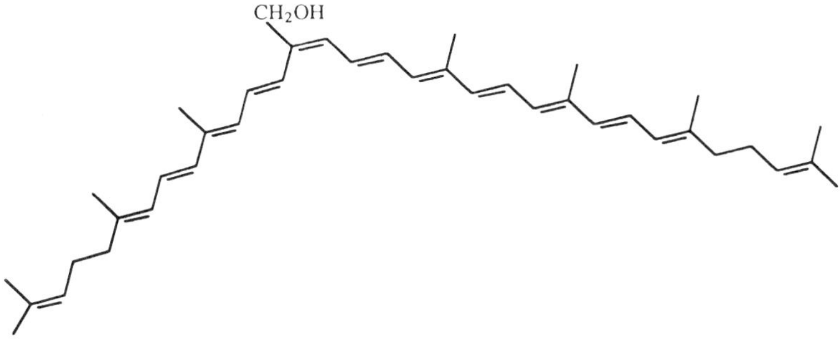

Lycopen-20-ol (13-*cis*-ψ,ψ-caroten-20-ol)
$C_{40}H_{56}O$. UV/Vis λ_{max} nm: *trans* – 345, 363, 450, 471, 500 (acetone); *neo U* –345, 362, 442, 466, 496 (acetone); *neo V* –345, 363, (446), 467, 495 (acetone). MS *m/z*: 552 $[M]^+$, 536 $[M\text{-}16]^+$, 534 $[M\text{-}18]^+$, 483 $[M\text{-}69]^+$, 460 $[M\text{-}92]^+$, 446 $[M\text{-}106]^+$, 444 $[M\text{-}108]^+$, 430 $[M\text{-}122]^+$, 394 $[M\text{-}158]^+$, 378 $[M\text{-}174]^+$.

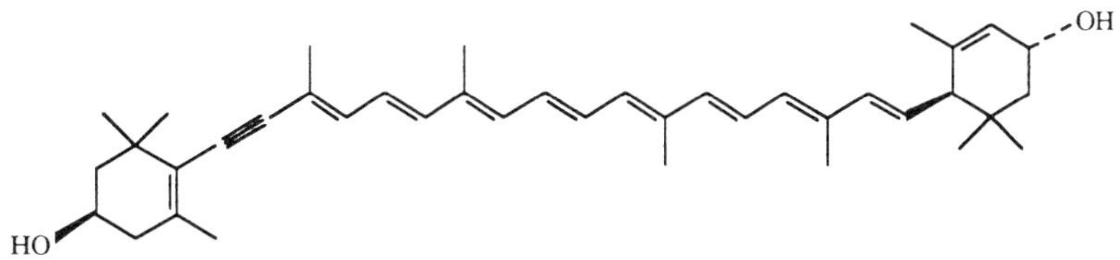

Monadoxanthin ((3*R*,3'*R*,6'*R*)-7,8-didehydro-ß,ε-carotene-3,3'-diol)
$C_{40}H_{54}O_2$. Red needles m.p. 163–165°C. UV/Vis λ_{max} nm: 456, 487 (benzene); 424, 447, 477 (ethanol); 422, 445, 475 (hexane). IR υ_{max} cm^{-1} (KBr): 2167*vw*. MS *m/z* (rel. int.): 566 $[M]^+$ (100), 548 $[M-18]^+$ (17), 533 $[M-18-15]^+$ (2), 530 $[M-18-18]^+$ (<1), 474 $[M-92]^+$ (4), 460 $[M-106]^+$ (2). NMR end-groups: 7-29.

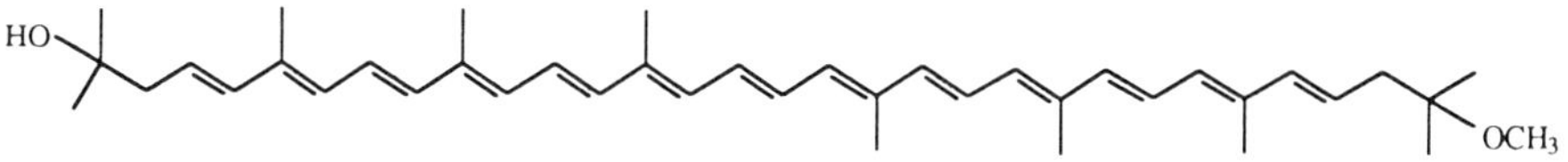

Monodemethylspirilloxanthin (1'-methoxy-3,4,3',4'-tetradehydro-1,2,1',2'-tetrahydro-ψ,ψ-caroten-1-ol)
$C_{41}H_{58}O_2$. UV/Vis as spirilloxanthin. MS *m/z* (rel. int.): 582. $[M]^+$ (6), 564 $[M-18]^+$ (1), 550 $[M-32]^+$ (1), 532 $[M-32-18]^+$ (0.5), 509 $[M-73]^+$ (1), 490 $[M-92]^+$ (0.5), 476 $[M-106]^+$ (4), 444 $[M-106-32]^+$ (1), 73 (100). NMR end-groups:38-39.

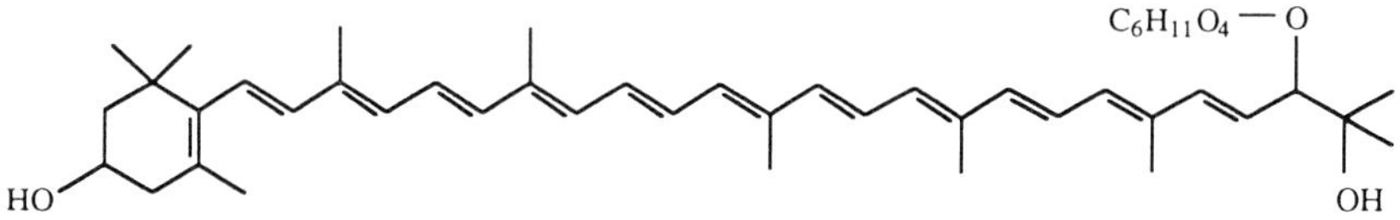

Myxoxanthophyll (2'-(ß-*L*-rhamnopyranosyloxy)-3',4';-didehydro-1',2'-dihydro-ß,ψ-carotene-3,1'-diol)
$C_{46}H_{66}O_7$. Red needles from acetone/petrol. m.p. 173°C. UV/Vis λ_{max} nm: (445), 471, 502 (methanol); (450), 478, 510 (acetone $A^{1\%}_{1cm}$ 2160); (460), 488, 522 ($CHCl_3$); 462, 488, 522 (benzene); (465), 493, 528 (pyridine). MS *m/z*: 730 $[M]^+$, 712 $[M-18]^+$, 694 $[M-36]^+$, 672 $[M-58]^+$, 638 $[M-92]^+$, 608 $[M-122]^+$, 584 $[M-146]^+$.
Tetra-acetate: MS *m/z* (rel. int.): 898 $[M]^+$ (20), 856 $[M-42]^+$ (2), 840 $[M-58]^+$ (3), 838 $[M-60]^+$ (2), 906 $[M-92]^+$ (3), 792 $[M-106]^+$ (10), 331 (100), 273 (60).

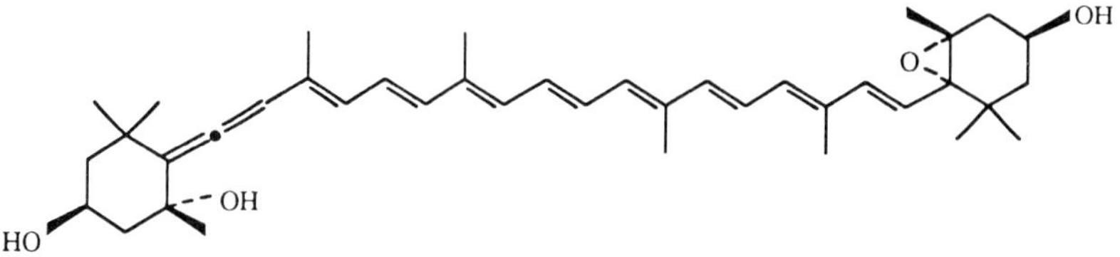

Neoxanthin ((3*S*,5*R*,6*R*,3'*S*,5'*R*,6'*S*)-5',6'-epoxy-6,7-didehydro-5,6,5',6'-tetrahydro-ß,ß-carotene-3,5,3'-triol)
$C_{40}H_{56}O_4$. m.p. 134°C. UV/Vis λ_{max} nm: 414, 436, 468 (diethyl ether); 415, 435, 462 (hexane); 415, 438, 467 (ethanol, $A^{1\%}_{1cm}$ 2243) (401, 424, 451, ethanol + HCl → neochrome); 423, 448, 476 ($CHCl_3$); 426, 453, 483 (benzene); 463, 493 (CS_2). CD (ethanol) nm (Δε): 222 (-2.0), 267 (-5.9), 299 (-0.5), 311 (-0.5), 328 (-0.1), 393 (+0.6), 420 (+1.3), 444 (+1.8), 475 (+2.0). MS *m/z* (rel. int.): 600 $[M]^+$ (50), 582 $[M-18]^+$ (10), 520 $[M-80]^+$(5), 5(3*S*,5*R*,6*S*,3'*S*,5'*R*,6'*R*)8 $[M-92]^+$ (3), 221 (5), 91 (100). IR υ_{max} cm^{-1}: 1931 (allene). NMR end-groups: 11-9.

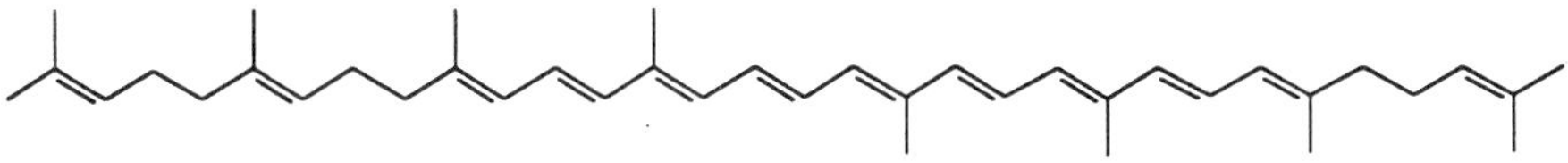

Neurosporene (7,8-dihydro-ψ,ψ-carotene)
$C_{40}H_{58}$. UV/Vis λ_{max} nm: 416, 440, 470 (hexane $A^{1\%}_{1cm}$ 2918); 414, 439, 467 (petrol); 416, 441, 470 (ethanol); 424, 451, 480 ($CHCl_3$). No optical activity. MS *m/z* (rel. int.): 538 $[M]^+$ (25), 469 $[M-69]^+$ (6), 446 $[M-92]^+$ (3), 401 $[M-137]^+$ (8), 69 (100). NMR end-groups 33-32.

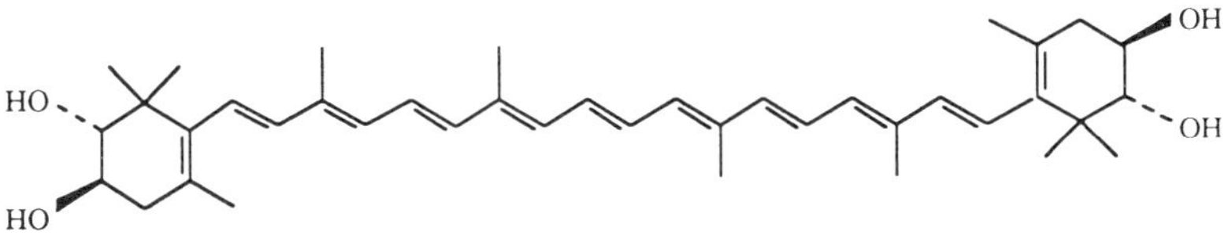

Nostoxanthin ((2*R*,3*R*,2'*R*,3'*R*)-ß,ß-carotene-2,3,2',3'-tetrol)
$C_{40}H_{56}O_4$. UV/Vis as zeaxanthin. MS *m/z* (rel.int.): 600 $[M]^+$ (100), 582 $[M-18]^+$ (12), 564 $[M-18-18]^+$ (1), 508 $[M-92]^+$ (11), 494 $[M-106]^+$ (2). CD (EPA) nm (Δε): 340 (+3.5), 313 (0), 282 (-16.7), 261 (0), 245 (+8.8), 232 (0), 220 (-11.9), 208 (0). NMR end-groups: 10-10.

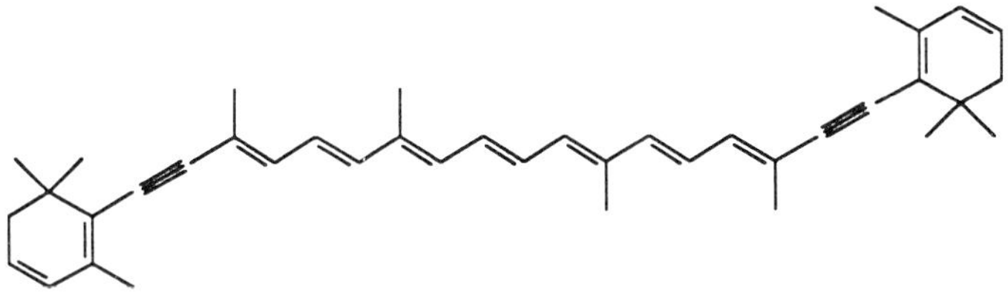

Octadehydro-ß,ß-carotene (3,4,7,8,3',4',7',8'-octadehydro-ß,ß-carotene) $C_{40}H_{48}$. UV/Vis λ_{max} nm: 9,9'-*dicis*; 327, (445), 472, 501 (hexane/ethanol); 9-*cis*; 380, (438), 466, 492 (hexane/ethanol); all-*trans*; 329, (382), 434, 458, 486 (hexane/ethanol). MS *m/z* (rel. int.): 528 [M]$^+$ (100), 513 [M-15]$^+$ (9), 508 (9), 480 (6), 410 (41), 302 (35), 247 (67). NMR end-groups: 2-2.

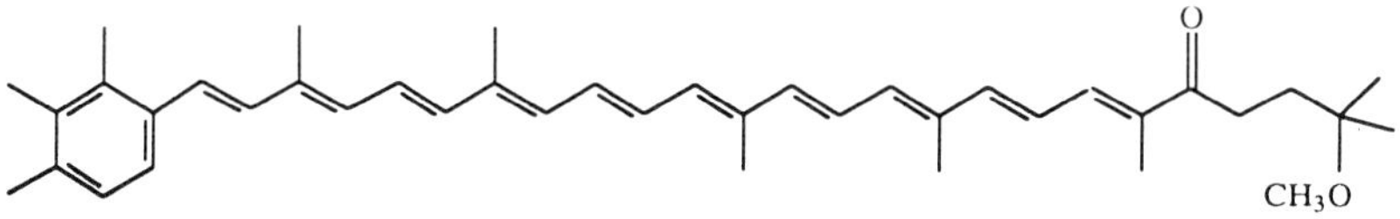

Okenone (1'-methoxy-1',2'-dihydro-χ,ψ-caroten-4'-one) $C_{41}H_{54}O_2$. Violet needles from acetone/CS_2. m.p. 161–163°C. UV/Vis λ_{max} nm: (360), 375, (460), 484, 516 (petrol, $A^{1\%}_{1cm}$ 2320); (360), 381, (465), 487, 518 (acetone). MS *m/z* (rel. int.): 578 [M]$^+$ (15), 546 [M-32]$^+$ (18), 486 [M-92]$^+$ (2), 472 [M-106]$^+$ (8), 440 [M-32-106]$^+$ (6), 133 (100). NMR end-groups: 31-41.

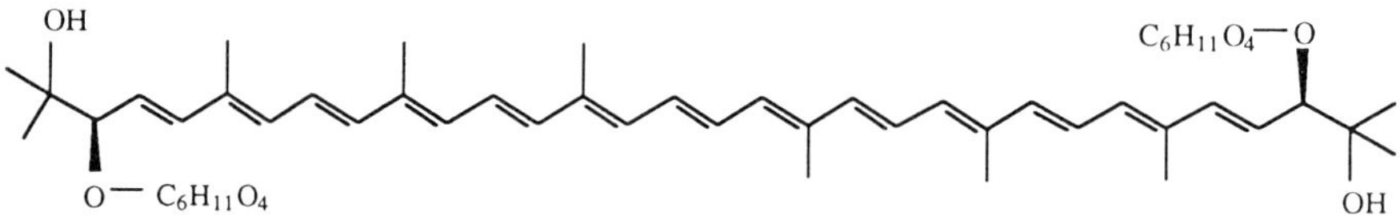

Oscillaxanthin ((2*R*,2'*R*)-2,2'-bis(ß-*L*-rhamnopyranosyloxy)-3,4,3',4'-tetradehydro-1,2,1',2'-tetrahydro-ψ,ψ-carotene-1,1'-diol) $C_{52}H_{76}O_{12}$. Red-black needles from pyridine-petrol. UV/Vis λ_{max} nm (as hexaacetate): 390, 470, 499, 534 (acetone); 397, 480, 510, 548 ($CHCl_3$); 465, 492, 527 (methanol, ethanol). MS (hexaacetate) *m/z*: 1144 [M]$^+$, 1068 [M-76]$^+$, 1052 [M-92]$^+$, 1038 [M-106]$^+$, 996 [M-106-42]$^+$, 980 [M-106-58]$^+$, 962 [M-76-106]$^+$, 946 [M-106-92]$^+$, 932 [M-106-106]$^+$, 872 [M-272]$^+$, 856 [M-288]$^+$, 838 [M-18-288]$^+$, 812 [M-290-42]$^+$, 766 [M-106-272]$^+$, 750 [M-106-288]$^+$, 749 [M-106-289]$^+$, 748 [M-106-290]$^+$.

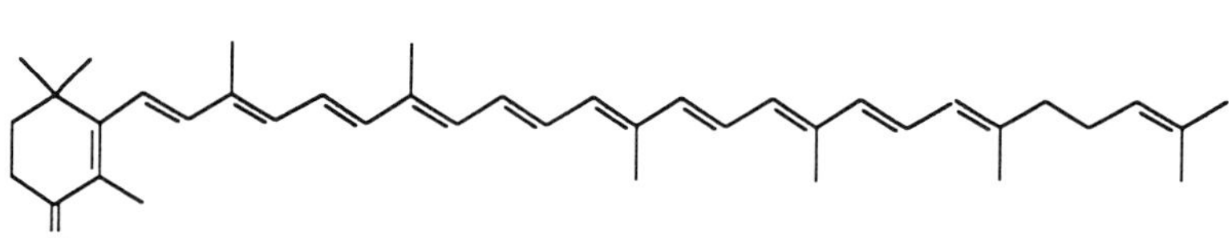

4-Oxo-γ-carotene (ß,ψ-caroten-4-one)
$C_{40}H_{54}O$. Red needles, m.p. 146°C. UV/Vis λ_{max} nm: 465 (490) (petrol);471,(490) (acetone $A^{1\%}_{1cm}$ 2380). IR (KBr) υ_{max} cm^{-1}: 1660 (conj. carbonyl). NMR end-groups: 17-32.

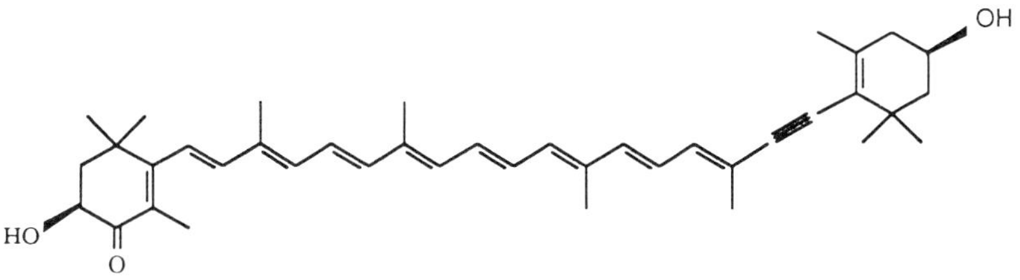

Pectenolone ((3*S*,3'*R*)-3,3'-dihydroxy-7'8'-didehydro-ß,ß-caroten-4-one)
$C_{40}H_{52}O_3$. Reddish needles, m.p. 162-163°C. UV/Vis λ_{max} nm: 459–462 (diethyl ether). MS *m/z*: 580 [M]$^+$, 562 [M-18]$^+$, 544 [M-18-18]$^+$, 488 [M-92]$^+$. IR (KBr) υ_{max} cm^{-1}: 2165 (acetylene). NMR end-groups: 18-7.

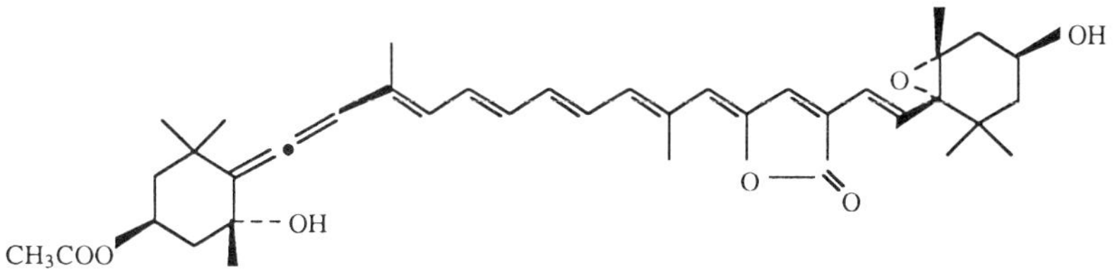

Peridinin ((3*S*,5*R*,6*R*,3'*S*,5'*R*,6'*S*)-5',6'-epoxy-3,5,3'-trihydroxy-6,7-didehydro-5,6,5',6'-tetrahydro-10,11,20-trinor-ß,ß-caroten-19',11'-olide 3-acetate)
$C_{39}H_{50}O_7$. Purple crystals from ether/hexane, m.p. 128–130°C. UV/Vis λ_{max} nm: 454, 484 (hexane); 466 (acetone); 475 (ethanol, $A^{1\%}_{1cm}$ 1350). IR (KBr) υ_{max} cm^{-1}: 1930 (allene). MS *m/z*: 630 [M]$^+$, 612 [M-18]$^+$, 594 [M-18-18]$^+$, 586 [M-44]$^+$, 570 [M-60]$^+$, 568 [M-18-44]$^+$ 552 [M-18-60]$^+$, 538 [M-92]$^+$, 534 [M-18-18-60]$^+$, 520 [M-18-92]$^+$, 508 [M-18-60-44]$^+$, 478 [M-92-60]$^+$. NMR end-groups: 12 (C-1 to C-10)-23.

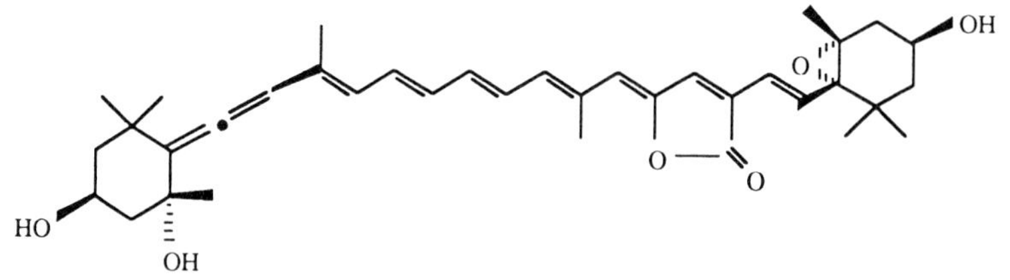

Peridininol (5',6'-epoxy-3,5,3'-trihydroxy-6,7-didehydro-5,6,5',6'-tetrahydro-10,11,20-trinor-ß,ß-caroten-19',11'-olide)
$C_{37}H_{48}O_6$. UV/Vis λ_{max} nm: 466 (acetone); 464 (methanol), 467.5, 493.5 (benzene). MS *m/z* (rel. int.): 588 $[M]^+$ (6.2), 570 $[M-18]^+$ (32.3), 552 $[M-18-18]^+$ (19.1), 544 $[M-44]^+$ (2.9), 526 $[M-62]^+$ (4.1), 508 $[M-80]^+$ (3.5), 496 $[M-92]^+$ (7.4), 479 $[M-110]^+$ (4.4), 181 (100). NMR end-groups: 11-23.

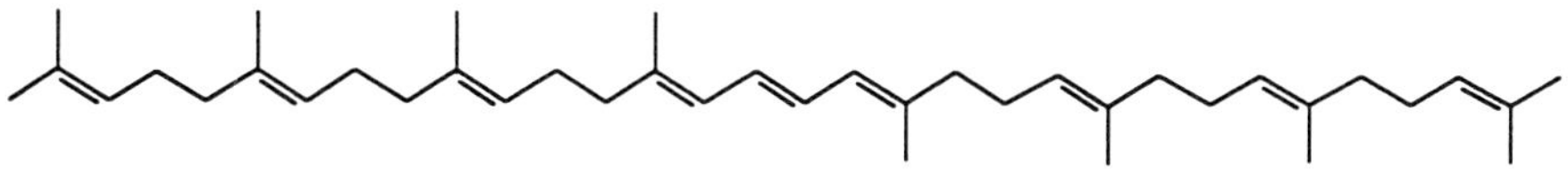

Phytoene (7,8,11,12,7',8',11',12'-octahydro-ψ,ψ-carotene)
$C_{40}H_{64}$. Colourless oil. UV/Vis λ_{max} nm: 276, 286, 297 (hexane, $A^{1\%}_{1cm}$ 757 (15*Z*), 915 (all-*E*)); 275, 285, 296 (15*Z*, petrol); 276, 286, 297 (all-*E*, petrol); 280, 291, 303 ($CHCl_3$). IR υ_{max} cm^{-1}: 767 (15*Z*), 950–990 (all-*E*). MS *m/z* (rel. int.): 544 $[M]^+$ (94), 450 $[M-94]^+$ (8), 339 $[M-205]^+$ (100), 69 (60). NMR end-groups: 35-35 (all-*E* and 15*Z*).

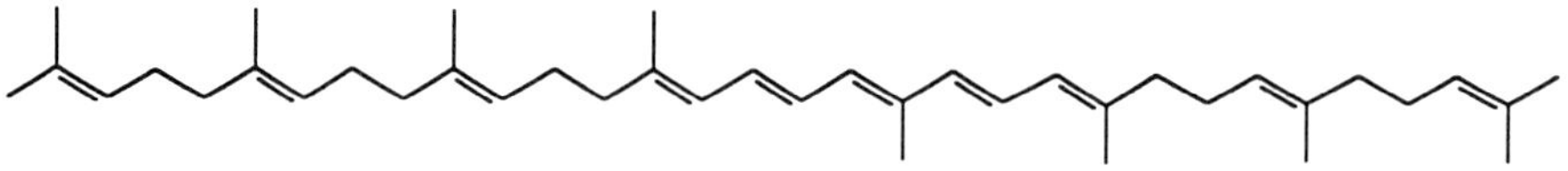

Phytofluene (7,8,11,12,7',8'-hexahydro-ψ,ψ-carotene)
$C_{40}H_{62}$. Colourless oil. UV/Vis λ_{max} nm: 331, 347, 366 (hexane, $A^{1\%}_{1cm}$ 1577); 331, 347, 367 (petrol); 332,348,366 (ethanol); 337, 354, 374 ($CHCl_3$). MS *m/z* (rel. int.): 542 $[M]^+$ (15), 405 $[M-137]^+$ (6), 337 $[M-205]^+$ (7), 69 (100). NMR end-groups 35-33.

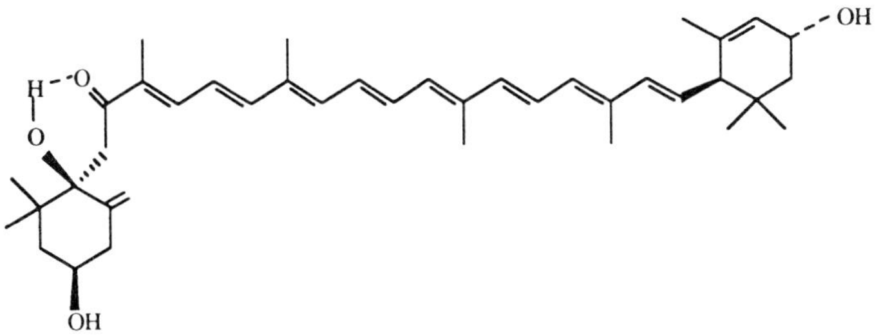

Prasinoxanthin ((3'*R*,6'*R*)-3,6,3'-trihydroxy-7,8-dihydro-γ,ε-caroten-8-one) $C_{40}H_{56}O_4$. UV/Vis λ_{max} nm: 446, (466) (ethanol). MS *m/z* (rel. int.): 600 $[M]^+$ (1), 564 $[M\text{-}18\text{-}18]^+$ (1), 446.3186 $[M\text{-}154]^+$, 428 $[M\text{-}154\text{-}18]^+$ (30). CD (EPA) nm. (Δε): 220 (+8), 230 (+2), 235 (+5), 245 (+4), 270 (+10). NMR end-groups: 27-29.

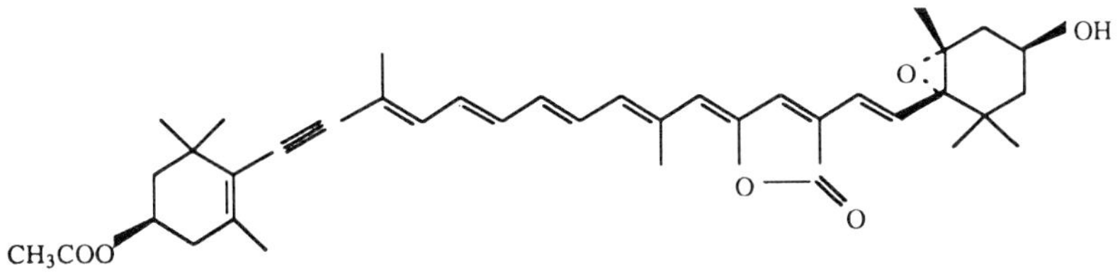

Pyrrhoxanthin (5',6'-epoxy-3,3'-dihydroxy-7,8-didehydro-5',6'-dihydro-10,11,20-trinor-ß,ß-caroten-19',11'-olide 3-acetate) $C_{39}H_{48}O_6$. UV/Vis λ_{max} nm: 457, 485 (petrol, $A^{1\%}_{1cm}$ 1260); 471 (ethanol); 475.4 (acetone); 512, 586 (CS_2); 459, 487 (hexane). IR (KBr) υ_{max} cm^{-1}: 2180 (acetylene). MS *m/z* (rel.int.): 612 $[M]^+$ (30), 610 $[M\text{-}2]^+$ (18), 594 $[M\text{-}18]^+$ (17), 592 $[M\text{-}2\text{-}18]^+$ (11), 576 $[M\text{-}18\text{-}18]^+$ (3), 568 $[M\text{-}44]^+$, 552 $[M\text{-}60]^+$ (19), 550 $[M\text{-}2\text{-}60; M\text{-}18\text{-}44]^+$ (16), 534 $[M\text{-}18\text{-}60]^+$ (22), 234 (98), 181 (100). NMR end-groups: 8 (C-1 to C-10)-23.

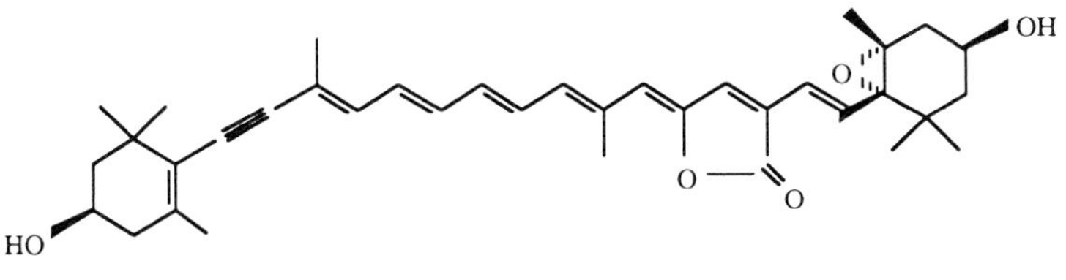

Pyrrhoxanthinol (5',6'-epoxy-3,3'-dihydroxy-7,8-didehydro-5',6'-dihydro-10,11,20-trinor-ß,ß-caroten-19',11'-olide)
$C_{37}H_{46}O_5$. UV/Vis λ_{max} . nm: 457.5 (acetone); 456.5 (484) (hexane); 459 (methanol). MS *m/z* (rel. int.): 570.3531 $[M]^+$ (11.6), 478 $[M-92]^+$ (7.9), 181 (100). NMR end-groups: 7 (C-1 to C-10)-23.

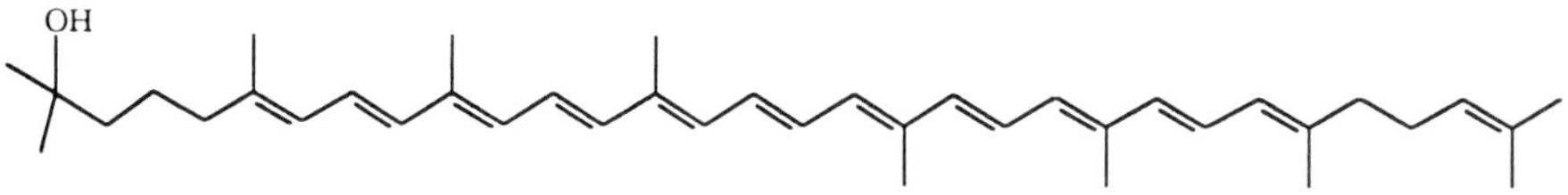

Rhodopin (1,2-dihydro-ψ,ψ-caroten-1-ol)
$C_{40}H_{58}O$. UV/Vis as lycopene. MS *m/z* (rel. int.): 554 $[M]^+$ (12), 536 $[M-18]^+$ (6), 485 $[M-69]^+$ (4), 467 $[M-87]^+$ (2), 462 $[M-92]^+$ (1), 448 $[M-106]^+$ (3), 430 $[M-106-18]^+$ (1), 69 (100). NMR end-groups: 36-32.

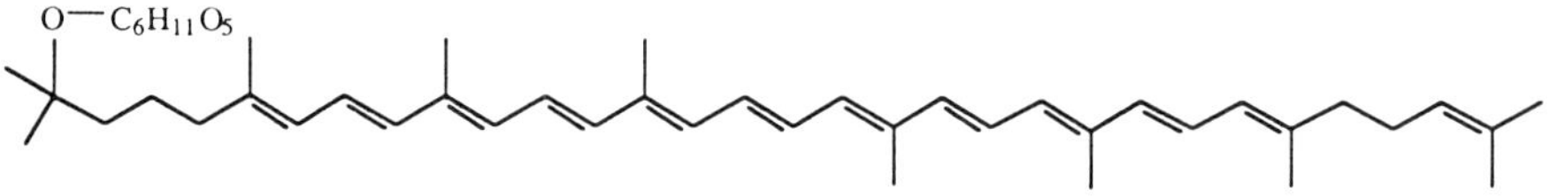

Rhodopin glucoside (1-(ß-*D*-glucopyranosyloxy)-1,2-dihydro-ψ,ψ-carotene)
$C_{46}H_{68}O_6$. m.p. 170–175°C. UV/Vis as rhodopin. MS (acetylated) *m/z*: 884 $[M]^+$.

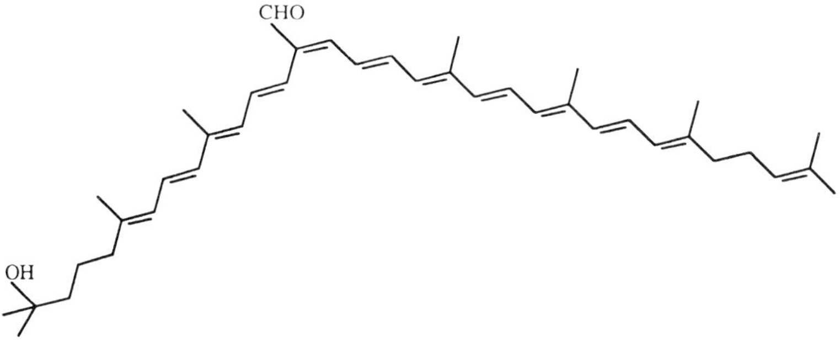

Rhodopin-20-al (13-*cis*-1-hydroxy-1,2-dihydro-ψ,ψ-caroten-20-al)
$C_{40}H_{56}O_2$. m.p. 195°C. UV/Vis λ_{max} nm: 365,505 (acetone, $A^{1\%}_{1cm}$ 2300); (358), 363,

398,495 (petrol).

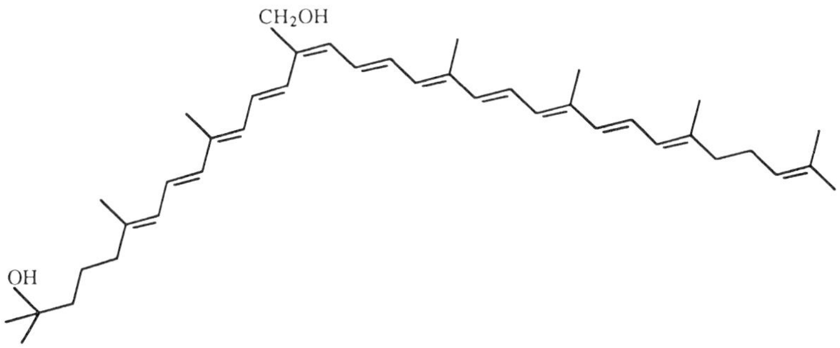

Rhodopin-20-ol (13-*cis*-1,2-dihydro-ψ,ψ-caroten-1,20-diol)
$C_{40}H_{58}O_2$. m.p. 182°C. UV/Vis λ_{max} nm: *trans* –367, 448, 472, 503 (acetone); *neo B* – 345, 362, (440), 466, 495 (acetone). MS *m/z*: 570 [M]$^+$, 568 [M-2]$^+$, 554 [M-16]$^+$, 552 [M-18]$^+$, 478 [M-92]$^+$, 464 [M-106]$^+$, 462 [M-108]$^+$, 412 [M-158]$^+$, 396 [M-174]$^+$.

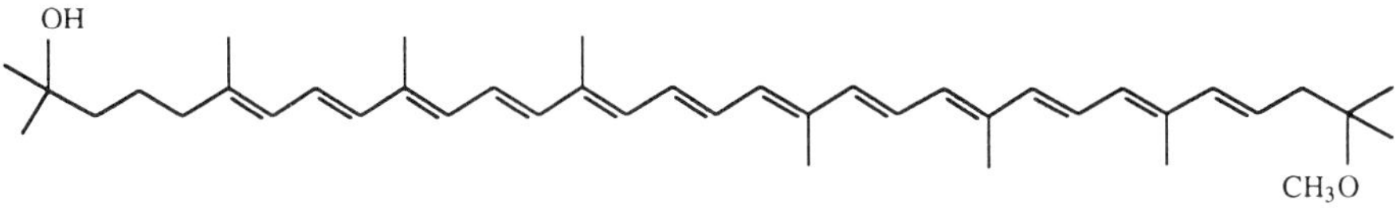

Rhodovibrin (1'-methoxy-3',4'-didehydro-1,2,1',2'-tetrahydro-ψ,ψ-caroten-1-ol)

$C_{41}H_{60}O_2$. UV/Vis λ_{max} nm: 455, 483, 516 (petrol); 460, 488, 522 (acetone); 469, 498, 532 ($CHCl_3$); 473, 503, 535 (benzene); 491, 522, 559 (CS_2). MS *m/z* (rel. int.): 584 [M]$^+$ (8), 566 [M-18]$^+$ (10), 552 [M-32]$^+$ (3), 511 [M-73]$^+$ (1), 492 [M-92]$^+$ (0.5), 478 [M-106]$^+$ (4), 460 [M-18-106]$^+$ (2), 73 (100). NMR end-groups: 36-39.

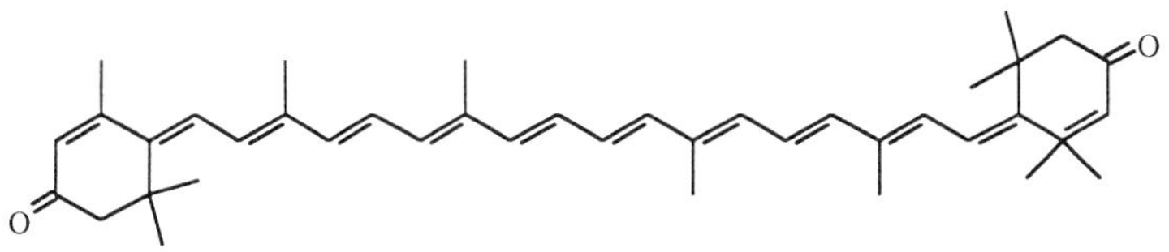

Rhodoxanthin (4',5'-didehydro-4,5'-*retro*-ß,ß-carotene-3,3'-dione)
$C_{40}H_{50}O_2$. Dark violet crystals from benzene/methanol, m.p. 219°C. UV/Vis λ_{max} nm: 456, 487, 521 (petrol); 458, 490, 524 (hexane $A^{1\%}_{1cm}$ 2500); 496, 538 (ethanol, on reduction with $NaBH_4$ 446, 472, 503); 474, 503.5,542 (benzene); 482, 510, 546 ($CHCl_3$); 491, 525, 564 (CS_2). MS *m/z* (rel. int.): 562 $[M]^+$ (100), 547 $[M-15]^+$ (5), 506 $[M-56]^+$ (1), 470 $[M-92]^+$ (24), 456 $[M-106]^+$(43), 91 (100). NMR end-groups 25-25 (6,7-*trans*), 29-29 (6,7-*cis*)

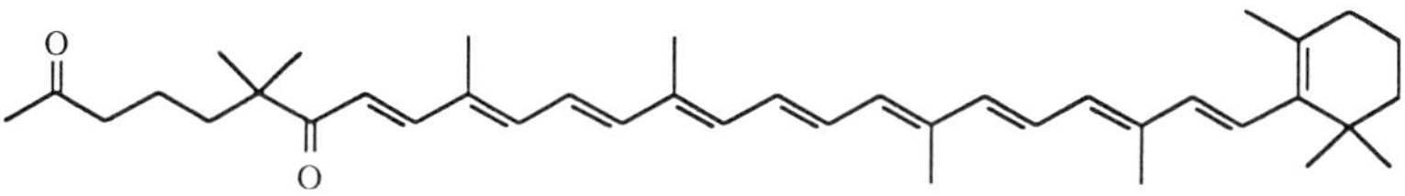

Semi-ß-carotenone (5,6-seco-ß,ß-carotene-5,6-dione)
$C_{40}H_{56}O_2$. Red plates from methanol, m.p. 118–119°C. UV/Vis λ_{max} nm: 440, 464, 495 (hexane $A^{1\%}_{1cm}$ 1850); 473 (ethanol); 480 ($CHCl_3$); 458, 486, 518 (benzene); 500, 525 (CS_2). MS *m/z* (rel.int.): 568 $[M]^+$ (10), 476 $[M-92]^+$ (1), 462 $[M-106]^+$ (6), 413 $[M-155]^+$(1), 69 (100). NMR end-groups: 24-1.

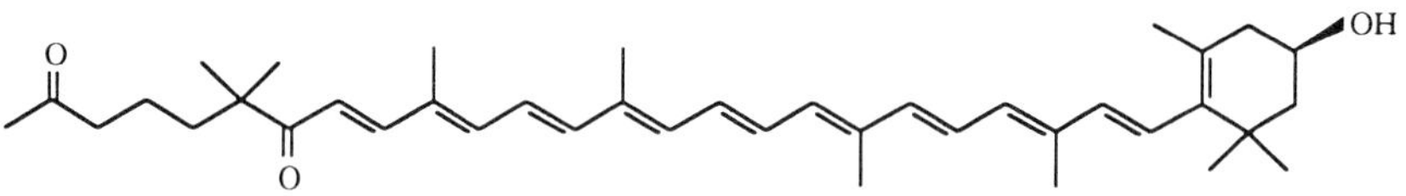

Similar compound: **Triphasiaxanthin** (3'-hydroxy-5,6-seco-ß,ß-carotene-5,6-dione)
$C_{40}H_{56}O_3$. UV/Vis as semi-ß-carotenone. MS *m/z* (rel.int.): 584 $[M]^+$ (20). NMR end groups 24-6.

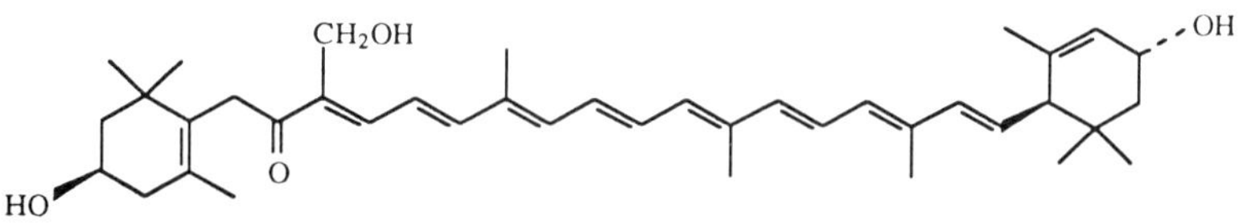

Siphonaxanthin ((3*R*,3'*R*,6'*R*)3,19,3'-trihydroxy-7,8-dihydro-ß,ε-caroten-8-one)
$C_{40}H_{56}O_4$. UV/Vis λ_{max} nm: (427), 450, 478 (petrol); 445–451 (ethanol,); 466 ($CHCl_3$); (375), 398, 421, 448 (reduced with $NaBH_4$, ethanol). MS *m/z*: 600[M]$^+$, 582 [M-18]$^+$, 564 [M-18-18]$^+$, 546 [M-54]$^+$, 429 [M-161]$^+$, 411 [M-189]$^+$. NMR end-groups: 15-29.

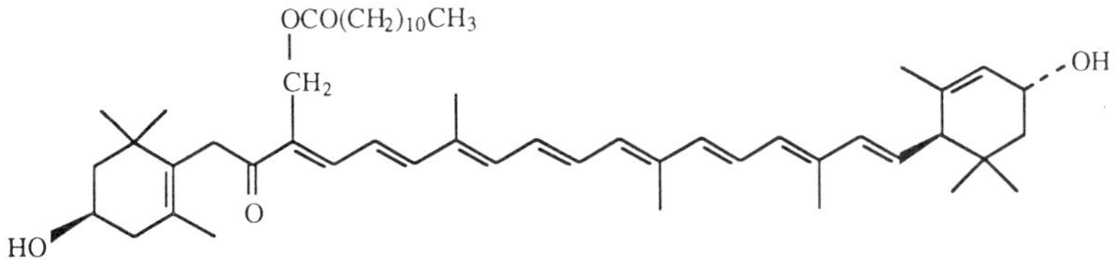

Siphonein ((3*R*,3'*R*,6'*R*)-3,19,3'-trihydroxy-7,8-dihydro-ß,ε-caroten-8-one 19-laurate).
$C_{52}H_{78}O_5$. UV/Vis λ_{max} nm: 448, (471) (ethanol). MS *m/z* (rel. int.): 782 [M]$^+$ (2), 764 [M-18]$^+$ (3), 600 [M-182]$^+$ (10), 582 [M-200]$^+$ (15).

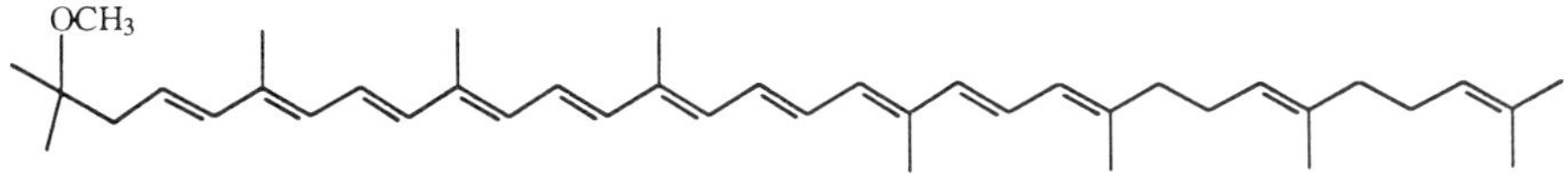

Spheroidene (1-methoxy-3,4-didehydro-1,2,7',8'-tetrahydro-ψ,ψ-carotene)
$C_{41}H_{60}O$. UV/Vis λ_{max} nm: 427, 454, 486 (petrol); 429, 454, 486 (ethanol); 440, 466, 500 ($CHCl_3$); 441, 468, 502 (benzene, $A^{1\%}_{1cm}$ 2785). MS *m/z* (rel. int.): 568 [M]$^+$ (10), 536 [M-32]$^+$ (10), 495 [M-73]$^+$ (2), 476 [M-92]$^+$ (2), 462 [M-106]$^+$ (6), 431 [M-137]$^+$(2), 430 [M-106-32]$^+$ (3), 73 (100), 69 (38). NMR end-groups: 39-33.

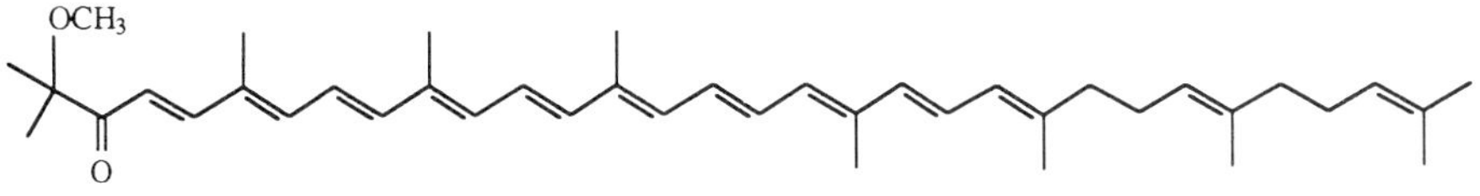

Spheroidenone (1-methoxy-3,4-didehydro-1,2,7',8'-tetrahydro-ψ,ψ-caroten-2-one)

$C_{41}H_{58}O_2$. UV/Vis λ_{max} nm: 460, 483, 515 (petrol); 488 (ethanol); 455, 484, 505 (acetone); 499 ($CHCl_3$); (475), 502, 530 (benzene); (490), 520, 553 (CS_2). MS *m/z* (rel. int.): 582 $[M]^+$ (8), 550 $[M\text{-}32]^+$ (1), 509 $[M\text{-}73]^+$ (3), 490 $[M\text{-}92]^+$ (3), 476 $[M\text{-}106]^+$ (12), 445 $[M\text{-}137]^+$ (4), 73 (100). NMR end-groups: 40-33.

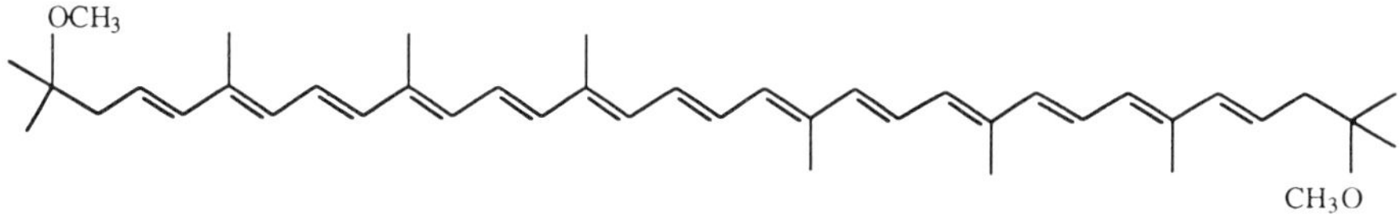

Spirilloxanthin (1,1'-dimethoxy-3,4,3',4'-tetradehydro-1,2,1',2'-tetrahydro-ψ,ψ-carotene)
$C_{42}H_{60}O_2$. m.p. 219.5–222°C. UV/Vis λ_{max} nm: 465, 492, 525 (petrol); 465, 491, 526 (ethanol); 475, 505, 543 ($CHCl_3$); 479, 510, 546 (benzene, $A^{1\%}_{1cm}$ 2470). MS *m/z* (rel. int.): 596 $[M]^+$ (15), 564 $[M\text{-}32]^+$ (3), 532 $[M\text{-}32\text{-}32]^+$ (1), 523 $[M\text{-}73]^+$ (1), 490 $[M\text{-}106]^+$ (12), 458 $[M\text{-}106\text{-}32]^+$ (5), 73 (100). NMR end-groups: 39-39.

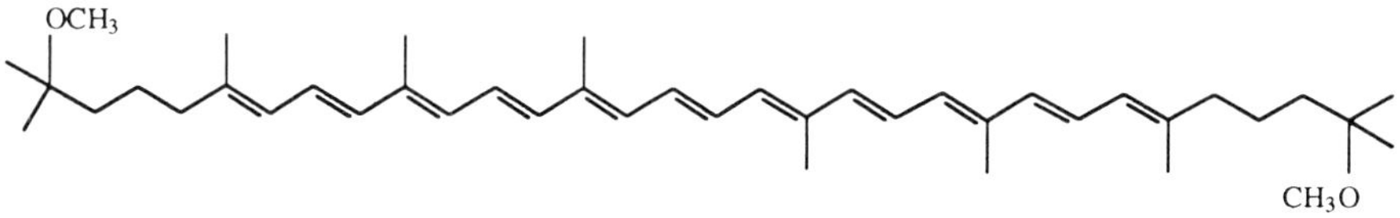

3,4,3',4'-Tetrahydrospirilloxanthin (1,1'-Dimethoxy-1,2,1',2'-tetrahydro-ψ,ψ-carotene)
$C_{42}H_{64}O_2$. UV/Vis λ_{max} nm: 443, 470, 501 (petrol). MS *m/z* (rel. int.): 592 $[M]^+$, 560 $[M\text{-}32]^+$, 528 $[M\text{-}32\text{-}32]^+$, 513 $[M\text{-}79]^+$, 500 $[M\text{-}92]^+$, 486 $[M\text{-}106]^+$, 468 $[M\text{-}92\text{-}32]^+$, 454 $[M\text{-}106\text{-}32]^+$, 73 (100).

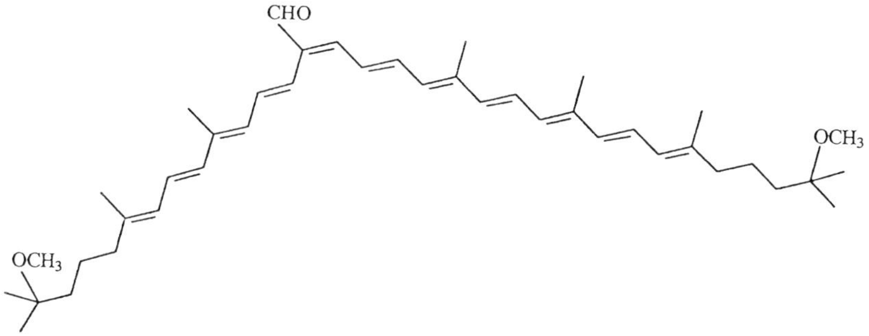

3,4,3',4'-Tetrahydrospirilloxanthinal (13-*cis*-1,1'-dimethoxy-1,2,1',2'-tetrahydro-ψ,ψ-caroten-20-al)
$C_{42}H_{62}O_3$. UV/Vis as lycopen-20-al and rhodopin-20-al.

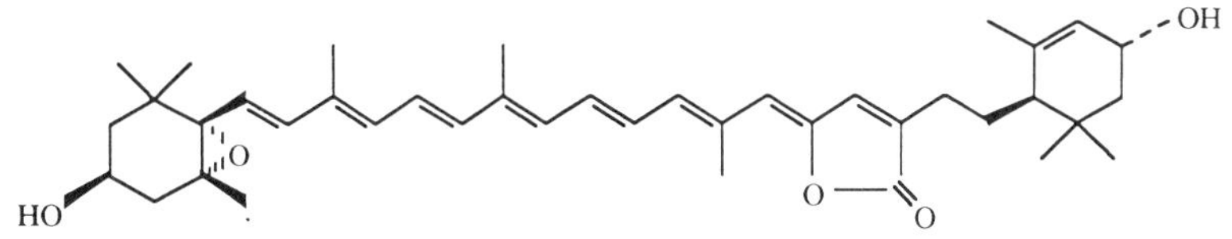

Uriolide (5,6-epoxy-3,3'-dihydroxy-5,6,7',8'-tetrahydro-ß,ε-caroten-11',19'-olide
$C_{40}H_{54}O_5$. UV/Vis λ_{max} nm: 448, 472 (Me_2CO); 448, 470 (MeOH); (427), 448, 478 (hexane). MS *m/z* (rel. int.): 614.3983 $[M]^+$ (30), 596 $[M-18]^+$ (50), 574 $[M-80]^+$ (25), 522 $[M-92]^+$ (35), 516 $[M-80-18]^+$ (25), 504 $[M-92-18]^+$ (15), 221 (100). CD (EPA) (Δε): 215 (0), 225 (-5), 240 (0), 250 (+1), 260 (0), 267 (-2), 278 (0), 315 (-2.5), 340 (0). NMR end-groups: 9-22.

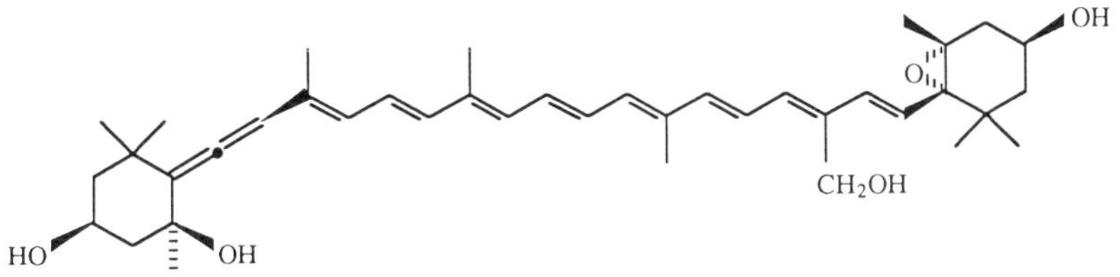

Vaucheriaxanthin (5',6'-epoxy-6,7-didehydro-5,6,5',6'-tetrahydro-ß,ß-carotene-3,5,3',19'-tetrol)
$C_{40}H_{56}O_5$. UV/Vis λ_{max} nm: 418, 436, 466 (ethanol); 426,451,480 ($CHCl_3$). MS *m/z* (rel. int.): 616 $[M]^+$ (10), 598 $[M-18]^+$ (21.7), 580 $[M-18-18]^+$ (14), 536 $[M-80]^+$, 524 $[M-92]^+$, 518 $[M-18-80]^+$, 506 $[M-92-18]^+$, 494 $[M-122]^+$, 488 $[M-92-18-18]^+$, 476 $[M-122-18]^+$. NMR end-groups: 11-14.

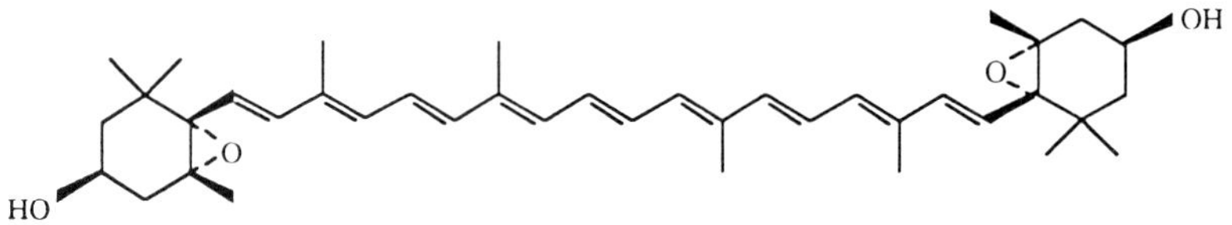

Violaxanthin ((3*S*,5*R*,6*S*,3'*S*,5'*R*,6'*S*)-5,6,5',6'-diepoxy-5,6,5',6'-tetrahydro-ß,ß-carotene-3,3'-diol)
$C_{40}H_{56}O_4$. Yellow-orange prisms (from methanol) m.p. 189°C. UV/Vis λ_{max} nm: 417,439,470 (petrol); 417,441,471 (ethanol, $A^{1\%}_{1cm}$ 2550); 426,453,483 (benzene, $A^{1\%}_{1cm}$ 2240); 426, 449, 478 ($CHCl_3$); 440, 470, 501 (CS_2). CD (ethanol) nm (Δε): 205 (-8.0), 230 (+11.6), 266 (-29.4), 316 (+3.6), 328 (+5.8), 395 (+0.9), 420 (+2.1), 444 (+3.4), 471 (+2.7). MS *m/z* (rel.int.): 600 $[M]^+$ (25), 582 $[M\text{-}18]^+$ (5), 564 $[M\text{-}18\text{-}18]^+$ (3), 520 $[M\text{-}80]^+$ (12), 508 $[M\text{-}92]^+$ (5), 502 $[M\text{-}18\text{-}80]^+$ (2), 440 $[M\text{-}80\text{-}80]^+$ (12), 221 (100), 181 (45). Details of several *Z*-isomers have been given by Molnar et al. 1986. NMR end-groups: 9-9.

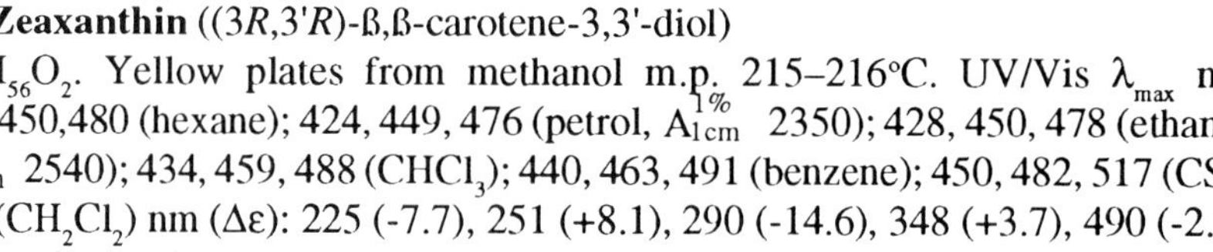

Zeaxanthin ((3*R*,3'*R*)-ß,ß-carotene-3,3'-diol)
$C_{40}H_{56}O_2$. Yellow plates from methanol m.p. 215–216°C. UV/Vis λ_{max} nm: 426,450,480 (hexane); 424, 449, 476 (petrol, $A^{1\%}_{1cm}$ 2350); 428, 450, 478 (ethanol, $A^{1\%}_{1cm}$ 2540); 434, 459, 488 ($CHCl_3$); 440, 463, 491 (benzene); 450, 482, 517 (CS_2). CD (CH_2Cl_2) nm (Δε): 225 (-7.7), 251 (+8.1), 290 (-14.6), 348 (+3.7), 490 (-2.5); (EPA): 224 (-18.0), 236 (0), 245 (+18.0), 260 (0), 284 (-24.8), 325 (0), 350 (+4.4). MS *m/z* (rel. int.): 568 $[M]^+$ (100), 550 $[M\text{-}18]^+$ (40), 532 $[M\text{-}18\text{-}18]^+$ (15), 476 $[M\text{-}92]^+$ (22), 462 $[M\text{-}106]^+$ (7). NMR end-groups: 6-6.

Index